Lecture Notes in Artificial Intelligence   4702

Edited by J. G. Carbonell and J. Siekmann

Subseries of Lecture Notes in Computer Science

T0180236

Joost N. Kok   Jacek Koronacki
Ramon Lopez de Mantaras   Stan Matwin
Dunja Mladenič   Andrzej Skowron (Eds.)

Knowledge Discovery in Databases: PKDD 2007

11th European Conference on Principles and Practice
of Knowledge Discovery in Databases
Warsaw, Poland, September 17-21, 2007
Proceedings

 Springer

Series Editors

Jaime G. Carbonell, Carnegie Mellon University, Pittsburgh, PA, USA
Jörg Siekmann, University of Saarland, Saarbrücken, Germany

Volume Editors

Joost N. Kok
Leiden University, The Netherlands
E-mail: joost@liacs.nl

Jacek Koronacki
Polish Academy of Sciences, Warsaw, Poland
E-mail: korona@ipipan.waw.pl

Ramon Lopez de Mantaras
Spanish National Research Council (CSIC), Bellaterra, Spain
E-mail: mantaras@iiia.csic.es

Stan Matwin
University of Ottawa, Ontario, Canada
E-mail: stan@site.uottawa.ca

Dunja Mladenič
Jožef Stefan Institute, Ljubljana, Slovenia
E-mail: dunja.mladenic@ijs.si

Andrzej Skowron
Warsaw University, Poland
E-mail: skowron@mimuw.edu.pl

Library of Congress Control Number: 2007934762

CR Subject Classification (1998): I.2, H.2, J.1, H.3, G.3, I.7, F.4.1

LNCS Sublibrary: SL 7 – Artificial Intelligence

ISSN   0302-9743
ISBN-10   3-540-74975-6 Springer Berlin Heidelberg New York
ISBN-13   978-3-540-74975-2 Springer Berlin Heidelberg New York

Springer is a part of Springer Science+Business Media

springer.com

© Springer-Verlag Berlin Heidelberg 2007
Printed in Germany

Typesetting: Camera-ready by author, data conversion by Scientific Publishing Services, Chennai, India
Printed on acid-free paper   SPIN: 12124534   06/3180   5 4 3 2 1 0

Preface

The two premier annual European conferences in the areas of machine learning and data mining have been collocated ever since the first joint conference in Freiburg, 2001. The European Conference on Machine Learning (ECML) traces its origins to 1986, when the first European Working Session on Learning was held in Orsay, France. The European Conference on Principles and Practice of Knowledge Discovery in Databases (PKDD) was first held in 1997 in Trondheim, Norway. Over the years, the ECML/PKDD series has evolved into one of the largest and most selective international conferences in machine learning and data mining. In 2007, the seventh collocated ECML/PKDD took place during September 17–21 on the central campus of Warsaw University and in the nearby Staszic Palace of the Polish Academy of Sciences.

The conference for the third time used a hierarchical reviewing process. We nominated 30 Area Chairs, each of them responsible for one sub-field or several closely related research topics. Suitable areas were selected on the basis of the submission statistics for ECML/PKDD 2006 and for last year's International Conference on Machine Learning (ICML 2006) to ensure a proper load balance among the Area Chairs. A joint Program Committee (PC) was nominated for the two conferences, consisting of some 300 renowned researchers, mostly proposed by the Area Chairs. This joint PC, the largest of the series to date, allowed us to exploit synergies and deal competently with topic overlaps between ECML and PKDD.

ECML/PKDD 2007 received 592 abstract submissions. As in previous years, to assist the reviewers and the Area Chairs in their final recommendation authors had the opportunity to communicate their feedback after the reviewing phase. For a small number of conditionally accepted papers, authors were asked to carry out minor revisions subject to the final acceptance by the Area Chair responsible for their submission. With very few exceptions, every full submission was reviewed by three PC members. Based on these reviews, on feedback from the authors, and on discussions among the reviewers, the Area Chairs provided a recommendation for each paper. The four Program Chairs made the final program decisions following a 2-day meeting in Warsaw in June 2007. Continuing the tradition of previous events in the series, we accepted full papers with an oral presentation and short papers with a poster presentation. We selected 41 full papers and 37 short papers for ECML, and 28 full papers and 35 short papers for PKDD. The acceptance rate for full papers is 11.6% and the overall acceptance rate is 23.8%, in accordance with the high-quality standards of the conference series. Besides the paper and poster sessions, ECML/PKDD 2007 also featured 12 workshops, seven tutorials, the ECML/PKDD Discovery Challenge, and the Industrial Track.

An excellent slate of Invited Speakers is another strong point of the conference program. We are grateful to Ricardo Bazea-Yates (Yahoo! Research Barcelona), Peter Flach (University of Bristol), Tom Mitchell (Carnegie Mellon University), and Barry Smyth (University College Dublin) for their participation in ECML/PKDD 2007. The abstracts of their presentations are included in this volume.

We distinguished four outstanding contributions; the awards were generously sponsored by the *Machine Learning Journal* and the *KD-Ubiq network*.

ECML Best Paper: Angela Kimming, Luc De Raedt and Hannu Toivonen: "Probabilistic Explanation-Based Learning"

PKDD Best Paper: Toon Calders and Szymon Jaroszewicz: "Efficient AUC-Optimization for Classification"

ECML Best Student Paper: Daria Sorokina, Rich Caruana, and Mirek Riedewald: "Additive Groves of Regression Trees"

PKDD Best Student Paper: Dikan Xing, Wenyuan Dai, Gui-Rong Xue, and Yong Yu: "Bridged Refinement for Transfer Learning"

This year we introduced the Industrial Track chaired by Florence d'Alché-Buc (Université d'Evry-Val d'Essonne) and Marko Grobelnik (Jožef Stefan Institute, Slovenia) consisting of selected talks with a strong industrial component presenting research from the area covered by the ECML/PKDD conference.

For the first time in the history of ECML/PKDD, the conference proceedings were available on-line to conference participants during the conference. We are grateful to Springer for accommodating this new access channel for the proceedings. Inspired by some related conferences (ICML, KDD, ISWC) we introduced videorecording, as we would like to save at least the invited talks and presentations of award papers for the community and make them accessible at http://videolectures.net/.

This year's Discovery Challenge was devoted to three problems: user behavior prediction from Web traffic logs, HTTP traffic classification, and Sumerian literature understanding. The Challenge was co-organized by Piotr Ejdys (Gemius SA), Hung Son Nguyen (Warsaw University), Pascal Poncelet (EMA-LGI2P) and Jerzy Tyszkiewicz (Warsaw University); 122 teams participated. For the first task, the three finalists were:

Malik Tahir Hassan, Khurum Nazir Junejo and Asim Karim from Lahore University, Pakistan

Krzysztof Dembczyński and Wojciech Kotłowski from Poznań University of Technology, Poland and Marcin Sydow from Polish-Japanese Institute of Information Technology, Poland

Tung-Ying Lee from National Tsing Hua University, Taiwan

Results for the other Discovery Challenge tasks were not available at the time the proceedings were finalized, but were announced at the conference.

We are all indebted to the Area Chairs, Program Committee members and external reviewers for their commitment and hard work that resulted in a rich but selective scientific program for ECML/PKDD. We are particularly grateful to those reviewers who helped with additional reviews at very short notice to assist us in a small number of difficult decisions. We further thank our Workshop and Tutorial Chairs Marzena Kryszkiewicz (Warsaw Technical University) and Jan Rauch (University of Economics, Prague) for selecting and coordinating the 12 workshops and seven tutorial events that accompanied the conference; the workshop organizers, tutorial presenters, and the organizers of the Discovery Challenge and the Industrial track; Richard van de Stadt and CyberChairPRO for competent and flexible support; Warsaw University and the Polish Academy of Sciences (Institute of Computer Science) for their local and organizational support. Special thanks are due to the Local Chair, Marcin Szczuka, Warsaw University (assisted by Michal Ciesiołka from the Polish Academy of Sciences) for the many hours spent making sure that all the details came together to ensure the success of the conference. Finally, we are grateful to the Steering Committee and the ECML/PKDD community that entrusted us with the organization of the ECML/PKDD 2007.

Most of all, however, we would like to thank all the authors who trusted us with their submissions, thereby contributing to the one of the main yearly events in the life of our vibrant research community.

September 2007

Joost Kok (PKDD Program co-Chair)
Jacek Koronacki (General Chair)
Ramon Lopez de Mantaras (General Chair)
Stan Matwin (ECML Program co-Chair)
Dunja Mladenič (ECML Program co-Chair)
Andrzej Skowron (PKDD Program co-Chair)

Organization

General Chairs

Ramon Lopez de Mantaras (Spanish Council for Scientific Research)
Jacek Koronacki (Polish Academy of Sciences)

Program Chairs

Joost N. Kok (Leiden University)
Stan Matwin (University of Ottawa and Polish Academy of Sciences)
Dunja Mladenič (Jožef Stefan Institute)
Andrzej Skowron (Warsaw University)

Local Chairs

Michał Ciesiołka (Polish Academy of Sciences)
Marcin Szczuka (Warsaw University)

Tutorial Chair

Jan Rauch (University of Economics, Prague)

Workshop Chair

Marzena Kryszkiewicz (Warsaw University of Technology)

Discovery Challenge Chair

Hung Son Nguyen (Warsaw University)

Industrial Track Chairs

Florence d'Alché-Buc (Université d'Evry-Val d'Essonne)
Marko Grobelnik (Jozef Stefan Institute)

Steering Committee

Jean-François Boulicaut
Rui Camacho
Johannes Fürnkranz
Fosca Gianotti
Dino Pedreschi
Myra Spiliopoulou

Pavel Brazdil
Floriana Esposito
João Gama
Alípio Jorge
Tobias Scheffer
Luís Torgo

Area Chairs

Michael R. Berthold
Olivier Chapelle
Kurt Driessens
Eibe Frank
Thomas Gärtner
Rayid Ghani
Eamonn Keogh
Mieczysław A. Kłopotek
Pedro Larranaga
Andreas Nürnberger
Bernhard Pfahringer
Luc De Raedt
Giovanni Semeraro
Myra Spiliopoulou
Luís Torgo

Hendrik Blockeel
James Cussens
Peter Flach
Johannes Fürnkranz
João Gama
Jerzy Grzymala-Busse
Kristian Kersting
Stefan Kramer
Claire Nedellec
George Paliouras
Enric Plaza
Tobias Scheffer
Władysław Skarbek
Hannu Toivonen
Paul Utgoff

Program Committee

Charu C. Aggarwal
Jesús Aguilar-Ruiz
David W. Aha
Nahla Ben Amor
Sarabjot Singh Anand
Annalisa Appice
Josep-Lluis Arcos
Walid G. Aref
Eva Armengol
Anthony J. Bagnall
Antonio Bahamonde
Sugato Basu
Bettina Berendt
Francesco Bergadano
Ralph Bergmann
Steffen Bickel

Concha Bielza
Mikhail Bilenko
Francesco Bonchi
Gianluca Bontempi
Christian Borgelt
Karsten M. Borgwardt
Daniel Borrajo
Antal van den Bosch
Henrik Boström
Marco Botta
Jean-François Boulicaut
Janez Brank
Thorsten Brants
Ulf Brefeld
Carla E. Brodley
Paul Buitelaar

Toon Calders
Luis M. de Campos
Nicola Cancedda
Claudio Carpineto
Jesús Cerquides
Kaushik Chakrabarti
Chien-Chung Chan
Amanda Clare
Ira Cohen
Fabrizio Costa
Susan Craw
Bruno Crémilleux
Tom Croonenborghs
Juan Carlos Cubero
Pádraig Cunningham
Andrzej Czyżewski
Walter Daelemans
Ian Davidson
Marco Degemmis
Olivier Delalleau
Jitender S. Deogun
Marcin Detyniecki
Belén Diaz-Agudo
Chris H.Q. Ding
Carlotta Domeniconi
Marek J. Druzdzel
Sašo Džeroski
Tina Eliassi-Rad
Tapio Elomaa
Abolfazl Fazel Famili
Wei Fan
Ad Feelders
Alan Fern
George Forman
Linda C. van der Gaag
Patrick Gallinari
José A. Gámez
Alex Gammerman
Minos N. Garofalakis
Gemma C. Garriga
Eric Gaussier
Pierre Geurts
Fosca Gianotti
Attilio Giordana
Robert L. Givan

Bart Goethals
Elisabet Golobardes
Pedro A. González-Calero
Marko Grobelnik
Dimitrios Gunopulos
Maria Halkidi
Mark Hall
Matthias Hein
Jose Hernandez-Orallo
Colin de la Higuera
Melanie Hilario
Shoji Hirano
Tu-Bao Ho
Jaakko Hollmen
Geoffrey Holmes
Frank Höppner
Tamás Horváth
Andreas Hotho
Jiayuan Huang
Eyke Hüllemeier
Masahiro Inuiguchi
Inaki Inza
Manfred Jaeger
Szymon Jaroszewicz
Rosie Jones
Edwin D. de Jong
Alípio Mário Jorge
Tamer Kahveci
Alexandros Kalousis
Hillol Kargupta
Andreas Karwath
George Karypis
Samuel Kaski
Dimitar Kazakov
Ross D. King
Frank Klawonn
Ralf Klinkenberg
George Kollios
Igor Kononenko
Bożena Kostek
Walter A. Kosters
Miroslav Kubat
Halina Kwasnicka
James T. Kwok
Nicolas Lachiche

Michail G. Lagoudakis
Niels Landwehr
Pedro Larranaga
Pavel Laskov
Mark Last
Dominique Laurent
Nada Lavrac
Quoc V. Le
Guy Lebanon
Ulf Leser
Jure Leskovec
Jessica Lin
Francesca A. Lisi
Pasquale Lops
Jose A. Lozano
Peter Lucas
Richard Maclin
Donato Malerba
Nikos Mamoulis
Suresh Manandhar
Stéphane Marchand-Maillet
Elena Marchiori
Lluis Marquez
Yuji Matsumoto
Michael May
Mike Mayo
Thorsten Meinl
Prem Melville
Rosa Meo
Taneli Mielikäinen
Bamshad Mobasher
Serafín Moral
Katharina Morik
Hiroshi Motoda
Toshinori Munakata
Ion Muslea
Olfa Nasraoui
Jennifer Neville
Siegfried Nijssen
Joakim Nivre
Ann Nowe
Arlindo L. Oliveira
Santi Ontañón
Miles Osborne
Martijn van Otterlo

David Page
Spiros Papadimitriou
Srinivasan Parthasarathy
Andrea Passerini
Jose M. Peña
Lourdes Peña Castillo
José M. Peña Sánchez
James F. Peters
Johann Petrak
Lech Polkowski
Han La Poutre
Philippe Preux
Katharina Probst
Tapani Raiko
Ashwin Ram
Sheela Ramanna
Jan Ramon
Zbigniew W. Ras
Chotirat Ann Ratanamahatana
Francesco Ricci
John Riedl
Christophe Rigotti
Celine Robardet
Victor Robles
Marko Robnik-Sikonja
Juho Rousu
Céline Rouveirol
Ulrich Rückert (TU München)
Ulrich Rückert (Univ. Paderborn)
Stefan Rüping
Henryk Rybiński
Lorenza Saitta
Hiroshi Sakai
Roberto Santana
Martin Scholz
Matthias Schubert
Michele Sebag
Sandip Sen
Jouni K. Seppänen
Galit Shmueli
Arno Siebes
Alejandro Sierra
Vikas Sindhwani
Arul Siromoney
Dominik Ślęzak

Carlos Soares
Maarten van Someren
Alvaro Soto
Alessandro Sperduti
Jaideep Srivastava
Jerzy Stefanowski
David J. Stracuzzi
Jan Struyf
Gerd Stumme
Zbigniew Suraj
Einoshin Suzuki
Roman Swiniarski
Marcin Sydow
Piotr Synak
Marcin Szczuka
Luis Talavera
Matthew E. Taylor
Yannis Theodoridis
Kai Ming Ting
Ljupco Todorovski
Volker Tresp
Shusaku Tsumoto
Karl Tuyls
Michalis Vazirgiannis
Katja Verbeeck
Jean-Philippe Vert

Michail Vlachos
Haixun Wang
Jason Tsong-Li Wang
Takashi Washio
Gary M. Weiss
Sholom M. Weiss
Shimon Whiteson
Marco Wiering
Slawomir T. Wierzchoń
Graham J. Williams
Stefan Wrobel
Ying Yang
JingTao Yao
Yiyu Yao
François Yvon
Bianca Zadrozny
Mohammed J. Zaki
Gerson Zaverucha
Filip Zelezny
ChengXiang Zhai
Yi Zhang
Zhi-Hua Zhou
Jerry Zhu
Wojciech Ziarko
Albrecht Zimmermann

Additional Reviewers

Rezwan Ahmed
Fabio Aiolli
Dima Alberg
Vassilis Athitsos
Maurizio Atzori
Anne Auger
Paulo Azevedo
Pierpaolo Basile
Margherita Berardi
Andre Bergholz
Michele Berlingerio
Kanishka Bhaduri
Konstantin Biatov
Jerzy Błaszczyński
Gianluca Bontempi
Yann-ael Le Borgne

Zoran Bosnic
Remco Bouckaert
Agnès Braud
Bjoern Bringmann
Emma Byrne
Olivier Caelen
Rossella Cancelliere
Giovanna Castellano
Michelangelo Ceci
Hyuk Cho
Kamalika Das
Souptik Datta
Uwe Dick
Laura Dietz
Marcos Domingues
Haimonti Dutta

Marc Dymetman
Stefan Eickeler
Timm Euler
Tanja Falkowski
Fernando Fernandez
Francisco J. Ferrer-Troyano
Cèsar Ferri
Daan Fierens
Blaz Fortuna
Alexandre Francisco
Mingyan Gao
Fabián Güiza
Anna Lisa Gentile
Amol N. Ghoting
Arnaud Giacometti
Valentin Gjorgjioski
Robby Goetschalckx
Derek Greene
Perry Groot
Philip Groth
Daniele Gunetti
Bernd Gutmann
Sattar Hashemi
Yann-Michael De Hauwere
Vera Hollink
Yi Huang
Leo Iaquinta
Alexander Ilin
Tasadduq Imam
Tao-Yuan Jen
Felix Jungermann
Andrzej Kaczmarek
Benjamin Haibe Kains
Juha Karkkainen
Rohit Kate
Chris Kauffman
Arto Klami
Jiri Klema
Dragi Kocev
Christine Koerner
Kevin Kontos
Petra Kralj
Anita Krishnakumar
Matjaž Kukar
Brian Kulis

Arnd Christian König
Christine Körner
Fei Tony Liu
Antonio LaTorre
Anne Laurent
Baoli Li
Zi Lin
Bin Liu
Yan Liu
Corrado Loglisci
Rachel Lomasky
Carina Lopes
Chuan Lu
Pierre Mahé
Markus Maier
Giuseppe Manco
Irina Matveeva
Nicola Di Mauro
Dimitrios Mavroeidis
Stijn Meganck
Ingo Mierswa
Mirjam Minor
Abhilash Alexander Miranda
João Moreira
Sourav Mukherjee
Canh Hao Nguyen
Duc Dung Nguyen
Tuan Trung Nguyen
Janne Nikkilä
Xia Ning
Blaž Novak
Irene Ntoutsi
Riccardo Ortale
Stanisław Osiński
Kivanc Ozonat
Aline Paes
Pance Panov
Thomas Brochmann Pedersen
Maarten Peeters
Ruggero Pensa
Xuan-Hieu Phan
Benjarath Phoophakdee
Aloisio Carlos de Pina
Christian Plagemann
Jose M. Puerta

Aritz Pérez
Chedy Raissi
M. Jose Ramirez-Quintana
Umaa Rebbapragada
Stefan Reckow
Chiara Renso
Matthias Renz
Francois Rioult
Domingo Rodriguez-Baena
Sten Sagaert
Luka Šajn
Esin Saka
Saeed Salem
Antonio Salmeron
Eerika Savia
Anton Schaefer
Leander Schietgat
Gaetano Scioscia
Howard Scordio
Sven Van Segbroeck
Ivica Slavkov
Larisa Soldatova
Arnaud Soulet
Eduardo Spynosa
Volkmar Sterzing
Christof Stoermann
Jiang Su
Piotr Szczuko

Alexander Tartakovski
Olivier Teytaud
Marisa Thoma
Eufemia Tinelli
Ivan Titov
Roberto Trasarti
George Tsatsaronis
Katharina Tschumitschew
Duygu Ucar
Antonio Varlaro
Shankar Vembu
Celine Vens
Marcos Vieira
Peter Vrancx
Nikil Wale
Chao Wang
Dongrong Wen
Arkadiusz Wojna
Yuk Wah Wong
Adam Woźnica
Michael Wurst
Wei Xu
Xintian Yang
Monika Zakova
Luke Zettlemoyer
Xueyuan Zhou
Albrecht Zimmermann

Sponsors

We wish to express our gratitude to the sponsors of ECML/PKDD 2007 for their essential contribution to the conference. We wish to thank Warsaw University, Faculty of Mathematics, Informatics and Mechanics, and Institute of Computer Science, Polish Academy of Sciences for providing financial and organizational means for the conference; the European Office of Aerospace Research and Developement, Air Force Office of Scientific Research, United States Air Force Research Laboratory, for their generous financial support.[1] KDUbiq European Coordination Action for supporting Poster Reception, Student Travel Awards, and the Best Paper Awards; Pascal European Network of Excellence for sponsoring the Invited Speaker Program, the Industrial Track and the video-recording of the invited talks and presentations of the four Award Papers; Jožef Stefan Institute, Slovenia, SEKT European Integrated project and Unilever R & D for their financial support; the *Machine Learning Journal* for supporting the Student Best Paper Awards; Gemius S.A. for sponsoring and supporting the Discovery Challenge. We also wish to express our gratitude to the following companies and institutions that provided us with data and expertise which were essential components of the Discovery Challenge: Bee Ware, l'École des Mines d'Alès, LIRMM - The Montpellier Laboratory of Computer Science, Robotics, and Microelectronics, and Warsaw University, Faculty of Mathematics, Informatics and Mechanics. We also acknowledge the support of LOT Polish Airlines.

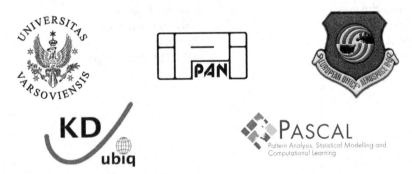

---

[1] AFOSR/EOARD support is not intended to express or imply endorsement by the U.S. Federal Government.

Table of Contents

Short Papers

Learning, Information Extraction and the Web[*]

Tom M. Mitchell

Machine Learning Department
Carnegie Mellon University, USA
tom.mitchell@cs.cmu.edu

Abstract. Significant progress has been made recently in semi-supervised learning algorithms that require less labeled training data by utilizing unlabeled data. Much of this progress has been made in the context of natural language analysis (e.g., semi-supervised learning for named entity recognition and for relation extraction). This talk will overview progress in this area, present some of our own recent research, and explore the possibility that now is the right time to mount a community-wide effort to develop a never-ending natural language learning system.

[*] Invited speakers at ECML/PKDD are supported by the PASCAL European network of excellence.

J.N. Kok et al. (Eds.): PKDD 2007, LNAI 4702, p. 1, 2007.

Putting Things in Order: On the Fundamental Role of Ranking in Classification and Probability Estimation⋆

Peter A. Flach

Department of Computer Science, University of Bristol, United Kingdom
Peter.Flach@bristol.ac.uk

Abstract. While a binary classifier aims to distinguish positives from negatives, a ranker orders instances from high to low expectation that the instance is positive. Most classification models in machine learning output some score of 'positive-ness', and hence can be used as rankers. Conversely, any ranker can be turned into a classifier if we have some instance-independent means of splitting the ranking into positive and negative segments. This could be a fixed score threshold; a point obtained from fixing the slope on the ROC curve; the break-even point between true positive and true negative rates; to mention just a few possibilities.

These connections between ranking and classification notwithstanding, there are considerable differences as well. Classification performance on n examples is measured by accuracy, an $O(n)$ operation; ranking performance, on the other hand, is measured by the area under the ROC curve (AUC), an $O(n \log n)$ operation. The model with the highest AUC does not necessarily dominate all other models, and thus it is possible that another model would achieve a higher accuracy for certain operating conditions, even if its AUC is lower.

However, within certain model classes good ranking performance and good classification performance are more closely related than suggested by the previous remarks. For instance, there is evidence that certain classification models, while designed to optimise accuracy, in effect optimise an AUC-based loss function [1]. It has also been known for some time that decision tree yield convex training set ROC curves by construction [2], and thus optimising training set accuracy is likely to lead to good training set AUC. In this talk I will investigate the relation between ranking and classification more closely.

I will also consider the connection between ranking and probability estimation. The quality of probability estimates can be measured by, e.g., mean squared error in the probability estimates (the Brier score). However, like accuracy, this is an $O(n)$ operation that doesn't fully take ranking performance into account. I will show how a novel decomposition of the Brier score into calibration loss and refinement loss [3] sheds light on both ranking and probability estimation performance. While previous decompositions are approximate [4], our decomposition is an exact one based on the ROC convex hull. (The connection between the ROC convex hull and calibration was independently noted by [5]). In the case of decision trees, the analysis explains the empirical evidence that probability estimation trees produce well-calibrated probabilities [6].

⋆ Invited speakers at ECML/PKDD are supported by the PASCAL European network of excellence.

J.N. Kok et al. (Eds.): PKDD 2007, LNAI 4702, pp. 2–3, 2007.

References

1. Rudin, C., Cortes, C., Mohri, M., Schapire, R.E.: Margin-based ranking meets boosting in the middle. In: Auer, P., Meir, R. (eds.) COLT 2005. LNCS (LNAI), vol. 3559, pp. 63–78. Springer, Heidelberg (2005)
2. Ferri, C., Flach, P.A., é Hernández-Orallo, J.: Learning decision trees using the area under the ROC curve. In: Sammut, C., Hoffmann, A.G. (eds.) Proceedings of the Nineteenth International Conference on Machine Learning (ICML 2002), pp. 139–146. Morgan Kaufmann, San Francisco (2002)
3. Flach, P.A., Matsubara, E.T.: A simple lexicographic ranker and probability estimator. In: Proceedings of the Eighteenth European Conference on Machine Learning (ECML (2007) (this volume) (2007)
4. Cohen, I., Goldszmidt, M.: Properties and benefits of calibrated classifiers. In: Boulicaut, J.-F., Esposito, F., Giannotti, F., Pedreschi, D. (eds.) PKDD 2004. LNCS (LNAI), vol. 3202, pp. 125–136. Springer, Heidelberg (2004)
5. Fawcett, T., Niculescu-Mizil, A.: PAV and the ROC convex hull. Machine Learning 68(1), 97–106 (2007)
6. Niculescu-Mizil, A., Caruana, R.: Predicting good probabilities with supervised learning. In: Proceedings of the Twenty-Second International Conference on Machine Learning (ICML 2005), pp. 625–632. ACM, New York (2005)

Mining Queries*

Ricardo Baeza-Yates

Yahoo! Research, Barcelona, Spain
and Yahoo! Research Latin America, Santiago, Chile
ricardo.baeza@upf.edu

Abstract. User queries in search engines and Websites give valuable information on the interests of people. In addition, clicks after queries relate those interests to actual content. Even queries without clicks or answers imply important missing synonyms or content. In this talk we show several examples on how to use this information to improve the performance of search engines, to recommend better queries, to improve the information scent of the content of a Website and ultimately to capture knowledge, as Web queries are the largest wisdom of crowds in Internet.

---

* Invited speakers at ECML/PKDD are supported by the PASCAL European network of excellence.

Adventures in Personalized Information Access*

Barry Smyth

Adaptive Information Cluster, School of Computer Science and Informatics,
University College Dublin, Ireland
barry.smyth@ucd.ie

Abstract. Access to information plays an increasingly important role in our
everyday lives and we have come to rely more and more on a variety of information access services to bring us the right information at the right time. Recently
the traditional one-size-fits-all approach, which has informed the development of
the majority of today's information access services, from search engines to portals, has been brought in to question as researchers consider the advantages of
more personalized services. Such services can respond to the learned needs and
preferences of individuals and groups of like-minded users. They provide for a
more proactive model of information supply in place of today's reactive models of
information search. In this talk we will consider the key challenges that motivate
the need for a new generation of personalized information services, as well as the
pitfalls that lie in wait. We will focus on a number of different information access
scenarios, from e-commerce recommender systems and personalized mobile portals to community-based web search. In each case we will describe how different
machine learning and data mining ideas have been harnessed to take advantage
of key domain constraints in order to deliver information access interfaces that
are capable of adapting to the changing needs and preferences of their users. In
addition, we will describe the results of a number of user studies that highlight
the potential for such technologies to significantly enhance the user experience
and the ability of users to locate relevant information quickly and reliably.

---

* Invited speakers at ECML/PKDD are supported by the PASCAL European network of excellence.

J.N. Kok et al. (Eds.): PKDD 2007, LNAI 4702, p. 5, 2007.

Experiment Databases: Towards an Improved Experimental Methodology in Machine Learning

Hendrik Blockeel and Joaquin Vanschoren

Computer Science Dept., K.U.Leuven, Celestijnenlaan 200A, 3001 Leuven, Belgium

Abstract. Machine learning research often has a large experimental component. While the experimental methodology employed in machine learning has improved much over the years, repeatability of experiments and generalizability of results remain a concern. In this paper we propose a methodology based on the use of *experiment databases*. Experiment databases facilitate large-scale experimentation, guarantee repeatability of experiments, improve reusability of experiments, help expliciting the conditions under which certain results are valid, and support quick hypothesis testing as well as hypothesis generation. We show that they have the potential to significantly increase the ease with which new results in machine learning can be obtained and correctly interpreted.

1 Introduction

Experimental assessment is a key aspect of machine learning research. Indeed, many learning algorithms are heuristic in nature, each making assumptions about the structure of the given data, and although there may be good reason to believe a method will work well in general, this is difficult to prove. In fact, it is impossible to theoretically prove that one algorithm is superior to another [15], except under specific conditions. Even then, it may be difficult to specify these conditions precisely, or to find out how relevant they are for real-world problems. Therefore, one usually verifies a learning algorithm's performance empirically, by implementing it and running it on (real-world) datasets.

Since empirical assessment is so important, it has repeatedly been argued that care should be taken to ensure that (published) experimental results can be interpreted correctly [8]. First of all, it should be clear how the experiments can be reproduced. This involves providing a complete description of both the experimental setup (which algorithms to run with which parameters on which datasets, including how these settings were chosen) and the experimental procedure (how the algorithms are run and evaluated). Since space is limited in paper publications, an online log seems the most viable option.

Secondly, it should be clear how generalizable the reported results are, which implies that the experiments should be general enough to test this. In time series analysis research, for instance, it has been shown that many studies were biased towards the datasets being used, leading to ill-founded or contradictory results [8]. In machine learning, Perlich et al. [10] describe how the relative

J.N. Kok et al. (Eds.): PKDD 2007, LNAI 4702, pp. 6–17, 2007.

performance of logistic regression and decision trees depends strongly on the size of dataset samples. Similarly, Hoste and Daelemans [6] show that in text mining, the relative performance of lazy learning and rule induction is dominated by the effect of parameter optimization, data sampling, feature selection, and their interaction. As such, there are good reasons for strongly varying the conditions under which experiments are run, and projects like Statlog [12] and METAL [11] made the first inroads into this direction.

In light of the above, it would be useful to have an environment for machine learning research that facilitates storage of the exact conditions under which experiments have been performed as well as large-scale experimentation under widely varying conditions. To achieve this goal, Blockeel [1] proposed the use of *experiment databases*. Such databases are designed to store detailed information on large numbers of learning experiments, selected to be highly representative for a wide range of possible experiments, improving reproducibility, generalizability and interpretability of experimental results. In addition, they can be made available online, forming "experiment repositories" which allow other researchers to query for and reuse the experiments to test new hypotheses (in a way similar to how dataset repositories are used to test the performance of new algorithms).

Blockeel introduced the ideas behind experiment databases and discussed their potential advantages, but did not present details on how to construct such a database, nor considered whether it is even realistic to assume this is possible. In this paper, we answer those questions. We propose concrete design guidelines for experiment databases, present a specific implementation consistent with these guidelines, and illustrate the use of this database. By querying it for specific experiments, we can directly test a wide range of hypotheses on the covered algorithms and verify or refine existing results. Finally, the database itself is a contribution to the machine learning community: this database, containing the results of 250,000 runs of well-known classification systems under varying conditions, is publicly accessible on the web to be queried by other researchers.

The remainder of this paper is structured as follows. In Sect. 2 we summarize the merits of experiment databases. In Sect. 3 we discuss the structure of such a database, and in Sect. 4 methods for populating it with experiments. Section 5 presents a case study: we implemented an experimental database and ran a number of queries in order to evaluate how easily it allows verification of existing knowledge and discovery of new insights. We conclude in Sect. 6.

2   Experiment Databases

An experiment database is a database designed to store a (large) number of experiments, containing detailed information on the datasets, algorithms, and parameter settings used, as well as the evaluation procedure and the obtained results. It can be used as a log of performed experiments, but also as a repository of experimental results that can be reused for further research.

The currently most popular experimental methodology in machine learning is to first come up with an hypothesis about the algorithms under study, then perform experiments explicitly designed to test this hypothesis, and finally interpret the results. In this context, experiment databases make it easier to keep an unambiguous log of all the performed experiments, including all information necessary to repeat the experiments.

However, experiment databases also support a new methodology: instead of designing experiments to test a specific hypothesis, one can design them to cover, as well as possible, the space of all experiments that are of interest in the given context. A specific hypothesis can then be tested by querying the database for those experiments most relevant for the hypothesis, and interpreting the returned results. With this methodology, many more experiments are needed to evaluate the learning algorithms under a variety of conditions (parameter settings, datasets,...), but the same experiments can be reused for many different hypotheses. For instance, by adjusting the query, we can test how much the observed performance changes if we add or remove restrictions on the datasets or parameter settings. Furthermore, as the query explictly mentions all restrictions, it is easy to see under which conditions the returned results are valid.

As an example, say Ann wants to test the effect of dataset size on the complexity of trees learned by C4.5. To do this, she selects a number of datasets of varying sizes, runs C4.5 (with default parameters) on those datasets, and interprets the results. Bob, a proponent of the new methodology proposed here, would instead build a large database of C4.5 runs (with various parameter settings) on a large number of datasets, possibly reusing a number of experiments from existing experiment databases. Bob then queries the database for C4.5 runs, selecting the dataset size and tree size for all runs with default parameter settings (explicitly mentioning this condition in his query), and plotting them against each other. If Ann wants to test whether her results on default settings for C4.5 are representative for C4.5 in general, she needs to set up new experiments. Bob, on the other hand, only has to ask a second query, this time not including the condition. This way, he can easily investigate under which conditions a certain effect will occur, and be more confident about the generality of his results.

The second methodology requires a larger initial investment with respect to experimentation, but may pay off in the long run, especially if many different hypotheses are to be tested, and if many researchers make use of experiments stored in such databases. For instance, say another researcher is more interested in the runtime (or another performance metric) of C4.5 on these experiments. Since this is recorded in the experiment database as well, the experiments will not have to be repeated. A final advantage is that, given the amount of experiments, Bob can train a learning algorithm on the available meta-data, gaining models which may provide further insights in C4.5's behavior.

Note that the use of experiment databases is not strongly tied to the choice of methodology. Although experiment databases are necessary for the second methodology, they can also be used with the first methodology, allowing experiments to be more easily reproduced and reused.

3   Database Structure

An experiment database should be designed to store experiments in such detail that they are perfectly repeatable and maximally reusable. In this section, we consecutively discuss how the learning algorithms, the datasets, and the experimental procedures should be described to achieve this goal. This discussion does not lead to a single best way to design an experiment database: in many cases several options remain, and depending on the purpose of the experiment database different options may be chosen.

3.1   Algorithm

In most cases, storing a complete symbolic description of the implementation of an algorithm is practically impossible. It is more realistic to store name and version of a system, together with a pointer to source code or an executable, so the experiment can be rerun under the same conditions. Some identification of the environment (e.g. the required operating system) completes this description.

As most algorithms have parameters that change their behavior, the values of these parameters must be stored as well. We call an algorithm together with specific values for its parameters an algorithm instantiation. For randomized algorithms, we store the seed for the random generator they use also as a parameter. As such, an algorithm instantiation is always a deterministic function.

Optionally, a characterization of the algorithm could be added, consisting of generally known or calculated properties [13,7]. Such a characterization could indicate, for instance, the class of approaches the algorithm belongs to (naive bayes, neural net, decision tree learner,...), whether it generally has high or low bias and variance, etc. Although this characterization is not necessary to ensure repeatability of the experiment, it may be useful when interpreting the results or when investigating specific types of algorithms.

3.2   Dataset

To describe datasets, one can store name, version and a pointer to a representation of the actual dataset. The latter could be an online text file (possibly in multiple formats) that the algorithm implementations can read, but it could also be a dataset generator together with its parameters (including the generator's random seed) or a data transformation function (sampling instances, selecting features, defining new features, etc.) together with its parameters and a pointer to the input dataset. If storage space is not an issue, one could also store the dataset itself in the database.

As with algorithms, an optional characterization of the dataset can be added: number of examples, number of attributes, class entropy, etc. These are useful to investigate how the performance of an algorithm is linked to properties of the training data. Since this characterization depends only on the dataset, not on the experiment, new features can be added (and computed for each dataset), and subsequently used in future analysis, without rerunning any experiments.

The same holds for the algorithm characterisation. This underlines the reusability aspect of experiment databases.

3.3 Experimental Procedure

To correctly interpret (and repeat) the outcome of the experiment, we need to describe exactly how the algorithm is run (e.g. on which machine) and evaluated. For instance, in case we use a cross-validation procedure to estimate the predictive performance of the algorithm on unseen data, this implies storing (a seed to generate) the exact folds[1]. Also the exact functions used to compute these estimates (error, accuracy,...) should be described. To make the experiments more reusable, it is advisable to compute a variety of much used metrics, or to store the information from which they can be derived. In the case of classifiers, this includes storing the full contingency table (i.e., for each couple of classes (i, j), the number of cases where class i was predicted as class j).[2]

Another important outcome of the experiment is the model generated by the algorithm. As such, we should at least store specific properties of these models, such as the time to learn the model, its size, and model-specific properties (e.g. tree depth) for further analysis. If storage space allows this, also a full representation of the model could be stored for later visualisation[3]. For predictive models, it might also be useful to store the individual (probabilities of) predictions for each example in the dataset. This allows to add and compute more evaluation criteria without rerunning the experiment.

4 Populating the Database

Next to storing experiment in a structured way, one also needs to select the right experiments. As we want to use this database to gain insight in the behavior of machine learning algorithms under various conditions, we need to have experiments that are as diverse as possible. To achieve this in practice, we first need to select the algorithm(s) of interest from a large set of available algorithms. To choose its parameter settings, one can specify a probability distribution for each different parameter according to which values should be generated (in the simplest case, this could be a uniformly sampled list of reasonable values).

Covering the dataset space is harder. One can select a dataset from a large number of real-world datasets, including for instance the UCI repository. Yet, one can also implement a number of data transformation methods (e.g., sampling the dataset, performing feature selection,...) and derive variants of real-world datasets in this way. Finally, one could use synthetic datasets, produced by

---

[1] Note that although algorithms should be compared using the same folds, these folds (seeds) should also be varied to allow true random sampling.

[2] Demšar [3] comments that it is astounding how many papers still evaluate classifiers based on accuracy alone, despite the fact that this has been advised against for many years now. Experiment databases may help eradicate this practice.

[3] Some recent work focuses on efficiently storing models in databases [4].

dataset generators. This seems a very promising direction, but the construction of dataset generators that cover a reasonably interesting area in the space of all datasets is non-trivial. This is a challenge, not a limitation, as even the trivial approach of only including publicly available datasets would already ensure a coverage that is equal to or greater than that of many published papers on general-purpose machine learning techniques.

At the same time however, we also want to be able to thoroughly investigate very specific conditions (e.g. very large datasets). This means we must not only cover a large area within the space of all interesting experiments[4], but also populate this area in a reasonably dense way. Given that the number of possible algorithm instantiations and datasets (and experimental procedures) is possibly quite large, the space of interesting experiments might be very high-dimensional, and covering a large area of such a high-dimensional space in a "reasonably dense" way implies running many experiments.

A simple, yet effective way of doing this is selecting random, but sensible, values for all parameters in our experiments. With the term parameter we mean any stored property of the experiment: the used algorithm, its parameters, its algorithm-independent characterization, the dataset properties, etc.

To imagine how many experiments would be needed in this case, assume that each of these parameter has on average v values (numerical parameters are discretized into v bins). Running $100v$ experiments with random values for all parameters implies that for each value of any single parameter, the average outcomes of about 100 experimental runs will be stored. This seems sufficient to be able to detect most correlations between outcomes and the value of this parameter. To detect n-th order interaction effects between parameters, $100v^n$ experiments would be needed. Taking, for example, $v = 20$ and $n = 2$ or $n = 3$, this yields a large number of experiments, but (especially for fast algorithms) not infeasible with today's computation power.

Note how this contrasts to the number of experimental runs typically reported on machine learning papers. Yet, when keeping many parameters constant to test a specific hypothesis, there is no guarantee that the obtained results generalize towards other parameter settings, and they cannot easily be reused for testing other hypotheses. The factor 100 is the price we pay for ensuring reusability and generalizability. Especially in the long run, these benefits easily compensate for the extra computational expense. The v^n factor is unavoidable if one wants to investigate n'th order interaction effects between parameters. Most existing work does not study effects higher than the second order.

Finally, experiments could in fact be designed in a better way than just randomly generating parameter values. For instance, one could look at techniques from active learning or Optimal Experiment Design (OED) [2] to focus on the most interesting experiments given the outcome of previous experiments.

---

[4] These are the experiments that seem most interesting in the studied context, given the available resources.

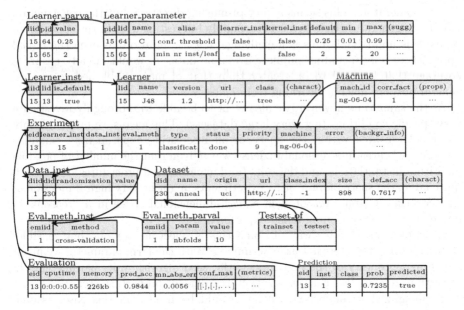

Fig. 1. A possible implementation of an experiment database

5   A Case Study

In this section we discuss one specific implementation of an experiment database. We describe the structure of this database and the experiments that populate it. Then, we illustrate its use with a few example queries. The experiment database is publicly available on `http://www.cs.kuleuven.be/~dtai/expdb`.

5.1   A Relational Experiment Database

We implemented an experiment database for classifiers in a standard RDBMS (MySQL), designed to allow queries about all aspects of the involved learning algorithms, datasets, experimental procedures and results. This leads to the database schema shown in Fig. 1. Central in the figure is a table of experiments listing the used *instantiations* of learning algorithms, datasets and evaluation methods, the experimental procedure, and the machine it was run on.

First, a *learner instantiation* points to a learning algorithm (**Learner**), which is described by the algorithm name, version number, a url where it can be downloaded and a list of characteristics. Furthermore, if an algorithm is parameterized, the parameter settings used in each learner instantiation (one of which is flagged as default) are stored in table **Learner_parval**. Because algorithms have different numbers and kinds of parameters, we store each parameter value assignment in a different row (in Fig. 1 only two are shown). The parameters are further described in table **Learner_parameter** with the learner it belongs to, its name and a specification of sensible values. If a parameter's value points to a learner instantiation (as occurs in ensemble algorithms) this is indicated.

Secondly, the used dataset, which can be instantiated with a randomization of the order of its attributes or examples (e.g. for incremental learners), is described in table Dataset by its name, download url(s), the index of the class attribute and 56 characterization metrics, most of which are mentioned in [9]. Information on the origin of the dataset can also be stored (e.g. whether it was taken from a repository or how it was preprocessed or generated).

Finally, we must store an evaluation of the experiments. The evaluation method (e.g. cross-validation) is stored together with its (list of) parameters (e.g. the number of folds). If a dataset is divided into a training set and a test set, this is defined in table Testset_of. The results of the evaluation of each experiment is described in table Evaluation by a wide range of evaluation metrics for classification, including the contingency tables[5]. The last table in Fig. 1 stores the (non-zero probability) predictions returned by each experiment.

5.2   Populating the Database

To populate the database, we first selected 54 classification algorithms from the WEKA platform[14] and inserted them together with all their parameters. Also, 86 commonly used classification datasets were taken from the UCI repository and inserted together with their calculated characteristics[6].

To generate a sample of classification experiments that covers a wide range of conditions, while also allowing to test the performance of some algorithms under very specific conditions, a number of algorithms were explored more thoroughly than others. In a first series of experiments, we ran all experiments with their default parameter settings on all datasets. In a second series, we defined at most 20 suggested values for the most important parameters of the algorithms SMO, MultilayerPerceptron, J48 (C4.5), 1R and Random Forests. We then varied each of these parameters one by one, while keeping all other parameters at default. In a final series, we defined sensible ranges for all parameters of the algorithms J48 and 1R, and selected random parameter settings (thus fully exploring their parameter spaces) until we had about 1000 experiments of each algorithm on each dataset. For all randomized algorithms, each experiment was repeated 20 times with different random seeds. All experiments (about 250,000 in total) where evaluated with 10-fold cross-validation, using the same folds on each dataset.

5.3   Querying and Mining

We will now illustrate how easy it is to use this experiment database to test a wide range of hypotheses on the behavior of these learning algorithms by simply writing the right queries and interpreting the results, or by applying data mining

---

[5] To help compare cpu times, a diagnostic test might be run on each machine and its relative speed stored as part of the machine description.

[6] As the database stores a 'standard' description of the experiments, other algorithm (implementations) or datasets can be used just as easily.

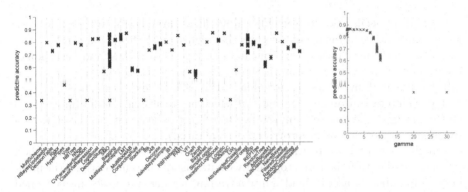

Fig. 2. Performance comparison of all algorithms on the `waveform-5000` dataset

Fig. 3. Impact of the γ-parameter on SMO

algorithms to model more complex interactions. In a first query, we compare the performance of all algorithms on a specific dataset:

```
SELECT l.name, v.pred_acc
FROM experiment e, learner_inst li, learner l, data_inst di, dataset d,
evaluation v
WHERE e.learner_inst = li.liid and li.lid = l.lid and e.data_inst =
di.diid and di.did = d.did and d.name='waveform-5000' and v.eid = e.eid
```

In this query, we select the algorithm used and the predictive accuracy registered in all experiments on dataset `waveform-5000`. We visualize the returned data in Fig. 2, which shows that most algorithms reach over 75% accuracy, although a few do much worse. Some do not surpass the default accuracy of 34%: besides SMO and ZeroR, these are ensemble methods that use ZeroR by default.

It is also immediately clear how much the performance of these algorithms varies as we change their parameter settings, which illustrates the generality of the returned results. SMO varies a lot (from default accuracy up to 87%), while J48 and (to a lesser extent) MultiLayerPerceptron are much more stable in this respect. The performance of RandomForest (and to a lesser extent that of SMO) seems to jump at certain points, which is likely bound to a different parameter value. These are all hypotheses we can now test by querying further.

For instance, we could examine which bad parameter setting causes SMO to drop to default accuracy. After some querying, a clear explanation is found by selecting the predictive accuracy and the gamma-value (kernel width) of the RBF kernel from all experiments with algorithm SMO and dataset `waveform-5000` and plotting them (Fig. 3). We see that accuracy drops sharply when the gamma value is set too high, and while the other modified parameters cause some variation, it is not enough to jeopardize the generality of the trend.

We can also investigate combined effects of dataset characteristics and parameter settings. For instance, we can test whether the performance 'jumps' of RandomForest are linked to the number of trees in a forest and the dataset size. Therefore, we select the dataset name and number of examples, the parameter value of the parameter named **nb of trees in forest** of algorithm **RandomForest**

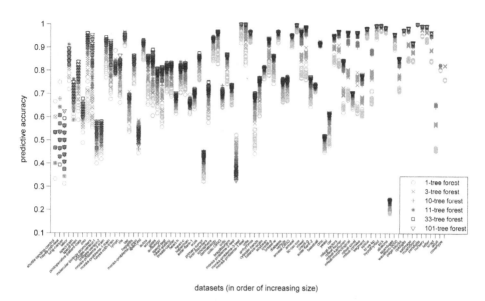

Fig. 4. The effect of dataset size and the number of trees for random forests

and the corresponding predictive accuracy. The results are returned in order of dataset size:

```
SELECT d.name, d.nr_examples, lv.value, v.pred_acc
FROM experiment e, learner_inst li, learner l, learner_parval lv,
learner_parameter p, data_inst di, dataset d, evaluation v
WHERE  e.learner_inst = li.liid and li.lid = l.lid and
l.name='RandomForest' and lv.liid = li.liid and lv.pid = p.pid and
p.alias='nb of trees in forest' and v.eid = e.eid
ORDER BY d.nr_examples
```

When plotted in Fig. 4, this clearly shows that predictive accuracy increases with the number of trees, usually leveling off between 33 and 101 trees, but with one exception: on the `monks-problems-2_test` dataset the base learner performs so badly (less than 50% accuracy, though there are only two classes) that the ensemble just performs worse when more trees are included. We also see that as the dataset size grows, the accuracies for a given forest size vary less, which is indeed what we would expect as trees become more stable on large datasets.

As said before, an experiment database can also be useful to verify or refine existing knowledge. To illustrate this, we verify the result of Holte [5] that very simple classification rules (like 1R) perform almost as good as complex ones (like C4, a predecessor of C4.5) on most datasets. We compare the average predictive performance (over experiments using default parameters) of J48 with that of OneR for each dataset. We skip the query as it is quite complex. Plotting the average performance of the two algorithms against each other yields Fig. 5.

We see that J48 almost consistently outperforms OneR, in many cases performing a little bit better, and in some cases much better. This is not essentially

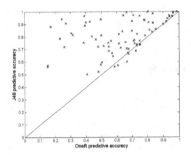
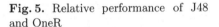

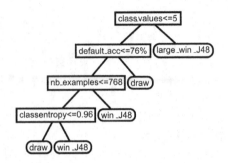

Fig. 5. Relative performance of J48 and OneR

Fig. 6. A meta-decision tree on dataset characteristics

different from Holte's results, though the average improvement does seem a bit larger here (which may indicate an improvement in decision tree learners and/or a shift towards more complex datasets).

We can also automatically learn under which conditions J48 clearly outperforms OneR. To do this, we queried for the difference in predictive accuracy between J48 and OneR for each dataset, together with all dataset characteristics. Discretizing the predictive accuracy yields a classification problem with 3 class values: "draw", "win_J48" (4% to 20% gain), and "large_win_J48" (20% to 70% gain). The tree returned by J48 on this meta-dataset is shown in Fig. 6, showing that a high number of class values often leads to a large win of J48 over 1R. Interestingly, Holte's study contained only one dataset with more than 5 class values, which might explain why smaller accuracy differences were reported.

Yet these queries only scratched the surface of all possible hypotheses that can be tested using the experiments generated for this case study. One could easily launch new queries to request the results of certain experiments, and gain further insights into the behavior of the algorithms. Also, one can reuse this data (possibly augmented with further experiments) when researching the covered learning techniques. Finally, one can also use our database implementation to set up other experiment databases, e.g. for regression or clustering problems.

6   Conclusions

We advocate the use of experiment databases in machine learning research. Combined with the current methodology, experiment databases foster repeatability. Combined with a new methodology that consists of running many more experiments in a semi-automated fashion, storing them all in an experiment database, and then querying that database, experiment databases in addition foster reusability, generalizability, and easy and thorough analysis of experimental results. Furthermore, as these databases can be put online, they provide a detailed log of performed experiments, and a repository of experimental results that can be used to obtain new insights. As such, they have the potential to speed up future research and at the same time make it more reliable, especially when supported by the development of good experimentational tools. We have discussed the construction of experiment databases, and demonstrated the feasibility and merits of this approach by presenting an publicly available experiment database containing 250,000 experiments and illustrating its use.

Acknowledgements

Hendrik Blockeel is Postdoctoral Fellow of the Fund for Scientific Research - Flanders (Belgium) (F.W.O.-Vlaanderen), and this research is further supported by GOA 2003/08 "Inductive Knowledge Bases".

References

1. Blockeel, H.: Experiment databases: A novel methodology for experimental research. In: Bonchi, F., Boulicaut, J.-F. (eds.) Knowledge Discovery in Inductive Databases. LNCS, vol. 3933, pp. 72–85. Springer, Heidelberg (2006)
2. Cohn, D.A.: Neural Network Exploration Using Optimal Experiment Design. Advances in Neural Information Processing Systems 6, 679–686 (1994)
3. Demšar, J.: Statistical Comparisons of Classifiers over Multiple Data Sets. Journal of Machine Learning Research 7, 1–30 (2006)
4. Fromont, E., Blockeel, H., Struyf, J.: Integrating Decision Tree Learning into Inductive Databases. In: Bonchi, F., Boulicaut, J.-F. (eds.) Knowledge Discovery in Inductive Databases. LNCS, vol. 3933, Springer, Heidelberg (2006)
5. Holte, R.: Very simple classification rules perform well on most commonly used datasets. Machine Learning 11, 63–91 (1993)
6. Hoste, V., Daelemans, W.: Comparing Learning Approaches to Coreference Resolution. There is More to it Than 'Bias'. In: Proceedings of the Workshop on Meta-Learning (ICML- 2005) pp. 20–27 (2005)
7. Kalousis, A., Hilario, M.: Building Algorithm Profiles for prior Model Selection in Knowledge Discovery Systems. Engineering Intelligent Systems 8(2) (2000)
8. Keogh, E., Kasetty, S.: On the Need for Time Series Data Mining Benchmarks: A Survey and Empirical Demonstration. In: Proceedings of the 8th ACM SIGKDD International Conference on Knowledge Discovery and Data Mining, pp. 102–111. ACM Press, New York (2002)
9. Peng, Y., et al.: Improved Dataset Characterisation for Meta-Learning. In: Lange, S., Satoh, K., Smith, C.H. (eds.) DS 2002. LNCS, vol. 2534, pp. 141–152. Springer, Heidelberg (2002)
10. Perlich, C., Provost, F., Siminoff, J.: Tree induction vs. logistic regression: A learning curve analysis. Journal of Machine Learning Research 4, 211–255 (2003)
11. METAL-consortium: METAL Data Mining Advisor, http://www.metal-kdd.org
12. Michie, D., Spiegelhalter, D.J., Taylor, C.C.: Machine Learning, Neural and Statistical Classification. Ellis Horwood, New York (1994)
13. Van Someren, M.: Model Class Selection and Construction: Beyond the Procrustean Approach to Machine Learning Applications. In: Paliouras, G., Karkaletsis, V., Spyropoulos, C.D. (eds.) Machine Learning and Its Applications. LNCS (LNAI), vol. 2049, pp. 196–217. Springer, Heidelberg (2001)
14. Witten, I.H., Frank, E.: Data Mining: Practical Machine Learning Tools and Techniques, 2nd edn. Morgan Kaufmann, San Francisco (2005)
15. Wolpert, D., Macready, W.: No free lunch theorems for search. SFI-TR-95-02-010 Santa Fe Institute (1995)

Using the Web to Reduce Data Sparseness in Pattern-Based Information Extraction

Sebastian Blohm and Philipp Cimiano

Institute AIFB, University of Karlsruhe, Germany
{blohm,cimiano}@aifb.uni-karlsruhe.de

Abstract. Textual patterns have been used effectively to extract information from large text collections. However they rely heavily on textual redundancy in the sense that facts have to be mentioned in a similar manner in order to be generalized to a textual pattern. Data sparseness thus becomes a problem when trying to extract information from hardly redundant sources like corporate intranets, encyclopedic works or scientific databases.

We present results on applying a weakly supervised pattern induction algorithm to Wikipedia to extract instances of arbitrary relations. In particular, we apply different configurations of a basic algorithm for pattern induction on seven different datasets. We show that the lack of redundancy leads to the need of a large amount of training data but that integrating Web extraction into the process leads to a significant reduction of required training data while maintaining the accuracy of Wikipedia. In particular we show that, though the use of the Web can have similar effects as produced by increasing the number of seeds, it leads overall to better results. Our approach thus allows to combine advantages of two sources: The high reliability of a closed corpus and the high redundancy of the Web.

1   Introduction

Techniques for automatic information extraction (IE) from text play a crucial role in all scenarios in which manually scanning texts for certain information is unfeasible or too costly. Nowadays, information extraction is thus for example applied on biochemical texts to discover unknown interactions between proteins (compare [13]) or to texts available in corporate intranets for the purpose of knowledge management (compare [16]). In many state-of-the-art systems, textual patterns are used to extract the relevant information. Textual patterns are in essence regular expressions defined over different levels of linguistic analysis. In our approach, we rely on simple regular expressions defined over string tokens. As the extraction systems should be easily adaptable to any domain and scenario, considerable research has been devoted to the automatic induction of patterns (compare [5,14,7]). Due to the fact that patterns are typically induced from a specific corpus, any such approach is of course affected by the problem of data sparseness, i.e. the problem that there will never be enough data to learn all relevant patterns. In the computational linguistics community, it has been shown that the Web can in some cases be effectively used to overcome data sparseness problems (compare [9]).

In this paper, we explore whether the Web can effectively help to overcome data sparseness as a supplementary data source for information extraction on limited corpora. In particular we build on a weakly-supervised pattern learning approach in which

J.N. Kok et al. (Eds.): PKDD 2007, LNAI 4702, pp. 18–29, 2007.

patterns are derived on the basis of a few seed examples. A bootstrapping approach then induces patterns, matches these on the corpus to extract new tuples and then alternates this process over several iterations. Such an approach has been investigated before and either applied only to the Web (see [3,1]) or only to a given (local) corpus [11]. We thus combine advantages of two sources: the high reliability of a closed corpus and the high redundancy of the Web.

The idea is as follows: given seed examples (e.g. $(Warsaw, Poland)$ and $(Paris, France)$) of a specific relation (e.g. *locatedIn*) to be extracted (appearing in the local corpus), we can consult the Web for patterns in which these examples appear. The newly derived patterns, which in essence are a generalization of plain string occurrences of the tuples, can then be matched on the Web in order to extract new examples which are taken into the next iteration as seeds. Then, we can search for patterns for the increased set of examples (coming from the Web) in the corpus, thus effectively leading to more patterns. Overall, we experiment with different variations of the basic pattern induction algorithm on seven different relation datasets. Our experiments show on the one hand that lack of redundancy can be definitely compensated by increasing the number of seeds provided to the system. On the other hand, the usage of the Web yields even better results and does not rely on the provision of more training data to the system in the form of seeds.

In this paper, we use Wikipedia[1] as local corpus and access the Web through the Google API. In the next Section, we motivate the need for an approach to overcome data sparseness both quantitatively and qualitatively by giving some examples. Then, in Section 3 we present our bootstrapping approach to pattern induction which alternates the usage of the (local) corpus and the Web and its implementation in our *Pronto* system. Section 4 presents our experiments and results. Before concluding, we discuss some related work in Section 5.

2 Motivation

Specialized text corpora such as company intranets or collections of scientific papers are non-redundant by design. Yet they constitute a valuable source for information extraction as they are typically more reliable and focussed than the general Web (cf. [8] for an analysis of structure and content of corporate intranets).

In our present experiments we use Wikipedia as a non-redundant, highly reliable and (somewhat) specialized text collection of limited size that is freely accessible to the entire community. As a first observation, we found that Wikipedia is hardly redundant. We computed the number of page co-occurrences of instances of four test relations taking the Google result count estimates for searches of individual relation instances limited to the Wikipedia site. As a result we found that most relation instances do not co-occur more than 100 times (median: 15). When doing the same counts on the entire Web, hardly any instance occurs less than 100 times, the median lies at 48000. The effect increases when considering that page co-occurrence does not suffice for a relation instance to be extracted. Patterns only match a limited context. In our case, we match 10 tokens around each link relating it to the document title. This reduces the number

---

[1] http://en.wikipedia.org

of times, a candidate relation instance occurs in the corpus dramatically to an average of 1.68 (derived by counting the number of times that the top 200 relation instances for each relation occur in the index).

It is thus our goal to assess how effectively Web-based extraction can serve as background knowledge to extraction on a smaller corpus. That is, we will not use the Web to extract additional information, but only to make up for a lack of redundancy in the small corpus. In particular no information found on the Web goes into the result set without being verified on the small corpus as otherwise the benefits of the smaller corpus (higher quality, domain specificity, availability of further background knowledge) would be lost. In what follows, we describe the approach in more detail.

3   Approach

Our Pronto system uses a generic pattern learning algorithm as it is typically applied on the Web. It works analogously to many of the approaches mentioned in the introduction implementing a similar bootstrapping-based procedure. The Pronto system has been previously described in further detail [2]. The algorithm starts with a set of initial tuples S' of the relation in question – so called *seeds* – and loops over a procedure which starts by acquiring occurrences of the tuples currently in S. Further, patterns are learned by abstracting over the text occurrences of the tuples. The new patterns are then evaluated and filtered before they are matched. From these matches, new tuples are extracted, evaluated and filtered. The process is stopped when the termination condition DONE is fulfilled (typically, a fixed number of iterations is set). The learning is thus inductive in nature abstracting over individual positive examples in a bottom-up manner. Learning essentially takes place in a generate-and-test manner.

Figure 1 describes our modification of the algorithm. It basically consists of a subsequent application of the loop body on the Web and the wiki. Web matching and wiki matching contribute to the same evolving set of tuples S but maintain separate pattern pools P_{web} and P_{wiki}. This separation is done to allow for different types of pattern representation for the different corpora.

An important novelty is checking each tuple t derived from the Web using PRESENT-IN-WIKI(t). This ensures that no knowledge that is actually not present in Wikipedia goes into the set of results. Otherwise, the extraction procedure would not be able to benefit from the higher quality in terms of precision that the wiki corpus can be assumed to present.

3.1   Extraction from the Web

Given a number of seeds at the start of each of the algorithm's iterations, occurrences of these seed tuples are searched on the Web. For example, given a tuple $(Stockholm, Sweden)$ for the *locatedIn* relation, the following query would be sent to the Google Web Search API:

```
"Stockholm" "Sweden"
```

For each instance of the *locatedIn* relation a fixed number $num_{matchTuples_{web}}$ of results is retrieved for a maximum of $num_{tupleLimit_{web}}$ instances. These occurrences

WEB-WIKI PATTERN INDUCTION($Patterns\,P', Tuples\,S'$)

```
 1   S ← S′
 2   P_web ← P′
 3   while not DONE
 4   do
 5       Occ_t ← WEB-MATCH-TUPLES(S)
 6       P_web ← P_web ∪ LEARN-PATTERNS(Occ_t)
 7       EVALUATE-WEB-PATTERNS(P_web)
 8       P_web ← {p ∈ P_web | WEB-PATTERN-FILTER-CONDITION(p)}
 9       Occ_p ← WEB-MATCH-PATTERNS(P_web)
10       S ← S + EXTRACT-TUPLES(Occ_p)
11       S ← {t ∈ S | PRESENT-IN-WIKI(t)}
12       EVALUATE-WEB-TUPLES(S)
13       S ← {t ∈ S | TUPLE-FILTER-CONDITION(t)}
14       Occ_t ← WIKI-MATCH-TUPLES(S)
15       P_wiki ← P_wiki ∪ LEARN-PATTERNS(Occ_t)
16       EVALUATE-WIKI-PATTERNS(P_wiki)
17       P_wiki ← {p ∈ P_wiki | WIKI-PATTERN-FILTER-CONDITION(p)}
18       Occ_p ← WIKI-MATCH-PATTERNS(P)
19       S ← S + EXTRACT-TUPLES(Occ_p)
20       EVALUATE-WIKI-TUPLES(S)
21       S ← {t ∈ S | TUPLE-FILTER-CONDITION(t)}
```

Fig. 1. Combined Web and wiki pattern induction algorithm starting with initial patterns P' and tuples S' maintaining two pattern pools P_{web} and P_{wiki}

serve as input to pattern learning if the arguments are at most $max_{argDist}$ tokens apart. For our experiments we chose $max_{argDist} = 4$, $num_{matchTuples_{web}} = 50$ and $num_{matchTuples_{web}} = 200$.

LEARN-PATTERNS generates more abstract versions of the patterns. We take a generate-and-test approach to learning. LEARN-PATTERNS produces a large amount of patterns by combining ("merging") sets of occurrences by keeping common tokens and replacing tokens in which the patterns differ by "*" wildcards. Thus, the generalization is effectively calculating the least general generalization (LGG) of two patterns as typically done in bottom-up ILP approaches (compare [10]).

To avoid too general patterns, a minimum number of non-wildcard tokens is enforced. To avoid too specific patterns, it is required that the merged occurrences reflect at least two different tuples.

EVALUATE-WEB-PATTERNS(P_{web}) assigns a confidence score to each pattern. The confidence score is derived as the number of different tuples from which the pattern has been derived through merging. This measure performs better than other strategies as shown in [2]. Evaluation is followed by filtering applying WEB-PATTERN-FILTER-CONDITION(p) which ensures that the top $pool_{web} = 50$ patterns are kept. Note that the patterns are kept over iterations but that old patterns compete against newly derived ones in each iteration.

EVALUATE-WEB-PATTERNS(P_{web}) matches the filtered pattern set on the Web retrieving $num_{matchPatterns_{web}}$ results per pattern. A pattern like "flights to ARG_1 , ARG_2 from ANY airport"

for the locatedIn relation would be translated into a Google-query as follows:

```
"flights to  *   * from * airport"
```

A subsequent more selective matching step enforces case and punctuation which are ignored by Google. All occurrences are stored in Occ_p from which EXTRACT-TUPLES(Occ_p) extracts the relevant relation instances by identifying what occurs at the positions of ARG_1 and ARG_2. For the present experiments we chose $num_{matchPatterns_{web}} = 200$.

The above-mentioned PRESENT-IN-WIKI(t) check ensures that Web extractions for which no corresponding link-title pair is present in the Wikipedia are eliminated. This way, the high quality of content of Wikipedia is used to filter Web results and only those instances are kept that could in principle have been extracted from Wikipedia. Yet, the Web results increase the yield of the extraction process.

All parameters employed have been determined through extensive initial tests.

3.2 Extraction from Wikipedia

This section presents our approach to pattern matching for relation extraction on Wikipedia. We describe pattern structure and index creation before going into detail on the individual steps of the algorithm in Figure 1.

For pattern matching on Wikipedia, we make use of the encyclopedic nature of the corpus by limiting focussing on pairs of hyperlinks and document titles. It is a common assumption when investigating the semantics in documents like Wikipedia (e.g. [17]) that key information on the entity described on a page p lies within the set of links on that page $l(p)$ and in particular that it is likely that there is a salient semantic relation between p and $p' \in l(p)$.

We therefore consider patterns consisting of the document title and a hyperlink within its context. The context of $2 * w$ tokens around the link is taken into account because we assume that this context is most indicative of the the nature of the semantic relation expressed between the entity described in the article and the one linked by the hyperlink. In addition, a flag is set to indicate whether the first or the second argument of the relation occurs in the title. Each token can be required to be equal to a particular string or hold a wildcard character. For our experiments we chose $w = 5$.

To allow for efficient matching of patterns and tuples we created an index of all hyperlinks within Wikipedia. To this end, we created a database table with one row for each title/link pair featuring one column for link, title and each context token position. The link was created from the Wiki-Syntax version of the document texts using a database dump from December 17th 2006. The table has over 42 Million records. We omitted another 2.3 Million entries for links lying within templates to maintain generality as templates are a special syntactic feature of Wikipedia that may not transfer to similar corpora. Tokenization has been done based on white space. Hyperlinks are considered one token. Punctuation characters and common sequences of punctuation characters as well as HTML markup sequences are considered separate tokens even if not separated by white space. HTML comments and templates were omitted.

Tuple Matching and Pattern Learning. For each of at most $num_{matchTuples_{wiki}} = 50$ tuples, WIKI-MATCH-TUPLES(S) sends two queries to the index. One for each possibility to map argument 1 and 2 to title and link. Like in the Web case there is a maximum limit for matches $num_{matchTuples_{wiki}} = 200$ but it is hardly ever enforced as virtually no tuple is mentioned more than three times as a link-title pair. The same LEARN-PATTERNS(Occ_t) method is applied as in the Web setting. Like in the Web setting, EVALUATE-WIKI-PATTERNS(P_{wiki}) takes into account the number of distinct tuples which participated in the creation of a pattern. Finally, WIKI-PATTERN-FILTER-CONDITION(p) retains the top $pool_{web} = 50$ patterns for matching.

Pattern Matching and Tuple Generation. WIKI-MATCH-PATTERNS(P) retrieves from the index a random sequence of $num_{matchPatterns_{wiki}} = 50$ matches of the pattern by selecting those entries for which the non-wildcard context tokens of the patterns are present in the correct positions. EXTRACT-TUPLES(Occ_p) then generates a tuple instance for each distinct title/link pair occurring in the selected index entries. EVALUATE-WIKI-TUPLES(S) and TUPLE-FILTER-CONDITION(t) are currently not enabled to maximize the yield from the wiki.

The termination condition DONE is currently implemented to terminate the processing after 10 iterations.

3.3 Summary

Extraction from both the Web and the wiki index follow the same basic procedure. Parameters have been adapted to the different levels of redundancy in the text collections. In addition, the pattern structure of the patterns have been chosen is different to allow link-title matches in the wiki and window co-occurrences for the Web. The PRESENT-IN-WIKI(t) check ensures that the Web only facilitates extraction but does not provide knowledge not present in the wiki.

4 Evaluation

The goal of this study is to show how information extraction from the Web can be used to improve extraction results on a smaller corpus, i.e. how extraction on a precise, specialized corpus can benefit from a noisy but redundant source. We do so by running our system in two configurations employing Web extraction and an additional baseline condition. As the assumption is that Web extraction can make up for the lack of redundancy which is particularly important in the beginning of the bootstrapping process, we compare how the different configurations behave when provided with smaller and bigger amounts of seed examples.

4.1 Datasets

For the selection of seed instances and for automatic evaluation of results, 7 data sets consisting of the extensions of relations have been created:

- *albumBy*: 19852 titles of music albums and their artists generated from the Wikipedia category "Albums by Artist".
- *bornInYear*: 172696 persons and their year of birth generated from the Wikipedia category "Births by Year".
- *currencyOf*: 221 countries and their official currency according to DAML export of the CIA World Fact Book[2]. Manual modifications were done to reflect the introduction of the Euro as official currency in many European countries.
- *headquarteredIn*: 14762 names of companies and the country they are based in generated from the Wikipedia category "Companies by Country".
- *locatedIn*: 34047 names of cities and the state and federal states they are located in generated from the Wikipedia category "Cities by Countries". Note that a considerable number of cities are contained in this data set with both their state and their federal state.
- *productOf*: 2650 vehicle product names and the brand names of their makers generated from the Wikipedia category "Vehicles by Brand".
- *teamOf*: 8307 soccer players and the national teams they were playing for between 1950 and 2006. [3]

It is important to note that also the Wikipedia collections have been compiled manually by authors who assigned the documents to the respective categories and checked by further community members. Thus, the datasets can be regarded to be of high quality. Further, due to the vast coverage of Wikipedia the extensions of the relations can be assumed to be relatively complete.

Most of the above described datasets have been obtained from Wikipedia by automatically resolving category membership with the help of the CatScan[4] Tool by Daniel Kinzler. CatScan was applied iteratively to also obtain members of sub-categories.

The data sets have been chosen to differ according to various dimensions, most notably in size. The *currencyOf* dataset, for example, is relatively small and constitutes a relation with clear boundaries. The other relations are likely not be reflected fully in the data sets.

Small samples (size 10, 50 and 100) of the datasets were taken as input seeds. With two exceptions[5], we took the number of in-links to the Wikipedia articles mentioned in each tuple as an indicator for their significance in the corpus and selected the top n samples with respect to the harmonic mean of these counts. Initial tests showed that taking prominent instances as seeds strongly increases the system performance over random seeds. It can be expected that in most real scenarios prominent seeds are available as they should be those best known to the users.

---

[2] http://www.daml.org/2001/12/factbook/

[3] This data set is a courtesy of the SmartWeb consortium (see http://www.smartweb-project.de/).

[4] http://meta.wikimedia.org/wiki/User:Duesentrieb/CatScan

[5] For cities we took the average living costs as an indicator to ensure that Athens Greece was ranked higher than Athens, New York. (Population would have skewed the sample towards Asian cities not prominently mentioned in the English Wikipedia.) For Albums we required titles to be at least 10 characters in length to discourage titles like "Heart" or "Friends"

4.2 Experimental Conditions

To assess the added value of Web extraction, we compare three configurations of the above algorithm.

Dual: Exactly as formalized in Figure 1, this condition iterates the bootstrapping performing both, Web and wiki extraction in every iteration.

Web once: The processing runs like in Figure 1 but the lines 5 to 12 are executed only in the first iteration. Thereby, the seed set is augmented once by a set of learned relation instances. After that, processing is left to Wikipedia extraction.

Wiki only: As a baseline condition, extraction is done on Wikipedia only. Thus line 5 to 12 in Figure 1 are omitted entirely.

Figure 1 is simplified in one respect. Initial tests revealed that performing the PRESENT-IN-WIKI(t) filter in every iteration was too strict so that bootstrapping was quenched. We therefore decided to apply the filter in every third iteration[6]. A considerable number of – also correct – instances were filtered out when applying the filter. Consequently we only present results after iteration 3, 6 and 9 for comparability reasons.

We performed extraction with each of the three configurations for 10 iterations while varying the size of the seed set. Presenting the 10, 50 and 100 most prominent relation instances as seed sets to test how the different configurations affect the system's ability to bootstrap the extraction process.

4.3 Evaluation Measures

In our experiments, we rely on the widely used P(recision) and R(ecall) measures to evaluate system output. These measures compute the ratio of correctly found instances to overall tuples extracted (precision) or all tuples to be found (recall).

As the fixed number of iterations in our experiments poses a fixed limit on the number of possible extractions we use a notion of *(R)elative (R)ecall* assuming maximally extracted number of tuples by any configuration in any iteration with the given relation. With $Y_r(i, m)$ being the Yield, i.e. number of extractions (correct and incorrect) at iteration i for relation r with method m and $p_r(i, m)$ the precision respectively, we can formalize relative recall as

$$RR_r(i, m) = \frac{Y_r(i, m) * P_r(i, m)}{\max_{i,m} Y_r(i, m)}$$

The F-measure (more precisely F_1-measure) is a combination of precision and recall by the harmonic mean.

4.4 Results

Figure 2 presents results of the extraction runs with the different configurations starting with seed sets of different sizes. The figures show precision, relative recall and

---

[6] As the filter is always applied to all tuples in S this does not lead to the presence of non-wiki patterns in the final results. Yet, the non-wiki patterns seem to help bootstrapping before they are eliminated.

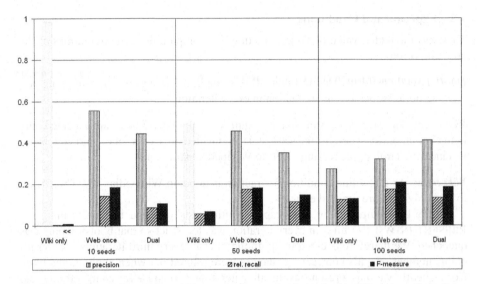

Fig. 2. Precision, relative recall and F-measure for results derived with different configurations and seed set sizes. Grayed columns are not very indicative as due to the low recall the results largely consist of the seed set. The mark $<<$ is to indicate that performance is statistically significantly worse than all other runs.

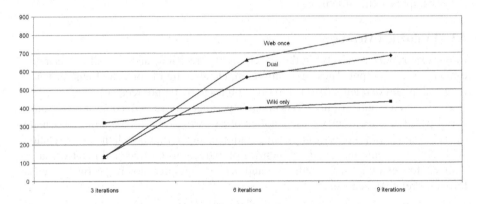

Fig. 3. Correct yield counts after 3, 6 and 9 iterations. Triangles mark Web once results, diamonds Dual and squares Wiki only. Strong lines indicate results with 50 seeds. Results with 10 seeds are higher 100 are lower.

F-measure after 9 iterations of the extraction algorithm. The scores are averaged over the performance on the seven relations from our testbed. Precision for the Web-supported configurations ranges between 0.32 and 0.55 depending on the configuration. We grayed the precision bars for the wiki only conditions with 10 and 50 seeds as the output contains largely seed tuples (95% for 10 seeds, 25% for 50 seeds) which accounts for the precision score.

We can observe that a purely wiki-based extraction performs very bad with 10 seeds and still far less from optimal with 50 seeds. A two-sided pairwise Student's t-test indicates in fact that the *Wiki only* strategy performs significantly worse than the other Web-based configurations at a seed set size of 10 ($\alpha = 0.05$) as well as for a seed set size of 50 ($\alpha = 0.1$). This clearly corroborates our claim that the integration of the Web improves results with respect to a Wiki-only strategy at 10 and 50 seeds.

Figure 3 shows the number of correctly extracted tuples averaged over the test relations after 3, 6 and 9 iterations. 50 seeds have been provided as training. In the wiki only configuration (square markers) the system is able to quickly derive a large number of instances but shows only slow increase of knowledge after iteration 3. The other configurations show a stronger incline between the iterations 3 and 9. This confirms the expected assumption that the low number of results when extracting solely from the wiki is due to an early convergence of the process. It is interesting to observe that the Web once condition slightly outperforms the Dual condition. This allows to assume that the major benefit of integrating the Web into the process lies in the initial extension of the seed set. Further investigation of this observation would require more iterations and further modifications of the configuration.

Overall, we can conclude that in this setting using the Web as background knowledge allows to produce more recall in hardly redundant corpora while maintaining the precision level. Yet, a larger seed set can also compensate for the lack of redundancy.

5 Related Work

The iterative induction of textual patterns is a method widely used in large-scale information extraction. Sergey Brin pioneered the use of Web search indices for this purpose [3]. Recent successful systems include KnowItAll which has been extended to automatic learning of patterns [7] and Espresso [11]. Espresso has been tested on the typical taxonomic *is-a* and *part-of* relations, but also the (political) *succession*, (chemical) *reaction* and *productOf* relations. Precision ranges between 49% and 85% for those relations. In a setup where it uses an algorithms similar to the one described above, KnowItAll is able to reach around 80% when limited to the task of named entity classification.

Apart from pattern-based approaches, a variety of supervised and semi-supervised classification algorithms has been applied to relation extraction. The methods include kernel-based methods [18,6] and graph-labeling techniques [4]. The advantage of such methods is that abstraction and partial matches are inherent features of the learning algorithm. In addition, kernels allow incorporating more complex structures like parse trees which cannot be reflected in text patterns. However, such classifiers require testing all possible relation instances while with text patterns extraction can be significantly speeded up using search indices. Classification thus requires linear-time processing of the corpus while search-patterns can lead to faster extraction.

In the present study, Wikipedia is used as a corpus. We used it to simulate an intranet scenario which shares with Wikipedia the properties of being more acurate, less spam-prone and much less redundant than the World Wide Web. Wikipedia is currently widely used as a corpus for information extraction from text. One example is a study by Suchanek et al. [15] who focus on high-precision ontology learning and population with methods specifically tailored to Wikipedia. Wikipedia's category system is exploited assuming typical namings and composition of categories that allow to deduce semantic relations from category membership. In [12] information extraction from Wikipedia text is done using hyperlinks as indicators for relations just like in the present study. As opposed to the work presented here it relies on WordNet as a hand-crafted formal taxonomy and is thus limited to relations for which such sources exist. Precision of 61-69% is achieved which is comparable to our results given the relative good extractability of the hyponomy and holonymy relations on which the tests have been performed.

6   Conclusion

The results we present here indicate that Web-based information extraction can help improving extraction results even if the task at hand requires extraction from a closed, non-redundant corpus. In particular, we showed that with extraction based on 10 seed examples and incorporating the Web as "background knowledge" better results could be achieved than using 100 seeds solely on Wikipedia. The potential of the approach lies in the fact that the additional information does not require formalization (like e.g. in WordNet) nor is it limited to a particular domain.

In future studies one can improve results by including additional techniques like part-of-speech tagging and named-entity tagging that have been omitted here to maintain generality of the study. In addition to the title-link pairs considered here, further indicators of relatedness can be considered to increase coverage.

We see applications of the derived results in domains like e-Science in particular fields in which research is focussed on some relations (e.g. protein interaction) and for which large non-redundant text collections are available.

Acknowledgements

The authors would like to thank Egon Stemle for technical assistance with our Wikipedia clone. This work was funded by the X-Media project (www.x-media-project.org) sponsored by the European Commission as part of the Information Society Technologies (IST) program under EC grant number IST-FP6-026978. Thanks to Google for giving enhanced access to their API.

References

1. Agichtein, E., Gravano, L.: Snowball: extracting relations from large plain-text collections. In: Proceedings of the fifth ACM conference on Digital Libraries (DL), pp. 85–94. ACM Press, New York (2000)
2. Blohm, S., Cimiano, P., Stemle, E.: Harvesting relations from the web -quantifiying the impact of filtering functions. In: Proceedings of the 22nd International Conference of the Association for the Advancement of Artificial Intelligence (AAAI) (to appear, 2007)
3. Brin, S.: Extracting patterns and relations from the world wide web. In: Schek, H.-J., Saltor, F., Ramos, I., Alonso, G. (eds.) EDBT 1998. LNCS, vol. 1377, Springer, Heidelberg (1998)

4. Chen, J., Ji, D., Tan, C.L., Niu, Z.: Relation extraction using label propagation based semi-supervised learning. In: Proceedings of the 21st International Conference on Computational Linguistics (COLING) and the 44th Annual Meeting of the Association for Computational Linguistics (ACL), pp. 129–136 (2006)
5. Ciravegna, F.: Adaptive information extraction from text by rule induction and generalisation. In: Proceedings of the International Joint Conference on Artificial Intelligence (IJCAI), pp. 1251–1256 (2001)
6. Culotta, A., Sorensen, J.: Dependency tree kernels for relation extraction. In: Proceedings of the 42nd Meeting of the Association for Computational Linguistics (ACL), pp. 423–429 (2004)
7. Downey, D., Etzioni, O., Soderland, S., Weld, D.: Learning text patterns for web information extraction and assessment. In: Proceedings of the AAAI Workshop on Adaptive Text Extraction and Mining (2004)
8. Fagin, R., Kumar, R., McCurley, K.S., Novak, J., Sivakumar, D., Tomlin, J.A., Williamson, D.P.: Searching the workplace web. In: Proceedings of the 12th International Conference on World Wide Web (WWW), pp. 366–375. ACM Press, New York (2003)
9. Kilgariff, A., Grefenstette, G.: Special Issue on the Web as a Corpus. Journal of Computational Linguistics 29 (2003)
10. Muggleton, S., Feng, C.: Efficient induction of logic programs. In: Proceedings of the 1st Conference on Algorithmic Learning Theory, pp. 368–381 (1990)
11. Pantel, P., Pennacchiotti, M.: Espresso: Leveraging generic patterns for automatically harvesting semantic relations. In: Proceedings of the 21st International Conference on Computational Linguistics (COLING) and the 44th Annual Meeting of the Association for Computational Linguistics (ACL), pp. 113–120 (2006)
12. Ruiz-Casado, M., Alfonseca, E., Castells, P.: Automatic extraction of semantic relationships for wordnet by means of pattern learning from wikipedia. In: Natural Language Processing and Information Systems, Springer, Berlin (2005)
13. Saric, J., Jensen, L., Ouzounova, R., Rojas, I., Bork, P.: Extraction of regulatory gene expression networks from pubmed. In: Proceedings of the Annual Meeting of the Association for Computational Linguistics (ACL), pp. 191–198 (2004)
14. Soderland, S.: Learning information extraction rules for semi-structured and free text. Machine Learning 34(1-3), 233–272 (1999)
15. Suchanek, F.M., Kasneci, G., Weikum, G.: Yago: A Core of Semantic Knowledge. In: Proceedings of the 16th International Conference on World Wide Web (WWW), pp. 697–706. ACM Press, New York (2007)
16. Uren, V., Cimiano, P., Iria, J., Handschuh, S., Vargas-Vera, M., Motta, E., Ciravegna, F.: Semantic annotation for knowledge management: Requirements and a survey of the state of the art. Journal of Web Semantics: Science, Services and Agents on the World Wide Web 4, 14–28 (2006)
17. Völkel, M., Krötzsch, M., Vrandecic, D., Haller, H., Studer, R.: Semantic wikipedia. In: Proceedings of the 15th International Conference on World Wide Web (WWW), pp. 585–594 (2006)
18. Zelenko, D., Aone, C., Richardella, A.: Kernel methods for relation extraction. Journal of Machine Learning Research 3, 1083–1106 (2003)

A Graphical Model for Content Based Image Suggestion and Feature Selection

Sabri Boutemedjet[1], Djemel Ziou[1], and Nizar Bouguila[2]

[1] Département d'informatique
Université de Sherbrooke, QC, Canada J1K 2R1
{sabri.boutemedjet,djemel.ziou}@usherbrooke.ca
[2] Concordia Institute for Information Systems Engineering
Concordia University, Montreal, QC, Canada H3G 1T7
bouguila@ciise.concordia.ca

Abstract. Content based image retrieval systems provide techniques for representing, indexing and searching images. They address only the user's short term needs expressed as queries. From the importance of the visual information in many applications such as advertisements and security, we motivate in this paper, the *Content Based Image Suggestion*. It targets the user's long term needs as a recommendation of products based on the user preferences in different situations, and on the visual content of images. We propose a generative model in which the visual features and users are clustered into separate classes. We identify the number of both user and image classes with the simultaneous selection of relevant visual features. The goal is to ensure an accurate prediction of ratings for multidimensional images. This model is learned using the minimum message length approach. Experiments with an image collection showed the merits of our approach.

1 Introduction

Information retrieval (IR) provides tools and techniques that help users to access, browse and summarize information stores efficiently. In the case of visual information, these techniques are addressed within content based image retrieval (CBIR) community. In retrieval, a user expresses the information need by formulating a search query generally in the form of image examples. The kind of information needs addressed in CBIR is short term. There is another kind of interests i.e. long term or permanent such as desires, tastes and preferences of each user. In today's e-market, products are described using both visual and textual information. From consumer psychology, the visual information has been recognized as an important factor that influences the consumer's decision making and has an important power of persuasion [18]. Indeed, images can convey meanings that cannot be expressed using words. Furthermore, it is well recognized [1] that the consumer choice is also influenced by the external environment or consumer's context defined by the time and location. For example, a consumer could express an information need during a travel that is different from the situation when she or he is working or even at home.

J.N. Kok et al. (Eds.): PKDD 2007, LNAI 4702, pp. 30–41, 2007.

In literature, user preferences are modeled within collaborative filtering (CF) and content based filtering (CBF) communities. CF approaches predict the relevance of a given product for the active user based on the preferences provided by a set of "like-minded" (similar tastes) users. Within the CF framework, each product is represented by its index considered as a categorical variable. The Aspect model [10] and the flexible mixture model (FMM) [24] are examples of some model-based CF approaches which involve the clustering as an underlying principle. The "correct" model order (number of parameters or clusters) was generally chosen "empirically" as a compromise between the model's complexity and the accuracy of recommendation. To the best of our knowledge, the issue of formally identifying the model order from the statistical properties of the data, was not addressed in CF literature. On the other hand, CBF approaches [19] [16], represent the user's profile using content descriptors and infer the relevance of unseen products based on the history of the active user. CBF approaches have targeted mainly textual data such as Web sites and newspapers [16]. Some hybrid approaches that combine CF and CBF [22] have been also proposed taking advantage of both methods.

In this paper, we motivate the *"Content Based Image Suggestion"* (CBIS). CBIS aims at the suggestion of products whose relevance is inferred from the history of users in different contexts on images of the previously consumed products. We try to make a direct "mapping" between products and their visual information described in terms of visual features and/or keywords extracted from images. In this work, we consider an image as a D-dimensional vector $v = (v_1, v_2, \ldots, v_D)$. The visual features may be local such as interest points [15] or global such as color, texture, or shape. The keywords can be automatically or semiautomatically extracted by annotation or recognition process. Therefore, text-based recommendations can be improved by capturing user preferences related to the added-value visual appearance of products. For example, figure 1 shows the list of products preferred by two users. Following a similar methodology in hybrid filtering approaches of text documents [22], the CBIS would consider the two users as "like-minded" since visually, they have preferred the same category of products ("motorbikes"). Then, the "camera" can be recommended to the user 2.

In order to predict the relevance of products for users in different contexts, we propose a probabilistic model which we call Visual Content Context-aware Flexible Mixture Model (VCC-FMM). In this model, users and visual documents are clustered separately into homogeneous groups as in FMM [24] except that images are grouped based on an additional visual information. The high dimensionality of visual documents v does not mean necessarily that the clustering structure is contained within the whole set of visual features. Indeed, it is common that high dimensional documents can be clustered based on an unknown set of few features. Moreover, the presence of many irrelevant features may deteriorate the performance of data modeling and increases the computational complexity [20]. The VCC-FMM defines the relevance of each visual feature as the degree of its dependence on class labels [26][21]. In literature [9], the process of feature selection in mixture models have not received as much attention as in supervised learning. The main reason is the absence of class labels that may guide the selection process in addition to the influence of the considered feature subset on the model order [7]. To address these issues, the VCC-FMM is learned from unlabeled data by minimizing a Minimum Message Length (MML) objective [27].

This paper is organized as follows. The next Section details the VCC-FMM model with an integrated feature selection. In Section 3, we discuss the identification of the model order using MML. Experimental results are presented in Section 4. Finally, we conclude this paper by a summary of the work.

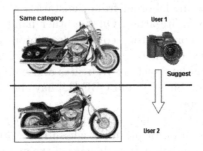

Fig. 1. The principle of Content Based Image Suggestion

2   The Visual Content Context Flexible Mixture Model

We consider a set of users $\mathcal{U} = \{1, 2, \ldots, N_u\}$, a set of visual documents $\mathcal{V} = \{v_1, v_2, \ldots, v_{N_v}\}$, and a set of possible contexts $\mathcal{E} = \{1, 2, \ldots, N_e\}$. We assume that a numeric rating r measures the relevance of a visual document for a given user and context. This rating r is defined on an ordered voting scale $\mathcal{R} = \{1, 2, \ldots, N_r\}$. First, we model the joint event $p(v, r, u, e)$ by equation (1) where two latent variables c and z label each observation $< u, e, v, r >$. The variable c carries the information about the visual similarity between images while the variable z denote "like-mindedness" of user preferences. The rating r for a given user u, context e and a visual document v can be predicted on the basis of probabilities $p(r|u, e, v)$ that can be derived by conditioning $p(u, e, v, r)$. The conditional independence assumptions among variables are illustrated by the graphical representation of the model in figure 2. The nodes denote random variables and edges (absence) denote conditional dependencies (independencies).

$$p(v, r, u, e) = \sum_{z=1}^{K} \sum_{c=1}^{M} p(z)p(c)p(u|z)p(e|z)p(v|c)p(r|z, c) \qquad (1)$$

where K and M denote the numbers of user classes and image classes, respectively. The quantities $p(z)$ and $p(c)$ denote the a priori weights of user and image classes. $p(u|z)$ and $p(e|z)$ denote the likelihood of a user and context to belong respectively to the user's class z. $p(r|z, c)$ is the probability to generate a rating for given user and image classes. We model $p(\boldsymbol{v}|c)$ using the Generalized Dirichlet distribution (GDD) [3][2] which is suitable for non Gaussian data such as images. This distribution has a more general covariance structure and provides multiple shapes. The distribution of the c-th component $\boldsymbol{\Theta}_c^*$ is given by:

$$p(\boldsymbol{v}|\boldsymbol{\Theta}_c^*) = \prod_{l=1}^{D} \frac{\Gamma(\alpha_{cl}^* + \beta_{cl}^*)}{\Gamma(\alpha_{cl}^*)\Gamma(\beta_{cl}^*)} v_l^{\alpha_{cl}^* - 1} (1 - \sum_{k=1}^{l} v_k)^{\gamma_{cl}^*} \qquad (2)$$

where $\sum_{l=l}^{D} v_l < 1$ and $0 < v_l < 1$ for $l = 1, \ldots, D$. $\gamma_{cl}^* = \beta_{cl}^* - \alpha_{cl+1}^* - \beta_{cl+1}^*$ for $l = 1, \ldots, D - 1$ and $\gamma_D^* = \beta_D^* - 1$. In equation (2) we have set $\boldsymbol{\Theta}_c^* = (\alpha_{c1}^*, \beta_{c1}^*, \ldots, \alpha_{cD}^*, \beta_{cD}^*)$. From the mathematical properties of the GDD, we can transform using a geometric transformation a data point $\boldsymbol{v}$ into another data point $\boldsymbol{x} = (x_1, \ldots, x_D)$ with independent features without loss of information [5][2]. In addition, each x_l of $\boldsymbol{x}$ generated by the c-th component, follows a Beta distribution $p_b(.|\theta_{cl}^*)$ with parameters $\theta_{cl}^* = (\alpha_{cl}^*, \beta_{cl}^*)$ which leads to the fact $p(\boldsymbol{x}|\boldsymbol{\Theta}_c^*) = \prod_{l=1}^{D} p_b(x_l|\theta_{cl}^*)$. Therefore, the estimation of the distribution of a D-dimensional GDD sample is indeed reduced to D-estimations of one-dimensional Beta distributions which is very interesting for multidimensional data sets. Since x_l are independent, we can extract *"relevant"* features in the representation space $\mathcal{X}$ as those that depend on class labels [26][21]. In other words, an irrelevant feature is independent of components θ_{cl}^* and follows another background distribution $p_b(.|\xi_l)$ common to all components. Let $\boldsymbol{\phi} = (\phi_1, \ldots, \phi_D)$ be a set of missing binary variables denoting the relevance of all features. ϕ_l is set to 1 when the l-th feature is relevant and 0 otherwise. The "ideal" Beta distribution θ_{cl}^* can be approximated as:

$$p(x_l|\theta_{cl}^*, \phi_l) \simeq \big(p_b(x_l|\theta_{cl})\big)^{\phi_l} \big(p_b(x_l|\xi_l)\big)^{1 - \phi_l} \qquad (3)$$

By considering each ϕ_l as Bernoulli variable with parameters $p(\phi_l = 1) = \epsilon_{l_1}$ and $p(\phi_l = 0) = \epsilon_{l_2}$ ($\epsilon_{l_1} + \epsilon_{l_2} = 1$) then, the distribution $p(x_l|\theta_{cl}^*)$ can be obtained after marginalizing over ϕ_l [14] as: $p(x_l|\theta_{cl}^*) \simeq \epsilon_{l_1} p_b(x_l|\theta_{cl}) + \epsilon_{l_2} p_b(x_l|\xi_l)$. The VCC-FMM model is given by equation (4). We notice that the work of [4] is special case of VCC-FMM.

$$p(\boldsymbol{x}, r, u, e) = \sum_{z=1}^{K} \sum_{c=1}^{M} p(z)p(u|z)p(e|z)p(c)p(r|z, c) \prod_{l=1}^{D} [\epsilon_{l_1} p_b(x_l|\theta_{cl}) + \epsilon_{l_2} p_b(x_l|\xi_l)]$$
$$(4)$$

3   Model Selection and Parameter Estimation Using MML

The variables U, E, R, Φ_l, Z and C are discrete and their distributions are assumed multinomial. We employ the following notation to simplify the presentation. The parameter vector of the multinomial distribution of a discrete variable A conditioned on its parent Π (predecessor) is denoted by θ_π^A (i.e.

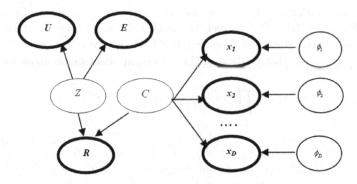

Fig. 2. Graphical representation of VCC-FMM

$A|_{\Pi=\pi} \sim Multi(1; \boldsymbol{\theta}_\pi^A))$ where $\theta_{\pi a}^A = p(A = a|\Pi = \pi)$ and $\sum_a \theta_{\pi a}^A = 1$. We have to estimate Θ defined by the parameters of multinomial distributions $\boldsymbol{\theta}_z^U, \boldsymbol{\theta}_z^E, \boldsymbol{\theta}^Z, \boldsymbol{\theta}^C, \boldsymbol{\theta}_z^R, \boldsymbol{\theta}_c^R, \boldsymbol{\theta}^{\phi_l}$ and the parameters of Beta distributions θ_{cl}, ξ_l. We employ the superscripts θ and ξ to denote the parameters of relevant and irrelevant Beta components, respectively $\left(\text{i.e.}\quad \theta_{cl} = (\alpha_{cl}^\theta, \beta_{cl}^\theta) \text{ and } \xi_l = (\alpha_l^\xi, \beta_l^\xi)\right)$. The log-likelihood of a data set of N independent and identically distributed observations $\mathcal{D} = \{< u^{(i)}, e^{(i)}, \boldsymbol{x}^{(i)}, r^{(i)} > |i = 1, \dots, N, u^{(i)} \in \mathcal{U}, e^{(i)} \in \mathcal{E}, \boldsymbol{x}^{(i)} \in \mathcal{X}, r^{(i)} \in \mathcal{R}\}$ is given by:

$$\log p(\mathcal{D}|\Theta) = \sum_{i=1}^N \log \sum_{z=1}^K \sum_{c=1}^M p(z)p(c)p(u^{(i)}|z)p(e^{(i)}|z)p(r^{(i)}|z,c) \times$$
$$\prod_{l=1}^D [\epsilon_{l_1} p_b(x_l^{(i)}|\theta_{cl}) + \epsilon_{l_2} p_b(x_l^{(i)}|\xi_l)] \tag{5}$$

The standard Expectation-Maximization (EM) algorithm for maximum likelihood estimation requires a good initialization and the knowledge of both M and K to converge to a good local optimum. Since both M and K are unknown, we employ the MML approach [27] for both estimation of the parameters and identification of K and M. In MML, a penalty term is introduced to the objective of $\mathcal{D}$ to penalize complex models as:

$$MML(K, M) = -\log p(\Theta) + \frac{1}{2} \log |I(\Theta)| + \frac{s}{2}(1 + \log \frac{1}{12}) - \log p(\mathcal{D}|\Theta) \tag{6}$$

where $|I(\Theta)|$, $p(\Theta)$, and s denote the Fisher information, prior distribution and the total number of parameters, respectively. The Fisher information of a parameter is the expectation of the second derivatives with respect to the parameter of the minus log-likelihood. It is common sense to assume an independence among the different groups of parameters which eases the computation of $|I(\Theta)|$ and $p(\Theta)$. Therefore, the joint prior is given by:

$$p(\Theta) = p(\boldsymbol{\theta}^Z)p(\boldsymbol{\theta}^C)[\prod_{z=1}^K p(\boldsymbol{\theta}_z^U)p(\boldsymbol{\theta}_z^E)p(\boldsymbol{\theta}_z^R)]\left(\prod_{l=1}^D [p(\xi_l)p(\epsilon_l)\prod_{c=1}^M p(\theta_{cl})]\right)\prod_{c=1}^M p(\boldsymbol{\theta}_c^R) \tag{7}$$

Besides, the Fisher information matrix is bloc-diagonal [8] which leads to

$$|I(\Theta)| = |I(\boldsymbol{\theta}^Z)||I(\boldsymbol{\theta}^C)| \prod_{c=1}^{M} |I(\boldsymbol{\theta}_c^R)| \left(\prod_{z=1}^{K} |I(\boldsymbol{\theta}_z^U)||I(\boldsymbol{\theta}_z^E)||I(\boldsymbol{\theta}_z^R)| \right) \left(\prod_{l=1}^{D} |I(\xi_l)| \right.$$

$$\left. |I(\epsilon_l)| \prod_{c=1}^{M} |I(\theta_{cl})| \right).$$ We approximate the Fisher information of VCC-FMM

from the complete likelihood which assumes the knowledge of z and c associated to each observation $< u^{(i)}, e^{(i)}, \boldsymbol{x}^{(i)}, r^{(i)} > \in \mathcal{D}$. The Fisher information of the parameters of multinomial distributions can be computed using the result found in [13]. Indeed, if the discrete variable A conditioned on its parent Π, has N_A different values $\{1, 2, \ldots, N_A\}$ in a data set of N observations, then $|I(\boldsymbol{\theta}_\pi^A)| = [(Np(\pi))^{N_A-1}]/[\prod_{a=1}^{N_A} \theta_{\pi a}^A]$, where $p(\pi)$ is the marginal probability of the parent Π. The proposed configuration of VCC-FMM does not involve variable ancestors (parents of parents). Therefore, the marginal probabilities $p(\pi)$ are simply the parameters of the multinomial distribution of the parent variable. Thus,

$$|I(\boldsymbol{\theta}_z^R)| = \frac{(N\theta_z^Z)^{N_r-1}}{\prod_{r=1}^{N_r} \theta_{zr}^R}, \quad |I(\boldsymbol{\theta}_c^R)| = \frac{(N\theta_c^C)^{N_r-1}}{\prod_{r=1}^{N_r} \theta_{cr}^R}, \quad |I(\boldsymbol{\theta}^Z)| = \frac{N^{K-1}}{\prod_{z=1}^{K} \theta_z^Z}$$

$$|I(\boldsymbol{\theta}^C)| = \frac{N^{M-1}}{\prod_{c=1}^{M} \theta_c^C}, \quad |I(\boldsymbol{\theta}_z^U)| = \frac{(N\theta_z^Z)^{N_u-1}}{\prod_{u=1}^{N_u} \theta_{zu}^U} \tag{8}$$

$$|I(\boldsymbol{\theta}_z^E)| = \frac{(N\theta_z^Z)^{N_e-1}}{\prod_{e=1}^{N_e} \theta_{ze}^E}, \quad |I(\boldsymbol{\theta}^\phi)| = N(\epsilon_{l_1}\epsilon_{l_2})^{-1}$$

The Fisher information of θ_{cl} and ξ_l can be computed by considering the log-likelihood of each feature taken separately [3]. After the second order derivations of this log-likelihood, we obtain:

$$|I(\theta_{cl})| = (N\theta_c^C \epsilon_{l_1})^2 \left| \left(\psi'(\alpha_{cl}^\theta)\psi'(\beta_{cl}^\theta) - \psi'(\alpha_{cl}^\theta + \beta_{cl}^\theta)(\psi'(\alpha_{cl}^\theta) + \psi'(\beta_{cl}^\theta)) \right) \right|$$

$$|I(\xi_l)| = (N\epsilon_{l_2})^2 \left| \left(\psi'(\alpha_l^\xi)\psi'(\beta_l^\xi) - \psi'(\alpha_l^\xi + \beta_l^\xi)(\psi'(\alpha_l^\xi) + \psi'(\beta_l^\xi)) \right) \right| \tag{9}$$

where Ψ is the trigamma function. In the absence of any prior knowledge on the parameters, we use the Jeffrey's prior for different groups of parameters as the square root of their Fisher information e.g. $p(\boldsymbol{\theta}^Z) \propto \prod_{z=1}^{K} (\theta_z^Z)^{-1/2}$. Replacing $p(\Theta)$ and $I(\Theta)$ in (6), and after discarding the first order terms, the MML objective of a data set $\mathcal{D}$ controlled by VCC-FMM is given by:

$$MML(K, M) = \frac{N_p}{2} \log N + M \sum_{l=1}^{D} \log \epsilon_{l_1} + \sum_{l=1}^{D} \log \epsilon_{l_2} + \frac{1}{2} N_p^Z \sum_{z=1}^{K} \log \theta_z^Z$$

$$+ \frac{1}{2}(N_r - 1) \sum_{c=1}^{M} \log \theta_c^C - \log p(\mathcal{D}|\Theta) \tag{10}$$

with $N_p = 2D(M+1) + K(N_u + N_e + N_r - 1) + MN_r$ and $N_p^Z = N_r + N_u + N_e - 3$. For fixed values of K, M and D, the minimization of the MML objective with respect to Θ is equivalent to a maximum a posteriori (MAP) estimate with the following improper Dirichlet priors [14]:

$$p(\boldsymbol{\theta}^C) \propto \prod_{c=1}^{M} (\theta_c^C)^{-\frac{N_r-1}{2}}, \quad p(\boldsymbol{\theta}^Z) \propto \prod_{z=1}^{K} (\theta_z^Z)^{-\frac{N_p^Z}{2}}, \quad p(\epsilon_1,\ldots,\epsilon_D) \propto \prod_{l=1}^{D} \epsilon_{l_1}^{-M} \epsilon_{l_2}^{-1}$$

(11)

3.1 Estimation of Parameters

We optimize the MML of the data set using the EM algorithm in order to estimate the parameters. The EM algorithm alternates between two steps. In the E-step, the joint posterior probabilities of the latent variables given the observations are computed as:

$$a_{lzc}^{(i)} = p(\phi_l = 1, u^{(i)}, e^{(i)}, x_l^{(i)}, r^{(i)}|z, c, \hat{\Theta}) = \hat{\theta}_{zu^{(i)}}^U \hat{\theta}_{ze^{(i)}}^E \hat{\theta}_{zr^{(i)}}^R \hat{\theta}_{cr^{(i)}}^R \epsilon_{l_1} p(x_l^{(i)}|\hat{\theta}_{cl})$$

$$b_{lzc}^{(i)} = p(\phi_l = 0, u^{(i)}, e^{(i)}, x_l^{(i)}, r^{(i)}|z, c, \hat{\Theta}) = \hat{\theta}_{zu^{(i)}}^U \hat{\theta}_{ze^{(i)}}^E \hat{\theta}_{zr^{(i)}}^R \hat{\theta}_{cr^{(i)}}^R \epsilon_{l_2} p(x_l^{(i)}|\hat{\xi}_l)$$

$$Q_{zci} = p(z, c|u^{(i)}, e^{(i)}, \boldsymbol{x}^{(i)}, r^{(i)}, \hat{\Theta}) = \frac{\hat{\theta}_z^Z \hat{\theta}_c^C \prod_l (a_{lzc}^{(i)} + b_{lzc}^{(i)})}{\sum_{z,c} \hat{\theta}_z^Z \hat{\theta}_c^C \prod_l (a_{lzc}^{(i)} + b_{lzc}^{(i)})}$$

(12)

In the M-step, the parameters are updated using the following equations:

$$\hat{\theta}_z^Z = \frac{\max\left(\sum_i \sum_c Q_{zci} - \frac{N_p^Z}{2}, 0\right)}{\sum_z \max\left(\sum_i \sum_c Q_{zci} - \frac{N_p^Z}{2}, 0\right)}, \quad \hat{\theta}_c^C = \frac{\max\left(\sum_i \sum_z Q_{zci} - \frac{N_r-1}{2}, 0\right)}{\sum_c \max\left(\sum_i \sum_z Q_{zci} - \frac{N_r-1}{2}, 0\right)}$$

(13)

$$\hat{\theta}_{zu}^U = \frac{\sum_{i:u^{(i)}=u} \sum_c Q_{zci}}{N\hat{\theta}_z^Z}, \quad \hat{\theta}_{ze}^E = \frac{\sum_{i:e^{(i)}=e} \sum_c Q_{zci}}{N\hat{\theta}_z^Z} \quad \hat{\theta}_{cr}^R = \frac{\sum_{i:r^{(i)}=r} \sum_z Q_{zci}}{N\hat{\theta}_c^C}$$

(14)

$$\hat{\theta}_{zr}^R = \frac{\sum_{i:r^{(i)}=r} \sum_c Q_{zci}}{N\hat{\theta}_z^Z} \quad \frac{1}{\epsilon_{l_1}} = 1 + \frac{\max\left(\sum_{z,c,i} \frac{Q_{zci}\epsilon_{l_2} p_b(x_l^{(i)}|\xi_l)}{\epsilon_{l_1} p_b(x_l^{(i)}|\theta_{cl})+\epsilon_{l_2} p_b(x_l^{(i)}|\xi_l)} - 1, 0\right)}{\max\left(\sum_{z,c,i} \frac{Q_{zci}\epsilon_{l_1} p_b(X_{il}|\theta_{cl})}{\epsilon_{l_1} p_b(x_l^{(i)}|\theta_{cl})+\epsilon_{l_2} p_b(x_l^{(i)}|\xi_l)} - M, 0\right)}$$

(15)

The parameters of Beta distributions θ_{cl} and ξ_l are updated using the Fisher scoring method based on the first and second order derivatives of the MML objective [3]. In order to avoid unfavorable local optimums, we use the deterministic EM annealing [25].

The update formulas of θ_c^C, θ_z^Z and ϵ_{l_1} involve a pruning behavior of components and features by forcing some weights to go to zero. It should be stressed that the speed of component pruning for θ_c^C during the first few iterations of the EM algorithm, depends on the size of the rating scale. For a large rating scale, the EM algorithm tends to remove quickly more components θ_{cl} during the first few iterations since the penalty term $\frac{N_r-1}{2}$ is high. On the other hand, for small rating scales such as "accept" or "reject" patterns (i.e. $N_r = 2$), the model tends to maintain more classes (i.e. penalty $= 1/2$) to explain variable user ratings.

4   Experiments

We consider I-VCC-FMM and D-VCC-FMM as two variants of VCC-FMM where the visual features are represented in $\mathcal{V}$ and $\mathcal{X}$, respectively. By this way, we evaluate the impact on the prediction accuracy of the naive Bayes assumption among visual features. Two additional variants are also considered: V-FMM and V-GD-FMM. The former does not handle the contextual information and assumes θ_{ze}^E constant for all $e \in \mathcal{E}$. In the latter, feature selection is not considered by setting $\epsilon_{l_1} = 1$ and pruning uninformative components ξ_l for $l = 1, \ldots, D$.

4.1   Data Set

We have mounted an ASP.NET Web site with SQL Server database in order to collect ratings from 27 subjects who participated in the experiment (i.e. $N_u = 27$) during a period of two months. The participating subjects are graduate students in faculty of science. Subjects received periodically (twice a day) a list of three images on which they assign relevance degrees expressed on a five star rating scale (i.e. $N_r = 5$). We define the context as a combination of two attributes: location $\mathcal{L} = \{in-campus, out-campus\}$ and time as $\mathcal{T} = (weekday, weekend)$ i.e $N_e = 4$. After the period of rating's acquisition, a data set $\mathcal{D}$ of 11050 ratings is extracted from the SQL Server database (i.e. $N = 11050$). We have used a collection of 4775 (i.e. $N_v = 4775$) images collected in part from Washington University [1] and another part from collections of free photographs on the Internet. We have categorized manually this collection into 41 categories. For visual content characterization, we have employed both local and global descriptors. For local descriptors, we use the Scale Invariant Feature Transform (SIFT) to represent image patches. This descriptor has been used with success in object recognition and has provided the best performance for matching. Then, we cluster SIFT vectors using K-Means which provides a visual vocabulary as the set of cluster centers or keypoints. After that, we generate for each image a normalized histogram of frequencies of each keypoint ("bag of keypoints") [6]. We have found that a number of 500 keypoints provided a good clustering for our collection. For global descriptors, we used the color correlogram [11] for image texture representation, and the edge histogram descriptor [12]. The color correlogram is built by considering the spatial arrangement of colors in the image for four displacements. A visual feature vector is represented in a 540-dimensional space ($D = 500 + 9 * 4 + 4 = 540$). We subdivide the data set $\mathcal{D}$ many times into two parts: for training and validation. Then, we measure the accuracy of the rating's prediction by the Mean Absolute Error (MAE) which is the average of the absolute deviation between the actual ratings (validation part) and the predicted ones.

---

[1] http://www.cs.washington.edu/research/imagedatabase.

4.2   First Experiment: Evaluating the Influence of Model Order on the Prediction Accuracy

This experiment investigates the influence of the assumed model order defined by K and M on the prediction accuracy of both I-VCC-FMM and D-VCC-FMM. While the number of image classes is known in the case of our collection, however, the number of user classes are not known in first sight. To validate the approach on a ground truth data $\mathcal{D}_{GT}$, we build a data set from preferences P_1 and P_2 of two most dissimilar subjects. We compute the dissimilarity in preferences on the basis of Pearson correlation coefficients. We sample ratings for 100 simulated users from the preferences P_1 and P_2 on images of four image classes. For each user, we generate 80 ratings ($\sim$ 20 ratings per context). Therefore, the ground truth model order is $K^* = 2$ and $M^* = 4$. The choice of image classes is purely motivated by convenience of presentation. Indeed, similar performance was noticed on the whole collection. We learn both I-VCC-FMM and D-VCC-FMM using one half of $\mathcal{D}_{GT}$ using different choices of training and validation data. The model order defined by $M = 15$ and $K = 15$ is used to initialize the EM algorithm for each partitioning of $\mathcal{D}_{GT}$.

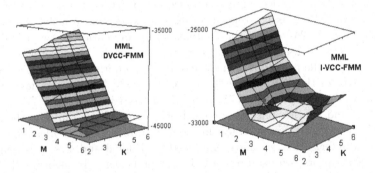

Fig. 3. MML criterion of the data set $\mathcal{D}_{GT}$ for D-VCC-FMM and I-VCC-FMM

Figure 3 shows that the MML approach has identified the correct number of user and image classes for both I-VCC-FMM and D-VCC-FMM on the synthetic data since the lowest MML was reported for the model order defined by $M = 4$ and $K = 2$. T The selection of the "correct" model order is important since it influences the accuracy of the prediction as illustrated by Figure 4. Furthermore, for $M > M^*$ the accuracy rating prediction is influenced more than the case of $K > K^*$. This experiment shows that the identification of the numbers of user and images classes is an important issue in CBIS.

4.3   Second Experiment: Comparison with State-of-the-Art

The aim of this experiment is to measure the contribution of the visual information and the user's context in making accurate predictions comparatively with some existing CF approaches. We make comparisons with the Aspect model [10],

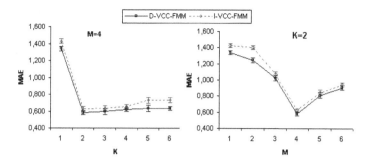

Fig. 4. Average MAE for different model orders

Pearson Correlation (PCC)[23], Flexible Mixture Model (FMM) [24], and User Rating Profile (URP) [17]. For accurate estimators, we learn the URP model using Gibs sampling. We retained for the previous algorithms, the model order that ensured the lowest MAE.

Table 1. Averaged MAE over 10 runs of the different algorithms on $\mathcal{D}$

	PCC(baseline)	Aspect	FMM	URP	V-FMM	V-GD-FMM	I-VCC	D-VCC
Avg MAE	1.327	1.201	1.145	1.116	0.890	0.754	0.712	0.645
Deviation	0.040	0.051	0.036	0.042	0.038	0.027	0.022	0.014
Improv.	0.00%	9.49%	13.71%	15.90%	32.94%	43.18%	51.62%	55.84%

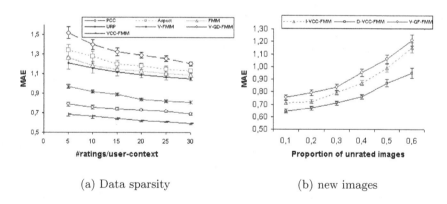

(a) Data sparsity   (b) new images

Fig. 5. MAE curves with error bars on the data set $\mathcal{D}$

The first five columns of table 1 show the added value provided by the visual information comparatively with pure CF techniques. For example, the improvement in the rating's prediction reported by V-FMM is 22.27% and 20.25% comparatively with FMM and URP, respectively. The algorithms (with context information) shown in the last three columns provide also an improvement (at least 15.28%) in the prediction accuracy comparatively with those which do not consider the context of the user. Also, we notice that feature selection is another important factor due to the improvement provided by I-VCC-FMM (5.57%) and D-VCC-FMM (14.45%) comparatively with V-GD-FMM. Furthermore, the naive Bayes assumption in I-VCC-FMM has increased (10.39%) MAE of D-VCC-FMM. In addition, it is reported in figure 5(a) that VCC-FMM is less sensitive to data sparsity (number of ratings per user) than pure CF techniques. Finally, the evolution of the average MAE provided VCC-FMM for different proportions of unrated images remains under 25% for up to 30% of unrated images as shown in Figure 5(b). We explain the stability of the accuracy of VCC-FMM for data sparsity and new images by the visual information since only cluster representatives need to be rated.

5   Conclusions

In this paper, we have motivated the content based image suggestion by modeling long term user needs to the visual information. We have proposed a graphical model by addressing two issues of unsupervised learning: the feature selection and the model order identification. Experiments showed the importance of the visual information and the user's context in making accurate predictions.

References

1. Belk, R.W.: Situational Variables and Consumer Behavior. Journal of Consumer Research 2, 157–164 (1975)
2. Bouguila, N., Ziou, D.: A Hybrid SEM Algorithm for High-Dimensional Unsupervised Learning Using a Finite Generalized Dirichlet Mixture. IEEE Trans. on Image Processing 15(9), 1785–1803 (2006)
3. Bouguila, N., Ziou, D.: High-dimensional unsupervised selection and estimation of a finite generalized dirichlet mixture model based on minimum message length. IEEE Trans. on PAMI (2007)
4. Boutemedjet, S., Ziou, D.: Content-based collaborative filtering model for scalable visual document recommendation. In: Proc. of IJCAI-2007 Workshop on Multimodal Information Retrieval (2007)
5. Connor, R.J., Mosimann, J.E.: Concepts of Independence for Proportions With a Generalization of the Dirichlet Distribution. Journal of the American Statistical Association 39, 1–38 (1977)
6. Csurka, G., Dance, C.R., Fan, L., Willamowski, J., Bray, C.: Visual categorization with bags of keypoints. In: Pajdla, T., Matas, J(G.) (eds.) ECCV 2004. LNCS, vol. 3024, Springer, Heidelberg (2004)
7. Dy, J.G., Brodley, C.E.: Feature selection for unsupervised learning. Journal of Machine Learning Research 5, 845–889 (2004)
8. Figueiredo, M.A.T., Jain, A.K.: Unsupervised learning of finite mixture models. IEEE Trans. on PAMI 24(3), 4–37 (2002)
9. Guyon, I., Elisseeff, A.: An Introduction to Variable and Feature Selection. Journal of Machine Learning Research 3, 1157–1182 (2003)
10. Hofmann, T.: Probabilistic Latent Semantic Indexing. In: Proc. of SIGIR (1999)

11. Huang, J., Kumar, S.R., Mitra, M., Zhu, W.J., Zabih, R.: Image indexing using color correlograms. In: Proc. of IEEE Conf, IEEE Computer Society Press, Los Alamitos (1997)
12. Jain, A., Vailaya, A.: Image retrieval using color and shape. Pattern Recognition 29(8), 1233–1244 (1996)
13. Kontkanen, P., Myllymki, P., Silander, T., Tirri, H., Grnwald, P.: On predictive distributions and bayesian networks. Statistics and Computing 10(1), 39–54 (2000)
14. Law, M.H.C., Figueiredo, M.A.T., Jain, A.K.: Simultaneous feature selection and clustering using mixture models. IEEE Trans. on PAMI, 26(9) (2004)
15. Lowe, D.G.: Distinctive image features from scale-invariant keypoints. International Journal of Computer Vision 60(2), 91–110 (2004)
16. Muramastsu, J., Pazzani, M., Billsus, D.: Syskill and Webert: Identifying Interesting Web Sites. In: Proc. of AAAI (1996)
17. Marlin, B.: Modeling User Rating Profiles For Collaborative Filtering. In: Proc. of NIPS (2003)
18. Messaris, P.: Visual Persuasion: The Role of Images in Advertising. Sage Pubns (1997)
19. Mooney, R.J., Roy, L.: Content-Based Book Recommending Using Learning for Text Categorization. In: Proc. 5th ACM Conf. Digital Libaries, ACM Press, New York (2000)
20. Ng, A.Y.: On feature selection: Learning with exponentially many irrelevant features as training examples. In: Proc. of ICML (1998)
21. Novovicova, J., Pudil, P., Kittler, J.: Divergence based feature selection for multi-modal class densities. IEEE Trans. on PAMI 18(2), 218–223 (1996)
22. Popescul, A., Ungar, L.H., Pennock, D.M., Lawrence, S.: Probabilistic Models for Unified Collaborative and Content-Based Recommendation in Sparse-Data Environments. In: Proc. of UAI (2001)
23. Resnick, P., Iacovou, N., Suchak, M., Bergstrom, P., Riedl, J.: Grouplens: An Open Architecture for Collaborative Filtering of Netnews. In: Proc. of ACM Conference on CSCW, ACM Press, New York (1994)
24. Si, L., Jin, R.: Flexible Mixture Model for Collaborative Filtering. In: Proc. of ICML, pp. 704–711 (2003)
25. Ueda, N., Nakano, R.: Deterministic Annealing EM Algorithm. Neural Networks 11(2), 271–282 (1998)
26. Vaithyanathan, S., Dom, B.: Generalized Model Selection for Unsupervised Learning in High Dimensions. In: Proc. of NIPS, pp. 970–976 (1999)
27. Wallace, C.: Statistical and Inductive Inference by Minimum Message Length. Information Science and Statistics. Springer, Heidelberg (2005)

Efficient AUC Optimization for Classification

Toon Calders[1] and Szymon Jaroszewicz[2]

[1] Eindhoven University of Technology, the Netherlands
[2] National Institute of Telecommunications, Warsaw, Poland

Abstract. In this paper we show an efficient method for inducing classifiers that directly optimize the area under the ROC curve. Recently, AUC gained importance in the classification community as a mean to compare the performance of classifiers. Because most classification methods do not optimize this measure directly, several classification learning methods are emerging that directly optimize the AUC. These methods, however, require many costly computations of the AUC, and hence, do not scale well to large datasets. In this paper, we develop a method to increase the efficiency of computing AUC based on a polynomial approximation of the AUC. As a proof of concept, the approximation is plugged into the construction of a scalable linear classifier that directly optimizes AUC using a gradient descent method. Experiments on real-life datasets show a high accuracy and efficiency of the polynomial approximation.

1 Introduction

In binary classification, often, the performance of classifiers is measured using the Area under the ROC Curve (AUC). Intuitively, the AUC of a classification function f expresses the probability that a randomly selected positive example gets a higher score by f than a randomly selected negative example. This measure has proven to be highly useful for evaluating classifiers, especially when class distributions are heavily skewed.

Recently, several new classifier training techniques have been developed that directly optimize the AUC. The main problem these algorithms face is that computing the AUC is a relatively costly operation: it requires sorting the database, a cost of order $n \log(n)$ for a database of size n. Also, in contrast to, e.g., the mean squared error, the AUC is not continuous on the training set, which makes the optimization task even more challenging. Therefore, often the algorithms optimize a slight variant of the AUC, that is differentiable. We denote this variant *soft-AUC*. The complexity of computing this soft-AUC, however, is even worse: it is of order n^2 for a database of size n. These high computational demands of AUC and soft-AUC seriously impact the scalability of these methods to large databases. Most of these algorithms therefore rely on sampling.

In this paper we present another option, namely the use of polynomial approximations for the AUC and the soft-AUC. The polynomial approximation has the advantage that it can be computed in only one scan over the database,

J.N. Kok et al. (Eds.): PKDD 2007, LNAI 4702, pp. 42–53, 2007.

and hence, it does not require resorting the database every time the AUC for a new or updated classification function is needed. Furthermore, when the classification function is only slightly changed, it is even possible to find the new AUC without a database scan, based on a small summary of the database.

We show experimentally that the polynomial approximation is very accurate and extremely efficient to compute; for soft-AUC, the traditional methods are already outperformed starting from a couple of hundred of tuples. Furthermore, the computation of the AUC can be plugged into all methods requiring repeated computations of the AUC. As a proof of concept, the approximation is plugged into a gradient descent method for training a linear classifier. It was implemented and tested on real-life datasets. With the approximation technique, similar AUC scores were reached as with existing techniques. The scalability and running times of the proposed approximation technique, however, were vastly superior.

To summarize, the main contribution of this paper is the development of an efficient procedure to approximate AUC and soft-AUC that scales very well with the size of the dataset. This method makes it possible to scale-up existing algorithms that optimize AUC directly.

2   Area Under the Curve and Classification

Consider the problem of assessing the quality of predictions for binary observations. Let $C(o)$ denote the class of an observation o. The *predicted* quantity might be a continuous quantity, e.g., a probability ranging from 0 to 1. This continuous quantity can be translated into a binary prediction by setting a threshold; if the predicted quantity is below the threshold, the result is a 0 prediction, otherwise, 1 is predicted. Depending on the threshold, there is a trade-off between precision and recall; on the one hand, if the threshold is low, recall of the 1-class will be high, but precision will be low, but on the other hand, if the threshold is high, precision will be high, but recall will be low. In order to characterize the quality of a predictor without fixing the threshold, *area under the ROC curve (AUC)* or its soft version *soft-AUC* can be used [2].

AUC. The AUC of a predictor f is defined as

$$AUC(f) \; := \; P(f(x) < f(y) | C(x) = 0, C(y) = 1) \; .$$

Given a set of negative examples $\mathcal{D}^0$, and a set of positive examples $\mathcal{D}^1$, the following *Wilcoxon-Man-Whitney* statistic [6], which we denote $auc(f, \mathcal{D}^0 \cup \mathcal{D}^1)$, is an unbiased estimator of $AUC(f)$:

$$auc(f, \mathcal{D}^0 \cup \mathcal{D}^1) \; := \; \frac{\sum_{t_0 \in \mathcal{D}^0} \sum_{t_1 \in \mathcal{D}^1} \mathbf{1}[f(t_0) < f(t_1)]}{|\mathcal{D}^0| \cdot |\mathcal{D}^1|} \; ,$$

where $\mathbf{1}[f(t_0) < f(t_1)]$ denotes the *indicator function* of $f(t_0) < f(t_1)$; that is, $\mathbf{1}[f(t_0) < f(t_1)]$ is 1 if $f(t_0) < f(t_1)$ is true, and otherwise it is 0.

Given a dataset $\mathcal{D} = \mathcal{D}^0 \cup \mathcal{D}^1$, the exact value of $auc(f, \mathcal{D})$ can be computed in time $\mathcal{O}(|\mathcal{D}| \log(|\mathcal{D}|))$ by sorting the tuples t in the database with respect to the value of $f(t)$ in ascending order, after which we scan the data and maintain a count of 0-examples which have the value of f less than the current tuple.

Soft-AUC. For some classification algorithms, such as gradient descent, however, it is problematic that the statistic $auc(f, \mathcal{D})$ is not continuous in f. Furthermore, another disadvantage of the AUC measure is that no weights are assigned to the difference in scores; $auc(f, \mathcal{D})$ fully takes into account a pair of a higher scoring positive example with a lower scoring negative example, even if the margin is small. Both problems: the non-differentiability and the insensitivity to the difference in scores between positive and negative examples, are solved by the introduction of the following *soft-AUC* statistic (parameterized by β):

$$s\_auc_\beta(f, \mathcal{D}^0 \cup \mathcal{D}^1) := \frac{\sum_{t_0 \in \mathcal{D}^0} \sum_{t_1 \in \mathcal{D}^1} sigmoid_\beta(f(t_1) - f(t_0))}{|\mathcal{D}^0| \cdot |\mathcal{D}^1|} ,$$

where $sigmoid_\beta(x)$ is the function $\frac{1}{1+e^{-\beta x}}$. This function approximates the step function, but smoothes out the region around 0. For $\beta \to \infty$, $sigmoid_\beta$ pointwise converges to the step function. The computational cost of the soft-AUC, however, is quadratic in the number of tuples. Similar measures have been introduced in the literature to deal with the non-differentiability issue. We believe, however, that soft-AUC has its own merits, and thus propose we it as a measure in its own right.

Optimizing the AUC Directly. Recently, many new classification algorithms have been proposed that directly optimize the AUC measure [7,3,1,5,10,9]. In these approaches, the AUC has to be computed repeatedly, as the classifier f is being changed during the training process. Because for large datasets $\mathcal{D}$, the cost of $\mathcal{O}(|\mathcal{D}| \log(|\mathcal{D}|))$ for every computation of the AUC-measure can be too high, it is often measured on only a small sample of the dataset.

In this paper, we propose another approach for optimizing the AUC directly. We propose the use of polynomial approximations of AUC and soft-AUC. These approximations have the advantage that they are more accurate than sampling, they can be computed in linear time, and it is possible to cache a concise summary of the dataset that allows, for small changes in the classification function, to compute the AUC without having to re-scan the dataset.

3   Polynomial Approximation of AUC and Soft-AUC

The key observation is that the indicator function $\mathbf{1}[f(t_0) < f(t_1)]$ (resp. sigmoid) can be approximated by a polynomial. We only give the derivation for the AUC, because the soft-AUC case is similar.

To approximate the indicator function, we actually approximate the function $H(x) = \mathbf{1}[x > 0]$, which is the well-known *Heaviside step-function*. The required indicator function is then $H(f(t_1) - f(t_0))$. In Figure 1, a polynomial (Chebyshev) approximation of $H(x)$ has been plotted.

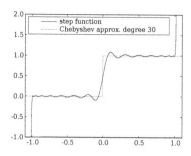

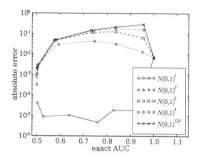

Fig. 1. Polynomial approximation of the Heaviside function

Fig. 2. Accuracy of the polynomial approximations

Let now $\sum_{k=0}^{d} c_k x^k$ be a polynomial approximation of $H(x)$ of degree d. Then,

$$H(f(t_1) - f(t_0)) \approx \sum_{k=0}^{d} c_k (f(t_1) - f(t_0))^k = \sum_{k=0}^{d} c_k \sum_{l=0}^{k} \binom{k}{l} f(t_1)^l (-f(t_0))^{k-l}$$

$$= \sum_{k=0}^{d} \sum_{l=0}^{k} \alpha_{kl} f(t_1)^l f(t_0)^{k-l}$$

where α_{kl} equals $c_k \binom{k}{l} (-1)^{k-l}$. This approximation of $H(x)$ leads directly to the following approximation for the *auc*. Let n_0 denote $|\mathcal{D}^0|$ and n_1 denote $|\mathcal{D}^1|$.

$$n_0 n_1 \, auc(f, \mathcal{D}) \approx \sum_{t_0 \in \mathcal{D}^0} \sum_{t_1 \in \mathcal{D}^1} \sum_{k=0}^{d} \sum_{l=0}^{k} \alpha_{kl} f(t_1)^l f(t_0)^{k-l}$$

$$= \sum_{k=0}^{d} \sum_{l=0}^{k} \alpha_{kl} \left(\sum_{t_1 \in \mathcal{D}^1} f(t_1)^l \right) \left(\sum_{t_0 \in \mathcal{D}^0} f(t_0)^{k-l} \right) \qquad (1)$$

Notice that in (1), the quantities $\sum_{t_1 \in \mathcal{D}^1} f(t_1)^l$ and $\sum_{t_0 \in \mathcal{D}^0} f(t_0)^{k-l}$ for $1 \leq l \leq k \leq d$ can all be computed in one scan, and then combined afterwards. Following a similar convention as [8], we introduce the notation $s(f, \mathcal{D}) := \sum_{t \in \mathcal{D}} f(t)$. Following this convention, the approximation becomes:

$$auc(f, \mathcal{D}) \approx \frac{\sum_{k=0}^{d} \sum_{l=0}^{k} \alpha_{kl} \, s(f^l, \mathcal{D}^1) s(f^{k-l}, \mathcal{D}^0)}{n_0 n_1} \qquad (2)$$

Hence, we get an approximation of the *auc* in one linear scan over the database.

4   Training a Linear Classifier with the Approximation

In this section we show how the polynomial approximation of the auc can be plugged into a gradient descent method for linear discriminative analysis that optimizes the area under the ROC curve. Notice that the approximation can be plugged into other classification inducers as well, in a very similar way.

Suppose a dataset $\mathcal{D} = \mathcal{D}^0 \cup \mathcal{D}^1$ with m numerical attributes has been given. We will represent the elements of $\mathcal{D}$ as vectors $\mathbf{x} = [x_1 \ldots x_m]$. The goal is now to find a vector of weights $\mathbf{w}$, such that the function $f_{\mathbf{w}}(\mathbf{x}) = \mathbf{w} \cdot \mathbf{x}$ maximizes AUC; that is, $auc(f_{\mathbf{w}}, \mathcal{D}^0 \cup \mathcal{D}^1)$ is maximal w.r.t. $\mathbf{w}$. To find such an optimal vector of weights $\mathbf{w}$, we use a gradient descent method. The gain of using the polynomial approximation for the AUC will be three-fold: first, a costly sorting operation in the computation of the AUC is avoided, second, based on the approximation we can estimate the gradient, and third, we do not have to re-scan the dataset every time the weights are adjusted by storing a small summary of the dataset.

Before presenting the complete algorithm, we explain its components.

Approximating the Gradient. To apply a gradient descent method, we need to compute the gradient. The AUC of $f_{\mathbf{w}}$ for a fixed set of examples $\mathcal{D}$, however, is not continuous in the weights $\mathbf{w}$. We assume that there is an underlying infinite distribution of which $\mathcal{D}$ is only a sample, and the gradient of the AUC is approximated by applying the derivative of the polynomial approximation of the AUC on the sample. Another way to interpret this approach is that we actually optimize the accurate polynomial approximation, instead of the AUC.

The gradient of the AUC w.r.t. the weights $\mathbf{w}$ is $\left[\frac{\partial auc(f)}{\partial w_1}, \ldots, \frac{\partial auc(f)}{\partial w_m} \right]$ and $\partial auc(f)/\partial w_i$ can be approximated by taking the partial derivatives of the polynomial approximation in Equation 2:

$$n_0 n_1 \frac{\partial auc(f)}{\partial w_i} \approx \sum_{k=0}^{d} \sum_{l=0}^{k} \alpha_{kl} \left(\frac{\partial s(f^l, \mathcal{D}^1)}{\partial w_i} s(f^{k-l}, \mathcal{D}^0) + s(f^l, \mathcal{D}^1) \frac{\partial s(f^{k-l}, \mathcal{D}^0)}{\partial w_i} \right) \tag{3}$$

In the case of a linear classifier, $f(x) = \mathbf{w} \cdot \mathbf{x} = \sum_{i=1}^{m} w_i x_i$, we get

$$\frac{\partial s(f^l, \mathcal{D}^1)}{\partial w_i} = \sum_{x \in \mathcal{D}^1} \frac{\partial f(x)^l}{\partial w_i} = \sum_{x \in \mathcal{D}^1} l f(x)^{l-1} x_i = l \cdot s(x_i f^{l-1}, \mathcal{D}^1) . \tag{4}$$

Combining (3) and (4), we get the following approximation for the derivative:

$$n_0 n_1 \frac{\partial auc(f)}{\partial w_i} \approx \sum_{k=0}^{d} \sum_{l=0}^{k} \alpha_{kl} \left(l \cdot s(x_i f^{l-1}, \mathcal{D}^1) \cdot s(f^{k-l}, \mathcal{D}^0) \right.$$
$$\left. + (k - l) \cdot s(f^l, \mathcal{D}^1) \cdot s(x_i f^{k-l-1}, \mathcal{D}^0) \right) . \tag{5}$$

Optimizing Along the Gradient. We now show how to choose an optimal value of the learning rate; i.e., optimize the weights along the gradient direction.

Update Rule. Suppose that the current weights are $\mathbf{w}$, and we have approximated the gradient $\mathbf{g}$. When updating the weights $\mathbf{w} \leftarrow \mathbf{w} + \gamma\mathbf{g}$, the optimal value of the learning rate γ needs to be determined. In our case, instead of minimizing along the gradient, it is better to find the optimal angle α between old and new weights in the plane spanned by the current weight vector and the gradient. The reason for this is that the AUC does not depend on the length of $\mathbf{w}$, only on its direction. Hence, the weight vectors under consideration when selecting the optimal α are given by $\cos(\alpha)\mathbf{w} + \sin(\alpha)\mathbf{g}$ with α between 0 and 2π. We show how we can avoid scanning the database to get an updated value for the AUC every time we change α in order to find the optimal value.

Avoiding re-scanning. We can approximate $n_0 n_1 \, auc(f_{\cos(\alpha)\mathbf{w}+\sin(\alpha)\mathbf{g}}, \mathcal{D})$ as follows (the scaling factor $\frac{1}{\sqrt{2}}$ is explained in the next paragraph)

$$\sum_{k=0}^{d}\sum_{l=0}^{k} \alpha_{kl} \left(\sum_{x\in\mathcal{D}^1} \left(\tfrac{\cos(\alpha)}{\sqrt{2}}\mathbf{w}\cdot\mathbf{x} + \tfrac{\sin(\alpha)}{\sqrt{2}}\mathbf{g}\cdot\mathbf{x} \right)^l \right) \cdot$$
$$\left(\sum_{x\in\mathcal{D}^0} \left(\tfrac{\cos(\alpha)}{\sqrt{2}}\mathbf{w}\cdot\mathbf{x} + \tfrac{\sin(\alpha)}{\sqrt{2}}\mathbf{g}\cdot\mathbf{x} \right)^{k-l} \right)$$
$$= \sum_{k=0}^{d}\sum_{l=0}^{k} \alpha_{kl} \left(\sum_{m=0}^{l} \beta_{l,m} s(f_\mathbf{w}^m f_\mathbf{g}^{l-m}, \mathcal{D}^1) \right) \cdot \left(\sum_{m=0}^{k-l} \beta_{k-l,m} s(f_\mathbf{w}^m f_\mathbf{g}^{k-l-m}, \mathcal{D}^0) \right)$$

where $\beta_{l,m}$ denotes $\binom{l}{m}2^{-\frac{l}{2}}\cos(\alpha)^m \sin(\alpha)^{l-m}$. Thus, after we have computed the gradient $\mathbf{g}$, one scan over the database is needed to compute $s(f_\mathbf{w}^m f_\mathbf{g}^{l-m}, \mathcal{D}^1)$ and $s(f_\mathbf{w}^m f_\mathbf{g}^{l-m}, \mathcal{D}^0)$, for all $1 \le l \le m \le d$. Based on this summary, the AUC of $f_{\cos(\alpha)\mathbf{w}+\sin(\alpha)\mathbf{g}}$ for all α can be computed without re-scanning the database.

Scaling of the Weights. One important problem we have to deal with in this application, is that the approximation of the Heaviside function is only accurate within the interval $[-1,1]$. Outside of this interval, the approximation quickly deteriorates, as can be seen in Figure 1. Therefore, we have to make sure that for all points $t_1 \in \mathcal{D}^1$ and $t_0 \in \mathcal{D}^0$, the difference $(f(t_1) - f(t_0))$ is in the interval $[-1,1]$. We show how this requirement can be met by re-scaling the weights vector $\mathbf{w}$. Obviously, re-scaling the weight vector only changes the magnitude of the scores of the classification function; the classifier and its AUC remain the same.

We need to re-scale the weights $\mathbf{w}$ in such a way that for all $\mathbf{x_0} \in \mathcal{D}^0$ and $\mathbf{x_1} \in \mathcal{D}^1$, the difference $f(x_1) - f(x_0) = \mathbf{w}\cdot\mathbf{x_1} - \mathbf{w}\cdot\mathbf{x_0}$ falls into the interval $[-1,1]$. A straightforward solution is as follows. Let m^1, M^1, m^0, M^0 be the following numbers:

$$m^0 = \min_{x_0\in\mathcal{D}^0} \mathbf{w}\cdot\mathbf{x_0}, \quad M^0 = \max_{x_0\in\mathcal{D}^0} \mathbf{w}\cdot\mathbf{x_0},$$
$$m^1 = \min_{x_1\in\mathcal{D}^1} \mathbf{w}\cdot\mathbf{x_1}, \quad M^1 = \max_{x_1\in\mathcal{D}^1} \mathbf{w}\cdot\mathbf{x_1}.$$

From these numbers it can be derived that $f(x_1) - f(x_0)$ always falls into the interval $[m^1 - M^0, M^1 - m^0]$. Based on this interval, $\mathbf{w}$ can be re-scaled appropriately. In our implementation we have opted to re-scale $\mathbf{w}$ by dividing it by $\max(M^0 - m^1, M^1 - m^0)$.

For the optimization along the gradient, we have to guarantee correct scaling for every α. Let $\mathbf{w}'$ denote $\frac{1}{\sqrt{2}}(\cos(\alpha)\mathbf{w} + \sin(\alpha)\mathbf{g})$. Observe now, for all $\mathbf{x}_0 \in \mathcal{D}^0, \mathbf{x}_1 \in \mathcal{D}^1$:

$$|f_{\mathbf{w}'}(\mathbf{x}_1) - f_{\mathbf{w}'}(\mathbf{x}_0)| = \frac{1}{\sqrt{2}}|(\cos(\alpha)(\mathbf{w}\mathbf{x}_1 - \mathbf{w}\mathbf{x}_0) + \sin(\alpha)(\mathbf{g}\mathbf{x}_1 - \mathbf{g}\mathbf{x}_0))|$$

$$\leq \frac{1}{\sqrt{2}}(\cos(\alpha) + \sin(\alpha)) \leq \frac{1}{\sqrt{2}}(\sqrt{2}) = 1$$

Notice that in the derivation we implicitly assume that $\mathbf{w}$ and $\mathbf{g}$ are appropriately scaled. Hence, when using update rule $\mathbf{w} \leftarrow \frac{1}{\sqrt{2}}(\cos(\alpha)\mathbf{w} + \sin(\alpha)\mathbf{g})$, we are guaranteed that the weights are scaled correctly.

Complete Algorithm. The complete algorithm is given in Algorithm 2. The number of iterations is fixed to *maxiter*. In every iteration, first the gradient is computed (lines 2 to 5). To this end, the database is scanned once to collect the necessary supports (lines 3 and 4). These supports are then combined to form the gradient (line 5). Once the gradient is found, the optimal angle α is computed. Again, first the necessary supports are counted in one scan over the database (lines 7 and 8). These supports suffice to find the optimal α without re-scanning the database. In the implementation, finding the optimal α is done by ranging over many different values of α evenly spread over $[0, 2\pi]$, and selecting the one that gives the highest AUC. The AUC scores for the different values of α can be computed without re-scanning the database. This optimization of α is performed by Algorithm 1 (line 4). The method is quite crude, but any other linear optimizer could be used instead. Once the optimal α has been found, the weights are updated (line 11), and the next iteration is entered.

It seems that our weight rescaling method requires an extra database scan. In our implementation, however, we combine it with support counting. Rescaling is done (if needed) continuously as records are read (this happens in lines 3,4 and 7,8). Thus, only two database scans per gradient descent iteration are required.

Soft-AUC. As we discussed earlier, the AUC does not take into account how close the points are to the decision boundary. Whether a pair of points $(\mathbf{x}_0, \mathbf{x}_1)$ contributes to the AUC solely depends on $f(\mathbf{x}_1)$ being larger than $f(\mathbf{x}_0)$, not on the magnitude of this difference. It would be more natural if small differences were counted less than large differences, like it is also the case in, e.g., mean squared error. This observation is the main motivation for the soft-AUC measure:

$$s\_auc_\beta(f, \mathcal{D}^0 \cup \mathcal{D}^1) := \frac{\sum_{t_0 \in \mathcal{D}^0} \sum_{t_1 \in \mathcal{D}^1} sigmoid_\beta(f(t_1) - f(t_0))}{|\mathcal{D}^0| \cdot |\mathcal{D}^1|}.$$

For optimizing soft-AUC, our method works perfectly well; having a good polynomial approximation is even easier, as the main difficulty, the steep step in the Heaviside, is avoided. There are, however, still some problems we have to take into account. First of all, re-scaling the weights no longer leaves the objective function unchanged. Therefore, the optimization problem actually becomes:

Algorithm 1. Optimize α

Input: $s(f_{\mathbf{w}}^{m} f_{\mathbf{g}}^{l-m}, \mathcal{D}^1)$, and $s(f_{\mathbf{w}}^{m} f_{\mathbf{g}}^{l-m}, \mathcal{D}^0)$ for all $1 \leq l \leq m \leq d$
Output: Optimal angle *ang*

1: $opt := 0$; $ang := 0$;
2: **for all** $\alpha := 0 \ldots 2\pi$ step .01 **do**
3:    Approx. AUC of $f_{\frac{1}{\sqrt{2}}(\cos(\alpha)\mathbf{w} + \sin \alpha \mathbf{g})}$, using Equation (6).
4:    **if** AUC $> opt$ **then**
5:      $opt :=$ AUC;
6:      $ang := \alpha$;
7: **return** *ang*

Algorithm 2. Learning a linear classifier

Input: Database $\mathcal{D} = \mathcal{D}^0 \cup \mathcal{D}^1$ with m attributes, initial weights $\mathbf{w}$, maximal number of iterations *maxiter*.
Output: Weights $\mathbf{w}$, reached via a gradient descent method

1: **for** $iter := 1 \ldots maxiter$ **do**
2:    {*Approximate gradient*}
3:    Count $s(x_i f^{l-1}, \mathcal{D}^1)$, $s(f^l, \mathcal{D}^1)$ for $l = 1 \ldots d$, $i = 1 \ldots m$ in one scan over $\mathcal{D}^1$.
4:    Count $s(x_i f^{l-1}, \mathcal{D}^0)$, $s(f^l, \mathcal{D}^0)$ for $l = 1 \ldots d$, $i = 1 \ldots m$ in one scan over $\mathcal{D}^0$.
5:    Approx. gradient $\mathbf{g}$ based on the supports counted in steps 1 and 2, using Equation (5).
6:    {*Approximate AUC of* $f_{\frac{1}{\sqrt{2}}(\cos(\alpha)\mathbf{w} + \sin(\alpha)\mathbf{g})}$}
7:    Count $s(f_{\mathbf{w}}^{m} f_{\mathbf{g}}^{l-m}, \mathcal{D}^1)$, for all $1 \leq l \leq m \leq d$ in one scan over $\mathcal{D}^1$.
8:    Count $s(f_{\mathbf{w}}^{m} f_{\mathbf{g}}^{l-m}, \mathcal{D}^0)$, for all $1 \leq l \leq m \leq d$ in one scan over $\mathcal{D}^0$.
9:    {*Update weights* $\mathbf{w}$}
10:    Find optimal α, with Algorithm 1
11:    $\mathbf{w} := \frac{1}{\sqrt{2}} \cos(\alpha)\mathbf{w} + \sin(\alpha)\mathbf{g}$
12: **return** $\mathbf{w}$

---

find optimal weights $\mathbf{w}$, with $\|\mathbf{w}\| = 1$, such that $s\_auc_\beta(f_w, \mathcal{D})$ is maximal. This requirement contradicts the scaling needed to keep the approximation accurate. Therefore, the re-scaling is kept, but, every time we need the approximations, the polynomial coefficients are recomputed, such that not $s\_auc_\beta(f_w, \mathcal{D})$ is approximated, but $s\_auc_{\beta/\|\mathbf{w}\|}(f_w, \mathcal{D})$. Put otherwise, instead of requiring that $\|\mathbf{w}\| = 1$, and $s\_auc_\beta(f_w, \mathcal{D})$ is optimal, we equivalently require that $s\_auc_{\beta/\|\mathbf{w}\|}(f_w, \mathcal{D})$ is optimal. We do not go into detail here due to lack of space.

5   Experimental Evaluation

We implemented the linear approximation of AUC and soft-AUC, and a linear classifier inducer based on these approximations. For both the approximation in isolation and the classifier inducer we test both the accuracy and the running times. For the polynomial approximations used in the experiments, a degree of

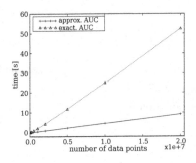 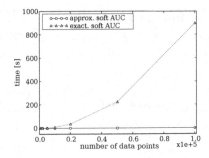

Fig. 3. Performance of the polynomial approximations

30 was chosen as a compromise. Higher values for the degree did not give a significant increase in accuracy, while decreasing performance. Numerical stability problems do become visible for high degrees, but for the degree of 30 no such problems occurred on any of the datasets used. It should be noted, however, that the optimal degree highly depends on the numerical precision of the computations and even on the architecture of the computer used.

Datasets. The characteristics of the datasets used for testing are given below. In case of the forest cover dataset only the two most frequent classes were kept. We tried two versions of the forest cover data, one with only 10 numerical attributes kept, and another with all attributes. This allowed us to see how binary attributes influence accuracy.

dataset	records	attrs
sonar	208	60
KDD Cup 04 physics	50000	78
forest cover 10 numeric attrs	495141	10
forest cover all attrs	495141	54
KDD Cup 98 all attrs	95412	464

All experiments in this section have been performed with 10 fold cross-validation.

Performance of the Polynomial Approximation. To test the accuracy of the polynomial approximation, we used synthetically generated data. The data was generated by randomly drawing positive examples with f-values with mean m_1 and standard deviation 1 and negative examples with mean m_2 and standard deviation 1 following a normal distribution, and raising this number to the power p. By varying the difference $m_1 - m_2$, different AUC values are obtained. The higher the value of p becomes, the smaller the average distances between the scores become, making the approximation difficult since many values will fall in the poorly approximated region of the Heaviside function. As can be seen in Figure 2, the accuracy of the approximation is very high in general, but deteriorates slightly when there are only small differences between the scores (high powers p). In the graphs in Figure 3 the running times for the approximations of the AUC and soft-AUC are given, showing significant performance gains.

Performance of the Linear Classifier. We begin by examining the performance and accuracy of training an AUC-maximizing linear classifier based on polynomial approximations. We used the maximum of 30 iterations, and the polynomial degree was set to 30. Figure 4 shows the results of comparing our approach with Linear Discriminant Analysis and SVM_perf, a version of Support Vector Machine minimizing AUC directly [9]. Since the SVM's performance depends on a parameter c we used three different values of this parameter.

Our approach achieves better values of AUC than the SVM and is often more than an order of magnitude faster. We were, e.g., not able to run the SVM on the KDD Cup'98 dataset. This is probably due to large number of attributes in this dataset. The main step of our method, the linear search, is totally independent of the number of attributes. Our approach minimized AUC directly without any performance problems. Forest cover gives worse results when all attributes are present. This is due to binary attributes which cause the occurrence of values of f very close to each other, thus causing significantly worse approximation.

In order to check the usefulness of direct AUC minimization we also compared it with Linear Discriminant Analysis, a standard linear classification technique. Due to the efficiency of our approach, we were able to perform direct AUC optimization on large datasets and thus obtain meaningful comparison. As it turns out, minimizing AUC directly does not give any visible improvement over classifiers built using LDA. This seems to confirm results presented in [4].

In [7] a method of fast AUC computation based on sampling is presented. We modified our algorithm to compute AUC directly on a small sample at each minimization step to obtain a similar approach. Figure 5 shows the results. It can be seen that polynomial approximations achieve higher accuracy in shorter time.

We now present some experiments on minimizing soft-AUC. We compared the method with a sampling based version. Figure 6 has the results. Again, it can be seen that our polynomial approximation gives better results than sampling. The experiment was extremely time consuming, since computing the exact soft-AUC for the final classifier took hours (quadratic time in number of records). At the same time, *building* the classifier using our approach took just seconds.

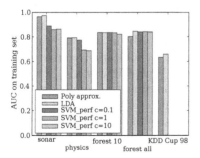 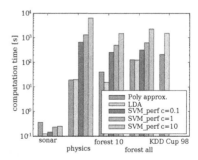

Fig. 4. AUC computation time for linear models built using polynomial approximation, Linear Discriminant Analysis and SVM_perf. Data on KDD Cup 98 is missing for SVM_perf due to excessive computation time.

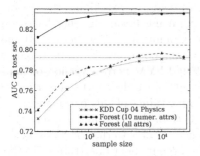
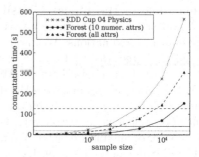

Fig. 5. Test set AUC and computation time for linear models built using exact AUC computation on samples. Horizontal lines denote test set AUC and computation time for respective models built using polynomial approximation.

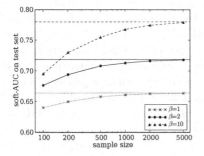
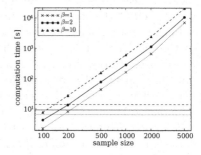

Fig. 6. Sampling vs. polynomial approximation for minimizing soft-AUC on KDD Cup 04 Physics datasets. Horizontal lines denote test set soft-AUC and computation time for respective models built using polynomial approximation.

Summary of Experimental Results. For the linear approximation, we tested the accuracy and the performance in comparison with the exact versions. The presented experiments support our claim that the approximation is very accurate and that there is a large performance gain in running time. For the linear classifier inducer, we compared both the performance w.r.t. running time and w.r.t. predictive power of the learned model, in comparison with sampling, Linear Discriminative Analysis (a linear model learner optimizing accuracy), and SVM_perf [9] (a version of Support Vector Machine learner, minimizing the AUC directly). The experiments show that the running times of our method are comparable to LDA, which is significantly lower than the time required by SVM_perf. On the other hand, sampling is not efficient as it requires too many examples to reach the same accuracy as our approximation. Hence, both in running time and predictive performance, our method is always comparable to the winner, hence combining the advantages of the different methods.

6   Summary and Conclusion

A polynomial approximation of the Area Under the ROC Curve, computable in linear time, has been presented, and was applied to inducing a classifier that optimizes AUC directly. We also proposed a soft-AUC measure which does not give simple $0/1$ scores to points close to the decision border.

Experimental evaluation has shown that the method is efficient and accurate compared to other methods for approximating the AUC. As a proof of concept, the method was plugged into the training of a linear classifier by optimizing the AUC or soft-AUC directly. With the approximation technique, similar AUC scores were reached as with existing techniques. The scalability and running times of the proposed approximation technique, however, are vastly superior.

Future work will include extending the approach to nonlinear classifiers.

References

1. Ataman, K., Street, W.N., Zhang, Y.: Learning to rank by maximizing auc with linear programming. In: IEEE International Joint Conference on Neural Networks (IJCNN 2006), pp. 123–129. IEEE Computer Society Press, Los Alamitos (2006)
2. Bradley, A.P.: Use of the area under the ROC curve in the evaluation of machine learning algorithms. Pattern Recognition 30(7), 1145–1159 (1997)
3. Brefeld, U., Scheffer, T.: AUC Maximizing Support Vector Learning. In: Proc. ICML workshop on ROC Analysis in Machine Learning (2005)
4. Cortes, C., Mohri, M.: Auc optimization vs. error rate minimization. In: Advances in Neural Information Processing Systems, vol. 16, MIT Press, Cambridge (2004)
5. Ferri, C., Flach, P., Hernandez-Orallo, J.: Learning decision trees using the area under the ROC curve. In: ICML, pp. 139–146 (2002)
6. Hanley, J.A., McNeil, B.J.: The meaning and use of the area under a receiver operating characteristic (ROC) curve. Radiology 143(1), 29–36 (1982)
7. Herschtal, A., Raskutti, B.: Optimising area under the roc curve using gradient descent. In: ICML, pp. 49–56. ACM Press, New York (2004)
8. Jaroszewicz, S.: Polynomial association rules with applications to logistic regression. In: KDD (2006)
9. Joachims, T.: A support vector method for multivariate performance measures. In: ICML (2005)
10. Rakotomamonjy, A.: Optimizing Area Under Roc Curve with SVMs. In: Workshop on ROC Analysis in Artificial Intelligence (2004)

Finding Transport Proteins
in a General Protein Database

Sanmay Das, Milton H. Saier, Jr., and Charles Elkan

University of California, San Diego, La Jolla, CA 92093, USA

Abstract. The number of specialized databases in molecular biology is growing fast, as is the availability of molecular data. These trends necessitate the development of automatic methods for finding relevant information to include in specialized databases. We show how to use a comprehensive database (SwissProt) as a source of new entries for a specialized database (TCDB, the Transport Classification Database). Even carefully constructed keyword-based queries perform poorly in determining which SwissProt records are relevant to TCDB; we show that a machine learning approach performs well. We describe a maximum-entropy classifier, trained on SwissProt records, that achieves high precision and recall in cross-validation experiments. This classifier has been deployed as part of a pipeline for updating TCDB that allows a human expert to examine only about 2% of SwissProt records for potential inclusion in TCDB. The methods we describe are flexible and general, so they can be applied easily to other specialized databases.

1   Introduction

The number of specialized databases in molecular biology is growing fast. The 2006 Database Issue of the journal *Nucleic Acids Research* (NAR) describes 94 new databases and updates of 68 existing databases [2]. The NAR Molecular Biology Collection Database is a compilation of 968 databases as of the 2007 update, 110 more than in the previous update [6]. The vast number of these databases and the number of high-throughput projects producing molecular biological information make it difficult for human curators to keep their databases up-to-date [12]. There has recently been much research on identifying documents containing information that should be included in specialized databases [8,5,13].

The traditional approach has been to apply text classification algorithms to the primary literature to determine whether or not a paper is relevant to the database. We propose an alternative approach: to leverage an existing, general protein database (namely SwissProt, http://www.expasy.org/sprot/) by directly screening its records for potential inclusion.

Previously, we investigated an approach in which an expert constructed detailed queries based on keywords and gene ontology terms to identify appropriate SwissProt records. The results of this approach were encouraging, but we hypothesized that a classifier trained on the content of SwissProt records could have higher precision and recall. This paper confirms this hypothesis, in the

J.N. Kok et al. (Eds.): PKDD 2007, LNAI 4702, pp. 54–66, 2007.

context of the Transport Classification Database (TCDB), a specialized protein database created and maintained at UCSD.

Our approach to automatically updating TCDB has three steps: (1) triage to filter out SwissProt records that are not relevant to the TCDB domain; (2) deciding which of the remaining proteins are novel enough to be included in TCDB (since TCDB is intended to be representative, not comprehensive, most transport proteins will not be included); and (3) actually incorporating data from SwissProt and other sources into TCDB.

This paper focuses on step (1) and briefly describes our approach to (2) and (3). We use a maximum-entropy classifier to select relevant proteins from Swiss-Prot. We demonstrate experimentally that this classifier discriminates effectively between transport-related proteins and others on the basis of text in SwissProt records. We show how to interleave training the classifier with creating a clean training set, which we have made publicly available. These results show that maintainers of specialized molecular biology databases can find records of interest with minimal preprocessing after the one-time effort required to create a clean training set.

Having an effective method for automatically selecting data from SwissProt is critical given the size of SwissProt and its growth rate (for example, 2015 new proteins were added to SwissProt between the releases of January 9 and January 23, 2007). Once the number of proteins to consider has been reduced, more complex analyses can then be applied in step (2). We discuss this process in the context of TCDB in Section 3. After the proteins to be included have been determined, the fact that we are working directly with SwissProt records is advantageous for step (3) because information can be transferred directly from one structured format to another.

2 A Pipeline to Identify Relevant and Novel Proteins

A transporter protein is one that imports or exports molecules through the membrane of a cell or organelle. The Transport Classification Database (TCDB) is a web-based resource (`http://www.tcdb.org`) that provides free access to data on proteins related to transmembrane transport. TCDB contains information compiled from over 3000 published papers and is intended to provide information on all recognized, functionally characterized, transmembrane molecular transport systems [11]. TCDB implements and defines the Transport Classification (TC) system, which was officially adopted by the International Union of Biochemistry and Molecular Biology in 2002 as the international standard for categorizing proteins involved in transmembrane transport.

As mentioned above, there are three well-defined steps in our pipeline for automatically updating TCDB. This section introduces terminology we use in the rest of the paper and explains our approach to each of the three steps.

The first step is a triage stage that selects only a subset of *relevant* proteins from SwissProt. We define any protein that is involved in any transmembrane transport process as relevant. Specialists in transport proteins estimate that approximately 10% of all proteins will meet this description of relevance. It does not automatically follow that 10% of all proteins in SwissProt will be relevant, since the distribution of proteins in SwissProt may be different from the distribution of proteins overall.

It might seem that a search for keywords like "transport" should be sufficient for this task, but even keyword searches constructed by experts are problematic in this domain. This paper describes how we use machine learning techniques to build a classifier that evaluates the relevance of protein records in SwissProt. It is not clear *a priori* that learning-based methods will have high precision and recall, because it is hard for an untrained human being who is not an expert in molecular biology to become skilled at finding relevant records without substantial expert coaching. However, experiments show that we can achieve 95% precision and 95% recall.

One of the major problems we encounter in the process of building a classifier is the absence of a reliable "negative" training set of protein records that are definitely *not* relevant. We know the SwissProt accession numbers of all proteins already in TCDB, but for any protein that is in SwissProt but not in TCDB, we do not know if it is irrelevant or whether it is relevant but not included in TCDB for any of many reasons. Therefore, the process of building a classifier has to involve the assembly of a reliable training set. Details on this process, including the construction of the training set, can be found in Section 3.

After the triage step, the second step is to decide which proteins should actually be included in TCDB. This is hard because TCDB is intended to be a representative database, not a comprehensive one. For example, a protein that is homologous to a protein already in TCDB should be included only if it performs a different function (for example, it acts upon a different substrate) or is in some other way sufficiently distinct (for example, it is from an organism in a different domain and has sufficient sequence dissimilarity). We will refer to proteins that should be included as *novel*. A major advantage of classifying SwissProt records, as opposed to published papers, is that once a record is found to be relevant, we can directly retrieve multiple kinds of information to perform further analysis. For example, we can analyze the sequences of proteins classified as relevant. At this stage we use a rule-based system to decide on the novelty of relevant proteins, largely because of its transparency and comprehensibility to biologists.

In the third stage of the pipeline, if a protein is relevant and novel, a human expert assigns its TC number and the protein information is entered into TCDB. We intend to automate this process as well in the future, but do not focus on the issues involved in this paper.

Since we make extensive use of expert judgments in what follows, it is important to characterize expertise in the TC domain. We define a "Level 1" expert to be a person who is an expert in molecular biology and the TC system. Such a person can definitively decide whether or not to enter a protein into TCDB, and assign TC numbers, using his/her knowledge of the TC system. A "Level 2" expert is someone who has substantial knowledge of transport proteins, but who cannot always decide whether a protein is novel, or what TC number is most appropriate. Note that even a Level 1 expert may not be able to make a final decision on relevance as defined above, because there may not be enough evidence about a protein. In these cases, the protein will not be included in TCDB.

3   Learning a Classifier to Determine Relevance

The first step of our pipeline consists of a classifier that uses text from certain fields of a SwissProt record to decide whether or not the record describes a protein involved in transmembrane transport. We do not consider TrEMBL records because SwissProt is carefully manually curated and we are largely interested in proteins that are sufficiently well-known and characterized for inclusion in SwissProt.

Evaluation Measures. Precision and recall are the primary measures of success for our classifier. Estimating precision and recall in the absence of known labels for all test examples is in itself a tricky problem [1,4]. We propose that the best way to estimate precision and recall in such circumstances is to perform two separate experiments. Precision is measured using an experiment in which randomly selected unseen test data are labeled as relevant or irrelevant by the classifier, and then all the examples labeled as relevant are manually checked. The proportion of examples labeled as relevant by the classifier that are also manually judged relevant gives a statistically valid estimate of precision. Unfortunately, performing a similar experiment for recall is impractical because it would require labeling the entire test set, not just the examples labeled as relevant by the classifier. Therefore, we use a ten-fold cross-validation experiment on the training data in order to measure recall, since the training set is fully labeled.

Task-specific utility functions are also sometimes used as measures of success, for example, in the document classification task of the 2005 TReC Genomics Track [7]. In our experimental results we present complete confusion matrices from which any function can be computed in addition to precision and recall numbers.

Choice of Features. We use a text representation of selected fields from Swiss-Prot records. The maximum-entropy classifier performs significantly better when using the chosen fields rather than the whole SwissProt record, according to preliminary experiments. This finding is consistent with the result of [3] that text classification algorithms (and, in particular, maximum-entropy methods) perform better when using only selected paragraphs from papers rather than the entire paper. Table 1 shows the fields we use.

The last feature mentioned in Table 1 is the reported number of transmembrane segments in the protein. This feature is derived from the "Features" field of the original SwissProt record, which contains position-wise annotations of the protein sequence. The number is alphabetized and concatenated with the letters "TM," so for example the word "TMFOUR" is added to the representation of the record for a protein annotated as having four transmembrane segments. Note that the number of transmembrane segments is not used until after the relabeling process described below, and we present final results with and without this feature. Also, the tokenization performed by the software package that we use [9] removes numeric strings and special characters, and only preserves the alphabetic parts of alphanumeric strings.

Table 1. Description of SwissProt fields used in the text classification process

SwissProt Field Title	Text Code	Description
Accession Number	AC	Accession number (unique)
Protein Name	DE	Full name of the protein
References	RT	Titles of papers referenced
Comments	CC	Human annotations
Keywords	KW	Assigned by curators
Ontologies (GO)	DR GO	Gene ontology (GO) terms
Features (Transmembrane Segments)	FT TRANSMEM	# transmembrane segments

One simple way to make the coding of SwissProt records more sophisticated, and possibly more useful to the learning algorithm, would be to add a tag to each word specifying which section of the SwissProt record it is found in. This tagging would, for example, treat the word "transmembrane" in the title of a paper differently from the same word in a functional annotation. While tagging words could be useful, based on experience with the feature encoding the number of transmembrane segments (see Table 2), a dramatic improvement in precision or recall is unlikely.

Selection of Training Data. The training set is created from the version of SwissProt dated September 23, 2006, which contains 232,345 records. The training set contains 2453 SwissProt records corresponding to proteins in TCDB. The features described above are extracted for each of these records, and these 2453 are labeled as positive examples. We select twice this number, that is, 4906 random records from SwissProt excluding the 2453 records known to be in TCDB. We expect the universe of SwissProt records to be significantly unbalanced with respect to our notion of relevance, and we want to reflect this fact in the training set. However, using a very unbalanced training set is problematic because it may lead to a loss in discriminative power in the learned classifier. The classification threshold can be adjusted after training if the classifier identifies too many records as positive, which is presumably the greatest risk of using a training set that is less unbalanced than the test data.

The 4906 records are initially assumed to be negative examples, but some of them are actually positive, since they are randomly selected from the entire universe of SwissProt proteins, which includes a significant number of proteins that are involved in transmembrane transport but are not in TCDB. These may be proteins that should be in TCDB (these are the ones that it is our ultimate goal to identify), or proteins that do not meet the criteria for inclusion in TCDB for a variety of reasons. For training a classifier to determine relevance as opposed to novelty, both these types of records should be labeled positive.

Below, we describe an iterative process for relabeling the negative examples as appropriate.

For final training and experiments after the relabeling process, we used the January 9, 2007 version of SwissProt for each of the 7359 records used, since modifications might have been made to the annotations or sequences.

Choosing a Classifier. We consider two classification algorithms, the naive Bayes classifier and the maximum-entropy classifier. We use the implementations provided in the Mallet toolkit [9]. Both algorithms use Bayesian probabilistic models. Space limitations preclude a detailed description of their properties in this paper; see [10] for a detailed comparison. The naive Bayes classifier builds a generative model using the strong assumption of conditional independence of the features given the class label, while the maximum-entropy model is discriminative, in that it directly maximizes the conditional log-likelihood of the data. The motivation behind maximum-entropy methods is to prefer the most uniform model that satisfies the constraints determined by the data [10].

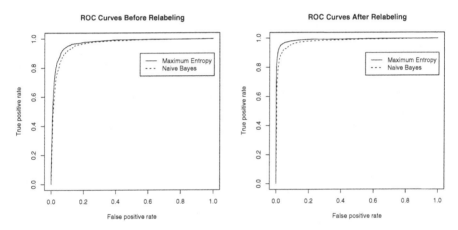

Fig. 1. ROC curves for the maximum-entropy (ME) and naive Bayes (NB) classifiers on the original (left) and relabeled (right) datasets. In the original dataset the random selections from SwissProt are assumed to all be negative examples. The areas under the curves are as follows: NB (original): 0.962, ME (original): 0.971, NB (relabeled): 0.979, ME (relabeled): 0.991. Curves are generated from ten-fold cross-validation experiments.

We compared the performance of the two classifiers on the original training data, assuming that all 4906 randomly selected proteins were, in fact, negative examples. In our application the ROC curve for the maximum-entropy method dominated that for Naive Bayes. We thus proceeded to relabel negative examples (described next) using the maximum entropy method. However, we repeated the ROC curve comparison after relabeling. Both ROC curves are shown in Figure 1. The success of the maximum-entropy method is likely related to its ability to adjust for redundant information, which is prevalent in our domain because the

Table 2. Confusion matrices (above) and precision and recall (below) after each iteration of relabeling and retraining. Performance improves with each iteration, both because incorrect labels are fixed, and because the correct labels are used in the next iteration of training.

Iteration 1			Iteration 2			Iteration 3			Iteration 4		
	Neg	Pos		Neg	Pos		Neg	Pos		Neg	Pos
Neg	4628	278	Neg	4530	183	Neg	4491	132	Neg	4456	130
Pos	260	2193	Pos	261	2385	Pos	141	2595	Pos	146	2627

(Rows represent ground truth and columns represent classifier predictions.
The row sums change between iterations due to relabeling.)

	Iteration 1	Iteration 2	Iteration 3	Iteration 4
Precision	88.75%	92.87%	95.16%	95.28%
Recall	89.40%	90.14%	94.85%	94.73%
"False" positives that	192/278	91/183	37/132	28/130
were actually positive				

same information can appear in SwissProt records in multiple places, including in the titles of papers, functional annotations, and keywords.

Updating the Negative Set. One major problem with the training data is that we have no assurance that the 4906 randomly selected proteins are, in fact, negative examples. There are a number of different approaches to learning with partially labeled data, but we focus on creating a clean training set.

Our approach relies on explicitly examining examples classified as positive and deciding on their relevance. We use this approach for two reasons. First, we need to examine these records in detail in any case for the second part of our task, determining whether or not an entry is novel, i.e. should be added to TCDB. Verifying relevance is a necessary part of evaluating novelty. Second, going through the relabeling process allows us to determine which factors are important and develop a screening process for the second part of the pipeline, which is described in detail in the next section. Third, once we have created a reliable negative set of records, it can be re-used by us and by others. We are making this database publicly accessible for researchers.

The process of relabeling is performed in several iterations as follows. Each iteration assumes that the relabeling from the previous iteration is the "ground truth" about relevance. The first iteration assumes that all 4906 randomly selected records are not relevant. Each iteration consists of the following steps:

1. All training examples are randomly divided into ten folds (sets of equal size).
2. For each fold F_i, a classifier is trained on the other nine folds and then applied to each example in F_i.
3. A confusion matrix and precision and recall numbers are computed using the predicted labels and the assumed true labels.
4. Each example in the "false positive" category of the confusion matrix is manually examined to determine relevance (details of this process are in the next subsection).

5. All "false positives" that are manually determined to be relevant are relabeled as positive for the next iteration.

The above process is repeated for four iterations. The process of relabeling becomes faster with each iteration, because proteins are often brought up as false positives repeatedly, but they only need to be examined for relevance once. The results from each iteration are shown in Table 2. The relabeled dataset is available at `http://people.csail.mit.edu/sanmay/data/`.

Table 3. Final training results are similar with and without the feature that codes the number of transmembrane segments

	Without TM feature		With TM feature
	Neg Pos		Neg Pos
Neg	4444 114	Neg	4438 120
Pos	139 2662	Pos	135 2666
Precision	95.89%	Precision	95.69%
Recall	95.04%	Recall	95.18%

Manually Determining Relevance. The process of determining whether or not a record is relevant proceeds in a series of steps that go from minimum to maximum expert intervention in the process. First, a number of features are derived to help in making the judgment.

1. The protein sequence is retrieved from SwissProt and then a BLAST search is performed against TCDB.
2. An indicator variable called "Interesting Papers" is defined for each Swiss-Prot record. A paper is thought to be interesting if it may have functional information about the protein. We use a heuristic to make this determination by eliminating any paper that contains two out of the three words "complete," "genome," and "sequence," or their cognates in the title.
3. Similar variables are defined for whether or not the protein may be hypothetical (if the name of the protein starts with "Hypothetical" or "Putative" or is of the form "Protein y⟨xxx⟩") and whether it is likely to be involved in bulk transport (indicated by the presence of the word "vacuole" or its cognates, or the term "endoplasmic reticulum," although transmembrane transporters can also be present in the vacuoles or the endoplasmic reticulum).

We then categorize many of the proteins as follows:

1. Those with best TC-BLAST e-value scores better than 10^{-50} (better means closer to 0, or smaller in magnitude) are automatically assumed to be relevant (TC-BLAST is NCBI blastp applied to all the proteins in TCDB).
2. Those with best TC-BLAST scores worse than 10^{-1} (further from 0, or greater in magnitude) that also have no interesting papers or are hypothetical are assumed to be irrelevant.

3. Proteins involved in bulk transport, which is indicated by the presence of the words mentioned above, but is verified by a human (who can be a non-expert) reading the functional annotation, are also assumed to be irrelevant.
4. Proteins with best TC-BLAST scores better than 10^{-10} which have functional annotations that indicate they perform the same function as their best BLAST hit in TCDB are assumed to be relevant.

The remaining proteins are analyzed by experts. Many of these can be judged by a Level 2 expert, but some need the judgment of a Level 1 expert.

Final Precision and Recall Results. The cross-validation experimental results in Table 2 show that a maximum-entropy classifier trained after the relabeling process achieves recall over 95%. This is a fair estimate of recall on unseen test data, and sufficiently high to be of excellent practical value. Figure 2 shows precision and recall when different proportions of the entire relabeled dataset are used for training and testing. The figure shows that performance continues to improve as more training data is used, perhaps indicating that further improvements may be achievable as we add more entries to TCDB that can serve as positive examples for future iterations of the classifier.

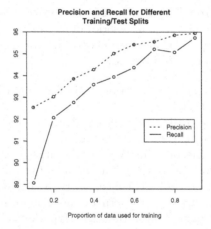

Fig. 2. Precision and recall of the maximum-entropy classifier when different fractions of the entire relabeled dataset are used for training. Results are averaged over ten different random splits for each data point.

To estimate precision on real test data, we use the same experiment that we use to judge the success of the second stage of the pipeline (described below). Out of 1000 randomly selected SwissProt records, the maximum-entropy classifier labels 99 as relevant. Of these 99, 82 are truly relevant, yielding an estimated "real-world" precision of 83%. This value is certainly high enough to be useful in practice.

It is fair to compare these results with the estimated precision and recall of rule sets for determining relevance that we had previously designed in consultation with experts. One set of less complex rules achieved 67% precision and 73.5% recall while another, more complex, set of rules achieved 78% precision and 71.5% recall. Therefore, it is clear that the classifier is learning non-obvious relations among words from different parts of the text of SwissProt records. The classifier can use combinations of terms appearing in the titles of papers, human annotations of the protein record, the gene ontology record of the protein, and the SwissProt keywords. Similar performance cannot be achieved from a rule-based search, even though the rule-based method focuses on terms thought to be the most important by human experts. For example, while the top twenty words ranked by information gain on the training set include terms like "transport," "transmembrane," and "transporter," which were included among the terms in the expert rulesets, they also include words like "multi-pass" that refer to the topology of the protein, which were not included in human rule-sets. The more important gain in performance is from the classifier's ability to combine and weight different terms in a flexible manner.

Rules for Deciding Novelty. A protein should be included in TCDB only if it is sufficiently novel to add value to the database. Many proteins that are relevant are not sufficiently novel. The most common reason for a relevant protein not to be included is that it is homologous with a high degree of sequence similarity and identical function to a protein already in TCDB. Another common reason for not adding a protein to TCDB is that it does not have sufficient experimental functional characterization.

We devised rules for recognizing when proteins identified as relevant are nevertheless not novel. The proteins not eliminated by these rules are analyzed by a Level 2 expert, and then if necessary by a Level 1 expert, who makes the final decision on which proteins to include in TCDB.

Measuring Performance. In order to estimate the precision of the classifier that predicts relevance (and the success of the rules that evaluate novelty), we train a final classifier on the entire relabeled training set. We then classify 1000 fresh records selected randomly from SwissProt. All fresh records classified as positive are then examined to determine whether they are genuinely relevant, and whether they are novel.

Of the 1000 fresh records, 99 are labeled positive by the final classifier. This is reasonable if approximately 10% of proteins are related to transmembrane transport. As expected, many of the 99 records classified as positive are eliminated by the rules for evaluating novelty and relevance. The rules label 67 of the 99 as relevant but not novel, and another 11 as not relevant. The remaining 21 records were presented to the Level 1 expert. In this set, 5 have been or will be added to TCDB, while 16 will not be included in TCDB: 3 are interesting but do not yet have sufficient functional data, 6 are irrelevant, 1 is a chaperone protein that is marginally relevant, and 6 are too similar to proteins already in TCDB.

Looking just at relevance, of the 99 records labeled as positive by the classifier, 82 are genuinely relevant. Precision is lower (83%) in this real-world experiment than in the cross-validation experiments (96%). There are at least three possible explanations for the decline in precision. One, the training set on which cross-validation experiments are performed is biased by virtue of containing many proteins from TCDB.[1] Two, the final classifier may simply overfit the training set. The third reason is drift over time in the concept of relevance: some proteins currently in TCDB are not relevant by the criteria used today, but were relevant according to criteria that were previously applied.

A Biological Success: Channel Toxins. A case study shows the benefits that automation can bring to databases like TCDB. While evaluating relevance and novelty, we came across several proteins that are toxins which specifically target transport mechanisms. Since the classifier repeatedly found these toxins, the Level 1 expert decided to introduce a new category into the TC system for them, named 8.B. (The broad Category 8 is for auxiliary proteins that are involved in regulating transport.) This experience shows that our automated methods can find a whole set of proteins that are sufficiently interesting to constitute a new category in TCDB. These proteins were not unknown prior to our experiments, but the expert never thought to include them until the classifier kept flagging them as relevant. The new 8.B category adds value to TCDB as new knowledge and also because channel toxins are important in medicine.

4   Discussion

Our work shows that it is possible to build a classifier that operates on an established general database like SwissProt to select records for potential inclusion in a more specialized database like TCDB, with high precision and recall. Similar classifiers should perform equally well for other specialized databases. Using the classifier to filter out about 90% of SwissProt makes it feasible to apply techniques like BLAST searches to the remaining records that are too expensive, or too inaccurate, to apply to all of SwissProt. The software described above is in real-world use by the biologists who maintain TCDB. For real-world use tools must be convenient, comprehensible, and transparent. Our pipeline meets these criteria.

It is important to consider two stages of evolution of the project for updating TCDB, or another specialized database. The first stage is to bring the database up-to-date, based on information already in SwissProt that was missed in the previous manual construction and updating. The second stage is to use the pipeline to screen fresh records continuously as they are entered into SwissProt.

SwissProt contained 270,778 protein entries as of June 12, 2007. Our experiments show that the maximum-entropy classifier can reduce the set of proteins

---

[1] For example, the words "escherichia," and "coli" have high information gain on the training set, because TCDB preferentially includes proteins from well-characterized species. A classifier trained primarily on records in TCDB might be too willing to include a protein from *E. coli* in the absence of words indicating a transport function. Similarly, transmembrane transport is overrepresented compared to bulk or intracellular transport; three of the false positives functioned in these types of transport.

we need to consider in more detail by a factor of 10, to around 27,000 proteins. The additional rules we have devised can be used by a combination of software and non-experts to eliminate perhaps 80% of these proteins, still leaving an additional 5400 for an expert to examine. The most critical direction to pursue next is to prioritize these records. The most useful criteria may be the expected quality of functional information present for a protein, which can be estimated from certain attributes of the papers cited in the SwissProt record. For example, prioritizing records that point to recent papers in particularly important journals is the approach currently preferred by the expert maintainers of TCDB.

We hope that this pipeline can continue to be used many years into the future so that experts can restrict the time they spend on updating the database manually. To achieve this, it will be necessary to screen SwissProt (new versions are released bi-weekly) for new proteins, as well as for proteins with updated functional annotations. Between the releases of January 9 and January 23, 2007, for example, 2015 records were added to SwissProt, and 102,269 entries had their annotations revised. Obviously, new records will have to be screened using the pipeline described in this paper, and this seems a feasible goal. While 102,269 is a daunting number, when screening revised SwissProt records, we will only be concerned with proteins that either are already in TCDB or have been marked as potentially interesting given more functional information. Therefore, we expect SwissProt to continue to serve effectively as a data source for updating TCDB.

Acknowledgments. This research is supported by NIH R01 grant number GM077402. The authors thank Aditya Sehgal, who was involved in designing and evaluating the keyword-based strategy for finding relevant SwissProt records.

References

1. Aslam, J.A., Pavlu, V., Yilmaz, E.: A statistical method for system evaluation using incomplete judgments. In: Proc. ACM SIGIR, pp. 541–548. ACM, New York (2006)
2. Bateman, A.: Editorial. Nucleic Acids Res. Database Issue, 34(D1) (2006)
3. Brow, T., Settles, B., Craven, M.: Classifying biomedical articles by making localized decisions. In: Proc. TReC 2005 (2005)
4. Craven, M., Kumlien, J.: Constructing biological knowledge bases by extracting information from text sources. In: Proc. 7th Intl. Conf. on Intelligent Systems for Molecular Biol. (1999)
5. Donaldson, I., et al.: PreBIND and Textomy–mining the biomedical literature for protein-protein interactions using a support vector machine. BMC Bioinformatics, 4(1) (2003)
6. Galperin, M.Y.: The molecular biology database collection: 2007 update. Nucleic Acids Res. Database Issue, 35 (2007)
7. William Hersh, A., Cohen, J., Yang, R.T., Roberts, B.P., Hearst, M.: Trec 2005 genomics track overview. In: Proc. TREC (2005)
8. Krallinger, M., Valencia, A.: Text-mining and information-retrieval services for molecular biology. Genome Biol. 6(7), 224–230 (2005)

9. McCallum, A.K.: Mallet: A machine learning for language toolkit (2002), http://mallet.cs.umass.edu
10. Nigam, K., Lafferty, J., McCallum, A.: Using maximum entropy for text classification. In: Proc. IJCAI-99 Workshop on Machine Learning for Inf. Filtering, pp. 61–67 (1999)
11. Saier Jr., M.H., Tran, C.V., Barabote, R.D.: TCDB: the Transporter Classification Database for membrane transport protein analyses and information. Nucleic Acids Res. 36(Database Issue) D181–D186 (2006)
12. Shatkay, H.: Hairpins in bookstacks: Information retrieval from biomedical text. Briefings in Bioinformatics 6(3), 222–238 (2005)
13. Yeh, A.S., Hirschman, L., Morgan, A.A.: Evaluation of text data mining for database curation: Lessons learned from the KDD Challenge Cup. Bioinformatics, 19(Suppl. 1) i331–i339 (2003)

Classification of Web Documents Using a Graph-Based Model and Structural Patterns

Andrzej Dominik, Zbigniew Walczak, and Jacek Wojciechowski

Warsaw University of Technology, Institute of Radioelectronics
Nowowiejska 15/19, 00-665 Warsaw, Poland
A.Dominik@elka.pw.edu.pl, Z.Walczak@elka.pw.edu.pl,
J.Wojciechowski@ire.pw.edu.pl

Abstract. The problem of classifying web documents is studied in this paper. A graph-based instead of traditional vector-based model is used for document representation. A novel classification algorithm which uses two different types of structural patterns (subgraphs): contrast and common is proposed. This approach is strongly associated with the classical emerging patterns techniques known from decision tables. The presented method is evaluated on three different benchmark web documents collections for measuring classification accuracy. Results show that it can outperform other existing algorithms (based on vector, graph, and hybrid document representation) in terms of accuracy and document model complexity. Another advantage is that the introduced classifier has a simple, understandable structure and can be easily extended by the expert knowledge.

1 Introduction

Classification problems have been deeply researched due to variety of applications. They appear in different fields of science and industry and may be solved using different techniques, e.g. neural networks, rough sets. One important research domain is automated categorization of text and web documents based on their content. Development of fast and accurate algorithms is required by web and corporate search engines to provide better quality of service (quality of search results in terms of accuracy and speed) for their users.

Algorithms for document categorization and classification operate on different data representation. The most popular document representation is based on vectors. According to this model, each term in a document becomes a feature (dimension). The value of each dimension in a vector is the weight of appropriate term in a given document. Weight usually denotes the frequency of a particular term inside a document or some other measure based on frequency.

The vector model has numerous advantages. It is simple and can be used with traditional classification algorithms that operate on vectors containing numerical values (e.g. k-NN algorithm, artificial neural networks, decision trees). The most important disadvantage of vector model is that it concentrates only on words frequency and ignores other sources of information. Additional information may

J.N. Kok et al. (Eds.): PKDD 2007, LNAI 4702, pp. 67–78, 2007.
© Springer-Verlag Berlin Heidelberg 2007

be obtained from both the structure of a text (e.g. the order in which the words appear or the location of a word within the document) and the structure of a document (e.g. markup elements (tags) inside HTML web document). Such information may be crucial and may greatly improve web documents classification accuracy.

Another document representation model is based on a graph. In this model terms refer to nodes. Nodes are connected by edges which provides both text and document structure information. This model overcomes major limitations of the vector model. On the other hand dealing with data represented by a graph is more complex than by vector (i.e. common graph operations such as graph isomorphism are NP-complete).

There are also mixed (hybrid) document models which use both representations: graph and vector. They were designed to overcome problems connected with simple representations. They capture structure information (using graph model) and represent relevant data using vector [13].

This paper makes a few important contributions. Firstly, the concepts of contrast and common subgraphs are extended and used for building a Contrast Common Patterns Classifier (CCPC). Secondly, some typical emerging patterns ideas are adapted to improve classification accuracy. Classification results for benchmark web document collections obtained by using the considered approach are provided and compared with existing algorithms using different document representations.

This paper is organized as follows. In Section 2 the state of the art in graph mining and document classification is briefly described. Graph based web document representation model is presented in Section 3. In Section 4 preliminary terminology on graph theory is introduced. The concept of our classifier and experiments results are described in Sections 5 and 6, respectively. Conclusions, final remarks and future work are in Section 7.

2   Related Work

In this section we review the state of the art in the areas associated with mining contrast and common graph patterns.

Contrast patterns are substructures that appear (appear frequently) in one class of objects and don't appear (appear infrequently) in other classes. In data mining patterns which uniquely identify certain class of objects are called jumping emerging patterns (JEP). Patterns common for different classes are called emerging patterns (EP). Concepts of jumping emerging patterns and emerging patterns have been deeply researched as a tool for classification purposes [7], [8], [11].

The concept of contrast subgraphs was studied in [18], [2]. Ting and Bailey [18] proposed an algorithm (containing backtracking tree and hypergraph traversal algorithm) for mining all disconnected contrast subgraphs from dataset.

Another relevant area to review is mining frequent structures. Frequent structure is a structure which appears in samples of a given dataset more frequently than the specified treshold. Agarwal and Srikant proposed an efficient algorithm for mining frequent itemsets in the transaction database called Apriori. Similar algorithms were later proposed for mining frequent subgraphs from graphs dataset: [10], [12]. They were also used for the classification purposes [4].

Mining patterns in graphs dataset which fulfil given conditions is a much more challenging task than mining patterns in decision tables (relational databases). The most computationally complex tasks are isomorphism and automorphism. The first is proved to be NP-complete while the complexity of the other one is still not known. All the algorithms for solving the isomorphism problem present in the literature, have an exponential time complexity in the worst case but polynomial solution has not been yet disproved. A universal exhaustive algorithm for both of these problems was proposed in [19]. It operates on the matrix representation of graphs and tries to find a proper permutation of nodes. Search space can be greatly reduced by using nodes invariants and iterative partitioning [9]. Moreover multiple graph isomorphism problems can be efficiently performed with canonical labelling [15], [9]. Canonical label is a unique representation (code) of a graph such that two isomorphic graphs have the same canonical label.

Another important issue is generating all non-isomorphic subgraphs of a given graph. The algorithm for generating DFS (Depth First Search) code [20] can be used to enumerate all subgraphs and reduce the number of required isomorphism checking.

One of the most popular approaches for document classification is based on k-NN (k-Nearest Neighbors) method. Different similarity measures where proposed for different document representations. For vector representation the most popular is cosine measure [17], [16] while for graph representation distance based on maximum common graph is widely used [17], [16]. Recently methods based on hybrid document representations became very popular. They are reported to provide better results than methods using simple representations. Markov and Last [13] proposed an algorithm that uses hybrid representation. It extracts subgraphs from a graph that represents document then creates vector with boolean values indicating relevant subgraphs.

3   Graph Representation of Web Documents

In this section we present basic information on graph based models for web document representation.

There are numerous methods for creating graphs from documents. In [16] six major algorithms were described: standard, simple, n-distance, n-simple distance, absolute frequency and relative frequency. All of these methods use adjacency of terms. Some of these methods were specially designed to deal with web documents by including markup elements information. In our case we used standard document representation (previously reported as being the most effective) with slight modifications (simplifications). We refer to this representation as standard simplified. This method produces a labeled (both nodes and edges) undirected multigraph. Detailed information on creating graph according to standard simplified model is provided below.

Before converting web document (HTML document) into a graph some pre-processing steps are taken. Firstly all words that do not provide any meaningful information about document's domain (e.g. "the", "and", "of") are removed from the text. Subsequently simple steaming method is performed in order to determine those word forms which should be considered to be identical (e.g. "graph" and "graphs"). Lastly, frequency of words appearing in document is calculated and document is divided into three major sections: title, which contains the text related to the documents title and any provided keywords; link, which is text appearing in clickable hyperlinks on the document; and text, which comprises any of the readable text in the document (this includes link text but not title and keyword text).

The final graph model is created on the previously preprocessed document. It contains only one parameter - N which refers to the number of nodes in a graph representing given document. This parameter is responsible for reducing computational complexity. N most frequently appearing terms in the text are extracted. Each such unique word becomes a node labeled with the term it represents. Note that there is only a single node for each word even if a word appears more than once in a text. If word a and word b are adjacent somewhere in a document section s, then there is an undirected edge from the node corresponding to a to the node corresponding to b with an edge label s. An edge is not added to the graph if the words are separated by certain punctuation marks (such as a period, comma).

4   Preliminary Terminology

In this section we introduce some basic concepts and definitions [18], [5], [6], [3] that are used in the subsequent sections.

Graphs are assumed to be undirected, connected (any two vertices are linked by a path), labelled (both vertices and edges posses labels) multigraphs (parallel edges and loops are allowed). By the size of a graph we mean the number of its edges. Capital letters (G, S, ...) denote single graphs while calligraphic letters ($\mathcal{G}$, $\mathcal{N}$, $\mathcal{P}$, ...) denote sets of graphs.

Definition 1. *Labelled graph G is a quadruple (V, E, α, β), where V is a non-empty finite set of vertices, E is non-empty finite multiset of edges $(E \subseteq V \cup [V]^2)$ and α, β are functions assigning labels to vertices and edges, respectively.*

Definition 2. *A graph $S = (W, F, \alpha, \beta)$ is a subgraph of $G = (V, E, \alpha, \beta)$ (written as $S \subseteq G$) if: (1) $W \subseteq V$ and(2) $F \subseteq E \cap (W \cup [W]^2)$.*

Definition 3. *Let $\mathcal{G}$ be a set of graphs and let $G' = (V', E', \alpha', \beta')$ and $G = (V, E, \alpha, \beta)$. We say that G' is isomorphic to G (written as $G' \simeq G$) if there is an injective function $f : V' \longrightarrow V$ such that: (1) $\forall e' = (u', v') \in E'$ $\exists e = (f(u'), f(v')) \in E$, (2) $\forall u' \in V', \alpha'(u') = \alpha(f(u'))$ and (3) $\forall e' \in E', \beta'(e') = \beta(f(e'))$. If $f : V' \longrightarrow V$ is a bijective function then G' is automorphic to G (written as $G' = G$). If G' is not isomorphic to G then we write $G' \not\simeq G$.*

A graph G' is $\mathcal{G}$-isomorphic (written as $G' \simeq \mathcal{G}$) if: (1) $\exists G \in \mathcal{G} : G' \simeq G$. A graph G' is not $\mathcal{G}$-isomorphic (written as $G' \not\simeq \mathcal{G}$) if: (1) $\forall G \in \mathcal{G} : G' \not\simeq G$.

Definition 4. *Given the set of graphs $\mathcal{G}_1, ..., \mathcal{G}_n$ and a graph $M_{\mathcal{G}_1,...,\mathcal{G}_n}$. $M_{\mathcal{G}_1,...,\mathcal{G}_n}$ is a common subgraph for $\mathcal{G}_1, ..., \mathcal{G}_n$ if: $\forall i \in \langle 1, n \rangle : M_{\mathcal{G}_1,...,\mathcal{G}_n} \simeq \mathcal{G}_i$. Set of all common subgraphs for $\mathcal{G}_1, ..., \mathcal{G}_n$ will be denoted by $\mathcal{M}_{\mathcal{G}_1,...,\mathcal{G}_n}$. Set of all minimal (with respect to size i.e. containing only one edge and either one or two vertices) common subgraphs for $\mathcal{G}_1, ..., \mathcal{G}_n$ will be denoted as $\mathcal{M}_{\mathcal{G}_1,...,\mathcal{G}_n}^{\mathrm{Min}}$.*

Definition 5. *Given the sets of graphs $\mathcal{N}$ and a graph P. A graph $C_{P \to \mathcal{N}}$ is a contrast subgraph of P with respect to $\mathcal{N}$ if: (1) $C_{P \to \mathcal{N}} \simeq P$ and (2) $C_{P \to \mathcal{N}} \not\simeq \mathcal{N}$. It is minimal (with respect to isomorphism) if all of $C_{P \to N}$'s strict subgraphs are not contrast subgraphs. Set of all minimal contrast subgraphs of P with respect to $\mathcal{N}$ will be denoted as $\mathcal{C}_{P \to \mathcal{N}}^{\mathrm{Min}}$.*

Definition 6. *Given the sets of graphs $\mathcal{P} = \{P_1, ..., P_n\}$ and $\mathcal{N}$. Let $\mathcal{C}_{P_i \to \mathcal{N}}^{\mathrm{Min}}$ be the set of all minimal contrast subgraphs of P_i with respect to $\mathcal{N}$, $i \in \langle 1, n \rangle$. $\mathcal{C}_{\mathcal{P} \to \mathcal{N}}^{\mathrm{Min}}$ is a set of all minimal contrast subgraphs of $\mathcal{P}$ with respect to $\mathcal{N}$ if: (1) $\forall C \in \mathcal{C}_{P_i \to \mathcal{N}}^{\mathrm{Min}} \ \exists J \in \mathcal{C}_{\mathcal{P} \to \mathcal{N}}^{\mathrm{Min}} : J \simeq C$, for $i \in \langle 1, n \rangle$, (2) $\forall J_1 \in \mathcal{C}_{\mathcal{P} \to \mathcal{N}}^{\mathrm{Min}} \ \neg \exists J_2 \in \mathcal{C}_{\mathcal{P} \to \mathcal{N}}^{\mathrm{Min}} \setminus J_1 : J_2 \simeq J_1$.*

$\mathcal{C}_{\mathcal{P} \to \mathcal{N}}^{\mathrm{Min}}$ contains all minimal subgraphs (patterns) which are in $\mathcal{P}$ (i.e. each subgraph in $\mathcal{C}_{\mathcal{P} \to \mathcal{N}}^{\mathrm{Min}}$ is isomorphic to at least one graph from $\mathcal{P}$) and are not present in $\mathcal{N}$ (i.e. each subgraph in $\mathcal{C}_{\mathcal{P} \to \mathcal{N}}^{\mathrm{Min}}$ is not isomorphic to any graph from $\mathcal{N}$). What is more $\mathcal{C}_{\mathcal{P} \to \mathcal{N}}^{\mathrm{Min}}$ contains only minimal (with respect to size and isomorphism) subgraphs.

Definition 7. *Given the sets of graphs $\mathcal{G}$, $\mathcal{N}$, $\mathcal{P}$ and a graph G. Let $\mathcal{S} = \{G' \in \mathcal{G} : G \simeq G'\}$. Support of graph G in $\mathcal{G}$ is defined as follows: $\mathrm{supp}_{\mathcal{G}}(G) = \frac{\mathrm{card}(\mathcal{S})}{\mathrm{card}(\mathcal{G})}$, where $\mathrm{card}(\mathcal{G})$ denotes the cardinal number of set $\mathcal{G}$. Growth rate of G in favour of $\mathcal{P}$ against $\mathcal{N}$ is expressed as follows: $\rho_{\mathcal{P} \to \mathcal{N}}(G) = \frac{\mathrm{supp}_{\mathcal{P}}(G)}{\mathrm{supp}_{\mathcal{N}}(G)}$.*

5   Web Document Classification Algorithm

In this section we propose a classification algorithm called: CCPC (Contrast Common Patterns Classifier). We present only the general concept without implementation details.

The concept of contrast subgraph is directly associated with the concept of jumping emerging pattern (JEP). They both define a pattern (either subgraph or set of items) exclusive for one class of objects. Similarly, the common subgraphs are associated with emerging patterns (EP), i.e. patterns that are present in both classes of objects. Measures for classical emerging patterns designed for classification purposes are mainly based on the support of a pattern in different classes of objects. This section adapts some classical scoring schemes to be used with contrast and common subgraphs.

Let $\mathcal{G}$ be a set of training graphs (graphs used for the learning of a classifier) and G be a test graph (graph to be classified). Let $\mathcal{G}$ be divided into n disjoint decision classes: $\mathcal{G}_1, ..., \mathcal{G}_n$; $\mathcal{G} = \bigcup_{i=1..n} \mathcal{G}_i$. Let $\mathcal{C}^{\mathrm{Min}}_{\mathcal{G}_i \to (\mathcal{G} \setminus \mathcal{G}_i)}$ be the set of all minimal contrast subgraphs of $\mathcal{G}_i$ with respect to graph set $(\mathcal{G} \setminus \mathcal{G}_i)$, where $i \in \langle 1, n \rangle$. Let $\mathcal{M}^{\mathrm{Min}}_{\mathcal{G}_1,...,\mathcal{G}_n}$ be the set of all minimal common subgraphs for $\mathcal{G}_1, ..., \mathcal{G}_n$.

Let us now define a few score routines used for classification. Score is obtained using contrast subgraphs according to the following equations ($i \in \langle 1, n \rangle$):

$$\mathrm{scConA}_{\mathcal{G}_i}(G) = \sum_{K \in \mathcal{K}} \mathrm{supp}_{\mathcal{G}_i}(K), \quad \mathcal{K} = \{K : K \in \mathcal{C}^{\mathrm{Min}}_{\mathcal{G}_i \to (\mathcal{G} \setminus \mathcal{G}_i)} \wedge K \simeq G\} \quad (1)$$

$$\mathrm{scConB}_{\mathcal{G}_i}(G) = \frac{1}{\lambda_{\mathcal{G}_i}} * \sum_{K \in \mathcal{K}} \mathrm{supp}_{\mathcal{G}_i}(K), \quad \mathcal{K} = \{K : K \in \mathcal{C}^{\mathrm{Min}}_{\mathcal{G}_i \to (\mathcal{G} \setminus \mathcal{G}_i)} \wedge K \simeq G\} \quad (2)$$

where $\lambda_{\mathcal{G}_i}$ is a scaling factor. Scaling factors are median values from statistics of the contrast scores (1) determined for each graph from classes: $\mathcal{G}_1, ..., \mathcal{G}_n$.

Score is also calculated using common subgraphs according to the following equations:

$$\mathrm{scComA}_{\mathcal{G}_i}(G) = \sum_{K \in \mathcal{K}} \mathrm{supp}_{\mathcal{G}_i}(K), \quad \mathcal{K} = \{K : K \in \mathcal{M}^{\mathrm{Min}}_{\mathcal{G}_1,...,\mathcal{G}_n} \wedge K \simeq G\} \quad (3)$$

$$\mathrm{scComB}_{\mathcal{G}_i}(G) = \sum_{K \in \mathcal{K}} \rho_{\mathcal{G}_i \to (\mathcal{G} \setminus \mathcal{G}_i)}(G), \quad \mathcal{K} = \{K : K \in \mathcal{M}^{\mathrm{Min}}_{\mathcal{G}_1,...,\mathcal{G}_n} \wedge K \simeq G\} \quad (4)$$

In (3) score depends directly on the support of the subgraphs, whereas in (4) it depends on the growth rate of certain patterns.

Classifier train process looks as follows. First all minimal (with respect to size and inclusion (non-isomorphic)) contrast subgraphs characteristic for each class are discovered ($\mathcal{C}^{\mathrm{Min}}_{\mathcal{G}_i \to (\mathcal{G} \setminus \mathcal{G}_i)}$, where $i \in \langle 1, n \rangle$). Then all minimal (with respect to size and inclusion (non-isomorphic)) common subgraphs for all classes are discovered ($\mathcal{M}^{\mathrm{Min}}_{\mathcal{G}_1,...,\mathcal{G}_n}$). Subgraph discovery is performed using DFS (Depth First Search) code [20] generation method and all necessary automorphism and isomorphism checking are performed using canonical labelling, nodes invariants, and iterative partitioning methods [15], [9]. Additional speed up can be achieved by limiting the size of discovered subgraphs i.e. instead of discover all contrast subgraphs only some of them are discovered up to given size (number of edges). This limitation may influence classification accuracy.

Classification process looks as follows. First scores based on contrast subgraphs are calculated for each class. We can choose between presented scoring schemes: scConA - we calculate $\mathrm{scConA}_{\mathcal{G}_i}(G)$ from eq. (1) for each decision class; scConB - we calculate $\mathrm{scConB}_{\mathcal{G}_i}(G)$ from eq. (2) for each decision class.

Test example G is assigned to a class with a highest score. If two or more decision classes have the highest score then G remains unclassified and scores based on common subgraphs are then calculated. Again we can choose one of the two approaches:scComA - we calculate $\mathrm{scComA}_{\mathcal{G}_i}(G)$ from eq. (3) for each decision class;scComB - we calculate $\mathrm{scComB}_{\mathcal{G}_i}(G)$ from eq. (4) for each decision class.

Test sample is assigned to a class with a higher score. If two or more decision classes have the highest score then G remains unclassified.

6   Experiments and Results

In order to evaluate the performance of the proposed classifiers we performed several experiments on three different benchmark collections of web documents, called F-series, J-series and K-series. The data comes from [1]. Each collection contains HTML documents which were originally news pages hosted at Yahoo (www.yahoo.com). Each document in every collection has a category (class) associated to the content of the document.

The original F-series collection contains 98 documents. Each of them is assigned to one or more (maximum three) of 17 subcategories of four major category areas. Some of the documents have conflicts subcategories (i.e. belonging to different major categories). We decided to remove those documents and simplify the problem by reducing the number of classes to major categories. The same operation was performed in [17], [13], [16], [14]. Final F-series collection contains 93 documents and four classes. Each document is assigned to one class.

The J-series collection contains 185 documents assigned to 10 categories while the K-series consists of 2340 documents belonging to 20 categories. In both cases each document is assigned to exactly one category.

We created one more web document collection, called K^{7th}-series. According to [17], this collection was created by selecting every 7th document from K-series. It contains 335 documents representing 19 classes.

Summary of the benchmark document collections is provided in Table 1.

Table 1. Detailed information on F-series, J-series, K-series, and K^{7th}-series document collections

Document	Number of		Percentage of documents in category		
collection	documents	categories	minimal	median	maximal
F-series	93	4	20.4	24.7	28.0
J-series	185	10	8.6	10.3	10.8
K-series	2340	20	0.4	3.0	21.1
K^{7th}-series	335	19	0.6	3.9	21.2

We basically concentrated our research on the following issues: performance of classifiers and influence of training and test data complexity. Performance of classifiers (ability to assign the correct class to a document) was evaluated using leave-one-out cross-validation procedure. Accuracy of a classifier is expressed as the percentage of correctly classified documents. By data complexity we mean number of nodes in graph (N) representing each document.

Figures 1, 2, and 3 show classification accuracy of our method using different scoring routines as a function of number of nodes in a graph representing each document for different document collections. Figure 4 shows percentage of documents classified by contrast graphs for different document collections using scConB scoring scheme.

Table 2. Comparison of classification accuracy

Doc. series	Document model	Algorithm description	Number of nodes	Classification accuracy
F	Vector	k-NN, cosine	NA	94.6
	Vector	k-NN, Jaccard	NA	94.6
	Graph	k-NN, MCS	30	96.8
	Hybrid	k-NN, Naive	100	95.7
	Hybrid	k-NN, Smart	100	95.7
	Graph	**CCPC**	**30/50/100**	**91.4/98.9/98.9**
J	Vector	k-NN, cosine	NA	74.6
	Vector	k-NN, Jaccard	NA	77.3
	Graph	k-NN, MCS	30/60	85.4/86.5
	Hybrid	k-NN, Naive	30	87.6
	Hybrid	k-NN, Smart	40	94.6
	Graph	**CCPC**	**30/45**	**86.5/91.4**
K	Vector	k-NN, cosine	NA	77.5
	Vector	k-NN, Jaccard	NA	80.4
	Graph	k-NN, MCS	40/100/150	78.2/84.6/85.7
	Hybrid	k-NN, Naive	100	86.0
	Hybrid	k-NN, Smart	120	86.3
	Hybrid	C4.5, Naive	100	78.0
	Hybrid	NBC, Smart	100	76.0
	Graph	**CCPC**	**40**	**86.3**
K^{7th}	Vector	k-NN, cosine	NA	67.5
	Graph	k-NN, MCS	30/70	65.3/77.0
	Graph	**CCPC**	**30/70/100**	**59.4/73.7/84.8**

For all data collections scoring routines based on contrast subgraphs (scConA, scConB) has dominant influence on accuracy. Scoring routines based on common subgraphs (scComA, scComB) have in most cases very little impact on final results. As far as scoring on contrast subgraphs are concerned obtained with the scConB scoring scheme are more accurate than those obtained with scConA.

For most collections (except for J-series) classification accuracy is increasing with increasing number of nodes in a graph.

Table 2 shows comparison of accuracy for different classifiers based on different document representation. We made a selection of best available results for the following methods: k-NN with the vector representation, cosine and Jaccard similarity measure [17], [16]; k-NN with the graph representation and maximum common subgraph similarity measure [17], [16]; k-NN with the hybrid representation, Manhattan similarity measure, naive and smart subgraph extraction [13], [14]; C4.5 with the hybrid representation, Manhattan similarity measure, naive and smart subgraph extraction [14].

Our classifier (CCPC) outperforms other methods for F-series (smallest document collection) and K-series (largest document collection). For K-series our method required only 40 nodes while the runner-up required 120 to provide similar accuracy. For J-series collection CCPC method provided second best

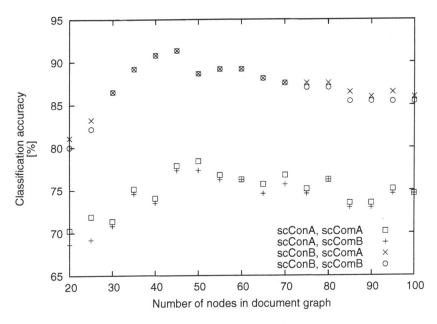

Fig. 1. Classification accuracy for J-series web document collection

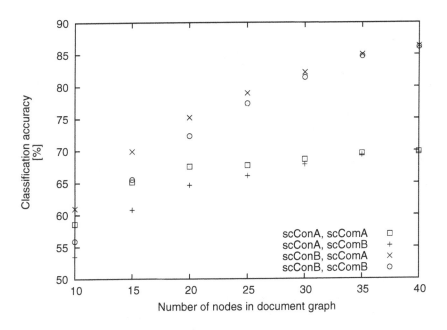

Fig. 2. Classification accuracy for K-series web document collection

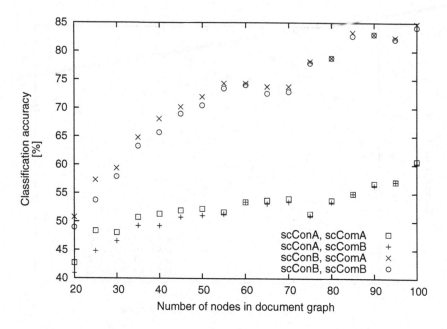

Fig. 3. Classification accuracy for K^{7th}-series web document collection

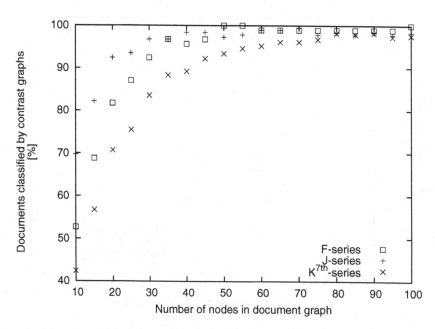

Fig. 4. Percentage of documents classified by contrast graphs for F-series, J-series and K^{7th}-series web document collections using scConB scoring scheme

result (after k-NN Smart). It is worth to mention that for all data collections we managed to achieve better results than methods based on vector document representation.

7   Conclusions

In this paper we presented a new approach for classifying web documents. Our algorithm (CCPC - Contrast Common Patterns Classifier) operates on a graph representation of a web document (HTML page) and uses concepts of contrast and common subgraphs as well as some ideas characteristic for emerging patterns technique.

The results show that our algorithm is competitive to existing schemes in terms of accuracy and data complexity (number of terms used for classification). For three document collections (F-series, K-series and K^{7th}) our method outperformed other approaches. What is more, construction and structure of our classifier is quite simple. This feature lets the domain expert to modify the classifier with professional knowledge by adding new patterns to contrast or common subgraphs sets or by modifying their supports.

The main concept of our classifier is domain independent so it can be used to solve classification problems in other areas (where data is represented by a graph) as well. It was already successfully applied in computational chemistry and chemical informatics for solving chemical compounds classification problems [6], i.e. detecting mutagenicity and carcinogenicity of chemical compounds for a given organism. The results shown that our algorithm outperformed the existing algorithms in terms of accuracy.

Our future research will concentrate on adapting the concept of contrast graphs to work with popular classification algorithm like k-NN. We will also try to apply our classifier in other research domains.

References

1. Datasets "pddpdata": ftp://ftp.cs.umn.edu/dept/users/boley/pddpdata/
2. Borgelt, C., Berthold, M.R.: Mining molecular fragments: Finding relevant substructures of molecules. In: ICDM '02: Proceedings of the 2002 IEEE International Conference on Data Mining (ICDM'02), Washington, DC, USA, pp. 51–58. IEEE Computer Society Press, Los Alamitos (2002)
3. Bunke, H., Shearer, K.: A graph distance metric based on the maximal common subgraph. Pattern Recognition Letters 19, 255–259 (1998)
4. Deshpande, M., Kuramochi, M., Karypis, G.: Frequent sub-structure-based approaches for classifying chemical compounds. In: ICDM '03: Proceedings of the 2002 IEEE International Conference on Data Mining (ICDM'03), pp. 35–42 (2003)
5. Diestel, R.: Graph Theory. Springer, New York (2000)
6. Dominik, A., Walczak, Z., Wojciechowski, J.: Classifying chemical compounds using contrast and common patterns. In: Beliczynski, B., Dzielinski, A., Iwanowski, M., Ribeiro, B. (eds.) ICANNGA 2007, vol. 4432, pp. 772–781. Springer-Verlag, Berlin Heidelberg (2007)

7. Dong, G., Li, J.: Efficient mining of emerging patterns: Discovering trends and differences. In: Knowledge Discovery and Data Mining, pp. 43–52 (1999)
8. Dong, G., Zhang, X., Wong, L., Li, J.: CAEP: Classification by aggregating emerging patterns. In: Discovery Science, pp. 30–42 (1999)
9. Fortin, S.: The graph isomorphism problem. Technical report, University of Alberta, Edmonton, Alberta, Canada (1996)
10. Inokuchi, A., Washio, T., Motoda, H.: An apriori-based algorithm for mining frequent substructures from graph data. In: Principles of Data Mining and Knowledge Discovery, pp. 13–23 (2000)
11. Kotagiri, R., Bailey, J.: Discovery of emerging patterns and their use in classification. In: Gedeon, T.D., Fung, L.C.C. (eds.) AI 2003. LNCS (LNAI), vol. 2903, pp. 1–12. Springer, Heidelberg (2003)
12. Kuramochi, M., Karypis, G.: Frequent subgraph discovery. In: ICDM '01: Proceedings of the 2001 IEEE International Conference on Data Mining (ICDM'01), pp. 313–320. IEEE Computer Society Press, Los Alamitos (2001)
13. Markov, A., Last, M.: Efficient graph-based representation of web documents. In: Proceedings of the Third International Workshop on Mining Graphs, Trees and Sequences (MGTS 2005), pp. 52–62 (2005)
14. Markov, A., Last, M., Kandel, A.: Model-based classification of web documents represenetd by graphs. In: Proceedings of WebKDD 2006: KDD Workshop on Web Mining and web Usage Analysis, iin conjunction with the 12th ACM SIGKDD International Conference on Knoowledge Discovery and Data Mining (KDD 2006), Philadelphia, PA, USA, ACM, New York (2006)
15. Read, R.C., Corneil, D.G.: The graph isomorph disease. Journal of Graph Theory 363, 339–363 (1977)
16. Schenker, A.: Graph-Theoretic Techniques for Web Content Mining. PhD thesis, University of South Florida (2003)
17. Schenker, A., Last, M., Bunke, H., Kandel, A.: Classification of web documents using a graph model. In: Proceedings of the Seventh International Conference on Document Analysis and Recognition (ICDAR 2003), vol. 01, pp. 240–244. IEEE Computer Society, Los Alamitos, CA, USA (2003)
18. Ting, R.M.H., Bailey, J.: Mining minimal contrast subgraph patterns. In: SIAM '06: Proceedings of the 2006 SIAM Conference on Data Mining, Maryland, USA (2006)
19. Ullmann, J.R.: An algorithm for subgraph isomorphism. J. ACM 23(1), 31–42 (1976)
20. Yan, X., Han, J.: gspan: Graph-based substructure pattern mining. In: ICML '02: Proceedings of the Nineteenth International Conference on Machine Learning (2002)

Context-Specific Independence Mixture Modelling for Protein Families

Benjamin Georgi [1], Jörg Schultz [2], and Alexander Schliep [1]

[1] Max Planck Institute for Molecular Genetics, Dept. of Computational Molecular
Biology, Ihnestrasse 73, 14195 Berlin, Germany
[2] Universität Würzburg, Dept. of Bioinformatics, 97074 Wuerzburg, Germany

Abstract. Protein families can be divided into subgroups with functional differences. The analysis of these subgroups and the determination of which residues convey substrate specificity is a central question
in the study of these families. We present a clustering procedure using
the *context-specific independence* mixture framework using a Dirichlet
mixture prior for simultaneous inference of subgroups and prediction of
specificity determining residues based on multiple sequence alignments
of protein families. Application of the method on several well studied
families revealed a good clustering performance and ample biological
support for the predicted positions. The software we developed to carry
out this analysis *PyMix - the Python mixture package* is available from
http://www.algorithmics.molgen.mpg.de/pymix.html.

1 Introduction

Proteins within the same family commonly fall into sub categories which differ by
functional specificity. The categorization and analysis of these subgroups is one of
the central challenges in the study of these families. In particular it is of interest
which residues determine functional specificity of a subgroup. These functional
residues are characterized by a strong signal of subgroup specific conservation.

A number of studies have focused on the question how to detect residues which
determine functional specificity based on prior knowledge of subtype membership. A review of these methods can be found in [14]. Among the approaches
taken were relative entropy based scores [12], classification based on similarity
to a data base of functional residue templates [4], contrasting position specific
conservation in orthologues and paralogues to predict functional residues [21].
In [26] the authors use known reference protein 3D structures to find conserved
discriminatory surface residues. One major limitation of these *supervised* approaches is the requirement of biological expert annotation of the number of
subtypes and subtype assignments for each sequence. Which then limits usefulness of these methods to cases where prior biological knowledge is abundant. In
the absence of such knowledge the inference of the subgroups becomes one central aspect of the prediction of functional residues. In many cases the subgroup
structure of a given family is a direct consequence of evolutionary divergence
of homologue sequences. As such it is not surprising that methods based on

J.N. Kok et al. (Eds.): PKDD 2007, LNAI 4702, pp. 79–90, 2007.

the phylogenetic tree of a family have been extensively and successfully used to study protein family subgroups [15,16,22,25]. However, the performance of these methods does degrade in cases where the evolutionary divergence between subgroups is large. Moreover phylogeny does not account for situations where functional relatedness of proteins arose from a process of convergent evolution. As such there is a need for additional methods for detection and analysis of the subgroups inherent in a set of related sequences. Here, we present the first unsupervised approach to simultaneously cluster related sequences and predict functional residues which does not rely on a phylogenetic tree. Prior work either relies on inference of phylogenetic trees or is unsupervised.

The clustering procedure employs the Bayesian *context-specific independence* mixture framework [9]. CSI mixtures have for instance been used for modeling of transcription factor binding sites [9], clustering of gene expression data [1] or the analysis of complex genetic diseases [10]. The central idea of the *context-specific independence* model is to adapt the number of model parameters to a level which is appropriate for a given data set. This notion of automatic adaption of a probabilistic model to the data has received considerable attention in the context of Bayesian networks [3,5,7].

One of the challenges of clustering protein families into subgroups based on the sequence is that the discriminating features one attempts to learn are a property of the structure rather than the sequence. As an example, consider three subgroups with perfect conservation of amino acids Leucine, Isoleucine and Tryptophan respectively at one position. A naive application of a clustering would consider said position to be highly discriminative for all three groups. Of course, this would be misleading due to the great similarity in chemical properties between Leucine and Isoleucine which makes them, to some extent, synonymous as far as structure is concerned. To adapt the CSI mixture model for this situation we apply a parameter prior in form of a mixture of Dirichlet distributions. These Dirichlet mixture priors have been successfully used to improve generalization properties of parameter estimates for probabilistic models for small sample sizes [23]. In the CSI framework a suitably chosen prior additionally acts to guide the structure learning towards distributions indicative of structural differences between the subgroups.

2   Methods

2.1   CSI Mixture Models

In this section we briefly introduce notation for conventional mixture models and our extension in the *context-specific independence* framework. For a more in depth coverage the reader is referred to [20] and [9] respectively. Let $X_1, ..., X_p$ be discrete random variables over the 20 amino acids and a gap symbol representing a multiple sequence alignment (MSA) with p positions (see Fig. 1a for an example). Given a data set D of N realizations, $D = x_1, ..., x_N$ with $x_i = (x_{i1}, ..., x_{ip})$ a conventional mixture density is given by

$$P(x_i) = \sum_{k=1}^{K} \pi_k \ f_k(x_i|\theta_k), \tag{1}$$

where the π_k are the mixture coefficients, $\sum_{k=1}^{K} \pi_k = 1$ and each component distribution f_k is a product distribution over $X_1, ..., X_p$ parameterized by parameters $\theta_k = (\theta_{k1}, ..., \theta_{kp})$

$$f_k(x_i|\theta_k) = \prod_{j=1}^{p} P_j(x_{ij}|\theta_{kj}). \tag{2}$$

The complete parameterization of the mixture is then given by $\theta = (\pi, \theta_1, ..., \theta_k)$. For a data set D of N samples the likelihood under mixture M is given by

$$P(D|M) = \prod_{i=1}^{N} P(x_i). \tag{3}$$

The way the mixture arises from a given MSA is visualized in Fig. 1; 1a) shows an example MSA with four positions and three subgroups $C_1 - C_3$ within the sequences. An abstract representation of the corresponding mixture model is shown in 1b). Here each position of the alignment is modelled by a discrete random variable $X_1 - X_4$ and each cell in the matrix represents a uniquely parameterized discrete distribution with parameters estimated from the amino acids of the sequences assigned to the subgroup at the respective positions.

The central quantity for both the parameter estimation with *Expectation Maximization* (EM) [6] as well as the subgroup assignment is the posterior of component membership given by

$$\tau_{ik} = \frac{\pi_k \ f_k(x_i|\theta_k)}{\sum_{k=1}^{K} \pi_k \ f_k(x_i|\theta_k)}, \tag{4}$$

i.e. τ_{ik} is the probability that a sample x_i was generated by component k.

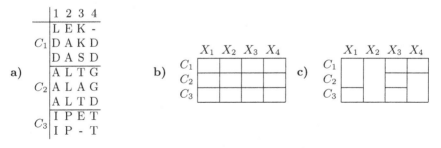

Fig. 1. a) Example input MSA. Eight sequences with four positions each divided into three subgroups. b) Model structure matrices for conventional mixture model and c) structure matrix for the CSI mixture model.

The basic idea of the CSI extension to the mixture framework is to automatically adapt model complexity to match the variability observed in the data. This is visualized in Fig. 1. In 1b) the structure matrix for a conventional mixture model is depicted. Each cell represents a uniquely parameterized distribution for each component and sequence position. In opposition to that a CSI model (Fig. 1 c) may assign the same distribution for a position to several components as indicated by the cell spanning multiple rows in the structure matrix. In example C_1 and C_2 share a distribution parameters for position X_1. For position X_2 all components have the same distribution and for position X_4 all components except C_1 have the same parameters. This not only yields a reduced model complexity, it also allows the direct characterization of protein subgroups by the model structure matrix. For instance it can be seen that position X_4 is uniquely characterizing component C_1. For a protein family data set this might indicate that position X_4 is a candidate for functional residue with respect to subgroup C_1.

Formally the CSI mixture model is defined as follows: Given the set $C = \{1, .., K\}$ of component indexes and sequence positions $X_1, ..., X_p$, let $G = \{g_j\}_{(j=1,...,p)}$ be the CSI structure of the model M. Then $g_j = (g_{j1}, ... g_{jZ_j})$ such that Z_j is the number of parameter subgroups for X_j and each $g_{jr}, r = 1, ..., Z_j$ is a subset of component indexes from C. Thus, each g_j is a partition of C into disjunct subsets such that each g_{jr} represents a subgroup of components with the same distribution for X_j. The CSI mixture distribution is then obtained by replacing $f_{kj}(x_{ij}|\theta_{kj})$ with $f_{kj}(x_{ij}|\theta_{g_j(k)j})$ in (2) where $g_j(k) = r$ such that $k \in g_{jr}$. Accordingly $\theta_M = (\pi, \theta_{X_1|g_{1r}}, ..., \theta_{X_p|g_{pr}})$ is the model parameterization. $\theta_{X_j|g_{jr}}$ denotes the different parameter sets in the structure for position j. The complete CSI model M is then given by $M = (G, \theta_M)$. Note that we have covered the structure learning algorithm in more detail in a previous publication [9].

2.2   Dirichlet Mixture Priors

In the Bayesian setting the fit of different models to the data is assessed by the model posterior $P(M|D)$ given by

$$P(M|D) \propto P(M)P(D|M),$$

where $P(D|M)$ is the likelihood of the data under M and $P(M)$ is the model prior. For $P(M) = P(K)P(G)$ a simple factored form was used with $P(K) = \gamma^K$ and $P(G) = \prod_{j=1}^{p} \alpha^{Z_j}$. $\gamma < 1$ and $\alpha < 1$ are hyperparameters which determine the strength of the bias for a less complex model introduced by the prior. The likelihood term $P(D|M)$ is given by

$$P(D|M) = P(D|\vec{\theta}_M)P(\vec{\theta}_M).$$

Here $P(D|\vec{\theta}_M)$ is simply the mixture likelihood (1) evaluated at the *maximum a posteriori* (MAP) parameters $\vec{\theta}_M$ and $P(\vec{\theta}_M)$ is a conjugate prior over the model parameters.

One choice of $P(\vec{\theta}_M)$ for discrete distributions θ is a mixture of Dirichlet distributions. A Dirichlet mixture prior (DMP) over a discrete distribution $\theta = (\theta_1, ..., \theta_Q)$ is given by

$$P(\theta) = \sum_{g=1}^{G} q_g \, D_g(\theta | \alpha_g), \tag{5}$$

where D_g is the Dirichlet density parameterized by $\alpha_g = (\alpha_{g1}, ..., \alpha_{gQ}), \alpha_{gz} > 0$,

$$D_g(\theta) = \frac{\Gamma(\sum_{z=1}^{Q} \alpha_{gz})}{\prod_{z=1}^{Q} \Gamma(\alpha_{gz})} \prod_{z=1}^{Q} \theta_z^{\alpha_{gz}-1}.$$

The DMP has a number of attractive properties for the modeling of protein families. Not only does the DMP retain conjugacy to the discrete distribution which guarantees closed form solutions for the parameter estimates, it also allows for a great degree of flexibility in the induced density over the parameter space. This allows for the integration of amino acid similarities in the structure learning procedure.

2.3   Parameter Estimation

As the Dirichlet distribution is conjugate to the multinomial distribution, the MAP estimates for θ can be computed conveniently.

To obtain the MAP for Dirichlet Mixture priors in case of a mixture of discrete distributions we extend the update rules in [23] where the formulas for the single distribution case have been derived in detail. The MAP solution for the distribution over position j in component k, $\theta_{kj} = (\theta_{kj1}, ..., \theta_{kjQ})$, where Q is the size of the alphabet Σ (21 for amino acids plus gap symbol) is given by

$$\theta_{kjz} = \sum_{g=1}^{G} q_g \frac{T_{kjz} + \alpha_{gz}}{T_k + |\alpha_g|} \tag{6}$$

where the $T_{kj} = (T_{kj1}, ..., T_{kjQ})$ are the expected sufficient statistics of mixture component k in feature j with

$$T_{kjz} = \sum_{i=1}^{N} \tau_{ki} \, \delta_{(x_{ij}=\Sigma_z)},$$

$T_{kj} = \sum_{z=1}^{Q} T_{kjz}$ and q_{gj} is the component membership posterior of θ_{kjz} under the DMP $P(\theta)$ computed according to (4).

2.4   Prior Parameter Derivation

In order to apply the DMP framework on the problem of regularizing the structure learning for protein families we have to specify the parameterization of $P(\theta)$. This includes the choice of G, the q_g and the α_g.

We considered three different approaches to arrive at choices for these parameters,

1. choice of parameters based on a PAM series amino acid substitution probability matrix,
2. use of previously published DMP regularizers [23] based on machine learning techniques and
3. heuristic parameter derivation based on basic chemical properties of the amino acids.

The latter approach proved to be most suitable for our purposes and therefore will be described in more detail below. It should be stressed however that the non-optimal performance of the DMPs from [23] may be caused by their focus on providing suitable regularization to compensate for small sample sizes. While this is certainly related, it is not quite the same as the kind of regularization we require for the CSI structure learning. Clearly a machine learning approach for specifying the prior parameters would be desirable. This however is not straightforward for two reasons: First, it is not clear how the training data for learning a DMP for this application would have to be assembled and secondly the optimization of DMPs is a difficult problem as many local minima exist [23]. In any case, it seems appealing to use a simple heuristically specified DMP in this first analysis in order to establish a baseline performance of the CSI mixtures in this application.

Table 1. The twenty amino acids can be characterized by nine chemical properties. A x in the table denotes the presence, a · the absence of a trait.

	A	R	N	D	C	Q	E	G	H	I	L	K	M	F	P	S	T	W	Y	V	-
Hydrophobic	x	·	·	·	x	·	·	x	x	x	x	x	x	x	·	·	x	x	x	x	·
Polar	·	x	x	x	·	x	x	·	x	·	·	x	·	·	·	x	x	x	x	·	·
Small	x	·	x	x	x	·	·	x	·	·	·	·	·	·	x	x	x	·	·	x	·
Tiny	x	·	·	·	·	·	·	x	·	·	·	·	·	·	·	x	·	·	·	·	·
Aliphatic	·	·	·	·	·	·	·	·	·	x	x	·	·	·	·	·	·	·	·	x	·
Aromatic	·	·	·	·	·	·	·	·	x	·	·	·	·	x	·	·	·	x	x	·	·
Positive	·	x	·	·	·	·	·	·	x	·	·	x	·	·	·	·	·	·	·	·	·
Negative	·	·	·	x	·	·	x	·	·	·	·	·	·	·	·	·	·	·	·	·	·
Charged	·	x	·	x	·	·	x	·	x	·	·	x	·	·	·	·	·	·	·	·	·

The impact of an amino acid substitution on the fold of a protein depends on the similarity of the chemical properties of the two amino acids. The more dissimilar the amino acids are, the more pronounced the effect on protein structure will be. The relevant chemical properties can be arranged into a hierarchy of more general and specific properties [18]. The nine properties we consider and the assignment of amino acids is summarized in Table 1. Here 'x' and '·' denote presence and absence of a property respectively. Note that the gap symbol '-' is negative for all properties.

Based on this characterization of the amino acids by their basic chemical properties we construct a DMP as follows: To each of the properties in Table 1 we assign a component D_g in the DMP. The parameters α_g are chosen such that α_{gj}

is larger if amino acid j has the property. This means we construct nine Dirichlet distributions which give high density to $\theta_{X_j|g_{jr}}$ with strong prevalence of amino acids with a certain property. The combination of all property specific D_g in the DMP then yields a density which allows the quantification of similarity between amino acids in the probabilistic framework. In order to arrive at a scheme to choose the parameters of the DMP the following constraints were taken into consideration:

- The strength of a Dirichlet distribution prior D_g is determined by the sum of its parameters $|\alpha_g|$. The size of $|\alpha_g|$ is also anti-proportional to the variance of D_g. To assign equal strength to all property specific Dirichlets D_g, all $|\alpha_g|$ are set to be identical.
- More general properties should receive greater weights q_g in the DMP.
- The strength of the prior, i.e. $|\alpha_g|$ should depend on the size of the data set N.

This leads to the following heuristics for choosing the DMP parameters: Let the strength of each D_g be one tenth of the data set size; i.e. $|\alpha_g| = \frac{N}{10}$ and $b = \frac{0.75\,|\alpha_g|}{21}$ the base value for the parameters α_g. Then $\alpha_{gj} = b$, for all amino acids were the property is absent and

$$\alpha_{gj} = b + \frac{0.25\,|\alpha_g|}{B_g},$$

for all amino acids where the property is present, where B_g denotes the number amino acid which have the property. Finally, the weights q_g are set to

$$q_g = \frac{B_g}{\sum_{g=1}^{G} B_g}$$

which means that more general properties receive proportionally higher weight in the prior. Thus, the priors in the model introduce two types of bias' into the structure learning. An unspecific preference for a less complex model given by $P(M)$ and a specific preference for parameters $\theta_{X_j|g_{jr}}$ that match the amino acid properties encoded in the prior $P(\theta)$.

2.5   Feature Ranking

To predict which features are functional residues for a given subgroup, it is necessary to refine the information in the CSI structure matrix by ranking the informative features. Since these features are distinguished by subgroup specific sequence conservation, the relative entropy is a natural choice to score for putative functional residues.

In order to quantify the relevance of X_j for subgroup i we assume a CSI structure in which X_j is uniquely discriminative for component i, i.e. $Z_j = 2$ with $g_{j1} = \{i\}$ and $g_{j2} = \{1, ..., K\} \setminus i$. Based on this structure a component-specific parameter set θ_{ji} and a parameter set for all other components θ_{other} are constructed by doing a single structural EM update.

The score for feature X_j in component i is then given by $S_{ij} = KL(\theta_{ji}, \theta_{other})$, where KL is the symmetric relative entropy. Note that this is somewhat similar to the setup used in [12]. The major difference being that in [12] subgroup assignments were assumed to be known and in this work the scoring is based on the posterior distribution of component membership and parameter estimates induced by the expected sufficient statistics in the structural EM framework.

3   Results

We evaluated the performance of CSI mixture models for protein subfamilies on a number of data set of different sizes from families with known subtype assignments and structural information. This allows for a validation of the clustering results. Any column in the alignment with more than 33% gaps was removed prior to the clustering. Model selection was carried out using the *Normalized entropy criterion* (NEC) [2]. To assess the impact of the DMP on model performance sensitivity and specificity of the clusterings with DMP were compared to mixtures with the same number of components but a simple uninformative single Dirichlet prior.

3.1   L-Lactate Dehydrogenase Family

We analyzed members of the L-lactate dehydrogenase family, which differ in their substrate specificity. We analyzed two subfamilies, malate and lactate dehydrogenases. In this family, despite substantial variance within the subfamilies and between them, a single position is responsible for defining substrate specificity. Taking PDB 1IB6 as reference sequence, an R in position 81 confers specificity for lactate whereas a Q in the same position would change the substrate to malate. Clusterings were computed for the 29 sequences in the PFAM seed alignment of that domain (PF00056). The alignment contained 16 lactate dehydrogenases (LDH) and 13 malate dehydrogenases (MDH). NEC model selection indicated 2 components to provide the best fit for the data. The two components separated the MDH/LDH groups without error for the DMP mixture.When using the uninformative prior, considerably lower sensitivities and specificities of around 75% were achieved. To assess the robustness of this result we repeatedly trained two component models with DMP and uninformative priors. Averaged over 10 models the DMP achieved sensitivity 95% (SD 1.8) and specificity 93% (SD 2.4), the uninformative prior yielded sensitivity 76% (SD 8.7) and specificity 75% (SD 9.3). Thus, our method was able to identify the two subfamilies correctly without any prior biological knowledge. The position identified as most informative for distinguishing the groups was indeed the one responsible for substrate specificity. Many of the other highly ranked residues were arranged around the NAD interaction site of the domain, which suggests they play a role in malate / lactate recognition.

3.2 Protein Kinase Family

The protein kinase super family is one of the largest and best studied protein families. The human genome contains more than 500 protein kinases [19] with many involved in different diseases like cancer or diabetes. The probably most prominent classification of this key players in signal transduction is between tyrosine and serine/threonine kinases. These can be further subdivided according to different regulatory mechanisms [13]. In our test case, we combined these levels of classifications by joining tyrosine kinases (TK) with two groups of serine threonine kinases, STE (Homologs of yeast Sterile 7, Sterile 11, Sterile 20 kinases) and AGC (Containing PKA, PKG and PKC families). An alignment of 1221 representative sequences of the subfamilies was obtained from the *Protein kinase resource* [24]. The three best NEC model selection scores were assigned to 2, 3 and 4 components. Since the scores were too similar for a clear choice of components, we will consider the clustering of all three models in the following. In the three component model each family acquired its own subgroup with a sensitivity of 79% and a specificity of 83%. Results for the uninformative prior were only slightly worse (about 1% in both sensitivity and specificity) for this data set. These results were highly robust in the repetitions with standard deviations of 0.1%-0.6% on the sensitivities and specificities of both prior types. In the following PDB 2cpk (cAMP-dependent protein kinase, alpha-catalytic subunit, Mus musculus) is used as reference sequence for residue numbering. A ranking of the informative features of the three component model with respect to the TK subgroup yielded within the top 20 positions a region of three residues (168-170) which has been experimentally shown to be important for kinase substrate specificity [11]. For the two component model the TK and STE sequences were collected in one subgroup and the second was almost exclusively AGC. The four component model finally yielded a high specificity clustering (98%) in which the AGC sequence got split over two components. The sensitivity was 76%.

3.3 Nucleotidyl Cyclase Family

Nucleotidyl cyclases play an important role in cellular signaling by producing the second messengers cAMP and cGMP which regulate the activity of many other signalling molecules. As cGMP and cAMP fulfill different biological roles, specificity of converting enzymes is imperative. Five residues have been experimentally confirmed to convey substrate specificity, namely 938, 1016, 1018, 1019, 1020 (numbering according to PDB 1AB8) [17]. We used this family as a test case for families with multiple sites involved in functional classification, complementing the L-lactate dehydrogenase family with a single site. We computed a MSA from 132 GC (EC 4.6.1.2) and AC (EC 4.6.1.1) sequences obtained from the ExPASy data base [8]. The NEC model selection indicated two components to provide the best fit. The model with optimal NEC produced a clustering with sensitivity of 83% and specificity 87% with respect to the GC / AC subgroups. For the uninformative prior these values decreased to 70% and 73% respectively. Averaged over 10 models the uninformative prior yielded a decreased performance of 59% (SD 5.3) sensitivity and 62% (SD 5.6) respectively. The averaged

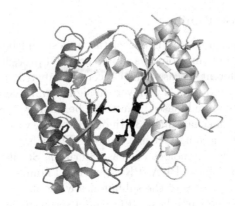

Fig. 2. Adenylyl cyclase with classifying sites highlighted - Subunit I in dark grey, subunit II in light grey. The 10 most informative sites were selected. Shown in black: experimentally validated identified sites, darkest grey: additional identified sites. A colored version of the figure is available from *http://algorithmics.molgen.mpg.de/pymix/Figure_Cyclase.png*

results for the DMP were sensitivity 73% (SD 4.3) and specificity 77% (SD 4.8). Figure 2 shows the three dimensional structure of 1AB8 with the 10 most informative sites highlighted. Indeed, these contain 4 of the sites involved in substrate specificity (1018 (ranked 2.), 1016 (3.), 938 (6.), 1019 (9.)). Further top ranking positions included sites which are part of the subunit I and II domain interface (919, 912, 911). Position 943 is right next to a forsoklin interaction site and position 891 interacts with magnesium. Residue 921 finally, is also a metal interacting site [27]. Thus, not only known substrate specific sites were identified, but also further functional sites. It would be interesting to experimentally test identified sites with no functional annotation.

4   Discussion

The results of CSI mixture-based clustering on a number of different protein families show that the approach is capable of simultaneously finding biological relevant subgroups, as well as predicting functional residues that characterize these groups. The functional residue prediction proved to be robust to some degree to imperfections in the clustering. This implies that our unsupervised approach to simultaneous clustering and determination of functional residues is feasible. Also note that our results for the functional residue prediction are strongly consistent with those reported by studies using supervised methods on the same families [12,26]. With regard to experimentally confirmed specificity determining residues found by these studies, we found 1/1 for L-lactate dehydrogenase, 3/3 for protein kinases and 4/5 for nucleotidyl cyclase.

The results also show that the DMP used in this analysis, in spite of being based on basic chemical properties and simple heuristics, consistently increases the performance of the mixture framework for the application on protein data, although the degree of improvement differs considerably between the families. This is not unexpected as one would expect differing amounts of synonymous substitutions within the various subgroups and that is the situation where the DMP makes the largest difference as compared to the uninformative prior. For comparison we also applied the tree-based method SPEL [22] to our data. The sources were obtained from the authors and run with default parameters. For the MDH/LDH data the true functional position 81 was not among the ten positions returned by SPEL. For the two larger data sets, there were implementation-issues and SPEL did unfortunately not produce any results.

For future work it might be worth investigating the impact of different DMPs on the clustering results and in particular whether customized DMPs for specific applications yield improvement over the more general purpose prior used in this work. Moreover, now that the usefulness of the method has been established on families with abundant prior knowledge about subgroups and structure, the next step must be to bring the method to bear to predict groups and functional residues on data sets where such knowledge does not exist yet. Finally, the software we developed to carry out this analysis *PyMix - the Python Mixture Package* is available from our home page *http://algorithmics.molgen.mpg.de/pymix.html*.

References

1. Barash, Y., Friedman, N.: Context-specific bayesian clustering for gene expression data. J. Comput. Biol. 9(2), 169–191 (2002)
2. Biernacki, C., Celeux, G., Govaert, G.: An improvement of the NEC criterion for assessing the number of clusters in a mixture model. Non-Linear Anal. 20(3), 267–272 (1999)
3. Boutilier, C., Friedman, N., Goldszmidt, M., Koller, D.: Context-specific independence in Bayesian networks. In: Uncertainty in Artificial Intelligence, pp. 115–123 (1996)
4. Chakrabarti, S., Lanczycki, C.J.: Analysis and prediction of functionally important sites in proteins. Protein Sci. 16(1), 4–13 (2007)
5. Chickering, D.M., Heckerman, D.: Efficient approximations for the marginal likelihood of bayesian networks with hidden variables. Mach. Learn. 29(2-3), 181–212 (1997)
6. Dempster, A., Laird, N., Rubin, D.: Maximum likelihood from incomplete data via the EM algorithm. Journal of the Royal Statistical Society, Series B, 1–38 (1977)
7. Friedman, N., Goldszmidt, M.: Learning bayesian networks with local structure. In: Proceedings of the NATO Advanced Study Institute on Learning in graphical models, pp. 421–459. Kluwer Academic Publishers, Norwell, MA, USA (1998)
8. Gasteiger, E., Gattiker, A., Hoogland, C., Ivanyi, I., Appel, D., Bairoch, A.: ExPASy: The proteomics server for in-depth protein knowledge and analysis. Nucleic Acids Res. 31(13), 3784–3788 (2003)
9. Georgi, B., Schliep, A.: Context-specific independence mixture modeling for positional weight matrices. Bioinformatics 22(14), e166–173 (2006)
10. Georgi, B., Spence, M.A., Flodman, P., Schliep, A.: Mixture model based group inference in fused genotype and phenotype data. In: Studies in Classification, Data Analysis, and Knowledge Organization, Springer, Heidelberg (2007)

11. Hanks, S.K., Quinn, A.M., Hunter, T.: The protein kinase family: conserved features and deduced phylogeny of the catalytic domains. Science 241(4861), 42–52 (1988)
12. Hannenhalli, S., Russell, R.B.: Analysis and prediction of functional sub-types from protein sequence alignments. J. Mol. Biol. 303(1), 61–76 (2000)
13. Hunter, T.: Protein kinase classification. Methods Enzymol 200, 3–37 (1991)
14. Jones, S., Thornton, J.M.: Searching for functional sites in protein structures. Curr. Opin. Chem. Biol. 8(1), 3–7 (2004)
15. Lazareva-Ulitsky, B., Diemer, K., Thomas, P.D.: On the quality of tree-based protein classification. Bioinformatics 21(9), 1876–1890 (2005) Comparative Study
16. Lichtarge, O., Bourne, H.R., Cohen, F.E.: An evolutionary trace method defines binding surfaces common to protein families. J. Mol. Biol. 257(2), 342–358 (1996)
17. Liu, Y., Ruoho, A.E., Rao, V.D., Hurley, J.H.: Catalytic mechanism of the adenylyl and guanylyl cyclases: modeling and mutational analysis. Proc. Natl. Acad. Sci. USA 94(25), 13414–13419 (1997)
18. Livingstone, C.D., Barton, G.J.: Protein sequence alignments: a strategy for the hierarchical analysis of residue conservation. Comput. Appl. Biosci. 9(6), 745–756 (1993)
19. Manning, G., Whyte, D.B., Martinez, R., Hunter, T., Sudarsanam, S.: The protein kinase complement of the human genome. Science 298(5600), 1912–1934 (2002)
20. McLachlan, G.J., Peel, D.: Finite Mixture Models. John Wiley & Sons, Chichester (2000)
21. Mirny, L.A., Gelfand, M.S.: Using orthologous and paralogous proteins to identify specificity-determining residues in bacterial transcription factors. J. Mol. Biol. 321(1), 7–20 (2002)
22. Pei, J., Cai, W., Kinch, L.N, Grishin, N.V.: Prediction of functional specificity determinants from protein sequences using log-likelihood ratios. Bioinformatics 22(2), 164–171 (2006)
23. Sjolander, K., Karplus, K., Brown, M., Hughey, R., Krogh, A., Mian, I.S, Haussler, D.: Dirichlet mixtures: A method for improving detection of weak but significant protein sequence homology. Technical report, University of California at Santa Cruz, Santa Cruz, CA, USA (1996)
24. Smith, C.M., Shindyalov, I.N., Veretnik, S., Gribskov, M., Taylor, S.S., Ten Eyck, L.F., Bourne, P.E.: The protein kinase resource. Trends Biochem. Sci. 22(11), 444–446 (1997)
25. Wicker, N., Perrin, G.R., Thierry, J.C., Poch, O.: Secator: a program for inferring protein subfamilies from phylogenetic trees. Mol. Biol. Evol. 18(8), 1435–1441 (2001)
26. Yu, G., Park, B.-H., Chandramohan, P., Munavalli, R., Geist, A., Samatova, N.F.: In silico discovery of enzyme-substrate specificity-determining residue clusters. J. Mol. Biol. 352(5), 1105–1117 (2005)
27. Zhang, G., Liu, Y., Ruoho, A.E., Hurley, J.H.: Structure of the adenylyl cyclase catalytic core. Nature 386(6622), 247–253 (1997)

An Algorithm to Find Overlapping Community Structure in Networks

Steve Gregory

Department of Computer Science
University of Bristol, BS8 1UB, England
steve@cs.bris.ac.uk

Abstract. Recent years have seen the development of many graph clustering algorithms, which can identify community structure in networks. The vast majority of these only find disjoint communities, but in many real-world networks communities overlap to some extent. We present a new algorithm for discovering overlapping communities in networks, by extending Girvan and Newman's well-known algorithm based on the *betweenness* centrality measure. Like the original algorithm, ours performs hierarchical clustering — partitioning a network into any desired number of clusters — but allows them to overlap. Experiments confirm good performance on randomly generated networks based on a known overlapping community structure, and interesting results have also been obtained on a range of real-world networks.

1 Introduction and Motivation

Many complex systems in the real world can be represented abstractly as networks (or graphs). Recently, with increasing availability of data about large networks and the need to understand them, the study of networks has become an important topic. A property that has been extensively studied is the existence of community structure in networks. A *cluster* (or *community* or *module*) is a subgraph such that the density of edges within it (*intracluster edges*) is greater than the density of edges between its vertices and those outside it (*intercluster edges*). A wide range of algorithms have been developed to discover communities in a network, including [4, 6, 11, 12, 13, 14].

Probably the best-known algorithm for finding community structure is Girvan and Newman's algorithm [6, 14], based on the betweenness centrality measure [5]. The *betweenness* (strictly, the *shortest-path betweenness*) of edge e, $c_B(e)$, is defined as the number of shortest paths, between all pairs of vertices, that pass along e. A high betweenness means that the edge acts as a bottleneck between a large number of vertex pairs and suggests that it is an intercluster edge. Although the algorithm is quite slow and is no longer the most effective clustering algorithm, it does give relatively good results. The algorithm works as follows:

1. Calculate edge betweenness of all edges in network.
2. Find edge with highest edge betweenness and remove it.
3. Recalculate edge betweenness for all remaining edges.
4. Repeat from step 2 until no edges remain.

J.N. Kok et al. (Eds.): PKDD 2007, LNAI 4702, pp. 91–102, 2007.
© Springer-Verlag Berlin Heidelberg 2007

This is a hierarchical, divisive, clustering algorithm. Initially, the n-vertex network (if connected) forms a single cluster. After one or more iterations, removing an edge (step 2) causes the network to split into two components (clusters). As further edges are removed, each component again splits, until n singleton clusters remain. The result is a *dendrogram*: a binary tree in which the distance of nodes from the root shows the order in which clusters were split. A cross-section of the dendrogram at any level represents a division of the network into any desired number of clusters.

In step 3, edge betweenness need not be recalculated for the whole network, but only for the component containing the edge removed in step 2, or for both components if removing the edge caused the component to split. (The edge betweenness of an edge depends only on the vertices and edges in the same component as it.)

Most algorithms assume that communities are disjoint, placing each vertex in only one cluster. However, in the real world, communities often overlap. For example, in collaboration networks an author might work with researchers in many groups, in biological networks a protein might interact with many groups of proteins, and so on.

In this paper we present a new algorithm to find overlapping community structure in networks. It is a hierarchical, divisive algorithm, based on Girvan and Newman's but extended with a novel method of splitting vertices. We describe the design of the algorithm in Section 2. In Section 3 we present some results on both artificial (computer-generated) and real-world networks. Section 4 compares our algorithm with a few others that can detect overlapping communities. Conclusions appear in Section 5.

2   Finding Overlapping Clusters

In any divisive hierarchical clustering algorithm, clusters are repeatedly divided into smaller (normally disjoint) clusters that together contain the same items. To allow overlapping clusters, there needs to be some way of splitting (copying) an item so that it can be included in more than one cluster when the cluster divides.

In the context of network clustering, assuming it is based entirely on the network structure, it seems reasonable to assume that each vertex should be in the same cluster as at least one of its neighbours, unless it is in a singleton cluster or no cluster at all. Therefore, a vertex v should be split into at most $d(v)$ copies, where $d(v)$ is the degree of v. We need to decide how many times a vertex should be split, and when a vertex should be split (e.g., at the beginning or when dividing a cluster).

Our algorithm extends Girvan and Newman's algorithm (the "GN algorithm") with a specific method of deciding *when* and *how* to split vertices. As in the original work, we only consider unipartite networks with undirected, unweighted edges. We name our new algorithm "CONGA" (Cluster-Overlap Newman Girvan Algorithm).

Splitting Vertices. In the GN algorithm, the basic operation is removing an edge. We introduce a second operation: splitting a vertex. If split, a vertex v always splits into *two* vertices v_1 and v_2: edges with v as an endvertex are redirected to v_1 or v_2 such that v_1 and v_2 each has at least one edge. By splitting repeatedly, a vertex v can eventually split into at most $d(v)$ vertices. Vertices are split incrementally during the clustering process. This binary splitting fits well into the GN algorithm because, like removing an edge, splitting a vertex may cause its cluster to split into two.

Split Betweenness. The key point of the CONGA algorithm is the notion of "split betweenness". This provides a way to decide (1) *when* to split a vertex, instead of removing an edge, (2) *which* vertex to split, and (3) *how* to split it. Clearly, v should only be split into v_1 and v_2 if these two vertices belong to different clusters. We could verify this by counting the number of shortest paths that would pass between v_1 and v_2 *if* they were joined by an edge. Then, if there were more shortest paths on $\{v_1, v_2\}$ than on any real edge, the vertex should be split; otherwise, an edge should be removed as usual. This is the basis of our method of splitting a vertex, which is as follows.

For any split of vertex v into v_1 and v_2, we add a new "imaginary" edge between v_1 and v_2. If u is a neighbour of v_1 and w is a neighbour of v_2, all shortest paths that passed through v along edges $\{u,v\}$, $\{v,w\}$ now pass along $\{u,v_1\}$, $\{v_1,v_2\}$, $\{v_2,w\}$. The imaginary edge has zero cost: the lengths of paths traversing it are unchanged, and no new shortest paths are created: paths beginning from v do not traverse this edge. We then calculate the betweenness $c_B(\{v_1,v_2\})$ of the imaginary edge. In general, there are $2^{d(v)-1}-1$ ways to split v into two. We call the split that maximizes $c_B(\{v_1,v_2\})$ the *best split* of v, and the maximum value of $c_B(\{v_1,v_2\})$ the *split betweenness* of v.

We modify the GN algorithm so that, at each step, it considers the split betweenness of every vertex as well as the edge betweenness of every edge. If the maximum split betweenness is greater than the maximum edge betweenness, the corresponding vertex is split, using its best split. (Note that imaginary edges are never actually added to the network, but are used only during the calculation of the split betweenness.)

Fig. 1(a) shows a network comprising two overlapping clusters: $\{a,b,c\}$ and $\{a,d,e\}$. Labels on the edges show edge betweennesses (with shortest paths counted in both directions). Fig. 1(b) shows a's best split into a_{bc} and a_{de}, with the imaginary edge (betweenness 8) shown as a dashed line. Fig. 1(c-d) shows the other two possible splits of a. In these, the imaginary edge has a lower betweenness, 4, proving that the split of Fig. 1(b) is the best split and the split betweenness of a is 8. Because this is greater than any edge betweenness, a should indeed be split.

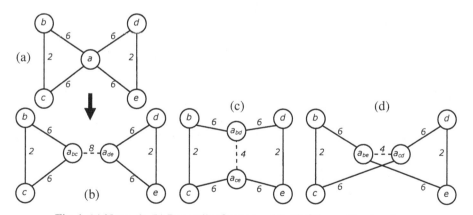

Fig. 1. (a) Network. (b) Best split of vertex a. (c), (d) Other splits of vertex a.

Fig. 2 shows a network which does not exhibit clustering. Here, any (2+2) split of a is a best split. The split betweenness of a is 8, which is the same as the betweenness of each edge. Therefore, by default, we remove any edge instead of splitting a.

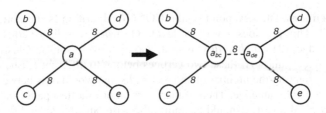

Fig. 2. Best split of vertex a: split betweenness of a is 8

Our method will never split a vertex into v_1 and v_2 such that v_1 has only one neighbour, u. This is because the betweenness of $\{v_1,v_2\}$ would be the same as that of $\{u,v_1\}$, as shown in Fig. 3, so removing edge $\{u,v\}$ would be preferred over splitting v. As a consequence of this, vertices with degree less than 4 are never split. In general, there are now only $2^{d(v)-1}-d(v)-1$ ways to split a vertex into two.

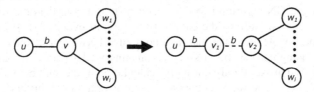

Fig. 3. Vertex will not split into vertices with degree 1

Vertex Betweenness and Split Betweenness. The split betweenness of a vertex v is the number of shortest paths passing between any member of n_1 and any member of n_2 via v, where n_1 and n_2 are disjoint sets containing all neighbours of v. By definition, this is no greater than the total number of shortest paths passing through v: the *vertex betweenness* of v [5]. It is simple to calculate vertex betweenness $c_B(v)$ from edge betweenness $c_B(e)$ [7]:

$$c_B(v) = \frac{1}{2} \sum_{e \in \Gamma(v)} c_B(e) - (n-1) \tag{1}$$

where $\Gamma(v)$ is the set of edges with v as an endvertex and n is the number of vertices in the component containing v. Therefore, as an optimization, we can use vertex betweenness as an upper bound on split betweenness: if the vertex betweenness of v is no greater than the maximum edge betweenness, there is no need to calculate v's split betweenness.

Calculating Split Betweenness. To calculate the split betweenness, and best split, of a vertex v, we first compute the pair betweennesses of v. The *pair betweenness* of v for $\{u,w\}$, where u and w are neighbours of v and $u \neq w$, is the number of shortest paths that traverse both edges $\{u,v\}$ and $\{v,w\}$. The vertex betweenness of v is the sum of all of its pair betweennesses.

We can represent the pair betweennesses of v, degree k, by a k-clique in which each vertex is labelled by one of v's neighbours and each edge $\{u,w\}$ is labelled by the pair betweenness "score" of v for $\{u,w\}$. Then, to find the best split of v:

1. Choose edge $\{u,w\}$ with minimum score.
2. Coalesce u and w to a single vertex, uw.
3. For each vertex x in the clique, replace edges $\{u,x\}$, score b_1, and $\{w,x\}$, score b_2, by a new edge $\{uw,x\}$ with score b_1+b_2.
4. Repeat from step 1 k-2 times (in total).

The labels on the remaining two vertices show the split, and the score on the remaining edge is the split betweenness.

This algorithm is not guaranteed to find the best split. To do that, we would need to try *all* edges in step 1 of each iteration, which would require exponential time. Our "greedy" method is much more efficient and, in practice, usually finds the best split or a close approximation to it. Fig. 4 shows how it finds the best split of vertex a of Fig. 1. There are k-2 = 2 phases; the edge chosen in step 1 of each phase is highlighted.

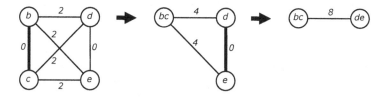

Fig. 4. Finding the best split of vertex a of Fig. 1

Calculating Pair Betweennesses. Pair betweennesses are computed while calculating edge betweenness, by a straightforward modification of the GN algorithm. The GN algorithm increments the betweenness of edge $\{i,j\}$ for all shortest paths beginning at each vertex s. CONGA does this *and* increments the pair betweennesses of i for all pairs $\{j,k\}$ such that k is a neighbour of i on a path between s and i.

There is some overhead, in both time and space, in computing pair betweennesses during the betweenness calculation. In most cases this information is not used because we can often determine, from the vertex betweenness, that a vertex should not be split. Therefore, our betweenness calculation is split into two phases, as shown below.

The CONGA Algorithm. Our complete algorithm is as follows:

1. Calculate edge betweenness of all edges in network.
2. Calculate vertex betweenness of vertices, from edge betweennesses, using Eq. (1).
3. Find candidate set of vertices: those whose vertex betweenness is greater than the maximum edge betweenness.
4. If candidate set is non-empty, calculate pair betweennesses of candidate vertices, and then calculate split betweenness of candidate vertices, using Eq. (1).
5. Remove edge with maximum edge betweenness or split vertex with maximum split betweenness (if greater).
6. Recalculate edge betweenness for all remaining edges in same component(s) as removed edge or split vertex.
7. Repeat from step 2 until no edges remain.

Complexity and Efficiency. The GN algorithm has a worst-case time complexity of $O(m^2n)$, where m is the number of edges and n is the number of vertices. In CONGA, each vertex splits into an average of up to $2m/n$ vertices, so the number of vertices after splitting is $O(m)$; the number of iterations is still $O(m)$ and the number of edges is unchanged. This makes the time complexity $O(m^3)$ in the worst case.

In practice, the speed depends on the number of vertices that are split. If more are split, more iterations are needed, the network becomes larger, and step 4 needs to be performed more frequently. Conversely, vertex splitting can cause the network to decompose into separate components more readily, which reduces the execution time.

3   Results

In this section we compare CONGA with the GN algorithm, to assess the effect of our extensions. We have tested both algorithms on computer-generated networks based on a known, possibly overlapping, community structure. Each network contains n vertices divided into c equally-sized communities, each containing nr/c vertices. Vertices are randomly and evenly distributed between communities such that each vertex is a member of r (≥ 1) communities on average. Edges are randomly placed between pairs of vertices with probability p_{in} if the vertices belong to the same community and p_{out} otherwise. In the special case where both r and p_{out} are 0, the network will be disconnected. Apart from this, all of our experiments use connected networks, constructed with a sufficiently high value of r or p_{out}, or both.

We measure how well each algorithm can recover the community structure from a network by using it to compute c clusters and comparing the result with the c known communities. Admittedly, c is not generally known for real-world networks, but this is still a useful and common way to assess clustering algorithms; e.g., [6, 14].

We calculate two values (all averaged over 10 graphs):

- *recall*: the fraction of vertex pairs belonging to the same community that are also in the same cluster.
- *precision*: the fraction of vertex pairs in the same cluster that also belong to the same community.

First (Fig. 5), we generated networks of 256 vertices divided into 32 communities, set $p_{out} = 0$ (i.e., no intercommunity edges) and $p_{in} = 0.5$, and increased the amount of overlap from $r = 1$ (i.e., no overlap) to $r = 3$. The number of edges (and hence the average degree) increases roughly quadratically with r, because the average community size is proportional to r and each vertex is a member of r communities. So the average degree is 4 for $r = 1$ but increases to approximately 15 for $r = 2$ and 32 for $r = 3$.

For the GN algorithm, as r increases, recall declines steadily because the (nonoverlapping) clusters are smaller than the communities; precision is quite high, though certainly not perfect, in this range. Suddenly, at around $r = 2$, recall increases and precision decreases, as most vertices are placed in a single cluster. In contrast, CONGA behaves very well up to about $r = 2$ and then deteriorates gradually.

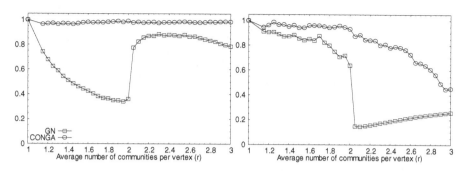

Fig. 5. recall (left), precision (right): $n=256$, $c=32$, $p_{in}=0.5$, $p_{out}=0$, various r

We have repeated this experiment for various values of c and p_{in}. The curves always show a similar shape, though the value of r at which precision drops varies.

To evaluate the algorithm on real-world networks, there is no correct solution with which to compare, so the quality of a clustering must be assessed in a different way. This is usually done by measuring the relative number of intracluster and intercluster edges, for example, by the *modularity* measure [13, 14]. However, there is no widely accepted alternative measure for use with overlapping clusters, but a promising candidate is the *average degree* measure [3]. We define the *vertex average degree* (*vad*) of a set of clusters S, where each cluster is a set of vertices, as:

$$vad(S) = \frac{2\sum_{C \in S} |E(C)|}{\sum_{C \in S} |C|} \qquad (2)$$

Another useful measure is the *overlap* of a set of clusters S: the sum of the cluster sizes divided by the size of the union of all clusters. (We do not claim that *vad* and overlap are mutually independent measures; that is outside the scope of this paper.)

We have run the CONGA and GN algorithms on several real-world examples, listed in Table 1. Execution times are shown for a 2.4GHz Pentium 4 processor.

Table 1. Algorithm's results on real-world networks

Name	Ref.	Vertices	Edges	Runtime (s)
Karate club	[19]	34	78	0.2
Dolphins	[9]	62	159	0.5
College football	[6]	115	613	7.8
Network science	[12]	379	914	12.5
Blogs	[18]	3982	6803	30411
Words	[10]	1000	3471	6767

"Karate club" [19], discussed in [6], represents a social network based on two disjoint communities. The communities are not reflected clearly in the network structure: there are eight intercommunity edges. GN finds an almost perfect (relative to the real-world situation) two-cluster solution, misclassifying one vertex. CONGA finds a different solution with a small overlap, 1.03. The *vad* is 4.45 for CONGA and 4.0 for GN, suggesting that the overlapping clustering is a good one (albeit incorrect).

"Dolphins" [9], discussed in [14], is a social network of dolphins, also based on two disjoint communities. Here there are only six intercommunity edges. GN finds the two communities correctly and CONGA finds the same division but with two vertices from the larger community included in both clusters: the overlap is 1.03. The *vad* is 4.91 for CONGA and 4.94 for GN.

"College football" [6] is a network based on games between teams that belong to 15 disjoint real-world communities. This network has many intercommunity edges. Neither algorithm finds a perfect 15-cluster solution; the one found by CONGA has a lower *vad* (5.87 *vs.* 7.18) and a large overlap: 1.75.

"Network science" [12] is a collaboration network of coauthorships. For such networks it is impossible to determine the number of real-world communities, and it seems reasonable to assume they might overlap. CONGA's solution has a higher *vad* than GN's for 14 or more clusters, and overlap increases with the number of clusters. CONGA's solution for 33 clusters is illustrated in Fig. 6: each cluster is identified by a letter or digit and each vertex is labelled with the cluster(s) to which it belongs.

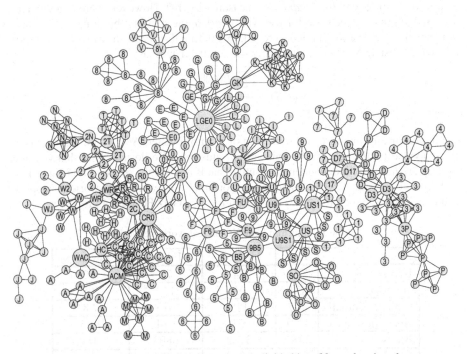

Fig. 6. Network science collaboration network divided into 33 overlapping clusters

"Blogs" [18] is a network of blogs on the MSN (now known as Windows Live™ Spaces) platform. An edge links people who have left two or more comments on each other's blog, and so are deemed to be acquainted. CONGA's solution has a consistently higher *vad* than GN's, especially for more than 90 clusters. The overlap increases with the number of clusters but levels off, reaching a maximum of 1.39.

"Words" is a non-social network: a contiguous 1000-vertex subgraph of a word association network from [15], converted from an original directed, weighted version [10]. CONGA successfully groups related words. For example, dividing it into 400 clusters, the word "form" appears in four: {contract, document, form, order, paper, signature, write}, {blank, entry, fill, form, up}, {compose, create, form, make}, {form, mold, shape}. (Related words in this network are not necessarily synonyms, as they are in this example.) Again, the *vad* for CONGA's solution is consistently higher than GN's; the overlap increases and tails off, reaching a maximum of 2.23.

4 Related Work

Pinney and Westhead [16, 17] have also proposed extending the GN algorithm with the ability to split vertices between clusters. The decision of whether to split a vertex or remove an edge is based entirely on edge betweenness and vertex betweenness. The highest-betweenness edge is removed only if its two endvertices have similar betweenness; i.e., if their ratio is between α and $1/\alpha$, where α is a parameter with suggested value 0.8 [16]. Otherwise the vertex with highest betweenness is temporarily removed. When a component splits into two or more subcomponents, each removed vertex is split and copied into each subcomponent, and all edges between the vertex copy and the subcomponent are restored, including any removed in previous steps. We have implemented this algorithm and compared it with CONGA; see below.

The clique percolation algorithm of Palla *et al.* [15], implemented in CFinder [1], finds overlapping clusters in a different way. Instead of dividing a network into its most loosely connected parts, it identifies the most densely connected parts. The parameter is not the number of clusters to be found but their density, k. A cluster is defined as the set of k-cliques that can all be reached from each other via a sequence of adjacent k-cliques; two k-cliques are *adjacent* if they share k-1 vertices. Each vertex may be in many clusters, or even none: e.g., degree-1 vertices are always ignored. We have run CFinder (v1.21) to compare its results with CONGA's; see below.

Baumes *et al.* [2, 3] present a collection of algorithms to find overlapping clusters. One algorithm iteratively improves a candidate cluster by adding vertices to and removing vertices from it while its density improves. Another removes vertices from a network until it breaks into disjoint components, forming initial clusters, and then replaces each removed vertex into one or more of the clusters, which might overlap.

Li *et al.* [8] form overlapping clusters using both the structure of the network *and* the content of vertices and edges. The first phase of their algorithm finds densely connected "community cores", similarly to the method of [15]. In the second phase, clusters are formed from cores by adding further triangles and edges whose content (assessed using keywords) is similar to that of the core.

Experiments. We have run the Pinney and Westhead ("P&W") and CFinder algorithms on computer-generated networks, to compare with CONGA. The number of communities c was input to both CONGA and P&W, but CFinder cannot make use of this information, so CFinder is clearly disadvantaged. To compensate for this, we show the CFinder results for all values of k (CFinder's only parameter). For each experiment we plot the F-measure: the harmonic mean of recall and precision.

Fig. 7(a) shows results on the networks of Section 3: p_{in} and p_{out} are fixed while r is varied. CONGA gives the best results of all algorithms tested, but performance declines for all algorithms for high r. CFinder gives its best performance for $r=2$, so in fairness to CFinder we use this value in subsequent experiments. In Fig. 7(b) we fix r and p_{out} and vary p_{in}. CONGA gives the best results, and they improve as p_{in} increases. In contrast, CFinder, for each k, reaches a peak at a different value of p_{in}; for smaller values its recall is reduced while for larger values its precision drops.

In Fig. 7(c) we fix r and p_{in} and vary p_{out}. This time, CONGA's performance suffers as p_{out} increases, because of reduced precision, while CFinder's performance is more stable. Finally, in Fig. 7(d), we test the hypothesis that CFinder should be more effective in cases where the number of communities is not known. We do this by generating networks in which a (varying) number, u, of the 256 individuals are placed in singleton communities and the remainder are divided between the 32 main communities; because $p_{out}>0$ these networks are still connected. In this experiment, CFinder with $k=4$ performs slightly better than CONGA. For both algorithms, recall decreases as u increases but CFinder's precision improves while CONGA's declines.

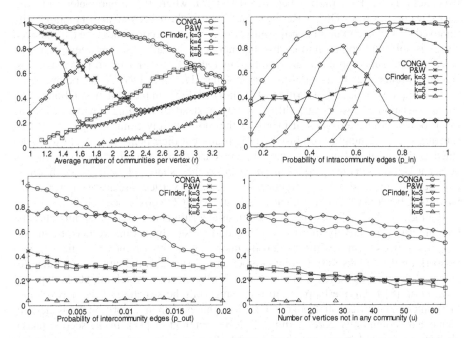

Fig. 7. F-measure for random networks with $n=256$, $c=32$. (a: upper left) $p_{in}=0.5$, $p_{out}=0$, various r; (b: upper right) $r=2$, $p_{out}=0$, various p_{in}; (c: lower left) $r=2$, $p_{in}=0.5$, various p_{out}; (d: lower right) $r=2$, $p_{in}=0.5$, $p_{out}=0.008$, various u.

Fig. 8 shows the execution times of all algorithms for the experiments of Fig. 7(a). For CONGA and P&W these times include the generation of the complete dendrogram, from which the solution for any number of clusters can be quickly extracted. The process is not stopped after the network is divided into 32 clusters. For CFinder,

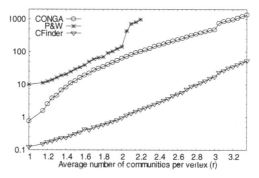

Fig. 8. Execution time (seconds) for $n=256$, $c=32$, $p_{in}=0.5$, $p_{out}=0$, various r

the times include the generation of solutions for all values of k. CONGA and P&W were implemented by the author in Java. Each experiment was run on a machine with dual AMD Opteron 250 CPUs (2.4GHz).

In summary, CONGA and CFinder seem to have complementary strengths and weaknesses: each may be better for a different application. CFinder is substantially faster than CONGA. P&W behaves in a similar way to CONGA but with worse results (for these networks); however, we have only tested it with one value of its parameter (α). The execution time of P&W is also the worst, but this may be because of the poor implementation rather than the algorithm itself.

5 Conclusions

We have presented an algorithm that seems to be effective in discovering overlapping communities in networks. Good results have been obtained for a range of random networks with overlap of more than 2, which is large relative to the number of communities: if a network has only 32 communities, an overlap of 3 means that each vertex is in the same community as ¼ of the whole network. As the number of communities is increased, the algorithm can cope with a larger overlap. The algorithm is not fast, but its speed is comparable with that of the GN algorithm from which it is derived.

Future work includes trying to improve the algorithm further and applying similar ideas to faster clustering algorithms than the GN algorithm. It is also worth investigating alternative ways of measuring the quality of an overlapping clustering; e.g., the *vad* measure. Finally, it would be interesting to study the overlapping nature of real-world networks, a subject that has received little attention (but see [15]). For example, it may be that the collaboration network of Fig. 6 naturally divides into a small number of disjoint clusters, possibly corresponding to research groups, but to decompose it further requires clusters to overlap.

Further information related to this paper, including the networks analysed and more results, can be found at http://www.cs.bris.ac.uk/~steve/networks/ .

Acknowledgements. I am very grateful to Peter Flach for his expert advice on several drafts of this paper. Thanks are also due to John Pinney for explaining his algorithm, and the four anonymous referees for their detailed comments.

References

1. Adamcsek, B., Palla, G., Farkas, I., Derényi, I., Vicsek, T.: CFinder: locating cliques and overlapping modules in biological networks. Bioinformatics 22, 1021–1023 (2006)
2. Baumes, J., Goldberg, M., Krishnamoorty, M., Magdon-Ismail, M., Preston, N.: Finding communities by clustering a graph into overlapping subgraphs. In: Proc. IADIS Applied Computing 2005, pp. 97–104 (2005)
3. Baumes, J., Goldberg, M., Magdon-Ismail, M.: Efficient identification of overlapping communities. In: Kantor, P., Muresan, G., Roberts, F., Zeng, D.D., Wang, F.-Y., Chen, H., Merkle, R.C. (eds.) ISI 2005. LNCS, vol. 3495, pp. 27–36. Springer, Heidelberg (2005)
4. Brandes, U., Gaertler, M., Wagner, D.: Experiments on graph clustering algorithms. In: Di Battista, G., Zwick, U. (eds.) ESA 2003. LNCS, vol. 2832, pp. 568–579. Springer, Heidelberg (2003)
5. Freeman, L.C.: A set of measures of centrality based on betweenness. Sociometry 40, 35–41 (1977)
6. Girvan, M., Newman, M.E.J.: Community structure in social and biological networks. Proc. Natl. Acad. Sci. USA 99, 7821–7826 (2002)
7. Koschützki, D., Lehmann, K.A., Peeters, L., Richter, S., Tenfelde-Podehl, D., Zlotowski, O.: Centrality indices. In: Brandes, U., Erlebach, T. (eds.) Network Analysis. LNCS, vol. 3418, Springer, Heidelberg (2005)
8. Li, X., Liu, B., Yu, P.S.: Discovering overlapping communities of named entities. In: Fürnkranz, J., Scheffer, T., Spiliopoulou, M. (eds.) PKDD 2006. LNCS (LNAI), vol. 4213, pp. 593–600. Springer, Heidelberg (2006)
9. Lusseau, D., Schneider, K., Boisseau, O.J., Haase, P., Slooten, E., Dawson, S.M.: The bottlenose dolphin community of Doubtful Sound features a large proportion of long-lasting associations. Behavioral Ecology and Sociobiology 54, 396–405 (2003)
10. Nelson, D.L., McEvoy, C.L., Schreiber, T.A.: The University of South Florida word association, rhyme and word fragment norms (1998), http://w3.usf.edu/FreeAssociation/
11. Newman, M.E.J.: Fast algorithm for detecting community structure in networks. Phys. Rev. E 69, 066133 (2004)
12. Newman, M.E.J.: Finding community structure in networks using the eigenvectors of matrices. Phys. Rev. E 74, 036104 (2006)
13. Newman, M.E.J.: Modularity and community structure in networks. Proc. Natl. Acad. Sci. USA 103, 8577–8582 (2006)
14. Newman, M.E.J., Girvan, M.: Finding and evaluating community structure in networks. Phys. Rev. E 69, 026113 (2004)
15. Palla, G., Derényi, I., Farkas, I., Vicsek, T.: Uncovering the overlapping community structure of complex networks in nature and society. Nature 435, 814–818 (2005)
16. Pinney, J.W.: Personal communication
17. Pinney, J.W., Westhead, D.R.: Betweenness-based decomposition methods for social and biological networks. In: Barber, S., Baxter, P.D., Mardia, K.V., Walls, R.E. (eds.) Interdisciplinary Statistics and Bioinformatics, pp. 87–90. Leeds University Press (2006)
18. Xie, N.: Social network analysis of blogs. MSc dissertation. University of Bristol (2006)
19. Zachary, W.W.: An information flow model for conflict and fission in small groups. Journal of Anthropological Research 33, 452–473 (1977)

Privacy Preserving Market Basket Data Analysis

Ling Guo, Songtao Guo, and Xintao Wu

University of North Carolina at Charlotte,
{lguo2,sguo,xwu}@uncc.edu

Abstract. Randomized Response techniques have been empirically investigated in privacy preserving association rule mining. However, previous research on privacy preserving market basket data analysis was solely focused on support/ confidence framework. Since there are inherent problems with the concept of finding rules based on their support and confidence measures, many other measures (e.g., correlation, lift, etc.) for the general market basket data analysis have been studied. How those measures are affected due to distortion is not clear in the privacy preserving analysis scenario.

In this paper, we investigate the accuracy (in terms of bias and variance of estimates) of estimates of various rules derived from the randomized market basket data and present a general framework which can conduct theoretical analysis on how the randomization process affects the accuracy of various measures adopted in market basket data analysis. We also show several measures (e.g., correlation) have monotonic property, i.e., the values calculated directly from the randomized data are always less or equal than those original ones. Hence, some market basket data analysis tasks can be executed on the randomized data directly without the release of distortion probabilities, which can better protect data privacy.

1 Introduction

The issue of maintaining privacy in association rule mining has attracted considerable attention in recent years [8,9,4,21]. Most of techniques are based on a data perturbation or Randomized Response (RR) approach [5], wherein the 0 or 1 (0 denotes absence of an item while 1 denotes presence of an item) in the original user transaction vector is distorted in a probabilistic manner that is disclosed to data miners.

However, previous research on privacy preserving market basket data analysis is solely focused on support/confidence framework. In spite of the success of association rules, there are inherent problems with the concept of finding rules based on their support and confidence. Various measures have been studied in market basket data analysis. In this paper we conduct theoretical analysis on the accuracy of various measures adopted previously in market data analysis. Our analysis is based on estimating the parameters of derived random variables. The estimated measure (e.g., Interest statistics) is considered as one derived variable. We present a general method, which is based on the Taylor series, for approximating the mean and variance of derived variables. We also derive interquantile ranges of those estimates. Hence, data miners are ensured that their estimates lie within these ranges with a high confidence.

There exists some scenario where data owners are reluctant to release the distortion probabilities since attackers may exploit those distortion probabilities to recover

J.N. Kok et al. (Eds.): PKDD 2007, LNAI 4702, pp. 103–114, 2007.

individual data. In this paper, we also show that some useful information can still be discovered directly from the randomized data without those distortion probabilities. Specifically, we show some market basket data analysis tasks (such as correlation analysis or independence hypothetical testing) can be conducted on the randomized data directly without distortion probabilities.

The remainder of this paper is organized as follows. In Section 2, we revisit the distortion framework and discuss how the Randomized Response techniques are applied to privacy preserving association rule mining. In Section 3, we conduct the theoretical analysis on how distortion process affects various other measures adopted in market basket data analysis. In Section 4, we show that some useful data mining results (e.g., dependence itemsets etc.) can be discovered even without the distortion values. We discuss the related work in Section 5 and conclude our work in Section 6.

2   Distortion Framework Revisited

The authors in [21,4,3] proposed the MASK scheme, which is based on Randomized Response, presented strategies of efficiently estimating the original support values of frequent itemsets from the randomized data. Their results empirically shown a high degree of privacy to the user and a high level of accuracy in the mining results can be simultaneously achieved. The privacy situation considered here is that perturbation is done at the level of individual customer records, without being influenced by the contents of the other records in the database. We also focus on a simple independent column perturbation, wherein the value of each attribute in the record is perturbed independently.

2.1   Randomization Procedure

Denoting the set of transactions in the database D by $\mathcal{T} = \{T_1, \cdots, T_n\}$ and the set of items in the database by $\mathcal{I} = \{A_1, \cdots, A_m\}$. Each item is considered as one dichotomous variable with 2 mutually exclusive and exhaustive categories ($0 = $ absence, $1 = $ presence). Each transaction can be logically considered as a fixed-length sequence of 1's and 0's. For item A_j, we use a 2×2 distortion probability matrix

$$P_j = \begin{pmatrix} \theta_0 & 1 - \theta_0 \\ 1 - \theta_1 & \theta_1 \end{pmatrix}$$

If the original value is in *absence* category, it will be kept in *absence* category with a probability θ_0 and changed to *presence* category with a probability $1 - \theta_0$. Similarly, if the original value is in *presence* category, it will be kept in *presence* with a probability θ_1 and changed to *absence* category with a probability $1 - \theta_1$. In this paper, we follow the original Warner RR model by simply setting $\theta_0 = \theta_1 = p_j$.

Let $\pi_{i_1, \cdots, i_k}$ denote the true proportion corresponding to the categorical combination $(A_{1i_1}, \cdots, A_{ki_k})$, where $i_1, \cdots, i_k \in \{0, 1\}$. Let π be vectors with elements $\pi_{i_1, \cdots, i_k}$, arranged in a fixed order. The combination vector corresponds to a fixed order of cell entries in the contingency table formed by the k-itemset. Table 1(a) shows one contingency table for a pair of two variables. We use the notation $\bar{A}$ ($\bar{B}$) to indicate that

A (B) is absent from a transaction. The vector $\pi = (\pi_{00}, \pi_{01}, \pi_{10}, \pi_{11})'$ corresponds to a fixed order of cell entries π_{ij} in the 2×2 contingency table. π_{11} denotes the proportion of transactions which contain both A and B while π_{10} denotes the proportion of transactions which contain A but not B. The row sum π_{1+} represents the support frequency of item A while the column sum π_{+1} represents the support frequency of item B.

The original database D is changed to D_{ran} after randomization. Assume $\lambda_{\mu_1,\cdots,\mu_k}$ is the probability of getting a response $(\mu_1, \cdots, \mu_k)$ and λ the vector with elements $\lambda_{\mu_1,\cdots,\mu_k}$ arranged in a fixed order(e.g., the vector $\lambda = (\lambda_{00}, \lambda_{01}, \lambda_{10}, \lambda_{11})'$ corresponds to cell entries λ_{ij} in the randomized contingency table as shown in Table 1(b)), we can get

$$\lambda = (P_1 \times \cdots \times P_k)\pi$$

where $\times$ stands for the Kronecker product. Let $P = P_1 \times \cdots \times P_k$, an unbiased estimate of π follows as

$$\hat{\pi} = P^{-1}\hat{\lambda} = (P_1^{-1} \times \cdots \times P_k^{-1})\hat{\lambda} \tag{1}$$

where $\hat{\lambda}$ is the vector of proportions observed from the randomized data corresponding to λ and P_j^{-1} denotes the inverse of the matrix P_j. Note that although the distortion matrices $P_1, \cdots, P_k$ are known, they can only be utilized to estimate the proportions of itemsets of the original data, rather than the precise reconstruction of the original 0-1 data.

2.2   Accuracy of Association Rule

Recently the authors in [11] investigated the accuracy of support and confidence measures for each individual association rule derived from the randomized data and presented an analytical formula for evaluating their accuracy in terms of bias and variance. From the derived variances, users can tell how accurate the derived association rules in terms of both support and confidence measures from the randomized data are.

Table 1. 2×2 contingency tables for two variables A,B

(a) Original				(b) After randomization		
	$\bar{B}$	B			$\bar{B}$	B

$$
\begin{array}{ccccc}
 & \bar{B} & B & & \\
\bar{A} & \pi_{00} & \pi_{01} & \pi_{0+} \\
A & \pi_{10} & \pi_{11} & \pi_{1+} \\
 & \pi_{+0} & \pi_{+1} & \pi_{++}
\end{array}
\qquad
\begin{array}{ccccc}
 & \bar{B} & B & & \\
\bar{A} & \lambda_{00} & \lambda_{01} & \lambda_{0+} \\
A & \lambda_{10} & \lambda_{11} & \lambda_{1+} \\
 & \lambda_{+0} & \lambda_{+1} & \lambda_{++}
\end{array}
$$

Assume item A and B are randomized using distortion matrix P_1 and P_2 respectively. For a simple association rule $A \Rightarrow B$ derived from the randomized data, it was shown in [11] that an unbiased estimate is $\hat{\pi} = P^{-1}\hat{\lambda} = (P_1^{-1} \times P_2^{-1})\hat{\lambda}$ with the covariance matrix as

$$c\hat{o}v(\hat{\pi}) = (n-1)^{-1}P^{-1}(\hat{\lambda}^\delta - \hat{\lambda}\hat{\lambda}')P'^{-1} \tag{2}$$

where $\hat{\lambda}^\delta$ is a diagonal matrix with the same diagonal elements as those of $\hat{\lambda}$ arranged in the same order. The last element of $\hat{\pi}$ corresponds to the estimated support value s

and the last element of $\hat{cov}(\hat{\pi})$ denotes its estimated variance. The estimated confidence c is

$$\hat{c} = \frac{\hat{s}_{AB}}{\hat{\theta}_A} = \frac{\hat{\pi}_{11}}{\hat{\pi}_{1|}}$$

and its variance as

$$\hat{var}(\hat{c}) \approx \frac{\hat{\pi}_{10}^2}{\hat{\pi}_{1+}^4}\hat{var}(\hat{\pi}_{11}) + \frac{\hat{\pi}_{11}^2}{\hat{\pi}_{1+}^4}\hat{var}(\hat{\pi}_{10}) - 2\frac{\hat{\pi}_{10}\hat{\pi}_{11}}{\hat{\pi}_{1+}^4}\hat{cov}(\hat{\pi}_{11}, \hat{\pi}_{10}) \tag{3}$$

The above results can be straightforwardly extended to the general association rule $\mathcal{X} \Rightarrow \mathcal{Y}$. Incorporating the derived estimate and variance, the $(1 - \alpha)100\%$ interquantile range for the estimated support and confidence is then derived. An $(1 - \alpha)100\%$ interquantile range, say $\alpha = 0.05$, shows the interval where the original value lies in with 95% probability. In other words, users shall have 95% confidence that the original value falls into this interquantile range.

3   Accuracy Analysis of Measures

The objective interestingness measure is usually computed from the contingency table. Table 2 shows various measures defined for a pair of binary variables [23]. Here we give results on how RR may affect the accuracy of those measures or analysis methods on market basket data.

Table 2. Objective measures for the itemset {A,B}

Measure	Expression	Measure	Expression
Support (s)	π_{11}	Confidence(c)	$\frac{\pi_{11}}{\pi_{1+}}$
Correlation (ϕ)	$\frac{\pi_{11}\pi_{00} - \pi_{01}\pi_{10}}{\sqrt{\pi_{1+}\pi_{+1}\pi_{0+}\pi_{+0}}}$	Cosine (IS)	$\frac{\pi_{11}}{\sqrt{\pi_{1+}\pi_{+1}}}$
Odds ratio (α)	$\frac{\pi_{11}\pi_{00}}{\pi_{10}\pi_{01}}$	Interest (I)	$\frac{\pi_{11}}{\pi_{1+}\pi_{+1}}$
Jaccard (ζ)	$\frac{\pi_{11}}{\pi_{1+}+\pi_{+1}-\pi_{11}}$	(PS)	$\pi_{11} - \pi_{1+}\pi_{+1}$
Mutual Info(M)	$\frac{\sum_i \sum_j \pi_{ij} log \frac{\pi_{ij}}{\pi_{i+}\pi_{+j}}}{-\sum_i \pi_{i+} log \pi_{i+}}$	Conviction (V)	$\frac{\pi_{1+}\pi_{+0}}{\pi_{10}}$
J-measure (J)	$\pi_{11}log\frac{\pi_{11}}{\pi_{1+}\pi_{+1}} + \pi_{10}log\frac{\pi_{11}}{\pi_{1+}\pi_{+0}}$	Certainty (F)	$\frac{\frac{\pi_{11}}{\pi_{1+}}-\pi_{+1}}{1-\pi_{+1}}$
Std. residues(e)	$\frac{\pi_{ij}-\pi_{i+}\pi_{+j}}{\sqrt{\pi_{i+}\pi_{+j}}}$	Likelihood (G^2)	$2\sum_i \sum_j \pi_{ij} log\frac{\pi_{ij}}{\pi_{i+}\pi_{+j}}$
Pearson (χ^2)	$\sum_i \sum_j \frac{\{\pi_{ij}-\pi_{i+}\pi_{+j}\}^2}{\pi_{i+}\pi_{+j}}$	Added Value(AV)	$\frac{\pi_{11}}{\pi_{1+}-\pi_{+1}}$

In this Section, we provide a general framework which can derive estimates of all measures using randomized data and the released distortion parameters. Furthermore, we present a general approach which can calculate the variance of those estimates in Section 3.1. By incorporating the Chebyshev Theorem, we show how to derive their interquantile ranges in Section 3.2.

3.1   Variances of Derived Measures

From Table 2, we can see that each measure can be expressed as one derived random variable (or function) from the observed variables (π_{ij} or their marginal totals π_{i+}, π_{+j}). Similarly, its estimate from the randomized data can be considered as another derived random variable from the input variables ($\hat{\pi}_{ij}, \hat{\pi}_{i+}, \hat{\pi}_{+j}$). Since we know how to derive variances of the input variables ($\hat{var}(\hat{\pi}_{ij})$) from the randomized data, our problem is then how to derive the variance of the derived output variable.

In the following, we first present a general approach based on the delta method [17] and then discuss how to derive the variance of chi-square statistics (χ^2) as one example.

Let z be a random variable derived from the observed random variables x_i ($i = 1, \cdots, n$): $z = g(x)$. According to the delta method, a Taylor approximation of the variance of a function with multiple variables can be expanded as

$$var\{g(x)\} = \sum_{i=1}^{k}\{g_i'(\theta)\}^2 var(x_i) + \sum_{i\neq j=1}^{k} g_i'(\theta)g_j'(\theta)cov(x_i, x_j) + o(n^{-r})$$

where θ_i is the mean of x_i, $g(x)$ stands for the function $g(x_1, x_2, \cdots, x_k)$, $g_i'(\theta)$ is the $\frac{\partial g(x)}{\partial x_i}$ evaluated at $\theta_1, \theta_2, \cdots, \theta_k$.

For market basket data with 2 variables, $\hat{\pi} = (\hat{\pi}_{00}, \hat{\pi}_{01}, \hat{\pi}_{10}, \hat{\pi}_{11})'$, the estimated chi-square is shown as

$$\hat{\chi}^2 = n(\frac{(\hat{\pi}_{00} - \hat{\pi}_{0+}\hat{\pi}_{+0})^2}{\hat{\pi}_{0+}\hat{\pi}_{+0}} + \frac{(\hat{\pi}_{01} - \hat{\pi}_{0+}\hat{\pi}_{+1})^2}{\hat{\pi}_{0+}\hat{\pi}_{+1}}$$
$$+ \frac{(\hat{\pi}_{10} - \hat{\pi}_{1+}\hat{\pi}_{+0})^2}{\hat{\pi}_{1+}\hat{\pi}_{+0}} + \frac{(\hat{\pi}_{11} - \hat{\pi}_{1+}\hat{\pi}_{+1})^2}{\hat{\pi}_{1+}\hat{\pi}_{+1}})$$

Let $x_1 = \hat{\pi}_{00}$, $x_2 = \hat{\pi}_{01}$, $x_3 = \hat{\pi}_{10}$ and $x_4 = \hat{\pi}_{11}$, we have

$$g(x_1, x_2, x_3, x_4) = \chi^2$$
$$= n[\frac{x_1^2}{(x_1 + x_2)(x_1 + x_3)} + \frac{x_2^2}{(x_1 + x_2)(x_2 + x_4)} +$$
$$\frac{x_3^2}{(x_3 + x_4)(x_3 + x_1)} + \frac{x_4^2}{(x_4 + x_3)(x_4 + x_2)} - 1]$$

Partial derivatives of the function $g()$ can be calculated respectively. By incorporating estimated expectations, variances and covariances of variables in function $g()$, the variance of function $g()$ can be estimated as

$$\hat{var}(g) \approx G_1^2\hat{var}(\hat{\pi}_{00}) + G_2^2\hat{var}(\hat{\pi}_{01}) + G_3^2\hat{var}(\hat{\pi}_{10}) + G_4^2\hat{var}(\hat{\pi}_{11})$$
$$+2G_1G_2\hat{cov}(\hat{\pi}_{00}, \hat{\pi}_{01}) + 2G_1G_3\hat{cov}(\hat{\pi}_{00}, \hat{\pi}_{10}) + 2G_1G_4\hat{cov}(\hat{\pi}_{00}, \hat{\pi}_{11})$$
$$+2G_2G_3\hat{cov}(\hat{\pi}_{01}, \hat{\pi}_{10}) + 2G_2G_4\hat{cov}(\hat{\pi}_{01}, \hat{\pi}_{11}) + 2G_3G_4\hat{cov}(\hat{\pi}_{10}, \hat{\pi}_{11})$$

where

$$G_1 = \frac{\partial g}{\partial x_1} = n\left[\frac{\hat{\pi}_{00}^2(\hat{\pi}_{01}+\hat{\pi}_{10})+2\hat{\pi}_{00}\hat{\pi}_{01}\hat{\pi}_{10}}{\hat{\pi}_{0+}^2\hat{\pi}_{+0}^2} - \frac{\hat{\pi}_{01}^2}{\hat{\pi}_{0+}^2\hat{\pi}_{+1}} - \frac{\hat{\pi}_{10}^2}{\hat{\pi}_{+0}^2\hat{\pi}_{1+}}\right]$$

$$G_2 = \frac{\partial g}{\partial x_2} = n\left[\frac{\hat{\pi}_{01}^2(\hat{\pi}_{00}+\hat{\pi}_{11})+2\hat{\pi}_{00}\hat{\pi}_{01}\hat{\pi}_{11}}{\hat{\pi}_{0+}^2\hat{\pi}_{+1}^2} - \frac{\hat{\pi}_{00}^2}{\hat{\pi}_{0+}^2\hat{\pi}_{+0}} - \frac{\hat{\pi}_{11}^2}{\hat{\pi}_{+1}^2\hat{\pi}_{1+}}\right]$$

$$G_3 = \frac{\partial g}{\partial x_3} = n\left[\frac{\hat{\pi}_{10}^2(\hat{\pi}_{11}+\hat{\pi}_{00})+2\hat{\pi}_{00}\hat{\pi}_{10}\hat{\pi}_{11}}{\hat{\pi}_{1+}^2\hat{\pi}_{+0}^2} - \frac{\hat{\pi}_{11}^2}{\hat{\pi}_{1+}^2\hat{\pi}_{+1}} - \frac{\hat{\pi}_{00}^2}{\hat{\pi}_{+0}^2\hat{\pi}_{0+}}\right]$$

$$G_4 = \frac{\partial g}{\partial x_4} = n\left[\frac{\hat{\pi}_{11}^2(\hat{\pi}_{01}+\hat{\pi}_{10})+2\hat{\pi}_{11}\hat{\pi}_{01}\hat{\pi}_{10}}{\hat{\pi}_{1+}^2\hat{\pi}_{+1}^2} - \frac{\hat{\pi}_{10}^2}{\hat{\pi}_{1+}^2\hat{\pi}_{+0}} - \frac{\hat{\pi}_{01}^2}{\hat{\pi}_{+1}^2\hat{\pi}_{0+}}\right]$$

Since $\chi^2 = n\phi^2$ where ϕ denotes correlation (A proof is given in Appendix A of [22]), $\phi = \sqrt{\chi^2/n} = \sqrt{g/n}$. As we know, $\frac{\partial \phi}{\partial x_i} = \frac{1}{2\sqrt{gn}}\frac{\partial g}{\partial x_i}$. Following the same procedure above, the variance of correlation ϕ can be approximated as

$$\hat{var}(\phi) \approx \frac{\hat{var}(g)}{4G_E}$$

where

$$G_E = n^2\left[\frac{\hat{\pi}_{00}^2}{\hat{\pi}_{0+}\hat{\pi}_{+0}} + \frac{\hat{\pi}_{01}^2}{\hat{\pi}_{0+}\hat{\pi}_{+1}} + \frac{\hat{\pi}_{10}^2}{\hat{\pi}_{1+}\hat{\pi}_{+0}} + \frac{\hat{\pi}_{11}^2}{\hat{\pi}_{1+}\hat{\pi}_{+1}} - 1\right].$$

Similarly we can derive variances of the estimated values of all measures shown in Table 2. Measures such as χ^2, interest factor, IS, PS, and Jaccard coefficient can be extended to more than two variables using the multi-dimensional contingency tables. We show the estimated chi-square statistics for k-itemset as one example.

$$\hat{\chi}^2 = n \sum_{u_1=0}^{1} \cdots \sum_{u_k=0}^{1} \frac{(\hat{\pi}_{u_1\cdots u_k} - \prod_{j=1}^{k}\hat{\pi}_{u_j}^{(j)})^2}{\prod_{j=1}^{k}\hat{\pi}_{u_j}^{(j)}} \quad (4)$$

It is easy to see $\hat{\chi}^2$ can be considered as one derived variable from the observed elements $\hat{\pi}_{u_1\cdots u_k}$ and the marginal totals $\hat{\pi}_{u_j}^{(j)}$ of the 2^k contingency table. Following the same delta method, we can derive its variance.

3.2 Interquantile Ranges of Derived Measures

To derive interquantile ranges of estimates, we need to explore the distribution of those derived variables. In [11], the authors have shown the estimate of support follows an approximate normal distribution and the estimate of confidence (i.e., a ratio of two correlated normal variables) follows a very complex $F(w)$ distribution. In general, we can observe that every element (e.g., $\hat{\pi}_{ij}$) in the derived measure expressions (shown in Table 2) has an approximate normal distribution, however, the derived measures usually do not have explicit distribution expressions. Hence we cannot calculate the critical values of distributions to derive the interquantile range. In the following, we provide an approximation to such range based on Chebyshev's theorem.

Theorem 1. *(Chebyshev's theorem) Let X be a random variable with expected value μ and finite variance σ^2. Then for any real $k > 0$, we have $Pr(|X - \mu| \geq k\sigma) \leq 1/k^2$.*

Chebyshev's Theorem gives a conservative estimate. It provides a lower bound to the proportion of measurements that are within a certain number of standard deviations from the mean. The theorem can be useful despite loose bounds because it can be applied to random variables of any distribution, and because these bounds can be calculated knowing no more about the distribution than the mean and variance. For example, the loose $(1 - \alpha)100\%$ interquantile range of correlation ϕ between A and B can be approximated as

$$[\hat{\phi} - \frac{1}{\sqrt{\alpha}}\sqrt{\hat{var}(\hat{\phi})}, \hat{\phi} + \frac{1}{\sqrt{\alpha}}\sqrt{\hat{var}(\hat{\phi})}]$$

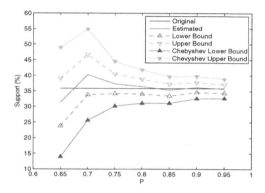

Fig. 1. Interquantile Range vs. varying p

From Chebyshev's theorem, we know for any sample, at least $(1 - (1/k)^2)$ of the observations in the data set fall within k standard deviations of the mean. When we set $\alpha = \frac{1}{k^2}$, we have $Pr(|X - \mu| \geq \frac{1}{\sqrt{\alpha}}\sigma) \leq \alpha$. Hence, $Pr(|X - \mu| \leq \frac{1}{\sqrt{\alpha}}\sigma) \geq 1 - \alpha$. The $(1 - \alpha)100\%$ interquantile range of the estimated measure is then derived.

Note that the interquantile range based on Chebyshev's Theorem is much larger than that based on known distributions such as normal distribution for support estimates. This is because that $\frac{1}{\sqrt{\alpha}} \geq z_{\alpha/2}$ where $z_{\alpha/2}$ is the upper $\alpha/2$ critical value for the standard normal distribution. In Figure 1, we show how the 95% interquantile ranges for the estimated support of one particular rule ($G \Rightarrow H$ from COIL data) change with varied distortion p from 0.65 to 0.95. We can see the interquantile range derived based on Chebyshev's theorem is wider than that derived from known normal distribution. As expected, we can also observe that the larger the p, the more accurate the estimate and the tighter the interquantile ranges.

4   Measures Derived from the Randomized Data Without p

Randomization still runs certain risk of disclosures. Attackers may exploit the released distortion parameter p to calculate the posterior probabilities of the original value based on the distorted data. It is considered as jeopardizing with respect to the original value if the posterior probabilities are significantly greater than the a-priori probabilities [5]. In this section, we consider the scenario where the distortion parameter p is not released to data miners. As a result individual privacy can be better preserved.

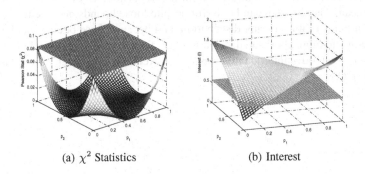

(a) χ^2 Statistics   (b) Interest

Fig. 2. measures from randomized data vs. varying p_1 and p_2

Result 1. *For any pair of items A, B perturbed with distortion parameter p_1 and p_2 ($p_1, p_2 \in [0, 1]$) respectively, we have*

$$\phi_{ran} \leq \phi_{ori} \quad M_{ran} \leq M_{ori}$$
$$G^2_{ran} \leq G^2_{ori} \quad \chi^2_{ran} \leq \chi^2_{ori}$$

where $\phi_{ori}, M_{ori}, G^2_{ori}, \chi^2_{ori}$ denote Correlation, Mutual Information, Likelihood Ratio, Pearson Statistics measures calculated from the original data respectively and $\phi_{ran}, M_{ran}, G^2_{ran}, \chi^2_{ran}$ correspond to measures calculated directly from the randomized data without knowing p_1 and p_2.

All other measures shown in Table 2 do not hold monotonic relations.

Proof. we include the proof of χ^2 in Appendix A and we skip proof of all other measures due to space limits.

Figure 2(a) and 2(b) show how the χ^2 Statistics (G and H) and Interest measures calculated from the randomized data varies with distortion parameters p_1 and p_2. We can easily observe that $\chi^2_{ran} \leq \chi^2_{ori}$ for all $p_1, p_2 \in [0, 1]$ and $I_{ran} \geq I_{ori}$ for some p_1, p_2 values.

We would emphasize that Result 1 is important for data exploration tasks such as hypothesis testing. It shows useful information can still be discovered from the randomized data even without knowing the distortion parameters. For example, testing pairwise independence between the original attributes is equivalent to testing pairwise independence between the corresponding distorted attributes. From the randomized data, if we discover an itemset which satisfies $\chi^2_{ran} \geq \chi^2_\alpha$, we can guarantee that dependence exists among the original itemset since $\chi^2_{ran} \leq \chi^2_{ori}$ holds for all p [1].

---

[1] The alternative hypothesis will be accepted if the observed data values are sufficiently improbable under the null hypothesis. Otherwise, the null hypothesis is not rejected.

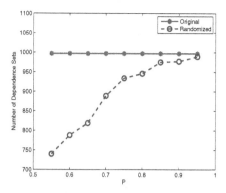

Fig. 3. The number of dependence itemsets vs. varying p

Figure 3 shows the number of dependence itemsets discovered from the randomized COIL data sets with varying p from 0.55 to 0.95. We can observe that the larger the distortion parameter p, the more dependence itemsets calculated directly from the randomized data. Even with $p = 0.55$, around 750 dependence sets can be discovered from the randomized data, which represents about 75% of 997 dependence itemsets derived from the original data.

5 Related Work

Privacy is becoming an increasingly important issue in many data mining applications. A considerable amount of work on privacy preserving data mining, such as additive randomization based [2,1] and projection based [6,19], has been proposed. Recently, a lot of research has focused on the privacy aspect of the above approaches and various point-wise reconstruction methods [16,15,13,12,18,14] have been investigated.

The issue of maintaining privacy in association rule mining has also attracted considerable studies [8,9,4,21,7,20]. Among them, some work [7,20] focused on sensitive association rule hiding where privacy is defined in terms of the output frequent itemsets or association rules. The work closest to our approach is that of [21,4,3] based on Randomization Response techniques. In [21,4], the authors proposed the MASK technique to preserve privacy for frequent itemset mining. In [4], the authors addressed the issue of providing efficiency in estimating support values of itemsets derived from the randomized data. Our paper focused on the issue of providing accuracy in terms of various reconstructed measures (e.g., support, confidence, correlation, lift, etc.) in privacy preserving market basket data analysis. Providing the accuracy of discovered patterns from randomized data is important for data miners. To the best of our knowledge, this has not been previously explored in the context of privacy preserving data mining although defining the significance of discovered patterns in general data mining has been studied (e.g., [10]).

6 Conclusion

In this paper, we have considered the issue of providing accuracy in privacy preserving market basket data analysis. We have presented a general approach to deriving variances of estimates of various measures adopted in market basket data analysis. We applied the idea of using interquantile ranges based on Chebyshev's Theorem to bound those estimates derived from the randomized market basket data. We theoretically show some measures (e.g., correlation) have monotonic property, i.e., the measure values calculated directly from the randomized data are always less than or equal to those original ones. As a result, there is no risk to introduce false positive patterns. Hence, some market basket data analysis tasks (such as correlation analysis or independence hypothetical testing) can be executed on the randomized data directly without the release of distortion probabilities. In the future, we are interested in exploring the tradeoff between the privacy of individual data and the accuracy of data mining results. We will also investigate how various measures are affected by randomization, e.g., which measures are more sensible to randomization.

Acknowledgments

This work was supported in part by U.S. National Science Foundation IIS-0546027.

References

1. Agrawal, D., Agrawal, C.: On the design and quantification of privacy preserving data mining algorithms. In: Proceedings of the 20th Symposium on Principles of Database Systems (2001)
2. Agrawal, R., Srikant, R.: Privacy-preserving data mining. In: Proceedings of the ACM SIGMOD International Conference on Management of Data, pp. 439–450. Dallas, Texas (May 2000)
3. Agrawal, S., Haritsa, J.: A framework for high-accuracy privacy-preserving mining. In: Proceedings of the 21st IEEE International Conference on Data Engineering, pp. 193–204. IEEE Computer Society Press, Los Alamitos (2005)
4. Agrawal, S., Krishnan, V., Haritsa, J.: On addressing efficiency concerns in privacy-preserving mining. Proc. of 9th Intl. Conf. on Database Systems for Advanced Applications (DASFAA), pp. 113–124 (2004)
5. Chaudhuri, A., Mukerjee, R.: Randomized Response Theory and Techniques. Marcel Dekker (1988)
6. Chen, K., Liu, L.: Privacy preserving data classification with rotation perturbation. In: Proceedings of the 5th IEEE International Conference on Data Mining. Houston,TX (November 2005)
7. Dasseni, E., Verykios, V., Elmagarmid, A.K., Bertino, E.: Hiding association rules by using confidence and support. In: Proceedings of the 4th International Information Hiding Workshop, pp. 369–383. Pittsburg, PA (April 2001)
8. Evfimievski, A., Gehrke, J., Srikant, R.: Limiting privacy breaches in privacy preserving data mining. In: Proceedings of the twenty-second ACM SIGMOD-SIGACT-SIGART symposium on Principles of database systems, pp. 211–222. ACM Press, New York (2003)

9. Evfimievski, A., Srikant, R., Agrawal, R., Gehrke, J.: Privacy preserving mining of association rules. In: Proceedings of the eighth ACM SIGKDD international conference on Knowledge discovery and data mining, pp. 217–228 (2002)

10. Gionis, A., Mannila, H., Mielikainen, T., Tsaparas, P.: Assessing data mining results via swap randomization. In: Proceedings of the 12th ACM International Conference on Knowledge Discovery and Data Mining (2006)

11. Guo, L., Guo, S., Wu, X.: On addressing accuracy concerns in privacy preserving association rule mining. Technical Report, CS Dept. UNC Charlotte (March 2007)

12. Guo, S., Wu, X.: On the lower bound of reconstruction error for spectral filtering based privacy preserving data mining. In: Proceedings of the 10th European Conference on Principles and Practice of Knowledge Discovery in Databases, pp. 520–527. Berlin, Germany (September 2006)

13. Guo, S., Wu, X.: On the use of spectral filtering for privacy preserving data mining. In: Proceedings of the 21st ACM Symposium on Applied Computing, pp. 622–626 (April 2006)

14. Guo, S., Wu, X.: Deriving private information from arbitraraily projected data. In: Proceedings of the 11th Pacific-Asia Conference on Knowledge Discovery and Data Mining, pp. 84–95. Nanjing, China, (May 2007)

15. Huang, Z., Du, W., Chen, B.: Deriving private information from randomized data. In: Proceedings of the ACM SIGMOD Conference on Management of Data, Baltimore, MA (2005)

16. Kargupta, H., Datta, S., Wang, Q., Sivakumar, K.: On the privacy preserving properties of random data perturbation techniques. In: Proceedings of the 3rd International Conference on Data Mining, pp. 99–106 (2003)

17. Kendall, M.G., Stuart, A.: The advanced theory of statistics. Hafner Pub. Co, New York (1969)

18. Liu, K., Giannella, C., Kargupta, H.: An attacker's view of distance preserving maps for privacy preserving data mining. In: Proceedings of the 10th European Conference on Principles and Practice of Knowledge Discovery in Databases. Berlin, Germany (September 2006)

19. Liu, K., Kargupta, H., Ryan, J.: Random projection based multiplicative data perturbation for privacy preserving distributed data mining. IEEE Transaction on Knowledge and Data Engineering 18(1), 92–106 (2006)

20. Oliveira, S., Zaiane, O.: Protecting sensitive knowledge by data sanitization. In: Proceedings of the 3rd IEEE International Conference on Data Mining, pp. 211–218. Melbourne, Florida (November 2003)

21. Rizvi, S., Haritsa, J.: Maintaining data privacy in association rule mining. In: Proceedings of the 28th International Conference on Very Large Data Bases (2002)

22. Silverstein, C., Brin, S., Motwani, R., Ullman, J.: Scalable techniques for mining causal structures. In: Proceedings of the 24th VLDB Conference. New York (1998)

23. Tan, P., Steinbach, M., Kumar, K.: Introduction to Data Mining. Addison-Wesley, Reading (2006)

A   Proof of Result 1

The chi-square calculated directly from the randomized data without knowing p is

$$\chi^2_{ran} = \frac{n(\lambda_{11} - \lambda_{1+}\lambda_{+1})^2}{\lambda_{1+}\lambda_{+1}\lambda_{0+}\lambda_{+0}} \tag{5}$$

The original chi-square can be expressed as

$$\chi^2_{ori} = \frac{n(\pi_{11} - \pi_{1+}\pi_{+1})^2}{\pi_{1+}\pi_{+1}\pi_{0+}\pi_{+0}}$$

$$= \frac{n(\lambda_{11} - \lambda_{1+}\lambda_{+1})^2}{f(p_1, p_2, \lambda_{0+}, \lambda_{1+}, \lambda_{+0}, \lambda_{+1}) + \lambda_{1+}\lambda_{+1}\lambda_{0+}\lambda_{+0}}$$

where $f(p_1, p_2, \lambda_{0+}, \lambda_{1+}, \lambda_{+0}, \lambda_{+1}) =$
$p_1 p_2 (p_1 - 1)(p_2 - 1) + p_1(p_1 - 1)\lambda_{+0}\lambda_{+1} + p_2(p_2 - 1)\lambda_{0+}\lambda_{1+}$

To prove $\chi^2_{ran} \leq \chi^2_{ori}$, we need $f(p_1, p_2, \lambda_{0+}, \lambda_{1+}, \lambda_{+0}, \lambda_{+1}) \leq 0$ holds for $\forall$ $\{p_1, p_2, \lambda_{0+}, \lambda_{1+}, \lambda_{+0}, \lambda_{+1}\}$.

As

$$\lambda_{+0} = \pi_{+0}p_2 + (1 - \pi_{+0})(1 - p_2) \quad \lambda_{+1} = 1 - \lambda_{+0}$$
$$\lambda_{0+} = \pi_{0+}p_1 + (1 - \pi_{0+})(1 - p_1) \quad \lambda_{1+} = 1 - \lambda_{0+}$$

$f()$ can be expressed as a function with parameters p_1 and p_2.

We can prove $f() \leq 0 \ \forall p_1, p_2 \in [1/2, 1]$ by showing 1) $f()$ is monotonically increasing with p_1 and p_2 and 2) $f(p_1 = 1, p_2 = 1) = 0$.

1) Since p_1 and p_2 are symmetric and independent, $f()$ can be expressed as

$$f(p_1) = ap_1^2 - ap_1 + \pi_{0+}(1 - \pi_{0+})$$
$$a = mp_2^2 - mp_2 + (1 - \pi_{+0})\pi_{+0} \tag{6}$$
$$m = -4\pi_{+0}^2 + 4\pi_{+0} - 4\pi_{0+}^2 + 4\pi_{0+} - 1$$

Note that $f(p_1)$ is monotonically increasing if $a \geq 0$. Since $0 \leq \pi_{+0}, \pi_{0+} \leq 1$, we have $-1 \leq m \leq 1$.

– When $0 \leq m \leq 1$,

$$\Delta = m^2 - 4m(1 - \pi_{+0})\pi_{+0} = -m(2\pi_{0+} - 1)^2 \leq 0$$

we have $a \geq 0$.

– When $-1 \leq m < 0$, since $\Delta = m^2 - 4m(1 - \pi_{+0})\pi_{+0} \geq m^2 > 0$ the roots for Equation 6 are $p_{21} = \frac{m - \sqrt{\Delta}}{2m} > 1$ and $p_{22} = \frac{m + \sqrt{\Delta}}{2m} < 0$, hence we have $a \geq 0$ for all $1/2 \leq p_2 \leq 1$.

Since $a \geq 0$, we have proved $f()$ is monotonically increasing with p_1. Similarly, we can prove $f()$ is monotonically increasing with p_2.

2) It is easy to check $f(p_1, p_2) = 0$ when $p_1 = p_2 = 1$.

Combining 1) and 2), we have proved $f() \leq 0 \ \forall p_1, p_2 \in [1/2, 1]$. Hence, we have $\chi^2_{ran} \leq \chi^2_{ori}$.

Feature Extraction from Sensor Data Streams for Real-Time Human Behaviour Recognition

Julia Hunter and Martin Colley

Department of Computer Science, University of Essex, Wivenhoe Park,
Colchester CO4 3SQ, U.K.
{jhunte,martin}@essex.ac.uk

Abstract. In this paper we illustrate the potential of motion behaviour analysis in assessing the wellbeing of unsupervised, vulnerable individuals. By learning the routine motion behaviour of the subject (i.e. places visited, routes taken between places) we show it is possible to detect unusual behaviours while they are happening. This requires the processing of continuous sensor data streams, and real-time recognition of the subject's behaviour. To address privacy concerns, analysis will be performed locally to the subject on a small computing device. Current data mining techniques were not developed for restricted computing environments, nor for the demands of real-time behaviour recognition. In this paper we present a novel, online technique for discretizing a sensor data stream that supports both unsupervised learning of human motion behaviours and real-time recognition. We performed experiments using GPS data and compared the results of Dynamic Time Warping.

Keywords: sensor data stream discretization; unsupervised learning; real-time behaviour recognition.

1 Introduction

There are many factors that can limit a person's ability to live a fully independent life, whether as a result of their age or due to physical or mental impairments. Yet often these people desire greater independence than they can safely be granted, such as a young child demanding greater freedom or an elderly person wishing to remain in their own home. The research that we describe here was conceived of with the needs of such people in mind. We are interested in the potential of human motion behaviour analysis in assessing the wellbeing of vulnerable individuals. We learn the routine motion behaviour of the subject (i.e. the places visited, and the routes taken between places) and then show it is possible to detect unusual behaviours while they are happening. An example of unusual behaviour could be as simple as taking a wrong turning and becoming lost. For a young child or a memory-impaired adult this could be a frightening and potentially dangerous situation; the quicker the responsible care-giver can be alerted, the better the outcome will likely be. This type of human motion behaviour analysis has not thus far received much attention within the research

J.N. Kok et al. (Eds.): PKDD 2007, LNAI 4702, pp. 115–126, 2007.

community. However, other kinds of human behaviour have been studied with the same goals of supporting vulnerable people in living independent lives. In particular, a large body of work investigates Activities of Daily Living (ADL) monitoring [1], where the execution of everyday tasks (such as food preparation, having a bath) is recorded and analyzed. It is widely accepted that analysis of the tasks performed can give an indication of a person's wellbeing, and that changes in ADL performance can indicate a change in the subject's condition. Of course, once the individual leaves their home the monitoring stops. We examine the potential of human motion behaviour analysis for extending this kind of monitoring beyond the home.

Section 2 considers some implementation practicalities. Section 3 reviews some data mining techniques that can be applied to the problem of learning and recognizing motion behaviour. Section 4 presents our solution to the same problem. Section 5 discusses the experiments performed to investigate the utility of our method. Section 6 draws conclusions and presents our ideas for further work.

2 Design and Definitions

Let us consider the scenario of a child who is allowed some travel independence yet whose parents would like to be kept aware of changes in their behaviour. If the child starts to take a forbidden shortcut to school the parents would like to find out, as they would if the child decided to go to the park instead of school. The nature of the application is that sensitive data is being collected and the monitored individual may feel that their privacy is being intruded upon. There is also concern that such data collection could be exploited in various ways in the case of a security breach. In order to address some of these concerns we decided that data collection and processing should both happen locally on a small, portable computing device. This means that the time and space complexity of the algorithms used must be kept to a minimum. Additionally, some potential users are computer illiterate; therefore behaviour learning should not require user interaction i.e. the learning algorithms should be unsupervised. Data is obtained from a GPS sensor, so we must work with the inherent coverage and accuracy constraints of the GPS system. These are all points that the design of the system must take into account.

Now let us consider some design issues that will guide implementation. We have decided that an individual's motion behaviour model consists of all the places visited and all the routes travelled; these are connected as a network. The resolution of the model is restricted by the accuracy and availability of GPS data; for example, we will distinguish between buildings but not between rooms in a building. A tested approach to the modelling of places is as a [latitude, longitude] point plus a radius [2] and we follow this example. The resolution of route modelling should support the differentiation between multiple routes connecting the same pair of places.

We define the recognition of an individual's behaviour to be the results of a comparison of what he is currently doing (whether in a place or on a journey) with his model. Real-time recognition means that this process is carried out straight away, rather than waiting for a convenient moment (such as the end of a journey).

3 Related Work

A large body of data mining literature is dedicated to the analysis of time series, and in particular *continuous* time series, where points in the series are real-numbered rather than discrete values (examples of discrete data are the binary values 0/1 or letters of the alphabet). Much work is devoted to the conversion of continuous data to a discrete representation so that discrete analysis can be carried out. This transformation is referred to as "symbolization" or "discretization". The efficiency of numerical computations is greatly increased when the data set is transformed from continuous to discrete [3]; this can lead to faster execution, which is important to real-time monitoring and control operations, and is often less sensitive to measurement noise [3]. It also allows the wealth of discrete data analysis techniques from fields such as bioinformatics and text processing to be applied [4]. Most of the time series data that is the object of data mining research consists of long-term observations of phenomena of a periodic nature, whether this is the performance of financial markets or a patient's heartbeat trace. Many techniques that are suited to this kind of data do not fit our needs. Rather, as explained by [5], the data we measure is the result of a person's intentions; conscious human choice affects every value collected. There are nonetheless some common transformations that we can apply:

Piecewise Approximation (Segmentation)
The time series is decomposed into n homogeneous pieces, segments, such that the data in each segment can be described accurately by a simple model (i.e. Piecewise Constant Approximation, Piecewise Linear Approximation) [6]. This is also referred to as *segmentation*. The distance measure used to compare two segmentations then uses the geometry of the representation [7]. Batch algorithms require n as an input parameter. Online versions, e.g. [13], also exist; here the input parameter is the maximum allowable error per segment. Unfortunately, where the data does not correspond well with the chosen model, over-fitting occurs [14].

Symbolization
SAX [4] takes as input the number of symbols and the number of sub-sequences to decompose each time series into. The initial reduction in dimensionality is obtained by finding the mean of each subsequence. The entire data set is then analysed in order to define the range of values mapping to each symbol, so that symbols are equiprobable overall. Finally the discrete representation is obtained by mapping subsequence means to symbols. Distance between two time series is obtained using a measure derived from Euclidean distance.

Feature Extraction
In order to extract features, some degree of prior knowledge about the data set is required. Domain experts may be used. In [8] motion is represented as a series of symbols, e.g. left turn, straight line. [9] quantify the maxima and minima in a data set and then use these points as a compressed representation.

No Transformation
It is also possible to compare time series using the original, raw data without this initial transformation. Two very common (and long-standing) approaches are [10]:

Euclidean Distance (ED) and Dynamic Time Warping (DTW). ED can only be calculated for equal-length time series, as every point in the candidate time series needs to be compared to its equivalent in the query time series. Thus it may be necessary to carry out interpolation first. ED has time complexity $O(n)$, with additional overhead attributable to interpolation. DTW [11] is able to compare time series of different lengths via a non-linear mapping, also utilizing a Euclidean distance. It is less sensitive to skewing on the time axis than ED and has time complexity $O(n^2)$. Both DTW and ED could have very large n since no dimensionality reduction is carried out.

These data representations and distance measures could be used with the GPS data to perform clustering of the journeys recorded and thus learn models for the routes followed. It is relatively straightforward to perform agglomerative hierarchical clustering [12] once every possible pair of journeys has been compared; any of the techniques described above could be applied to this learning task. However, we wish to use the same techniques to then perform real-time recognition; this requires that data transformation and comparison should be performed online, i.e. while the journey is taking place. The data mining techniques discussed here emphasise accuracy (and in some cases scalability). They are often implemented using batch algorithms that were not intended to give real-time responses of the type we seek. In the context of our application, this would mean waiting for a particular journey to be completed before searching for it in the model. Considering the example of a confused person becoming lost and walking for an hour in the wrong direction, this delay is not acceptable. Real-time recognition demands immediate responses, even if this means basing the response on incomplete information.

In summary, many data mining algorithms implement batch processing, an approach that is unsuited to real-time recognition. Additionally, algorithms have been designed for accurate processing of large data sets; these are not adapted to the processing and storage limitations of small computing devices.

The remainder of this paper presents our solution to the online processing and analysis of sensor data so that it can be used for unsupervised learning and real-time recognition on a device with limited computing resources.

4   Online Data Stream Symbolization (ODSS)

Information relating to the subject's position is obtained from a GPS receiver. The GPS receiver can generate *track angle* data as well as [latitude, longitude] coordinates. The track angle is the angle of travel in degrees with respect to true north, generated by the GPS receiver from the change in [latitude, longitude]. We decided to use track angle as the input to our behaviour model instead of [latitude, longitude] because it efficiently encodes changes in two-dimensional position using one-dimensional data. Fig. 1 and fig. 2 show [latitude, longitude] and track angle plots for a single journey. Inspection of track angle data generated by a person moving about shows frequent areas of steady (or near-steady) state corresponding to periods of straight-line travel. The associated value is the real-numbered track angle. These steady state sections vary in length and frequencies according to the route travelled, but are always present to some degree.

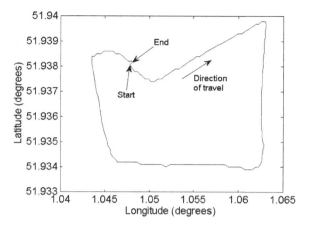

Fig. 1. Plot of latitude against longitude for a circular route travelled clockwise

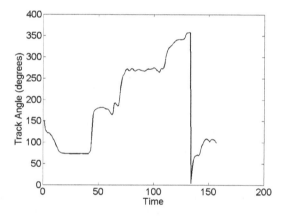

Fig. 2. Plot of track angle against time step for the journey in fig. 1

This characteristic is interesting because it has the potential to uniquely characterize a journey. It can be treated as a representation of the continuous time series to which discrete analysis techniques can be applied. We refer to this representation as a set of symbols, where the symbols are defined in an online fashion as the analysis proceeds. Each symbol consists of a real-numbered value and a tolerance (this is one input to the feature extraction process) describing the degree of variation tolerated with the region of steady state. The second input parameter is the minimum duration of a state for it to be extracted as a symbol. This is the essence of our Online Data Stream Symbolization (ODSS) approach. The results of applying ODSS to the journey in fig. 1 are shown in fig. 3. The input parameters used here are: tolerance of +/- 4 degrees and minimum duration of 12s; these values were selected empirically and had previously been found to work well on a range of data. The overall set of symbols is not restricted, and as more journeys are processed the total number of symbols used by all the journeys could become quite large. In this sense it is different from other symbol definition processes such as that used by [4], which fix the number and definition of the symbols in advance of performing the transformation. The ordering of the symbols is preserved, as this reflects the information in the time dimension of the original data. The symbols are considered to belong to an *ordered* set.

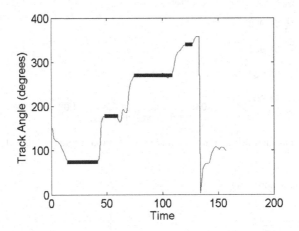

Fig. 3. ODSS results (tolerance = +/- 4 degrees, minimum duration = 12s) superimposed on the track angle data seen in fig. 2. The symbol set can be expressed as the ordered set {73, 178, 270, 340}.

The ODSS approach is similar to Piecewise Constant Approximation (PCA) in that the real-numbered input data maps to a few real-numbered symbols, where these output values are unknown a priori. The difference lies in the fact that PCA fits a constant model to the whole data set, even to those sections that are not constant. Fig. 4 shows a 6-segmentation PCA that has been applied to the track angle data in fig. 2; ODSS and PCA agree about the central 4 segments, while the poorer-fitting 2 outer segments are not identified by ODSS.

The ODSS approach also relates to feature extraction such as [9], in that the feature sought (constant value) is established with prior knowledge of the data set, but can take any value.

Having obtained an ODSS representation for two separate journeys (p and q), a means of comparing the two journeys is required. This in turn depends on being able to compare a pair of symbols. We take one symbol from each journey and compare them using the tolerance, t. The symbols are considered to be a match if equation 1 holds true.

$$(p_i - q_j) \leq 2*t \tag{1}$$

A discrete similarity measure can then be used to compare a pair of journeys. We chose to use the Jaccard similarity coefficient (J). J takes a value in the range [0,1], where high J indicates a high degree of similarity, and vice versa. Equation 2 defines J for two sets A and B. We use equation 1 to work out which symbols are common to both sets.

$$J(A,B) = |A \cap B| / |A \cup B| \tag{2}$$

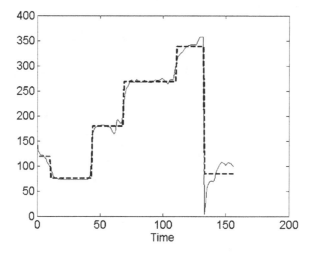

Fig. 4. The 6-segmentation produced by PCA, superimposed onto the track angle data of fig. 2

Our symbol sets are ordered sets, so the set notation in equation 2 is considered here to apply to ordered sets. This important distinction is illustrated in table 1 using the example sets: A={1,3,5,7}, B={7,1,5} (N.B. for the purposes of this example only, symbol equality corresponds to standard mathematical equality).

Given a similarity measure it is then possible to proceed to clustering. The similarity of all possible pairs of journeys is calculated, and used as the input to an agglomerative hierarchical clustering algorithm. The output is used to identify groups of similar journeys.

Table 1. Example showing how ordering affects J, where A={1,3,5,7}, B={7,1,5}

Ordering	A ∩ B	A ∪ B	$J = |A \cap B| / |A \cup B|$
No	{1,5,7}	{1,3,5,7}	3/4 = 0.75
Yes	{1,5}	{7,1,3,5,7}	2/5 = 0.4

5 Experiments

5.1 Experimental Scenario

We envisaged a scenario that would provide a framework for the experiments. A young child walks alone through the park (fig. 5) to school in the morning and is accompanied home by a group of friends in the afternoon. In the afternoon the child is allowed to take the direct route home through the park (C – A) that is obscured by trees. In the morning the parents prefer the child to take the longer route to school (A – B – C) that avoids the treed area by a large margin. In this scenario the parents would like to know about changes in their child's behaviour, but would like to be

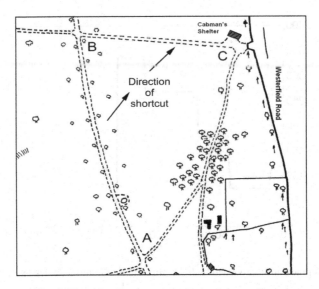

Fig. 5. Map of park where experimental data collected

flexible in setting boundaries for the child. For example, they don't mind if the child cuts the corner at B a bit to shorten their journey but they would like to know if the child is starting to deviate a significant amount from the prescribed route.

5.2 Data Collection and Preparation

Data was collected while walking around the park shown in fig. 5. The data was collected over two visits, with several journeys being walked on each occasion and stationary pauses used to demarcate the journeys. The routes travelled are defined in table 2.

Table 2. Definition of routes travelled

Name of route	Description
Permitted (P)	A – B – C
Return (R)	C – A
Forbidden (F)	A – C
Shortcut (S)	A – (cut the corner at B) – C. The degree to which the corner is cut varies from a few meters to a route running almost parallel with Forbidden

Data was downloaded from the PDA to the PC for processing. Journeys were isolated by identifying velocity = 0 (corresponding to pauses in data collection). All the data was used and there was no attempt to exclude any noisier data from the analysis.

5.3 Experimental Method and Results

Aims. Obtain clusters identifying the main routes travelled. First do this using ODSS, then perform a comparative analysis based on DTW. Investigate a working value for the dissimilarity measure in each case.

Clustering using ODSS. ODSS was applied to each journey (31 in total) in turn. A similarity matrix was then obtained by calculating J for every pair of journeys. This matrix was transformed into a dissimilarity matrix by subtracting all elements from one and then used as the input to the hierarchical agglomerative clustering *linkage* routine available in Matlab. The linkage method used was *average*, meaning that the inter-cluster distance between two clusters x and y is calculated to be the average distance between all pairs of objects in cluster x and cluster y. Results are presented graphically as a dendrogram in fig. 6.

ODSS Results. Dimensionality reduction is significant: for example, a journey consisting of 200 data points might be reduced to 5 symbols using this method. Each tick on the x-axis of the dendrogram represents a single journey; the tick label consists of a letter (e.g. 's' for shortcut) and a number indicating the journey instance (e.g. 's5' is the 5th shortcut). Please refer to table 2 for the meanings of other letters. There is no hard threshold that can divide similar from dissimilar journeys; the dendrogram can be used to ascertain a working value for the application in question. We begin by placing a threshold at 0.5 and consider that all cluster joins below this line are valid. This gives us: a cluster of 'r' journeys; a cluster of 'f' journeys (incorporating the most extreme shortcut that strongly resembles the forbidden route); a cluster of 'p' journeys (incorporating the least extreme shortcut that strongly resembles the permitted route); and several small clusters containing the remaining 's' journeys. Journeys r2 and f2 both join their respective clusters, but above the chosen threshold. Inspection of plots of the raw data shows that these two journeys are noisier than the remainder of the dataset, which explains the greater apparent dissimilarity.

Clustering using DTW. Pairs of continuous time series are used as the input to the DTW algorithm, which uses Euclidean distance to calculate the dissimilarity between the two series. In order to generate results for comparison with ODSS, DTW was used to calculate the distance between all pairs of journeys. This resulted in a dissimilarity matrix that could be clustered in the same way as the one produced by ODSS. The dendrogram is shown in fig. 7. The dissimilarity values have been rescaled so that the values vary between 0 and 1 for this set of results; this is to facilitate comparison with the ODSS results.

DTW Results. Again we select a threshold of 0.5 and then explore the results in the same way as the ODSS clustering: the 'r' journeys form a tight cluster; the 'f' journeys do the same, and pull in the most similar 's' journey too (s10). The next identifiable cluster contains two 'p' journeys plus a fairly similar 's' journey. The remainder of the clusters are more difficult to explain since the 's' and 'p' journeys show strong heterogeneous links; whereas we know that some of the 's' journeys are very different from the 'p' journeys, and very different from each other too.

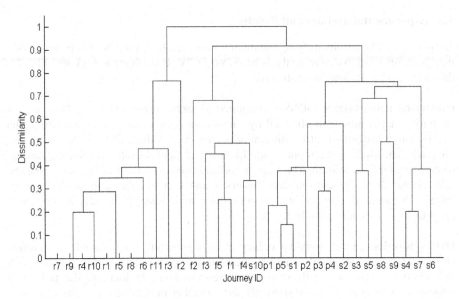

Fig. 6. Results of clustering using ODSS

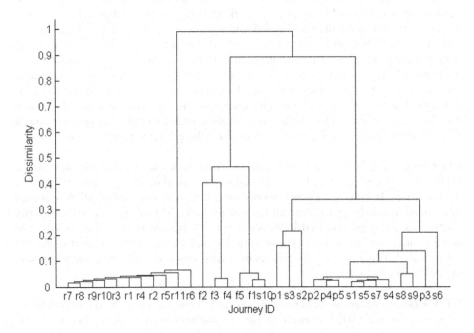

Fig. 7. Results of clustering using DTW

6 Conclusions and Further Work

The results show that ODSS is able to cluster a set of unlabelled routes; it does this at least as well as the equivalent clustering based on DTW for the real GPS data set used. Continuous data is transformed to a discrete representation, with dimensionality reduction of the order of 40:1 for this data set. Transformation can be performed online and requires no batch processing, nor prior definition of symbol sets. A well-established similarity measure, the Jaccard coefficient, can be applied to this discrete representation; this in turn allows hierarchical clustering to be performed. ODSS succeeds in identifying quite small changes in path followed, despite its low-resolution representation. When analysing the behaviour of vulnerable individuals, the ability to detect such small changes could be important.

ODSS forms the basis of our real-time recognition system, which compares partial (ongoing) behaviours with the model of previous behaviours, again using the Jaccard coefficient. A sliding window approach is used to adapt the methods described in this paper to a real-time context. This allows behaviour assessments to be carried out repeatedly, as symbols are extracted from the data stream. The significance of this is the ability to rapidly detect unusual behaviour without waiting, for example, for a journey to end. Initial real-time recognition results are encouraging and are the main focus of ongoing experiments.

References

1. Patterson, D., Fox, D., Kautz, H., Philipose, M.: Expressive, Tractable and Scalable Techniques for Modeling Activities of Daily Living UbiHealth 2003, Seattle, WA (2003)
2. Ashbrook, D., Starner, T.: Learning Significant Locations and Predicting User Movement with GPS. In: ISWC 2002, IEEE, Los Alamitos (2002)
3. Daw, C.S., Finney, C.E.A., Tracy, E.R: A Review of Symbolic Analysis of Experimental Data. Review of Scientific Instruments 74, 916–930 (2003)
4. Lin, J., Keogh, E., Lonardi, S., Chiu, B.A: Symbolic Representation of Time Series, with Implications for Streaming Algorithms. In: Proc. 8th ACM SIGMOD workshop on Research issues in data mining and knowledge discovery, pp. 2–11. ACM Press, New York (2003)
5. Vasquez, D., Large, F., Fraichard, T., Laugier, C.: Intentional Motion, Online Learning and Prediction. In: Proc. Int. Conf. on Field and Service Robotics, Port Douglas (AU) (2005)
6. Bingham, E., Gionis, A., Haiminen, N., Hiisila, H., Mannila, H., Terzi, E.: Segmentation and dimensionality reduction. In: SIAM Data Mining Conference (SDM) (2006)
7. Keogh, E., Pazzani, M.: An enhanced representation of time series which allows fast and accurate classification and relevance feedback. In: proceedings of the 4th Int'l Conference on Knowledge Discovery and Data Mining. New York, NY, pp. 239–241 (1998)
8. Xiaolei Li, X., Han, J., Kim, S.: Motion-Alert: Automatic Anomaly Detection in Massive Moving Objects. In: Proc. 2006 IEEE Int. Conf. on Intelligence and Security Informatics (ISI'06), San Diego, CA (2006)
9. Pratt, K., Fink, E.: Search for Patterns in Compressed Time Series. International Journal of Image and Graphics 2(1), 89–106 (2002)
10. Keogh, E., Kasetty, S.: Data Mining and Knowledge Discovery. 7(4), 349–371 (2003)

11. Keogh, E., Ratanamahatana, A.: Everything you know about Dynamic Time Warping is Wrong. In: 3rd Workshop on Mining Temporal and Sequential Data, in conjunction with 10th ACM SIGKDD Int. Conf. Knowledge Discovery and Data Mining (KDD-2004), Seattle, WA (2004)
12. Sneath, P.H.A, Sokal, R.R: Numerical taxonomy; the principles and practice of numerical classification. W. H. Freeman, San Francisco (1973)
13. Palpanas, T., Vlachos, M., Keogh, E., Gunopulos, D., Truppel, W.: Online Amnesic Approximation of Streaming Time Series. In: ICDE. Boston, MA, USA (2004)
14. Lemire, D.: A Better Alternative to Piecewise Linear Segmentation. SIAM Data Mining 2007 (2007)

Generating Social Network Features for Link-Based Classification

Jun Karamon[1], Yutaka Matsuo[2], Hikaru Yamamoto[3], and Mitsuru Ishizuka[1]

[1] The University of Tokyo, 7-3-1 Hongo, Bunkyo-ku, Tokyo, Japan
karamon@mi.ci.i.u-tokyo.ac.jp,ishizuka@i.u-tokyo.ac.jp
[2] National Institute of Advanced Industrial Science and Technology,
1-18-13 Soto-kanda, Chiyoda-ku, Tokyo, Japan
y.matsuo@aist.go.jp
[3] Seikei University, 3-3-1 Kichijoji Kitamachi, Musashino-shi, Tokyo, Japan
yamamoto@econ.seikei.ac.jp

Abstract. There have been numerous attempts at the aggregation of attributes for relational data mining. Recently, an increasing number of studies have been undertaken to process social network data, partly because of the fact that so much social network data has become available. Among the various tasks in link mining, a popular task is *link-based classification*, by which samples are classified using the relations or links that are present among them. On the other hand, we sometimes employ traditional analytical methods in the field of social network analysis using e.g., centrality measures, structural holes, and network clustering. Through this study, we seek to bridge the gap between the aggregated features from the network data and traditional indices used in social network analysis. The notable feature of our algorithm is the ability to invent several indices that are well studied in sociology. We first define general operators that are applicable to an adjacent network. Then the combinations of the operators generate new features, some of which correspond to traditional indices, and others which are considered to be new. We apply our method for classification to two different datasets, thereby demonstrating the effectiveness of our approach.

1 Introduction

Recently, increasingly numerous studies have been undertaken to process network data (e.g., social network data and web hyperlinks), partly because of the fact that such great amounts of network data have become available. *Link mining* [6] is a new research area created by the intersection of work in link analysis, hypertext and web mining, relational learning, and inductive logic programming and graph mining. A popular task in link mining is *link-based classification*, classifying samples using the relations or links that are present among them. To date, numerous approaches (e.g. [8]) have been proposed for link-based classification, which are often applied to social network data.

A social network is a social structure comprising nodes (called *actors*) and relations (called *ties*). Prominent examples of recently studied social networks

J.N. Kok et al. (Eds.): PKDD 2007, LNAI 4702, pp. 127–139, 2007.

are online social network services (SNS), weblogs (e.g., [1]), and social book-marks (e.g., [7]). As the world becomes increasingly interconnected as a "global village"[18], network data have multiplied. For that reason, among others, the needs of mining social network data are increasing. A notable feature of social network data is that it is a particular type of relational data in which the target objects are (in most cases) of a single type, and relations are defined between two objects of the type. Sometimes a social network consists of two types of objects: a network is called an affiliation network or a two-mode network.

Social networks have traditionally been analyzed in the field of social net-work analysis (SNA) in sociology [16,14]. Popular modes of analysis include centrality analysis, role analysis, and clique and cluster analyses. These analyses produce indices for a network, a node, or sometimes for an edge, that have been revealed as effective for many real-world social networks over the half-century history of social studies. In complex network studies [17,3], which is a much younger field, analysis and modeling of scale-free and small world networks have been conducted. Commonly used features of a network are clustering coefficients, characteristic path lengths, and degree distributions.

Numerous works in the data mining community have analyzed social networks [2,13]. For example, L. Backstrom et al. analyzed the social groups and commu-nity structure on LiveJournal and DBLP data [2]. They build eight community features and six individual features, and subsequently report that one feature is unexpectedly effective: for moderate values of k, an individual with k friends in a group is significantly more likely to join if these k friends are themselves mutual friends than if they are not. Apparently, greater potential exists for such new features using a network structure, which is the motivation of this research. Although several studies have been done to identify which features are useful to classify entities, no comprehensive research has been undertaken so far to generate the features effectively, including those used in social studies.

In this paper, we propose an algorithm to generate the various network fea-tures that are well studied in social network analysis. We define primitive oper-ators for feature generation to create structural features. The combinations of operators enable us to generate various features automatically, some of which cor-respond to well-known social network indices (such as centrality measures). By conducting experiments on two datasets, the Cora dataset and @cosme dataset, we evaluate our algorithm.

The contributions of the paper are summarized as follows:

- Our research is intended to bridge a gap between the data mining community and the social science community; by applying a set of operators, we can effectively generate features that are commonly used in social studies.
- The research addresses link-based classification from a novel approach. Be-cause some features are considered as novel and useful, the finding might be incorporated into future studies for improving performance for link-based classification.
- Our algorithm is applicable to social networks (or one-mode networks). Be-cause of the increasing amount of attention devoted to social network data,

especially on the Web, our algorithm can support further analysis of the network data, in addition to effective services such as recommendations of communities.

This paper is organized as follows. Section 2 presents related works of this study. In Section 3, we show details of the indices of social network analysis. In Section 4, we propose our method for feature generation by defining nodesets, operators, and aggregation methods. Section 5 describes experimental results for two datasets, followed by discussion and conclusions.

2   Related Work

Various models have been developed for relational learning. A notable study is that of *Probabilistic Relational Models (PRMs)* [5]. Such models provide a language for describing statistical models over relational schema in a database. They extend the Bayesian network representation to enable incorporation of a much richer relational structure and are applicable to a variety of situations. However, the process of feature generation is decoupled from that of feature selection and is often performed manually. Alexandrin et al. [11] propose a method of statistical relational learning (SRL) with a process for systematic generation of features from relational data. They formulated the feature generation process as a search in the space of a relational database. They apply it to relational data from Citeseer, including the citation graph, authorship, and publication, in order to predict the citation link, and show the usefulness of their method.

C. Perlich et al. [10] also propose aggregation methods in relational data. They present the hierarchy of relational concepts of increasing complexity, using relational schema characteristics and introduce target-dependent aggregation operators. They evaluate this method on the noisy business domain, or IPO domain. They predict whether an offer was made on the NASDAQ exchange and draw conclusions about the applicability and performance of the aggregation operators.

L. Backstrom et al. [2] analyzes community evolution, and shows that some *structural features* characterizing individuals' positions in the network are influential, as well as some *group features* such as the level of activity among members. They apply a decision-tree approach to LiveJournal data and DBLP data, which revealed that the probability of joining a group depends in subtle but intuitively natural ways not just on the number of friends one has, but also on the ways in which they are mutually related. Because of the relevance to our study, we explain the individuals' features used in their research in Table 1; they use eight community features and six individual features. Our purpose of this research can be regarded as generating such features automatically and comprehensively to the greatest degree possible.

Our task is categorized into link-based object classification in the context of link mining. Various methods have been used to address tasks such as loopy belief propagation and mean field relaxation labeling [15]. Although these models are useful and effective, we do not attempt to generate such probabilistic or

Table 1. Features used in [2]

Features related to an individual u and her set S of friends in community C
Number of friends in community ($|S|$).
Number of adjacent pairs in $S(|(u,v)|u,v \in S \land (u,v) \in E_C|)$.
Number of pairs in S connected via a path in E_C.
Average distance between friends connected via a path in E_C.
Number of community members reachable from S using edges in E_C.
Average distance from S to reachable community members using edges in E_C.

statistical models in this study because it is difficult to compose such models using these basic operations.

3   Social Network Features

In this section, we overview commonly-used indices in social network analysis and complex network studies. We call such attributes *social network features* throughout the paper.

One of the simplest features of a network is its *density*. It describes the general level of linkage among the network nodes. The graph density is defined as the number of edges in a (sub-)graph, expressed as a proportion of the maximum possible number of edges.

Within social network analysis, the centrality measures are an extremely popular index of a node. They measure the structural importance of a node, for example, the power of individual actors. There are several kinds of centrality measures [4]; the most popular ones are as follows:

Degree. The degree of a node is the number of links to others. Actors who have more ties to other actors might be advantaged positions. It is defined as $C_i^D = \frac{k_i}{N-1}$, where k_i is the degree of node i and N is the number of nodes.

Closeness. Closeness centrality emphasizes the distance of an actor to all others in the network by focusing on the distance from each actor to all others. It is defined as $C_i^C = (L_i)^{-1} = \frac{N-1}{\sum_{j \in G} d_{ij}}$, where L_i is the average geodesic distance of node i, and d_{ij} is the distance between nodes i and j.

Betweenness. Betweenness centrality views an actor as being in a favored position to the extent that the actor falls on the geodesic paths between other pairs of actors in the network. It measures the number of all the shortest paths that pass through the node. It is defined as $C_i^B = \frac{\sum_{j<k \in G} n_{jk}(i)/n_{jk}}{(N-1)(N-2)}$, where n_{jk} denotes the number of the shortest paths between nodes j and k, and $n_{jk}(i)$ is the number of those running through node i.

A popular variation of centrality measure is the eigenvector centrality (also known as PageRank or stationary probability). Because we do not target the eigenvector centrality in this paper, we do not explain it here but we will discuss it in Section 6.

Another useful set of network indices is the characteristic path length (sometimes denoted as L) and clustering coefficient (denoted as C), which are the most important and frequently-invoked characteristics of complex network studies.

Characteristic path length. The characteristic path length L is the average distance between any two nodes in the network (or a component).

Clustering coefficient. The clustering for a node is the proportion of edges between the nodes within its neighborhood divided by the number of edges that could possibly exist between them. The clustering coefficient C is the average of clustering of each node in the network.

There are other groups of indices such as structural equivalence (defined on a pair of nodes), and structural holes (defined on a node). We do not explain all the indices but readers can consult literature on social network analysis [16,14].

4   Methodology

In this section, we define the elaborate operators that generate social-network features. Using our model, we attempt to generate features that are often used in social science. Our intuition is simple; recognizing that traditional studies in social science have shown the usefulness of several indices, we can assume that feature generation toward the indices is also useful.

Then, how can we design the operators so that they can effectively construct various types of social network features? Through trial and error, we can come up with the feature generation in three steps; we first select a set of nodes. Then the operators are applied to the set of nodes to produce a list of values. Finally, the values are aggregated into a single feature value. Eventually, we can construct indices such as characteristic path length L, clustering coefficient C, and centralities. Below, we explain each step in detail.

4.1   Defining a Node Set

First, we define a node set. We consider two types of node sets: one is based on a network structure; the other is based on the category of a node.

Distance-based node set. Most straightforwardly, we can choose the nodes that are adjacent to node x. The nodes are, in other words, those of distance one from node x. The nodes with distance two, three, and so on can be defined as well. We define a set of nodes as follows.

- $C_x^{(k)}$: a set of nodes within distance k from x.

Note that $C_x^{(k)}$ does not include node x itself. $C_x^{(\infty)}$ means a set of nodes that are reachable from node x.

Table 2. Operator list

Notation	Input	Output	Description	Stage
$C_x^{(1)}$	node x	a nodeset	adjacent nodes to x	1
$C_x^{(\infty)}$	node x	a nodeset	reachable nodes from x	2
$N_p \cap C_x^{(1)}$	node x	a nodeset	all positive nodes adjacent to x	3
$N_p \cap C_x^{(\infty)}$	node x	a nodeset	all positive nodes reachable from x	3
$s^{(1)}$	a nodeset	a list of values	1 if connected, 0 otherwise	1
t	a nodeset	a list of values	distance between a pair of nodes	1
t_x	a nodeset	a list of values	distance between node x and other nodes	2
u_x	a nodeset	a list of values	1 if the shortest path includes node x, 0 otherwise	2
Avg	a list of values	a value	average of values	1
Sum	a list of values	a value	summation of values	1
Min	a list of values	a value	minimum of values	1
Max	a list of values	a value	maximum of values	1
$Ratio_p$	two values	value	ratio of value on positive nodes($N_p \cap C_x^{(k)}$) by all nodes ($C_x^{(k)}$)	4

Category-based node set. We can define a set of nodes with a particular value of some attribute. Although various attributes can be targeted, for link-based classification, we specifically examine the value of the category attribute of a node to be classified. We denote a set of positive nodes as N_p.

Considering both distance-based and category-based node sets, we can define the conjunction of the sets, e.g., $C_x^{(1)} \cap N_p$.

4.2   Operation on a Node Set

Given a nodeset, we can conduct several calculations to the node set. Below, we define operators to two nodes, and then expand it to a nodeset with an arbitrary number of nodes.

The most straightforward operation for two nodes is to check whether the two nodes are adjacent or not. A slight expansion is performed to check whether the two nodes are within distance k or not. Therefore, we define the operator as follows:

$$s^{(k)}(x, y) = \begin{cases} 1 & \text{if nodes } x \text{ and } y \text{ are connected within } k \\ 0 & \text{otherwise} \end{cases}$$

Another simple operation for two nodes is to measure the geodesic distance between the two nodes on the graph. We can define an operator as follows:

$$t(x, y) = \text{distance between } x \text{ to } y = \arg\min_k\{s^{(k)}(x, y) = 1\}$$

If given a set of more than two nodes (denoted as N), these two operations are applied to each pair of nodes in N. For example, if we are given a node set

$\{n_1, n_2, n_3\}$, we calculate $s^{(1)}(n_1, n_2)$, $s^{(1)}(n_1, n_3)$, and $s^{(1)}(n_2, n_2)$ and return a list of three values, e.g. $(1, 0, 1)$. We denote this operation as $s^{(1)} \circ N$.

In addition to s and t operations, we define two other operations. One is to measure the distance from node x to each node, denoted as t_x. Instead of measuring the distance of two nodes, $t_x \circ N$ measures the distance of each node in N from node x. Another operation is to check the shortest path between two nodes. Operator $u_x(y, z)$ returns 1 if the shortest path between y and z includes node x. Consequently, $u_x \circ N$ returns a set of values for each pair of $y \in N$ and $z \in N$. Operations t_x and u_x focus on node x in terms of the distance and the shortest path, and can be considered fundamental.

4.3   Aggregation of Values

Once we obtain a list of values, several standard operations can be added to the list. Given a list of values, we can take the summation (Sum), average (Avg), maximum (Max), and minimum (Min). For example, if we apply Sum aggregation to a value list $(1, 0, 1)$, we obtain a value of 2. We can write the aggregation as e.g., $Sum \circ s^{(1)} \circ N$. Although other operations can be performed, such as taking the variance or taking the mean, we limit the operations to the four described above.

Additionally, we can take the difference or the ratio of two obtained values. For example, if we obtain 2 by $Sum \circ s^{(1)} \circ N$ and 1 by $Sum \circ s^{(1)} \circ C_x$, the ratio is $2/1 = 2.0$.

We can thereby generate a feature by subsequently defining a nodeset, applying an operator, and aggregating the values. Because the number of possible combinations is enormous, we apply some constraints on the combinations. First, when defining a nodeset, k is an arbitrary integer theoretically; however, we limit k to be 1 or infinity for simplicity. Operator $s^{(k)}$ is used only as $s^{(1)}$. We also limit taking the ratio only to those two values with and without a positive nodeset.

The nodesets, operators, and aggregations are shown in Table 2. We have $4(nodesets) \times 4(operators) \times 4(aggregations) = 64$ combinations. If we consider the ratio, there are ratios for $C_x^{(1)}$ to $N_p \cap C_x^{(1)}$, and for $C_x^{(\infty)}$ to $N_p \cap C_x^{(\infty)}$. In all, there are $4 \times 4 \times 2$ more combinations, and 96 in total. Each combination corresponds to a feature of node x. Note that some combinations produce the same value; for example, $Sum \circ t_x \circ C_x^{(1)}$ is the same as $Sum \circ s \circ C_x^{\infty}$, representing the degree of node x.

The resultant value sometimes corresponds to a well-known index as we intend in the design of the operators. For example, the network density can be denoted as $Avg \circ s^{(1)} \circ N$. It represents the average of edge existence among all nodes; it therefore corresponds to the density of the network. Below, we describe other examples that are used in the social network analysis literature.

- diameter of the network: $Min \circ t \circ N$
- characteristic path length: $Avg \circ t \circ N$
- degree centrality: $Sum \circ s_x^{(1)} \circ N_x^{(1)}$
- node clustering: $Avg \circ s^{(1)} \circ N_x^{(1)}$

- closeness centrality: $Avg \circ t_x \circ C_x^{(\infty)}$
- betweenness centrality: $Sum \circ u_x \circ C_x^{(\infty)}$,
- structural holes: $Avg \circ t \circ N_x^{(1)}$

We can generate several features that have been shown to be effective in existing studies [2]. A couple of examples are the following

- Number of friends in community $= Sum \circ S_x^{(1)} \circ (C_x^{(1)} \cap N_p)$ and
- Number of adjacent pairs $= Sum \circ s^{(1)} \circ (N_x^{(1)} \cap N_p)$.

These features represent some of the possible combinations. Some lesser-known features might actually be effective.

5   Experimental Result

In this section, we describe empirical results obtained using our social network feature generation. Through the experiment, we show the usefulness of the generated features toward link-based classification problems. We classify a node into categories using the relations around the node.

5.1   Datasets and Task

After generating features, we investigate which features are better to classify the entities. We employ a decision tree technique following [2] to generate the decision tree (using C4.5 algorithm [12]). We use two datasets: Cora database and @cosme. We first explain the characteristics of these datasets, and then describe the results and findings.

Cora dataset. This dataset, contributed by A. McCallum [9], contains 300,000 scientific papers related to computer science classified into 69 research areas. About 10,000 papers include detailed information about properties such as the title, author names, a journal name, and the year of publication. In addition, each paper has information about its cited literature. We therefore have a citation network in which a node is a paper and an (undirected) edge is a citation. We do not use direction information on edges.

Training and test data are created as follows: we randomly select nodes from among those in the target category and those which cite or are cited by a paper in the target category. We randomly select one-fifth of the whole 69 categories as target categories. For example, in the case of the category *Neural networks in Machine Learning in Artificial Intelligence*, the number of all nodes is 1682; the number of positive nodes (in this category) is 781. Because the negative examples are the nodes which are not in the category but which have a direct relation with other nodes in the category, the settings are more difficult than those used when we select negative examples randomly.

Fig. 1. Top three levels of the decision tree in using up to Stage 2 operators

Fig. 2. Top three levels of the decision tree using all operators

@cosme dataset. @cosme (www.cosme.net) is the largest online community site of "for-women" communities in Japan. It provides information and reviews related to cosmetic products. Users of @cosme can post their reviews of cosmetic products (100.5 thousand items of 11 thousand brands) on the system. Notable characteristics of @cosme are that a user can register other users who can be trusted, thereby creating a social network of users.

Because a user of @cosme can join various communities on the site, we can classify users into communities, as was done with the Cora dataset. The nodes are selected from among those who are the members of the community, or those who have a relation with a user in the community. Here we target popular communities with more than 1000 members[1]. In case of *I love Skin Care* communities, the number of nodes is 5730 and the number of positive nodes is 2807.

5.2 Experimental Results

We generate features defined in Table 2 for each dataset. To record the effectiveness of operators, we first limit the operators of Stage 1, as shown in Table 2; then we include the operators of Stage 2, those of Stage 3, and one of Stage 4.

Table 4 shows the values of recall, precision, and F-value for the Cora dataset. The performance is measured by 10-fold cross validation. As we use more operators, the performance improves. Figures 1 and 2 show the top three levels of the decision tree when using operators of Stage 1 and 2, and all the operators. We can see in Fig. 1 that the top level node of the decision tree is $Sum \circ s^{(1)} \circ C_x^{(\infty)}$, which is the number of edges that node x has, or the degree centrality. The second top node is $Sum \circ t_x \circ C_x^{(1)}$, which also corresponds to a degree centrality (in a different expression).

If we add operators in Stage 3 and Stage 4, we obtain a different decision tree as in Fig. 2. The top node is the ratio of the number of positive and all nodes neighboring node x. It means that if the number of neighboring nodes in the category is larger, the node is more likely to be in the category, which can be reasonably understood. We can see in the third level the ratio of $Avg \circ s^{(1)} \circ C_x^{(\infty)}$, which corresponds to the density of the subgraph including node x. There are

---

[1] Such as *I love Skin Care* community, *Blue Base* community and *I love LUSH* community.

Table 3. Recall, precision, and F-value in the @cosme dataset as adding operators

	Recall	Precision	F-value
Stage 1	0.429	0.586	0.494
Stage 2	0.469	0.593	0.523
Stage 3	0.526	0.666	0.586
Stage 4	0.609	0.668	0.636

Table 4. Recall, precision, and F-value in Cora dataset as adding operators

	Recall	Precision	F-value
Stage 1	0.427	0.620	0.503
Stage 2	0.560	0.582	0.576
Stage 3	0.724	0.696	0.709
Stage 4	0.767	0.743	0.754

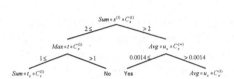

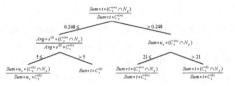

Fig. 3. Top three levels of the decision tree using up to Stage 2 operators in @cosme dataset

Fig. 4. Top three levels of the decision tree using all operators in @cosme dataset

also features calculating the ratio of $Sum \circ t_x \circ C_x^\infty$, which is a closeness centrality, and $Sum \circ u_x \circ C_x^{(1)}$, which corresponds to a betweenness centrality.

The results of @cosme dataset are shown in Table. 3. The trend is the same as that for the Cora dataset; if we use more operators, the performance improves. The decision trees when using up to Stage 2 operators and all operators are shown in Figs. 3 and 4. The top level node of Fig. 3 is $Sum \circ t_x \circ C_x^{(1)}$, which is the number of edges among nodes adjacent to node x. The top level node in Fig. 4 is the ratio of the summation of the path length of reachable positive nodes from node x to the summation of the path length of all reachable nodes. In the third level, we can find $Sum \circ t \circ C_x^{(1)}$. This value is not well known in social network analysis, but it measures the distance among neighboring nodes of node x. The distance is 1 if the nodes are connected directly, and 2 if the nodes are not directly connected (because the nodes are connected via node x). Therefore, it is similar to clustering of node x. Table 5 shows the effective combinations of operators (which appear often in the obtained decision trees) in Cora dataset[2].

In summary, various features have been shown to be important for classification, some of which correspond to well-known indices in social network analysis such as degree centrality, closeness centrality, and betweenness centrality. Some indices seem new, but their meanings resemble those of the existing indices. Nevertheless, the ratio of values on positive nodes to all nodes is useful in many cases. The results support the usefulness of the indices that are commonly used in the

---

[2] The score $1/r$ is added to the combination if it appears in the r-th level of the decision tree, and we sum up the scores in all the case. (Though other feature weighting is possible, we maximize the correspondence to the decision trees explained in the paper.)

Table 5. Effective combinations of operators in Cora dataset

Rank	Combination	Description
1	$Sum \circ t_x \circ (C_x^{(1)} \cap N_p)$	The number of positive nodes adjacent to node x.
2	$Sum \circ t_x \circ C_x^{(1)}$	The number of nodes adjacent to node x.
3	$Sum \circ s^{(1)} \circ (C_x^{(\infty)} \cap N_p)$	The density of the positive nodes reachable from node x.
4	$Sum \circ s^{(1)} \circ (C_x^{(1)} \cap N_p)$	The number of edges among positive nodes adjacent to node x.
5	$Max \circ t \circ (C_x^{(1)} \cap N_p)$	Whether there is a triad including node x and two positive nodes.
6	$Sum \circ s^{(1)} \circ C_x^{(1)}$	The number of edges among nodes adjacent to node x.
7	$Sum \circ s^{(1)} \circ C_x^{(\infty)}$	The number of edges among nodes reachable from node x.
8	$Max \circ u_x \circ (C_x^{(\infty)} \cap N_p)$	Whether the shortest path includes node x.
9	$Max \circ s^{(1)} \circ (C_x^{(1)} \cap N_p)$	Whether there is a triad including node x and two positive nodes.
10	$Ave \circ s^1 \circ C_x^{(\infty)}$	The Density of the component.

social network literature, and illustrate the potential for further composition of useful features.

6   Discussion

We have determined the operators so that they remain simple but cover a variety of indices. There are other features that can not be composed in our current setting, but which are potentially composable. Examples include

- centralization: e.g., $Max_{n \in N} \circ Sum \circ s^{(1)} \circ C_x^{(\infty)} - Avg_{n \in N} \circ Sum \circ s^{(1)} \circ C_x^{(\infty)}$
- clustering coefficient: $Avg_{n \in N} \circ Avg \circ s^{(1)} \circ N$,

both need additional operators. There are many other operators; for example, we can define the distance of two nodes according to the probability of attracting a random surfer. Eigenvector centrality is a difficult index to implement using operators because it requires iterative processing (or matrix processing). We do not argue that the operators that we define are optimal or better than any other set of operators; we show the first attempt for composing network indices. Elaborate analysis of possible operators is an important future task.

One future study will compare the performance with other existing algorithms for link-based classification, i.e., *approximate collective classification algorithms* (ACCA) [15]. Our algorithm falls into a family of models proposed in Inductive Logic Programming (ILP) called *propositionalization* and *upgrade*. More detailed discussion of the relations to them is available in a longer version of the paper.

7   Conclusions

In this paper, we proposed an algorithm to generate various network features that are well studied in social network analysis. We define operators to generate the features using combinations, and show that some of which are useful for node classification. Both the Cora dataset and @cosme dataset show similar trends. We can find empirically that commonly-used indices such as centrality measures and density are useful ones among all possible indices. The ratio of values, which has not been well investigated in sociology studies, is also sometimes useful.

Although our analysis is preliminary, we believe that our study shows an important bridge between the KDD research and social science research. We hope that our study will encourage the application of KDD techniques to social sciences, and vice versa.

References

1. Adamic, L., Glance, N.: The political blogosphere and the 2004 u.s. election: Divided they blog. In: LinkKDD-2005 (2005)
2. Backstrom, L., Huttenlocher, D., Lan, X., Kleinberg, J.: Group formation in large social networks: Membership, growth, and evolution. In: Proc. SIGKDD'06 (2006)
3. Barabási, A.-L.: LINKED: The New Science of Networks. Perseus Publishing, Cambridge, MA (2002)
4. Freeman, L.C.: Centrality in social networks: Conceptual clarification. Social Networks 1, 215–239 (1979)
5. Friedman, N., Getoor, L., Koller, D., Pfeffer, A.: Learning probabilistic relational models. In: Proc. IJCAI-99, pp. 1300–1309 (1999)
6. Getoor, L., Diehl, C.P.: Link mining: A survey. SIGKDD Explorations, 2(7) (2005)
7. Golder, S., Huberman, B.A.: The structure of collaborative tagging systems. Journal of Information Science (2006)
8. Lu, Q., Getoor, L.: Link-based classification using labeled and unlabeled data. In: ICML Workshop on the Continuum from Labeled to Unlabeled Data in Machine Learning and Data Mining (2003)
9. McCallum, A., Nigam, K., Rennie, J., Seymore, K.: Automating the construction of internet portals with machine learning. Information Retrieval Journal 3, 127–163 (2000), http://www.research.whizbang.com/data.
10. Perlich, C., Provost, F.: Aggregation based feature invention and relational concept classes. In: Proc. KDD 2003 (2003)
11. Popescul, A., Ungar, L.: Statistical relational learning for link prediction. In: IJCAI03 Workshop on Learning Statistical Models from Relational Data (2003)
12. Quinlan, J.R.: C4.5: Programs for Machine Learning. Morgan Kaufmann, California (1993)
13. Sarkar, P., Moore, A.: Dynamic social network analysis using latent space models. SIGKDD Explorations: Special Edition on Link Mining (2005)
14. Scott, J.: Social Network Analysis: A Handbook, 2nd edn. SAGE publications (2000)

15. Sen, P., Getoor, L.: Link-based classification. In: Technical Report CS-TR-4858, University of Maryland (2007)
16. Wasserman, S., Faust, K.: Social network analysis. Methods and Applications. Cambridge University Press, Cambridge (1994)
17. Watts, D.: Six Degrees: The Science of a Connected Age. W. W. Norton & Company (2003)
18. Wellman, B.: The global village: Internet and community. The Arts & Science Review, University of Toronto 1(1), 26 30 (2006)

An Empirical Comparison of Exact Nearest Neighbour Algorithms

Ashraf M. Kibriya and Eibe Frank

Department of Computer Science
University of Waikato
Hamilton, New Zealand
{amk14,eibe}@cs.waikato.ac.nz

Abstract. Nearest neighbour search (NNS) is an old problem that is of practical importance in a number of fields. It involves finding, for a given point q, called the query, one or more points from a given set of points that are nearest to the query q. Since the initial inception of the problem a great number of algorithms and techniques have been proposed for its solution. However, it remains the case that many of the proposed algorithms have not been compared against each other on a wide variety of datasets. This research attempts to fill this gap to some extent by presenting a detailed empirical comparison of three prominent data structures for exact NNS: KD-Trees, Metric Trees, and Cover Trees. Our results suggest that there is generally little gain in using Metric Trees or Cover Trees instead of KD-Trees for the standard NNS problem.

1 Introduction

The problem of nearest neighbour search (NNS) is old [1] and comes up in many fields of practical interest. It has been extensively studied and a large number of data structures, algorithms and techniques have been proposed for its solution. Although nearest neighbour search is the most dominant term used, it is also known as the best-match, closest-match, closest-point and the post office problem. The term similarity search is also often used in the information retrieval field and the database community. The problem can be stated as follows:

Given a set of n points S in some d-dimensional space X and a distance (or dissimilarity) measure M, the task is to preprocess the points in S in such a way that, given a query point $q \in X$, we can quickly find the point in S that is nearest (or most similar) to q.

A natural and straightforward extension of this problem is k-nearest neighbour search (k-NNS), in which we are interested in the k ($\leq |S|$) nearest points to q in the set S. NNS then just becomes a special case of k-NNS with $k=1$.

Any specialized algorithm for NNS, in order to be effective, must do better than simple linear search (the brute force method). Simple linear search, for n d-dimensional data points, gives $\mathrm{O}(dn)$ query time[1] and requires no preprocessing.

---

[1] Time required to return the nearest neighbour(s) of a given query.

J.N. Kok et al. (Eds.): PKDD 2007, LNAI 4702, pp. 140–151, 2007.

Ideal solutions exist for NNS for $d \leq 2$, that give $O(d\log n)$ query time, and take $O(dn)$ space and $O(dn\log n)$ preprocessing time. For $d = 1$ it is the binary search on a sorted array, whereas for $d = 2$ it is the use of Voronoi diagrams and a fast planar point location algorithm [2]. For $d > 2$, all the proposed algorithms for NNS are less than ideal. Most of them work well only in the expected case and only for moderate d's (≤ 10). At higher d's all of them suffer from the curse-of-dimensionality [3], and their query time performance no longer improves on simple linear search. Algorithms that give better query time performance at higher d's exist but only for relaxations of NNS, i.e. for approximate NNS [4,5], near neighbour search [6,7], and approximate near neighbour search [7].

KD-Trees are among the most popular data structures used for NNS. Metric Trees are newer and more broadly applicable structures, and also used for NNS. Recently a new data structure, the Cover Tree, has been proposed [8], which has been designed to work well even at higher dimensions provided the data has a low intrinsic dimensionality. This paper presents an empirical comparison of these three data structures, as a review of the literature shows that they have not yet been compared against each other. The comparison is performed on synthetic data from a number of different distributions to cover a broad range of possible scenarios, and also on a set of real-world datasets from the UCI repository.

The rest of the paper is structured as follows. Section 2 contains a brief overview of the three data structures that are compared. Section 3 presents the experimental comparison. It outlines the evaluation procedure employed, and also presents the empirical results. The paper concludes with some final remarks in Section 4.

2   Brief Overview of the NNS Data Structures

The following sub-sections give a brief overview of KD-Trees, Metric Trees and Cover Trees. Particular emphasis has been given to Cover Trees, to provide an intuitive description of the technique.

2.1   KD-Trees

KD-Trees, first proposed by Bentley [9], work by partitioning the point-space into mutually exclusive hyper-rectangular regions. The partitioning is achieved by first splitting the point-space into two sub-regions using an axis-parallel hyperplane, and then recursively applying the split process to each of the two sub-regions. For a given query q, only those regions of the partitioned space are then inspected that are likely to contain the kth nearest neighbour. Recursive splitting of the sub-regions stops when the number of data points inside a sub-region falls below a given threshold. To handle the degenerate case of too many collinear data points, in some implementations the splitting also stops when the maximum relative width of a rectangular sub-region (relative to the whole point-space) falls below a given threshold. KD-Trees require points in vector form, and use this representation very efficiently.

Each node of a KD-Tree is associated with a rectangular region of the point-space that it represents. Internal nodes, in addition to their region, are also associated with an axis-parallel hyperplane that splits their region. The hyperplane is represented by a dimension and a value for that dimension, and it conceptually sits orthogonal to that selected dimension at the selected value, dividing the internal node's region.

A number of different strategies have been proposed in the literature for the selection of the dimension and the value used to split a region in KD-Trees. This paper uses the Sliding Midpoint of Widest Side splitting strategy, which produces good quality trees—trees that adapt well to the distribution of the data and give good query time performance. This strategy, given in [10], splits a region along the midpoint of the dimension in which a region's hyper-rectangle is widest. If, after splitting, one sub-region ends up empty, the selected split value is slid towards the non-empty sub-region until there is at least one point in the empty sub-region. For a detailed description, and a comparison of Sliding Midpoint of Widest Side to other splitting strategies, see [11].

The search for the nearest neighbours of a given query q is carried out by recursively going down the branch of the tree that contains the query. On reaching a leaf node, all its data points are inspected and an initial set of k-nearest neighbours is computed and stored in a priority queue. During backtracking only those regions of the tree are then inspected that are closer than the kth nearest neighbour in the queue. The queue is updated each time a closer neighbour is found in some region that is inspected during backtracking. At the start, the queue is initialized with k null elements and their distance to q set to infinity.

2.2   Metric Trees

Metric Trees, also known as Ball Trees, were proposed by Omohundro [12] and Uhlmann [13]. The main difference to KD-Trees is that regions are represented by hyper-spheres instead of hyper-rectangles. These regions are not mutually exclusive and are allowed to overlap. However, the points inside the regions are not allowed to overlap and can only belong to one sub-region after a split. A split is performed by dividing the current set of points into two subsets and forming two new hyper-spheres based on these subsets. As in KD-Trees, splitting stops when for some sub-region the number of data points falls below a given threshold. A query is also processed as in KD-Trees, and only those regions are inspected that can potentially contain the kth nearest neighbour. Metric Trees are more widely applicable than KD-Trees, as they only require a distance function to be known, and do not put any restriction on the representation of the points (i.e. they do not need to be in vector form, as in KD-Trees).

Each node of a Metric Tree is associated with a ball comprising the hyper-spherical region that the node represents. The ball is represented by its centre, which is simply the centroid of the points it contains, and its radius, which is the distance of the point furthest from the centre.

A number of different construction methods for Metric Trees can be found in the literature. This paper uses the Points Closest to Furthest Pair method proposed by Moore [14]. This method first finds the point that is furthest from the centre of a spherical region (centre of the whole point-space in the beginning), and then finds another point that is furthest from this furthest point. The method, thus, tries to find the two points in a region that are furthest from each other. Then, points that are closest to one of these two points are assigned to one child ball, and the points closest to the other one are assigned to the other child ball. The method produces good quality Metric Trees that adapt well to the distribution of the data. A detailed comparison of this method with other construction methods for Metric Trees can be found in [11].

2.3   Cover Trees

Cover Trees [8] try to exploit the intrinsic dimensionality of a dataset. They are based on the assumption that datasets exhibit certain restricted or bounded growth, regardless of their actual number of dimensions.

Cover Trees are N-ary trees, where each internal node has an outdegree $\leq N$. Each node of the tree contains a single point p, and a ball which is centred at p. The points are arranged in levels, such that each lower level acts as a cover for the previous level, and each lower level has balls half the radius than the ones at the previous level. The top level consists of a single point with a ball centred at it that has radius $2^{i'}$, with an i' big enough to cover the entire set of data points. The next level consists of points with balls of half the radius than the top-most ball $(2^{i'-1})$, which cover the points at a finer level. The bottom-most level consists of points that have balls covering only those single points. A point at any level i in the tree is also explicitly present in all the lower levels.

The structure is built by arbitrarily selecting a point from the list of data points and creating the top-level ball. This same point is then used to build a smaller ball at the next lower level. This creation of smaller balls from the same point is repeated until we reach a level where a ball covers only that single point. Then the procedure backtracks to the last higher-level cover ball that still has unprocessed points, arbitrarily picks the next available point, and then recursively builds cover balls for this point at lower levels. The procedure is illustrated graphically in Figure 1.

When searching for the nearest neighbours of a given query q, we go down the levels of the tree, inspecting nodes at each level. At each level i we add only those nodes for further inspection whose centre points are inside the query ball (i.e. the ball centered at the query). The radius of the query ball is set to the distance of the current best k^{th} nearest neighbour (found from among the centre points of the nodes so far inspected) plus the radius of the balls at the current level i (which is 2^i). This amounts to shrinking the query ball as we go down the levels, and inspecting children of only those nodes whose ball centres are within the query ball. The search stops when at some level the inspected nodes are all leaf nodes with no children. At this stage the k-nearest neighbours in our priority queue are the exact k-nearest neighbours of the query. The procedure is illustrated graphically in Figure 2. Note that the figure shows the final shrunken query ball at each level.

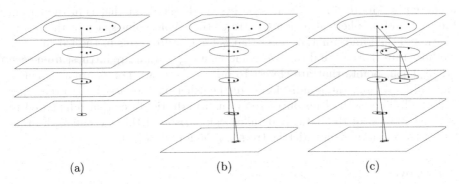

Fig. 1. Illustration of the construction method for Cover Trees. Tree at the end of (a) the first branch of recursion, (b) the second branch of recursion, and (c) the third and final branch of recursion.

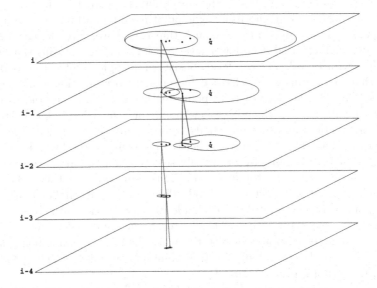

Fig. 2. Illustration of Cover Tree query. The query ball shrinks as the search proceeds.

3   Empirical Comparison of the Data Structures

The comparison of the data structures is performed on synthetic as well as real-world data. Synthetic data was used to experiment in controlled conditions, to assess how they behave for increasing n (no. of data points) and increasing d (no. of dimensions), while keeping the underlying distribution constant.

On synthetic data, the evaluation of the data structures was carried out for $d = 2, 4, 8, 16, 32, 80$ and $n = 1000, 2000, 4000, 8000, 16000, 100000$. For each combination of n and d, data points were generated from the following distributions: uniform, Gaussian, Laplace, correlated Gaussian, correlated Laplace, clustered

Gaussian, clustered ellipsoids, straight line (not parallel to any axis), and noisy straight line. Most of these distribution are provided in the ANN library [10], the rest were added for this research. The correlated Gaussian and correlated Laplacian distributions are designed to model data from speech processing, the line distributions were added to test extreme cases, and the remaining distributions, especially the clustered ones, model data that occurs frequently in real-world scenarios. More details on these distributions can be found in [11].

The data structures were built for each generated set of data points, and were evaluated first on 1000 generated query points that had the same distribution as the data, and then on another 1000 generated query points that did not follow the distribution of the data, but had uniform distribution. In other words, results were obtained for increasing d for a fixed n, and for increasing n for a fixed d, when the query did and when it did not follow the distribution of the data. Moreover, each of these evaluations were repeated 5 times with different random number seeds and the results were averaged. Note that for each dataset the dimensions were normalized to the $[0, 1]$ range.

To obtain results for real-world data, we selected datasets from the UCI repository that had at least 1000 examples. In each case, the class attribute was ignored in the distance calculation. Nominal attributes were treated as integer-valued attributes, and all attributes were normalized. Missing values were replaced by the mean for numeric attributes, and the mode for nominal ones. On each dataset, the data structures were evaluated 5 times using a random 90/10 data/query set split, and the results reported are averages of those 5 runs. Also, the evaluations for both the artificial and the real-world data were repeated for $k = 1, 5$, and 10 neighbours.

All three data structures compared have a space requirement of $O(n)$. For Cover Trees though, the exact space is comparatively higher since it has maximum leaf size 1, but for KD-Trees and Metric Trees it is very similar as they both have maximum leaf size 40. The construction time for Cover Trees is $O(c^6 n \log n)$ [8] (where c is the expansion constant of the dataset [8]), but for KD-Trees and Metric Trees, with their chosen construction methods, it is not guaranteed. However, in the expected case they do construct in $O(n \log n)$ time. The query time of Cover Trees is $O(c^{12} \log n)$ [8], whereas for KD-Trees and Metric Trees it is $O(\log n)$ in the expected case for lower d's. Note that the constant c for Cover Trees is related to the assumption of restricted growth of a dataset, and can sometimes vary largely even within a dataset [8]. Hence, for all the data structures the space is guaranteed, but the construction and query times can best be observed empirically.

For the comparison of query time, linear search is also included in the experiments as a baseline. All compared techniques, including the linear search, were augmented with Partial Distance Calculation [15,11], which skips the complete distance calculation of a point if at some stage the distance becomes larger than the distance of the best k^{th} nearest neighbour found so far.

For all the experiments the leaf size of KD-Trees and Metric Trees was set to 40. The threshold on a node's maximum relative width in KD-Trees was set to 0.01. All algorithms were implemented in Java and run under the same experimental conditions.[2] The Cover Tree implementation we used is a faithful port of the original C implementation provided by the authors. Note that the base for the radii of the balls in the algorithm was set to 1.3 instead of 2 in the C implementation, and thus also in ours.

3.1 Results

We present construction times and query times for the synthetic data. For the real-world datasets, only the query times are given, to support the main conclusions observed from the synthetic data.

Figure 3 shows the construction times of the structures on synthetic data for increasing n, for $d = 4$, and also for increasing d, for $n = 16000$. Figures 4 and 5 show the query times, Figure 4 for increasing n for $k = 5$ and $d = 8$, and Figure 5 for increasing d for $k = 5$ and $n = 16000$. All axes are on log scale.

It can be observed from Figure 3 that KD-Trees exhibit the best construction time overall. On all but the line distribution, their construction time grows at the same rate as for the other techniques, but is a constant times faster. The construction time of Cover Trees is very similar to that of Metric Trees on distributions other than the line, but for $d > 16$ it grows exponentially and becomes worst overall.

Considering query time, Figures 4 and 5 show that all three tree methods suffer from the curse-of-dimensionality, and generally become worse than linear search for $d > 16$. At higher d's they are only better than linear search if the points are clustered or lie on a line. KD-Trees are the best method if the query points have the same distribution as the data used to build the trees, otherwise KD-Trees are best for low d's, but for higher d's Cover Trees are best. Metric trees generally perform somewhat better than Cover Trees when the query points have the same distribution as the original data, and somewhat worse otherwise. However, their query times are generally quite close. When the query distribution is not changed to be uniform, KD-Trees, in terms of both construction and query time, are worse than the others only for points lying on a line, a case that is uncommon in practice. Trends like the ones in Figures 3, 4 and 5 were also observed for $k = 1$ and $k = 10$, and other values of d and n.

Table 1 shows the query time of the data structures on the UCI data. All the techniques are compared against KD-Trees, and the symbols ○ and ● denote respectively, whether the query time is significantly worse or better compared to KD-Trees, according to the corrected resampled paired t-test [16]. It can be observed that KD-Trees are significantly better than the rest on most datasets. In some cases they are still better than linear search even at higher d's (the dimensions are given in brackets with the name of a dataset). It can also be observed that, in most cases, and in contrast to the results on the artificial data, Cover Trees outperform Metric Trees.

---

[2] All implementations are included in version 3.5.6 of the Weka machine learning workbench, available from http://www.cs.waikato.ac.nz/ml/weka.

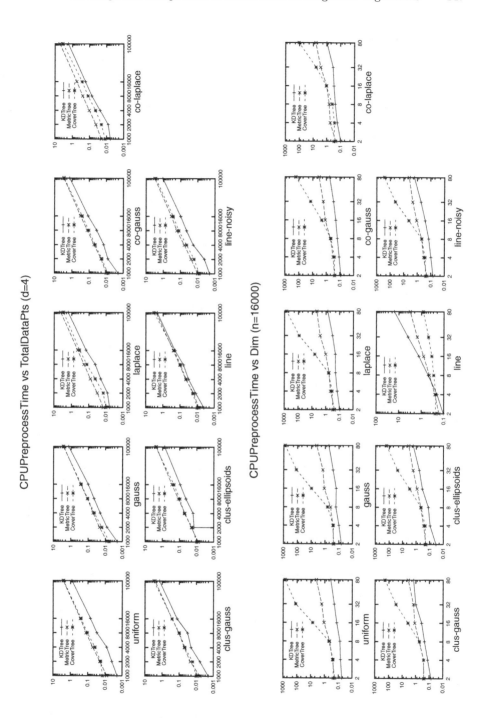

Fig. 3. CPU construction time of the data structures for increasing n, for $d = 4$, and for increasing d, for $n = 16000$

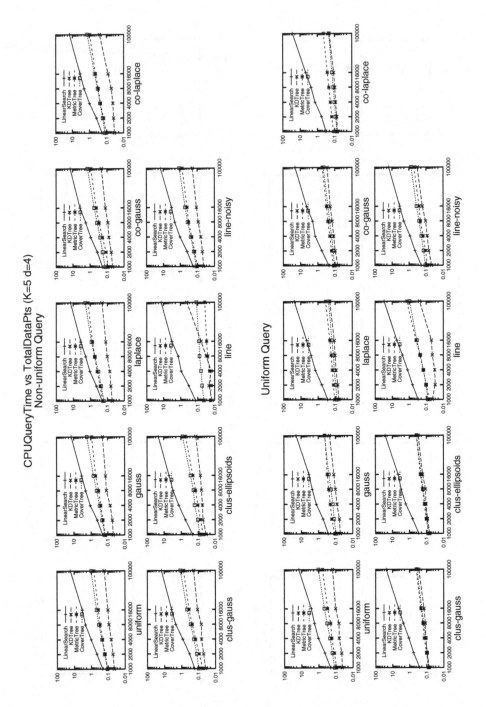

Fig. 4. CPU query time of the data structures for increasing n, for $d = 4$

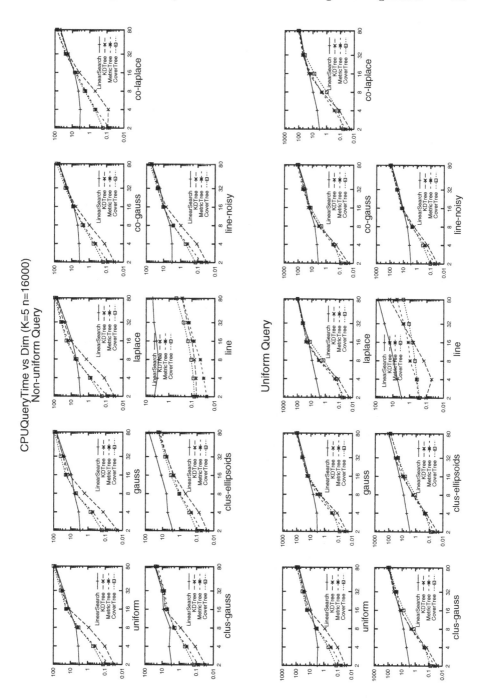

Fig. 5. CPU query time of the data structures for increasing d, for $n = 16000$

Table 1. Query time of the data structures on UCI data

Dataset	KD-Trees	Linear Search	Metric Trees	Cover Trees
car(7)	0.03	0.07 ○	0.08 ○	0.07 ○
mfeat(7)	0.02	0.11 ○	0.03	0.04 ○
cmc(10)	0.02	0.05 ○	0.07 ○	0.04 ○
german-credit(21)	0.06	0.06	0.09 ○	0.09 ○
segment(20)	0.03	0.13 ○	0.08 ○	0.08 ○
page-blocks(11)	0.04	0.76 ○	0.17 ○	0.18 ○
sick(30)	0.15	0.60 ○	0.78 ○	0.21 ○
hypothyroid(30)	0.21	0.82 ○	1.06 ○	0.27 ○
kr-vs-kp(37)	0.40	0.57 ○	1.03 ○	0.54 ○
nursery(9)	0.76	2.91 ○	7.31 ○	4.54 ○
mushroom(23)	0.34	2.44 ○	4.05 ○	1.04 ○
pendigits(17)	0.44	3.01 ○	1.22 ○	1.07 ○
splice(62)	2.10	1.93 ●	2.53 ○	2.29 ○
waveform(41)	4.67	4.35 ●	6.05 ○	6.00 ○
letter(17)	4.00	11.42 ○	8.20 ○	6.16 ○
optdigits(65)	4.50	4.79 ○	5.52 ○	4.13 ●
ipums-la-97-small(61)	4.91	4.60 ●	6.27 ○	5.53 ○
ipums-la-98-small(61)	4.48	4.00 ●	5.77 ○	5.25 ○
ipums-la-99-small(61)	6.42	5.60 ●	8.22 ○	7.63 ○
internet-usage(72)	26.90	23.90 ●	35.73 ○	32.45 ○
auslan2(27)	23.71	660.73 ○	100.33 ○	101.62 ○
auslan(14)	28.54	2162.14 ○	297.02 ○	123.70 ○
ipums-la-98(61)	474.78	364.63 ●	602.31 ○	580.48 ○
census-income-test(42)	189.06	556.99 ○	976.03 ○	624.07 ○
ipums-la-99(61)	666.84	513.60 ●	862.59 ○	839.27 ○
abalone(9)	0.06	0.27 ○	0.20 ○	0.12 ○
ailerons(41)	4.35	8.57 ○	11.20 ○	10.47 ○
bank32nh(33)	11.06	9.82 ●	13.84 ○	14.56 ○
2dplanes(11)	12.81	42.56 ○	39.08 ○	23.05 ○
bank8FM(9)	0.68	1.76 ○	1.52 ○	1.51 ○
cal-housing(9)	1.33	7.60 ○	2.60 ○	2.70 ○
cpu-act(22)	0.54	3.32 ○	1.91 ○	1.79 ○
cpu-small(13)	0.23	2.52 ○	1.02 ○	0.92 ○
delta-ailerons(6)	0.10	0.81 ○	0.39 ○	0.40 ○
delta-elevators(7)	0.21	1.48 ○	1.00 ○	0.94 ○
elevators(19)	3.28	7.69 ○	8.55 ○	7.71 ○
fried(11)	16.08	45.07 ○	61.27 ○	47.43 ○
house-16H(17)	3.53	25.79 ○	12.93 ○	10.06 ○
house-8L(9)	1.30	16.79 ○	4.57 ○	3.91 ○
CorelFeatures-ColorHist(33)	16.67	157.90 ○	155.29 ○	75.05 ○
CorelFeatures-ColorMoments(10)	23.64	90.38 ○	54.72 ○	50.14 ○
CorelFeatures-CoocTexture(17)	20.83	110.80 ○	32.76 ○	32.56 ○
CorelFeatures-LayoutHist(33)	35.01	177.49 ○	120.31 ○	104.83 ○
el-nino(12)	173.40	481.58 ○	2056.06 ○	1000.63 ○
kin8nm(9)	0.89	1.93 ○	2.20 ○	1.85 ○
mv(11)	8.56	36.28 ○	21.60 ○	12.11 ○
pol(49)	1.20	14.98 ○	9.61 ○	6.02 ○
puma32H(33)	9.86	8.42 ●	12.21 ○	12.96 ○
puma8NH(9)	0.94	1.97 ○	2.33 ○	2.02 ○
quake(4)	0.01	0.07 ○	0.02	0.02 ○

○/● Statistically worse/better at 95% confidence level.

4   Conclusions

Most of the data structures and techniques proposed since the initial inception of the NNS problem have not been extensively compared with each other, making it hard to gauge their relative performance.

KD-Trees are one of the most popular data structures used for NNS for moderate d's. Metric Trees are more widely applicable, and also designed for moderate d's. The more recently proposed Cover Trees have been designed to exploit the low intrinsic dimensionality of points embedded in higher dimensions. This paper has presented an extensive empirical comparison of these three techniques on artificial and real-world data. It shows that Metric Trees and Cover Trees do not perform better than KD-Trees in general on the standard NNS problem. On our synthetic data, Cover Trees have similar query time to Metric Trees, but they outperform Metric Trees on real-world data. However, Cover Trees have a higher construction cost than the other two methods when the number of dimensions grows.

References

1. Minsky, M., Papert, S.: Perceptrons, pp. 222–225. MIT Press, Cambridge (1969)
2. Aurenhammer, F.: Voronoi diagrams–A survey of a fundamental geometric data structure. ACM Computing Surveys 23(3), 345–405 (1991)
3. Hastie, T., Tibshirani, R., Friedman, J.: The Elements of Statistical Learning: Data Mining, Inference, and Prediction. Springer, Heidelberg (2001)
4. Liu, T., Moore, A.W., Gray, A.G.: Efficient exact k-NN and nonparametric classification in high dimensions. In: Proc. of NIPS 2003, MIT Press, Cambridge (2004)
5. Indyk, P., Motwani, R.: Approximate nearest neighbors: towards removing the curse of dimensionality. In: Proc. 13th Annual ACM symposium on Theory of Computing, pp. 604–613. ACM Press, New York (1998)
6. Nene, S.A., Nayar, S.K.: A simple algorithm for nearest neighbor search in high dimensions. IEEE Trans. Pattern Anal. Mach. Intell. 19(9), 989–1003 (1997)
7. Datar, M., Immorlica, N., Indyk, P., Mirrokni, V.S.: Locality-sensitive hashing scheme based on p-stable distributions. In: Proc. 20th Annual Symposium on Computational Geometry, pp. 253–262. ACM Press, New York (2004)
8. Beygelzimer, A., Kakade, S., Langford, J.: Cover trees for nearest neighbor. In: Proc. 23rd International Conference on Machine learning, pp. 97–104. ACM Press, New York (2006)
9. Bentley, J.L.: Multidimensional binary search trees used for associative searching. Commun. ACM 18(9), 509–517 (1975)
10. Mount, D.M., Arya, S.: ANN: A library for approximate nearest neighbor searching. In: CGC 2nd Annual Fall Workshop on Computational Geometry (1997) Available from http://www.cs.umd.edu/~mount/ANN
11. Kibriya, A.M.: Fast algorithms for nearest neighbour search. Master's thesis, Department of Computer Science, University of Waikato, New Zealand (2007)
12. Omohundro, S.M.: Five balltree construction algorithms. Technical Report TR-89-063, International Computer Science Institute (December 1989)
13. Uhlmann, J.K.: Satisfying general proximity / similarity queries with metric trees. Information Processing Letters 40(4), 175–179 (1991)
14. Moore, A.W.: The anchors hierarchy: Using the triangle inequality to survive high dimensional data. In: Proc. 16th Conference on Uncertainty in Artificial Intelligence, pp. 397–405. Morgan Kaufmann, San Francisco (2000)
15. Bei, C.D., Gray, R.M.: An improvement of the minimum distortion encoding algorithm for vector quantization. IEEE Transactions on Communications 33(10), 1132–1133 (1985)
16. Nadeau, C., Bengio, Y.: Inference for the generalization error. Machine Learning 52(3), 239–281 (2003)

Site-Independent Template-Block Detection

Aleksander Kołcz[1] and Wen-tau Yih[2]

[1] Microsoft Live Labs, Redmond WA, USA
[2] Microsoft Research, Redmond WA, USA

Abstract. Detection of template and noise blocks in web pages is an important step in improving the performance of information retrieval and content extraction. Of the many approaches proposed, most rely on the assumption of operating within the confines of a single website or require expensive hand-labeling of relevant and non-relevant blocks for model induction. This reduces their applicability, since in many practical scenarios template blocks need to be detected in arbitrary web pages, with no prior knowledge of the site structure. In this work we propose to bridge these two approaches by using within-site template discovery techniques to drive the induction of a site-independent template detector. Our approach eliminates the need for human annotation and produces highly effective models. Experimental results demonstrate the usefulness of the proposed methodology for the important applications of keyword extraction, with relative performance gain as high as 20%.

1 Introduction

The Web has become the most important information resource. Numerous applications and services for information retrieval and content extraction have been created to analyze web documents automatically in order to serve high-quality, relevant information to millions of users daily. The effectiveness of these applications relies on their ability to identify the "important" portions of a web page and separate them from other items that may be present, such as navigation bars, advertisements, etc. Detecting these *template* or *noise* blocks has thus become a major preprocessing step in several practical web applications [14][4]. In this paper, we propose a robust and effective approach to detect the template blocks. In contrast to many existing approaches, our method operates in a site-independent fashion, where information such as site structure is not needed. It is trained using a large collection of data sampled from various websites. Reliable labels are gathered automatically based on statistics of the characteristics of in-domain pages, which completely eliminates the expensive process of human annotation. This makes it very attractive for many application scenarios. Our approach not only achieves high accuracy in its own task – template block detection, but also brings significantly positive impact on the application of *keyword extraction*.

J.N. Kok et al. (Eds.): PKDD 2007, LNAI 4702, pp. 152–163, 2007.
© Springer-Verlag Berlin Heidelberg 2007

2   Prior and Related Work

When analyzing the content of a web-page, the hypertext is usually first represented as a DOM tree. The goal of web-page cleaning is then to postprocess it
by either removing certain element nodes or by assigning different importance
weights to different blocks. Both block elimination [14] and block weighting [13]
have been shown to improve the accuracy of document clustering and classification. The differences between various solutions proposed in the literature lie in
the areas outlined below.

2.1   Segmentation Heuristics

Not all HTML tags impact the block appearance of a web page. Domain knowledge is typically used to identify the subset of possible HTML tags (e.g., table
elements) that are considered as possible block separators [4], although not all
such nodes correspond to clearly identifiable rectangular regions of a page. [2] restrict valid block nodes to such that, when rendered, are visually separated from
other regions by horizontal and vertical lines and, in addition, have content that
is sufficiently cohesive. Although in most cases a block will contain some text
and clickable links, in applications such as advertisement removal, a block may
be defined as an area in the HTML source that maps to a clickable image [9].

2.2   The Same-Site Assumption

When approaching the problem of noise block detection, many researchers have
taken the "same-site" assumption, i.e., restricted the problem to the confines
of the same website, where it is likely that web pages were generated using a
set of common templates or layout styles [1,4,14,7]. In particular, whether a
block should be considered as part of the template can be naturally defined by
its relative block frequency. In [1], a shingling algorithm was used to compute
robust fingerprints of block content. These fingerprints were then applied to
identify near-duplicate blocks on pages belonging to the same site with some
additional constraints on the nature of a block and the set of pages for which
block occurrences were counted. In [7], rather than computing block similarity
via text shingling, a direct MD5 digest of the HTML snippet corresponding to
a block was computed and the relative frequency of such fingerprints in the
collection of pages of a site were counted.

Judging block similarity by a fingerprint match may be brittle. A number
of authors considered comparing pairs of blocks directly. [14] compressed pages
from the same site into a site tree, where nodes from different pages are considered equivalent if they have the same position in the DOM tree and the same
attributes. However, nodes can be merged also when the textual content is sufficiently similar even if the HTML tags/attributes are different. Block-to-block
similarity computations are thus limited to nodes in related positions in their
DOM trees. A similar approach was taken by [4], where all blocks of all pages are
compared for similarity using both textual and structural features. The computational complexity is reduced by keeping track of corresponding blocks in pages

that had already been processed. [3] proposed an approach that scales well to large data collections. Here, blocks are first grouped into clusters of individual styles and then further grouped by content similarity within each cluster.

2.3 Site-Independent Relevant/Noise Block Identification

In practical applications, simple site-independent heuristics are often applied to remove top/bottom banner ads and boiler-plates. In [9], a rule-based system is proposed to extract banner ads, although this is a problem narrower than noise-block detection. In [4], blocks are categorized as text-based, image/link based etc., depending on the ratio of one particular type of content within a block to all others, which leads to different weights of different features. The feature weights are then compared among the blocks belonging to a single page to identify, for example, the largest text block. Block attributes were also used in [4] in conjunction with labeled data to train an inductive model which showed good performance in identifying the relevant blocks using a small set of 250 pages. Song *et al.* [10] took an approach relying on a two-dimensional layout of a web page. It was shown that the combination of positional and non-positional features can lead to very accurate predictive models given a classifier such as an SVM and a quantity of labeled data.

3 Bridging the Gap Between Site-Dependent and Site-Independent Approaches

The results of published research in the noise-block detection area can be summarized as follows:

- It is possible to automatically detect template blocks within pages of the same site.
- Using hand-labeled data corresponding to noise and non-noise blocks in pages from multiple sites, it is possible to build accurate inductive models estimating the importance of a block on a page.
- Removal or down-weighting of content in template blocks tends to increase the accuracy of data mining tasks such as classification or clustering.

One notes that although the template discovery process for a single site is automatic, building site independent noise detection models has so far relied on expensive hand labeling of all blocks within a page for a training corpus. Our contribution is to apply automatic template discovery to generate large quantities of labeled data for use in building the inductive models. We hypothesize that to achieve high accuracy one does not need to acquire labels for all blocks within a training web-page. Given a large quantity of web pages, we propose to pool only those blocks from individual pages that can be confidently identified as either a template or a non-template. The lack of a hand-labeled truth set might be seen as preventing evaluation of the quality of the resulting model. We take the position, however, that in practical applications, the quality and impact of web-page cleaning should be primarily measured by the secondary metrics of how the accuracy of the data mining tasks is affected.

Our approach can be described as follows (the algorithm settings used in our experiments are further detailed in Section 4):

1. A sample of pages is collected from sites that have substantial number of pages, which enables application of site-specific template discovery techniques. A variety of sampling regimens could be considered. Traditionally, uniform sampling of pages produced by a webcrawl of a particular site tends to be used. However, this does not guarantee fair access to all site content, due to the implementation details of any particular crawler used and the fact that many sites contain dynamic content. An alternative might be to sample from a web-log pertaining to an actual site, which focuses on pages that are actually getting accessed.

2. A block segmentation algorithm is applied to each page and block frequency statistics within each site are computed. Since our goal is to identify template blocks within the set of pages sampled from a site rather than building a template detection model for that site, a simple fingerprint-based technique (adapted from [7]) is used. The block identification algorithm is constrained to HTML elements that are typically rendered in the form of a visual block. Additionally, a valid block needs to contain sufficient amount of textual content.

3. Blocks having sufficiently high document frequency are considered as templates. Unique blocks are considered as non-templates. The remaining blocks are ignored. This provides a labeled set of blocks for a particular site.

4. The feature vectors for labeled blocks extracted from the collection of sites are aggregated into a dataset used to train an inductive template detection model. We consider features describing a block in isolation from the remainder of the page (e.g., its textual content including length and word count, punctuation and formatting of text, the presence and frequency of HTML tags, with separate counting of image and anchor tags), features capturing the relative difference of the block from the remainder of the page (KL divergence between textual contents, relative word and tags counts), as well as features capturing the position of the block node within the DOM tree (e.g., line number, distance from the root, distance from the left-most sibling). The feature generation process aims to introduce a rich representation of potentially correlated attributes.

5. Given the training data, a model is induced and subsequently applied in a site-independent manner on a page-by-page basis. A number of machine learning techniques can be used in modeling. Section 4 discusses the techniques chosen in this work in more detail.

While block frequency counting can lead to fairly confident identification of template blocks, assuming that the ones that are very infrequent are non-templates may sometimes be inaccurate. An infrequent block could for example be a block in a new format that has not yet been widely deployed, or a result of a copy and paste of the block, or the entire page, from somewhere else. To limit the effect of class-noise in our system, blocks with frequency greater than one but smaller than a frequent-threshold cutoff point are ignored. Since the template blocks are repetitive, for any web site many more examples of non-template blocks can be acquired. To curb the amount of imbalance, in the hierarchical processing of DOM trees we retain only the top-most non-template block that decomposes in non-template blocks only (i.e., all of its sub-blocks have the site frequency of 1).

4   Experiment Details

We used two independent sets of data (one that was used in [7] and one derived from the Open Directory Project (ODP)[1]) to learn template detection models. The motivation for using two alternative sets was to assess the importance of a particular sampling regimen on the usefulness of the resulting model. The two datasets differ drastically in the sampling philosophy. One uses a uniform sample from the data generated by a large-scale web-crawl and one uses a human generated collection of links, with unknown biases as to what areas of a site are being accounted for.

4.1   Template Detection Using the Crawl-Sample Dataset

The data for this experiment corresponded to the results published in [7], where a random sample of domains was extracted from a large web crawl (of approx. 2,000,000,000 pages) and subsequently pruned to retain only those sites having at least 200 pages. This resulted in a dataset containing 109 sites and 21,349 pages. For each page, the HTML content was parsed into a DOM tree from which nodes corresponding to invalid HTML elements or invalid attributes of valid elements were removed. Comment and script sections were also eliminated. A potential block node had to satisfy the following two requirements:

- It had to correspond to a predefined set of HTML tags that typically affect visual block layout {blockquote, dd, div, dl, dt, h1, h2, h3, h4, h5, h6, li, ol, pre, small, table, td, th, tr, ul}.
- It had to have sufficient text content. For each node, the underlying text was extracted and processed to remove the leading and trailing space as well as to collapse consecutive whitespace into one. A block was then considered to be a candidate if the length of the normalized text content was at least 40 characters and at least 3 unique words.

When processing web pages belonging to the same site, potential block nodes on each page were selected and for each one the page count of the node was recorded. A node was identified by an MD5 digest of its textual content, so that nodes having different markups but resolving to the same text were treated as

---

[1] http://www.dmoz.org

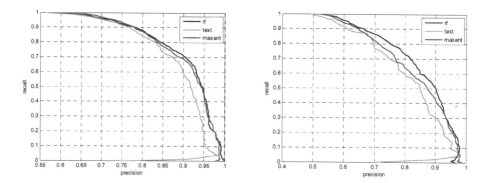

Fig. 1. Template detection performance over the web-crawl dataset (left) and the ODP dataset (right) in terms of the precision-recall curve. Using text-only features performs well (using Naive Bayes, labeled as *text*), but further improvements are possible by considering non-textual attributes (using logistic regression, labeled as *maxent*). Random forest (labeled as *rf*) has an advantage over logistic regression, which is more pronounced for the ODP dataset.

synonymous. In the hierarchical DOM structure, if a node was found to have too little textual content, it and all its descendants were subsequently ignored. Similarly, duplicates were also ignored. In the end, for each site a frequency distribution of the textual fingerprints was obtained and, following [7], a node was declared to be a template if it was present in at least 10% of the site's pages. A node was declared a definite non-template candidate if it occurred only on a single page.

Given the identity of template and non-template nodes, the data corresponding to each site were processed again to extract feature sets out of each node. Additionally, in this pass, a node was considered a non-template if it did not contain any template nodes or frequent nodes underneath its node hierarchy. This prevented, for example, the full web-page to be considered as an example of a non-template. The essence was to capture the positional information of a block in the DOM tree and in the page source, as well as the content and markup properties of a block.

The positive and negative example feature vectors due to individual sites were grouped together. In a 10-fold *cross-site* cross-validation experimental procedure, in 10 runs the data corresponding to 10% of sites were used for testing with the remaining data used to build a predictive model. Note that we *did not* use data from the same site for both training and testing, and there was no known relationship between different sites.

For each cross-validation round the training set was deduplicated and then the training data underwent MDL-based discretization [5] and was converted to binary features. Again, training nodes with the same feature vectors were deduplicated. The same discretization model was applied to the test data, although no deduplication was performed in this case.

The feature vectors consisted of two main components: the subset corresponding to word features and the subset corresponding to all other features, with text attributes being much more numerous. Given that natural sparsity of text, we decided to split the modeling into building a text based classifier first and using its output score as a feature to be combined with other features in inducing the final template-block classifier. For text classification we used a version of Naive Bayes (labeled as *text*). For the overall classifier, we considered comparing logistic regression (labeled as *maxent*) as the linear model and random forest (labeled as *rf*) as the non-linear model. Figure 1 (left) shows the precision-recall curves for the classifiers considered. At the relatively high precision of 90%, it is possible to achieve recall of 70%, but since the non-template data did not undergo hand-labeling it is possible that precision is actually higher. This can be considered as indicative of high effectiveness, especially given that the data from the sites over which the classifiers are tested were not used in their induction. Note that text features alone tend to be a good predictor, but combination with other attributes leads to substantial improvement, particularly in the area of high precision. Logistic regression performs comparably to a random forest, although the latter appears to have an advantage in the region of precision between 80% and 95%.

4.2   Template Detection Using the ODP Dataset

A uniform sample of a web crawl is not always easy to obtain. On the other hand, various biased samples of the web (collections of links, directories, etc.) are readily available. A question arises to what extent a useful template block detection model can be defined using such biased samples. In order to answer this question, we used the Oct 6, 2006 snapshot of the Open Directory Project (ODP). We sorted the sites according to their ODP page presence and considered the ones having at least 200 pages in the directory. In cases where the number of pages listed was greater, we took a random sample of 200. After retaining only valid unique HTML documents, the dataset consisted of 21,869 web pages corresponding to 131 different sites. This dataset was used to derive a template-block model (the ODP template model) using methodology and settings identical as in the experiment described in Section 4.1. The accuracy of template detection using these data is summarized in Figure 1 (right). Qualitatively, the results are similar to those obtained using the web-crawl data, although the detection rates at any given precision appear to be lower. This could be attributed to the bias of ODP "sampling", which might have resulted in more true-template blocks being considered as non-templates. Also, note that for this dataset the benefit of using non-linear random forest rather than logistic regression appears to be more pronounced. All in all, even with non-uniform sampling, the proposed template detection methodology appears to be quite effective.

5   Application to Cross-Domain Keyword Extraction

Noise-block detection has previously been found beneficial to traditional information retrieval tasks such as document ranking. Here we examine if it can help in the relatively recent application of *keyword extraction* [6,11,12,8,15], the fundamental technology for content-targeted advertising such as Google's AdSense program and Yahoo's Contextual Match product.

Given a target web document, the goal is to identify keywords that are relevant to the content. The output keywords can be used to match keywords advertisers bid on. Corresponding advertisements are then displayed on the target page, which is supposed to be relevant to the content. Keywords contained just in the template components of a web page are of questionable value to the advertising system. The removal of template blocks should thus help the system to focus on the relevant content.

The general approach to extracting keywords from documents is to first consider each phrase or word in the target document as candidate. The importance of the candidate keyword to the document is evaluated based on various properties or features of the candidate, while the weights of those features are usually determined by machine learning algorithms such as Naive Bayes [6] or logistic regression [8,15]. No matter what learning algorithm is used, the quality of a keyword extraction system still heavily depends on the features.

Although many different attributes have been shown to be useful in this task, previous work has identified the three most important features capturing different properties of a candidate keyword: term frequency, document frequency and search query log. Term frequency provides a rough idea on how important this term is, relative to the target document. Document frequency downweights stopwords and common terms in the document collection. Finally, search query log provides an additional source of term importance.

5.1   Impact of Template-Blocks on Keyword Extraction

Intuitively, template blocks should be removed from the target page before a keyword extraction system is applied. After all, the extracted keywords are expected to be relevant to the content of the page, not the template blocks such as navigational side bars. In addition, terms that occur in one template block can often be found in several other template blocks on the same page. If template blocks are not removed before further processing, these terms will have higher term frequency and can be mistakenly judged as more relevant than terms that only appear in the real content block. These phenomena can be magnified if the terms happen to be frequently queried keywords. Nevertheless, good results were reported without first "de-noising" the documents by removing template blocks [15]. We hypothesize that the negative effect of template blocks is alleviated by the document frequency feature in their work. When the target document is similar to the document collection where document frequency is derived from, de-noising is implicitly done by the document frequency feature. In this case, terms that appear in the template block will not only have higher term frequency but also higher document frequency. As a result, the importance of these template terms will not be overly emphasized.

However, in practice, a machine learning based system is often faced with the problem of *domain adaptation* – the system is developed or trained in one domain but deployed in another. Due to the different distribution of the data, performance degradation of the system is commonly observed. In this scenario, since the template in the training data may not appear in the testing data, document frequency derived using the training document collection can no longer downweight the importance of the new template terms, which may therefore be judged as relevant keywords.

5.2 Experiments

In order to validate the assumption that our template-block detection method is crucial to keyword extraction when faced with the cross-domain problem, we collected two sets of documents from different sources for experiments. The first set of documents, denoted as **IA**, is a random sample of 828 pages from the Internet Archive[2]. Since this dataset was used mainly for training the keyword extraction system, we intentionally made it more diversified. In particular, no two pages are originally from the same site. The second set of documents, denoted as **MSN**, is a collection of 477 pages from MSN.com, which covers topics such as *autos, health, finance* and etc. This dataset was used for testing the cross-domain behavior of the keyword extractor. Therefore, data set **IA** did not include any pages from **MSN.com**. Each page in these two datasets was annotated with relevant keywords for training and evaluation.

We followed the approach described in [15] to build a state-of-the-art web keyword extraction using dataset **IA**. We preserved all the features described in [15] except the linguistic features and the URL feature, which do not seem to change the performance much. Depending on the experiment setting, the document frequency may be derived from different collections of documents and our template-block detector may be used as a preprocessing step to remove template blocks. In the experiments, we used the template detector trained over the ODP data, using the random forest classifier and set to produce the expected template recall of 50%.

As suggested in [15], to measure the performance of keyword extraction we used the *top-N* score, calculated as follows. For a given document, we count how many of the top N keywords output by the system are "correct" (i.e., they were also picked by the annotator), and then divide this number by the maximum achievable number of correct predictions. Mathematically, if the set of top-N keywords output by the system is K_n and the set of keywords picked by the annotator for this document is A, then the *top-N* score for this test document is $|A \cap K_n| / \min(|A|, N)$. We report the average *Top-1,-2,-3,-5,-10* scores over the documents in the test set.

Table 1 presents the results of our first four different experimental settings, with datasets **IA** and **MSN** used for training and testing, respectively. The *ORD* configuration is the baseline, where the trained system is applied on the testing documents directly. The document frequency features used here were derived using dataset **IA** only. Assume that we do not know *a priori* where the test pages

---

[2] http://www.archive.org

Table 1. The results of keyword extraction: *ORD* – trained and tested as usual; *DN* – removing template-blocks in test pages before extracting keywords; *DF* – similar to *ORD* but the document frequency list is derived using both document sets **IA** and **MSN**; *DN-DF* – the combination of *DN* and *DF*.

	Top1	Top2	Top3	Top5	Top10
ORD	34.80	27.16	24.28	19.23	17.06
DN	37.74	28.11	25.48	22.14	20.55
DF	41.09	31.68	26.04	22.44	19.17
DN-DF	40.46	32.00	27.66	23.71	21.28

come from, i.e., we have no knowledge on which web site the keyword extractor will be deployed at training time. To reduce the negative influence brought by the template blocks, the *DN* (de-noising) configuration first applies our template detector to the *testing* pages to remove the template blocks. The same keyword extractor is then applied to the preprocessed pages. In Section 5.1, we assumed that if document frequency is derived using a document collection that is similar to the evaluation domain, then including the new document frequency features implicitly de-noises the data. This assumption is validated in the experimental setting *DF* (new document frequency). In this configuration, the document frequency statistic was derived using data sets **IA** and **MSN** together, and was used in both training and testing. Finally, the *DN-DF* configuration tests whether removing template-blocks in the testing documents can still enhance the results after using the new document frequency statistic.

From Table 1, we can clearly see that de-noising the data using our method does improves the results of keyword extraction. Comparing *DN* with *ORD*, this preprocessing step improves the *top-N* scores consistently (the differences are all statistically significant except for Top-2)[3]. As we can see, having a better document frequency list is still a more powerful solution *if the testing domain is known in advance*. The *top-N* scores of *DF* are consistently better than *ORD* and *DN* (statistically significant compared to all Top-*N* scores of *ORD*, but only statistically significant on Top-2 and Top-10 compared to *DN*). Interestingly, the result can be further improved with the preprocessing step of de-noising. Except the *Top-1* score, all other *top-N* scores of *DN-DF* are even better than *DF* (the differences on Top-3,-5,-10 are statistically significant).

In all the aforementioned experiments, de-noising was used only on the testing data but not on the training data. It is therefore interesting to examine whether using de-noised training data can further improve the results. Table 2 presents the results of this set of experiments. The *DN_both* configuration applies de-noising in both training and testing. *DN-DF_both* is similar to *DN_both* except that document frequency is derived using datasets **IA** and **MSN** together. Comparing these two results with the *DN* and *DN-DF* rows in Table 1, we can see that de-noising the training data does not seem to help. This may be due

---

[3] Statistical significance test was done using 2-tail paired-t test on the top-*N* scores of each test document. Significance was tested at the 95% level.

Table 2. The results of keyword extraction when (1) de-noising is applied on both training and testing data, and (2) a different document frequency static is used: *DN_both* – removing template-blocks in training and test pages before extracting keywords; *DN-DF_both* – similar to *DN_both* but the document frequency list is derived using both document sets **IA** and **MSN**; *ODP-DF* – when document frequency is derived using the pages in the ODP dataset (see text for details)

	Top1	Top2	Top3	Top5	Top10
DN_both	36.48	27.68	24.84	21.71	20.47
DN-DF_both	39.20	31.79	26.96	23.44	21.04
ODP-DF	29.35	25.58	23.57	19.14	16.82

to the fact that the document frequency derived from the in-domain document collection shadows the positive effect that de-noising can bring.

Finally, we want to emphasize that although using document frequency from the target domain documents has the implicit effect of de-noising, a DF table based on large collection of diversified documents cannot provide similar results. This is due to the fact that words in template blocks can no longer appear in most documents. To validate this point, we generated a DF table using the ODP dataset (see Section 4.2) and used it in both training and testing. As shown in the last row of Table 2, this result is the worst among all the settings we tested, probably because the distribution difference between the target domain (MSN dataset) and the ODP dataset.

6   Conclusions

We considered the problem of building a web-site independent template-block detection. Our contribution was to acknowledge the effectiveness of within-site template-block detection and use it to provide data for building cross-site template-block classifiers. The proposed methodology is accurate in identifying true template blocks as training examples. With the possible contamination of non-template data taken into account, our cross-site template detectors were able to achieve recall of as high as 70% at the precision of 90%. Given that the training and test data originated in different sites, this represents high level of accuracy of practical importance. We were also able to show that useful template detection models can be learnt with biased samples of individual websites. This further increases the flexibility of creating template detection models using readily available data.

Our methodology of template-block removal was assessed by its impact on the target application of keyword extraction. Template removal proved to be universally beneficial in this task, with relative increases in performance of as much as 20%, pointing at information extraction tasks as the class of applications where noise filtering is likely to improve performance.

Acknowledgement. We would like to thank David Gibson for sharing the dataset used in [7]. We are also grateful to Chris Meek for the helpful discussion on applying the work to keyword extraction.

References

1. Bar-Yossef, Z., Rajagopalan, S.: Template detection via data mining and its applications. In: Proc. of the 11th World Wide Web Conference (2002)
2. Cai, D., Yu, S., Wen, J., Ma, W.: VIPS: a vision-based page segmentation algorithm. Technical Report MSR-TR-2003-79, Microsoft Research Asia (2003)
3. Chen, L., Ye, S., Li, X.: Template detection for large scale search engines. In: Proceedings of the 21st Annual ACM Symposium on Applied Computing (SAC'06), pp. 1094–1098. ACM Press, New York (2006)
4. Debnath, S., Mitra, P., Pal, N., Giles, C.: Automatic identification of informative sections of web pages. IEEE Transactions on Knowledge and Data Engineering 17(9), 1233–1246 (2005)
5. Fayyad, U., Irani, K.: Multi-interval discretization of continuousvalued attributes for classification learning. In: Proceedings of the 13th International Joint Conference on Artificial Intelligence, pp. 1022–1029 (1993)
6. Frank, E., Paynter, G.W., Witten, I.H., Gutwin, C., Nevill-Manning, C.G.: Domain-specific keyphrase extraction. In: Proc. of IJCAI-99, pp. 668–673 (1999)
7. Gibson, D., Punera, K., Tomkins, A.: The volume and evolution of web page templates. In: Proc. of the 14th World Wide Web Conference, pp. 830–839 (2005)
8. Goodman, J., Carvalho, V.R.: Implicit queries for email. In: CEAS-05 (2005)
9. Kushmerick, N.: Learning to remove internet advertisements. In: Proceedings of AGENTS-99 (1999)
10. Song, R., Liu, H., Wen, J., Ma, W.: Learning block importance models for web pages. In: Proc. of the 13th World Wide Web Conference, pp. 203–211 (2004)
11. Turney, P.D.: Learning algorithms for keyphrase extraction. Information Retrieval 2(4), 303–336 (2000)
12. Turney, P.D.: Coherent keyphrase extraction via web mining. In: Proc. of IJCAI-03, pp. 434–439 (2003)
13. Yi, L., Liu, B.: Web page cleaning for web mining through feature weighting. In: Proc. of 18th International Joint Conference on Artificial Intelligence (2003)
14. Yi, L., Liu, B., Li, X.: Eliminating noisy information in web pages for data mining. In: Proceedings of the ACM SIGKDD International Conference on Knowledge Discovery and Data Mining (KDD-2003), ACM Press, New York (2003)
15. Yih, W., Goodman, J., Carvalho, V.: Finding advertising keywords on web pages. In: Proceedings of the 15th World Wide Web Conference (2006)

Statistical Model for Rough Set Approach to Multicriteria Classification

Krzysztof Dembczyński[1], Salvatore Greco[2], Wojciech Kotłowski[1],
and Roman Słowiński[1,3]

[1] Institute of Computing Science, Poznań University of Technology,
60-965 Poznań, Poland
{kdembczynski,wkotlowski,rslowinski}@cs.put.poznan.pl
[2] Faculty of Economics, University of Catania, 95129 Catania, Italy
salgreco@unict.it
[3] Institute for Systems Research, Polish Academy of Sciences, 01-447 Warsaw, Poland

Abstract. In order to discover interesting patterns and dependencies in data, an approach based on rough set theory can be used. In particular, Dominance-based Rough Set Approach (DRSA) has been introduced to deal with the problem of multicriteria classification. However, in real-life problems, in the presence of noise, the notions of rough approximations were found to be excessively restrictive, which led to the proposal of the Variable Consistency variant of DRSA. In this paper, we introduce a new approach to variable consistency that is based on maximum likelihood estimation. For two-class (binary) problems, it leads to the isotonic regression problem. The approach is easily generalized for the multi-class case. Finally, we show the equivalence of the variable consistency rough sets to the specific risk-minimizing decision rule in statistical decision theory.

1 Introduction

In decision analysis, a multicriteria classification problem is considered that consists in assignment of objects to m *decision classes* Cl_t, $t \in T = \{1, \ldots, m\}$. The classes are preference ordered according to an increasing order of class indices, i.e. for all $r, s \in T$, such that $r > s$, the objects from Cl_r are strictly preferred to objects from Cl_s. Objects are evaluated on a set of *condition criteria*, i.e. attributes with preference ordered value sets. It is assumed that a better evaluation of an object on a criterion, with other evaluations being fixed, should not worsen its assignment to a decision class. In order to construct a preference model, one can induce it from a *reference (training)* set of objects U already assigned to decision classes. Thus, multicriteria classification problem resembles typical classification problem considered in machine learning [6,11] under monotonicity constraints: the expected decision value increases with increasing values on condition attributes. However, it still may happen that in U, there exists an object x_i not worse than another object x_k on all condition attributes, however, x_i is assigned to a worse class than x_k; such a situation violates the

J.N. Kok et al. (Eds.): PKDD 2007, LNAI 4702, pp. 164–175, 2007.
© Springer-Verlag Berlin Heidelberg 2007

monotone nature of data, so we shall call objects x_i and x_k *inconsistent with respect to dominance principle*.

Rough set theory [13] has been adapted to deal with this kind of inconsistency and the resulting methodology has been called *Dominance-based Rough Set Approach* (DRSA) [7,8]. In DRSA, the classical indiscernibility relation has been replaced by a dominance relation. Using the rough set approach to the analysis of multicriteria classification problem, we obtain lower and upper (rough) approximations of unions of decision classes. The difference between upper and lower approximations shows inconsistent objects with respect to the dominance principle. It can happen that due to the presence of noise, the data is so inconsistent, that too much information is lost, thus making the DRSA inference model not accurate. To cope with the problem of excessive inconsistency the *variable consistency* model within DRSA has been proposed (VC-DRSA) [9].

In this paper, we look at DRSA from a different point of view, identifying its connections with statistics and statistical decision theory. Using the maximum likelihood estimation we introduce a new variable consistency variant of DRSA. It leads to the statistical problem of isotonic regression [14], which is then solved by the optimal object reassignment problem [5]. Finally, we explain the approach as being a solution to the problem of finding a decision minimizing the empirical risk [1].

Notation. We assume that we are given a set $U = \{(x_1, y_1), \ldots, (x_\ell, y_\ell)\}$, consisting of ℓ training objects, with their decision values (class assignments), where each $y_i \in T$. Each object is described by a set of n condition criteria $Q = \{q_1, \ldots, q_n\}$ and by $\mathrm{dom} q_i$ we mean the set of values of attribute q_i. For each i, $\mathrm{dom} q_i$ is ordered by some weak preference relation, here we assume for simplicity $\mathrm{dom} q_i \subseteq \mathbb{R}$ and the order relation is a linear order $\geq$. We denote the evaluation of object x_i on attribute q_j by $q_j(x_i)$. Later on we will abuse a bit the notation, identifying each object x with its evaluations on all the condition criteria, $x \equiv (q_1(x), \ldots q_n(x))$ and denote $X = \{x_1, \ldots, x_\ell\}$. By *class* $Cl_t \subset X$, we mean a set of objects, such that $y_i = t$, i.e. $Cl_t = \{x_i \in X : y_i = t, 1 \leq i \leq \ell\}$.

2 Classical Variable Precision Rough Set Approach

The classical rough set approach [13] (which does not take into account any monotonicity constraints) is based on the assumption that objects having the same description are indiscernible (similar) with respect to the available information [13,8]. The indiscernibility relation I is defined as:

$$I = \{(x_i, x_j) \in X \times X : \ q_k(x_i) = q_k(x_j) \ \forall q_k \in Q\} \tag{1}$$

The equivalence classes of I (denoted $I(x)$ for some object $x \in X$) are called *granules*. The lower and upper approximations of class Cl_t are defined, respectively, by:

$$\underline{Cl_t} = \{x_i \in X : I(x_i) \subseteq Cl_t\} \qquad \overline{Cl_t} = \bigcup_{x_i \in Cl_t} I(x_i) \tag{2}$$

For application to the real-life data, a less restrictive definition was introduced under the name of *variable precision rough set model* (VPRS) [16] and is expressed in the probabilistic terms. Let $\Pr(Cl_t|I(x))$ be a probability that an object x_i from granule $I(x)$ belongs to the class Cl_t. The probabilities are unknown, but are estimated by frequencies $\Pr(Cl_t|I(x)) = \frac{|Cl_t \cap I(x)|}{|I(x)|}$. Then, the lower approximation of class Cl_t is defined as:

$$\underline{Cl_t} = \bigcup_{I(x):x \in X} \{I(x): \Pr(Cl_t|I(x)) \geq u\} \tag{3}$$

so it is the sum of all granules, for which the probability of class Cl_t is at least equal to some threshold u.

It can be shown that frequencies used for estimating probabilities are the maximum likelihood (ML) estimators under assumption of common class probability distribution for every object within each granule. The sketch of the derivation is the following. Let us choose some granule $G = I(x)$. Let n_G be the number of objects in G, and for each class Cl_t, let n_G^t be the number of objects from this class in G. Then the decision value y has a multinomial distribution when conditioned on granule G. Let us denote those probabilities $\Pr(y = t|G)$ by p_G^t. Then, the conditional probability of observing $n_G^1, \ldots n_G^t$ objects in G (conditional likelihood) is given by $L(p; n_G|G) = \prod_{t=1}^{m}(p_G^t)^{n_G^t}$, so that the log-likelihood is given by $\mathcal{L}(p; n_G|G) = \ln L(n; p, G) = \sum_{t=1}^{m} n_G^t \ln p_G^t$. The maximization of $\mathcal{L}(p; n_G|G)$ with additional constraint $\sum_{t=1}^{m} p_G^t = 1$ leads to the well-known fomula for ML estimators $\hat{p}_G^t$ in multinomial distribution:

$$\hat{p}_G^t = \frac{n_G^t}{n_G} \tag{4}$$

which are exactly the frequencies used in VPRS. This observation will lead us in section 4 to the definition of the variable consistency for dominance-based rough set approach.

3   Dominance-Based Rough Set Approach (DRSA)

Within DRSA [7,8], we define the *dominance* relation D as a binary relation on X in the following way: for any $x_i, x_k \in X$ we say that x_i *dominates* x_k, $x_i D x_k$, if on every condition criterion from Q, x_i has evaluation not worse than x_k, $q_j(x_i) \geq q_j(x_k)$, for $j = 1, \ldots, n$. The dominance relation D is a partial pre-order on X, i.e. it is reflexive and transitive. The *dominance principle* can be expressed as follows:

$$x_i D x_j \implies y_i \geq y_j \tag{5}$$

for any $x_i, x_j \in X$. We say that two objects $x_i, x_j \in X$ are consistent if they satisfy the dominance principle. We say that object x_i is consistent, if it is consistent with every other object from X.

The rough approximations concern granules resulting from information carried out by the decisions. The decision granules can be expressed by upward and downward unions of decision classes, respectively:

$$Cl_t^{\geq} = \{x_i \in X : y_i \geq t\} \qquad Cl_t^{\leq} = \{x_i \in X : y_i \leq t\} \qquad (6)$$

The condition granules are dominating and dominated sets defined, respectively, for each $x \in X$, as:

$$D^+(x) = \{x_i \in X : x_i D x\} \qquad D^-(x) = \{x_i \in X : x D x_i\} \qquad (7)$$

Lower approximations of $Cl_t^{\geq}$ and $Cl_t^{\leq}$ are defined as:

$$\underline{Cl_t^{\geq}} = \{x_i \in X : D^+(x_i) \subseteq Cl_t^{\geq}\} \qquad \underline{Cl_t^{\leq}} = \{x_i \in X : D^-(x_i) \subseteq Cl_t^{\leq}\} \quad (8)$$

Upper approximations of $Cl_t^{\geq}$ and $Cl_t^{\leq}$ are defined as:

$$\overline{Cl_t^{\geq}} = \{x_i \in X : D^-(x_i) \cap Cl_t^{\geq} \neq \emptyset\} \quad \overline{Cl_t^{\leq}} = \{x_i \in X : D^+(x_i) \cap Cl_t^{\leq} \neq \emptyset\} \quad (9)$$

4   Statistical Model of Variable Consistency in DRSA

In this section, we introduce a new model of variable consistency DRSA (VC-DRSA), by miming the ML estimation shown in section 2. The name *variable consistency* instead of *variable precision* is used in this chapter only to be consistent with the already existing theory [9].

In section 2, although it was not mentioned straightforward, while estimating the probabilities, we have made the assumption that in a single granule $I(x)$, each object $x \in G$ has the same conditional probability distribution, $\Pr(y = t|I(x)) \equiv p_G^t$. This is due to the property of indiscrenibility of objects within a granule. In case of DRSA, indiscernibility is replaced by a dominance relation, so that a different relation between the probabilities must hold. Namely, we conclude from the dominance principle that:

$$x_i D x_j \implies p_i^t \geq p_j^t \qquad \forall t \in T, \ \forall x_i, x_j \in X \qquad (10)$$

where p_i^t is a probability (conditioned on x_i) of decision value at least t, $\Pr(y \geq t|x_i)$. In other words, if object x_i dominates object x_j, probability distribution conditioned at point x_i *stochastically dominates* probability distribution conditioned at x_j. Equation (10) will be called *stochastic dominance principle*.

In this section, we will restrict the analysis to two-class (binary) problem, so we assume $T = \{0, 1\}$ (indices start with 0 for simplicity). Notice, that $\underline{Cl_0^{\geq}}$ and $\underline{Cl_1^{\leq}}$ are trivial, so that only $\underline{Cl_1^{\geq}}$ and $\underline{Cl_0^{\leq}}$ are used and will be denoted simply by $\underline{Cl_1}$ and $\underline{Cl_0}$, respectively. We relax the definition of lower approximations for $T = \{0, 1\}$ in the following way (in analogy to the classical variable precision model):

$$\underline{Cl_t} = \{x_i \in X : p_i^t \geq \alpha\}, \qquad (11)$$

where $\alpha \in (0.5, 1]$ is a chosen *consistency level*. Since we do not know probabilities p_i^t, we will use instead their ML estimators $\hat{p}_i^t$. The conditional likelihood function (probability of decision values with X being fixed) is a product of binomial distributions and is given by $\prod_{i=1}^{\ell} (p_i^1)^{y_i} (p_i^0)^{1-y_i}$, or using $p_i \equiv p_i^1$ (since $p_i^0 = 1 - p_i$), is given by $\prod_{i=1}^{\ell} (p_i)^{y_i} (1 - p_i)^{1-y_i}$. The log-likelihood is then

$$\mathcal{L}(p; y|X) = \sum_{i=1}^{\ell} (y_i \ln(p_i) + (1 - y_i) \ln(1 - p_i)) \tag{12}$$

The stochastic dominance principle (10) simplifies to:

$$x_i D x_j \implies p_i \geq p_j \qquad \forall x_i, x_j \in X \tag{13}$$

To obtain probability estimators $\hat{p}_i$, we need to maximize (12) subject to constraints (13). This is exactly the problem of statistical inference under the order restriction [14]. Before investigating properties of the problem, we state the following theorem:

Theorem 1. *Object $x_i \in X$ is consistent with respect to the dominance principle if and only if $\hat{p}_i = y_i$.*

Using Theorem 1 we can set $\hat{p}_i = y_i$ for each consistent object $x_i \in X$ and optimize (12) only for inconsistent objects, which usually gives a large reduction of the problem size (number of variables). In the next section, we show that solving (12) boils down to the isotonic regression problem.

5   Isotonic Regression

For the purpose of this paper we consider the simplified version of the *isotonic regression problem* (IRP) [14]. Let $X = \{x_1, \ldots, x_\ell\}$ be a finite set with some pre-order relation $D \subseteq X \times X$. Suppose also that $y: X \to \mathbb{R}$ is some function on X, where $y(x_i)$ is shortly denoted y_i. A function $y^*: X \to \mathbb{R}$ is an *isotonic regression* of y if it is the optimal solution to the problem:

$$\text{minimize} \sum_{i=1}^{\ell} (y_i - p_i)^2$$
$$\text{subject to } x_i D x_j \implies p_i \geq p_j \qquad \forall 1 \leq i, j \leq \ell \tag{14}$$

so that it minimizes the squared error in the class of all *isotonic* functions p (where we denoted $p(x_i)$ as p_i in (14)). In our case, the ordering relation D is the dominance relation, the set X and values of function y on X, i.e. $\{y_1, \ldots, y_\ell\}$ will have the same meaning as before. Although squared error in (14) seems to be arbitrarily chosen, it can be shown that minimizing many other error functions leads to the same function y^* as in the case of (14). Suppose that Φ is a convex function, finite on an interval I, containing the range of function y on X, i.e. $y(X) \subseteq I$ and Φ has value $+\infty$ elsewhere. Let ϕ be a nondecreasing function on

I such that, for each $u \in I$, $\phi(u)$ is a subgradient of Φ. For each $u, v \in I$ define the function $\Delta_\Phi(u, v) = \Phi(u) - \Phi(v) - (u - v)\phi(v)$. Then the following theorem holds:

Theorem 2. *[14] Let y^* be an isotonic regression of y on X, i.e. y^* solves (14). Then it holds:*

$$\sum_{x_i \in X} \Delta_\Phi(y_i, f(x_i)) \geq \sum_{x_i \in X} \Delta_\Phi(y_i, y^*(x_i)) + \sum_{x_i \in X} \Delta_\Phi(y^*(x_i), f(x_i)) \qquad (15)$$

for any isotonic function f with the range in I, so that y^* minimizes

$$\sum_{x_i \in X} \Delta_\Phi(y_i, f(x_i)) \qquad (16)$$

in the class of all isotonic functions f with range in I. The minimizing function is unique if Φ is strictly convex.

It was shown in [14] that by using the function:

$$\Phi(u) = \begin{cases} u \ln u + (1 - u) \ln(1 - u) & \text{for } u \in (0, 1) \\ 0 & \text{for } u \in \{0, 1\} \end{cases} \qquad (17)$$

in Theorem 2, we end up with the problem of maximizing (12) subject to constraints (13). Thus, we can find solution to the problem (12) subject to (13) by solving the IRP (14).

Suppose A is a subset of X and $f \colon X \to \mathbb{R}$ is any function. We define $Av(f, A) = \frac{1}{|A|} \sum_{x_i \in A} f(x_i)$ to be an average of f on a set A. Now suppose y^* is the isotonic regression of y. By a *level set* of y^*, $[y^* = a]$ we mean the subset of X, on which y^* has constant value a, i.e. $[y^* = a] = \{x \in X \colon y^*(x) = a\}$. The following theorem holds:

Theorem 3. *[14] Suppose y^* is the isotonic regression of y. If a is any real number such that the level set $[y^* = a]$ is not empty, then $a = Av(y, [y^* = a])$.*

Theorem 3 states, that for a given x, $y^*(x)$ equal to the average of y over all the objects having the same value $y^*(x)$. Since there is a finite number of divisions of X into level sets, we conclude there are only finite number of values that y^* can possibly take. In our case, since $y_i \in \{0, 1\}$, all values of y^* must be of the form $\frac{r}{r+s}$, where r is the number of objects from class Cl_1 in the level set, while s is the number of objects from Cl_0.

6   Minimal Reassignment Problem

In this section we briefly describe the *minimal reassignment problem* (MRP), introduced in [5]. We define the reassignment of an object $x_i \in X$ as changing its decision value y_i. Moreover, by minimal reassignment we mean reassigning the smallest possible number of objects to make the set X consistent (with respect to the dominance principle). One can see, that such a reassignment of objects corresponds to indicating and correcting possible errors in the dataset. To find minimal reassignment, one can formulate a linear program. Such problems were already considered in [3] (under the name *isotonic separation*, in the context of binary and multi-class classification) and also in [2] (in the context of boolean regression).

Assume $y_i \in \{0, 1\}$. For each $x_i \in X$ we introduce a binary variable d_i which is to be a new decision value for x_i. The request that the new decision values must be consistent with respect to the dominance principle implies:

$$x_i D x_j \implies d_i \geq d_j \qquad \forall 1 \leq i, j \leq \ell \qquad (18)$$

Notice, that (18) has the form of the stochastic dominance principle (13). The reassignment of an object x_i takes place if $y_i \neq d_i$. Therefore, the number of reassigned objects (which is also the objective function for MRP) is given by $\sum_{i=1}^{\ell} |y_i - d_i| = \sum_{i=1}^{\ell} (y_i(1 - d_i) + (1 - y_i)d_i)$, where the last equality is due to the fact, that both $y_i, d_i \in \{0, 1\}$ for each i. Finally notice that the matrix of constraints (18) is totally unimodular, so we can relax the integer condition for d_i reformulating it as $0 \leq d_i \leq 1$, and get a linear program [3,12]. Moreover, constraint $0 \leq d_i \leq 1$ can be dropped, since if there were any $d_i > 1$ (or $d_i < 0$) in any feasible solution, we could decrease their values down to 1 (or increase up to 0), obtaining a new feasible solution with smaller value of the objective function. Finally, for the purpose of the paper, we rewrite the problem in the following form:

$$\text{minimize} \sum_{i=1}^{\ell} |y_i - d_i|$$
$$\text{subject to } x_i D x_j \implies d_i \geq d_j \qquad \forall 1 \leq i, j \leq \ell \qquad (19)$$

Comparing (19) with (14), we notice that, although both problems emerged in different context, they look very similar and the only difference is in the objective function (L_1-norm in MRP instead of L_2-norm in IRP). In fact, both problems are closely connected, which will be shown in the next section.

7   Connection Between IRP and MRP

To show the connection between IRP and MRP we consider the latter to be in more general form, allowing the cost of reassignment to be different for different classes. The *weighted* minimal reassignment problem (WMRP) is given by

$$\text{minimize} \sum_{i=1}^{\ell} w_{y_i} |y_i - d_i|$$
$$\text{subject to } x_i D x_j \implies d_i \geq d_j \qquad \forall 1 \leq i, j \leq \ell \qquad (20)$$

where w_{y_i} are arbitrary, positive weights associated with decision classes. The following results hold:

Theorem 4. *Suppose $\hat{p} = \{\hat{p}_1, \ldots, \hat{p}_\ell\}$ is an optimal solution to IRP (14). Choose some value $\alpha \in [0,1]$ and define two functions:*

$$l(p) = \begin{cases} 0 & \text{if } p \le \alpha \\ 1 & \text{if } p > \alpha \end{cases} \tag{21}$$

and

$$u(p) = \begin{cases} 0 & \text{if } p < \alpha \\ 1 & \text{if } p \ge \alpha \end{cases} \tag{22}$$

Then the solution $\hat{d}^l = \{\hat{d}^l_1, \ldots, \hat{d}^l_\ell\}$ such that $\hat{d}^l_i = l(\hat{p}_i)$ for each $i \in \{1, \ldots, \ell\}$, and the solution $\hat{d}^u = \{\hat{d}^u_1, \ldots, \hat{d}^u_\ell\}$ such that $\hat{d}^u_i = u(\hat{p}_i)$ for each $i \in \{1, \ldots, \ell\}$, are the optimal solutions to WMRP (20) with weights:

$$
\begin{aligned}
w_0 &= p \\
w_1 &= 1 - p
\end{aligned}
\tag{23}
$$

Moreover, if $\hat{d} = \{\hat{d}_1, \ldots, \hat{d}_\ell\}$ is an optimal integer solution to WMRP with weights (23), it must hold $\hat{d}^l_i \le \hat{d}_i \le \hat{d}^u_i$, for all $i \in \{1, \ldots, \ell\}$. In particular, if $\hat{d}^l \equiv \hat{d}^u$, the solution to the WMRP is unique.

Theorem 4 clearly states, that if the optimal value for a variable $\hat{p}_i$ in IRP (14) is greater (or smaller) than α, then the optimal value for the corresponding variable $\hat{d}_i$ in the WMRP (20) with weights (23) is 1 (or 0). In particular, for $\alpha = \frac{1}{2}$ we have $w_0 = w_1 = 1$, so we obtain MRP (19). It also follows from Theorem 4, that if α cannot be taken by any $\hat{p}_i$ in the optimal solution $\hat{p}$ to the IRP (14), the optimal solution to the WMRP (20) is unique. It follows from Theorem 3 (and discussion after it), that $\hat{p}$ can take only finite number of values, which must be of the form $\frac{r}{r+s}$, where $r < \ell_1$ and $s < \ell_1$ are integer (ℓ_0 and ℓ_1 are numbers of objects from class, respectively, 0 and 1). Since it is preferred to have a unique solution to the reassignment problem, from now on, we always assume that α was chosen not to be of the form $\frac{r}{r+s}$ (in practice it can easily be done by choosing α to be some simple fraction, e.g. 2/3 and adding some small number ϵ). We call such value of α to be *proper*.

It is worth noticing that WMRP is easier to solve than IRP. It is linear, so that one can use linear programming, it can also be transformed to the network flow problem [3] and solved in $O(n^3)$. In the next section, we show, that to obtain lower and upper approximations for the VC-DRSA, it is enough to solve IRP and solves two reassignment problems instead.

8 Summary of the Statistical Model for DRSA

We begin with reminding the definitions of lower approximations of classes (for two-class problem) for consistency level α:

$$\underline{Cl}_t = \{x_i \in X : p^t_i \ge \alpha\} \tag{24}$$

for $t \in \{0, 1\}$. The probabilities p^t are estimated using the ML approach and from the previous analysis it follows that the set of estimators $\hat{p}$ is the optimal solution to the IRP.

As it was stated in the previous section we choose α to be proper, so that the definition (24) can be equivalently stated as:

$$\underline{Cl}_1 = \{x_i \in X : \hat{p}_i > \alpha\}$$
$$\underline{Cl}_0 = \{x_i \in X : 1 - \hat{p}_i > \alpha\} = \{x_i \in X : \hat{p}_i < 1 - \alpha\} \tag{25}$$

where we replaced the probabilities by their ML estimators. It follows from Theorem 4, that to obtain $\underline{Cl}_0$ and $\underline{Cl}_1$ we do not need to solve IRP. Instead we solve two weighted minimal reassignment problems (20), first one with weights $w_0 = \alpha$ and $w_1 = 1 - \alpha$, second one with $w_0 = 1 - \alpha$ and $w_1 = \alpha$. Then, objects with new decision value (optimal assignment) $\hat{d}_i = 1$ in the first problem form $\underline{Cl}_1$, while objects with new decision value $\hat{d}_i = 0$ in the second problem form $\underline{Cl}_0$. It is easy to show that the boundary between classes (defined as $X - (\underline{Cl}_1 \cup \underline{Cl}_0)$) is composed of objects, for which new decision values are different in those two problems.

9   Extension to the Multi-class Case

Till now, we focused on binary classification problems considered within DRSA. Here we show, how to solve the general problem with m decision classes.

We proceed as follows. We divide the problem into $m - 1$ binary problems. In tth binary problem, we estimate the lower approximations of upward union for class $t+1$, $\underline{Cl}_{t+1}^{\geq}$, and the lower approximation of downward union for class t, $\underline{Cl}_t^{\leq}$ using the theory stated in the section 8 for two-class problem with $Cl_0 = Cl_t^{\leq}$ and $Cl_1 = Cl_{t+1}^{\geq}$. Notice, that for the procedure to be consistent, it must hold if $t' > t$ than $\underline{Cl}_{t'}^{\geq} \subseteq \underline{Cl}_t^{\geq}$ and $\underline{Cl}_t^{\leq} \subseteq \underline{Cl}_{t'}^{\leq}$. In other words, the solution has to satisfy the property of inclusion that is one of the fundamental properties considered in rough set theory. Fortunately, we have:

Theorem 5. *For each $t = 1, \ldots, m - 1$, let $\underline{Cl}_t^{\leq}$ and $\underline{Cl}_{t+1}^{\geq}$ be the sets obtained from solving two-class isotonic regression problem with consistency level α for binary classes $Cl_0 = Cl_t^{\leq}$ and $Cl_1 = Cl_{t+1}^{\geq}$. Then, we have:*

$$t' \geq t \Longrightarrow \underline{Cl}_t^{\leq} \subseteq \underline{Cl}_{t'}^{\leq} \tag{26}$$
$$t' \geq t \Longrightarrow \underline{Cl}_{t'+1}^{\geq} \subseteq \underline{Cl}_{t+1}^{\geq} \tag{27}$$

10   Decision-Theoretical View

In this section we look at the problem of VPRS and VC-DRSA from the point of view of statistical decision theory [1,11]. A decision-theoretic approach has already been proposed in [15] (for VPRS) and in [10] (for DRSA). The theory presented here for VPRS is slightly different than in [15], while the decision-theoretic view for DRSA proposed in this section is completely novel.

Suppose, we seek for a function (classifier) $f(x)$ which, for a given input vector x, predicts value y as well as possible. To assess the goodness of prediction, the *loss function* $L(f(x), y)$ is introduced for penalizing the prediction error. Since x and y are random variables, the overall measure of the classifier $f(x)$ is the *expected loss* or *risk*, which is defined as a functional:

$$R(f) = E[L(y, f(x))] = \int L(y, f(x))dP(y, x) \tag{28}$$

for some probability measure $P(y, x)$. Since $P(y, x)$ is unknown in almost all the cases, one usually minimize the *empirical risk*, which is the value of risk taken for the points from a training sample U:

$$R_e(f) = \sum_{i=1}^{\ell} L(y_i, f(x_i)). \tag{29}$$

Function f is usually chosen from some restricted family of functions. We now show that the rough set theory leads to the classification procedures, which are naturally suited for dealing with problems when the classifiers are allowed to abstain from giving answer in some cases.

Let us start with VPRS. Assume, that we allow the classifier to give no answer, which is denoted as $f(x) =?$. The loss function suitable for the problem is the following:

$$L_c(f(x), y) = \begin{cases} 0 & \text{if } f(x) = y \\ 1 & \text{if } f(x) \neq y \\ a & \text{if } f(x) =? \end{cases} \tag{30}$$

There is a penalty a for giving no answer. To be consistent with the classical rough set theory, we assume, that any function must be constant within each granule, i.e. for each $G = I(x)$ for some $x \in X$, we have:

$$x_i, x_j \in G \Longrightarrow f(x_i) = f(x_j) \qquad \forall x_i, x_j \in X \tag{31}$$

which is in fact the principle of indiscernibility. We now state:

Theorem 6. *The function f^* minimizing the empirical risk (29) with loss function (30) between all functions satisfying (31) is equivalent to the VPRS in the sense, that $f^*(G) = t$ if and only if granule G belongs to the lower approximation of class t with the precision threshold $u = 1 - a$, otherwise $f^*(G) =?$.*

Concluding, the VPRS can be derived by considering the class of functions constant in each granule and choosing the function f^*, which minimizes the empirical risk (29) for loss function (30) with parameter $a = 1 - u$. As we see, classical rough set theory suits well for considering the problems when the classification procedure is allowed not to give predictions for some x.

We now turn back to DRSA. Assume, that to each point x, the classifier f assigns the interval of classes, denoted $[l(x), u(x)]$. The lower and upper ends of each interval are supposed to be consistent with the dominance principle:

$$x_i D x_j \implies l(x_i) \geq l(x_j) \qquad \forall x_i, x_j \in X$$
$$x_i D x_j \implies u(x_i) \geq u(x_j) \qquad \forall x_i, x_j \in X \qquad (32)$$

The loss function $L(f(x), y)$ is composed of two terms. First term is a penalty for the size of the interval (degree of imprecision) and equals to $a(u(x) - l(x))$. Second term measures the accuracy of the classification and is zero, if $y \in [l(x), u(x)]$, otherwise $f(x)$ suffers additional loss equal to distance of y from the closer interval range:

$$L(f(x), y) = a(u(x) - l(x)) + I(y \notin [l(x), u(x)]) \min\{|y - l(x)|, |y - u(x)|\} \quad (33)$$

where $I(\cdot)$ is an indicator function. We now state:

Theorem 7. *The function f^* minimizing the empirical risk (29) with loss function (33) between all interval functions satisfying (32) is equivalent to the statistical VC-DRSA with consistency level $\alpha = 1 - a$ in the sense, that for each $x \in X$, $x \in \underline{Cl}_t^{\geq}$ or $x \in \underline{Cl}_t^{\leq}$ if and only if $t \in f^*(x)$.*

Concluding, the statistical VC-DRSA, can be derived by considering the class of interval functions, for which the lower and upper ends of interval are isotonic (consistent with the dominance principle) and choosing the function f^*, which minimizes the empirical risk (29) with loss function (33) with parameter $a = 1 - \alpha$.

11   Conclusions

The paper introduced a new variable consistency theory for Dominance-based Rough Set Approach. Starting from the general remarks about the estimation of probabilities in the classical rough set approach (which appears to be maximum likelihood estimation), we used the same statistical procedure for DRSA, which led us to the isotonic regression problem. The connection between isotonic regression and minimal reassignment solutions was considered and it was shown that in the case of the new variable consistency model, it is enough to solve minimal reassignment problem (which is linear), instead of the isotonic regression problem (quadratic). The approach has also been extended to the multi-class case by solving $m - 1$ binary subproblems for the class unions. The proposed theory has an advantage of basing on well investigated maximum likelihood estimation method – its formulation is clear and simple, it unites seemingly different approaches for classical and dominance-based case.

Finally notice that a connection was established between statistical decision theory and rough set approach. It follows from the analysis that rough set theory can serve as a tools for constructing classifiers, which can abstain from assigning a new object to a class in case of doubt (in classical case) or give imprecise prediction in the form of interval of decision values (in DRSA case). However, rough set theory itself has a rather small generalization capacity, due to its nonparametric character, which was shown in section 10. The plans for further research are to investigate some restricted classes of functions which would allow to apply rough set theory directly for classification.

References

1. Berger, J.: Statistical Decision Theory and Bayesian Analysis. Springer, New York (1993)
2. Boros, E., Hammer, P.L., Hooker, J.N.: Boolean regression. Annals of Operations Research 58, 3 (1995)
3. Chandrasekaran, R., Ryu, Y.U., Jacob, V., Hong, S.: Isotonic separation. INFORMS J. Comput. 17, 462–474 (2005)
4. Dembczyński, K., Greco, S., Kotłowski, W., Słowiński, R.: Quality of Rough Approximation in Multi-Criteria Classification Problems. In: Greco, S., Hata, Y., Hirano, S., Inuiguchi, M., Miyamoto, S., Nguyen, H.S., Słowiński, R. (eds.) RSCTC 2006. LNCS (LNAI), vol. 4259, pp. 318–327. Springer, Heidelberg (2006)
5. Dembczyński, K., Greco, S., Kotłowski, W., Słowiński, R.: Optimized Generalized Decision in Dominance-based Rough Set Approach. LNCS. Springer, Heidelberg (2007)
6. Duda, R., Hart, P.: Pattern Classification. Wiley-Interscience, New York (2000)
7. Greco, S., Matarazzo, B., Słowiński, R.: Rough approximation of a preference relation by dominance relations. European Journal of Operational Research 117, 63–83 (1999)
8. Greco, S., Matarazzo, B., Słowiński, R.: Rough sets theory for multicriteria decision analysis. European Journal of Operational Research 129(1), 1–47 (2001)
9. Greco, S., Matarazzo, B., Słowiński, R., Stefanowski, J.: In: Ziarko, W., Yao, Y. (eds.) RSCTC 2000. LNCS (LNAI), vol. 2005, pp. 170–181. Springer, Heidelberg (2001)
10. Greco, S., Słowiński, R., Yao, Y.: Bayesian Decision Theory for Dominance-based Rough Set Approach. Lecture Notes in Computer Science 4481, 134–141 (2007)
11. Hastie, T., Tibshirani, R., Friedman, J.: The Elements of Statistical Learning. Springer, Heidelberg (2003)
12. Papadimitriou, C.H., Steiglitz, K.: Combinatorial Optimization. Dover Publications, New York (1998)
13. Pawlak, Z.: Rough sets. International Journal of Information & Computer Sciences 11, 341–356 (1982)
14. Robertson, T., Wright, F.T., Dykstra, R.L.: Order Restricted Statistical Inference. John Wiley & Sons, Chichester (1998)
15. Yao, Y., Wong, S.: A decision theoretic Framework for approximating concepts. International Journal of Man-machine Studies 37(6), 793–809 (1992)
16. Ziarko, W.: Probabilistic Rough Sets. In: Ślęzak, D., Wang, G., Szczuka, M., Düntsch, I., Yao, Y. (eds.) RSFDGrC 2005. LNCS (LNAI), vol. 3641, pp. 283–293. Springer, Heidelberg (2005)

Classification of Anti-learnable Biological and Synthetic Data

Adam Kowalczyk

National ICT Australia and
Department of Electrical & Electronic Engineering,
The University of Melbourne,
Parkville, Vic. 3010, Australia

Abstract. We demonstrate a binary classification problem in which standard supervised learning algorithms such as linear and kernel SVM, naive Bayes, ridge regression, k-nearest neighbors, shrunken centroid, multilayer perceptron and decision trees perform in an unusual way. On certain data sets they classify a randomly sampled training subset nearly perfectly, but systematically perform worse than random guessing on cases unseen in training. We demonstrate this phenomenon in classification of a natural data set of cancer genomics microarrays using cross-validation test. Additionally, we generate a range of synthetic datasets, the outcomes of 0-sum games, for which we analyse this phenomenon in the i.i.d. setting.

Furthermore, we propose and evaluate a remedy that yields promising results for classifying such data as well as normal datasets. We simply transform the classifier scores by an additional 1-dimensional linear transformation developed, for instance, to maximize classification accuracy of the outputs of an internal cross-validation on the training set. We also discuss the relevance to other fields such as learning theory, boosting, regularization, sample bias and application of kernels.

1 Introduction

Anti-learning is a non-standard phenomenon involving both dataset and classification algorithms, which has been encountered in some important biological classification tasks. In specific binary classification tasks, a range of standard supervised learning algorithms, such as linear and kernel SVM, naive Bayes, ridge regression, k-nearest neighbors, shrunken centroid, multilayer perceptron and decision trees behave in an unusual way. While they easily learn to classify a randomly sampled training subset nearly perfectly, they *systematically and significantly* perform worse than random guessing if tested on cases unseen in training. Thus reversing the classifier scores can deliver an accurate predictor, far more accurate than the original machine. In such a case we say that the dataset is *anti-learnable* by our classifier.

In this paper we shall demonstrate this phenomenon on a natural data set, a cancer genomics microarray dataset generated for classification of response

J.N. Kok et al. (Eds.): PKDD 2007, LNAI 4702, pp. 176–187, 2007.
© Springer-Verlag Berlin Heidelberg 2007

to treatment in esophageal cancer [1,2] and a synthetic dataset introduced in this paper. For the esophageal dataset, the previous analysis points towards a biological origin of a specific anti-learnable signal in the data [3], although the exact nature of such a mechanism is unclear at this stage.

We start with analysis of synthetic anti-learnable datasets, which are the outcomes of specific 0-sum games (Section 2). For such data we can use analytical methods and prove that anti-learning is the logical consequence of a specific configuration of dataset (Section 2.3). Further, for such datasets we can generate samples of arbitrary size, hence we can use the standard independently identically distributed (*i.i.d.*) setting rather than cross-validation for experimental evaluation. This leads to generation of non-conventional learning curves (Section 2.1) showing a continuum of behavior modes, starting with anti-learning for small size samples to classical, consistent generalization (asymptotic) bounds in the large size training samples limit.

In order to build a bridge to the esophageal data, we have used our synthetic model to generate a dataset of similar size (50 samples, split evenly between two labels and each represented by 10000 features). Then we classified the original and synthetic datasets using a range of classifiers combined with aggressive feature selection (t-test filter). We observe a strong similarity between learning curves for both datasets, which indirectly supports the hypothesis of deterministic origins of an "anti-learnable signal" in the esophageal dataset.

Independently, we demonstrate and evaluate some algorithms, which can successfully classify such non-standard data as well as standard datasets seamlessly. The idea here is to combine the classifiers scores with a module trained to "interpret" them accordingly. In our case, this is exclusively a simple 1-dimensional linear transformation developed to maximize a chosen objective function of the scores from internal cross-validation on the training set (Section 2.2). We show analytically and empirically, that such modified algorithms can perform well in Sections 2.1, 2.3 and 3.

Links to related research. There is a direct link to previous papers on perfect anti-learning [4,5] as follows. A specific cases of WL-game introduced in Section 2 (the magnitude $\mu \equiv$ const and single case per mode) generate the "class symmetric" kernel data studied in those papers. As mentioned before, the paper [3] studied significance of anti-learning in esophageal cancer dataset. A form of anti-learning is in KDD'02, Task 2 data: the anti-learning occurs for standard SVM and persists for the aggressive feature selection [6,7]. Finally, the existence of anti-learning is compatible with predictions of "No Free Lunch Theorems" [8].

2   Anti-learnable Signature of a 0-Sum Game

An individual outcome of the game is represented by a d_0-dimensional *state vector* $s = (s_1, ..., s_{d_0}) \in \mathbb{R}^{d_0}$, with each dimension corresponding to a "player". The players split into three groups: potential winners, indexed 1 to d_0^+; potential losers, index $d_0^+ + 1$ to $d_0^+ + d_0^-$; and remaining $d_0 - d_0^+ - d_0^- \geq 0$ neutral players.

The outcome is uniquely determined by two parameters, the *magnitude* $\mu_s > 0$ and *mode*, $M_s \in \{1, ..., d_0^+ + d_0^-\}$, which here is the index of the player, as follows:

$$s_i = \begin{cases} y_s \mu_s, & \text{for } i = M_s; \\ -y_s \mu_s/(d_0 - 1), & \text{for } M_s \neq i \leq d_0^+ + d_0^-; \\ 0, & \text{otherwise,} \end{cases} \tag{1}$$

for $i = 1, 2, ..., d_0$, where the *label* y_s is defined as 1 if $1 \leq M_s \leq d_0^+$ and -1, otherwise. Thus if $y_s = +1$, the M_sth player is a big-time winner, while the reaming, non-neutral players are uniformly worse-off. The opposite holds for $y_s = -1$, hence the name *Win-Loss game* or shortly *WL-game*. Note that s as above satisfies the 0-sum constraint:

$$\sum_{i=1}^{d_0} s_i = 0. \tag{2}$$

The subspace $S \subset \mathbb{R}^{d_0}$ of all such possible state vectors is called the *state space*. In general the state vector s is observed indirectly, via the measurement vector $\boldsymbol{x} = (x_1, ..., x_d) \in \mathbb{R}^d$ which is a linear mixture of state variables

$$\boldsymbol{x} = A\boldsymbol{s}, \tag{3}$$

where A is a $d \times d_0$ matrix. If rank(A) = d_0, then the label classes in both $S \subset \mathbb{R}^{d_0}$ and its image

$$X := AS = \{A\boldsymbol{s} \; ; \; \boldsymbol{s} \in S\} \subset \mathbb{R}^d$$

are *linearly separable*. Indeed, any hyperplane defined by the equation $s_i = 0$ for $i > d_0^+ + d_0^-$ always separates these datasets in $\mathbb{R}^{d_0}$, hence its image separates the data in span$(X) \subset \mathbb{R}^d$ and could be easily extended to the whole $\mathbb{R}^d$.

In general we shall consider $d \geq d_0$. In the particular case of $d = d_0$ and $A = I$ being the identity matrix, we say the game is directly observable. Another special case of interest, due to ease of analytical analysis, is *orthogonal mixing* with columns of A are composed of orthogonal vectors of equal length, i.e.

$$A^T A = CI, \tag{4}$$

where $C > 0$. We shall refer to this game as *orthogonal WL-game*. The above condition ensures that the following relations hold for dot-products:

$$C^{-1}\boldsymbol{x} \cdot \boldsymbol{x}' = \boldsymbol{s} \cdot \boldsymbol{s}' = \begin{cases} \mu_s \mu_{s'} d_0/(d_0 - 1), & \text{if } M_s = M_{s'}; \\ -\mu_s \mu_{s'} d_0/(d_0 - 1)^2 < 0, & \text{if } y_s = y_{s'} \text{ but } M_s \neq M_{s'} \\ \mu_s \mu_{s'} d_0/(d_0 - 1)^2 > 0, & \text{otherwise, i.e. if } y_s \neq y_{s'}, \end{cases} \tag{5}$$

for any two state vectors $\boldsymbol{s}, \boldsymbol{s}' \in S$, $\boldsymbol{x} = A\boldsymbol{s}$ and $\boldsymbol{x}' = A\boldsymbol{s}'$.

The equation (5) is the crucial relation for the theoretical understanding of anti-learning in this dataset. It states for instances of different modes: any two of the opposite label are more correlated than any two of the same label.

2.1   Empirical Learning Curves for Orthonormal WL-Game

Dataset. We have used WL-game to generate finite dataset $(\boldsymbol{x}_j, y_j) \in \mathbb{R}^d \times \{\pm 1\}$, $j = 1, ..., n$ as follows. First, we selected a random sample of states $(\boldsymbol{s}_j) \in S^n$ and generated a mixing $d \times d_o$ matrix A by Gramm-Schmidt orthonormalisation of columns of a random matrix; then we defined $y_j := y_{\boldsymbol{s}_j}$ and $\boldsymbol{x}_j := A\boldsymbol{s}_j$.

Performance metrics. We use the Area under Receiver Operating Characteristics ($AROC$ or AUC) [9,10], the plot of the True Positive versus False Negative error rates, as our main performance metric. Additionally, we also use Accuracy (ACC) defined as the average of the True Positive and the True Negative rates. Both metrics are insensitive to the class distribution in the test set. For both the value of 0.5 represents performance of trivial classifiers, be it random guessing or allocation of all example to one class; value 1 will be allocated to the perfect classifier and value 0 to the perfectly wrong one.

2.2   The i.i.d. Learning Curves

This experiment has been designed to demonstrate that anti-learning is a phenomenon of learning from a low size sample that disappears in the large size sample limit. We have used a synthetic orthogonal WL-game ($d_0^+ = d_0^- = 100$, $d_0 = 250$ and $d = 300$) to generate 2000 sample data set for re-sampling of a training set from, and then for testing classifiers (on the whole dataset). The results are plotted in Figure 1. We discuss the selected classifiers first.

Centroid. The *centroid* (*Cntr*), our basic (linear) classifier, is defined as follows:

$$f(\boldsymbol{x}) := \frac{2}{\|\boldsymbol{w}\|^2} \boldsymbol{w} \cdot \boldsymbol{x} - \frac{\|\boldsymbol{w}_+\|^2 - \|\boldsymbol{w}_-\|^2}{\|\boldsymbol{w}\|^2}$$

where $\boldsymbol{x} \in \mathbb{R}^d$, $\boldsymbol{w}_y := \sum_{i, y_i = y} \boldsymbol{x}_i / n_y$, $y = \pm 1$, and $\boldsymbol{w} = (w_j) := \boldsymbol{w}_+ - \boldsymbol{w}_-$. It is a very simple machine, does not depend on tuning parameters, is the "high regularisation" limit of SVMs and ridge regression [11], and performs well on microarray classification tasks [11]. (The scaling and the bias b are such that the scores of "class" centers are equal to class labels, i.e. $f(\boldsymbol{w}_y) = y$ for $y = \pm 1$.)

Cross-Validation Learners. In Figure 1 we observe that for small size training samples, both $AROC$ and ACC for primary classifiers such as SVM can reach values close to those for a classifier perfectly misclassifying the data. In such a case, the classifier $-f$ will classify data nearly perfectly. Obviously, for larger training samples, the reverse is true and f is preferred. Can such a decision to reverse the classifier or not be made in a principled way? The obvious way to address this issue is to perform additional cross validation on the training data in order to detect the "mode" of the classifier. A short reflection leads to the conclusion that there are a few possible strategies which can be used to insure that the proper detection of the mode actually happen. Perhaps the most straightforward one is as follows.

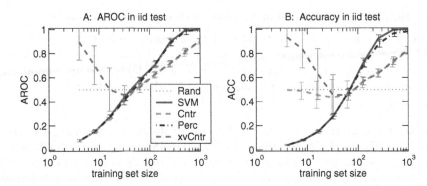

Fig. 1. Area under Reciver Operating Characteristic (AROC) and accuracy (ACC) as functions of increasing random training subset size for synthetic orthogonal WL-game data. We plot the averages of 100 tests on the whole dataset of 2000 instances with standard deviation marked by bars. We have used the following classifiers: Centroid (Cntr), hard margin support vector machine (SVM), Rosenblatt's perceptron [12] and xvCntr generated by Algorithm 1.

Algorithm 1 (xvL_1). **Given: a training set** $Tr = (\boldsymbol{x}_i, y_i) \in (\mathbb{R}^d \times \{\pm 1\})^n$, **an algorithm** $f = \mathcal{A}(Tr)$;

1. Calculate cross-validation results, e.g. for LOO: $f_{vx}(i) := f^{\backslash i}(\boldsymbol{x}_i)$ for $f^{\backslash i} := \mathcal{A}(Tr \backslash (\boldsymbol{x}_i, y_i))$, $i = 1, ..., n$;
2. Calculate $aroc_{vx} := AROC((f_{vx}(i), y_i)_{i=1}^n)$;

Output classifier: $f = \phi \circ f' := \mathrm{sgn}(aroc_{xv} - 0.5) \times f'$, where $f' = \mathcal{A}(Tr)$.

Obviously one can use cross-validation schemes other than the leave-one-out (LOO) and can optimize other measures than $AROC$ in designing moderation of the output scores or just train an additional classifier. An example follows.

Algorithm 2 (xvL_2). **Use cross validation scores to train an additional 1-dimensional classifier** $\phi := \mathcal{A}_2((f_{xv}(i), y_i)_{i=1}^n)$ **and then use the superposition** $\phi \circ f$ **instead of** $f = \mathcal{A}(Tr)$.

The function ϕ as in the above two Algorithms will be called a *reverser*.

Discussion of Figure 1. We clearly see disappearance of anti-learning phenomena in the large size sample limit. Note the poor performance of Cntr in Figure 1.B. This is due to the poor selection of the bias and is compatible with results of Theorem 2 and Corollary 1. The large variance for xvCntr is caused by a few cases of small size training samples which had the duplicate examples from the same mode, causing the miss-detection of the anti-learnable mode. Note that such single occurrence in 100 trails could result in std $\approx \sqrt{1/100} = 0.1$.

2.3   IID Anti-learning Theorem

In this section we generalise the analysis of the WL-game in Section 2 to more general kernel machines and prove a formal result on anti-learning for small size samples in the i.i.d. setting observed in Figure 1. We consider a *kernel function* $k : X \times X \to \mathbb{R}$, on the measurements space $X = AS \subset \mathbb{R}^d$, although we do not need to assume that it is symmetric or positive definite, which are typical assumptions in the machine learning field [12,13,14]. Further, we assume probability distribution Pr on the state space S and consider an i.i.d. training n-sample $(s_i) \in S^n$, with associated n-tuple of measurement-label pairs:

$$Tr := \big((x_i, y_i)\big)_{i=1}^n := \big((As_i, y_{s_i})\big)_{i=1}^n \subset \big(\mathbb{R}^d \times \{\pm 1\}\big)^n$$

and modes $M_i := M_{s_i}$, for $i = 1, ..., n$. We assume we are given an algorithm that produces a *kernel machine* $f = \mathcal{KM}(k, Tr) : X \to \mathbb{R}$ of the form

$$f(x) = \sum_{i=1}^n y_i \alpha_i k(x_i, x) + b \not\equiv \text{const}, \tag{6}$$

$$\alpha_i \geq 0 \quad \& \quad \sum_{i=1}^n y_i \alpha_i = 0. \tag{7}$$

for every $x \in X$. Many algorithms, including popular flavors of SVM [13,12], the centroid (see Section 2.2) and Rosenblatt's perceptron [12], generate solutions satisfying the above conditions. If $b = 0$, we say that f is a *homogenous* machine.

We recall here a re-formulation of our metrics in terms of a probability distribution Pr on S and the order statistic U [10], for convenience:

$$AROC(f, S) = Pr\big[f(x_s) < f(x_{s'}) \mid y_s = -1 \ \& \ y_{s'} = +1\big]$$
$$- \frac{1}{2} Pr\big[f(x_s) = f(x_{s'}) \mid y_s \neq y_{s'}\big], \tag{8}$$

$$ACC(f, S) = \frac{1}{2} Pr\big[f(x_s) < 0 \mid y_s = -1\big] + \frac{1}{2} Pr\big[f(x_s) > 0 \mid y_s = -1\big]. \tag{9}$$

Let $P_{\max} := \max_M Pr[M_s = M]$ denote the maximum probability of a mode and by $\pi_y := Pr[y_s = y]$ be the prior probability of label y for $y = \pm 1$.

Theorem 1. *Assume that the kernel function k satisfies the condition*

$$y_s y_{s'} k(x_s, x_{s'}) < y_s y_{s'} b_0, \tag{10}$$

for every $s, s' \in S$ such that $M_{s'} \neq M_s$, where $b_0 \in \mathbb{R}$ is a constant. Let function $\psi : \mathbb{R} \to \mathbb{R}$ be monotonically increasing on the range of k, i.e. for $\xi \in k(X \times X)$. Then for any kernel machine f trained for kernel $\psi \circ k$ on the n-sample Tr:

$$y_s \sum_{i=1}^n y_i \alpha_i \ \psi \circ k(x_i, x_s) < 0, \tag{11}$$

for every $s \in S$ such that $M_s \notin \{M_1, .., M_n\}$. Moreover, there exists B such that

$$AROC(f, S) \leq n P_{\max} / \min_y \pi_y, \tag{12}$$

$$ACC(f + B, S) \leq \frac{n}{2} P_{\max} / \min_y \pi_y. \tag{13}$$

Thus the homogenous kernel machine $s \mapsto \sum_{i=1}^{n} y_i \alpha_i \ \psi \circ k(x_i, x_s)$ misclassifies every instance with mode unseen in training (see Eqn. 11).

Remark 1. The significance of the monotonic function ψ is that it allows extension of results automatically to many classes of practical kernels which can be represented as a monotonic function of the dot-product kernel. These include the polynomial kernels and, under the additional assumption of fixed magnitude of measurement vectors, the radial basis kernels.

Proof. First, let us note that if assumption (10) holds, then it also holds for the kernel $k \leftarrow \psi \circ k$ and constant $b_0 \leftarrow \psi(b_0)$. This reduces the proof to the special case of $\psi(\xi) = \xi$ for every $\xi \in \mathbb{R}$, assumed from now on.

For a proof of (11) let us assume that (10) holds. Then

$$y_s f(x_s) = y_s \sum_{i=1}^{n} y_i \alpha_i k(x_i, x) < y_s b_0 \sum_{i=1}^{n} y_i \alpha_i = 0.$$

Now we proceed to the proof of (12). Denote by $P := Pr[s, \ M_s \in \{M_1, ..., M_n\}]$ the probability of an instance s having its mode present in the training set; by P_y the probability of such an instance s with label y. By (11) any two instances with modes not in the training sets are miss-ordered by f, hence

$$AROC(f, S) \leq 1 - \left(1 - \frac{P_-}{\pi_-}\right)\left(1 - \frac{P_+}{\pi_+}\right)$$

$$\leq 1 - \left(1 - \frac{P_-}{\min(\pi_-, \pi_+)}\right)\left(1 - \frac{P_+}{\min(\pi_-, \pi_+)}\right)$$

$$\leq \max_{0 \leq x \leq P} 1 - \left(1 - \frac{x}{\min(\pi_-, \pi_+)}\right)\left(1 - \frac{P - x}{\min(\pi_-, \pi_+)}\right)$$

$$= \frac{P}{\min(\pi_-, \pi_+)} \leq \frac{n P_{\max}}{\min(\pi_-, \pi_+)}.$$

This completes the proof of (12). The proof of (13) follows

$$ACC(f + B, S) \leq \frac{1}{2}\left(\frac{P_+}{\pi_+} + \frac{P_-}{\pi_-}\right) \leq \frac{P_+ + P_-}{2\min(\pi_-, \pi_+)} = \frac{P}{2\min(\pi_-, \pi_+)} \leq \frac{n P_{\max}}{2\min(\pi_-, \pi_+)}.$$

$\square$

Corollary 1. *Let $\phi : \mathbb{R} \to \mathbb{R}$ be a reverser generated by either Algorithm 1 or 2 for the homogeneous kernel machines. Then there exists a bias $B \in \mathbb{R}$ such that*

$$AROC(\phi \circ f, S) \geq 1 - n\frac{P_{\max}}{\min_y Pr[y_s = y]}, \tag{14}$$

$$ACC(\phi \circ f + B, S) \geq 1 - n\frac{P_{\max}}{2\min_y Pr[y_s = y]}, \tag{15}$$

with confidence $> \prod_{i=1}^{n-1}(1 - iP_{\max}) > 1 - \frac{(n-1)n}{2}P_{\max}.$

Note the "paradoxical" meaning of this result, compatible with experiments in Figure 1. The smaller the sample, the more accurate generalisation, provided the anti-learnable mode is detected and dealt with accordingly.

A simple proof (omitted) uses two observations: (i) that $\prod_{i=1}^{n-1}(1 - iP_{\max})$ is the lower bound on the probability of drawing n-different modes in that many samples, and (ii) that the assumptions insure that the inequality (11) holds for every kernel machine, hence also for f_{xv}, the pooled results of the cross-validation.

Note that for the orthogonal WL-game the dot product kernel $k(\boldsymbol{x}, \boldsymbol{x}') := \boldsymbol{x}\cdot\boldsymbol{x}'$ satisfies the assumption (10) of Theorem 1, see Eqn. 5. Thus we have

Corollary 2. *Corollary 1 holds for the linear kernel and orthogonal WL-game.*

3   Examples of Anti-learning in Natural Data

The esophageal adenocarcinoma dataset (AC) consists of 25 expressions of 9857 genes measured by cDNA microarrays in cancer biopsies collected from esophageal adenocarcinoma patients [1,2], prior to chemo-radio-therapy (CRT) treatment[1]. The binary labels were allocated according to whether the patient responded to the subsequent treatment (11 cases) or not (14 cases). The aim of the experiment was to assess the feasibility of developing a predictor of the response to treatment for clinical usage (an open problem, critical for clinical treatment).

We have also generated another synthetic data set, the output of the WL-game, but with $10,000 * 1000$ mixing matrix A drawn from the standard normal distribution (we have used $d_0^+ = d_0^- = 75$ and $d = 1000$). The data set consisted of 25 instances of each of the two labels. Back-to-back comparison of the classification of these two datasets in Figure 2 shows very similar trends indirectly linking the non-standard properties of AC-data to the anti-learning as understood in Section 2. Here we plot AROC as a function of number of features selected by t-test applied to the training set data only. In Figures 2.A & B we have used the following classifiers: Centroid (Cntr), hard margin support vector machine (SVM), shrunken centroid (PAM) [15] and 5-nearest neighbours (5-NN). In Figures C & D we have used various versions of xv-learner generated by the Algorithm 2 with and $\mathcal{A}_2$ generating the 1-dimensional linear reverser $\psi(\xi) := A\xi + B$ maximizing accuracy of the internal 2-fold cross-validation.

In Figure 3 we plot results for the additional test of 8 supervised learning algorithms on the natural AC-data. We observe that all averages are clearly below random guessing level of 0.5. These results show that the anti-learning persists for a number of standard classifiers, including multilayer algorithms such as decision trees or multilayer neural networks.

---

[1] Raw array data and protocols used are available at http://www.ebi.ac.uk/ arrayexpress/Exp. The processed data used in this paper is available from http://nicta.com.au/people/kowalczyka

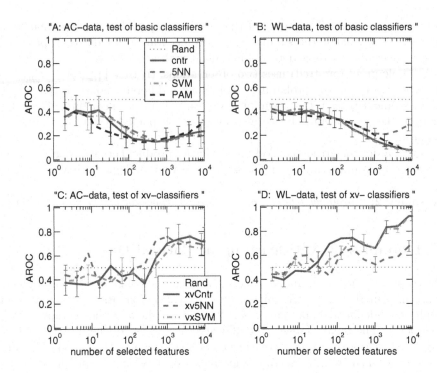

Fig. 2. Comparison of classification of natural adenocarcinoma (AC-data) and synthetic WL-game dataset for selected classifiers. We plot average of 20 repeats of 5-fold cross-validation. For all classifiers but PAM, the genes were selected using t-test applied to the training subset only. Note that PAM has built-in feature selection routine.

4   Discussion

The crux of anti-learning in our synthetic model is the inequality (5) stating that two examples of the opposite label are more "similar" to each other than two of the same label. This is a direct consequence of the 0-sum game constraints (2) combined with the "winner take all" paradigm. Such a simple "Darwinian" mechanism makes it plausible to argue that anti-learning signatures can arise in the biological datasets. However, there are also many other models generating anti-learnable signature, for instance a model of mimicry, which we shall cover elsewhere.

Anti-learning and esophageal adenocarcinoma. There are at least two reasons why research into anti-learning is currently critical for the project on prediction of CRT response in esophageal adenocarcinoma. Firstly, we need to prove that the measurements of gene expressions contain signal suitable for the prediction, so continuation of this expensive line of research is warranted. Secondly, apart from direct utility of CRT response prediction, there is a secondary, perhaps ultimate goal of this research, which is the determination of biology (say

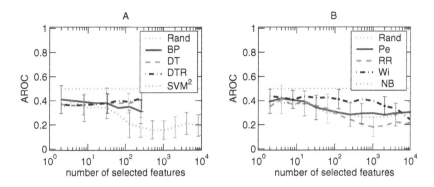

Fig. 3. Anti-learning performance of 8 selected classifiers on the natural AC-dataset (Figures A & C) and synthetic WL-game datasets (Figures B & D). The setting is similar to that in Figure 2, except that in Figure A we have tested for a smaller number of (preselected) features only, ≤ 256, as some of the implementations used did not run for the high dimensional input. For the following three algorithms: BP - Back-Propagation neural network, 5 hidden notes and 1 output, DT - Decision Trees and DTR - Regression Trees, we have used standard Matlab toolbox implementations, newff.m and treefit.m, respectively. For the remaining five algorithms, i.e. NB - Naive Bayes, Pe - Perceptron, RR - Ridge Regression [13], SVM^2 - SVM with the second order polynomial kernel and Wi - Winnow [16], we have used local custom implementations.

pathways) governing the CRT response, which could lead to a new treatment. As supervised learning signature of those precesses is most likely "anti-learnable" in view of our research, its proper interpretation and analysis is possible only from the position of anti-learning, since otherwise the data makes no sense and cannot lead to satisfactory conclusions.

Regularization. High regularization [12,13] is not an answer to the anti-learning challenge. In particular, the centroid, which is a "high regularization" limit of SVM and ridge regression [11], is systematically anti-learning on AC- and WL-datasets. Moreover, according to Theorem 2, for some datasets such as WL-game outcomes, SVM will anti-learn independently of how much regularization is used in its generation.

Kernels. Now let us consider the case of non-linear transformation of data via application of a kernel k [12,14,13]. Our Theorem 2, Remark 2 and Figure 3 argue that for some datasets the anti-learning extends to the popular kernels including the polynomial the radial basis kernels. This is compatible with a common sense observation that the anti-learning is not an issue of the too-poor hypothesis class, which is the main intuitive justification for kernel application.

Boosting. Now we turn to boosting, another heuristic for improving generalization of hard to learn data. The observation is that (ada)boosting [17] weak learners satisfying conditions (6), (7) and (11) outputs a convex combination of them, which again satisfies these conditions, hence the conclusion of Theorem 2 (see [5] for the similar argument line). Thus here the boosting does not change much at all. Intuitively, this is what one should expect: the boosting is effective in some cases where training data is difficult to classify. However, in the case in question, the training data is deceptively easy to deal with, but gives no clues of the performance on an independent test set.

Anti-learning and Overfitting. Overfitting is a deficiency of an algorithm with excessive capacity [12] which fits a model to idiosyncracies and noise of the training data. However, the anti-learning we are concerned with here is essentially different issue. Firstly, we prove that it is possible for a predictor to operate well below the accuracy of random guessing and still be a reliable forecaster. Secondly, we have shown that the anti-learning can be a signature of a deterministic phenomena (see the WL-game definition in Section 2).

The large sample limit and VC bounds. It follows clearly from Figure 1 that there is no contradiction between anti-learning and predictions of the learning theory such as VC-bounds [12,13]. Anti-learning occurs for a small size training set, where the asymptotic predictions of VC-theory are vacuous, and disappear in the large size sample limit, where VC-bounds hold.

5   Conclusions

We have demonstrated the existence of strong anti-learning behavior by a number of supervised learning algorithms on natural and synthetic data. Moreover, we have shown that a simple addition of an extra decision step, a reverser, can exploit this systematic tendency and lead to accurate predictor. Thus anti-learning is not a manifestation of over-fitting classifiers to the noise, but a systematic though usual, mode of operation of a range of supervised learning algorithms exposed to a non-standard dataset. Such a phenomenon, whenever encountered, should be systematically investigated rather than labelled as failure and forgotten. On a level of datamining we can offer a rough explanation of anti-learning by a specific geometry in the dataset, though this surely does not account for all of the phenomena encountered in nature. More research is needed into handling such datasets in practice as well as into the natural processes capable of generating such signatures.

Acknowledgements

We thank Justin Bedo and Garvesh Raskutti of NICTA, and Danielle Greenawalt and Wayne Phillips of Peter MacCallum Cancer Centre for help in preparation of this paper.

National ICT Australia is funded through the Australian Government's *Backing Australia's Ability* initiative, in part through the Australian Research Council.

This work was supported in part by the IST Programme of the European Community, under the PASCAL Network of Excellence, IST-2002-506778. This publication only reflects the authors' views.

References

1. Greenawalt, D., Duong, C., Smyth, G., Ciavarella, M., Thompson, N., Tiang, T., Murray, W., Thomas, R., Phillips, W.: Gene Expression Profiling of Esophageal Cancer: Comparative analysis of Barrett's, Adenocarcinoma and Squamous Cell Carcinoma. Int J. Cancer 120, 1914–1921 (2007)
2. Duong, C., Greenawalt, D., Kowalczyk, A., Ciavarella, M., Raskutti, G., Murray, W., Phillips, W., Thomas, R.: Pre-treatment gene expression profiles can be used to predict response to neoadjuvant chemoradiotherapy in esophageal cancer. Ann Surg Oncol (accepted, 2007)
3. Kowalczyk, A., Greenawalt, D., Bedo, J., Duong, C., Raskutti, G., Thomas, R., Phillips, W.: Validation of Anti-learnable Signature in Classification of Response to Chemoradiotherapy in Esophageal Adenocarcinoma Patients. Proc. Intern. Symp. on Optimization and Systems Biology, OSB (to appear, 2007
4. Kowalczyk, A., Chapelle, O.: An analysis of the anti-learning phenomenon for the class symmetric polyhedron. In: Jain, S., Simon, H.U., Tomita, E. (eds.) Proceedings of the 16th International Conference on Algorithmic Learning Theory, Springer, Heidelberg (2005)
5. Kowalczyk, A., Smola, A.: Conditions for antilearning. Technical Report HPL-2003-97(R.1), NICTA, NICTA, Canberra (2005)
6. Kowalczyk, A., Raskutti, B.: One Class SVM for Yeast Regulation Prediction. SIGKDD Explorations 4(2) (2002)
7. Raskutti, B., Kowalczyk, A.: Extreme re-balancing for svms: a case study. SIGKDD Explorations 6(1), 60–69 (2004)
8. Wolpert, D.H.: The lack of a priori distinctions between learning algorithms. Neural Computation 8(7), 1341–1390 (1996)
9. Provost, F., Fawcett, T.: Robust classification for imprecise environments. Machine Learning 42(3), 203–231 (2001)
10. Bamber, D.: The area above the ordinal dominance graph and the area below the receiver operating characteristic graph. J. Math. Psych. 12, 387–415 (1975)
11. Bedo, J., Sanderson, C., Kowalczyk, A.: An efficient alternative to svm based recursive feature elimination with applications in natural language processing and bioinformatics. In: Australian Conf. on Artificial Intelligence, pp. 170–180 (2006)
12. Vapnik, V.: Statistical Learning Theory. John Wiley and Sons, New York (1998)
13. Cristianini, N., Shawe-Taylor, J.: An Introduction to Support Vector Machines. Cambridge University Press, Cambridge (2000)
14. Schölkopf, B., Smola, A.: Learning with Kernels. MIT Press, Cambridge, MA (2002)
15. Tibshirani, R., Hastie, T., Narasimhan, B., Chu, G.: Class prediction by nearest shrunken centroids, with applicaitons to dna microarrays. Stat. Sci. 18, 104–117 (2003)
16. Kivinen, J., Warmuth, M.K.: Additive versus exponentiated gradient updates for linear prediction. In: Proc. 27th Annual ACM Symposium on Theory of Computing, pp. 209–218. ACM Press, New York (1995)
17. Freund, Y., Schapire, R.E.: A decision-theoretic generalization of on-line learning and an application to boosting. Journal of Computer and System Sciences 55(1), 119–139 (1997)

Improved Algorithms for Univariate Discretization of Continuous Features

Jussi Kujala and Tapio Elomaa

Institute of Software Systems
Tampere University of Technology
P.O. Box 553, FI-33101 Tampere, Finland
jussi.kujala@tut.fi elomaa@cs.tut.fi

Abstract. In discretization of a continuous variable its numerical value range is divided into a few intervals that are used in classification. For example, Naïve Bayes can benefit from this processing. A commonly-used supervised discretization method is Fayyad and Irani's recursive entropy-based splitting of a value range. The technique uses MDL as a model selection criterion to decide whether to accept the proposed split.

We argue that theoretically the method is not always close to ideal for this application. Empirical experiments support our finding. We give a statistical rule that does not use the ad-hoc rule of Fayyad and Irani's approach to increase its performance. This rule, though, is quite time consuming to compute. We also demonstrate that a very simple Bayesian method performs better than MDL as a model selection criterion.

1 Introduction

A common way of handling continuous information — such as *weight* and *volume* of an object — in classifiers is to discretize the variable's value range. Discretization produces typically disjoint intervals that mutually cover the continuous value range of the attribute. Some classifiers, like Naïve Bayes (NB), actually prefer information that composes of parts that have only few possible values [1,2]. We consider the supervised setting; i.e., a learning algorithm has access to a labeled *training set* $S = \{ (x_1, y_1), \ldots, (x_n, y_n) \}$, where instance x_i is composed of the feature values and y_i is the class label of example i. *Univariate* approaches consider one independently measured attribute at a time, while *multivariate* approaches take several (usually all) attributes into account simultaneously.

The literature on discretization algorithms is vast (see e.g., [1,3,4,5] and the references therein). Many univariate and multivariate discretization algorithms have been proposed. Fayyad and Irani's [6] entropy-based discretization algorithm is arguably the most commonly used supervised discretization approach. In addition to entropy calculation the method also takes advantage of the *minimum description length* (MDL) principle, so we will call this algorithm ENT-MDL. The main reasons for the success of ENT-MDL are probably its comprehensibility and quite good performance. The other most popular discretization techniques are unsupervised approaches equal-width binning (EWB) and equal-frequency binning [7,8,1].

J.N. Kok et al. (Eds.): PKDD 2007, LNAI 4702, pp. 188–199, 2007.
© Springer-Verlag Berlin Heidelberg 2007

Fayyad and Irani's approach is based on recursive binary splitting of the (sub)interval at the point that appears the most promising according to the entropy measure. Whether to actually implement the suggested split is tested using a MDL model selection criterion. In this paper we show that Fayyad and Irani's MDL rule is not optimal in discretization and that it is not sound. Replacing it with a Bayesian criterion leads to an algorithm that work as well, if not better. In addition, we propose a well-founded test statistic that performs very well in practice without any ad-hoc rules attached to it. This test statistic, though, is expensive to compute.

The remainder of this paper is organized as follows. Section 2 reviews the background of this work—Naïve Bayesian classifier and discretization of continuous features. In Section 3 we recapitulate Fayyad and Irani's [6] ENT-MDL algorithm in more detail and consider its theoretical and practical properties. We then propose to replace the MDL model selection criterion with a simple Bayesian one. Section 5 puts forward a test statistic to decide on splitting. This approach does not need any ad hoc techniques to support it. In Section 6 we report on an empirical evaluation of the techniques discussed in this work. The experiments confirm that the straightforward Bayesian rule slightly outperforms the MDL rule and the test statistic can match the performance of both of these heuristics. Finally, Section 7 presents the concluding remarks of this paper.

2 Related Work and Approaches to Discretization

In general, a classifier associates a feature vector x with a class label y. Values in x are information measured from an object and y is the identity of the object that we are interested in. A *discrete* feature has a finite number of possible values, while a *continuous* feature can attain values in some infinite totally ordered set. For example, the weight of an object can attain values in the set of positive real numbers $\mathbb{R}^+$.

In this section we first recapitulate the Naïve Bayes classifier. It is a simple and effective classifier for discrete features. Naïve Bayes gives us a motivation for discretization of continuous features. We, then, briefly review previous work on discretization.

2.1 Our Motivation: Naïve Bayes Classifier

Naïve Bayes classifier uses the training set to infer from the given features x the label y we want to know. It assumes that the feature-label pairs (x, y) in the training set have been generated independently from some distribution D. NB takes advantage of *Bayesian inference* in labeling:

$$\mathbf{P}(y \mid x) \propto \mathbf{P}(x \mid y) \, \mathbf{P}(y) \, .$$

The naïvity in NB is to assume that different features in $x = (x_1, \ldots, x_d)$ are statistically independent given the class:

$$\mathbf{P}(x \mid y) = \mathbf{P}(x_1, \ldots, x_d \mid y) \approx \mathbf{P}(x_1 \mid y) \cdots \mathbf{P}(x_d \mid y) \, .$$

This simplification enables it to avoid *the curse of dimensionality*, the fact that the number of samples needed to estimate a joint distribution of several features grows exponentially in their number. Under the independence assumption we only estimate the marginal distribution of each feature, and these densities do not depend on the number of features. The trade-off is that the independence is unlikely to hold which may lead to decreased accuracy in classification.

The empirical performance of Naïve Bayes classifier has, nevertheless, been shown to be good in several experiments [1,9]. It appears that the assumption that features are independent does not necessarily hinder the performance even when false [10]. Domingos and Pazzani [11] have argued why this is so.

2.2   Related Work on Discretization

Naïve Bayes needs to know for each feature x_i the probability of attaining a particular value v, $\mathbf{P}(x_i = v \mid y)$. For discrete features the conditional probability can be easily estimated from the training set by counting the number of labels y for which it holds that $x_i = v$. For continuous features it is an interesting question how to choose these probabilities given the training set. This problem has attained significant attention. For a comprehensive survey of the associated research see [3]. Here we only review work that is most related to ours.

For a classifier the most fundamental aim of discretization is to place the interval borders so that its predictive power is good on yet unseen examples. In discretization we could consider all features simultaneously and, for example, minimize the empirical error rate on the training set. Unfortunately multidimensional empirical error minimization is NP-complete [12,13,14] although polynomial time *approximation algorithm* exists [14,15]. In general the methods for multivariate discretization are computationally expensive.

Hence simpler univariate discretization methods are actually used. Moreover, Naïve Bayes is in some sense inherently univariate, because of the assumption of the statistical independence between features. For example, Figure 1(b) demonstrates a situation in which neither of the available attributes can clarify class distribution and multivariate discretization would be beneficial. However, Naïve Bayes cannot take advantage of multivariate discretization because the marginal distributions are mixed.

Early continuous feature handling in NB assumed that each feature conforms separately to some probability distribution — e.g., normal distribution [16]. The necessary parameters were then estimated from the training set. However, sometimes features are not distributed as assumed and then the performance suffers. A continuous feature can be binned to intervals of equal width, reducing the continuous-valued estimation to a discrete one. From a statistical point of view this models a continuous feature with a piecewise uniform distribution, where each uniform distribution corresponds to an interval. This is more flexible than using a more limited distribution, especially if the number of intervals can depend on the training set.

Figure 1(a) demonstrates that the unsupervised EWB is sometimes suboptimal. There is a slight performance drop if the label distribution suddenly changes

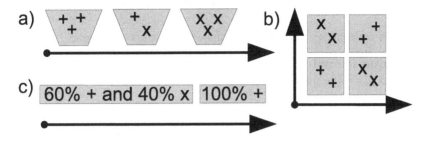

Fig. 1. The class labels are + and x. Subfigure a) shows how EWB can make suboptimal choices and b) depicts a case where univariate methods fail. Naïve Bayes cannot either take advantage of the best split in this case. In situation c) empirical error minimization fails to distinguish between two adjacent distributions, because their majority class is the same.

"within an interval". Hence several methods have been invented to place the interval borders in a more intelligent way [7,17,5].

Catlett [8] proposed to apply recursive partitioning based on entropy of the observed label distribution of a discretized feature. Intuitively the entropy measures the amount of randomness of a source producing random items. In this approach an interval is split at a point that results in minimum entropy. Formally, let $\widehat{P}_I(y_i)$ be the *empirical probability* of observing the label y_i on interval I; i.e., the ratio of labels y_i to all labels in the interval I. Then the entropy of the label distribution on I is defined as:

$$H(I) = \sum_{y_i} \widehat{P}_I(y_i) \log_2 \frac{1}{\widehat{P}_I(y_i)},$$

where the sum is over all labels. The entropy of the label distribution on a feature is the sum over all intervals:

$$H(S) = \sum_{I} \frac{|I|}{|S|} H(I),$$

where $|I|$ is the number of examples in the interval I and $|S|$ is the total number of examples in the training set S.

Several heuristic rules were used to decide when to stop the recursive partitioning in Catlett's [8] approach. In ENT-MDL Fayyad and Irani [6] proposed to use a single MDL-based stopping rule. We proceed to review ENT-MDL in more detail. It is based on modeling the assumed true distribution on a feature as accurately as possible. This is in contrast to the empirical error minimization, which must be regulated, e.g., by restricting the number of final intervals and can lose information of the distribution. The problem with error-based discretization is that it cannot separate two adjacent intervals that have the same majority class, even though it might be beneficial for further processing in a classifier (see Figure 1(c)) [18].

3 Fayyad and Irani's Recursive Discretization: ENT-MDL

For a given training set we have two somewhat distinct problems:

1. How many intervals to use?
2. How to place the intervals?

ENT-MDL uses a MDL criterion to answer the first question and entropy to answer the second one. Minimizing the entropy of the label distribution for a fixed number of intervals yields a discretization in which, intuitively, the empirical label distribution is as unsurprising as possible. However, no efficient method for minimizing entropy for a feature is known. ENT-MDL uses a *heuristic*: given an interval it splits it at the point that minimizes the joint entropy of the two resulting subintervals. This heuristic is applied recursively. To address the issue of the total number of intervals Fayyad and Irani suggest that a test be done whether to actually execute a split.

This test solves a model selection problem where the candidate models are:

M_0 labels on the interval are generated independently from the same distribution.

M_i there is a distribution for the instances up to the index i, $i > 0$, and a separate one for the instances after that. The labels are generated independently.

In this case a model M_i that splits the data has always more explanatory power on the training set than M_0 which refrains from splitting. This behavior is an example of overfitting, because a more complicated model can fit the training data very well, but may not have any predictive power on instances it has not seen. ENT-MDL uses MDL to choose the model. In short, MDL selects the model that makes it possible to compress the data — in this case the class labels — the most. The compression used in ENT-MDL is a *two part code*. The first part codes the model and the second part codes the data.

More precisely, Fayyad and Irani encode the data on any interval with approximately $|I| H(I)$ bits using an optimal code [19]. Then the model M_0 is encoded with $k H(I)$ bits, where k is the number of labels on the interval I. Thus, the total bit length of the data and the model is

$$(|I| + k)H(I).$$

Similarly the models $M_i, i > 0$, are encoded with

$$k_1 H(I_1) + k_2 H(I_2) + \log_2(3^k - 2) + \log_2(|I| - 1)$$

bits, where I_1 is the first subinterval and k_1 is the number of distinct labels on it. I_2 and k_2 are defined similarly. The additional terms follow from the fact that there are two intervals.

3.1   Theoretical and Empirical Properties of ENT-MDL

The code lengths proposed by Fayyad and Irani [6] do not derive from real codes, because for example we cannot encode the model M_0 within the bits given. They suggest that M_0 is sent for example as a codebook for a Huffman code that codes each label individually. First, the expected code word length $H(I)$ for a label is different from the usual arithmetic average of the code word lengths. For example, if we have two labels with codes 1 and 0, i.e., two bits, and the probability of the former is 0.9, then $kH(I) \approx 0.94$. Second, if the item labels are coded individually, then the sum of the code word lengths for the data can be $|I|$ bits greater than $|I|H(I)$, because there are no fractional bits.

Instead we need to use a non-universal nearly optimal code for sequences, like the arithmetic code or a Huffman code that encodes sequences, and these codes in general need to know the probabilities on the labels. For $|I|$ items and k labels there are

$$P = \binom{|I| + k - 1}{k - 1}$$

different sets of probabilities (the number of ways we can allocate $|I|$ items to k bins). Hence, on average a single set of probabilities takes $\log_2 P$ bits. This, then, is a lower bound for the length of the model M_0 unless we have some additional *a priori* knowledge on the probabilities or use an approximation of the model M_0.

In general, a problem with the accurate use of two part MDL is that the user is relied on giving the code, and an optimal code for the application may be difficult to come up with.

The performance of ENT-MDL increases if it splits the range of a given feature at least once. We call this property *autocutting* and denote this method by ENT-MDL-A. It is unclear to us whether Fayyad and Irani [6] meant that autocutting should be always done with ENT-MDL. Clearly, if the rule used to determine whether to split were approximately optimal, then this kind of behavior would be unnecessary. In Section 6 (Tables 1 and 2) we see that autocutting is empirically beneficial, because it increases the average prediction accuracy and the increase is statistically significant in four test domains. Furthermore, the average accuracy does not decrease significantly in any of our test domains.

Let us give an example of a situation in which ENT-MDL makes a wrong choice.

Example 1. Let an interval I have n examples that have binary class — either 0 or 1 — and there are equal numbers of instances from both classes. The first half I_1 of I contains 30% of the 1s and the second half I_2 contains 70% of them. The entropies of these intervals are $H(I) = 1$ and $H(I_1) = H(I_2) \approx 0.88$. Let $\mathcal{H}_0$ be the hypothesis that the labels I are generated uniformly and $\mathcal{H}_1$ the hypothesis that the distribution changes at some point; i.e., we should split. ENT-MDL chooses $\mathcal{H}_0$ if the following holds:

$$\left(1 + \frac{2}{n}\right) n < 2 \cdot 0.88 \cdot \frac{n}{2} \left(1 + \frac{2}{n/2}\right) + \log_2(n - 1) + \log_2(7).$$

If we use accurate values then if $n > 91$ ENT-MDL chooses to split, and does not split when $n = 91$. The probability that there is at least this discrepancy between entropies $H(I)$ and $H(I_1), H(I_2)$ is approximately of the order 0.35% *if* $\mathcal{H}_0$ is true with the label probabilities approximated from the empirical label frequencies. We estimated this probability by generating 100 000 intervals of the length n from $\mathcal{H}_0$ and computing the entropies of the intervals according to $\mathcal{H}_0$ and $\mathcal{H}_1$. Note that $\mathcal{H}_1$ always chooses the best split for the given generated labels. We then compared the difference of these entropies to those from the original data. In only 353 cases the difference was larger. Hence we should not choose $\mathcal{H}_0$, because the labels are not typical for it when compared to the hypothesis $\mathcal{H}_1$. This example is also valid if we consider MDL where the rule is based on a real code, as discussed above.

Note that we implicitly assume that we can approximate the real $\mathcal{H}_0$ with the one estimated from the item labels on the interval I. This does not affect our results to a great extent, because $\mathcal{H}_0$ is a simple hypothesis, hence it overfits only slightly. We also tested altering the frequencies for $\mathcal{H}_0$ and observed that the results were the same.

In experiments this kind of case appears to happen for example in the UCI Bupa Liver Database. For this domain a decrease of 6.2 percentage units in prediction accuracy results when ENT-MDL is used instead of ENT-MDL-A. We verified this behavior by manually checking for this domain that MDL did not split when it really should have.

4   Simple Bayesian Methods in Discretization: ENT-BAY

Let us study replacing the MDL criterion used in ENT-MDL with a Bayesian method. Bayesian model selection is a well known and much used tool. It unifies formal reasoning and intuitive prior knowledge of the user in a convenient manner. Given models $M_0, \ldots, M_n$ the Bayesian approach selects the model M_i that maximizes the posterior probability of having generated the data S:

$$\mathbf{P}(S \mid M_i)\,\mathbf{P}(M_i)\,,$$

where $\mathbf{P}(M_i)$ is the *a priori* probability of the model M_i given by the user. Two part MDL can be seen as a special case of the Bayesian approach in which $\mathbf{P}(M_i)$ is obtained from the code length for the model M_i.

In our case, the model M_0 corresponds to the no-split decision and models M_i with $i > 0$ correspond to the cases where the interval is split at instance with index i. We can, of course, set the priors in several ways. In subsequent empirical evaluation we study the following straightforward way. We assign a prior 0.95 to M_0 and the remaining probability mass is divided evenly to the other models. We call this method ENT-BAY95.

Having to assign the priors is both an advantage and a drawback. Priors offer flexibility, because they are intuitive and user can set them according to the needs of the problem. On the other hand, there are no true priors and selecting them can be a nuisance. It is worth noting that Fayyad and Irani [6] too consider a Bayesian test, but prefer MDL, because they view the selection of priors to be too arbitrary. We show in our empirical studies that the simple prior given above performs well in all tested problems. Thus, it can be used if the user does not wish to select the prior himself. If the user chooses to customize the prior distribution, then we presume that the results would be even better.

5   Using a Test Statistic to Decide on the Splits

A problem with the discretization schemes described above is that they can be improved with the ad-hoc technique of autocutting. This means that when used as such the schemes do not work as well as they should.

We demonstrate an alternative approach to decide whether to split: using a *test statistic* derived from the data. In statistics using such an approach is a standard method. In Section 6 we see that this approach works better than the previous discretization methods without autocutting. In discretization a χ^2-distributed test statistic has been used in the ChiMerge algorithm to decide whether to merge adjacent intervals together [20]. We give a test statistic which, when $\mathcal{H}_0$ is approximately true, tells us how likely it is that the best split produces $\mathcal{H}_1$. If we find the situation unlikely, then we can reject $\mathcal{H}_0$ and execute the split. We call this method ENT-TEST.

The test statistic is derived as follows. Denote the labels on the current interval with a vector $\boldsymbol{y}$. Let $\mathbf{P}(\boldsymbol{y} \mid \mathcal{H}_0(\boldsymbol{y}))$ be the probability of generating $\boldsymbol{y}$ according to hypothesis $\mathcal{H}_0$ when the parameters are estimated from $\boldsymbol{y}$ itself. Similarly let $\mathbf{P}(\boldsymbol{y} \mid \mathcal{H}_1(\boldsymbol{y}))$ be the probability according to $\mathcal{H}_1$. Now we need to know the probability of obtaining a pair $\langle \mathbf{P}(\boldsymbol{y}' \mid \mathcal{H}_0(\boldsymbol{y}')), \mathbf{P}(\boldsymbol{y}' \mid \mathcal{H}_1(\boldsymbol{y}')) \rangle$, where $\boldsymbol{y}' \sim \mathcal{H}_0$, that is less likely than the actual pair $\langle \mathbf{P}(\boldsymbol{y} \mid \mathcal{H}_0(\boldsymbol{y})), \mathbf{P}(\boldsymbol{y} \mid \mathcal{H}_1(\boldsymbol{y})) \rangle$. We have two problems:

1. How to generate $\boldsymbol{y}' \sim \mathcal{H}_0$ given that we do not know the *exact* probabilities of the labels under the null hypothesis $\mathcal{H}_0$?
2. How to define "less likely"?

We answer these questions by approximating that $\mathcal{H}_0$ is a permutation on the class labels that we have seen. Because $\mathcal{H}_0$ is a very simple hypothesis this estimation from the empirical data is likely to be close enough to the "truth" for our purposes and additionally $\mathbf{P}(\boldsymbol{y}' \mid \mathcal{H}_0(\boldsymbol{y}'))$ becomes a constant. Then we only need to compute $\sum_{\boldsymbol{y}' \in Y'} \mathbf{P}(\boldsymbol{y}')$, where Y' is the set of $\boldsymbol{y}'$s such that $\mathbf{P}(\boldsymbol{y}' \mid \mathcal{H}_1(\boldsymbol{y}')) < \mathbf{P}(\boldsymbol{y} \mid \mathcal{H}_1(\boldsymbol{y}))$ and $\mathbf{P}(\boldsymbol{y}')$ is the probability of $\boldsymbol{y}'$ according to $\mathcal{H}_0$. Unfortunately we do not know how to solve this problem efficiently. We resort to sampling from $\mathcal{H}_0$, i.e., generating vectors of data $\boldsymbol{y}'$ from $\mathcal{H}_0$. This is expensive, because we need to generate many vectors if we want to remove the effect of randomness from sampling.

Of course, this method gives a likelihood value and in empirical experiments we need to decide how small the likelihood value can be before splitting. In experiments we chose to split if the likelihood was below 10%. The number of samples drawn from $\mathcal{H}_0$ was fifty. The results of these experiments are given in Table 1 and Table 2. We can see that the significance value of 10% gives a good performance with respect to ENT-MDL. It is worth noticing is that ENT-TEST does not depend on autocutting to improve the performance. However, unless we can do the significance test efficiently, this method is limited to cases in which enough computational power is available to handle the sampling. It is an interesting open question whether a more efficient method to calculate the likelihood exists.

Why do we use such a complicated distribution? We could assume the number of a particular label in a partitioned interval to be normally distributed. Its parameters could, then, be taken from the unpartitioned interval. We can use the normal distribution, because it approximates quite well the multinomial one, which is the real distribution for the number of labels when the number of trials is fixed. Then these normally distributed values for both subintervals could be joined to form a variable that is χ^2-distributed; i.e., it is a sum of normalized normally distributed values squared. However, there is a flaw in this approach. The problem is that this works for a split that is in a fixed location on the interval, but in our case the hypothesis $\mathcal{H}_1$ selects the one that is the best according to its criteria. Hence, the numbers of the different labels do not conform to our assumption on their distribution.

An alternative approach to a test statistic is to simply use a test set. Unfortunately, in empirical tests this approach did not perform well. As the small number of samples in small intervals is probably to blame, the k-set validation could be more useful. However, we have not experimented with this approach yet.

6   Empirical Evaluation

We evaluate EWB, ENT-MDL, ENT-BAY, and ENT-TEST on 16 domains from the UCI machine learning repository. Also versions of ENT-MDL and ENT-BAY that carry out autocutting are included in this comparison. For each domain we randomly split the data to a training set and test set, with two-thirds being in the training set and the rest in the test set. We iterate the procedure thirty times for each domain. For an interval I the probability $\mathbf{P}(I \mid y)$ was estimated using Laplacian correction; i.e., each interval has one additional training example with label y.

The average prediction accuracies are given in Table 1 and statistical significance tests using t-test[1] with confidence level 0.95 are in Table 2. From these results we see that autocutting benefits both ENT-MDL and ENT-BAY. The resulting increase in average accuracy over all 16 test domains is 0.8 percentage units for ENT-MDL and 0.6 percentage units for ENT-BAY. Unsupervised EWB is the clear loser in these experiments, but still it is able to win in some domains. In these ones the numerical values of attributes are probably important. The

---

[1] The assumptions behind the t-test are violated and as Dietterich [21] argues this can result in inaccurate significance measurements. However, we also used the Wilcoxon signed-rank test, which has fewer assumptions, and the results were identical.

Table 1. Performance of discretization algorithms on Naïve Bayes. The average classification accuracy over 30 repetitions of randomized training set selection for 16 UCI domains is shown. Also the average over all 16 domains is given.

	EWB	MDL	MDL-A	BAY95	BAY95-A	TEST-10%
Iris	94.5	94.0	93.5	93.9	94.0	94.7
Glass	60.7	63.7	67.3	68.8	69.4	69.0
Bupa	61.6	57.1	63.3	57.4	62.2	60.0
Pima	75.0	74.7	74.1	75.5	74.5	74.2
Ecoli	83.5	84.9	85.0	85.9	85.5	84.9
Segmentation	79.0	81.3	84.0	81.6	83.2	82.3
Wine	97.1	98.3	98.3	98.3	98.3	98.1
Australian	85.2	85.3	85.0	85.2	85.7	85.5
German	71.8	71.9	73.2	71.4	73.9	74.7
Iono	85.9	89.8	89.2	90.3	88.3	89.8
Sonar	74.4	75.1	75.4	74.4	77.8	75.6
Wisconsin	97.4	97.6	97.4	97.6	97.5	97.4
Letter	61.2	73.6	73.6	73.5	73.6	73.5
Abalone	58.0	58.7	58.3	58.4	58.2	58.4
Vehicle	60.1	58.4	59.2	61.7	61.4	62.0
Page	92.3	93.4	93.4	93.4	93.5	93.2
Average	77.4	78.6	79.4	79.2	79.8	79.6

Table 2. Number of statistically significant wins using the t-test with 0.95 confidence level. The figure in a cell denotes the number of wins (out of 16) that the discretization algorithm mentioned on the row obtains with respect to the one on the column.

	EWB	MDL	MDL-A	BAY95	BAY95-A	TEST-10%
EWB	•	2	2	1	0	0
MDL	5	•	0	0	1	0
MDL-A	8	4	•	3	0	1
BAY95	8	2	2	•	1	1
BAY95-A	9	5	2	3	•	0
TEST-10%	9	4	2	2	1	•

inefficient ENT-TEST is better than pure ENT-MDL or ENT-BAY, and performs approximately the same when autocutting is factored in. It also has the least number of statistically significant losses against the other algorithms. The two entropy-based approaches ENT-MDL and ENT-BAY have quite similar overall performance. However, ENT-BAY is slightly better than ENT-MDL and wins more often against EWB.

7   Conclusions

In this paper we discussed the flaws in the theoretical justification of Fayyad and Irani's [6] entropy-based recursive discretization algorithm. The MDL criterion used to stop the recursive partitioning is not based on real codes. We proposed to replace the MDL criterion with an extremely simple Bayesian model selection criterion. In empirical evaluation the Bayesian approach has similar, though, slightly better overall performance than the MDL approach. Of course, the success of discretization algorithms varies from domain to domain. The Bayesian approach has the advantage of being simpler than the MDL approach and, furthermore, can be easily customized by the user.

We also put forward a test statistic to decide on partitioning. This approach does not need heuristic techniques to improve its performance like the other entropy-based techniques do. Empirical evaluation shows this approach to have a comparative performance with the heuristic approaches, but unfortunately it is expensive to compute.

In this work we have demonstrated that better working new efficient heuristic approaches to discretization and (inefficient) well-founded approaches can be developed. In the long run would be interesting to find solutions to the discretization problem that are at the same time *efficient* and *theoretically justified*.

Acknowledgments

This work has been financially supported by Tampere Graduate School in Information Science and Engineering (TISE), Academy of Finland, and Nokia Foundation.

References

1. Dougherty, J., Kohavi, R., Sahami, M.: Supervised and unsupervised discretization of continuous features. In: Prieditis, A., Russell, S. (eds.) Proc. 12th International Conference on Machine Learning, pp. 194–202. Morgan Kaufmann, San Francisco, CA (1995)
2. Hsu, C.N., Huang, H.J., Wong, T.T.: Implications of the Dirichlet assumption for discretization of continuous variables in naive Bayesian classifiers. Machine Learning 53, 235–263 (2003)
3. Liu, H., Hussain, F., Tan, C.L., Dash, M.: Discretization: An enabling technique. Data Mining and Knowledge Discovery 6, 393–423 (2002)
4. Yang, Y., Webb, G.I.: A comparative study of discretization methods for naive-Bayes classifiers. In: Proc. Pacific Rim Knowledge Acquisition Workshop (PKAW), pp. 159–173 (2002)
5. Elomaa, T., Rousu, J.: Efficient multisplitting revisited: Optima-preserving elimination of partition candidates. Data Mining and Knowledge Discovery 8, 97–126 (2004)
6. Fayyad, U.M., Irani, K.B.: Multi-interval discretization of continuous-valued attributes for classification learning. In: Proc. 13th International Joint Conference on Artificial Intelligence, pp. 1022–1027. Morgan Kaufmann, San Francisco, CA (1993)
7. Wong, A., Chiu, D.: Synthesizing statistical knowledge from incomplete mixed-mode data. IEEE Transactions on Pattern Analysis 9, 796–805 (1987)

8. Catlett, J.: On changing continuous attributes into ordered discrete attributes. In: Kodratoff, Y. (ed.) Machine Learning - EWSL-91. LNCS, vol. 482, pp. 164–178. Springer, Heidelberg (1991)

9. Hand, D.J., Yu, K.: Idiot Bayes? not so stupid after all. International Statistical Review 69, 385–398 (2001)

10. Rish, I.: An empirical study of the naive Bayes classifier. In: IJCAI-01 workshop on "Empirical Methods in AI" (2001)

11. Domingos, P., Pazzani, M.: On the optimality of the simple Bayesian classifier under zero-one loss. Machine Learning 29, 103–130 (1997)

12. Chlebus, B.S., Nguyen, S.H.: On finding optimal discretizations for two attributes. In: Polkowski, L., Skowron, A. (eds.) RSCTC 1998. LNCS (LNAI), vol. 1424, pp. 537–544. Springer, Heidelberg (1998)

13. Elomaa, T., Rousu, J.: On decision boundaries of naïve Bayes in continuous domains. In: Lavrač, N., Gamberger, D., Todorovski, L., Blockeel, H. (eds.) PKDD 2003. LNCS (LNAI), vol. 2838, pp. 144–155. Springer, Heidelberg (2003)

14. Călinescu, G., Dumitrescu, A., Karloff, H., Wan, P.J.: Separating points by axis-parallel lines. International Journal of Computational Geometry & Applications 15, 575–590 (2005)

15. Elomaa, T., Kujala, J., Rousu, J.: Approximation algorithms for minimizing empirical error by axis-parallel hyperplanes. In: Gama, J., Camacho, R., Brazdil, P.B., Jorge, A.M., Torgo, L. (eds.) ECML 2005. LNCS (LNAI), vol. 3720, pp. 547–555. Springer, Heidelberg (2005)

16. John, G., Langley, P.: Estimating continuous distributions in Bayesian classifiers. In: Proc. 11th Annual Conference on Uncertainty in Artificial Intelligence, pp. 338–345. Morgan Kaufmann, San Francisco (1995)

17. Fayyad, U.M., Irani, K.B.: On the handling of continuous-valued attributes in decision tree generation. Machine Learning 8, 87–102 (1992)

18. Kohavi, R., Sahami, M.: Error-based and entropy-based discretization of continuous features. In: Simoudis, E., Han, J.W., Fayyad, U. (eds.) Proc. 2nd International Conference on Knowledge Discovery and Data Mining, pp. 114–119. AAAI Press, Menlo Park, CA (1996)

19. Cover, T.M., Thomas, J.A.: Elements of Information Theory. John Wiley & Sons, New York (1991)

20. Kerber, R.: Chimerge: Discretization of numeric attributes. In: Proc. 10th National Conference on Artificial Intelligence, pp. 123–128. MIT Press, Cambridge (1992)

21. Dietterich, T.G.: Approximate statistical test for comparing supervised classification learning algorithms. Neural Computation 10, 1895–1923 (1998)

Efficient Weight Learning for Markov Logic Networks

Daniel Lowd and Pedro Domingos

Department of Computer Science and Engineering
University of Washington, Seattle WA 98195-2350, USA
{lowd,pedrod}@cs.washington.edu

Abstract. Markov logic networks (MLNs) combine Markov networks and first-order logic, and are a powerful and increasingly popular representation for statistical relational learning. The state-of-the-art method for discriminative learning of MLN weights is the voted perceptron algorithm, which is essentially gradient descent with an MPE approximation to the expected sufficient statistics (true clause counts). Unfortunately, these can vary widely between clauses, causing the learning problem to be highly ill-conditioned, and making gradient descent very slow. In this paper, we explore several alternatives, from per-weight learning rates to second-order methods. In particular, we focus on two approaches that avoid computing the partition function: diagonal Newton and scaled conjugate gradient. In experiments on standard SRL datasets, we obtain order-of-magnitude speedups, or more accurate models given comparable learning times.

1 Introduction

Statistical relational learning (SRL) focuses on domains where data points are not i.i.d. (independent and identically distributed). It combines ideas from statistical learning and inductive logic programming, and interest in it has grown rapidly in recent years [6]. One of the most powerful representations for SRL is Markov logic, which generalizes both Markov random fields and first-order logic [16]. Representing a problem as a Markov logic network (MLN) involves simply writing down a list of first-order formulas and learning weights for those formulas from data. The first step is the task of the knowledge engineer; the second is the focus of this paper.

Currently, the best-performing algorithm for learning MLN weights is Singla and Domingos' voted perceptron [19], based on Collins' earlier one [3] for hidden Markov models. Voted perceptron uses gradient descent to approximately optimize the conditional likelihood of the query atoms given the evidence. Weight learning in Markov logic is a convex optimization problem, and thus gradient descent is guaranteed to find the global optimum. However, convergence to this optimum may be extremely slow. MLNs are exponential models, and their sufficient statistics are the numbers of times each clause is true in the data. Because this number can easily vary by orders of magnitude from one clause to another, a learning rate that is small enough to avoid divergence in some weights is too

J.N. Kok et al. (Eds.): PKDD 2007, LNAI 4702, pp. 200–211, 2007.

small for fast convergence in others. This is an instance of the well-known problem of ill-conditioning in numerical optimization, and many candidate solutions for it exist [13]. However, the most common ones are not easily applicable to MLNs because of the nature of the function being optimized. As in Markov random fields, computing the likelihood in MLNs requires computing the partition function, which is generally intractable. This makes it difficult to apply methods that require performing line searches, which involve computing the function as well as its gradient. These include most conjugate gradient and quasi-Newton methods (e.g., L-BFGS). Two exceptions to this are scaled conjugate gradient [12] and Newton's method with a diagonalized Hessian [1]. In this paper we show how they can be applied to MLN learning, and verify empirically that they greatly speed up convergence. We also obtain good results with a simpler method: per-weight learning rates, with a weight's learning rate being the global one divided by the corresponding clause's empirical number of true groundings.

Voted perceptron approximates the expected sufficient statistics in the gradient by computing them at the MPE state (i.e., the most likely state of the non-evidence atoms given the evidence ones, or most probable explanation). Since in an MLN the conditional distribution can contain many modes, this may not be a good approximation. Also, using second-order methods requires computing the Hessian (matrix of second-order partial derivatives), and for this the MPE approximation is no longer sufficient. We address both of these problems by instead computing expected counts using MC-SAT, a very fast Markov chain Monte Carlo (MCMC) algorithm for Markov logic [15].

The remainder of this paper is organized as follows. In Section 2 we briefly review Markov logic. In Section 3 we present several algorithms for MLN weight learning. We compare these algorithms empirically on real-world datasets in Section 4, and conclude in Section 5.

2 Markov Logic

A Markov logic network (MLN) consists of a set of first-order formulas and their weights, $\{(w_i, f_i)\}$. Intuitively, a formula represents a noisy relational rule, and its weight represents the relative strength or importance of that rule. Given a finite set of constants, we can instantiate an MLN as a Markov random field (MRF) in which each node is a grounding of a predicate (atom) and each feature is a grounding of one of the formulas (clauses). This leads to the following joint probability distribution for all atoms:

$$P(\mathrm{X} = \mathrm{x}) = \frac{1}{Z} \exp \left(\sum_i w_i n_i(x) \right)$$

where n_i is the number of times the ith formula is satisfied by the state of the world x and Z is a normalization constant, required to make the probabilities of all worlds to sum to one.

The formulas in an MLN are typically specified by an expert, or they can be obtained (or refined) by inductive logic programming or MLN structure learning [10]. Many complex models, and in particular many non-i.i.d. ones, can be very compactly specified using MLNs.

Exact inference in MLNs is intractable. Instead, we can perform approximate inference using Markov chain Monte Carlo (MCMC), and in particular Gibbs sampling [7]. However, when weights are large convergence can be very slow, and when they are infinite (corresponding to deterministic dependencies) ergodicity breaks down. This remains true even for more sophisticated alternatives like simulated tempering. A much more efficient alternative, which also preserves ergodicity in the presence of determinism, is the MC-SAT algorithm, recently introduced by Poon and Domingos [15]. MC-SAT is a "slice sampling" MCMC algorithm that uses a modified satisfiability solver to sample from the slice. The solver is able to find isolated modes in the distribution very efficiently, and as a result the Markov chain mixes very rapidly. The slice sampling scheme ensures that detailed balance is (approximately) preserved. In this paper we use MC-SAT for inference.

3   Weight Learning for MLNs

Given a set of formulas and a database of atoms, we wish to find the formulas' maximum *a posteriori* (MAP) weights, i.e., the weights that maximize the product of their prior probability and the data likelihood. In this section, we describe a number of alternative algorithms for this purpose.

Richardson and Domingos [16] originally proposed learning weights generatively using pseudo-likelihood [2]. Pseudo-likelihood is the product of the conditional likelihood of each variable given the values of its neighbors in the data. While efficient for learning, it can give poor results when long chains of inference are required at query time. Singla and Domingos [19] showed that pseudo-likelihood is consistently outperformed by discriminative training, which maximizes the conditional likelihood of the query predicates given the evidence ones. Thus, in this paper we focus on this type of learning.[1]

3.1   Voted Perceptron

Gradient descent algorithms use the gradient, $\mathbf{g}$, scaled by a learning rate, η, to update the weight vector $\mathbf{w}$ in each step:

$$\mathbf{w}_{t+1} = \mathbf{w}_t + \eta \mathbf{g}$$

In an MLN, the derivative of the conditional log-likelihood with respect to a weight is the difference between the number of true groundings of the corresponding clause in the data, and the expected number according to the model:

$$\frac{\partial}{\partial w_i} \log P(Y\!=\!y|X\!=\!x) = n_i - E_w[n_i]$$

---

[1] For simplicity, we omit prior terms throughout; in our experiments, we use a zero-mean Gaussian prior on all weights with all algorithms.

where y is the state of the non-evidence atoms in the data, and x is the state of the evidence.

The basic idea of the voted perceptron (VP) algorithm [3] is to approximate the intractable expectations $E_w[n_i]$ with the counts in the most probable explanation (MPE) state, which is the most probable state of non-evidence atoms given the evidence. To combat overfitting, instead of returning the final weights, VP returns the average of the weights from all iterations of gradient descent.

Collins originally proposed VP for training hidden Markov models discriminatively, and in this case the MPE state is unique and can be computed exactly in polynomial time using the Viterbi algorithm. In MLNs, MPE inference is intractable but can be reduced to solving a weighted maximum satisfiability problem, for which efficient algorithms exist such as MaxWalkSAT [9]. Singla and Domingos [19] use this approach and discuss how the resulting algorithm can be viewed as approximately optimizing log-likelihood. However, the use of voted perceptron in MLNs is potentially complicated by the fact that the MPE state may no longer be unique, and MaxWalkSAT is not guaranteed to find it.

3.2   Contrastive Divergence

The contrastive divergence (CD) algorithm is identical to VP, except that it approximates the expectations $E_w[n_i]$ from a small number of MCMC samples instead of using the MPE state. Using MCMC is presumably more accurate and stable, since it converges to the true expectations in the limit. While running an MCMC algorithm to convergence at each iteration of gradient descent is infeasibly slow, Hinton [8] has shown that a few iterations of MCMC yield enough information to choose a good direction for gradient descent. Hinton named this method *contrastive divergence*, because it can be interpreted as optimizing a difference of Kullback-Leibler divergences. Contrastive divergence can also be seen as an efficient way to approximately optimize log-likelihood.

The MCMC algorithm typically used with contrastive divergence is Gibbs sampling, but for MLNs the much faster alternative of MC-SAT is available. Because successive samples in MC-SAT are much less correlated than successive sweeps in Gibbs sampling, they carry more information and are likely to yield a better descent direction. In particular, the different samples are likely to be from different modes, reducing the error and potential instability associated with choosing a single mode.

In our experiments, we found that five samples were sufficient, and additional samples were not worth the time: any increased accuracy that 10 or 100 samples might bring was offset by the increased time per iteration. We avoid the need for burn-in by starting at the last state sampled in the previous iteration of gradient descent. (This differs from Hinton's approach, which always starts at the true values in the training data.)

3.3   Per-weight Learning Rates

VP and CD are both simple gradient descent procedures, and as a result highly vulnerable to the problem of ill-conditioning. Ill-conditioning occurs when the *condition number*, the ratio between the largest and smallest absolute eigenvalues of the Hessian, is far from one. On ill-conditioned problems, gradient descent is very slow, because no single learning rate is appropriate for all weights. In MLNs, the Hessian is the negative covariance matrix of the clause counts. Because some clauses can have vastly greater numbers of true groundings than others, the variances of their counts can be correspondingly larger, and ill-conditioning becomes a serious issue.

One solution is to modify both algorithms to have a different learning rate for each weight. Since tuning every learning rate separately is impractical, we use a simple heuristic to assign a learning rate to each weight:

$$\eta_i = \frac{\eta}{n_i}$$

where η is the user-specified global learning rate and n_i is the number of true groundings of the ith formula. (To avoid dividing by zero, if $n_i = 0$ then $\eta_i = \eta$.) When computing this number, we ignore the groundings that are satisfied by the evidence (e.g., $A \Rightarrow B$ when A is false). This is because, being fixed, they cannot contribute to the variance.

We refer to the modified versions of VP and CD as VP-PW and CD-PW.

3.4 Diagonal Newton

When the function being optimized is quadratic, Newton's method can move to the global minimum or maximum in a single step. It does so by multiplying the gradient, $\mathbf{g}$, by the inverse Hessian, $\mathbf{H}^{-1}$:

$$\mathbf{w}_{t+1} = \mathbf{w}_t - \mathbf{H}^{-1}\mathbf{g}$$

When there are many weights, using the full Hessian becomes infeasible. A common approximation is to use the *diagonal* Newton (DN) method, which uses the inverse of the diagonalized Hessian in place of the inverse Hessian. DN typically uses a smaller step size than the full Newton method. This is important when applying the algorithm to non-quadratic functions, such as MLN conditional log likelihood, where the quadratic approximation is only good within a local region.

The Hessian for an MLN is simply the negative covariance matrix:

$$\frac{\partial}{\partial w_i \partial w_j} \log P(Y\!=\!y|X\!=\!x) = E_w[n_i]E_w[n_j] - E_w[n_i n_j]$$

Like the gradient, this can be estimated using samples from MC-SAT. In each iteration, we take a step in the diagonalized Newton direction:

$$w_i = w_i - \alpha \frac{n_i - E_w[n_i]}{E_w[n_i^2] - (E_w[n_i])^2}$$

The step size α could be computed in a number of ways, including keeping it fixed, but we achieved the best results using the following method. Given a search direction $\mathbf{d}$ and Hessian matrix $\mathbf{H}$, we compute the step size as follows:

$$\alpha = \frac{\mathbf{d}^T \mathbf{g}}{\mathbf{d}^T \mathbf{H} \mathbf{d} + \lambda \mathbf{d}^T \mathbf{d}}$$

where $\mathbf{d}$ is the search direction. For a quadratic function and $\lambda = 0$, this step size would move to the minimum function value along $\mathbf{d}$. Since our function is not quadratic, a non-zero λ term serves to limit the size of the step to a region in which our quadratic approximation is good. After each step, we adjust λ to increase or decrease the size of the so-called *model trust region* based on how well the approximation matched the function. Let Δ_{actual} be the actual change in the function value, and let Δ_{pred} be the predicted change in the function value from the previous gradient and Hessian and our last step, $\mathbf{d}_{t-1}$:

$$\Delta_{pred} = \mathbf{d}_{t-1}^T (\mathbf{g}_{t-1} + \mathbf{H}_{t-1} \mathbf{g}_{t-1})/2$$

A standard method for adjusting λ is as follows [5]:

$$\text{if } (\Delta_{actual}/\Delta_{pred} > 0.75) \text{ then } \lambda_{t+1} = \lambda_t/2$$
$$\text{if } (\Delta_{actual}/\Delta_{pred} < 0.25) \text{ then } \lambda_{t+1} = 4\lambda_t$$

Since we cannot efficiently compute the actual change in log-likelihood, we approximate it as the product of the step we just took and the gradient after taking it: $\Delta_{actual} = \mathbf{d}_{t-1}^T \mathbf{g}_t$. Since the log-likelihood is a convex function, this product is a lower bound on the improvement in the actual log-likelihood. When this value is negative, the step is rejected and redone after adjusting λ.

In models with thousands of weights or more, storing the entire Hessian matrix becomes impractical. However, when the Hessian appears only inside a quadratic form, as above, the value of this form can be computed simply as:

$$\mathbf{d}^T \mathbf{H} = (E_w[\textstyle\sum_i d_i n_i])^2 - E_w[(\textstyle\sum_i d_i n_i)^2]$$

The product of the Hessian by a vector can also be computed compactly [14]. Note that α is computed using the full Hessian matrix, but the step direction is computed from the diagonalized approximation which is easier to invert.

Our per-weight learning rates can actually be seen as a crude approximation of the diagonal Newton method. The number of true groundings not satisfied by evidence is a heuristic approximation to the count variance, which the diagonal Newton method uses to rescale each dimension of the gradient. The diagonal Newton method, however, can adapt to changes in the second derivative at different points in the weight space. Its main limitation is that clauses can be far from uncorrelated. The next method addresses this issue.

3.5 Scaled Conjugate Gradient

Gradient descent can be sped up by, instead of taking a small step of constant size at each iteration, performing a line search to find the optimum along the chosen descent direction. However, on ill-conditioned problems this is still inefficient, because line searches along successive directions tend to partly undo the effect of each other: each line search makes the gradient along its direction zero, but the next line search will generally make it non-zero again. In long narrow valleys, instead of moving quickly to the optimum, gradient descent zigzags.

A solution to this is to impose at each step the condition that the gradient along previous directions remain zero. The directions chosen in this way are called *conjugate*, and the method *conjugate gradient* [18]. We used the Polak-Ribiere method for choosing conjugate gradients since it has generally been found to be the best-performing one. Conjugate gradient methods are some of the most efficient available, on a par with quasi-Newton ones. Unfortunately, applying them to MLNs is difficult, because line searches require computing the objective function, and therefore the partition function Z, which is highly intractable. (Computing Z is equivalent to computing all moments of the MLN, of which the gradient and Hessian are the first two.)

Fortunately, we can use the Hessian instead of a line search to choose a step size. This method is known as *scaled conjugate gradient* (SCG), and was originally proposed by Møller [12] for training neural networks. In our implementation, we choose a step size the same way as in diagonal Newton.

Conjugate gradient is usually more effective with a preconditioner, a linear transformation that attempts to reduce the condition number of the problem (e.g., [17]). Good preconditioners approximate the inverse Hessian. We use the inverse diagonal Hessian as our preconditioner. We refer to SCG with the preconditioner as PSCG.

4   Experiments

4.1   Datasets

Our experiments used two standard relational datasets representing two important relational tasks: Cora for entity resolution, and WebKB for collective classification.

The Cora dataset consists of 1295 citations of 132 different computer science papers, drawn from the Cora Computer Science Research Paper Engine. This dataset was originally labeled by Andrew McCallum[2]. We used a cleaned version from Singla and Domingos [20], with five splits for cross-validation.

The task on Cora is to predict which citations refer to the same paper, given the words in their author, title, and venue fields. The labeled data also specifies which pairs of author, title, and venue fields refer to the same entities. In our experiments, we evaluated the ability of the model to deduplicate fields as well as citations. Since the number of possible equivalances is very large, we used the canopies found by Singla and Domingos [20] to make this problem tractable.

The MLN we used for this is very similar to the "MLN(B+C+T)" model used by Singla and Domingos [20]. Its formulas link words to citation identity, words to field identity, and field identity to citation identity. In this way, word co-occurrence affects the probability that two citations are the same both

---

[2] http://www.cs.umass.edu/~mccallum/data/cora-refs.tar.gz

indirectly, through field similarities, and directly. These rules are repeated for each word appearing in the database so that individualized weights can be learned, representing the relative importance of each word in each context. This model also features transitive closure for all equivalence predicates.

We did two things differently from Singla and Domingos. First, we added rules that relate words to field identity but apply equally to all words. Because these rules are not specific to particular words, they can potentially improve generalization and reduce overfitting. Secondly, we learned weights for all rules. Singla and Domingos set the weights for all word-specific rules using a naive Bayes model, and only learned the other rules' weights using VP. Our learning problem is therefore much harder and more ill-conditioned, but our more powerful algorithms enabled us to achieve the best results to date on Cora.

In our version, the total number of weights is 6141. During learning, the number of ground clauses exceeded 3 million.

The WebKB dataset consists of labeled web pages from the computer science departments of four universities. We used the relational version of the dataset from Craven and Slattery [4], which features 4165 web pages and 10,935 web links, along with the words on the webpages, anchors of the links, and neighborhoods around each link.

Each web page is marked with some subset of the categories: person, student, faculty, professor, department, research project, and course. Our goal is to predict these categories from the web pages' words and link structures.

We used a very simple MLN for this model, consisting only of formulas linking words to page classes, and page classes to the classes of linked pages. The "word-class" rules were of the following form:

$$\texttt{Has(page,word)} \Rightarrow \texttt{Class(page,class)}$$
$$\neg\texttt{Has(page,word)} \Rightarrow \texttt{Class(page,class)}$$

We learned a separate weight for each of these rules for each (**word, class**) pair. Classes of linked pages were related by the formula:

$$\texttt{Class(page1,class1)} \wedge \texttt{LinksTo(page1,page2)} \Rightarrow \texttt{Class(page2,class2)}$$

We learned a separate weight for this rule for each pair of classes. When instantiated for each word and class, the model contained 10,891 weights. While simple to write, this model represents a complex, non-i.i.d. probability distribution in which query predicates are linked in a large graph. During learning, the number of ground clauses exceeded 300,000.

We estimated the condition number for both Cora and WebKB at the point where all weights are zero. (Because our learning problem is not quadratic, the condition number depends on the current weights.) The size of these problems makes computing the condition number of the full Hessian matrix difficult, but we can easily compute the condition number of the diagonalized Hessian, which is simply the largest ratio of two clause variances. For Cora, this was over 600,000, while for WebKB it was approximately 7000. This indicates that both learning problems are ill-conditioned, but Cora is much worse than WebKB.

4.2   Metrics

To score our models, we ran MC-SAT for 100 burn-in and 1000 sampling iterations on the test data. The marginal conditional probability of each query
atom is the fraction of samples in which the atom was true with a small prior
to prevent zero counts.

From these marginal probabilities, we estimate conditional log-likelihood
(CLL) by averaging the log marginal probabilities of the true values of the query
predicates. CLL is the metric all of the algorithms attempt to optimize. However,
in cases such as entity resolution where the class distribution is highly skewed,
CLL can be a poor metric. For this reason, we also look at AUC, the area under
the precision-recall curve. The disadvantage of AUC is that it ignores calibration: AUC only considers whether true atoms are given higher probability than
false atoms.

4.3   Methodology

We ran our experiments using five-way cross-validation for Cora and four-way
cross-validation for WebKB. For each train/test split, one of the training datasets
was selected as a validation set and the remaining ones formed the tuning set.
The tuning procedure consisted of training each algorithm for four hours on
the tuning sets with various values of the learning rate. For each algorithm on
each split, we chose the learning rates that worked best on the corresponding
validation set for each evaluation metric.

We used the implementation of voted perceptron for MLNs in the Alchemy
package [11], and implemented the other algorithms as extensions of Alchemy.
For DN, SCG, and PSCG, we started with $\lambda = 1$ and let the algorithm adjust it
automatically. For algorithms based on MC-SAT, we used 5 samples of MC-SAT
for each iteration of the learning algorithm. The width of the Gaussian prior was
set for each dataset based on preliminary experiments.

After tuning all algorithms, we reran them for 10 hours with their respective
training sets, including the held-out validation data. For the gradient descent
algorithms, we averaged the weights from all iterations.

4.4   Results

Our results for the Cora and WebKB datasets are shown in Figure 1. Error
bars are omitted for clarity; at the final data point, all differences exceed twice
the standard error. For AUC, we computed the standard deviation using the
technique of Richardson and Domingos [16].

PSCG is the most accurate of all the algorithms compared, obtaining the best
CLL and AUC on both Cora and WebKB. It converges relatively quickly as well:
on WebKB, the PSCG learning curve dominates all others after 2 minutes; on
Cora, it dominates after 15 minutes. DN is consistently close behind PSCG in
CLL and AUC, briefly doing better when PSCG starts to overfit. In contrast,
VP and CD consistently converge more slowly to worse AUC and CLL.

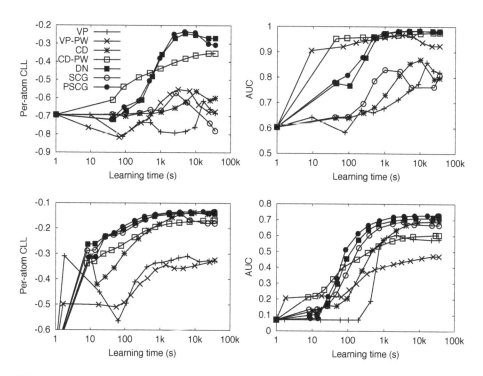

Fig. 1. CLL and AUC for Cora (above) and WebKB (below). Learning times are shown on a logarithmic scale.

On Cora, the algorithms that adjust the search direction using true clause counts or count variance do much better than those that do not. This suggests that these techniques help greatly in cases of extreme ill-conditioning. Without a preconditioner, even SCG does poorly. This is because, like VP and CD, the first step it takes is in the direction of the gradient. On a very ill-conditioned dataset like Cora, the gradient is a very poor choice of search direction.

The AUC results we show for Cora are for all query predicates—SameAuthor, SameVenue, SameTitle, and SameBib. When computing the AUC for just the SameBib predicate, PSCG reaches a high of 0.992 but ends at 0.990 after overfitting slightly. DN and CD-PW do about the same, ending at AUCs of 0.992 and 0.991, respectively. All of these algorithms exceed the 0.988 AUC reported by Singla and Domingos [20], the best previously published result on this dataset, and they do so by more than twice the standard error.

On WebKB, the ill-conditioning is less of an issue. PSCG still does better than SCG, but not drastically better. VP-PW and CD-PW actually do worse than VP and CD. This is because the per-weight learning rates are much smaller for the relational rules than the word-specific rules. This makes the relational rules converge much more slowly than they should.

The performance of some of the algorithms sometimes degrades with additional learning time. For some of the algorithms, such as PSCG, DN, and VP-PW on Cora, this is simply a symptom of overfitting. More careful tuning or a better prior could help correct this. But for other algorithms, such as SCG and VP on Cora, the later models perform worse on training data as well. For SCG, this seems to be the result of noisy inference and very ill-conditioned problems, which cause even a slight error in the step direction to potentially have a significant effect. Our lower bound on the improvement in log-likelihood prevents this in theory, but in practice a noisy gradient may still cause us to take bad steps. PSCG suffers much less from this effect, since the preconditioning makes the learning problem better behaved. For VP and CD, the most likely cause is learning rates that are too high. Our tuning experiments selected the learning rates that worked best after four hours on a smaller set of data. The increased amount of data in the test scenario increased the magnitude of the gradients, making these learning rates less stable than they were in the tuning scenario. This extreme sensitivity to learning rate makes learning good models with VP and CD much more difficult. We also experimented with the stochastic meta-descent algorithm [21], which automatically adjusts learning rates in each dimension, but found it to be too unstable for these domains.

In sum, the MLN weight learning methods we have introduced in this paper greatly outperform the voted perceptron. Given similar learning time, they learn much more accurate models; and, judging from the curves in Figure 1, running VP until it reaches the same accuracy as the better algorithms would take an extremely long time.

5   Conclusion

Weight learning for Markov logic networks can be extremely ill-conditioned, making simple gradient descent-style algorithms very slow to converge. In this paper we studied a number of more sophisticated alternatives, of which the best-performing one is preconditioned scaled conjugate gradient. This can be attributed to its effective use of second-order information. However, the simple heuristic of dividing the learning rate by the true clause counts for each weight can sometimes give very good results. Using one of these methods instead of gradient descent can yield a much better model in less time.

Acknowledgments. This research was funded by a Microsoft Research fellowship awarded to the first author, DARPA contract NBCH-D030010/02-000225, DARPA grant FA8750-05-2-0283, NSF grant IIS-0534881, and ONR grant N-00014-05-1-0313. The views and conclusions contained in this document are those of the authors and should not be interpreted as necessarily representing the official policies, either expressed or implied, of DARPA, NSF, ONR, or the United States Government.

References

1. Becker, S., Le Cun, Y.: Improving the convergence of back-propagation learning with second order methods. In: Proc. 1988 Connectionist Models Summer School, pp. 29–37. Morgan Kaufmann, San Francisco (1989)
2. Besag, J.: On the statistical analysis of dirty pictures. Journal of the Royal Statistical Society, Series B 48, 259–302 (1986)
3. Collins, M.: Discriminative training methods for hidden Markov models: Theory and experiments with perceptron algorithms. In: Proc. CEMNLP-2002 (2002)
4. Craven, M., Slattery, S.: Relational learning with statistical predicate invention: Better models for hypertext. Machine Learning 43(1/2), 97–119 (2001)
5. Fletcher, R.: Practical Methods of Optimization, 2nd edn. Wiley-Interscience, New York, NY (1987)
6. Getoor, L., Taskar, B.: Introduction to Statistical Relational Learning. MIT Press, Cambridge (2007)
7. Gilks, W.R., Richardson, S., Spiegelhalter, D.J. (eds.): Markov Chain Monte Carlo in Practice. Chapman and Hall, London, UK (1996)
8. Hinton, G.E.: Training products of experts by minimizing contrastive divergence. Neural Computation 14(8), 1771–1800 (2002)
9. Kautz, H., Selman, B., Jiang, Y.: A general stochastic approach to solving problems with hard and soft constraints. In: Du, D., Gu, J., Pardalos, P.M. (eds.) The Satisfiability Problem: Theory and Applications, pp. 573–586. American Mathematical Society, New York (1996)
10. Kok, S., Domingos, P.: Learning the structure of Markov logic networks. In: Proc. ICML-2005, pp. 441–448. ACM Press, New York (2005)
11. Kok, S., Singla, P., Richardson, M., Domingos, P.: The Alchemy system for statistical relational AI. Technical report, Department of Computer Science and Engineering, University of Washington, Seattle, WA (2005), http://alchemy.cs.washington.edu/
12. Møller, M.: A scaled conjugate gradient algorithm for fast supervised learning. Neural Networks 6, 525–533 (1993)
13. Nocedal, J., Wright, S.: Numerical Optimization. Springer, New York (2006)
14. Pearlmutter, B.: Fast exact multiplication by the Hessian. Neural Computation 6(1), 147–160 (1994)
15. Poon, H., Domingos, P.: Sound and efficient inference with probabilistic and deterministic dependencies. In: Proc. AAAI-2006, pp. 458–463. AAAI Press (2006)
16. Richardson, M., Domingos, P.: Markov logic networks. Machine Learning 62, 107–136 (2006)
17. Sha, F., Pereira, F.: Shallow parsing with conditional random fields. In: Proc. ACL-2003 (2003)
18. Shewchuck, J.: An introduction to the conjugate gradient method without the agonizing pain. Technical Report CMU-CS-94-125, School of Computer Science, Carnegie Mellon University (1994)
19. Singla, P., Domingos, P.: Discriminative training of Markov logic networks. In: Proc. AAAI-2005, pp. 868–873. AAAI Press (2005)
20. Singla, P., Domingos, P.: Entity resolution with Markov logic. In: Proc. ICDM-2006, pp. 572–582. IEEE Computer Society Press, Los Alamitos (2006)
21. Vishwanathan, S., Schraudolph, N., Schmidt, M., Murphy, K.: Accelerated training of conditional random fields with stochastic gradient methods. In: Proc. ICML-2006 (2006)

Classification in Very High Dimensional Problems with Handfuls of Examples

Mark Palatucci and Tom M. Mitchell

School of Computer Science
Carnegie Mellon University, Pittsburgh, Pennsylvania 15213, USA
{mpalatuc,tom.mitchell}@cs.cmu.edu

Abstract. Modern classification techniques perform well when the number of training examples exceed the number of features. If, however, the number of features greatly exceed the number of training examples, then these same techniques can fail. To address this problem, we present a hierarchical Bayesian framework that shares information between features by modeling similarities between their parameters. We believe this approach is applicable to many sparse, high dimensional problems and especially relevant to those with both spatial and temporal components. One such problem is fMRI time series, and we present a case study that shows how we can successfully classify in this domain with 80,000 original features and only 2 training examples per class.

1 Introduction

There are many interesting domains that have high dimensionality. Some examples include the stream of images produced from a video camera, the output of a sensor network with many nodes, or the time series of functional magnetic resonance images (fMRI) of the brain. Often we want use this high dimensional data as part of a classification task. For instance, we may want our sensor network to classify intruders from authorized personnel, or we may want to analyze a series of fMR images to determine the cognitive state of a human subject.

Unfortunately, for many of these high dimensional classification tasks, the number of available training examples is far fewer than the number of dimensions. Using regularization can certainly help, and classifiers like logistic regression with L_1 penalized weights have been shown to scale to many thousands of dimensions. There are other techniques like PCA, ICA, and manifold learning that explicitly try to reduce the data dimension. These methods, however, are unlikely to help when the amount of training data is only a few examples per class.

For many of these sparse, high dimensional problems the features are not truly independent. This is easy to imagine for time series data as features may not change much from one time point to the next. If we assumed that our data were temporally continuous, we could imagine smoothing each feature by other features nearby in time. This smoothing could remove noise and improve our estimate of the feature.

J.N. Kok et al. (Eds.): PKDD 2007, LNAI 4702, pp. 212–223, 2007.

Any assumption that we make *a priori* introduces inductive bias into our learning task. If the assumption is accurate then the bias will help the learning task when the number of training examples is very limited. Thus, to build a classifier that will perform well with small numbers of examples, we desire a way to incorporate any inductive biases (i.e. domain knowledge) we might have about the relationships between features.

We present such a classifier based on a hierarchical Bayesian model. Our model is both parametric and generative, and allows us to encode assumptions about the features *a priori*. We demonstrate this classifier on fMRI time series data and show that it scales tractably (even with 80,000 features). The classifier is robust to noise and extraneous features, and can classify with only 2 examples per class as compared to a standard Gaussian Naive Bayes classifier that fails completely on the same data.

1.1 Case Study: Cognitive State Classification Using Functional Magnetic Resonance Images

Recent work has shown that it is possible to classify cognitive states from fMRI data. For example, researchers have been able to determine the category of words that a person is reading (e.g. fruits, buildings, tools, etc.) [10] by analyzing fMR images of their neural activity. Others have shown that is is possible to classify between drug addicted persons and non-drug using controls [15]. One study even used fMRI data to classify whether participants were lying or telling the truth [4].

Classification in this domain is tricky. The data are very high dimensional and noisy, and training examples are sparse. A typical experiment takes a 3D volumetric image of the brain every second. Each image has roughly 5,000 voxels[1], each of which measures the neural activity at a specific location in the brain[2].

The experiments considered here are often divided into trials, with each lasting approximately 60 seconds. A trial is repeated several times within an experiment to collect multiple samples of the subject's neural response to some stimulus. A classifier may treat each voxel-timepoint as a feature, and each trial would be one example of that voxel-timepoint. Thus, an experiment with V voxels, T images per trial, and N trials would have $V * T$ features, with only N examples per feature. A typical experiment may have $V = 5,000$, $T = 60$, and $N = 20$, yielding 300,000 features with only 20 training examples per feature (per class). With this much data and such few examples, it is amazing that classification is even possible.

Why reducing the number of examples for fMRI experiments is important
Although others have shown classification methods that work for this domain, even these methods fail as we further reduce the number of training examples (to say 2-3 examples per class). Human subjects can get fatigued after long periods

---

[1] The total number of voxels depends on the fMR scanner and particular subject.
[2] fMRI technically measures blood oxygenation level which is believed to be correlated with underlying neural activity.

in the fMR scanner, and any movements they make reduce the usability of their data. Reducing the number of trials needed for an experiment would improve participant comfort, and would allow the testing of more varied stimuli within a given allotment of time.

1.2   Related Work

Hierarchical Bayesian methods have been used for quite some time within the statistics community for combining data from similar experiments. A good introduction is given in [6]. In general, these methods fall under "shrinkage" estimators. If we want to estimate several quantities that we believe are related, then in the absence of large sample sizes for the individual quantities, these methods shrink (smooth) the estimate toward some statistic of the combined quantities. For example, if we want to compute the mean for each of several random variables, we could shrink the sample mean for each variable toward the sample mean over all the variables. If the samples sizes are small and the variables related, this can provide a better estimate of the individual means.

Shrinkage estimators are very similar in spirit to multi-task learning algorithms within the machine learning community. With multi-task[3] or "lifelong" learning, the goal is to leverage "related" information from similar tasks to help the current learning task [14]. The overall goal in both these communities is to learn concepts with fewer data. A good example of using hierarchical Bayes in a multi-task learning application is [7]. There has also been some interesting theoretical work to explain why these methods are beneficial [2,1].

Hierarchical Bayesian methods have been applied successfully within the fMRI domain to the task of multiple subject classification. [13] demonstrates a hierarchical model that improves classification of a single human subject by combining data from multiple subjects within the same study. Our model, by contrast, focuses on sharing information between features of a single subject.

The most similar work to ours within the fMRI domain is [11,12]. This work demonstrates that sharing parameters between voxels can lead to more accurate models of the fMRI data. Our work by comparison, does not directly couple the parameters of shared features, but rather shares information through hyperparameters.

2   Models

2.1   Gaussian Naive Bayes

The hierarchical model we describe below is based on the Gaussian Naive Bayes (GNB) classifier. The classifier is popular for fMRI classification tasks because it scales to thousands of features, and is robust to noise and irrelevant features. The model is based on Bayes rule:

$$\mathbb{P}\left(Y|X\right) \propto \mathbb{P}\left(X|Y\right)\mathbb{P}\left(Y\right)$$

where $X \in \Re^J$ represents the example and $Y \in \{0,1\}$ is the class label. We treat each component X_j of the vector X as a feature in the classifier where

$1 \leq j \leq J$. The classifier makes the assumption that the X_j are independent given the class variable Y. We can then model the likelihood of the ith example for a feature j using a normal Gaussian:

$$X_{ij}|Y = c \sim N(\theta_j^{(c)}, \sigma_j^{2(c)}) \quad i = 1 \ldots N$$

where N is the total number of examples. The joint likelihood then becomes the product over all the features:

$$\mathbb{P}(X|Y = c) = \prod_{j=1}^{J} \mathbb{P}(X_j|Y = c)$$

The classification rule is therefore:

$$\text{Predicted Class} = \underset{c}{\operatorname{argmax}} \, \mathbb{P}(Y = c) \prod_{j=1}^{J} \mathbb{P}(X_j|Y = c)$$

$$= \underset{c}{\operatorname{argmax}} \, \mathbb{P}(Y = c) \prod_{j=1}^{J} N(\widehat{\theta}_j^{(c)}, \widehat{\sigma}_j^{2(c)}) \tag{1}$$

Here, $\widehat{\theta}_j^{(c)}$ and $\widehat{\sigma}_j^{2(c)}$ are just the sample mean and variance for each feature j and class c, and the prior $\mathbb{P}(Y = c)$ is given simply by the relative class frequencies in the training data. Here $\delta(\cdot)$ is the indicator function:

$$\widehat{\theta}_j^{(c)} = \frac{1}{\sum_{i=1}^{N} \delta(Y_i = c)} \sum_{i=1}^{N} \delta(Y_i = c) X_{ij}$$

$$\widehat{\sigma}_j^{2(c)} = \frac{1}{\sum_{i=1}^{N} \delta(Y_i = c)} \sum_{i=1}^{N} \delta(Y_i = c)(X_{ij} - \widehat{\theta}_j^{(c)})^2$$

$$\mathbb{P}(Y = c) = \frac{1}{N} \sum_{i=1}^{N} \delta(Y_i = c).$$

2.2   Standard Hierarchical Bayesian Model

If we believe that the individual θ_j are related by some distribution, then we can incorporate that belief using a hierarchical model. For example, if we thought that the θ_j were all drawn from a common normal distribution, then we could model that as:

$$X_{ij}|\theta_j \sim N(\theta_j, \sigma^2)$$
$$\theta_j \sim N(\mu, \tau^2)$$

Here μ and τ^2 are called *hyperparameters* for the model. (Note that for notational simplicity, we'll leave out mention of the class c until we return to the subject of classification.) Now in this hierarchical model, we want to know the posterior distribution of $\theta_j | \mu, \tau^2, X$. Intuitively, we want to know our best estimate of θ_j given not only the data, but also our prior belief about its distribution. If we assume for the moment the sampling variance is common across features, that is $\forall j, \sigma_j^2 = \sigma^2$, then we can obtain the MAP estimate of θ_j as:

$$\widehat{\theta}_j = \frac{\frac{N}{\sigma^2} \bar{X}_{\bullet j} + \frac{1}{\tau^2} \mu}{\frac{N}{\sigma^2} + \frac{1}{\tau^2}} \tag{2}$$

Equation (2) is surprisingly intuitive. It is just the weighted average of the sample mean $\bar{X}_{\bullet j} = \frac{1}{N} \sum_{i=1}^{N} X_{ij}$ and the hypermean μ, where N is the sample size. The weights are given by the inverse of the respective variances since σ^2 / N is the variance of the sample mean. If the number of samples N is large, then we see more weight being placed on the sample mean. Similarly, if the number of samples is small, the variance of the sample mean may be larger than that of the hypermean. More weight would be placed on the hypermean. The beauty of this estimator is that it automatically balances the estimate with the number of available samples. As N grows large, the weight on the hypermean grows smaller.

Of course there are a few difficulties that we must address. One problem is that usually we do not know the variance σ^2. This quantity must somehow be estimated from the data. Another problem is how to choose the hyperparameters μ and τ^2. We could perform a fully Bayesian approach and apply a non-informative hyperprior distribution to μ and τ^2. This would then require simulation to calculate the posterior for θ_j. Another, more tractable approach is to estimate the hyperparameters directly from the data. This technique is often called *empirical Bayes*[9] and uses point estimates for the hyperparameters:

$$\widehat{\mu} = \frac{1}{J} \sum_{j=1}^{J} \bar{X}_{\bullet j} \qquad \widehat{\tau}^2 = \frac{1}{J} \sum_{j=1}^{J} (\bar{X}_{\bullet j} - \widehat{\mu})^2$$

Here we are just taking the sample mean and variance for all the individual sample means. We use a similar empirical approach in the method we now describe.

2.3   Feature Sharing Empirical Bayesian Model

One problem with the standard hierarchical model is the assumption that all the parameters θ_j are drawn from the same distribution. To demonstrate this, consider two variables that are perfectly correlated while the parameters that characterize their distributions are wildly different. Assuming the parameters for these two variables are drawn from a common normal distribution would lead to poor estimates of the hyperparameters μ and τ^2 and subsequently the smoothed parameter θ_j. Nonetheless, the variables certainly contain information about each other that we want to leverage when making an estimate about either one.

We address this problem by allowing each θ_j to be drawn from a different distribution. We propose an approach that uses the parameters of other related variables, say θ_k and θ_i, to estimate the hyperparameters of the distribution for

θ_j. We define this formally as follows: assume we have two random variables, X and Y, parameterized by θ_X and θ_Y. Let $m_{X \to Y}(\theta_X)$ be a *parameter transformation* function that maps parameters of variable X to those of variable Y. Let $\mathcal{G}_j$ be the index set of all other variables that we believe contain information about variable j. Let $G_j = |\mathcal{G}_j|$ be the number of variables in that set.

We define a new smoothing estimator based on the normal model in Equation (2). Rather than assume all θ_j come from a common distribution, we assume that each θ_j has its own variance and hyperparameters μ_j and τ_j.

$$\widehat{\theta}_j = \frac{\frac{N}{\widehat{\sigma}_j^2}\bar{X}_{\bullet j} + \frac{1}{\widehat{\tau}_j^2}\widehat{\mu}_j}{\frac{N}{\widehat{\sigma}_j^2} + \frac{1}{\widehat{\tau}_j^2}} \tag{3}$$

These hyperparameters are calculated by point estimates of the transformed parameters of the variables in $\mathcal{G}_j$:

$$\widehat{\mu}_j = \frac{1}{G_j}\sum_{g=1}^{G_j} m_{g \to j}(\bar{X}_{\bullet g}) \tag{4}$$

$$\widehat{\tau}_j^2 = \frac{1}{G_j}\sum_{g=1}^{G_j}\left(m_{g \to j}(\bar{X}_{\bullet g}) - \widehat{\mu}_j\right)^2 \tag{5}$$

Intuitively, we first compute estimates of the variable j's parameters from the other variables, and use these to estimate the hyperparameter μ_j. We then smooth the sample mean using this hypermean as before.

Note that we still need estimates for the variances σ_j^2. Let $m'_{g \to i}(\cdot)$ be the parameter transformation function for the variance parameters. We could take the mean of these transformed parameters as before:

$$\widehat{\sigma}_j^2 = \frac{1}{G_j}\sum_{g=1}^{G_j} m'_{g \to j}(S_g^2) \tag{6}$$

where S_g^2 is sample variance for feature g. Empirically, we have found that pooling the variance parameters together $m'_{g \to i}(\sigma_g^2) = \sigma_g^2$ and taking the median (vs. mean) gives a estimator that is robust to extremely noisy variables:

$$\forall j, \widehat{\sigma}_j^2 = \text{median}\{S_1^2, S_2^2, \ldots, S_{G_j}^2\} \tag{7}$$

Given sets of sharing groups and parameter transformation functions, we can define a feature sharing classifier using the new estimators defined in Equations (3),(4),(5), and (7). The classifier is still based on the Gaussian Naive Bayes rule defined in Equation (1). Only now, for each class c we replace the estimate for $\widehat{\theta}_j$ with that from Equation (3) and $\widehat{\sigma}_j^2$ with either Equation (6) or Equation (7).

3   Case Study of Feature Sharing with fMRI Data

We now demonstrate this feature sharing model on a real fMRI classification task. We first show how to formulate the problem into the feature sharing framework described above, and then compare the feature sharing classifier against a standard Gaussian Naive Bayes classifier for the same task.

3.1 Notation

Since fMRI data are a time series we consider each voxel-timepoint as a feature. We index a particular example for a feature as X_{ivt} where i is the trial(example), v is the voxel, and t is the timepoint. The sample mean for a particular feature would then be $\bar{X}_{\bullet vt} = \frac{1}{N} \sum_{i=1}^{N} X_{ivt}$ (where N is the number of trials) and the sample variance would be $S_{vt}^2 = \frac{1}{N-1} \sum_{i=1}^{N} (X_{ivt} - \bar{X}_{\bullet vt})^2$.

3.2 Feature Sharing Empirical Bayesian Model for fMRI

There are two important questions we need to answer to formulate this problem into the feature sharing framework:

1. **Which of the features are related?** Specifically, for a feature j, what is the index set $\mathcal{G}_j$ of features that share information (parameters)?
2. **How are the features related?** Specifically, what are the *parameter transformation* functions $m_{k \to j}(\cdot)$ that map the parameters from feature k to feature j?

To answer these questions for the fMRI domain we consider a key observation made in [11]: *the time courses for neighboring voxels are often similar up to a scaling factor.* We can see this effect by observing several correlated neighborhoods (4-5 voxels) in Figure 1. We use this domain knowledge to define a feature sharing scheme for fMRI:

1. For feature j, let the index set of shared features $\mathcal{G}_j$ be the immediate spatial neighbors of a voxel. Since a voxel is indexed by integer {x,y,z} locations, there can be a maximum of 26 neighbors per voxel.
2. We define $m_{k \to j}(\cdot)$ to be the mean *parameter transformation* function from feature k to feature j. We define the function as a linear scaling factor $m_{k \to j}(\bar{X}_{\bullet k}) = \beta_{k \to j}\bar{X}_{\bullet k}$. We must remember, however, that each voxel-timepoint is a feature. To simplify, we'll assume that the parameter transformation function is the same for each pair of voxels, regardless of the time-point. Therefore, for voxels j and k at any time t we have $m_{kt \to jt}(\bar{X}_{\bullet kt}) = \beta_{k \to j}\bar{X}_{\bullet kt}$. We also define the variance parameter transformation $m'_{kt \to jt}(\cdot)$ to be the median pooling estimator described in Equation (7) [3].

We can solve for the $\beta_{k \to j}$ constants by assuming a linear regression model:

$$\bar{X}_{\bullet jt} = \beta_{k \to j}\bar{X}_{\bullet kt} + \epsilon$$

$$\widehat{\bar{X}}_{\bullet jt} = \widehat{\beta}_{k \to j}\bar{X}_{\bullet kt}$$

---

[3] We have found empirically that for the variance parameter it is advantageous to share over all the voxel-timepoints rather than just the immediate neighbors.

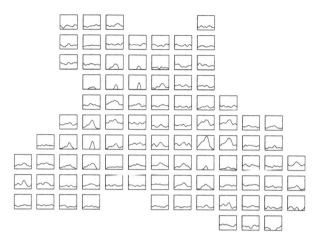

Fig. 1. Time series of the voxels in the visual cortex averaged over all trials. We see that several local neighborhoods (4-5 voxels) are similar but have different amplitudes.

This allows us to find estimates $\widehat{\beta}_{k \to j}$ using the usual method of least squares:

$$\widehat{\beta}_{k \to j} = \min_{\beta} \sum_{t=1}^{T} (\bar{X}_{\bullet jt} - \beta \bar{X}_{\bullet kt})^2$$

which is given by:

$$\widehat{\beta}_{k \to j} = \frac{\sum_{t=1}^{T} \bar{X}_{\bullet jt} \bar{X}_{\bullet kt}}{\sum_{t=1}^{T} \bar{X}_{\bullet kt}^2} \tag{8}$$

Now that we have our sharing groups and parameter transformation functions we can define a hierarchical model for fMRI:

$$X_{ivt}|\theta_{vt} \sim N(\theta_{vt}, \sigma^2)$$
$$\theta_{vt} \sim N(\mu_{vt}, \tau_{vt}^2)$$

Combining all these equations together, we can now define a feature sharing classifier for fMRI:

A Feature Sharing Classifier for fMRI:

   For each class c, compute:

1. $\widehat{\sigma}^{2(c)} = \text{median}(S_{11}^{2(c)}, \ldots, S_{1T}^{2(c)}, S_{21}^{2(c)}, \ldots, S_{2T}^{2(c)}, \ldots, S_{VT}^{2(c)})$

2. $\widehat{\beta}_{k \to j}^{(c)} = \dfrac{\sum_{t=1}^{T} \bar{X}_{\bullet jt}^{(c)} \bar{X}_{\bullet kt}^{(c)}}{\sum_{t=1}^{T} \bar{X}_{\bullet kt}^{2(c)}}$ For any pairs of voxels j, k that share features

$$3.\ \widehat{\mu}_{vt}^{(c)} = \frac{1}{G_{vt}} \sum_{j=1}^{G_{vt}} \widehat{\beta}_{j\to v}^{(c)} \bar{X}_{\bullet jt}^{(c)} \qquad \widehat{\tau}_{vt}^{2(c)} = \frac{1}{G_{vt}} \sum_{j=1}^{G_{vt}} (\widehat{\beta}_{j\to v}^{(c)} \bar{X}_{\bullet jt}^{(c)} - \widehat{\mu}_{vt}^{(c)})^2$$

$$4.\ \widehat{\theta}_{vt}^{(c)} = \frac{\frac{N^{(c)}}{\widehat{\sigma}^{2(c)}} \bar{X}_{\bullet vt}^{(c)} + \frac{1}{\widehat{\tau}_{vt}^{2(c)}} \widehat{\mu}_{vt}^{(c)}}{\frac{N^{(c)}}{\widehat{\sigma}^{2(c)}} + \frac{1}{\widehat{\tau}_{vt}^{2(c)}}}$$

The predicted class is then:

$$\operatorname*{argmax}_{c} \mathbb{P}(Y = c) \prod_{v=1}^{V} \prod_{t=1}^{T} N(\widehat{\theta}_{vt}^{(c)}, \widehat{\sigma}^{2(c)})$$

3.3   Experimental Results

Classification Task

We consider the task of classifying whether a subject in an fMRI experiment is "viewing a picture" or "reading a sentence". In this fMRI dataset[4], functional images of the brain were taken every 500ms (for 8 seconds). Each image recorded the neural activity at approximately 5,000 different locations (voxels) in the brain. We consider each *voxel-timepoint* as a feature, thus there were approximately $5,000 * 16 = 80,000$ features per trial. There were 20 "viewing a picture" trials and 20 "reading a sentence" trials. This experiment was repeated for 13 different human subjects.

Test Method

We performed the following testing method to estimate the error of the classifiers:

1. Split the dataset randomly in half. One half is used for training and one half is used for testing. We enforce an equal number of examples per class. Therefore, our training and test sets each have 20 examples total (10 per class).
2. Sample, at random, 2 examples per class from the training set. These are the training examples for this round.
3. Train on the sampled training examples in (2) and test on all examples in the test set.
4. Repeat 1-3 ten times and report the average error.

Discussion

In Figure 2, we show the results of the Feature Sharing classifier compared to a standard Gaussian Naive Bayes classifier for the 13 human subjects available in this study. In this experiment we used all available voxels in the brain ($\approx$ 5,000 per subject) yielding $\approx$80,000 features. Notice that there were *only two training examples per class*. The standard Gaussian Naive Bayes (GNB) classifier

---

[4] The dataset used is available at: http://www.cs.cmu.edu/afs/cs/project/theo-73/www/index.html

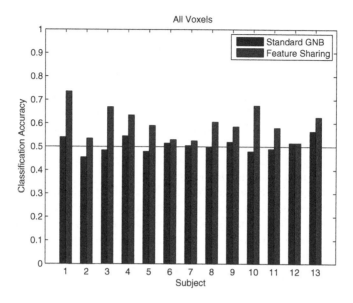

Fig. 2. Accuracies of the standard Gaussian Naive Bayes classifier and the Feature Sharing classifier for 13 human subjects with two training examples per class. The classifier uses all voxels in the brain. Since there are two classes, random accuracy is 0.5.

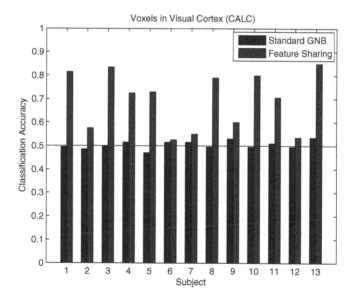

Fig. 3. Accuracies of the standard Gaussian Naive Bayes classifier and the Feature Sharing classifier for 13 human subjects with two training examples per class. The classifier uses only voxels in the Visual Cortex (CALC).

performed with near random accuracy for all subjects. The Feature Sharing classifier we described above shows considerable improvement, demonstrating that it is possible to classify even with an extremely small number of training examples.

In Figure 3, we show the results of the same experiment, except now we use only the voxels located in the visual cortex of the brain (≈ 300 per subject). These voxels are known to contain highly discriminating signal for this particular classification task. The interesting thing to note here is that the standard GNB classifier still fails with random accuracy on all subjects. The Feature Sharing classifier, however, is able to capitalize on the extra signal in these voxels, showing dramatic improvements for many subjects.

In the Feature Sharing classifier, we achieved the best results by sharing both the mean and variance parameters between features. We have found empirically, however, that sharing the variance parameter plays the larger role in improving overall classification accuracy. While this might seem surprising at first, some interesting theoretical work [5] shows that in the bias/variance decomposition under 0/1 loss, the variance dominates the bias. This may suggest why sharing the variance parameters caused the larger increase in performance.

4   Conclusion and Future Work

We have shown a feature sharing framework for classifying in very high dimensional problems with only a small number of training examples. This classifier is based on empirical Bayes and allows us to model relationships between features by assuming each class conditional parameter has its own hyperdistribution. The parameters for these hyperdistributions are estimated by sharing information between related groups of features.

We demonstrated this model on a fMRI classification task and showed how we can successfully classify in a problem with 80,000 spatially and temporally related features and only two training examples per class. We used domain knowledge of fMRI to specify feature sharing over local neighborhoods with a linear scaling factor.

An interesting future direction would be to automatically determine groups of features that share information rather than defining each group by the set of immediate neighbors. We could imagine learning a metric between features directly from the data, and then using that metric to define the parameter transformation functions.

Acknowledgements

We would like to thank Indra Rustandi and Francisco Pereira for useful discussions.

Mark Palatucci is supported by a NSF Graduate Research Fellowship and by a grant from the W.M. Keck Foundation.

References

1. Baxter, J.: A bayesian/information theoretic model of learning to learn via multiple task sampling. Machine Learning 28, 7–39 (1997)
2. Ben-David, S., Schuller, R.: Exploiting task relatedness for multiple task learning. In: Sixteenth Annual Conference on Learning Theory COLT (2003)
3. Caruana, R.: Multitask learning. Machine Learning 28(1), 41–75 (1997)
4. Davatzikos, C., et al.: Classifying spatial patterns of brain activity with machine learning methods: application to lie detection. Neuroimage 28(1), 663–668 (2005)
5. Friedman, J.H.: On bias, variance, 0/1 loss, and the curse-of-dimensionality. Data Mining and Knowledge Discovery 1(1), 55–77 (1997)
6. Gelman, A., Carlin, J., Stern, H., Rubin, D.: Bayesian Data Analysis, 2nd edn. Chapman and Hall/CRC Press, Boca Raton, NY (2003)
7. Heskes, T.: Solving a huge number of similar tasks: a combination of multi-task learning and a hierarchical bayesian approach. In: International Conference of Machine Learning ICML (1998)
8. Hutchinson, R.A., Mitchell, T.M., Rustandi, I.: Hidden process models. In: International Conference of Machine Learning ICML (2006)
9. Lee, P.M.: Bayesian Statistics, 3rd edn. Hodder Arnold, London, UK (2004)
10. Mitchell, T.M., Hutchinson, R., Niculescu, R.S., Pereira, F., Wang, X., Just, M., Newman, S.: Learning to decode cognitive states from brain images. Machine Learning 57(1-2), 145–175 (2004)
11. Niculescu, R.S.: Exploiting Parameter Domain Knowledge for Learning in Bayesian Networks. Carnegie Mellon Thesis: CMU-CS-05-147, Pittsburgh, PA (2005)
12. Niculescu, R.S., Mitchell, T.M.: Bayesian network learning with parameter constraints. Journal of Machine Learning Research 7, 1357–1383 (2006)
13. Rustandi, I.: Hierarchical gaussian naive bayes classifier for multiple-subject fmri data. In: NIPS Workshop: New Directions on Decoding Mental States from fMRI Data (2006)
14. Thrun, S.: Learning to learn: Introduction. In: Learning To Learn (1996)
15. Zhang, L., Samaras, D., Tomasi, D., Alia-Klein, N., Leskovjan, L.C.A., Volkow, N., Goldstein, R.: Exploiting temporal information in functional magnetic resonance imaging brain data. In: MICCAI Conference Proceedings, pp. 679–687 (2005)

Domain Adaptation of Conditional Probability Models Via Feature Subsetting

Sandeepkumar Satpal and Sunita Sarawagi*

IIT Bombay
sunita@iitb.ac.in

Abstract. The goal in domain adaptation is to train a model using labeled data sampled from a domain different from the target domain on which the model will be deployed. We exploit unlabeled data from the target domain to train a model that maximizes likelihood over the training sample while minimizing the distance between the training and target distribution. Our focus is conditional probability models used for predicting a label structure $\mathbf{y}$ given input $\mathbf{x}$ based on features defined jointly over $\mathbf{x}$ and $\mathbf{y}$. We propose practical measures of divergence between the two domains based on which we penalize features with large divergence, while improving the effectiveness of other less deviant correlated features. Empirical evaluation on several real-life information extraction tasks using Conditional Random Fields (CRFs) show that our method of domain adaptation leads to significant reduction in error.

1   Introduction

Most statistical learning techniques are based on the assumption that the training data is representative of the distribution on which the trained model is deployed. This assumption gets routinely broken in applications like information extraction, speech recognition, text classification, and opinion mining that are being increasingly used at large scales. In most such applications, an offline phase is used to collect carefully labeled data for training. However, the settings during deployment could be highly varied with little or no labeled data for that setting. For example, it is easy to find plenty of labeled data for named entity recognition in news articles but our goal might be to recognize person names from blogs. It is not easy to find labeled data for blogs but there is no dearth of unlabeled data.

Our goal in domain adaptation is to use labeled data from some domain to train a model that maximizes accuracy in a target domain for which we only have unlabeled data available. We concentrate on adapting structured learning tasks that model the conditional probability of a predicted structure $\mathbf{y}$ given input $\mathbf{x}$ as a linear exponential function of features defined over $\mathbf{x}$ and $\mathbf{y}$. A logistic classifier is a special case of such models where the predicted structure is a single discrete class label. Such conditional models allow users the flexibility of defining features without bothering about whether they are correlated or not.

---

* Contact author.

J.N. Kok et al. (Eds.): PKDD 2007, LNAI 4702, pp. 224–235, 2007.

Therefore, most real-life applications of these models involve a large number of features, contributing in varying strengths to the prediction task. With overfitting avoided using a suitable regularizer, these models provide state-of-the-art accuracy values in settings where features behave the same way in the training and target domain [1,2,3]. However, we observed that such models are rather brittle in that they perform very poorly on target data with even a small subset of features distorted in spite of other highly correlated features remaining intact.

We show how to detect features with large divergence in the two domains and penalize the more distorted features so that other less deviant correlated features start exerting a larger influence. A challenge is designing a reliable measure of divergence given only unlabeled data from the target domain whereas our features are defined over function of both labels $\mathbf{y}$ and input $\mathbf{x}$. We propose a measure of distortion as a function of the difference in expectation over the target samples and the trained conditional model. We formulate this as an optimization problem and present efficient algorithms for solving it. On seven real-life datasets, we show that our domain adapted classifier provides much higher accuracy than an unadapted model.

The rest of the paper is organized as follows. We discuss related work in Section 2. We describe our basic learning model in Section 3 and present our approach to domain adaptation in Section 4. We report results of an empirical evaluation of our model in Section 5.

2   Related Work

Transfer learning: In transfer learning [4,5,6,7] the goal is to use available training data from a related domain, along with training data from the target domain, to train the target classifier. A popular technique is to use the classifier in the related domain to define a prior [4,6,7] for the classifier trained using the in-domain data. For example, [7] proposes to first create a classifier using training data from the related domain. The output parameters are used as the mean of a Gaussian prior for the second classifier trained using labeled data of the target domain. A different type of prior is defined in [8] where the prior is used to give more importance to features that are useful across domains. Another interesting approach is based on replicating features so that shared features exploit labeled data from both domains whereas domain-specific features are trained only using in-domain data [5]. Our goal is different in that we do not have any labeled data from the target domain. Transfer learning is supervised domain adaptation whereas we are interested in unsupervised domain adaptation.

Structural correspondence learning: A recent proposal [9,10] for unsupervised domain adaptation is to define new features that capture the correspondence between features in the two domains. The new features are weights of "mini" classifiers that predict value of user-chosen anchor features that remain invariant across the domains. Successful domain adaptation will require both addition and deletion of features. Deletion is required for features that are missing or severely distorted, whereas when features are substituted, for example, the inter-author separator is changed from "comma" to a "new line", addition of features that capture their correspondence is more useful. Given that most structured learning tasks involve many correlated features, careful feature subsetting could lead to significant accuracy gains, as we show in this paper.

Robust learning: A different approach to handling features that are distorted in the test data is to learn classifiers that are robust to limited amounts of distortion. For example, [11] shows how to create SVM classifiers that provide good worst case performance with the deletion of any subset of features of size no more than k. In robust learning a model is trained once unlike in the case of domain adaptation where the model is *retrained* to adapt to any systematic difference between the two domains.

Correcting sample selection bias: In some cases, the training distribution fails to be representative of the test distribution because of a selection bias in the training instances, for example due to active learning. A popular strategy to correct for the bias [12,13] is to weight training examples differentially. Such methods are not likely to be useful for domain adaptation because all instances from the train domain could have very small probability in the target domain and the real issue is that of choosing the right representation through feature reweighting rather than instance reweighting.

In summary, the problem of unsupervised domain adaptation is related to, but distinct, from many problems in machine learning. To the best of our knowledge, domain adaptation via feature subsetting has not been addressed before in the literature.

3   Background

3.1   The Basic Learning Model

We consider conditional models of structure learning where the goal is to predict a label $\mathbf{y}$ from a structured space $\mathcal{Y}$ given an input $\mathbf{x}$. We assume a feature vector representation $\mathcal{F} : (\mathbf{x}, \mathbf{y}) \mapsto \mathcal{R}^K$ that maps any $(\mathbf{x}, \mathbf{y})$ pair to a vector of K reals. The conditional probability model is a log-linear function over these features. Thus, $\Pr(\mathbf{y}|\mathbf{x})$ is this Gibbs distribution

$$\Pr(\mathbf{y}|\mathbf{x}, \mathbf{w}) = \frac{1}{z_{\mathbf{w}}(\mathbf{x})} \exp \mathbf{w} \cdot \mathbf{f}(\mathbf{x}, \mathbf{y}) \tag{1}$$

where $\mathbf{w}$ is the parameter vector of the model where the kth component w_k is called the weight of feature f_k. The term $z_{\mathbf{w}}(\mathbf{x}) = \sum_{\mathbf{y}'} \exp \mathbf{w} \cdot \mathbf{f}(\mathbf{x}, \mathbf{y}')$ is a normalizing constant.

In practice, each feature $f_k(\mathbf{x}, \mathbf{y})$ is defined as a sum of local features that apply over smaller subsets of variables. When the features decompose over cliques of an undirected graph on labels $\mathbf{y}$, we get Conditional Random Fields [1]. This decomposition is exploited for efficient inference over the space of variables $\mathbf{y}$. For example, in information extraction, the underlying graph is a linear chain where features decompose over pairs of adjacent variables.

During training the goal is to maximize log-likelihood over a given training set $D = \{(\mathbf{x}_\ell, \mathbf{y}_\ell)\}_{\ell=1}^N$ expressed as

$$L(\mathbf{w}) = \sum_\ell \log \Pr(\mathbf{y}_\ell | \mathbf{x}_\ell, \mathbf{w}) = \sum_\ell (\mathbf{w} \cdot \mathbf{f}(\mathbf{x}_\ell, \mathbf{y}_\ell) - \log z_\mathbf{w}(\mathbf{x}_\ell)) \qquad (2)$$

We wish to find a $\mathbf{w}$ that maximizes $L(\mathbf{w})$. In practice, the norm of $\mathbf{w}$ is not allowed to grow too large to avoid overfitting. This is achieved by subtracting a regularization term $R(\mathbf{w}) = \frac{||\mathbf{w}||^\gamma}{\sigma^2}$ with $\gamma = 1$ or 2 and a user-provided variance σ^2. The resultant objective is convex, and can thus be maximized by gradient ascent, or one of many related methods.

During deployment, given an input $\mathbf{x}$, we predict a $\mathbf{y}$ for which $\Pr(\mathbf{y}|\mathbf{x})$ is maximum. The justification for this step is that the test data follows the same distribution as the training data, using which we learnt a $\mathbf{w}$ so as to maximize the probability of the correct prediction.

3.2   Train and Target Data Distributions

In domain adaptation we need to deploy a model in a domain where the distribution of $(\mathbf{x}, \mathbf{y})$ is different from the distribution from which the training data was obtained. Let $\mathcal{D}$ denote the distribution from which the training sample D was taken. Let $\mathcal{D}'$ denote the target distribution on which we wish to deploy the model. We do not have any labeled data from $\mathcal{D}'$, instead we have lots of unlabeled data D'. Let $D' = \{(\mathbf{x}_\ell)\}_{\ell=1}^{N'}$.

In domain adaptation our goal is to use both the labeled samples D from $\mathcal{D}$ and the unlabeled samples D' from distribution $\mathcal{D}'$ to train a model that maximizes accuracy on $\mathcal{D}'$. The accuracy in the $\mathcal{D}$ distribution is of no interest to us. Therefore the normal goal during CRF training of maximizing likelihood of D is not justified anymore because D is not representative of the distribution on which the model will be deployed. This is also what makes the problem different from semi-supervised learning where the labeled and unlabeled data come from the same distribution.

4   Domain Adaptation

Our approach to domain adaptation is to choose a representation where the training and test distributions are close, and once that is achieved we can justify training a model to maximize accuracy on the labeled training domain. Our starting representation is the user provided feature vector $\mathbf{f}(\mathbf{x}, \mathbf{y})$. During domain adaptation we select the subset S of features such that the distance between the train and target distributions is small in the projected space while maximizing likelihood on the training data. Our ideal objective of maximizing likelihood of the target distribution $\mathcal{D}$ for which we have no labeled samples

$$\text{argmax}_{\mathbf{w}} \sum_{(\mathbf{x},\mathbf{y})\in\mathcal{D}'} \sum_{k} w_k f_k(\mathbf{x},\mathbf{y}) - \log z_{\mathbf{w}}(\mathbf{x}) \qquad (3)$$

is replaced with the achievable objective

$$\text{argmax}_{\mathbf{w},S} \sum_{(\mathbf{x},\mathbf{y})\in D} \sum_{k\in S} w_k f_k(\mathbf{x},\mathbf{y}) - \log z_{\mathbf{w}}(\mathbf{x}) \qquad (4)$$

$$\text{such that } \text{dist}(\mathcal{D},\mathcal{D}'|S,D,D') \le \epsilon.$$

where $\text{dist}(\mathcal{D},\mathcal{D}'|S,D,D')$ is a suitable measure of distance between the two domains in a representation corresponding to the features in set S and as estimated from the labeled samples D from $\mathcal{D}$ and unlabeled samples D' from $\mathcal{D}'$.

4.1   Distance Function

We next discuss how to measure the distance between the two distributions. A direct approach is to first estimate their full $(\mathbf{x},\mathbf{y})$ distributions using sample data and then measure the distance between the two distributions using some function like KL distance. This is often difficult and requires a lot of training data. One of the main reasons for the success of the conditional approach for structured learning tasks is that they do not require the modeling of the distribution over $\mathbf{x}$.

Recently, [13] proposed to correct for sample selection bias in the training data by reducing the difference in the mean of the $\mathbf{x}$ features in the training and target distribution. There are several reasons why this method will not work well in our setting. First, in structured learning settings, the feature representation is in terms of both $\mathbf{x}$ and $\mathbf{y}$. Even if, we consider the scalar classification problem where we simplify the feature representation to be a cross product of features defined over $\mathbf{x}$ and labels $\mathbf{y}$, we can obtain more accurate distance measures by comparing the $\mathbf{x}$ means of each $\mathbf{y}$ separately rather than collapsing them on single means. Also, the method proposed in [13] assumes that $\Pr(\mathbf{y}|\mathbf{x})$ is the same in the training and test distribution. In our case, we assume that there exist some representation under which the two distributions are the same, but this is not true for all representations. In particular, this is not true for the starting representation used during normal training.

We propose to compare the two distributions by comparing component-wise the means of the features in their $(\mathbf{x},\mathbf{y})$ space. Let E_D^k and $E_{D'}^k$ denote the expected value of the kth feature under distributions $\mathcal{D}$ and $\mathcal{D}'$ respectively. For the training distribution, we estimate it empirically from the sample D as $E_D^k = \sum_{(\mathbf{x}_\ell,\mathbf{y}_\ell)\in D} \frac{f_k(\mathbf{x}_\ell,\mathbf{y}_\ell)}{N}$. For the target distribution $\mathcal{D}'$ since in the sample D' we have only $\mathbf{x}$ values, we use the expected value of the feature as calculated under the $\Pr(\mathbf{y}|\mathbf{x},\mathbf{w})$ distribution. Thus,

$$E_{D'}^k = \frac{1}{N'} \sum_{\mathbf{x}_\ell\in D'} \sum_{\mathbf{y}} f_k(\mathbf{x}_\ell,\mathbf{y}) \Pr(\mathbf{y}|\mathbf{x}_\ell,\mathbf{w}) \qquad (5)$$

Using $\mathbf{E}_D$ and $\mathbf{E}_{D'}$, we replace $\mathrm{dist}(\mathcal{D}, \mathcal{D}'|S, D, D')$ with the distance between the above sample means as $\sum_{k \in S} d(E_D^k, E_{D'}^k)$. The precise form of the distance function will depend on the nature of the specific features. For example, for sparse binary features, it is useful to interpret the mean values as probability of occurrence of a binomial distribution. In such cases, distance measures like cross-entropy and the log-odds ratio seem meaningful [14]. When the features are arbitrary real values, a $L1$ or square distance would be more appropriate.

4.2   Overall Objective

In terms of the new distance function, we can rewrite the objective as

$$\mathrm{argmax}_{\mathbf{w}, S} \sum_{(\mathbf{x}, \mathbf{y}) \in D} \sum_{k \in S} w_k f_k(\mathbf{x}, \mathbf{y}) - \log z_{\mathbf{w}}(\mathbf{x})$$
$$\text{such that } \sum_{k \in S} d(E_D^k, E_{D'}^k) \leq \epsilon. \tag{6}$$

The above objective presents a difficult combinatorial optimization problem over the exponentially many subsets of features. We convert the discrete feature selection problem to a soft selection problem by rewriting the constraint $\sum_{k \in S} d(E_D^k, E_{D'}^k) \leq \epsilon$ as $\sum_{k=1}^{K} |w_k|^\gamma d(E_D^k, E_{D'}^k) \leq \epsilon'$. Also, using the Lagrange dual formulation, we push the constraints into the objective and get the equivalent objective for an appropriate value of λ as

$$\mathrm{argmax}_{\mathbf{w}} \sum_{(\mathbf{x}, \mathbf{y}) \in D} \sum_k w_k f_k(\mathbf{x}, \mathbf{y}) - \log z_{\mathbf{w}}(\mathbf{x}) - \lambda \sum_k |w_k|^\gamma d(E_D^k, E_{D'}^k) \tag{7}$$

The above formulation has several intuitive interpretations. We can treat this as a standard accuracy-regularized training method with the only difference that the w_k are weighted in proportional to the distance between the training and target distribution along the k-th feature component. A feature with a large distance should get a smaller weight. Another interpretation is in terms of prior distributions over the parameters where the variance is not constant over all features, as is normally the case, but is inversely proportional to the divergence of the feature over the two distributions. When γ is 1 the prior is a Laplace distribution and when $\gamma = 2$ the prior is a Gaussian distribution with variance of the kth parameter as $\frac{1}{d(E_D^k, E_{D'}^k)}$. So when the distance is large, the parameter is likely to stay close to its mean value of 0.

4.3   Training Algorithm

We now discuss how we solve the optimization problem in Equation 7. For concreteness, we assume that $\gamma = 2$ and the distance function is the square distance defined as $d(E_D^k, E_{D'}^k) = (E_D^k - E_{D'}^k)^2$. The final objective then becomes.

$$L(\mathbf{w}) = \mathrm{argmax}_{\mathbf{w}} \sum_{(\mathbf{x}, \mathbf{y}) \in D} \left(\sum_k w_k f_k(\mathbf{x}, \mathbf{y}) - \log z_{\mathbf{w}}(\mathbf{x}) \right) - \lambda \sum_k w_k^2 (E_D^k - E_{D', \mathbf{w}}^k)^2$$

where $E_{D',\mathbf{w}}^k = \frac{1}{|D'|} \sum_{\mathbf{x}_i \in D'} \sum_{\mathbf{y}} f_k(\mathbf{x}_i, \mathbf{y}) \frac{\exp \mathbf{wf}(\mathbf{x}_i, \mathbf{y})}{z_{\mathbf{w}}(\mathbf{x}_i)}$. The above is a smooth differentiable function of $\mathbf{w}$. We can use standard gradient descent approaches to solve it. The gradient with respect to the kth parameter is

$$\frac{\partial L}{\partial w_k} = \sum_{(\mathbf{x},\mathbf{y}) \in D} f_k(\mathbf{x}, \mathbf{y}) - N E_{D,\mathbf{w}}^k - 2\lambda(w_k(E_D^k - E_{D',\mathbf{w}}^k)^2 - \sum_j w_j^2 (E_D^j - E_{D',\mathbf{w}}^j)\frac{\partial E_{D',\mathbf{w}}^j}{\partial w_k})$$

where

$$\frac{\partial E_{D',\mathbf{w}}^j}{\partial w_k} = \frac{1}{N'} \sum_{\mathbf{x}_i \in D'} \sum_{\mathbf{y}} f_j(\mathbf{x}_i, \mathbf{y}) \frac{\exp \mathbf{wf}(\mathbf{x}_i, \mathbf{y})}{z_{\mathbf{w}}(\mathbf{x}_i)}(f_k(\mathbf{x}_i, \mathbf{y}) - \sum_{\mathbf{y}'} f_k(\mathbf{x}_i, \mathbf{y}') \frac{\exp \mathbf{wf}(\mathbf{x}_i, \mathbf{y}')}{z_{\mathbf{w}}(\mathbf{x}_i)})$$

$$= \frac{1}{N'} \sum_{\mathbf{x}_i \in D'} \sum_{\mathbf{y}} f_j(\mathbf{x}_i, \mathbf{y}) \Pr(\mathbf{y}|\mathbf{x}_i)(f_k(\mathbf{x}_i, \mathbf{y}) - \sum_{\mathbf{y}'} f_k(\mathbf{x}_i, \mathbf{y}') \Pr(\mathbf{y}'|\mathbf{x}_i))$$

$$= (E_{D',\mathbf{w}}^{jk} - E_{D',\mathbf{w}}^j E_{D',\mathbf{w}}^k)$$

where $E_{D'}^{jk}$ is the expectation of the product of features j and k with respect to the empirical $\mathbf{x}$ distribution from D' and $\Pr(\mathbf{y}|\mathbf{w}, \mathbf{x})$. With respect to these distributions, the term $(E_{D',\mathbf{w}}^{jk} - E_{D',\mathbf{w}}^j E_{D',\mathbf{w}}^k)$ represents the covariance between features j and k. As in normal CRF training [1], we have to exploit the decomposability of the label space to evaluate these terms tractably.

There are two problem with the above objective.

1. The function is not convex, unlike the normal CRF objective with constant weighting of the regularizers.
2. The gradient is expensive to compute since the covariance terms are quadratic in the number of features. In typical structured learning tasks, for example in information extraction, the number of features tend to be very large.

We address both these issues by following a nested iterative approach to training. In each iteration, we fix feature distances with respect to the current values of the parameters and find the optimum value of the parameters treating the distance values as constant. This makes the inner optimization problem convex and linear in the number of features. We found that in practice with two or three iterations we get most of the benefit of complete training at significantly reduced cost.

5   Experiments

We evaluate the effectiveness of our proposed method on seven domain adaptation tasks constructed from the following four entity extraction benchmarks.

CoNLL 2003 dataset. The ConLL 2003 dataset[1] is a well-known benchmark for Named Entity Recognition where the goal is to extract entities like persons, organizations, and locations from news articles.

---

[1] http://cnts.uia.ac.be/conll2003/ner/

Cora citations. Cora citations [3] consists of citations collected from the reference section of several academic papers. The extraction task is to find author names, titles, venue, and year.

Cora headers. Cora headers [3] consists of headers of research papers covering fields like the title, author names, affiliations, and abstract of a paper. Even though headers and citations come from the same repository, the way authors and titles appear in paper headers is very different from the way they appear in paper citations, making it interesting for domain adaptation.

Citeseer citations. This dataset consists of journal articles we collected from Citeseer and therefore formatted slightly differently from the Cora dataset. Also, unlike Cora it consists only of journal entries. The dataset is available at http://www.it.iitb.ac.in/~sunita/data/personalBib.tar.gz.

Table 1. Description of domain adaptation tasks used in our experiments

Task	Train domain	Target domain	Label	Train		Target	
				#train	#test	#train	#test
Cite_Cora	Citeseer citations	Cora citations	Author	35	62	205	294
Cora_Cite	Cora citations	Citeseer citations	Author	155	294	39	62
Title_Caps	Citeseer citations	All-Caps	Title	35	62	39	62
Author_Caps	Citeseer citations	All-Caps	Author	35	62	39	62
Cite_Conll	Citeseer citations	CoNLL	Person	35	62	808	1191
Conll_Cite	CoNLL	Citeseer citations	Person	304	1191	39	62
Hdr_Cite	Cora headers	Citeseer citations	Title	45	87	39	62

In Table 1 we provide details of seven domain adaptation tasks created using various combination of these four datasets as the train and target domains and the extracted label. In tasks Title_Caps and Author_Caps the target domain differs from the train domain only in one respect: all words are fully capitalized in the target domain whereas in the train domain they are normal text records with a mix and capital and small letters. The last four columns specify for each of the two domains, the number of records used during training and testing respectively. For the target domain, the training documents are unlabeled.

We used a sequential CRF [1,2] with L2 regularization as our baseline model for information extraction. The package that we used is downloadable from [15]. We used the BCEU encoding of the entities where an entity like person name is decomposed into four labels: Begin-person, Continue-person, End-person, and Unique-person. Each token contributed two types of features: (1) the token itself if it was encountered in the training set and, (2) the set of regular expressions like digit or not, capitalized or not that the token matches. For each label i, these features where fired for the ith word and two words to the left and right of the word.

We evaluated our methods using F1 accuracy[2] at the level of individual tokens. We do not report span-level accuracy because the lack of standardization in what defines the boundaries of an entity, makes it difficult to get useful cross-domain comparison at the span-level. For example, in Citeseer the last punctuation ("."") is outside the title entity whereas in Cora it is inside. In each experiment performance was averaged over four runs obtained by varying the subset of instances used for training and testing. Unless otherwise stated, our default method of domain adaptation uses $\gamma = 1$, $\lambda = 1$ and the square log-odd distance function $(\log E_D^k - \log E_{D'}^k)^2$. This distance function has been shown to work well [14] for sparse indicator features commonly found in information extraction tasks. We used the ϵ-approximation trick proposed in [16] for handling the discontinuity of the objective when $\gamma = 1$.

5.1   Overall Improvement with Domain Adaptation

In Table 2 we show the accuracy of the original unadapted model and the adapted model trained using our method respectively called "Original" and "Adapted". Along with the accuracy on the target domain, for comparison we also show accuracy on the train domain. In all cases, we find that the accuracy of the target domain improves with domain adaptation. In some cases, the accuracy improvement is very dramatic, for example increasing from 26% to 69% on the second task.

Table 2. F1 Accuracy before and after domain adaptation

Dataset-Name	Train domain		Target domain	
	Original	Adapted	Original	Adapted
Cite_Cora	97.4	95.9	30.7	62.7
Cora_Cite	98.2	97.6	26.0	68.6
Title_Caps	94.4	93.2	41.8	90.1
Author_Caps	97.4	94.3	85.8	94.0
Cite_Conll	97.4	95.8	40.1	45.0
Conll_Cite	90.5	85.8	40.9	41.9
Hdr_Cite	85.3	76.0	12.0	27.8

For Title_Caps and Author_Caps where the target domain is just a fully capitalized version of the train domain, we find that the unadapted model performs very poorly whereas with adaptation we get accuracy comparable to the accuracy on the train domain. This illustrates the importance of adaptation even in domains that differ only slightly from the training domain. The top few features of the original model whose weight reduces almost to zero in the adapted model are: IsInitCapital, IsInitCapital.left-2, IsInitCapital.right+2, W_Extract, IsAllSmallCase, IsAllSmallCase.left-2, IsAllSmallCase. right+2. Most of these are case related features which have no importance in

---

[2] F1 is defined as 2*precision*recall/(precision+recall.)

the target domain. In contrast, the top few features whose weight increases significantly are `Punctuation`, `Punctuation.left-1`, `Punctuation.right+1`, `W_ACM.right+2`. These features remain invariant in the two domains since they are related to punctuation or fully capitalized words.

Another interesting observation from these tables is that on the train domain while the accuracy does drop after adapting to a different domain, the drop is only slight. This shows that in most cases, the model has other redundant features that start playing a role when some subset of its features are penalized.

5.2   Comparison with Other Methods

In Table 3 we compare our default method of domain adaptation to a number of other alternatives.

We compare with the recently proposed structural correspondence learning (SCL) [9] (described in Section 2). We find that SCL also shows significant accuracy improvements beyond the original unadapted model but the gain is lower than our method in all except the last dataset. Since our method of feature deletion is orthogonal to the SCL approach of feature addition, we also report results with both methods combined in the "SCL+Our" column of Table 3. In most cases, the combined method is better than either of the two.

We also compare our method to semi-supervised learning (SSL) proposed in [17] which adds to the training objective an additional goal of minimizing entropy labels for the unlabeled documents. In column SSL of Table 3 we show the results for the weight settings for which we obtained highest accuracy. Quite predictably, SSL is not competitive as a method of domain adaptation. We show

Table 3. Comparison of our method of domain adaptation with alternatives

Task	Original	Adapted	SCL	SCL+Our	SSL	x-dist	$\gamma = 2$	Square-dist
Cite_Cora	30.7	62.7	47.3	63.3	31.5	27.6	63.6	32.8
Cora_Cite	26.0	68.6	68.6	67.8	26.0	76.2	75.9	46.0
Title_Caps	41.8	90.1	80.1	90.9	46.8	90.3	77.3	46.0
Author_Caps	85.8	94.0	87.1	94.7	86.4	94.3	94.2	86.4
Cite_Conll	40.1	45.0	52.1	45.1	40.4	40.9	45.7	32.2
Conll_Cite	40.9	41.9	43.9	41.1	43.0	36.8	43.6	44.1
Hdr_Cite	12.0	27.8	57.9	38.9	19.7	24.3	23.7	18.5

the importance of comparing the mean of features in the joint $(\mathbf{x}, \mathbf{y})$ space instead of means along the projected $\mathbf{x}$ space as proposed in [13]. The latter is simpler to optimize because the distance function is independent of $\mathbf{w}$ and we get a simple convex objective. The results shown in column $\mathbf{x}$-dist of Table 3 indicate that in almost all cases the performance of the $\mathbf{x}$-only distance function is significantly worse than our method.

We vary our choice of γ from 1 to 2, that is using weighted L2 regularizer instead of L1 in column $\gamma = 2$ of Table 3. We find that our default of L1 distance performs much better than L2. This observation agrees with earlier reports on the efficacy of feature selection using L1 instead of L2 regularizers. Next, we vary our default choice of the distance function. We chose log-odds ratio because it has been found to perform better on sparse Bernoulli features. Instead, if we use a regular square distance between the expected values of features, we find that the accuracy is much worse as shown in the column marked Square-dist.

5.3 Effect of Training Data

Another interesting aspect of domain adaptation is the performance of the adapted model with increasing training data. In Figure 1 we show the accuracy of the adapted model on the target domain and the unadapted model on the train domain with increasing labeled training data. The y axis is the change in error compared to the error with 10% training data. As expected with statistical learners, with increasing training data, the error within the domain decreases. In contrast, the error of the adapted model either stays almost the same or increases slightly with more out-of-domain training data.

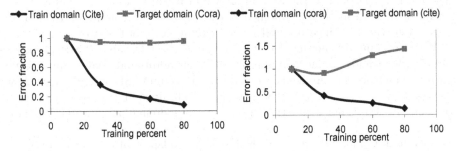

Fig. 1. Effect of increasing labeled training data on train and target domains for tasks Cite_Cora (left) and Cora_Cite (right)

6   Conclusion

In this paper we proposed a new method of unsupervised domain adaptation that selects a subset of features for which the distance between the train and target distribution is minimized while maximizing likelihood of the labeled data. The main challenge in this task is estimating distribution distance in the $(\mathbf{x}, \mathbf{y})$ space in which the model features are defined given only unlabeled samples from the target domain. We defined a distance measure and a method for solving the combined optimization problem that is both efficient and leads to significant accuracy improvements. In future, we would like to develop a theoretical analysis of this algorithm.

Acknowledgments. The work reported here was supported by grants from Microsoft Research and an IBM Faculty award.

References

1. Lafferty, J., McCallum, A., Pereira, F.: Conditional random fields: Probabilistic models for segmenting and labeling sequence data. In: Proceedings of the International Conference on Machine Learning (ICML-2001), Williams, MA (2001)
2. Sha, F., Pereira, F.: Shallow parsing with conditional random fields. In: Proceedings of HLT-NAACL (2003)
3. Peng, F., McCallum, A.: Accurate information extraction from research papers using conditional random fields. In: HLT-NAACL, pp. 329–336 (2004)
4. Li, X., Bilmes, J.: A Bayesian Divergence Prior for Classifier Adaptation. In: Eleventh International Conference on Artificial Intelligence and Statistics (AISTATS-2007) (2007)
5. Daumé, III H.: Frustratingly easy domain adaptation. In: Conference of the Association for Computational Linguistics (ACL), Prague, Czech Republic (2007)
6. Ando, R., Zhang, T.: A framework for learning predictive structures from multiple tasks and unlabeled data. Journal of Machine Learning Research 6, 1817–1853 (2005)
7. Chelba, A.: Adaptation of maximum entropy capitalizer: Little data can help a lot. In: EMNLP (2004)
8. Jiang, J., Zhai, C.: Exploiting domain structure for named entity recognition. In: HLT-NAACL, pp. 74–81 (2006)
9. Blitzer, J., McDonald, R., Pereira, F.: Domain Adaptation with Structural Correspondence Learning. In: Proceedings of the Empirical Methods in Natural Language Processing (EMNLP) (2006)
10. Ben-David, S., Blitzer, J., Crammer, K., Pereira, F.: Analysis of representations for domain adaptation. In: Advances in Neural Information Processing Systems 20, MIT Press, Cambridge, MA (2007)
11. Globerson, A., Rowels, S.: Nightmare at test time: robust learning by feature deletion. In: ICML, pp. 353–360 (2006)
12. Zadrozny, B.: Learning and evaluating classifiers under sample selection bias. In: ACM International Conference Proceeding Series, ACM Press, New York (2004)
13. Huang, J., Smola, A., Gretton, A., Borgwardt, K., Schölkopf, B.: Correcting Sample Selection Bias by Unlabeled Data. In: Advances in Neural Information Processing Systems 20, MIT Press, Cambridge, MA (2007)
14. Mladenic, D., Grobelnik, M.: Feature selection for unbalanced class distribution and naive bayes. In: ICML '99: Proceedings of the Sixteenth International Conference on Machine Learning, pp. 258–267 (1999)
15. Sarawagi, S.: The crf project: a java implementation (2004), http://crf.sourceforge.net
16. Lee, S.I., Lee, H., Abbeel, P., Ng, A.Y.: Efficient l1 regularized logistic regression. In: AAAI (2006)
17. Jiao, F., Wang, S., Lee, C.H., Greiner, R., Schuurmans, D.: Semi-supervised conditional random fields for improved sequence segmentation and labeling. In: ACL (2006)

Learning to Detect Adverse Traffic Events from Noisily Labeled Data

Tomáš Šingliar and Miloš Hauskrecht

Computer Science Dept, University of Pittsburgh, Pittsburgh, PA 15260
{tomas,milos}@cs.pitt.edu

Abstract. Many deployed traffic incident detection systems use algorithms that require significant manual tuning. We seek machine learning incident detection solutions that reduce the need for manual adjustments by taking advantage of massive databases of traffic sensor network measurements. First, we show that a rather straightforward supervised learner based on the SVM model outperforms a fixed detection model used by state-of-the-art traffic incident detectors. Second, we seek further improvements of learning performance by correcting misaligned incident times in the training data. The misalignment is due to an imperfect incident logging procedure. We propose a label realignment model based on a dynamic Bayesian network to re-estimate the correct position (time) of the incident in the data. Training on the automatically realigned data consistently leads to improved detection performance in the low false positive region.

1 Introduction

The cost of highway accidents is significantly reduced by prompt emergency response. With real-time traffic flow data, automated incident detection systems promise to detect accidents earlier than human operators. Earlier response and accident impact mitigation lead to significant savings of money and life.

The most widely deployed traffic incident detection models are fixed-structure models that combine and threshold a set of signals such as volumes, speed and speed derivatives [1]. Tuning of these thresholds requires extensive involvement of traffic experts. What is worse, as the settings extracted for one site typically do not transfer to a new site, the tuning costs are multiplied by the number of sites in the network. Transferability of detection algorithms is a central concern in automatic incident detection [2]. We investigate how machine learning can help design transferable detection algorithms.

Machine learning approaches to automatic incident detection are made possible by the wealth of data collected by networks of traffic sensors installed nowadays on many highways. Models that can be automatically tuned from data could reduce or eliminate costly recalibrations and improve performance [3,4,5,6]. We experiment with SVM-based detection and show it easily outperforms the optimally calibrated standard model (California 2).

J.N. Kok et al. (Eds.): PKDD 2007, LNAI 4702, pp. 236–247, 2007.

However, the learning framework can be further improved. In particular, the labels for incident data are imperfect; the initial time of incidents is logged with a variable delay. Consequently, the incident label may be misaligned with the onset of the observed changes in traffic flow caused by the incident. Training a learner with such badly aligned data yields a suboptimal detector.

We approach the alignment problem using machine learning methods as well. We propose a new dynamic Bayesian network [7] that models the misalignment problem probabilistically with respect to traffic flow quantities and the label position. We train the model on the manually realigned data from a *single* highway segment. Once learned, the model can be transferred to other highway segments to correct the incident labeling. The realignment model generates new incident labels in temporal data that are then used to train a supervised classifier such as a SVM to obtain the detection algorithm. This approach allows us to learn, with a *limited* amount of human effort, a more reliable detection model. We demonstrate the improvement in detector quality on traffic flow and incident data collected in the Pittsburgh highway network.

2   The Data and Detection Task

In this section, we look at the available data and define the incident detection task together with the relevant performance metrics.

2.1   Traffic and Incident Data

The data are collected by a network of sensors that use a number of physical principles to detect passing vehicles. Three traffic quantities are normally observed and aggregated over a time period: the average *speed*, *volume* (number of passing vehicles) and *occupancy* (the percentage of road taken up by cars). Incidents that the metropolitan Traffic Management Center (TMC) was aware of are noted in the data: their approximate location, time of accident and time of clearing by emergency responders (Figure 1). Short free-text accident descriptions are also available.

The detection task is to continuously observe the data stream and raise an alarm when the readings indicate an incident[1]. An incident restricts the capacity of the roadway by blocking one or more lanes, forcing drivers to slow down to navigate around it. This will result at a temporary drop in the number and density of vehicles passing the downstream sensor. Upstream of the accident, a jam forms. When the tail end of the jam approaches the nearest sensor, it will cause a drop in measured speed and and increase in vehicle density. The time when the sensors detect the anomaly depends on the utilization of the highway, distance to the sensors and severity of the incident.

---

[1] The term *incident* includes vehicular accidents as well as unscheduled emergency roadwork, debris on the road and many other hazards. Most incidents are accidents and we will use the terms interchangeably.

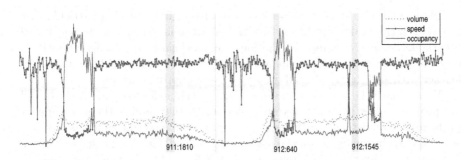

Fig. 1. A section of the raw data. The red (solid), green (solid with markers) and blue (dotted) lines represent average occupancy, average speed and total volume, respectively. Time is on the horizontal axis. The vertical (orange) stripes represent the reported accidents durations. A thin grey vertical line is drawn at midnight of each day. The numbers at the bottom encode accident time as recorded by TMC. Some accidents square with congestions perfectly (912:640 – September 12th, 6:40am), some are slightly shifted (912:1545) and some even have no observable effect on traffic (911:1810). The incident at 912:640 is mostly obscured by morning peak traffic – compare to the morning traffic on the previous day.

2.2   Performance Metrics

A false alarm occurs when the system raises an alarm, but no accident is present. The false alarm rate (FAR) is the number of false alarms divided by the number of detector invocations. The detection rate (DR) is the number of correctly detected incidents divided by the number of incidents that actually occurred. Receiver operating characteristic (ROC) curves [8] are the standard metric designed to quantify detection of one-off binary events. Because accidents affect the traffic for a longer span of time and the detections are not equally valuable around the beginning and the end of the span, we instead prefer the activity monitor operating characteristic (AMOC) curve as the primary performance metric. AMOC curves are used for evaluation of rare event detection performance, such as detection of disease outbreaks [9]. AMOC curves relate false alarm rate (FAR) to time-to-detection (TTD). TTD is defined here as the difference between the time of the start of the first data interval that was labeled as "accident" and the reported incident time. Note that this number can be negative because of the delayed incident recording. As we cannot guarantee to detect all accidents, we introduce a two-hour time-to-detection limit for accidents that remain undetected. When a scalar metric is desired, we compare detectors on $AUC_{1\%}$, the area under the curve restricted to the $(0, 0.01)$ sub-interval of FAR. This is a better indicator of detector performance in the usable low-false-positive region than the area under the entire curve.

The target performance at which a system is considered useful depends chiefly on its users. A study [5] surveying traffic managers found that they would seriously consider using an algorithm that achieves a DR over 88% and FAR under

2%. For any rare event detection system, a low FAR is absolutely essential. A system with a high FAR subjects its users to "alarm fatigue", causing them to ignore it.

3 The Detection Models

In this section, we present the detection models that operate on the original data supplied by the TMC.

3.1 The California 2 Model

"California #2" is a popular model against which new detection algorithms are often compared and runs in most deployed incident detection systems [2]. California #2 (abbreviated CA2) is a fixed-structure model that proceeds as follows:

- Let $Occ(s_{up})$ denote occupancy at the upstream sensor s_{up} and $Occ(s_{down})$ the same at the downstream sensor. If $Occ(s_{up}) - Occ(s_{down}) > T_1$, proceed to the next step.
- If $(Occ(s_{up}) - Occ(s_{down}))/Occ(s_{up}) > T_2$, proceed to the next step. The rationale behind this step is while a capacity-reducing accident will always produce large absolute differences in occupancy, these may also be produced under almost stalled traffic conditions.
- If $(Occ(s_{up}) - Occ(s_{down}))/Occ(s_{down}) > T_3$, wait until the next reading. If T_3 is still exceeded, flag an alarm. The wait is introduced to cut down on false alarms.

Thresholds T_1, T_2, T_3 need to be calibrated manually for each sensor site. Without access to an expert, but with plenty of data, we resorted to an exhaustive parameter grid-search as described in Section 5.

3.2 Model Learning and Features

The CA2 algorithm uses a surprisingly limited set of features. Could a better detection performance be achieved if the detector took advantage of multiple features? And which features? Clearly, the readings at the upstream sensor s_{up} and the downstream sensor s_{down} at the time of detection should be included. Sharp changes in traffic flow may also indicate an accident. Therefore, we include features computed as differences and proportions of the traffic variates to their previous value. Finally, unlike a benign congestion, an accident should cause radically different flow characteristics at the upstream and downstream sensors. This motivates the inclusion of features that correlate the measurements spatially, as differences and proportions of the respective measurements at upstream and downstream sensors.

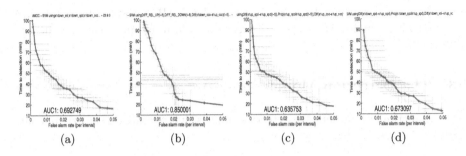

Fig. 2. Performance of the SVM model, for different feature sets. The features are: (a) All readings for the sensor. (b) California 2 features (the occupancy ratios). (c) All of current and previous step measurements. (d) All current measurements together with differences and proportions of the corresponding readings at the upstream and downstream sensors. For drawing the curves, the intercept of the SVM hyperplane is varied in the (-1,1) range, giving a lower bound on the true performance [10]. For each value of the detection threshold, we compute the average FAR and TTD over 10 train/test splits and draw the graph point as well as both of the corresponding error bars. The area under the portion of the curve up to 1% FAR is reported as AUC1.

3.3   SVM Detector

Having defined the potentially informative features, we pick a learner from the palette of learning tools. We had two reasons for choosing the SVM model [11]. First, in preliminary experiments it outperformed logistic regression and several variations of dynamic Bayesian network detectors [12]. Second, the SVM is fairly robust to irrelevant features, allowing us to include features that are weakly informative individually, but perhaps become strong predictors in aggregate. The SVM was learned in the straightforward way. Datapoints falling into the intervals labeled as "incident" in the data were assigned class 1, the remaining datapoints class -1. Misclassification cost was selected as to balance for unequal class sizes: if there are N instances of class 1 and M instance of class -1, then the misclassification of "-1" as "1" costs N/M and 1 vice versa.

The performance of the SVM detector using different feature sets can be seen in the curves and the associated $AUC_{1\%}$ values in Figure 2. It appears that for our data, the direct sensor readings (speed, volume, occupancy) provide most of the detection leverage. Addition of the spatial difference (and proportion) features affects the performance minimally. The temporal difference features do bring a small improvement, albeit one that fails to confirm the perception of temporal difference as an important feature [1]. This could be in part explained by the fact that our data are 5 minute averages and the sharp temporal effects important for detection are somewhat averaged out. A detector using the features of the CA2 algorithm is included for comparison. The results confirm our intuition: the SVM detectors using multiple features outperform that using only CA2 features (the comparison to CA2 itself follows later).

3.4   Persistence Checks

False alarm rate can be traded for detection time with the alarm signal post-processing technique known as persistence check. k-persistence check requires that the alarm condition persist for k additional time periods before the alarm is raised. Note that CA2 already has a built-in 1-persistence check in its last step. We experimented with the optimal (in the sense of minimizing $AUC_{1\%}$) choice of k for the SVM detector with the basic measurement features (on the training site data). Best performance is attained at $k = 1$ and all SVM experiments are therefore conducted using that persistence value.

4   Label Realignment Model

Our objective is to detect the accident as soon as its impact *manifests* in sensor readings. This time will always lag the time the accident actually *happened*. The lag amount depends, among other factors, on the capacity utilization of the roadway and the relative position of the sensor and accident locations. The time that the incident is *reported* to the TMC and logged in the database may precede or follow the time of manifestation. Differences between these times lead to label misalignment.

There are two things that the detector can latch onto; the short period when the accident's impact builds up (e.g. speed falls) around the sensor, and the longer steady state condition with lowered speeds or jammed traffic. To optimize detection time, we should focus the detector at the transient period. The transient period is very short and any misalignment will cause the accident start label to fall outside of it. It is therefore crucial for supervised learning that the label is precisely aligned with the observed impact of the accident. The end-of-incident labels are less important: by the time the incident is cleared, the emergency response has already taken place. We do not attempt to align incident clearing times.

By definition, misalignment can only occur in positive instances, i.e. those sequences that contain an incident. We need a method to correct the alignment of incident labels in the training set so that the learned model accuracy may improve.

4.1   A Model of Incident Sequences

Consider a single positive sequence S of traffic feature vectors. An incident start label r denotes the point in sequence S where the single incident is reported to occur. The label realignment task is to output the label ℓ pointing where the incident truly began to manifest in S. For label realignment, we model the sequence of feature vectors with a special dynamic Bayesian network model, shown in Figure 3. In the model, A represents the true accident time and takes on a value from $\{1, \ldots, L\}$, where L is the sequence length. Each *impact* variable $I^{(k)}$ is a binary indicator of incident impacting the traffic flow at time k. Each I is a part of the intra-slice Bayesian network that models the interaction between the

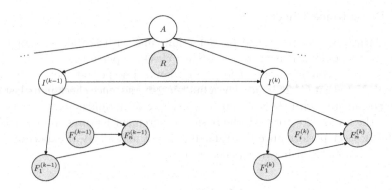

Fig. 3. Two slices of the temporal probabilistic model for realignment. As usual, shaded nodes represent observed random variables; unshaded nodes correspond to latent variables. There are a total of L slices; the superscript (k) denotes the variable's instantiation in the k-the time slice.

traffic measurement features $F_1, \ldots, F_n$. We place no restrictions on the within-slice network in general. In order to keep the model presentation simple, we do not draw arrows between the features in subsequent slices. However, features may depend on values at other nearby timepoints; for instance the "occupancy derivative" $F(t) = Occ(t) - Occ(t-1)$ depends on the previous measurement.

The variables A and $I^{(k)}$ have a special relationship, expressed in their probability distributions. First, we express the lack of prior knowledge about A by defining the prior $P(A)$ to be the uniform distribution on the set $\{1, \ldots, L\}$. Second, the conditional distribution is also fixed, expressing that once an incident starts impacting traffic, it continues to do so:

$$P(I^{(k)} = 1 | A = k', I^{(k-1)} = 1) = 1$$
$$P(I^{(k)} = 1 | A = k', I^{(k-1)} = 0) = 1 \quad \text{if } k = k',$$
$$0 \quad \text{otherwise.} \tag{1}$$

We can afford this simplification because we only want to model the accident onset and disregard the accident clearing event.

The report time R depends on the true accident time and is assumed to obey a conditional Gaussian distribution: $P(R|A = k) \sim N(k + \mu, \sigma^2)$, with μ, σ identical for all k. Equivalently, the amount of misalignment has a stationary Gaussian distribution: $R - A \sim N(\mu, \sigma^2)$.

4.2   Inference for Realignment

We perform inference in this model in its unrolled form. Basic variable elimination is the best suited inference method for the realignment model. It deals well with the unusual distributions and is also efficient for this model, because the special form of the inter-slice probability distribution simplifies the inference task – there are only L indicator sequences with $p(I_1, \ldots, I_L) > 0$ to sum over.

Using the probability distribution p defined by the above model, the label alignment task can be formulated as a posterior mode problem. Given that the data places the incident start label at r, we reassign the label to ℓ, so that

$$\ell = \underset{k}{\operatorname{argmax}}\, p(A = k | R = r, \mathbf{F}^{(1)}, \ldots, \mathbf{F}^{(L)}), \qquad (2)$$

where $\mathbf{F}^{(t)} = (F_1^{(t)}, \ldots, F_n^{(t)})$ is the t-th observation vector.

4.3 Learning and Transfer to New Locations

Now, we must parameterize a separate model for each sensor pair defining a highway segment (site). Let A denote the single calibration (training) site and let B_j, $j = 1, \ldots, S$ be the test sites. While one could learn the model in a fully unsupervised manner with the EM algorithm [13], there is little guarantee that the learning would converge to the intended interpretation. Instead, we first learn p^A, the sequence model for A, from manually aligned data. Manual alignment gives us a fully observed dataset $\mathbf{X}^A = (\mathbf{F}^A, R^A, I^A, A^A)$ and maximum likelihood learning becomes trivial:

$$\Theta_{ML}^A = \underset{\Theta}{\operatorname{argmax}}\, p(\mathbf{X}^A | \Theta) \qquad (3)$$

Then, inference in the model parameterized with Θ_{ML}^A can be applied to realign the labels for the B-sites where the manual annotation is unavailable. Of course, accident impact at each site B_j differs from that of the site A. The resulting labeling will be imperfect, but it still provides a good initial estimate. The EM-algorithm for estimation of Θ^{B_j} can proceed from there with a smaller risk of converging to an undesirable local optimum. Additionally, the sufficient statistics obtained in the estimation of Θ^A are stored and used to define the conjugate prior over Θ^{B_j}. Thus the resulting parameterization of a testing site model is a maximum a posteriori (MAP) estimate:

$$\Theta_{MAP}^{B_j} = \underset{\Theta}{\operatorname{argmax}}\, p(\mathbf{X}^{B_j} | \Theta) p(\Theta | \Theta_{ML}^A). \qquad (4)$$

In the EM algorithm that estimates $\Theta_{MAP}^{B_j}$, the expectation step corresponds to inference of the unobserved labels A^{B_j} and I^{B_j}. The maximization step re-estimates the parameters of the conditional distributions $p(R|A)$ and $p(F_i|I)$. We consider the EM converged if the labeling (the posterior modes, see Equation 2) does not change in two consecutive iterations. For our dataset, the EM always converges in less than 5 iterations.

5 Experimental Evaluation

In this section we describe the experimental setup and report the results. All statistics reported are averages and standard deviations across 10 cross-validation splits, even where error bars were dropped for sake of readability. Error bars in all graphs represent one standard deviation.

Table 1. Sites included in the evaluation, with number of incidents

Site	S_{Train}	S_{Test1}	S_{Test2}	S_{Test3}
# incidents	145	100	97	92

5.1 Evaluation Framework

We evaluated our model on four sites with the highest numbers of accident reports in the area. The incident reports at S_{Train} were manually aligned to the incident manifestations in the data. The manual realignment was also aided by the free-text incident descriptions from the TMC database.

We evaluate the models under the cross-validation framework. The dataset consists of three long sequences per sensor, one for each of the three traffic variates. We divide the data into train/test splits by incidents, making sure an entire incident sequence makes it into one and only one of the sets. To create the training set, we first select I_{train} "incident" sequences of preset length L so that the reported time of the incident falls in the middle of the incident sequence. In the rare case that more than one incident should occur in or in the vicinity of a sequence, we exclude such sequence from the data. C "control" sequences without an incident are selected so that no incident is recorded within additional $L/2$ datapoints before and after the control sequence. This safeguards against the imprecise accident recording. By choosing I_{train} and C, the class prior in the training set can be biased towards incident occurrences. The testing set consists of the $I_{test} = I_{all} - I_{train}$ incident sequences that were not selected for the training set. Additional sequences without accidents are added so that the testing set has class prior equal to that in the entire dataset.

To obtain the experimental statistics, we use 10 different train/test splits using the above method, with $I_{train} : I_{test} \approx 70 : 30$, sequence length $L = 100$ and $C = 50$ for training. For testing, instead of choosing a fixed C, we make sure the proportion of positive (incident) instances approximates the proportion observed in the entire dataset.

In each cross-validation fold, the positive training sequences are realigned and the quality of the detection is evaluated on the testing set, *using the original incident labeling*. While testing on the original labeling will result in a measurement that is somewhat off, the skew will be consistent across detectors and relative comparisons remain valid. If we evaluated on the realigned data, we would run the risk of having both the realignment algorithm and the detector make the same mistake in lockstep, losing touch with the data.

5.2 Detection and Alignment Model Specifics

To represent incident impact on traffic, we use a Naive Bayes intra-slice model with binary indicator I and two features, F_1: the difference in occupancy at the upstream sensor in the previous and following interval and F_2: the same difference in speed. Both features are assumed to follow a conditional Gaussian distribution.

Table 2. Summary of the $AUC_{1\%}$ performance statistics for the three detection algorithms and four evaluation sites. Some sites are more amenable to automatic detection, but consistent improvement is noted from CA2 to SVM on original data to SVM on realigned data.

Detector	Site			
	S_{Train}	S_{Test1}	S_{Test2}	S_{Test3}
CA2	0.838	0.451	1.177	1.180
SVM/orig	0.682	0.179	0.807	0.474
SVM/realign	0.547	0.149	0.763	0.389

The CA2 algorithm is normally tuned by experts who choose the three thresholds. Since we did not have services of an expert, we found the parameterization by an exhaustive procedure trying all possible settings of the three parameters on a discrete grid covering a wide range of parameter values. The "best performance" for the purpose of parameterization was defined as the best DR at a fixed FAR of 1%. This was an extremely time-consuming procedure that is impractical for a metropolitan network with hundreds of sensors, not to mention uninteresting from the learning perspective.

5.3   Experimental Results

The root of the mean squared difference between the hand-labeled incident manifestations and the recorded events is approximately 8.5 intervals. After automatically re-aligning the recorded events with the incidents, the RMS difference decreases to approximately 2.2 intervals. The decrease in training error affirms that the model indeed picks up the accident effect.

The average amount of misalignment at the training site is only 2.2 minutes (incidents are on average logged 2.2 minutes after they become observable in data), but with a standard deviation of more than 43 minutes. This is a serious amount of misalignment, it implies that the label position is on average off by 8 or 9 time steps.

The quality of the resulting labels is most relevantly measured by the improvement in the $AUC_{1\%}$ performance metric of a classifier learned on the realigned data. The $AUC_{1\%}$ values for the three methods (CA2, SVM, SVM after relabeling) are summarized in Table 2. The standard deviation of TTD and FAR obtained together with the 10-fold cross-validated averages are represented by the vertical and horizontal bars, respectively, around each operating point on the curves in Figure 4. The table shows that the SVM detector learned on the original data consistently improves over the CA2 method for every testing site. Similarly, the SVM detector learned on the label-realigned data realizes an improvement over the original SVM detector. The absolute performance varies significantly between testing sites as it depends on a number of site specifics: the distance between the accident site and the upstream sensor, volume of traffic, the presence of a shoulder lane where the vehicles may be removed from the flow of traffic, etc.

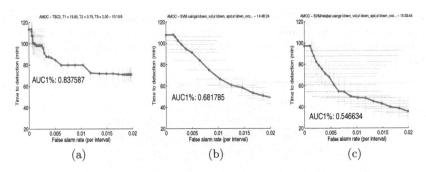

Fig. 4. Train site A with human-labeled data. Detection performance of (a) California 2 (b) SVM learned on original labeling, (c) SVM learned on the relabeled data.

6   Conclusions

Learning is a viable approach to construction of incident detection algorithms. It easily leads to detectors that outperform traditional hand-crafted detectors. With sufficient data now available, it can do away with the problem of manual tuning and re-tuning of the detectors to adapt to new deployment locations and changing traffic patterns.

However, the data obtained from such complex systems is inherently noisy. We proposed an algorithm that deals successfully with noise in event label timing and demonstrated that it improves the data quality to allow more successful learning of incident detectors. Of course, a number of specific questions about our approach remain open. One could devise finer incident models and offset distributions; relax the assumption of independence of time-to-recording and incident impact severity – a more severe accident is perhaps more easily noticed. Explicitly modeling time-of-day and the expected traffic pattern looks especially promising as it permits the definition of an "unexpected" congestion, presumably more indicative of an accident.

While the realignment algorithm was motivated by and presented in context of incident detection, it is generally applicable to situations where events are marked noisily in data streams. For instance, similar uncertainty in labeling alignment accompanies detection of intonation events in speech recognition [14].

References

1. Martin, P., Perrin, H., Hansen, B.: Incident detection algorithm evaluation. Technical Report UTL-0700-31, Utah Traffic Laboratory (July (2000)
2. Stephanedes, Y., Hourdakis, J.: Transferability of freeway incident detection algorithms. Technical Report Transportation Research Record 1554, Transportation Research Board, National Research Council (1996)
3. Dia, H., Rose, G., Snell, A.: Comparative performance of freeway automated incident detection algorithms (1996)

4. Rose, G., Dia, H.: Freeway automatic incident detection using artificial neural networks. In: Proceedings of the International Conference on Application of New Technology to Transport Systems, vol. 1, pp. 123–140 (1995)
5. Ritchie, S.G., Abdulhai, B.: Development, testing and evaluation of advanced techniques for freeway incident detection. Technical Report UCB-ITS-PWP-97-22, California Partners for Advanced Transit and Highways (PATH) (1997)
6. Parkanyi, E., Xie, C.: A complete review of incident detection algorithms and their deployment: What works and what doesn't. Technical Report NETCR37, New England Transportation Consortium (2005)
7. Dean, T., Kanazawa, K.: A model for reasoning about persistence and causation. Computational Intelligence 5, 142–150 (1989)
8. Provost, F.J., Fawcett, T.: Analysis and visualization of classifier performance: Comparison under imprecise class and cost distributions. In: Knowledge Discovery and Data Mining, pp. 43–48 (1997)
9. Cooper, G., Dash, D., Levander, J., Wong, W.K., Hogan, W., Wagner, M.: Bayesian biosurveillance of disease outbreaks. In: Proceedings of the 20th Annual Conference on Uncertainty in Artificial Intelligence (UAI-04), pp. 94–103. AUAI Press, Arlington, Virginia (2004)
10. Bach, F., Heckerman, D., Horvitz, E.: On the path to an ideal ROC curve: Considering cost asymmetry in learning classifiers. In: Cowell, R.G., Ghahramani, Z. (eds.) Proceedings of AISTATS05, pp. 9–16 (2005)
11. Mangasarian, O.L., Musicant, D.R.: Lagrangian support vector machine classification. Technical Report 00-06, Data Mining Institute, Computer Sciences Department, University of Wisconsin, Madison, Wisconsin (June 2000), ftp://ftp.cs.wisc.edu/pub/dmi/tech-reports/00-06.ps
12. Singliar, T., Hauskrecht, M.: Towards a learning incident detection system. In: ICML 2006 Workshop on Machine Learning Algorithms for Surveillance and Event Detection (2006)
13. Dempster, A., Laird, N., Rubin, D.: Maximum likelihood for incomplete data via the EM algorithm. Journal of Royal Statistical Society 39, 1–38 (1977)
14. Taylor, P.A.: Analysis and synthesis of intonation using the Tilt model. Journal of the Acoustical Society of America 107(3), 1697–1714 (2000)

IKNN: Informative K-Nearest Neighbor Pattern Classification

Yang Song[1], Jian Huang[2], Ding Zhou[1], Hongyuan Zha[1,2], and C. Lee Giles[1,2]

[1] Department of Computer Science and Engineering
[2] College of Information Sciences and Technology,
The Pennsylvania State University,
University Park, PA 16802, U.S.A.

Abstract. The K-nearest neighbor (KNN) decision rule has been a ubiquitous classification tool with good scalability. Past experience has shown that the optimal choice of K depends upon the data, making it laborious to tune the parameter for different applications. We introduce a new metric that measures the informativeness of objects to be classified. When applied as a query-based distance metric to measure the closeness between objects, two novel KNN procedures, Locally Informative-KNN (LI-KNN) and Globally Informative-KNN (GI-KNN), are proposed. By selecting a subset of most informative objects from neighborhoods, our methods exhibit stability to the change of input parameters, number of neighbors(K) and informative points (I). Experiments on UCI benchmark data and diverse real-world data sets indicate that our approaches are application-independent and can generally outperform several popular KNN extensions, as well as SVM and Boosting methods.

1   Introduction

The K-nearest neighbor (KNN) classifier has been both a workhorse and benchmark classifier [1,2,4,11,14]. Given a query vector x_0 and a set of N labeled instances $\{x_i, y_i\}_1^N$, the task of the classifier is to predict the class label of x_0 on the predefined P classes. The K-nearest neighbor (KNN) classification algorithm tries to find the K nearest neighbors of x_0 and uses a majority vote to determine the class label of x_0. Without prior knowledge, the KNN classifier usually applies Euclidean distances as the distance metric. However, this simple and easy-to-implement method can still yield competitive results even compared to the most sophisticated machine learning methods.

The performance of a KNN classifier is primarily determined by the choice of K as well as the distance metric applied [10]. However, it has been shown in [6] that when the points are not uniformly distributed, predetermining the value of K becomes difficult. Generally, larger values of K are more immune to the noise presented, and make boundaries more smooth between classes. As a result, choosing the same (optimal) K becomes almost impossible for different applications.

Since it is well known that by effectively using prior knowledge such as the distribution of the data and feature selection, KNN classifiers can significantly

J.N. Kok et al. (Eds.): PKDD 2007, LNAI 4702, pp. 248–264, 2007.

improve their performance, researchers have attempted to propose new approaches to augmenting the performance of KNN method. e.g., Discriminant Adaptive NN [9] (DANN), Adaptive Metric NN [6] (ADAMENN), Weight Adjusted KNN [8] (WAKNN), Large Margin NN [13] (LMNN) and etc. Despite the success and rationale of these methods, most have several constraints in practice. Such as the effort to tune numerous parameters (DANN introduces two new parameters, K_M and ϵ; ADAMENN has six input parameters in total that could potentially cause overfitting), the required knowledge in other research fields (LMNN applies semidefinite programming for the optimization problem), the dependency on specific applications (WAKNN is designed specifically for text categorization) and so on. Additionally, in spite of all the aforementioned constraints, choosing the proper value of K is still a crucial task for most KNN extensions, making the problem further compounded.

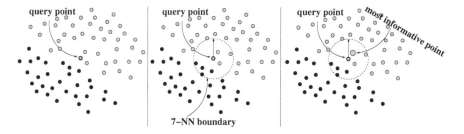

Fig. 1. A toy classification problem. (Left) The original distribution of two classes. (Middle) Results of $k = 7$ NN method where the query point is misclassified. (Right) One of our proposed methods, (LI-KNN), chooses one informative point for prediction.

Therefore, it is desirable to augment the performance of traditional KNN without introducing much overhead to this simple method. We propose two KNN methods that are ubiquitous and their performances are insensitive to the change of input parameters. Figure 1 gives an example that shows the motivation of our approach, in which the traditional KNN method fails to predict the class label of the query point with $K = 7$. One of our proposed method (LI-KNN) takes the same value of K, finds the most *informative* point ($I = 1$) for the query point according to the new distance metric, and makes a correct prediction.

1.1 Our Contribution

In this paper, we propose two novel, effective yet easy-to-implement extensions of KNN method whose performances are relatively insensitive to the change of parameters. Both of our methods are inspired by the idea of *informativeness*. Generally, a point(object) is treated to be *informative* if it is *close to the query point and far away from the points with different class labels*. Specifically, our paper makes the following contributions:

(1) We introduce a new concept named informativeness to measure the importance of points, which can be used as a distance metric for classification. (2)

Based on the new distance metric, we propose an efficient *locally informative* KNN (LI-KNN) method. (3) By learning a weight vector from training data, we propose our second method that finds the *globally informative* points for KNN classification (GI-KNN). (4) We perform a series of experiments on real world different data sets by comparing with several popular classifiers including KNN, DANN, LMNN, SVM and Boosting. (5) We discuss the optimal choice of the input parameters (K and I) for LI-KNN and GI-KNN and demonstrate that our methods are relatively insensitive to the change of parameters.

The rest of the paper is organized as follows: Section 2 presents related work about different approaches to improve KNN pattern classification; section 3 introduces the definition of informativeness and our first algorithm LI-KNN; section 4 continues to propose the second learning method GI-KNN; we apply the proposed methods to both synthetic and real-world data sets in section 5 for evaluation; finally we conclude in section 6.

2   Related Work

The idea of nearest neighbor pattern classification was first introduced by Cover and Hart in [4], in which the decision rule is to assign an unclassified sample point to the classification of the nearest of a collection of predetermined classified points. The authors proved that when the amount of data approaches infinity, the one nearest neighbor classification is bounded by twice the asymptotic error rate as the Bayes rule, independent of the distance metric applied.

Hastie and Tibshirani [9] developed an adaptive method of nearest neighbor classification (DANN) by using local discrimination information to estimate a subspace for global dimension reduction. They estimate between (B) and within (W) the sum-of-squares matrices, and use them as a local metric such as $\sum = W^{-1}BW^{-1}$. They showed that their work can be generalized by applying specialized distance measures for different problems.

Weinberger et al. [13] learned a Mahanalobis distance metric for KNN classification by using semidefinite programming, a method they call large margin nearest neighbor (LMNN) classification. Their method seeks a large margin that separates examples from different classes, while keeping a close distance between nearest neighbors that have the same class labels. The method is novel in the sense that LMNN does not try to minimize the distance between all examples that share the same labels, but only to those that are specified as *target neighbors*. Experimental results exhibit great improvement over KNN and SVM.

By learning locally relevant features from nearest neighbors, Friedman [7] introduced a flexible metric that performs recursively partitioning to learn local relevances, which is defined as $I_i^2(z) = (Ef - E[f|x_i = z])^2$, where Ef denotes the expected value over the joint probability density $p(x)$ of an arbitrary function $f(x)$. The most informative feature is recognized as the one giving the largest deviation from $P(x|x_i = z)$.

Han et al. [8] proposed an application of KNN classification to text categorization by using adjusted weight of neighbors (WAKNN). WAKNN tries to learn the best weight for vectors by measuring the cosine similarity between documents. Specifically, the similarity function is defined as $cos(X, Y, W) = \frac{\sum_{t \in T}(X_t \times W_t) \times (Y_t \times W_t)}{\sqrt{\sum_{t \in T}(X_t \times W_t)^2} \times \sqrt{\sum_{t \in T}(Y_t \times W_t)^2}}$, where X and Y are two documents, W the weight vector and T the set of features (terms). Optimizations are also performed to speed up WAKNN. The experiments on benchmark data sets indicate that WAKNN consistently outperforms KNN, C4.5 and several other classifiers.

3   Locally Informative KNN (LI-KNN)

Without prior knowledge, most KNN classifiers apply Euclidean distances as the measurement of the *closeness* between examples. Since it has already been shown that treating the neighbors that are of low relevance as the same importance as those of high relevance could possibly degrade the performance of KNN procedures [7], we believe it to be beneficial to further explore the information exhibited by neighbors. In this section, we first propose a new distance metric that assesses the informativeness of points given a specific query point. We then proceed to use it to augment KNN classification and advocate our first method, LI-KNN.

3.1   Definition of Informativeness

We use the following naming conventions. Q denotes the query point, K indicates the K nearest neighbors according to a distance metric, and I denotes most informative points based on equation (1). For each point, x_i denotes the i's feature vector, x_{ij} its j's feature and y_i its class label. Let N represent the total number of training points, where each point has P features.

Definition 1. *Specify a set of training points $\{x_i, y_i\}_1^N$ with $x_i \in \mathbb{R}^P$ and $y_i \in \{1, ...m\}$. For each query point x_i, the **informativeness** of each of the remaining N-1 points $\{x_j, y_j\}_1^N$ is defined as:*

$$\mathcal{I}(x_j|Q = x_i) = -\log(1 - \mathcal{P}(x_j|Q = x_i)) * \mathcal{P}(x_j|Q = x_i), \ j = 1, ...N, j \neq i \ (1)$$

where $\mathcal{P}(x_j|Q = x_i)$ is the probability that point x_j is informative (w.r.t. Q), defined as:

$$\mathcal{P}(x_j|Q = x_i) = \frac{1}{Z_i} \left\{ \mathrm{Pr}(x_j|Q = x_i)^\eta \left(\prod_{n=1}^N \left(1 - \mathrm{Pr}(x_j|Q = x_n)\mathbb{I}_{[y_j \neq y_n]}\right) \right)^{1-\eta} \right\} (2)$$

The first term $\mathrm{Pr}(x_j|Q = x_i)^\eta$ in equation (2) can be interpreted as the likelihood that point x_j is close to the Q, while the second part indicates the probability that x_j far apart from dissimilar points. The indicator $\mathbb{I}[.]$ equals to 1 if the condition is met and 0 otherwise. Z_i is a normalization factor and η is introduced as a balancing factor that determines the emphasis of the first term. Intuitively, η is set to $\frac{N_{x_j}}{N}$, where N_{x_j} represents the number of points in the same class of x_j.

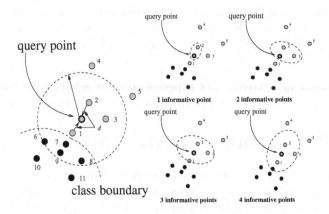

Fig. 2. An illustration of 7-NN and the corresponding i informative points for the query point. (Left) 7-NN classification and the real class boundary. (Right) $i(i = \{1, 2, 3, 4\})$ informative points for the same query.

The rationale of informativeness is that two points are likely to share the same class label when their distance is sufficiently small, assuming the points have a uniform distribution. This idea is the same as KNN classification. On the other hand, compared to traditional KNN classifiers that measures pairwise distances between the query point and neighbors, our metric also calculates the closeness between neighbor points, i.e., the informative points should also have a large distance from dissimilar points. This further guarantees that the locations of other informative points have the same class label maximum likelihood.

Figure 2(left) gives an example for clarification, in which point 1 and point 2 (with the same class label) both have the same distance d from Q but point 1 is closer to the real class boundary. Thus, point 1 is more likely to be closer to the points in other classes. As such we claim that point 1 is less informative than point 2 for Q by DEFINITION 1. Again, assuming the distribution over the concept location is uniform, it is more likely that points (e.g., 3 & 4) have the same label as points 1 & 2 and will more likely distribute around point 2 than point 1.

3.2   Informativeness Implementation

To define $\Pr(x_j|Q = x_i)$ in equation (2), we can model the causal probability of an individual point on Q as a function of the distance between them:

$$\Pr(x_j|Q = x_i) = f(\|x_i - x_j\|_p) \tag{3}$$

where $\|x_i - x_j\|_p$ denotes the p-norm distance between x_i and x_j. To achieve higher probability when two points are close to each other, we require $f(.)$ to be a function inverse to the distance between two points. The generalized Euclidean distance metric satisfies this requirement. Thus, equation (3) can be defined as

$$\Pr(x_j|Q = x_i) = \exp(-\frac{||x_i - x_j||^2}{\gamma}) \qquad \gamma > 0 \qquad (4)$$

In practice, it is very likely that the features have different importance, making it desirable to find the best weighting of the features. Specifically, we define $||x_i - x_j||^2 = \sum_p w_p(x_{ip} - x_{jp})^2$, where w_p is a scaling that reflects the relative importance of feature p. Although there exists numerous functions for calculating w_p, here we specify it as follows:

$$w_p = \frac{1}{m}\sum_{k=1}^{m} w_{pk} = \frac{1}{m}\sum_{k=1}^{m} \text{Var}_{\mathbf{x}_k}(\mathbf{x}_{pk}) \qquad (5)$$

We obtain w_p by averaging over all classes' weights w_{pk}, which is calculated using the variance of all points in each class k at feature p, denoted by $\text{Var}_{\mathbf{x}_k}(\mathbf{x}_{pk})$.

The normalization factor Z_i in equation (2) ensures the well-defined probabilities of neighbors for a given query point x_i. Specifically,

$$Z_i = \sum_{j=1}^{N} \Pr(x_j|Q = x_i), \quad \sum_{j=1}^{N} \mathcal{P}(x_j|Q = x_i) = 1 \qquad (6)$$

so that the normalization is guaranteed.

Based on the implementation, we have the following proposition:

Proposition 1. *Given a specific query x_0, $\forall\, x_i, x_j$ that satisfies $||x_i - x_0||^2 = kd$ and $||x_j - x_0||^2 = d$ with $d \in \mathbb{R}^+, k > 1$, $\mathcal{I}(x_i|x_0) < \exp((1-k)d)^\eta \mathcal{I}(x_j|x_0)$.*

Proof. For simplicity, we only consider the case that x_i and x_j are in the same class, i.e., $y_i = y_j$. Without loss of generality, we let $\gamma = 1$ for equation (4). We have

$$\begin{aligned}
\frac{\mathcal{P}(x_j|Q = x_0)}{\mathcal{P}(x_i|Q = x_0)} &= \frac{\Pr(x_j|Q = x_0)^\eta H(x_j)^{1-\eta}}{\Pr(x_i|Q = x_0)^\eta H(x_i)^{1-\eta}} \\
&= \frac{\exp(-d)^\eta H(x_j)^{1-\eta}}{\exp(-kd)^\eta H(x_i)^{1-\eta}} \\
&= \exp((k-1)d)^\eta \frac{H(x_j)^{1-\eta}}{H(x_i)^{1-\eta}} \qquad (7)
\end{aligned}$$

where $H(\mathbf{x}) = \left(\prod_{n=1}^{N}\left(1 - \Pr(\mathbf{x}|Q = x_n)\mathbb{I}_{[y \neq y_n]}\right)\right)$. Since $H(\cdot)$ is independent of the query point, its expected value (taken over $\mathbf{x}$ and each x_n) can be defined as

$$\begin{aligned}
E(H(\mathbf{x})) &= E\left(\prod_{n=1}^{N}\left(1 - \Pr(\mathbf{x}|Q = x_n)\mathbb{I}_{[y \neq y_n]}\right)\right) \\
&= \prod_{n=1}^{N}\left(E\left(1 - \Pr(\mathbf{x}|Q = x_n)\mathbb{I}_{[y \neq y_n]}\right)\right)
\end{aligned}$$

$$= \prod_{n=1}^{N} \left(E(1 - \exp(-\|x - x_n\|^2) \mathbb{I}_{[y \neq y_n]}) \right)$$

$$= \prod_{n=1}^{N} \left((1 - E \exp(-\|\mathbf{x} - x_n\|^2) \mathbb{I}_{[y \neq y_n]}) \right)$$

Recall that x_i and x_j are in the same class, thus the set of dissimilar points (say $\{x'_n, y'_n\}_1^q$) should be the same. The above equation can then be simplified by removing the indicator variables:

$$E(H(\mathbf{x})) = \prod_{n=1}^{q} \left((1 - E \exp(-\|\mathbf{x} - x'_n\|^2)) \right)$$

$$= \prod_{n=1}^{q} \left(1 - \int_1^N \exp(-\|\mathbf{x} - x'_n\|^2) d\mathbf{x} \right)$$

with $N \to \infty$, it is easy to verify that $E(H(x_i)) = E(H(x_j))$. Applying the results to equation (7), we have

$$\frac{\mathcal{P}(x_j | Q = x_0)}{\mathcal{P}(x_i | Q = x_0)} = \exp((k - 1)d)^\eta > 1 \quad (\text{with } k > 1) \tag{8}$$

Applying equation (8) to equation (1), we finally have:

$$\frac{\mathcal{I}(x_j | Q = x_0)}{\mathcal{I}(x_i | Q = x_0)} = \frac{\log(1 - \mathcal{P}(x_j | Q = x_0))}{\log(1 - \mathcal{P}(x_i | Q = x_0))} \cdot \exp((k - 1)d)^\eta$$

$$= \log_{(1 - \mathcal{P}(x_i | Q = x_0))}(1 - \mathcal{P}(x_j | Q = x_0)) \cdot \exp((k - 1)d)^\eta$$
$$> \exp((k - 1)d)^\eta \qquad \qquad \square$$

3.3 LI-KNN Classification

So far we have proposed to compute the informativeness of points in the entire data distribution for a specific Q. However, considering the high dimensionality and large number of data points in practice, the computational cost could be prohibitively high. We propose to make use of the new distance metric defined in equation (1) by restricting the computation between the nearest neighbors in an augmented *query-based* KNN classifier.

Algorithm 1 gives the pseudo-code of LI-KNN classification. Instead of finding the informative points for each x_i by going over the entire data set, LI-KNN retrieves I *locally* informative points by first getting the K nearest neighbors (we consider the Euclidean distance here). It then applies equation (1) to the K local points and the majority label between the I points are assigned to x_i. Specifically, when $I = 1$, LI-KNN finds only the most informative point, i.e., $y_i = \arg\max_{y_k, k \in \{1, \dots, K\}} \mathcal{I}(x_k | Q = x_i)$. In this way the computational cost of finding the most informative points is reduced to a local computation. Noticeably, when

---

Algorithm 1. LI-KNN Classification

1: **Input:** (S, K, I)
     target matrix: $S = \{x_i, y_i\}_1^N$
     number of neighbors: $K \in \{1, ..., N-1\}$
     number of informative points: $I \in \{1, ..., K\}$
2: Initialize $err \leftarrow 0$
3: **for** each query point x_i ($i = 1$ to N) **do**
4:     find K nearest neighbors $\mathcal{X}_K$ using Euclidean distance
5:     find I most informative points among K neighbors (equation (1))
6:     majority vote between the I points to determine the class label of x_i
7:     **if** x_i is misclassified
8:         $err \leftarrow err + 1/N$
9:     **end if**
10: **end for**
11: **Output:** err

K equals to N, the locally informative points are exactly the optimal informative points for the entire data distribution as in DEFINITION 1. Likewise, when I equals to K, LI-KNN performs exactly the same as KNN rule.

At the first glance, it seems that LI-KNN introduces one more parameter I for the KNN method. However, by carefully checking the requirement for points to be informative, it is not hard to figure out that LI-KNN is relatively insensitive to both K and I. (1) Regardless of the choice of K, the points that are closest (in Euclidean distance) to Q are always selected as neighbors, which by equation (2) have a high probability to be informative. (2) On the other hand, given a fixed number of K, the informativeness of the local points are fixed which insures that the most informative ones are always chosen. For example, in Figure 2(left), point 2 & 3 are selected as the neighbors for Q with K increasing from 3 to 7. Meanwhile, when K equals to 7 and I ranges from 1 to 3, the informative sets (Figure 2(right)) are {2},{2, 3} and {2, 3, 1} respectively, which include the most informative points in all cases that ensures Q is classified correctly. In practice, cross-validation is usually used to determine the best value of K and I.

4   GI-KNN Classification

The LI-KNN algorithm classifies each individual query point by learning informative points separately, however, the informativeness of those neighbors are then discarded without being utilized for other query points. Indeed, in most scenarios, different queries Q may yield different informative points. However, it is reasonable to expect that some points are more informative than others, i.e., they could be informative neighbors for several different points. As a result, it would seem reasonable to put more emphasis on those points that are *globally* informative. Since it has been shown that KNN classification [13] can be improved by learning from training examples a distance metric, in this section we enhance the power of the informativeness metric and propose a boosting-like iterative method, namely a *globally informative* KNN (GI-KNN) that aims to learn the best weighting for points within the entire training set.

4.1   Algorithm and Analysis

The goal of GI-KNN is to obtain an optimum weight vector A from all training points. The algorithm iterates M predefined steps to get the weight vector, which was initially set to be uniform. In each iteration, an individual point is classified in the same way as LI-KNN by finding I informative neighbors, with the only exception that in GI-KNN the distance metric is a weighted Euclidean distance whose weight is determined by A (line 5 & 6 in Algorithm 2, where $D(x_i, \mathbf{x})$ denotes the Euclidean distance between x_i and all the remaining training points, and $D_A(x_i, \mathbf{x})$ is the weighted distance). We use $\epsilon_m^i \in (0, 1)$ to denote the *weighted* expected weight loss of x_i's informative neighbors during step m. The cost function C_m^i is a smooth function of ϵ_m^i, which guarantees it to be in the range of (0,1) and positively related with ϵ_m^i. Here we use tanh function as the cost function, depicted in Figure 3[1]. The weight vector A is updated in the manner that if x_i is classified incorrectly, the weights of its informative neighbors which have different labels from x_i are decreased exponentially to the value of C_m^i (line 9, $e(x_i, x_\ell) = C_m^i$ if $y_i \neq y_\ell$; line 13, $A(x_\ell) \leftarrow A(x_\ell) \cdot \exp(-e(x_i, x_\ell))$). Meanwhile, the weights remain the same for neighbors in the same class with x_i even if x_i is misclassified (line 9, $e(x_i, x_\ell) = 0$ if $y_i = y_\ell$). Clearly, the greater the weight the query point is, the higher the penalty of misclassification will be for the selected neighbors. The vector A is then normalized before the next iteration.

Instead of rewarding those points that classify Q correctly by increasing their weights, the weights of neighbors remain unchanged if Q is classified correctly. This could potentially cause accumulative effects to points whose weights that once increased will always increase in the following steps, ending up with dominant large weights. As a result, we penalize those points that give the wrong prediction and have different labels with Q. Therefore, by updating the weight vector before the next iteration, they will be less likely to be selected as the neighbors for the same Q.

While GI-KNN has several parallels to Boosting such as the structure of the algorithm, GI-KNN differs from Boosting in the way weights are updated. Specifically, Boosting assigns high weights to points that are misclassified in the current step, so that the weak learner can attempt to fix the errors in future iterations. In GI-KNN classification, the objective is to find globally informative points, thus higher weights are given to the neighbors that seldom makes wrong predictions. Notice that the weight of the query point remains unchanged at that time, because the weight is updated for a specific point if and only if it is chosen to be one of the informative points for Q.

Another difference from Boosting is that the objective of the Boosting training process is to find a committee of discriminant classifiers that combines the

---

[1] In practice, we did not find much difference in performance for different τ. Therefore, we choose $\tau = 1$ for our implementation.

Algorithm 2. GI-KNN Training

1: **Input:** (T, K, I, M)
      training set: $T = \{\mathbf{x}, \mathbf{y}\} \in \mathbb{R}^{N \times P}$
      number of neighbors: $K \in \{1, ..., N-1\}$
      number of informative points: $I \in \{1, ..., K\}$
      number of iterations: $M \in \mathbb{R}$
2: **Initialization:** $A = \{1, ..., 1\} \in \mathbb{R}^{N \times 1}$ [*the weight vector*]
3: **for** $m = 1$ to M **do**
4:     **for** each query point x_i ($i = 1$ to N) **do**
5:        $D_A(x_i, \mathbf{x}) = \frac{D(x_i, \mathbf{x})}{A}$ [*calculate the weighted distance*]
6:        $\mathcal{N}_m(x_i) \leftarrow I$ most informative points according to $D_A(x_i, \mathbf{x})$
7:        $\epsilon_m^i = A(x_i) \cdot E_A[\mathcal{N}_m(x_i)] = A(x_i) \cdot \frac{1}{I} \sum_{i-1}^{I} A(\mathcal{N}_m(i))$
8:        $C_m^i = \frac{1}{2}(1 + tanh(\tau * (\epsilon_m^i - \frac{1}{2})))$
9:

$$e(x_i, x_\ell) = \begin{cases} C_m^i & \text{if } y_i \neq y_\ell; \\ 0 & \text{if } y_i = y_\ell. \end{cases}$$

10:       **if** point x_i is classified incorrectly [*update the neighbors' weights*]
11:          $err_m \leftarrow err_m + \frac{1}{N}$
12:          **for** each x_ℓ ($\ell \in \mathcal{N}_m(x_i)$) **do**
13:             $A(x_\ell) \leftarrow A(x_\ell) \cdot \exp(-e(x_i, x_\ell))$
14:          **end for**
15:       renormalizes A so that $\sum_{i=1}^{N} A(i) = N$
16:     **end for**
17:     $\xi_m \leftarrow err_m - err_{m-1}$
18: **end for**
19: **Output:** the weight vector A

---

weak learners, while GI-KNN tries to learn a query-based distance metric by focusing on finding the best weight vector for each training instance so that the misclassification rate of training examples could be minimized.

4.2   Learning the Weight Vector

At completion, the learned vector A can be used along with L_2 distance metric for KNN classification at each testing point $\mathbf{t_0}$. Specifically, given the training set $T = \{x_i, y_i\}_1^N$, the distance between $\mathbf{t_0}$ and each training point x_i is defined as

$$D(\mathbf{t_0}, x_i) = \|\mathbf{t_0} - x_i\|_{A_i} = \frac{\sqrt{(\mathbf{t_0} - x_i)^T (\mathbf{t_0} - x_i)}}{A_i} \tag{9}$$

By adding weights to data points, GI-KNN in essence is similar in effect to learning a Mahalanobis distance metric $D(x_i, x_j)$ for k-nearest neighbor classification. i.e., $D(x_i, x_j) = D_{\mathcal{A}}(x_i, x_j) = \|x_i - x_j\|_{\mathcal{A}} = \sqrt{(x_i - x_j)^T \mathcal{A}(x_i - x_j)}$, where $\mathcal{A}$ determines the similarity between features. In our case, A measures the importance of each training point rather than their features.

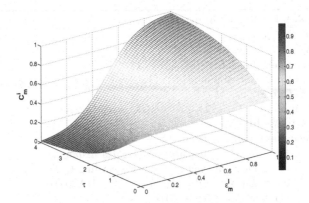

Fig. 3. Cost function (C) used for GI-KNN

In practice, we make two modifications to Algorithm 2 to reduce the computational cost. First, the L_2 distances between all training points are computed and stored (say in matrix D) at the beginning of the program. Then, instead of updating D real-time for each query point (line 5 in Algorithm 2), we do it after each external iteration. In another words, for each point, we update the weight vector if necessary, but use the same A for all points in the same iteration. After each round, D is updated with the new weight vector. Similarly, rather than normalizing A after each classification of Q (line 15 in Algorithm 2), the normalization is performed only after each external iteration. We discover that empirically these two modifications do not degrade the performance in most scenarios.

4.3   Complexity of GI-KNN Training

The major overhead of GI-KNN training phase is the time needed to find the informative neighbors for each point (line 5 of Algorithm 2). Specifically, $\mathcal{N}_m(i)$ keeps the indices of informative neighbors for point x_i, whose length is controlled by the input parameter I. Given K and I, the effort to find the nearest neighbors is bounded by $O(KP)$, where P denotes dimension of the input data. Calculating and ranking the informativeness of K nearest neighbors involves computing the pairwise distances between them and thus costs $O(K^2P)$ time to train. Thus the total time is bounded by $O(KP)+O(K^2P) = O(K^2P)$ for each point. Therefore, the training process requires approximately $O(K^2PMN)$ time for N training examples and M iterations. Remember that the traditional KNN classification costs $O(KPN)$ for the same setting, while LI-KNN requires $O(K^2PN)$. When the K is not very large, the computational complexity is nearly the same for KNN and LI-KNN, both of which are equal to one iteration time for GI-KNN.

5   Experiments

In this section, we present experimental results with benchmark and real-world data that demonstrate the different merits of LI-KNN and GI-KNN. LI-KNN and GI-KNN are first rigorously tested by several standard UCI data sets. Then our proposed methods are applied to text categorization using the real-world data from CiteSeer Digital Library[2]. Finally, we investigate their performance on images by applying to image categorization on the COIL-20[3] bench-marked data sets.

For performance evaluation, several classifiers are used for comparison. The classic KNN [4] classifier is used as the baseline algorithm. We implemented DANN [9] as an extension of KNN[4]. To be more convincing, we also compare with one of the newest KNN extensions – Large Margin Nearest Neighbor Classification (LMNN)[5]. Two discriminant classifiers are also compared: a Support Vector Machine (SVM) and a Boosting classifier. We use the AdaBoost.MH [12] and the Multi-class SVM [5] software (K.Crammer et al.[6]) for multi-class classification.

5.1   UCI Benchmark Corpus

We evaluate our algorithms by using 10 representative data sets from UCI Machine Learning Repository[7]. The size of the data sets ranges from 150 to 20,000 with dimensionality between 4 and 649, including both two classes and multi-class data ($C = 3, 26$ etc). For evaluation, the data sets are split into training sets and testing sets with a fixed proportion of 4:1. Table 1 reports the best testing error rates for these methods, averaged over ten runs. Our methods on these data sets exhibit competitive results in most scenarios.

Figure 4(a) shows the stability of LI-KNN on the testing errors rates of the Iris data set. KNN always incurs higher error rates than our algorithms. The performance of LI-KNN is depicted for four different values of I. It is obvious that even with different values of I (given the same K), the results are similar, indicating that the performance of LI-KNN does not degrade when the number of informative points changes. In addition, with the change of K, LI-KNN is relatively stable regarding the error rate. The variation of LI-KNN is roughly 3%, meaning that K does not have a large impact on the results of LI-KNN.

Figure 4(b) compares Boosting and GI-KNN on the Breast Cancer data for the first 1,000 iterations. Overall, GI-KNN incurs lower error rates. From 620 to about 780 iterations GI-KNN's error rates are slightly higher than Boosting. However, the error rates of Boosting fluctuate quite a bit from 0.048 to 0.153, while GI-KNN is relatively stable and the performance varies only between (0.043, 0.058). Moreover, our algorithm obtains the optimal results significantly earlier.

---

[2] http://citeseer.ist.psu.edu

[3] http://www1.cs.columbia.edu/CAVE/software/softlib/coil-20.php

[4] During the experiment, we set $K_M = max(N/5, 50)$ and $\epsilon = 1$ according to the original paper.

[5] The code is available at http://www.seas.upenn.edu/~kilianw/lmnn/

[6] See http://www.cis.upenn.edu/~crammer/code-index.html

[7] http://www.ics.uci.edu/~mlearn/MLRepository.html

Table 1. Testing error rates for KNN, DANN, LMNN, SVM, Boosting, LI-KNN and GI-KNN of 10 UCI Benchmark data sets. N, D and C denote the number of instances, dimensionality and number of classes respectively. Numbers in the parentheses indicate the optimal neighbors K for KNN, DANN and LMNN, (K, I) for LI-KNN, and number of iterations M for GI-KNN and Boosting.

Data Sets	N	D	C	KNN	DANN	LMNN	**LI-KNN**	**GI-KNN**	SVM	Boosting
Iris	150	4	3	0.044 (9)	0.040 (5)	0.053 (3)	0.013 (9, 5)	**0.010** (25)	0.042	0.038 (45)
Wine	178	13	3	0.263 (3)	0.250 (7)	**0.031** (5)	0.137 (15, 1)	0.137 (13)	0.205	0.192 (135)
Glass	214	10	2	0.372 (5)	0.436 (5)	0.356 (3)	**0.178** (7, 3)	0.198 (202)	0.222	**0.178** (304)
Ionosphere	351	34	2	0.153 (5)	0.175 (7)	0.100 (5)	0.127 (5, 3)	0.127 (8)	**0.090**	0.092 (156)
Breast	699	9	2	0.185 (7)	0.120 (11)	0.927 (5)	0.080 (4, 1)	**0.045** (48)	0.052	0.048 (657)
Heart	779	14	5	0.102 (3)	0.117 (5)	0.092 (5)	**0.078** (7, 1)	**0.078** (192)	**0.078**	0.080 (314)
Digit	2000	649	10	0.013 (3)	0.010 (3)	0.009 (3)	**0.005** (19, 1)	**0.005** (137)	0.010	**0.005** (175)
Isolet	7797	617	26	0.078 (11)	0.082 (11)	0.053 (5)	0.048 (13, 3)	**0.042** (175)	0.044	**0.042** (499)
Pendigits	10992	16	10	0.027 (3)	0.021 (5)	**0.020** (3)	**0.020** (9, 1)	**0.020** (42)	0.033	0.038 (482)
Letter	20000	16	10	0.050 (5)	0.045 (3)	0.042 (5)	0.045 (5, 3)	0.040 (22)	**0.028**	0.031 (562)

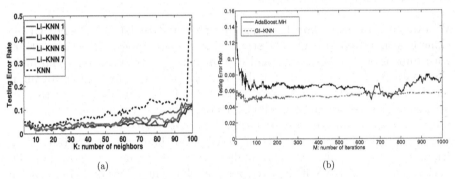

(a)   (b)

Fig. 4. (a) Results on Iris for K from 1 to 100. LI-KNN chooses the number of informative points (I) to be 1, 3, 5 and 7. (b) Results on Breast Cancer for AdaBoost.MH and GI-KNN (with $K = 5$ and $I = 1$). The best result for GI-KNN is slightly better (0.045) than that of AdaBoost.MH (0.048).

5.2 Application to Text Categorization

For text categorization experiments we use the CiteSeer data set consisting of nearly 750,000 documents primarily in the domain of computer science. Several types of data formats are indexed concurrently (*txt, pdf, ps, compressed files, etc.*). For the purpose of text categorization, we only make use of plain text files. For convenience, the metadata of the documents, i.e., the titles, abstracts and keyword fields are used in our experiments.

Document class labels are obtained from the *venue impact page*[8] which lists 1,221 major venues whose titles are named according to the DBLP[9] format. We make use of the top 200 publication venues listed in DBLP in terms of impact

---

[8] http://citeseer.ist.psu.edu/impact.html

[9] http://dblp.uni-trier.de/

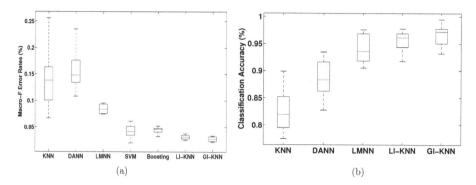

Fig. 5. (a) Box plots of macro-F error rates (100% minus Macro-F scores) on CiteSeer data set summarizes the average F scores on 193 classes. Our algorithms have very low error rates on average, with very small deviations. Plus(+) signs indicate the outliers. **(b)** Classification Accuracies on CiteSeer data set for KNN, LI-KNN and GI-KNN with different number of neighbors ($K = \{1, ..., 50\}$). Our algorithms generally demonstrate more stable results and higher accuracies.

rates, each of which was referred as a class label. Furthermore, we intentionally filtered those classes that contain too few examples (i.e., less than 100 documents). Overall, the total number of documents used for the experiments is 118,058, which are divided into a training set and testing set by 10-fold cross-validation. Meanwhile, we keep the imbalance of the classes, i.e., some classes have more training examples than others. Documents are treated as *bag-of-words* and *tf-idf* weights of terms are used to generate the feature space.

Figure 5(a) shows the box plots *macro-F* error rates. The optimal parameters (e.g., the number of iterations M and so on) are estimated by 10-fold cross-validation on the training set. It is evident that the spread of the error distribution for our algorithms are very close to zero, which clearly indicates that LI-KNN and GI-KNN obtain robust performance over different classes. Meanwhile, our algorithms incur lower error rates even for small classes, making them potentially good choices for imbalanced data set classification.

We further show the stability of our algorithms by box plots of the classification accuracies for different number of neighbors. Figure 5(b) depicts the results of KNN, DANN and our algorithms for K from 1 to 50 with a fixed number of $I = 1$ (i.e., only the most informative neighbor). The mean accuracies are higher for LI-KNN and GI-KNN than KNN, and the variations are almost half as that of KNN and DANN.

5.3   Object Recognition on COIL-20

We use the processed version of COIL-20 database for object recognition. The database is made up with 20 gray-scale objects, each of which consists 72 images with size 128×128. Figure 6(a) shows a sample image of each of the 20 objects.

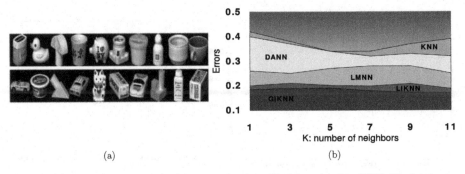

(a)   (b)

Fig. 6. (a) Randomly generated images from each object in the COIL-20 database.
(b) Results on COIL-20 with different number of neighbors.

We treat each object as one class, splitting the data into training and testing set
with the proportion of 3:1. Figure 6(b) shows the classification errors regarding
the 5 algorithms, where K ranges from 1 to 11. GI-KNN and LI-KNN generally
outperform others with the best parameters, while both show stable results with
the change of K.

5.4 Discussion

Our I-KNN procedures introduce two adjustable tuning parameters K and I, it is
then desirable to automate the choice of them. Theoretically we did not prove the
optimal choices for either K or I, however, empirical studies with different ranges
of values on several data sets allow us to draw a rough conclusion. Basically, the
value of K should be reasonably big. The bigger K is, the more information can
be gathered to estimate the distribution of neighborhood for the query point.
However, with the increase of K, the effort to compute the informativeness of
neighbors (equation (2)) grows exponentially as well. In practice, we figured out
that $K \in (7, 19)$ could be a good trade-off regardless of data size. Meanwhile, a
smaller I is preferable to give the best predictions. Experimental results indicate
that $I = 1$ and 3 usually achieve the best results, and the performance generally
degrades with the increase of I. There is potentially another parameter to tune,
i.e., η in equation (2), to balance the contribution of the first term. However, we
only use $\eta = \frac{N_{x_j}}{N}$ here.

We have observed that most influential on the running time on both algorithms is the computation cost of the informativeness metric, of which the normalization factor (equation (2) and (6)) takes most of the time. To further
improve the performance, we remove the normalization part in our experiments,
i.e., equation (2) and (6). This significantly reduced the complexity of our model
and did not jeopardize the performance very much.

Regarding the choice of the cost function C_m^i for GI-KNN training (line 8 in
Algorithm 2), since GI-KNN has a different objective (to find the best weight
vector) than boosting and other machine learning algorithms (to minimize a
smooth convex surrogate of the 0-1 loss function), we did not compare the

performance between different loss functions like exponential loss, hinge loss and so on. Since we believe that the performance change will not be significant by exhaustively searching for the best loss function. The choice of different loss functions has already been extensively studied, interested readers can refer to [3] for details.

6 Conclusion and Future Work

This paper presented two approaches namely LI-KNN and GI-KNN to extending KNN method with the goal of improving classification performance. Informativeness was introduced as a new concept that is useful as a query-based distance metric. LI-KNN applied this to select the most informative points and predict the label of a query point based on the most numerous class with the neighbors; GI-KNN found the globally informative points by learning a weight vector from the training points. Rigorous experiments were done to compare the performance between our methods and KNN, DANN, LMNN, SVM and Boosting. The results indicated that our approaches were less sensitive to the change of parameters than KNN and DANN, meanwhile yielded comparable results to SVM and Boosting. Classification performance on UCI benchmark corpus, CiteSeer text data, and images suggests that our algorithms were application-independent and could possibly be improved and extended to diverse machine learning areas.

Questions regarding the GI-KNN algorithm are still open for discussion. Can we possibly prove the convergence of GI-KNN, or is there an upper-bound for this algorithm given specific K and I? More practically, is it possible to stop earlier when the optimum results are achieved? As a boosting-like algorithm, can we replace the 0-1 loss function with a smooth convex cost function to improve the performance? Furthermore, it will be interesting to see whether the informativeness metric can be applied to semi-supervised learning or noisy data sets.

References

1. Athitsos, V., Alon, J., Sclaroff, S.: Efficient nearest neighbor classification using a cascade of approximate similarity measures. In: CVPR '05, pp. 486–493. IEEE Computer Society, Washington, DC, USA (2005)
2. Athitsos, V., Sclaroff, S.: Boosting nearest neighbor classifiers for multiclass recognition. In: CVPR '05, IEEE Computer Society, Washington, DC, USA (2005)
3. Bartlett, P., Jordan, M., McAuliffe, J.: Convexity, classification and risk bounds. J. Amer. Statist. Assoc. 101, 138–156 (2006)
4. Cover, T., Hart, P.: Nearest neighbor pattern classification. IEEE Transactions on Information Theory 13(1), 21–27 (1967)
5. Crammer, K., Singer, Y.: On the algorithmic implementation of multiclass kernel-based vector machines. J. Mach. Learn. Res. 2, 265–292 (2002)
6. Domeniconi, C., Peng, J., Gunopulos, D.: Locally adaptive metric nearest-neighbor classification. IEEE Trans. Pattern Anal. Mach. Intell. 24(9), 1281–1285 (2002)

7. Friedman, J.: Flexible metric nearest neighbor classification. technical report 113, stanford university statistics department (1994)
8. Han, E.-H.S., Karypis, G., Kumar, V.: Text categorization using weight adjusted k -nearest neighbor classification. In: 5th Pacific-Asia Conference on Knowledge Discovery and Data Mining (PAKDD), pp. 53–65 (2001)
9. Hastie, T., Tibshirani, R.: Discriminant adaptive nearest neighbor classification. IEEE Trans. Pattern Anal. Mach. Intell. 18(6), 607–616 (1996)
10. Latourrette, M.: Toward an explanatory similarity measure for nearest-neighbor classification. In: ECML '00: Proceedings of the 11th European Conference on Machine Learning, London, UK, pp. 238–245. Springer-Verlag, Heidelberg (2000)
11. Peng, J., Heisterkamp, D.R., Dai, H.K.: LDA/SVM driven nearest neighbor classification. In: CVPR '01, p. 58. IEEE Computer Society, Los Alamitos, CA, USA (2001)
12. Schapire, R.E., Singer, Y.: Improved boosting algorithms using confidence-rated predictions. In: COLT' 98, pp. 80–91. ACM Press, New York (1998)
13. Weinberger, K.Q., Blitzer, J., Saul, L.K.: Distance metric learning for large margin nearest neighbor classification. In: NIPS (2005)
14. Zhang, H., Berg, A.C., Maire, M., Svm-knn, J.M.: Discriminative nearest neighbor classification for visual category recognition. In: CVPR '06, pp. 2126–2136. IEEE Computer Society, Los Alamitos, CA, USA (2006)

Finding Outlying Items in
Sets of Partial Rankings

Antti Ukkonen[1,3] and Heikki Mannila[1,2,3]

[1] Helsinki University of Technology
[2] University of Helsinki
[3] Helsinki Institute for Information Technology

Abstract. Partial rankings are totally ordered subsets of a set of items. For example, the sequence in which a user browses through different parts of a website is a partial ranking. We consider the following problem. Given a set D of partial rankings, find items that have strongly different status in different parts of D. To do this, we first compute a clustering of D and then look at items whose average rank in the cluster substantially deviates from its average rank in D. Such items can be seen as those that contribute the most to the differences between the clusters. To test the statistical significance of the found items, we propose a method that is based on a MCMC algorithm for sampling random sets of partial rankings with exactly the same statistics as D. We also demonstrate the method on movie rankings and gene expression data.

1 Introduction

Partial rankings are totally ordered subsets of a set of items. For example, the set of items might contain all products available at an Internet store, while a partial ranking contains only products viewed by one user, ranked in the order the user clicked on their descriptions. Partial rankings can be found for example in clickstream analysis, collaborative filtering and different scientific applications, such as analysis of microarray data.

Given a set of partial rankings we can construct a clustering so that similar rankings are assigned to the same cluster [6]. The rankings belonging to the same cluster can be aggregated to form a condensed representation of the cluster. This representation can be for example a total or partial order on the complete set of items. However, comparing the aggregate representations between clusters can sometimes be difficult. Especially if the number of items is very large, it can be hard to quickly identify features that separate the clusters from each other.

For example, consider microarray data where the expression levels of a number of genes have been measured in different tissues or under different conditions. Typically this kind of data is represented as a matrix, where the rows correspond to different genes and columns to different tissues/conditions. This data can be converted to partial rankings by sorting the tissues separately for each gene in decreasing order of the level of expression. These rankings are indeed partial, due to missing data. The partial rankings can be clustered to find out in what

J.N. Kok et al. (Eds.): PKDD 2007, LNAI 4702, pp. 265–276, 2007.
© Springer-Verlag Berlin Heidelberg 2007

tissues the expression of genes belonging to a cluster is exceptionally strong or weak. This type of analysis is relevant in cases where one wants to identify the tissues or conditions in which a certain set of genes is more active than the rest.

As another example, consider movie ratings given by viewers. These ratings can be converted to partial rankings as well. In general people tend to prefer the same movies: if a movie is very good (or bad), then it is likely that the vast majority of all viewers considers it is good (or bad). But some titles divide the opinions of the viewers more than others. One of such films is for example Pulp Fiction by Quentin Tarantino. People either think it is a very good movie – maybe because of the distinct style of the director – or are appalled by the violence and use of language. What we might thus expect, is that when movie rating data is divided to, say, two clusters, the titles that end up as discriminative are movies that have a fairly strong fan base, but are frowned upon or otherwise disliked by others.

This leads to the idea of an alternative representation for a cluster. Instead of using a total or partial order as the aggregate representation, we can list those items that are ranked either distinctively high or low by the partial rankings in a cluster when compared to an aggregate representation of the entire data set. This provides a way of characterizing a cluster of partial rankings in terms of the items that separate it from the complete data. We call such items *outlying items*. A similar approach was proposed in [3], but it uses a different definition for outlyingness.

The second question concerns the statistical significance of the found items. For evaluating the "outlyingness" of an item we need a way to generate artificial sets of partial rankings with exactly the same statistics as the real input data. To this end we propose an MCMC algorithm for sampling sets of partial rankings from an equivalence class that is defined by the following statistics: number and length distribution of partial rankings, occurrence and co-occurrence frequencies of the items and the probabilities $Pr(u \prec v)$ that an item u precedes another item v in the partial rankings for all u and v. Especially important are the probabilities $Pr(u \prec v)$, as they play a major part in identifying the outlying items. We consider an item a significant outlier if it behaves differently in real data compared to random data sampled from the same equivalence class.

The rest of this paper is organized as follows. The definition of an outlying item and significance testing is discussed in Section 2. The MCMC algorithm used for generating random data is described in Section 3. Experiments and results are presented in Section 4 while Section 5 provides a short conclusion.

2   Problem Statement

2.1   Basic Definitions

Let M be a set of items to be ranked. In the following when we discuss an item u, it is always assumed to belong to the set M. A *partial ranking* ϕ is a totally ordered subset of M. Note that this is not the same as a *partial order* that concerns all of M but leaves the mutual ranking between some items unspecified. A set of partial rankings is denoted D.

Given the set D, we can compute a number of *statistics* that describe it. Some simple examples of such statistics are the size of D, length distribution of the partial rankings, occurrence-, and co-occurrence frequencies of the items. As the data contains rankings, the most important statistic is related to the mutual order of the items. For each pair (u, v), $u, v \in M$, we consider the probability that item u precedes the item v in D, denoted $Pr(u \prec v)$. We estimate $Pr(u \prec v)$ with the fraction of partial rankings in D that place u before v. Note that u and v do not need to be adjacent in the partial ranking. If u and v never occur together in a partial ranking, we set $Pr(u \prec v) = Pr(v \prec u) = 0.5$, and in general we always have $Pr(u \prec v) + Pr(v \prec u) = 1$. The probabilities are arranged in a $|M| \times |M|$ matrix C_D, so that $C_D(u, v) = Pr(u \prec v)$. We call C_D the *pair order matrix* associated with the set of partial rankings D.

All of the statistics discussed above can be used to define *equivalence classes* over the set of all possible sets of partial rankings. We denote with $\mathcal{C}(D)$ the class of sets of partial rankings that all have exactly the same statistics as the set D. Later, in Section 3, we will discuss an algorithm for sampling uniformly from $\mathcal{C}(D)$.

2.2   Finding Outlying Items

Let D' be a subset of D. Typically D' is obtained by computing a clustering of D and letting D' correspond to one cluster. Our aim in this paper is to discover items that behave differently in D' when compared to D. To do this we use the pair order matrices C_D and $C_{D'}$. More specifically, we are interested in the quantities

$$S_D(u) = \sum_v C_D(u, v) \quad \text{and} \quad S_{D'}(u) = \sum_v C_{D'}(u, v),$$

which are simply the row sums of the pair order matrices corresponding to D and D' for item u. These are indicators of the *global rank* of an item in a set of partial rankings . For example, if $S_D(u)$ is very large, then the item u should be placed before most of the other items in a global ranking based on D, as u tends to precede them in the partial rankings in D. Likewise, if $S_D(u)$ is small the item should be placed after most of the other items in the global ranking.

Given the subset D' of D, we consider an item *outlying* in D' if the difference

$$X(u) = S_{D'}(u) - S_D(u) \tag{1}$$

is significantly above (or below) zero. This would mean the rankings that belong to D' tend to place u more often before (or after) the other items than the rankings belonging to D in general. Hence, the subset D' is at least partially characterized by the way in which it ranks the item u. See [3] for a slightly different approach for finding outlying items.

Given the sets D and D' we can sort the items in decreasing order of their *outlyingness* $|X(u)|$ and pick the h topmost items as interesting representatives for the set D'. These items may contain both items that are ranked unusually high or unusually low in D' when compared to D.

Thus, the definition of an outlying item is very simple. Computing $X(u)$ is almost trivial, as we only need to build the pair order matrices and compute their row sums, but doing this alone may lead to incorrect conclusions. Consider the case where all partial rankings in D are completely random, the probabilities $Pr(u \prec v)$ are all approximately 0.5 for all u and v. If we use some clustering algorithm to divide D to two non-overlapping clusters, the result will be arbitrary. However, in one of the clusters some of the items may have a slightly higher global rank than in the entire data D. These will be identified as outlying, even though they were found largely by coincidence. To prevent us from finding such items in real data sets, we propose methods for evaluating the significance of the outlyingness of an item.

2.3   Testing the Significance of Outlying Items

There are two possible pitfalls when using $|X(u)|$ to find outlying items for a subset D'. First, we must address the reliability of $|X(u)|$ as the indicator of outlyingness for a fixed item u. Especially we want to determine if the deviation of $X(u)$ from zero for a specific u is only caused by the values in the pair order matrix C_D. If this were the case then we should observe high deviations from zero for $X(u)$ also with sets of partial rankings that differ from D but have the same pair order matrix.

The second issue is related to the validity of the set of outlying items in general. Suppose that we generate a random set $\hat{D}$ of partial rankings so that $C_D = C_{\hat{D}}$ and compute a clustering of $\hat{D}$ to k clusters. The meaningfulness of the set of outlying items found in the real data can be questioned if roughly the same number of (possibly different) items u have equally high deviations of $X(u)$ from zero in a cluster of the random data.

The basic approach for using random data is the following:

1. Compute a clustering of the real data D and find the sets of outlying items for each cluster using $|X(u)|$.
2. Pick a set $\hat{D}$ of partial rankings from $\mathcal{C}(D)$ (the equivalence class of sets of partial rankings with same statistics as D) uniformly at random.
3. Compute a clustering of $\hat{D}$ and record the $|X(u)|$ values for each item in each cluster. If enough samples of $|X(u)|$ have been obtained, go to next step, otherwise go to step 2.
4. Estimate $E[|X(u)|]$ and $Var[|X(u)|]$ for all $u \in M$ from the samples.
5. Compute a significance measure for $|X(u)|$ based on $E[|X(u)|]$ and $Var[|X(u)|]$.

Hence, we assume the $X(u)$s are normally distributed. The significance is measured by the distance of $|X(u)|$ from $E[|X(u)|]$ in standard deviations. That is, we let

$$Z(u) = \frac{||X(u)| - E[|X(u)|]|}{\sqrt{Var[|X(u)|]}}.$$

For example, if $Z(u) > 3$, it is fairly safe to assume that u indeed is a significantly outlying item.

To address the significance of the entire set of outlying items, we compute the quantity $Y(D') = \sum_u X(u)^2$, where the $X(u)$s are deviations in subset D'. This is done both for the original set of partial rankings and each random data. The significance of the deviations in the original data is again expressed as the distance of $Y(D')$ from $E[\sum_u X(u)^2]$ computed over all the clusters in random data sets. Denote this by Q. Large values of Q indicate that the set of outlying items in D' is more significant, as it means that in random data the $X(u)$s deviate on the average less from zero.

3   Sampling from $\mathcal{C}(D)$

In order to test for the outlyingness $X(u)$ of an item u we must have a way of generating random sets of partial rankings with exactly the same features as the original data D. Features we want to preserve are the size of D, length distribution of the partial rankings, occurrence and co-occurrence frequencies of all items, and most importantly the pair order matrix C_D. Recall, that data sets with the same statistics as a given data set D belong to the equivalence class $\mathcal{C}(D)$. We present a simple MCMC algorithm for sampling sets of partial rankings uniformly from $\mathcal{C}(D)$.

3.1   The SWAP-PAIRS Algorithm

The basic idea of the algorithm is to perform swaps of adjacent items in the partial rankings of D. Suppose that u and v are adjacent in the partial ranking ϕ with u before v. Swapping u and v in ϕ has no effect on the length of ϕ or the frequencies of u or v in general, but only on $C_D(u, v)$ which is decremented and $C_D(v, u)$ which is incremented, both by the same amount. To preserve the values of $C_D(u, v)$ and $C_D(v, u)$ we must perform another swap with the opposite effect, i.e., we must find a partial ranking ψ where u and v are adjacent, but v precedes u, and swap those. Combining these two swaps results in a new set of partial rankings $\hat{D}$ with ϕ and ψ changed, but having the pair order matrix $C_{\hat{D}}$ equal to the original C_D. The algorithm is called SWAP-PAIRS and it simply starts from D and performs a sequence of such swaps at random.

To do the swaps efficiently we preprocess the data and compute the set A of *swappable pairs*. More formally we let

$$A = \{\{u, v\} | uv \in \phi \wedge vu \in \psi \wedge \phi, \psi \in D\}, \tag{2}$$

where $uv \in \phi$ denotes that items u and v are adjacent in ϕ with u before v. Note that if the pair $\{u, v\}$ is swappable and u and v are swapped, then $\{u, v\}$ remains swappable. However, as a consequence of a swap some other pairs may become swappable and some other unswappable. For example, consider the following set of partial rankings:

SWAP-PAIRS:
1. Pick the swap (u, v, ϕ, ψ) uniformly at random from all possible swaps in the current state.
2. Swap the positions of u and v both in ϕ and ψ with probability $\min(1, \frac{N(current)}{N(swapped)})$. Update the set of possible swaps accordingly.
3. If we have done enough swaps, output $\hat{D}$, otherwise go to step 1.

Fig. 1. A high level description of the SWAP-PAIRS algorithm

$$\phi_1\colon 1\ 2\ 3\ 4\ 5$$
$$\phi_2\colon 7\ 8\ 4\ 3\ 6$$
$$\phi_3\colon 3\ 2\ 6\ 4\ 1$$

It is quickly seen that 3 and 4 are a swappable pair as they are adjacent in both ϕ_1 and ϕ_2 but in different order. Also note that 2 and 3 form a swappable pair with partial rankings ϕ_1 and ϕ_3. In this case we have thus $A = \{\{2,3\}, \{3,4\}\}$. Lets say we decide to swap 3 and 4 and obtain:

$$\phi_1'\colon 1\ 2\ 4\ 3\ 5$$
$$\phi_2'\colon 7\ 8\ 3\ 4\ 6$$
$$\phi_3\colon 3\ 2\ 6\ 4\ 1$$

Now $\{2,3\}$ is no longer swappable, as 2 and 3 are no longer adjacent in ϕ_1' and we must remove this pair from A. However, now 4 and 6 are adjacent in both ϕ_2' and ϕ_3 and their order is different, so we can add $\{4,6\}$ to the set of swappable pairs and are left with $A = \{\{3,4\}, \{4,6\}\}$.

In addition to the list of swappable pairs we use a data structure, denoted S, that quickly returns the set of relevant partial rankings when given a swappable pair. We let $S(u,v) = \{\phi \in D | uv \in \phi\}$. The structure S is also computed during preprocessing and updated during the execution of the algorithm.

Finally, to sample uniformly from $\mathcal{C}(D)$ we must address one additional issue. We discuss some notation first. A *swap* is the tuple (u, v, ϕ, ψ), where u and v are items and ϕ and ψ are partial rankings in D. The swap (u, v, ϕ, ψ) means that items u and v are swapped in rankings ϕ and ψ. Denote the number of different possible swaps at the current state of the Markov chain with $N(current)$, and with $N(swapped)$ the same number for a state reachable by one swap from the current state. It is easy to see that $N(current) = \sum_{\{u,v\} \in A} |S(u,v)||S(v,u)|$. As it is possible that $N(current) \neq N(swapped)$, a simple algorithm that just picks possible swaps at random doesn't converge to the uniform distribution over $\mathcal{C}(D)$. To remedy this we use the Metropolis-Hastings step when performing a swap. That is, first the swap (u, v, ϕ, ψ) is picked uniformly at random from the set of all possible swaps at the current state, and we accept (u, v, ϕ, ψ) with probability $\min(1, \frac{N(current)}{N(swapped)})$. Intuitively the chain always moves to states with fewer possible swaps than the current state, and if the next state has a larger number of possible swaps, the transition is accepted with a probability less than 1. Pseudocode for the SWAP-PAIRS algorithm is given in Figure 1.

The number of possible swaps at a given state is of order $O(|M|^2|D|^2)$ in the worst case, which can get prohibitively large. In practice our implementation never stores the set of possible swaps explicitly, but only uses the A and S structures. The swap (u, v, ϕ, ψ) is computed (step 1 of algorithm) by first picking the pair $\{u, v\}$ from A with probability $\frac{|S(u,v)||S(v,u)|}{N(current)}$ and then picking ϕ and ψ uniformly at random from $S(u, v)$ and $S(v, u)$, respectively. This can be done in time $O(|M|^2)$ when elements of S are accessible in constant time. Complexity of the swap (step 2 of the algorithm) depends on the type of data structure used for A and $S(u, v)$ as they must be modified as a result of the swap. Our simple implementation uses sorted random access lists that can be updated in time $O(\log |M|^2)$ and $O(\log |D|)$ in case of A and $S(u, v)$, respectively. Hashing would provide a constant time solution, but might make step 1 more complicated as we would need to sample uniformly from the values of a hash table. In practice, however, the biggest bottleneck of the algorithm is sampling the swappable pair $\{u, v\}$ from A, because the probabilities need to be updated on every iteration for each pair.

3.2   Theoretical Questions and Convergence Diagnostics

The problem of creating sets of partial rankings with the same features as a given initial set by performing swaps of adjacent items is interesting in its own right. The problem is very similar to the one discussed in [4] and more recently in [2] and [1] in the context of 0-1 matrices. There the problem is to generate 0-1 matrices with exactly the same row and column sums as a given initial matrix.

It is easy to see that SWAP-PAIRS preserves the statistics used to define the class $\mathcal{C}(D)$. However, it is not obvious that $\mathcal{C}(D)$ is connected with respect to the swap operation. In [4] it is shown that the set of 0-1 matrices with same row and column sums is connected with respect to a certain local transformation of the matrix values. Whether this holds also with $\mathcal{C}(D)$ is an interesting open question. Moreover, estimating the size of $\mathcal{C}(D)$ is another task for future work.

A more practical problem concerns the convergence of the Markov chain defined using the swap operation. To use the algorithm for sampling sets of partial rankings we must know how many swaps to make to be sure that the resulting rankings are uncorrelated with the initial state D. In general analyzing the mixing times of Markov chains formally is nontrivial. We can, however, empirically evaluate the sequence of sampled sets $\hat{D}$ of partial rankings in terms of their distance from D.

To do this, we define the function δ as measure of the distance between two sets of partial rankings generated by the swap randomization algorithm. As SWAP-PAIRS only swaps items within a partial ranking, the items belonging to each ranking stay the same. Hence, the i:th partial ranking in the swapped data set, denoted $\hat{D}(i)$, is in fact a permutation of the i:th partial ranking of the original data set, denoted $D(i)$. We define $\delta(D, \hat{D}) = |D|^{-1} \sum_i d(D(i), \hat{D}(i))$, where d is some distance function between permutations. We use Kendall's tau, which is the number of pairs of items that are ordered differently by D and $\hat{D}$. The measure δ is thus the average permutation distance between the partial rankings in D and $\hat{D}$.

To see when the chain has converged we must see how δ behaves as the swapping progresses. Denote by $\hat{D}_j$ the set of partial rankings obtained from D by performing j swaps. We say the chain has converged after r swaps when $\delta(D, \hat{D}_r) \approx \delta(D, \hat{D}_\infty)$. In practice we can determine r by starting the chain from D and stopping when $\delta(D, \hat{D}_j)$ no longer increases. This way of measuring convergence can be questioned as it does not directly use any of the estimated parameters ($X(u)$ in this case), but it is sufficient for our immediate concern of keeping consecutive samples uncorrelated.

4   Experiments

4.1   Data Sets

In the following we briefly discuss the data sets used in the experiments. Table 2 summarizes some of their statistics.

Movielens data. The MovieLens data[1] was originally collected by the GroupLens research group at University of Minnesota. It contains 10^6 ratings for about 3900 movies from over 6000 users. The ratings are given on a scale of 1-5.

Before turning the ratings into partial rankings we preprocess the data as follows. First we discard movies that have been ranked by less than 1000 users. This is done to reduce the number of different movies to 207. As many movies have been seen by only very few users the data does not contain enough information about their relation to the other movies. Next we prune users who have not used the entire scale of five stars in their ratings. This way the resulting partial rankings are more useful as they all reflect the entire preference spectrum from "very bad" to "excellent". This leaves us with 2191 users. For each user we create a partial ranking by picking uniformly at random at most three movies with the same number of stars and ordering them according to the number of stars so that better movies are ranked before the worse ones. The mutual order between two movies with the same number of stars is arbitrary. We call the resulting data set MOVIELENS.

Microarray data. We use a publicly available[2] microarray data from [5]. The data contains expression levels of 1375 genes in 60 cell lines associated to different types of cancer. This data is converted to partial rankings by first sorting the cell lines in decreasing order of expression, separately for each gene. If the expression of a gene for some cell line is unavailable, then this cell line is omitted from the ranking for that gene. Finally we select a random sample of 20 cell lines, again separately for each gene, to have one partial ranking of 20 items for each gene. We call this data set NCI60.

---

[1] http://www.movielens.org/ (24.4.2007)

[2] http://discover.nci.nih.gov/datasetsNature2000.jsp (24.4.2007)

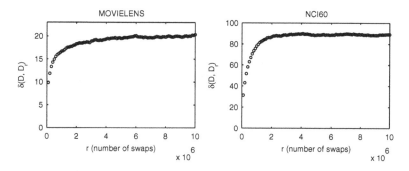

Fig. 2. Convergence of the SWAP-PAIRS algorithm with a movie ranking data (left) and a gene expression data (right). The measure $\delta(D, D_r)$ is the average Kendall distance between rankings in the original data D and the permuted data D_r after r swaps.

4.2   Convergence of SWAP-PAIRS

To make sure that the samples we obtain from $\mathcal{C}(D)$ are not correlated, we use the measure δ discussed above. Before sampling random sets of partial rankings to be used in the actual experiments, we ran the SWAP-PAIRS algorithm on both NCI60 and MOVIELENS data sets for ten million swaps and measured the distance $\delta(D, \hat{D}_r)$ every 10^5 steps. Running time of this test was a little over 14 minutes with the MOVIELENS data and about five minutes with the NCI60 data using a simple Java implementation of SWAP-PAIRS with JRE 1.5 on a 1.8GHz CPU. This difference is due to the different number of items in the data sets. The results are shown in Figure 2.

It is immediate that with the NCI60 data the algorithm seems to have converged after roughly four million swaps. The subsequent samples are all at approximately equal distance from the original set of partial rankings. With MOVIE-LENS the convergence is not as rapid. Even after ten million swaps it appears that $\delta(D, \hat{D}_r)$ is still slightly increasing, albeit extremely slowly.

For our purposes we considered it is enough to make 5 million swaps between samples with both data sets. In case of NCI60 this should yield very uncorrelated samples and also with MOVIELENS the samples should be usable.

4.3   Results

With both data sets we first computed a clustering to three clusters, and then determined the list of outlying items for each cluster. The validity of the found items was tested using the method discussed. We used 100 random data sets sampled with the SWAP-PAIRS algorithm.

When MOVIELENS is clustered to three clusters, we obtain one group with 195, one with 985 and a third one with 1011 partial rankings. The topmost part of Table 3 shows the five movies with largest positive and negative deviations of $X(u)$ from zero in cluster 1. As it contains so few items, the deviations are not very significant in terms of the $Z(u)$ measure. We report the items nonetheless,

Table 1. Outlying items found in clusters 2 and 3 computed from the NCI60 data

Cell line ID	$X(u)$	$E[\|X(u)\|]$	$Std[\|X(u)\|]$	$Z(u)$
Cluster 2: (412 rankings), $Q = 18.92$				
CO:COLO205	14.34	3.9	2.76	3.78
CO:HT29	13.74	3.93	2.72	3.61
CO:SW-620	13.45	3.79	2.57	3.76
CO:KM12	12.14	4.92	3.18	2.27
CO:HCC-2998	11.15	3.47	2.3	3.34
LE:HL-60	10.87	3.66	2.66	2.71
CO:HCT-15	10.71	3.89	2.98	2.29
Cluster 3: (434 rankings), $Q = 12.33$				
BR:MDA-N	19.31	3.8	2.57	6.04
BR:MDA-MB-435	19.05	3.59	2.64	5.86
ME:SK-MEL-5	17	4.71	3.24	3.79
ME:SK-MEL-28	13.83	4.71	3.11	2.93
ME:M14	13.48	3.25	2.27	4.5
ME:UACC-62	13.48	3.96	2.76	3.45
ME:SK-MEL-2	12.22	4.5	3.07	2.52
ME:MALME-3M	11.74	4.15	2.98	2.55
ME:UACC-257	11.17	3.65	2.48	3.03

Table 2. Statistics of the data sets used in the experiments

	MOVIELENS	NCI60
number of partial rankings	2191	1375
total number of items to rank	207	60
average length of partial ranking	13.25	20
length of shortest partial ranking	6	20
length of longest partial ranking	15	20

as they have a nice interpretation. Items that have a high value of $X(u)$ in cluster 1 are successful mainstream action titles, while those that have a low value (are disliked by the viewers) are older, maybe movies for a more mature audience. Clusters 2 and 3 are even more interesting. Table 3 shows the five most preferred and disliked movies for both clusters that are also significant in terms of $Z(u)$. One immediately notices that three movies (Being John Malkovich, Fargo and Reservoir Dogs) that are preferred by cluster 2 have a large negative value of $X(u)$ in cluster 3. In fact, both Pulp Fiction and Election almost made the list as well, as $X = -31.73$ (with $Z = 2.94$) for Pulp Fiction and $X = -35.09$ (with $Z = 4.43$) for Election in cluster 3. These movies are titles that viewers typically either love or hate. This result suggests that the outlying items can be used to identify "causes of controversy" in partial rankings. Also, using the randomization for identifying significant items proved useful, as for instance in cluster 3 the movie having the highest $X(u)$ was The Blair Witch Project with $X = 70.47$ but as its Z was only 0.92, we omitted it from Table 3. There were

Table 3. Outlying items found in three clusters computed from the MOVIELENS data

Movie title	$X(u)$	$E[\|X(u)\|]$	$Std[\|X(u)\|]$	$Z(u)$
Cluster 1: (195 rankings), $Q = 1.10$				
Twister	40.55	18.35	14.78	1.5
The Lost World: Jurassic Park	34.68	13.87	11.29	1.84
Robocop	33.51	16.03	12.2	1.43
The Fifth Element	31.82	14.39	12	1.45
Mad Max 2	31.25	17.64	13.36	1.02
Glory	-37.88	17.52	14.94	1.36
The Godfather	-37.92	14.74	11.31	2.05
Amadeus	-39.47	17.79	13.76	1.58
The Sting	-44.54	19.98	15.92	1.54
North by Northwest	-50.41	25.64	20.24	1.22
Cluster 2: (985 rankings), $Q = 0.85$				
Being John Malkovich	50.27	10.66	9.8	4.04
Fargo	46	10.27	10.37	3.45
Pulp Fiction	32.63	8.9	7.76	3.06
Election	31.91	7.65	6.19	3.92
Reservoir Dogs	30.82	8.02	7.13	3.2
Star Trek: First Contact	-21.67	6.96	5.23	2.81
Mary Poppins	-21.84	6.17	5.6	2.8
Apollo 13 (1995)	-24.24	7.03	5.63	3.06
Forrest Gump (1994)	-29.41	7.7	7.04	3.08
Star Wars: Episode V	-45.31	11.37	9.94	3.41
Cluster 3: (1011 rankings), $Q = 1.46$				
Independence Day	47.15	10.95	9.47	3.82
Ghost	34.69	9.16	7.73	3.3
The Rock	30.91	9.47	7.1	3.02
Men in Black	26.62	8.01	7.37	2.52
Big	24.26	5.5	4.84	3.87
Young Frankenstein	-38.71	7.92	5.94	5.18
Reservoir Dogs	-45.71	8.02	7.13	5.28
Taxi Driver	-46.28	10.62	7.86	4.53
Being John Malkovich	-53.64	10.66	9.8	4.39
Fargo	-60.22	10.27	10.37	4.82

several other examples as well, such as American Beauty with $X = 50.87$ and $Z = 0.94$ in cluster 2. Maybe somewhat unfortunately, in MOVIELENS the $X(u)$s tend to deviate by the same amount also in random data, as shown by the small Q values.

Results for NCI60 are not interpreted as easily. Table 1 shows the cell lines with a large positive deviation of $X(u)$ in clusters 2 and 3 as well as their $Z(u)$ values. In both clusters the cell lines with high values of $X(u)$ come from the same type of cancer. In cluster 2 all but one cell line is a sample from a colon cell, while in cluster 3 all but two are samples associated to melanoma. Interestingly, in this case all the Q values are very high indicating that in random data the deviations of $X(u)$ from zero are in general far smaller.

5   Conclusions and Future Work

We have presented a method for finding items from sets of partial rankings that have different status in different parts of the input. We called such items outlying items. The method is based on clustering of the partial rankings and subsequently computing a simple statistic $X(u)$ for each item. Those items u with $X(u)$ deviating from zero are considered outlying. We also discussed how to generate random data to evaluate the $X(u)$ statistic. The results indicate that the methods can be used to discover items that "divide the opinions" of the partial rankings.

Some interesting questions for future work include properties of the equivalence class $\mathcal{C}(D)$ defined by a set of partial rankings. Another direction for is sampling from $\mathcal{C}(D)$ in a more efficient way than what is possible with the SWAP-PAIRS algorithm. For example, replacing the swap with another kind of transformation that result in shorter mixing times of the Markov chain would be very useful. Another strategy would be to construct a model that can be used to sample the sets of partial rankings more easily. One could, for example, find a partial order P, such that the pair order matrix $C_{\hat{D}}$ obtained, when $\hat{D}$ is generated from uniformly sampled linear extensions of P, would approximate C_D as good as possible.

References

1. Cobb, G., Chen, Y.: An application of markov chain monte carlo to community ecology. American Mathematical Monthly 110, 264–288 (2003)
2. Gionis, A., Mannila, H., Mielikäinen, T., Tsaparas, P.: Assessing data mining results via swap randomization. In: Proceedings of the 12th ACM SIGKDD International Conference on Knowledge Discovery and Data Mining, pp. 167–176. ACM Press, New York (2006)
3. Kamishima, T., Akaho, S.: Efficient clustering for orders. In: Proceedings of the Sixth IEEE International Conference on Data Mining - Workshops (ICDMW'06), pp. 274–278. IEEE Computer Society Press, Los Alamitos (2006)
4. Ryser, H.J.: Combinatorial properties of matrices of zeros and ones. Canad. J. Math. 9, 371–377 (1957)
5. Scherf, U., Ross, D.T., Waltham, M., Smith, L.H., Lee, J.K., Tanabe, L., Kohn, K.W., Reinhold, W.C., Myers, T.G., TAndrews, D., Scudiero, D.A., Eisen, M.B., Sausville, E.A., Pommier, Y., Botstein, D., Brown, P.O., Weinstein, J.N.: A gene expression database for the molecular pharmacology of cancer. Nature Genetics 24(3), 236–244 (2000)
6. Ukkonen, A., Mannila, H.: Finding representative sets of bucket orders from partial rankings (submitted for review)

Speeding Up Feature Subset Selection Through Mutual Information Relevance Filtering

Gert Van Dijck and Marc M. Van Hulle

Katholieke Universiteit Leuven, Computational Neuroscience Research Group, bus 1021,
B-3000 Leuven, Belgium
Gert.VanDijck@mtm.kuleuven.be,
Marc.VanHulle@med.kuleuven.be

Abstract. A relevance filter is proposed which removes features based on the mutual information between class labels and features. It is proven that both feature independence and class conditional feature independence are required for the filter to be statistically optimal. This could be shown by establishing a relationship with the conditional relative entropy framework for feature selection. Removing features at various significance levels as a preprocessing step to sequential forward search leads to a huge increase in speed, without a decrease in classification accuracy. These results are shown based on experiments with 5 high-dimensional publicly available gene expression data sets.

1 Introduction

With the ever increasing feature dimensionality in pattern recognition problems such as in gene expression data [1], text categorization problems or image classification problems, it is important to develop algorithms which increase the speed of the search for optimal feature subsets. Nowadays, feature dimensionalities often reach 10000 to 100000 features.

Besides the speed of a feature selection approach it is important to prove under which conditions optimality will be achieved. In this paper we prove the optimality of an information theoretic (mutual information-based) relevance filter. It can be proven under which conditions it will be optimal by establishing a formal relationship with the *conditional relative entropy* framework from [2]. This relationship will be derived in section 2. Special care in this article is taken towards statistical significance of feature relevance by means of permutation testing, this is explored in section 2.

Experiments in section 4 show that an increase in speed in the wrapper search [3] is obtained if it is preceded by the relevance filter. Although optimality of the filter requires conditions, such as feature independence and class conditional feature independence, the experiments show that the filter does not decrease the classification performance of the feature sets found in the wrapper search.

2 Relevance Filtering

Several other authors have been proposing mutual information as a feature selection criterion: see e.g. MIFS [4], for classification problems and [5] for regression

J.N. Kok et al. (Eds.): PKDD 2007, LNAI 4702, pp. 277–287, 2007.
© Springer-Verlag Berlin Heidelberg 2007

problems. Here, we show that a mutual information criterion for classification problems can be derived from the conditional relative entropy framework for feature selection [2]. This relationship between conditional relative entropy framework for feature selection and mutual information for feature selection has not been established before.

2.1 Optimality of Marginal Feature Relevance

Following lemma establishes a link between the conditional relative entropy framework and marginal mutual information.

Lemma 1: If features are both independent and class conditional independent, then the *conditional relative entropy* (a special case of the Kullback-Leibler divergence, KL) between 2 posterior probabilities is equivalent to a sum of mutual information (MI) contributions of the class label with the omitted features individually.
More formally this can be written as follows.

$$if\ independence:\ \forall F_1, F_2, ..., F_n : p(F_1, F_2, ..., F_n) = \prod_{i=1}^{n} p(F_i)$$

$$and\ class\ conditional\ independence:\ \forall F_1, F_2, ..., F_n, C : p(F_1, F_2, ..., F_n \mid C) = \prod_{i=1}^{n} p(F_i \mid C)$$

$$then\ KL(P(C \mid F_1, F_2, ..., F_n) \parallel P(C \mid F_{l_1}, F_{l_2}, ..., F_{l_{n1}})) = \sum_{m_i} MI(C; F_{m_i})$$

$$with\ \left\{ F_{l_1}, F_{l_2}, ..., F_{l_{n1}} \right\} \subset \left\{ F_1, F_2, ..., F_n \right\}\ and$$

$$\left\{ F_{m_1}, F_{m_2}, ..., F_{m_{n2}} \right\} = \left\{ F_1, F_2, ..., F_n \right\} \setminus \left\{ F_{l_1}, F_{l_2}, ..., F_{l_{n1}} \right\}$$

Where, $F_1, ...F_n$ refer to the full set of features, indices $l_1, ...l_{n1}$ refer to the included features and indices $m_1, ...m_{n2}$ refer to the omitted features. Note that the full feature set consists of the union of the set of included and omitted features and hence n = n1 + n2. Firstly, we remark that feature independence (condition 1) is not a sufficient condition, also class conditional independence (condition 2) is required. This should not come as a surprise, because the Naïve Bayes (NB) classifier also requires class conditional independence. Despite these assumptions, experiments have shown that the NB outperforms many other classifiers [6].

Proof
Starting from the definition of the conditional relative entropy between 2 posterior probabilities $P(C \mid F_1, F_2, ..., F_n)$ and $P(C \mid F_{l_1}, F_{l_2}, ..., F_{l_{n1}})$, and using the convention that $0\ln(0) = 0$:

$$\sum_c \iiint_f P(c, f_1, f_2, ..., f_n) \ln \left[\frac{P(c \mid f_1, f_2, ..., f_n)}{P(c \mid f_{l_1}, f_{l_2}, ..., f_{l_{n1}})} \right] df_1 df_2 ... df_n \tag{1}$$

Using Bayes' theorem and dividing both the numerator and the denominator by P(c) within the logarithm this is equivalent to:

$$= \sum_c \iiint_f P(c, f_1, f_2, ..., f_n) \ln \left[\frac{\dfrac{P(f_1, f_2, ..., f_n \mid c) P(c)}{p(f_1, f_2, ..., f_n) P(c)}}{\dfrac{P(f_{l_1}, f_{l_2}, ..., f_{l_{n_1}} \mid c) P(c)}{p(f_{l_1}, f_{l_2}, ..., f_{l_{n_1}}) P(c)}} \right] df_1 df_2 ... df_n \qquad (2)$$

Using the assumption of both independence and class conditional independence this can be further written as:

$$= \sum_c \iiint_f P(c, f_1, f_2, ..., f_n) \ln \frac{\prod_{i=1}^{n} P(f_i \mid c) P(c)}{\prod_{i=1}^{n} P(f_i) P(c)} df_1 df_2 ... df_n$$

$$\qquad (3)$$

$$- \sum_c \iiint_f P(c, f_1, f_2, ..., f_n) \ln \frac{\prod_{i=1}^{n_1} P(f_{l_i} \mid c) P(c)}{\prod_{i=1}^{n_1} P(f_{l_i}) P(c)} df_1 df_2 ... df_n$$

Using the definition of conditional probabilities this is equal to:

$$= \sum_{i=1}^{n} \sum_c \iiint_f P(c, f_1, f_2, ..., f_n) \ln \frac{P(f_i, c)}{P(f_i) P(c)} df_1 df_2 ... df_n$$

$$\qquad (4)$$

$$- \sum_{i=1}^{n_1} \sum_c \iiint_f P(c, f_1, f_2, ..., f_n) \ln \frac{P(f_{l_i}, c)}{P(f_{l_i}) P(c)} df_1 df_2 ... df_n$$

Integrating out variables not appearing within the logarithms and applying the definition of mutual information: $MI(F_i; C) = \sum_c \int_{F_i} P(c, f_i) \ln \frac{P(f_i, c)}{P(f_i) P(c)} df_i$, we finally obtain:

$$= \sum_{i=1}^{n} MI(F_i; C) - \sum_{i=1}^{n_1} MI(F_{l_i}; C) \qquad (5)$$

Which is equivalent to:

$$= \sum_{i=1}^{n_2} MI(F_{m_i}; C) \qquad (6)$$

This ends the proof. $\square$

This means that in going from the full set $\{F_1, F_2, ..., F_n\}$ to a subset $\{F_{l_1}, F_{l_2}, ..., F_{l_{n_1}}\} \subset \{F_1, F_2, ..., F_n\}$, the distance between the full set and the subset, i.e.

$KL(P(C \mid F_1, F_2, ..., F_n) \parallel P(C \mid F_{l_1}, F_{l_2}, ..., F_{l_{n1}}))$, consists of the individual mutual information contributions between the omitted features and the class variable. As shown in the derivation, this holds under the assumption of both feature independence and class conditional feature independence. Knowing that the mutual information is always larger than or equal to 0, see [7], we can remove the features for which $MI(F_i;C) = 0$. Hence, features for which the mutual information with the class variable C is 0 can be omitted, without an increase in distance to full posterior probability. The effect of the filter on the performance of feature subset selection will be investigated in the experimental section of this article. However, there are 2 very important reasons for making the assumptions: accuracy and speed. As the result of the theorem shows, we can limit ourselves to 1 dimensional estimators of the mutual information. In general, the accuracy of mutual information estimators decreases with increasing dimensionality and fixed number of data points. Secondly, the mutual information between the class label and each feature can be computed independently from every other feature. Hence, there is a large potential in the increase in speed, while all computations can be scheduled in parallel.

2.2 Permutation Testing

As shown in lemma 1, under appropriate conditions, features for which the mutual information with the class variable, $MI(F_i;C)$, equals 0 can be removed. However, testing whether $MI(F_i;C) = 0$, is not straightforward. Mutual information [7] is defined based on the true underlying density $P(F_i,C)$ and this density is in general inaccessible. Therefore, one needs to rely on finite sample estimators: $\widehat{MI}(F_i;C)$ of the mutual information. Estimators of the mutual information based on histograms are biased [8], due to the discretization in intervals and thus need to be avoided. In this article the mutual information is estimated by a recent k-nearest neighbor approach to entropy estimation which does not need a discretization of the features [9]:

$$\hat{H}(F_i) = -\psi(k) + \psi(N) + \log c_d + \frac{d}{N} \sum_{i=1}^{N} \log \varepsilon(i) \qquad (7)$$

Here $\psi(.)$ is the psi-function, N the number of data points, $\varepsilon(i)$ is twice the distance from the i'th data point to its k'th neighbor, d the dimensionality and c_d the volume of the d-dimensional unit ball. For the maximum norm one has $c_d = 1$, for the Euclidean norm $c_d = \pi^{d/2}/\Gamma(1 + d/2)$, with $\Gamma(.)$ the gamma-function. Because features are considered individually, d is equal to 1.

The mutual information can then be computed from the entropy by means of:

$$\widehat{MI}(F_i;C) = \hat{H}(F_i) - \sum_{j=1}^{k} \hat{H}(F_i \mid c_j)P(c_j) \qquad (8)$$

Here $\hat{H}(F_i \mid c_j)$ is the entropy of the feature F_i conditioned on the class c_j, $P(c_j)$ is the probability for the j'th class. We take 'k' equal to 6 in the experiments.

An intelligent approach to test whether the data is likely to be a realization of a certain null-hypothesis can be obtained by manipulating the data such that the null-hypothesis is fulfilled and computing the test-statistic under these manipulations. Hereto, class labels are randomly permuted; this causes the class variable to be independent of the feature F_i. Any statistical dependence that is left is contributed to coincidence. This procedure is known as permutation testing. We restrict ourselves to 1000 permutations.

3   Feature Subset Selection

Here, we propose the wrapper subset selection that will be used in the experiments. The wrapper selection consists of following components: search procedure, induction algorithm and the criterion function to evaluate a particular subset.

3.1   Search Procedure

Many feature selection algorithms have been proposed and explored in the literature. The most well-known search procedures are: sequential search algorithms [10], Branch and bound [11], Genetic algorithms [12], [13]. Sequential forward search (SFS) is the algorithm being used here. We discuss it in more detail here, because it will allow us to explain some of the observations made in the experiments.

Suppose we have a set $\mathbf{Y}$ of D available features: $\mathbf{Y} = \{F_1, F_2, \dots F_D\}$; denote the set of k selected features by $\mathbf{X}_k = \{x_i: 1 \leq i \leq k, x_i \in \mathbf{Y}\}$. Suppose that we also dispose of a feature selection criterion function $J(.)$. The particular choice of $J(.)$ will be addressed in 3.2. The following pseudo-code defines the SFS algorithm. We implicitly assume in the input that we have feature realizations $\mathbf{f}^j = [f_1^j, f_2^j, \dots f_D^j]$ and class labels c^j to our disposal, where index 'j' refers to a particular data point. This is written in shorthand notation by the variable notation $F_1, F_2, \dots F_D$ and C in the input. In the initialization, we set the performance of the empty set $J(\mathbf{X}_{opt} = \varnothing)$ equal to 0. It is assumed that 0 is the minimum value of $J(.)$. In each iteration of the body of the pseudo-code, the most significant feature is searched for, until some predefined dimension d.

$\underline{Input:}$ features $F_1, F_2, \dots F_D$; class labels C; dimension d
$\underline{Initialize:}$ $\mathbf{Y} = \{F_1, F_2, \dots F_D\}$; $\mathbf{X}_0 = \varnothing$; $\mathbf{X}_{opt} = \varnothing$; $J(\mathbf{X}_{opt}) = 0$
$For\ k = 0\ to\ d\text{-}1$
   $F_i' = argmax_{Fi \in Y} J(\mathbf{X}_k \cup \{F_i\})$;determine the most significant feature F_i' in $\mathbf{Y}$.
   $\mathbf{X}_{k+1} = \mathbf{X}_k \cup F_i'$;update the selected feature set.
   $Y = Y\text{-}\{F_i'\}$;update the set of available features.
   $if\ (J(\mathbf{X}_{k+1}) > J(\mathbf{X}_{opt}))$; update the optimal feature set.
     $\mathbf{X}_{opt} = \mathbf{X}_{k+1}$
   end
end
$\underline{Output:}$ $\mathbf{X}_{opt}$

Here it is assumed that $d \leq D$. The most significant feature F'_i is added to the existing subset (X_k) to form a new subset of features (X_{k+1}). Subsequently this most significant feature is removed from the set of available features Y. Only when the performance of the new subset is better than the optimal set found so far (X_{opt}), the optimal subset is updated. Finally, the feature set with the highest performance is given as an output. If this performance is obtained for several subsets, the smallest subset will be returned, this is due to the strict inequality constraint in the update of X_{opt}.

The SFS is among the feature selection algorithms with the lowest time complexity: $O(D^2)$, see [14], with 'D' the number of features in the full feature set. BAB, SFFS, SBFS have time complexities in the order of $O(2^D)$.

A second motivation for the use of SFS can be found in table 3 of [12]. Here, we observe that the performances of SFS on different data sets are often only a few percent smaller than SFFS and the hybrid genetic algorithms (HGA's). Moreover, the reported results are obtained with leave-one-out cross-validation (LOO-CV) without considering separate training and test sets. It may be that, with the use of separate training sets and test sets, the advantage of the more complex search algorithms, when tested on independent test sets and when taking statistical significance into account, will vanish.

We work with independent test and training sets. In this case the problem of model selection (feature subset selection) and model evaluation is performed on different data sets and this allows the assessing of the overfitting behavior of feature subset selection. Moreover, when using several test sets, the statistical significance of the combined hybrid filter/wrapper approach can be compared to the stand-alone wrapper approach. The creation of several training and test sets is further explained in the experimental section of this paper.

3.2 Performance Estimation

As the performance measure $J(.)$, we use the leave-one out cross-validation (LOO-CV) criterion. Note, that this measure is in general non-monotonic: adding a feature to an existing set X_k, can either increase ($J(X_{k+1}) \geq J(X_k)$) or decrease ($J(X_{k+1}) \leq J(X_k)$) the LOO-CV performance. If LOO-CV is used as a criterion the lowest value is 0 and hence the initialization ($J(X_{opt} = \varnothing) = 0$) in the pseudo-code is correct. If the LOO-CV performance is used, the maximum value that can be achieved is equal to 1. Once this optimal performance is achieved, the loop in the pseudo-code does not need to run until d -1, because no subset with more features can achieve a higher-performance. As the induction algorithm we use the k nearest neighbor approach (k-NN), with k equal to 1. The use of this induction algorithm is motivated by: the ease of use and important feature subset selection articles [12], [14] have been using k-NN as well.

4   Experiments

The data sets used in the experiments are 5 gene-expression data sets: ALL-AML leukemia [15], central nervous system (CNS) [16], colon tumor [17], diffuse large B-cell lymphoma (DLBCL) [18] and MLL leukemia [19]. These data sets are summarized in table 1. These are challenging data sets due to the number of dimensions and the limited number of samples available. Some preprocessing steps have been applied.

Firstly, features are normalized by subtracting the mean and dividing by the standard deviation. Secondly, if missing values exist, these are replaced by sampling from a normal distribution with the same mean and standard deviation as the samples from the same class for which values are available. If no separate training sets and tests were available, approximately two-third of the samples of each class was assigned to the training set and the rest to the test set. The number of samples assigned to the training ad test sets for each data set are summarized in table 1. Note that the first 4 data sets form a binary classification problem, and the fifth a ternary one. For the sake of repeatability and comparability of our results, we do not consider the performance on a single training and test set, but also on reshuffled versions.

Table 1. Summary of properties of the data sets: TR, training set; TE, test set; C1, class 1; C2, class 2; C3, class 3; N, number of features

DATA SET	TR	TR C1	TR C2	TE	TE C1	TE C2	N
ALL	38	27	11	34	20	14	7129
CNS	39	25	14	21	14	7	7129
COLON	42	27	15	20	13	7	2000
DLBCL	32	16	16	15	8	7	4026
		20	17		4	3	
MLL		C3			C3		
	57	20		15	8		12582

Each data set is reshuffled 20 times and has the same number of samples from each class assigned to the training and test set, as shown in table 1. Where possible, these reshufflings are stratified: they contain the same number of examples of each class in the training and the test set as in the original data set. The reshuffling strategy implies that a feature selection procedure can be run 20 times on a different training set of the same data set. The outcome of each run can than be used to compute the test result on the associated test set. This also allows one to assess the variability of the test result and to test the statistical significance of the test results.

4.1 Discussion

In table 2, the results on the 5 data sets are shown for different p-values of the relevance filter. For p = 0, no features are removed and thus the sequential forward search is run on the full dimensionality of the data sets. Hence, this allows to compare the proposed hybrid approach to the stand-alone wrapper approach. For p = 0.1, 0.2 and so on, only those features are retained which exceed the [p*1000] order statistic of the mutual information. The testing performances are clearly much lower and change only slightly with changing p. However, due to the large standard deviations, these changes cannot be shown to be significant with changing 'p'. The first row for each data set in table 2 contains the training performances averaged over the 20 different training sets.

The second row contains the average test performances averaged over the 20 different test sets. The standard deviations of the performances for the training and test sets are shown as well.

The third row of each data set contains the speed-up factor. This has been obtained by summing the time needed to run the SFS algorithm on the 20 training sets for a particular p-value, divided by the same sum without filtering (p = 0). This explains the '1' values in the p = 0 column. The time needed for the filter is ignored: the computations of the MI for the 1000 permutations can be run in parallel; moreover the MI computations for all features can be run in parallel as well, because feature dependencies are ignored. This causes the time of the filter not more than tens of seconds to a few seconds.

From the training performances in table 2, we observe that the wrapper search finds in many cases subsets for which the LOO-CV performances on the training sets is 100% or very close to 100%. The filter has no effect on these training performances, except for the CNS and the colon data set filtered at the p = 1 level. Here considerable lower training performances are obtained: 82.2% and 89.0% respectively.

The average test performances are considerable lower than their corresponding training performances. This shows that it is very important to consider separate training and test sets in these small sample size settings. The SFS procedure tends to overfit heavily. In order to assess whether the filter improves the test performance compared to the p = 0 (unfiltered) test performance, hypothesis testing has been applied. This is indicated by the symbols '0' and '+' beneath the testing performances. A '0' indicates no change and the '+' indicates an improvement. The first symbol is the result of applying a t-test at the $\alpha = 0.05$ significance level, this tests whether the mean of the p = 0 level is different from the mean at the other levels. The second symbol is the result of applying the Wilcoxon rank-sum test at the $\alpha = 0.05$ level. The t-test assumes Gaussianity of the average performances and is therefore sensitive to outliers in the test performance.

An interesting observation is that by filtering at subsequent p-levels, e.g. p = 0, 0.1 and 0.2, sometimes exactly the same subsets are obtained in SFS. This can be explained as follows. Suppose that 5 features were sufficient to obtain a 100% training performance. If at the next higher p-level none of these features are removed, then they will be selected again as the optimal subset. This is due to the deterministic

Table 2. Result of the sequential forward search with various levels of relevance filtering. Average recognition rate results with standard deviations are shown for training and test sets. The speed-up factors for the filtered data sets are shown as well.

Data Set	P = 0.0	P = 0.1	P = 0.2	P = 0.4	P = 0.6	P = 0.8	P = 0.9	P = 1.0
ALL train	100.0 ± 0.0	100.0 ± 0.0	100.0 ± 0.0	100.0 ± 0.0	100.0 ± 0.0	100.0 ± 0.0	100.0 ± 0.0	100.0 ± 0.0
ALL test	84.0 ± 10.0	82.9 ± 10.0 0 0	83.8 ± 10.6 0 0	82.9 ± 10.2 0 0	83.1 ± 10.9 0 0	85.1 ± 9.6 0 0	86.0 ± 8.4 0 0	86.9 ± 5.7 0 0
ALL speed up	1	1.42	1.54	1.84	2.39	3.72	4.17	63.69
CNS train	99.5 ± 1.1	99.5 ± 1.3	99.7 ± 0.8	99.6 ± 0.9	99.4 ± 1.1	99.1 ± 1.7	99.2 ± 1.2	82.2 ± 4.9
CNS test	55.2 ± 10.1	56.7 ± 8.9 0 0	54.8 ± 8.5 0 0	54.8 ± 10.3 0 0	57.1 ± 10.1 0 0	57.6 ± 8.7 0 0	59.3 ± 9.9 0 0	66.2 ± 9.0 + +
CNS speed up	1	1.15	2.12	2.62	2.61	4.67	10.94	2352
Colon train	99.9 ± 0.5	99.9 ± 0.5	100.0 ± 0	99.8 ± 0.7	99.8 ± 0.7	99.8 ± 0.7	99.5 ± 1.0	89.0 ± 4.3
Colon test	69.5 ± 9.6	72.7 ± 7.7 0 0	73.8 ± 9.4 0 0	73.0 ± 7.5 0 0	74.8 ± 8.5 0 0	70.3 ± 9.1 0 0	75.5 ± 10.5 0 0	76.0 ± 10.6 + 0
Colon speed up	1	1.07	1.91	1.38	1.59	3.63	4.88	309.7
DLBCL train	100.0 ± 0.0	100.0 ± 0.0	100.0 ± 0.0	100.0 ± 0.0	100.0 ± 0.0	100.0 ± 0.0	100.0 ± 0.0	100.0 ± 0.0
DLBCL test	75.7 ± 13.2	75.0 ± 12.6 0 0	75.0 ± 12.6 0 0	75.7 ± 12.5 0 0	78.3 ± 10.3 0 0	81.3 ± 7.7 0 0	82.3 ± 7.9 0 0	80.3 ± 11.5 0 0
DLBCL speed up	1	1.40	1.50	1.38	2.44	3.0	4.36	46.05
MLL train	100.0 ± 0.0	100.0 ± 0.0	100.0 ± 0.0	100.0 ± 0.0	100.0 ± 0.0	100.0 ± 0.0	100.0 ± 0.0	100.0 ± 0.0
MLL test	84.0 ± 11.7	84.0 ± 11.7 0 0	84.0 ± 11.7 0 0	82.7 ± 10.9 0 0	82.7 ± 10.9 0 0	84.7 ± 8.9 0 0	81.0 ± 9.7 0 0	88.0 ± 9.3 0 0
MLL speed up	1	1.03	1.05	1.37	1.67	1.83	1.76	6.70

behavior of the SFS procedure. This explains the occurrence of the same test performances for subsequent p-values: this can be observed in the MLL test performances (for p = 0, 0.1 and 0.2) and to a smaller extent in the DLBCL data set (for p = 0.1 and 0.2).

From table 2, it can be observed that in almost all cases no significant change in the test performance is obtained after filtering and that both the t-test and Wilcoxon rank-sum test agree on this. A significant improvement is obtained for the CNS data set and the colon data set both filtered at the p = 1 level, although in the latter data set the t-test and the Wilcoxon rank-sum test disagree on this. It is also for these cases that the LOO-CV training performance is lower. Hence, it seems that filtering at a high significance level reduces somewhat the overfitting behavior.

The most interesting results are obtained with the increase of the speed of the SFS algorithm on the filtered data sets. Especially the gain in speed for the p = 0.9 and p = 1 are large. For p = 1 the speed up factors are respectively: 63.69, 2352, 309.7, 46.05, 6.70. The large differences in some of these gains can be explained by the number of features left compared to the full dimensionality. These are for the different data sets respectively: 169 out of 7129, 10 out of 7129, 10 out of 2000, 85 out of 4026 and 2011 out of 12582. In some cases, it is possible that the gain in speed does not increase with a small increase in p. This can be explained by the fact that when initially in the search a few good genes are found with 100% LOO-CV performance, the SFS can be stopped. However, if a data set is filtered at a higher p-level, some of the genes may be removed. The SFS procedure might then need to search a longer time to find a 100% LOO-CV subset, or even worse it needs to execute all iterations of the loop. This can be observed in the DLBCL data set (when p is changed from p = 0.2 to 0.4) and in the colon data set (when p is changed from p = 0.2 to 0.4 and 0.6).

5   Conclusions

It is proven that features, for which the mutual information is equal to 0, can be removed without loss of information, if features are both independent and class conditional independent. This result has been obtained by establishing the link with the conditional relative entropy framework for feature subset selection. This allowed motivating mutual information as an optimal criterion in feature selection, rather than a heuristic one. Removing features at various significance levels of the mutual information statistic prior to sequential forward search (SFS) with the 1-NN method does not decrease the average performance of the test sets. A huge increase in speed can be obtained by first filtering features at the 0.9 or 1.0 level.

Acknowledgements. The first author is supported by the Institute for the Promotion of Innovation through Science and Technology in Flanders (IWT Vlaanderen).

The second author is supported by the Excellence Financing program of the K.U. Leuven (EF 2005), the Belgian Fund for Scientific Research Flanders (G.0248.03, G.0234.04), the Flemish Regional Ministry of Education (Belgium) (GOA 2000/11), the Belgian Science Policy (IUAP P5/04), and the European Commission (NEST-2003-012963, STREP-2002-016276, and IST-2004-027017). This work made use of the HPC (High Performance Computing) infrastructure of the K. U. Leuven.

References

1. Quackenbush, J.: Microarray Analysis and Tumor Classification. The New England Journal of Medicine 354, 2463–2472 (2006)
2. Koller, D., Sahami, M.: Toward Optimal Feature Selection. In: Proceedings of the Thirteenth International Conference on Machine Learning, pp. 284–292. Morgan Kaufmann, San Francisco (1996)
3. Kohavi, R., John, G.H.: Wrappers for Feature Subset Selection. Artificial Intelligence 97, 273–324 (1997)
4. Battiti, R.: Using Mutual Information for Selecting Features in Supervised Neural Net Learning. IEEE Transactions on Neural Networks 5, 537–550 (1994)
5. Van Dijck, G., Van Hulle, M.: Speeding-up the Wrapper Feature Subset Selection in Regression by Mutual Information Relevance and Redundancy Analysis. In: Proceedings of the 16th International Conference on Artificial Neural Networks, pp. 31–40 (2006)
6. Domingos, P., Pazzani, M.: On the Optimality of the Simple Bayesian Classifier under Zero-one Loss. Machine Learning 29, 103–130 (1997)
7. Cover, T.M., Thomas, J.A.: Elements of Information Theory, 2nd edn. John Wiley & Sons, Hoboken New Jersey (2006)
8. Paninski, L.: Estimation of Entropy and Mutual Information. Neural Computation 15, 1191–1253 (2003)
9. Kraskov, A., Stögbauer, H., Grassberger, P.: Estimating Mutual Information. Physical Review E 69, 066138-1 – 066138-16 (2004)
10. Pudil, P., Novovičová, J., Kittler, J.: Floating Search Methods in Feature Selection. Pattern Recognition Letters 15, 1119–1125 (1994)
11. Narendra, P.M., Fukunaga, K.: A Branch and Bound Algorithm for Feature Subset Selection. IEEE Transactions on Computers C- 26, 917–922 (1977)
12. Oh, I.-S., Lee, J.-S., Moon, B.-R.: Hybrid Genetic Algorithms for Feature Selection. IEEE Transactions on Pattern Analysis and Machine Intelligence 26, 1424–1437 (2004)
13. Siedlecki, W., Sklansky, J.: A Note on Genetic Algorithms for Large-scale Feature Selection. Pattern Recognition Letters 10, 335–347 (1989)
14. Kudo, M., Sklansky, J.: Comparison of Algorithms that Select Features for Pattern Classifiers. Pattern Recognition 33, 25–41 (2000)
15. Golub, T.R., et al.: Molecular Classification of Cancer: Class Discovery and Class Prediction by Gene Expression Monitoring. Science 286, 531–537 (1999)
16. Pomeroy, S.L., et al.: Prediction of Central Nervous System Embryonal Tumour Outcome Based on Gene Expression. Nature 415, 436–442 (2002)
17. Alon, U., et al.: Broad Patterns of Gene Expression Revealed by Clustering Analysis of Tumor and Normal Colon Tissues Probed by Oligonucleotide Arrays. In: Proceedings of the National Academy of Sciences of the United States of America vol. 96, pp. 6745–6750 (1999)
18. Alizadeh, A.A., et al.: Distinct Types of Diffuse Large B-cell Lymphoma Identified by Gene Expression Profiling. Nature 403, 503–511 (2000)
19. Armstrong, S.A., et al.: MLL Translocations Specify a Distinct Gene Expression Profile that Distinguishes a Unique Leukemia. Nature Genetics 30, 41–47 (2002)

A Comparison of Two Approaches to Classify with Guaranteed Performance

Stijn Vanderlooy[1] and Ida G. Sprinkhuizen-Kuyper[2]

[1] MICC-IKAT, Universiteit Maastricht, P.O. Box 616, 6200 MD Maastricht,
The Netherlands
s.vanderlooy@micc.unimaas.nl

[2] NICI, Radboud University Nijmegen, P.O. Box 9104, 6500 HE Nijmegen,
The Netherlands
i.kuyper@nici.ru.nl

Abstract. The recently introduced transductive confidence machine approach and the ROC isometrics approach provide a framework to extend classifiers such that their performance can be set by the user prior to classification. In this paper we use the k-nearest neighbour classifier in order to provide an extensive empirical evaluation and comparison of the approaches. From our results we may conclude that the approaches are competing and promising generally applicable machine learning tools.

1 Introduction

In the past decades supervised learning algorithms have been applied to solve various classification tasks with growing success. However, it remains difficult to apply the learned classifiers in domains where incorrect classifications have high costs. Examples of such domains are medical diagnosis and law enforcement.

Recently two approaches have been introduced that extend an existing classifier such that the performance can be set by the user prior to classification. The approaches are called the transductive confidence machine (TCM) approach [1] and the receiver operating characteristic (ROC) isometrics approach [2]. They can construct reliable classifiers since a desired performance can be guaranteed. The key idea of both approaches is to identify instances for which there is uncertainty in the true label. These uncertain instances are inherent to the classification task and they lead to (unaffordable) incorrect classifications. The TCM approach assigns multiple labels to uncertain instances and the ROC isometrics approach leaves uncertain instances unclassified. The approaches are novel with respect to existing reject rules since they are classifier independent and they can guarantee any preset performance. This is not the case when, for example, thresholding the posterior probabilities of a naive Bayes classifier: an upper bound on the error is only guaranteed when the prior is correct [1].

In this paper we use the k-nearest neighbour classifier in combination with the TCM approach and the ROC isometrics approach. In this way we obtain two new classifiers. We use benchmark datasets to verify if the classifiers indeed guarantee a preset performance. We also compare them by analyzing which and

J.N. Kok et al. (Eds.): PKDD 2007, LNAI 4702, pp. 288–299, 2007.

how many instances are considered as uncertain. The best classifier deals with as few uncertain instances as possible in order to guarantee the preset performance.

The remainder of this paper is as follows. Section 2 defines the problem statement that we address. Sections 3 and 4 explain the TCM approach and the ROC isometrics approach, respectively. Section 5 investigates empirically how well both approaches perform and a comparison between them is given. Section 6 provides a discussion. Finally, Section 7 concludes that the approaches are competing and promising generally applicable machine learning tools.

2 Problem Statement

We consider the supervised machine learning setting. The instance space is denoted by $\mathcal{X}$ and the label space by $\mathcal{Y}$. An example is of the form $z = (x, y)$ with $x \in \mathcal{X}$ and $y \in \mathcal{Y}$. Training data are considered as a sequence of iid examples:

$$S = (x_1, y_1), \ldots, (x_n, y_n) = z_1, \ldots, z_n \ . \tag{1}$$

We desire that the performance of a classifier can be set by the user prior to classification. For example, if the user specifies an accuracy of 95%, then the percentage of incorrect classifications may not exceed 5%. Preset performance and empirical performance are used to denote the user-specified performance and the performance of the classifier, respectively. Thus, empirical performance should be at least preset performance. We also desire that the classifier is efficient in the sense that the number of instances with a reliable and useful classification is high. A classifier is not efficient when it outputs many possible labels for most instances or when it refuses to classify most instances.

3 Transductive Confidence Machine Approach

The TCM approach allows classifiers to assign a set of labels to instances. These sets are called prediction sets. A prediction set contains more than one label if there is uncertainty in the true label of the instance.

Subsection 3.1 explains the construction of prediction sets. Subsection 3.2 shows how to use the k-nearest neighbour classifier (k-NN) as a TCM and Subsection 3.3 shows how to assess the quality of a TCM.

3.1 Algorithm

To construct a prediction set for an unlabeled instance x_{n+1}, TCMs operate in a transductive way [1,3]. Each possible label $y \in \mathcal{Y}$ is tried as a label for instance x_{n+1}. In each try, the example $z_{n+1} = (x_{n+1}, y)$ is formed and added to the training data. Then, each example in the extended sequence:

$$(x_1, y_1), \ldots, (x_n, y_n), (x_{n+1}, y) = z_1, \ldots, z_{n+1} \ , \tag{2}$$

is assigned a nonconformity score by means of a nonconformity measure. This measure defines how nonconforming an example is with respect to other available examples.[1] The nonconformity score of example z_i is denoted by α_i.

Since nonconformity scores can be scaled arbitrary, the nonconformity score α_{n+1} is compared to all other α_i in order to know how nonconforming the artificially created example z_{n+1} is in the extended sequence.

Definition 1. *Given a sequence of nonconformity scores $\alpha_1, \ldots, \alpha_{n+1}$ with $n \geq 1$, the p-value of label y assigned to an unlabeled instance x_{n+1} is defined as:*

$$p_y = \frac{|\{i = 1, \ldots, n+1 : \alpha_i \geq \alpha_{n+1}\}|}{n+1} . \tag{3}$$

If the p-value is close to its lower bound $1/(n+1)$, then example z_{n+1} is very nonconforming. The closer the p-value is to its upper bound 1, the more conforming example z_{n+1} is. Hence, the p-value indicates how likely it is that the tried label is in fact the true label of the unlabeled instance. A TCM outputs the set of labels with p-values above a preset significance level ϵ.

Definition 2. *A transductive confidence machine determined by some nonconformity measure is a function that maps each sequence of examples $z_1, \ldots, z_n$ with $n \geq 1$, unlabeled instance x_{n+1}, and significance level $\epsilon \in [0, 1]$ to the prediction set:*

$$\Gamma^\epsilon(z_1, \ldots, z_n, x_{n+1}) = \{y \in \mathcal{Y} \mid p_y > \epsilon\} . \tag{4}$$

Given a preset significance level ϵ, the performance of a TCM measured by means of accuracy is $100(1 - \epsilon)\%$ [1]. For instance, if the user specifies a significance level of 0.05, then 5% of the computed prediction sets do not contain the true label of the corresponding instances. Thus, the performance is 95%.

3.2   TCM-kNN

Any classifier can be applied as a TCM when the used nonconformity measure identifies an example as nonconforming when its classification is uncertain. This subsection reviews TCM-kNN, a TCM based on k-NN [5].

According to k-NN, an example is nonconforming when it is far from nearest neighbours of the same class and close to nearest neighbours of different classes. So, given example $z_i = (x_i, y_i)$, define an ascending ordered sequence $D_i^{y_i}$ with distances from instance x_i to its nearest neighbours with label y_i. Similarly, let $D_i^{-y_i}$ contain ordered distances from instance x_i to its nearest neighbours with label different from y_i. The nonconformity score is then defined as:

$$\alpha_i = \frac{\sum_{j=1}^{k} D_{ij}^{y_i}}{\sum_{j=1}^{k} D_{ij}^{-y_i}} , \tag{5}$$

---

[1] The next subsection provides a nonconformity measure for the k-nearest neighour classifier. Measures for other classifiers are found in [1,4] and references therein.

with subscript j representing the j-th element in a sequence [5]. Clearly, the nonconformity score is monotonically increasing when distances to nearest neighbours of the same class increase and/or distances to nearest neighbours of different classes decrease.

3.3 Quality Assessment

The quality of TCMs is assessed by two key statistics. The first statistic is the percentage of prediction sets that contain the true label. This is the empirical performance and its value should be at least the preset performance. The second statistic is efficiency. Efficiency indicates how useful the prediction sets are and it is measured as the percentage of prediction sets with exactly one label. Prediction sets with multiple labels indicate that each of these labels may be correct.[2]

4 ROC Isometrics Approach

The key idea of the ROC isometrics approach is to leave some instances unclassified if there is uncertainty in the true label of those instances. This is done in such a way that a preset performance is guaranteed.

Subsection 4.1 outlines the approach and Subsection 4.2 shows how it can be used for k-NN. Subsection 4.3 shows how to perform quality assessment. For simplicity, we assume a classification task with a positive and negative class.

4.1 Algorithm

The ROC isometrics approach requires two positive values $l(x \mid p)$ and $l(x \mid n)$ that indicate the likelihood that an instance x is positive and negative, respectively. The likelihood values are combined into a score.

Definition 3. *The score of instance x is defined as:*

$$l(x) = \frac{l(x \mid p)}{l(x \mid n)} \ , \tag{6}$$

or ∞ if $l(x \mid n) = 0$.

Scores are used to rank instances from most likely positive to most likely negative [6]. An unlabeled instance is classified as positive when its score is at least the value of some threshold.

Definition 4. *An ROC curve is a plot with false positive rate (fpr) on the horizontal axis and true positive rate (tpr) on the vertical axis. It shows fpr and tpr values for each possible threshold on scores [7].*

---

[2] Prediction sets can also be empty. However, most empty prediction sets occur when the preset significance level results in a TCM with lower performance than is achieved by the conventional classifier [1,4]. This is clearly not an interesting situation.

Theorem 1. *For any point on and below the convex hull of an ROC curve (ROCCH), a classifier can be constructed by thresholding the scores in such a way that it achieves the fpr and tpr values represented by that point [7].*

In the ROC isometrics approach, the user has to preset a desired performance for each class. Positive class performance is defined as the fraction of positive classifications that are correct. Negative class performance is defined analogously. For each preset performance, an ROC isometric is constructed.

Definition 5. *ROC isometrics are curves that connect points with the same performance in the (fpr, tpr) plane [8].*

The intersection point of a positive (negative) class isometric and the ROCCH represents a classifier with the positive (negative) class performance as preset by the user. The intersection point of the two isometrics themselves represents the classifier that guarantees the preset performance on both classes (henceforth simply called performance). We distinguish three cases as shown in Fig. 1.

- **Case 1:** the isometrics intersect on the ROCCH.
  The classifier that guarantees the preset performance lies on the ROCCH. From Theorem 1 it follows that the classifier can be constructed.
- **Case 2:** the isometrics intersect below the ROCCH.
  Theorem 1 also applies in this case. However, classifiers corresponding to points on the ROCCH between the intersection points of the isometrics and the ROCCH have higher performance.[3]
- **Case 3:** the isometrics intersect above the ROCCH.
  Theorem 1 does not apply in this case. The proposed solution is to identify two thresholds $a > b$ that correspond with the intersection points of the positive class isometric and the ROCCH, and the negative class isometric and the ROCCH, respectively. A new instance x_{n+1} is classified as positive if $l(x_{n+1}) \geq a$ and as negative if $l(x_{n+1}) \leq b$. Otherwise, the instance is left unclassified since its classification is uncertain. It has been shown that the resulting classifier guarantees the preset performance (an unclassified instance is not counted as an error) [2].

4.2   ROC-kNN

Any classifier can be used to construct an ROC curve. Some classifiers such as naive Bayes and neural networks naturally provide likelihood values. For other classifiers, such as k-NN, a post-processing technique is needed.

We computed the likelihood value of instance x_i according to k-NN by computing the inverse of the nonconformity score (5). Our result of using k-NN in the ROC isometrics approach is denoted by ROC-kNN.

---

[3] This case is analogous to the case of choosing a significance level in the TCM approach such that empty prediction sets occur (i.e., the preset performance is lower than can be achieved by the conventional classifier).

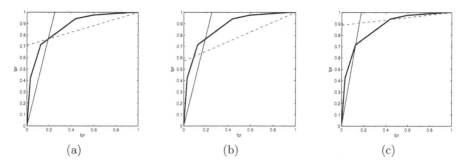

           (a)   (b)   (c)

Fig. 1. Location of the intersection point of a positive class isometric (———) and a negative class isometric (– – –): (a) Case 1, (b) Case 2, and (c) Case 3

4.3 Quality Assessment

Analogously to the TCM approach, quality assessment of the ROC isometrics approach is performed by reporting empirical performance (accuracy) and efficiency. Efficiency is now measured by the percentage of instances for which a label is predicted.

5 Experiments

This section provides the experimental results. We performed experiments with TCM-kNN and ROC-kNN on a number of benchmark datasets. Subsection 5.1 describes these datasets and Subsection 5.2 outlines the experimental setup. Subsection 5.3 evaluates and compares the two classifiers in terms of empirical performance and efficiency.

5.1 Benchmark Datasets

We tested TCM-kNN and ROC-kNN on six well-known binary datasets from the UCI benchmark repository [9] and four binary datasets from a recent machine learning competition [10]. The datasets vary greatly in size and in class ratio. The classes are denoted by positive class (p) and negative class (n). As a preprocessing step, all instances with missing feature values are removed as well as duplicate instances. Features are standardized to have zero mean and unit variance. Finally, linear discriminant analysis is used to project the data into a linear one-dimensional space. This allows for a post-hoc visualization (see Section 5.3). Table 1 summarizes the main characteristics of the datasets.

5.2 Experimental Setup

Nearest neighbours are found with Euclidean distances. The number of nearest neighbours is restricted to $k = 1, 2, \ldots, 10$ and chosen such that the average accuracy of k-NN on the test folds is maximized (using 10-fold cross validation).

Table 1. Benchmark datasets: name, number of examples, minority class, percentage of examples in the minority class, and the accuracy of k-NN (10-fold cross validation). The first six datasets are from the UCI repository, the last four are from the competition.

name	size	min. class	% min. class	accuracy
heart statlog	270	p	44.44	84.81
house votes	342	p	34.21	94.13
ionosphere	350	n	35.71	90.57
monks3	432	n	48.15	78.70
sonar	208	p	46.63	89.90
spect	219	n	12.79	88.12
ada	4560	p	24.81	83.55
gina	3460	p	49.16	93.31
hiva	4220	p	3.52	97.22
sylva	14390	p	6.15	97.57

Once the value of k is chosen for each dataset, the classifiers TCM-kNN and ROC-kNN are applied using 10-fold cross validation. For TCM-kNN, each test instance is used in combination with the training fold to compute its prediction set. For ROC-kNN, each training fold is used to construct an ROCCH in order to find the threshold(s). We report on the average performance (accuracy) and average efficiency over the test folds. To compare both classifiers on efficiency, we say that TCM-kNN leaves an instance unclassified when the corresponding prediction set contains multiple labels. To construct ROC isometrics, we set both the positive class performance and the negative class performance equal to the preset performance. In this way, a preset performance (accuracy) is obtained [2]. We consider five preset performances that we believe to be of interest in many classification tasks: 95%, 96%, 97%, 98%, and 99%.[4]

5.3   Quality Assessment of TCM-kNN and ROC-kNN

In this section we report on the experimental results. Results on the `hiva` and `sylva` datasets are omitted for preset performances below 98% since the conventional classifier has higher performance. No results of ROC-kNN on the `spect` dataset are obtained since the negative class isometric did not intersect the (badly structured) ROCCH.

Table 2 shows that the empirical performance equals the preset performance up to statistical fluctuations, even for the small datasets. These results verify that the performance of both classifiers can be preset. Since some datasets have a highly unbalanced class distribution, it is desired that the empirical performances on each class are approximately equal. Table 3 shows the differences between the positive class performances and the negative class performances. A positive value indicates that the positive class performance is higher than the

---

[4] We do not consider 100% preset performance since a TCM will leave all instances unclassified. A comparison with the ROC isometrics approach is then inappropriate.

negative class performance. The differences are large for TCM-kNN when it is applied on datasets with an unbalanced class distribution. The sign of these differences shows that the classifier uses easy-to-classify instances from the majority class to mask bad performance on the minority class, except for `ionosphere` and `monks3`. The large differences for `spect` are explained by a positive class performance of 100% for all preset performances. As expected, the differences for ROC-kNN show that the preset performance is guaranteed for both classes, with the exception of `ionosphere`, `monks3`, and `ada`. For these datasets, the classifier seems to suffer more from statistical fluctuations. In [2] it is proved that the ROC isometrics approach guarantees a preset performance on each class.

Table 4 shows the efficiency (i.e., the percentage of classified instances) of TCM-kNN and ROC-kNN. In general, efficiency declines exponentially when the preset performance is increased. There is no clear relation between dataset characteristics and efficiency, e.g., datasets with a highly unbalanced class distribution such as `hiva` and `sylva` can still have relatively few unclassified instances. In addition, neither of the two classifiers can be claimed as the most efficient. Noteworthy is the bad efficiency of ROC-kNN on `ionosphere`, `ada`, and `hiva` compared to that of TCM-kNN. This is due to a large number of positive instances and negative instances for which the likelihood values did not result in good scores to discriminate both classes. However, ROC-kNN is the most efficient on the majority of the remaining datasets.

An unclassified instance should be an instance for which there is uncertainty in the true label. We verified this by checking visually if TCM-kNN and ROC-kNN leave instances unclassified that lie in or close to the overlap of the class data histograms. Figure 2 gives an example. Since the number of unclassified instances is limited, we expect that many instances left unclassified by TCM-kNN and ROC-kNN are identical. Table 5 (left part) verifies a large percentage of identical unclassified instances that, in general, increases when the preset performance increases.[5] However, focusing on identical unclassified instances underestimates the resemblance of the two classifiers since unclassified instances very close to each other should also be considered as identical. Therefore, Table 5 (right part) shows the percentage of unclassified instances that are approximately identical, i.e., unclassified instances of both classifiers that are among the ten nearest neighbours of each other. The results clearly show that the classifiers are similar in terms of identifying instances with uncertainty in the true label.

6 Discussion

Our experimental results show that TCM-kNN and ROC-kNN are competing classifiers in terms of guaranteed performance and efficiency. This section provides a discussion on these results by elaborating on some noteworthy differences between the TCM approach and the ROC isometrics approach.

---

[5] Reported values are computed as the fraction of the number of equal unclassified instances and the minimum number of unclassified instances of the two classifiers. The fraction is then converted to a percentage.

Table 2. Empirical performances of TCM-kNN (left part) and ROC-kNN (right part): the preset performances are guaranteed up to statistical fluctuations, even for small datasets

dataset	95%	96%	97%	98%	99%	95%	96%	97%	98%	99%
heart	95.2	96.0	96.7	98.5	99.3	94.5	95.6	97.0	98.0	98.8
house votes	95.3	95.9	97.1	97.9	99.1	95.8	96.1	97.2	97.7	98.7
ionosphere	95.4	96.3	97.1	98.3	98.7	95.1	95.0	96.1	97.5	98.6
monks3	95.1	95.8	97.4	98.1	99.3	95.6	96.0	96.9	97.7	99.0
sonar	95.5	96.5	97.5	98.5	99.5	94.8	95.9	97.0	98.1	98.8
spect	95.2	96.2	97.6	98.6	99.5	-	-	-	-	-
ada	95.0	96.0	97.0	98.0	99.0	96.0	96.9	97.8	98.5	99.3
gina	95.0	96.0	97.0	98.1	99.1	95.0	96.0	97.0	98.0	99.0
hiva	-	-	-	98.0	99.0	-	-	-	98.0	99.0
sylva	-	-	-	98.0	99.0	-	-	-	98.0	99.0

Table 3. Differences between the empirical performances on each class of TCM-kNN (left part) and ROC-kNN (right part): TCM-kNN gives empirical performances that, in general, are far less balanced over the classes than is the case with ROC-kNN

dataset	95%	96%	97%	98%	99%	95%	96%	97%	98%	99%
heart	1.0	1.0	-0.4	-0.7	-1.9	0.2	1.8	-2.5	-2.5	-2.0
house votes	-7.9	-6.2	-5.3	-4.2	-2.4	-2.1	-1.7	-1.5	-0.1	-2.1
ionosphere	-5.7	-4.6	-4.3	-2.8	-1.9	-1.7	-1.9	-7.4	-10.0	-8.3
monks3	7.4	7.0	5.2	3.7	1.6	5.0	7.8	6.9	1.3	1.2
sonar	1.7	0.0	-1.0	-0.8	-1.0	-0.2	-0.1	-2.2	1.1	-1.3
spect	32.5	26.2	19.5	7.0	3.33	-	-	-	-	-
ada	-12.1	-9.8	-7.4	-5.4	-2.6	-5.0	-6.5	-4.5	-5.0	-6.9
gina	0.5	0.6	0.1	-0.5	-0.3	0.0	0.0	-0.2	-0.5	-0.6
hiva	-	-	-	-33.7	-19.7	-	-	-	-0.6	-1.3
sylva	-	-	-	-24.0	-13.3	-	-	-	-0.2	0.1

The TCM approach extends a classifier such that any preset performance (accuracy) can be guaranteed when a consistent nonconformity measure is used [1,4,5]. In contrast, the highest performance that can be achieved using the ROC isometrics approach depends on the ROCCH. Any preset performance can be guaranteed when the first line segment is vertical and the last line segment is horizontal. This is the case when the highest score and the lowest score are assigned exclusively to positive instances and negative instances, respectively. A few outliers or noise in the data can therefore cause that some preset performances cannot be guaranteed. In addition, randomization of two thresholds on scores is needed when an isometric does not intersect the ROCCH in an endpoint of two adjacent line segments [7]. For small datasets this can result in a deviation from preset performance, although in our experiments the deviations are negligible, as seen in Table 6. Finally, limited amount of data can yield an ROC curve that is a bad estimate of the true curve.

Table 4. Efficiency of TCM-kNN (left part) and ROC-kNN (right part): no classifier is clearly more efficient than the other classifier

dataset	95%	96%	97%	98%	99%	95%	96%	97%	98%	99%
heart	65.7	55.9	52.0	46.0	31.7	55.6	54.0	47.0	40.0	35.0
house votes	99.0	97.6	95.8	93.7	79.0	99.4	99.4	97.6	95.9	87.9
ionosphere	83.9	81.2	78.3	72.3	50.5	84.0	78.0	52.0	35.7	30.0
monks3	71.2	69.6	33.2	31.3	28.9	57.2	53.3	52.8	51.2	44.2
sonar	89.2	85.9	78.6	73.6	64.4	90.0	84.5	79.5	77.5	72.5
spect	43.0	38.4	34.7	27.4	18.7	-	-	-	-	-
ada	57.3	53.6	48.9	41.6	28.6	42.6	37.8	31.9	25.2	19.6
gina	96.5	93.8	90.3	84.7	74.0	96.5	93.8	89.5	83.8	67.2
hiva	-	-	-	98.2	92.0	-	-	-	80.0	73.6
sylva	-	-	-	97.9	95.0	-	-	-	99.5	96.6

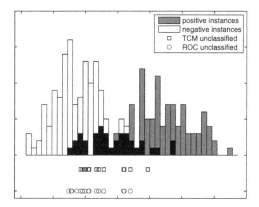

Fig. 2. Unclassified instances of TCM-kNN ($\square$) and ROC-kNN (O) for `heart statlog` dataset with a preset performance of 95%: the dark region in the middle of the class histograms shows the overlap of the histograms (the region of uncertainty).

On the other hand, the ROC isometrics approach has a significant computational advantage since a classifier is only trained once and subsequently used to construct an ROCCH. In addition, the approach is able to incorporate different costs of incorrect classifications, e.g., one can specify that classifying a negative instance incorrectly is more severe than classifying a positive instance incorrectly. The ability to incorporate costs is important for two reasons: (1) most application domains have a non-uniform cost distribution and (2) the cost distribution often changes in time. Costs are incorporated via the isometrics since the (fpr, tpr) plane is independent of the cost distribution. Finally, isometrics can be constructed for a variety of performance metrics such as the m-estimate and the F-measure [2].

The TCM approach can be applied without modification to multi-class classification problems. This is not the case for the ROC isometrics approach: given C classes, the search space has dimension $C^2 - C$ and isometrics are not yet

Table 5. Comparison of unclassified instances: identical unclassified (left part) and approximately identical unclassified (right part)

dataset	95%	96%	97%	98%	99%	95%	96%	97%	98%	99%
heart	84.6	84.8	93.0	92.4	95.4	97.8	93.2	100.0	100.0	98.9
house votes	0.0	33.3	71.4	81.3	73.9	100.0	100.0	100.0	100.0	82.6
ionosphere	45.5	55.6	65.8	69.9	80.0	64.5	66.7	75.3	82.8	85.9
monks3	86.2	87.7	94.2	94.9	90.2	97.6	95.4	96.6	97.7	91.0
sonar	100.0	92.3	90.2	95.6	100.0	100.0	96.2	100.0	100.0	100.0
spect	-	-	-	-	-	-	-	-	-	-
ada	100.0	100.0	100.0	100.0	98.7	100.0	100.0	100.0	100.0	100.0
gina	82.5	89.5	99.4	98.1	93.5	84.2	95.2	100.0	100.0	100.0
hiva	-	-	-	86.8	100.0	-	-	-	100.0	100.0
sylva	-	-	-	79.4	94.4	-	-	-	81.0	96.6

Table 6. Standard deviations of the empirical performance by applying ROC-kNN ten times: randomization of thresholds has minor influence, even for small datasets

dataset	95%	96%	97%	98%	99%
heart	0.77	0.44	1.08	0.95	1.03
house votes	0.44	0.36	0.26	0.51	0.23
ionosphere	0.33	0.33	0.42	1.35	0.40
monks3	0.76	0.62	0.65	0.44	0.44
sonar	0.34	0.65	0.45	0.60	0.43
spect	-	-	-	-	-
ada	0.08	0.12	0.25	0.13	0.17
gina	0.07	0.06	0.03	0.04	0.06
hiva	-	-	-	0.09	0.02
sylva	-	-	-	0.02	0.01

investigated in this space. (See [6] for an overview of approaches to multi-class ROC analysis.) We are currently extending the ROC isometrics approach to multi-class classification problems.

7   Conclusions

In this paper we used the k-nearest neighbour classifier in combination with the TCM approach and the ROC isometrics approach. The two resulting classifiers are applied on ten benchmark datasets in order to provide an extensive empirical evaluation and comparison in terms of performance and efficiency.

We review our contributions and formulate three conclusions. First, we verified that ROC-kNN guarantees a preset performance. This is also the case for TCM-kNN, as pointed out by earlier results [1,4]. Experiments with a naive Bayes classifier gave similar results. We may conclude that, dependent on the classification task, the user can preset the performance such that the number of incorrect classifications that may still occur is acceptable. Second, the approaches can identify instances for which there is uncertainty in the true label. These instances are difficult to classify and analyzing them can result in a better understanding of the problem at hand. Therefore, we may conclude that the approaches provide valuable feedback to the user. Third, we discussed the advantages and disadvantages of each approach. We conclude that the ROC isometrics approach is preferred in the following four situations: (1) fast processing of instances, (2) a balanced performance over the classes, (3) non-uniform cost distribution, and (4) choice of performance metric. The TCM approach is preferred in the following two situations: (1) low performance of a conventional classifier (badly structured ROCCH), and (2) limited amount of data. Clearly, the approaches are generally applicable and promising machine learning tools that should find their way into practice.

Acknowledgments

We thank the reviewers for useful suggestions. The first author is supported by the Dutch Organization for Scientific Research (NWO), grant nr: 634.000.435.

References

1. Vovk, V., Gammerman, A., Shafer, G.: Algorithmic Learning in a Random World. Springer, New York (2005)
2. Vanderlooy, S., Sprinkhuizen-Kuyper, I., Smirnov, E.: An analysis of reliable classifiers through ROC isometrics. In: Lachiche, N., Ferri, C., Macskassy, S., ICML (eds.) 2006 Workshop on ROC Analysis (ROCML 2006), Pittsburgh, USA, pp. 55–62 (2006)
3. Gammerman, A., Vovk, V.: Prediction algorithms and confidence measures based on algorithmic randomness theory. Theoretical Computer Science 287, 209–217 (2002)
4. Vanderlooy, S., van der Maaten, L., Sprinkhuizen-Kuyper, I.: Off-line learning with transductive confidence machines: an empirical evaluation. Technical Report 07-03, Universiteit Maastricht, Maastricht, The Netherlands (2007)
5. Proedrou, K., Nouretdinov, I., Vovk, V., Gammerman, A.: Transductive confidence machines for pattern recognition. Technical Report 01-02, Royal Holloway University of London, London, UK (2001)
6. Lachiche, N., Flach, P.: Improving accuracy and cost of two-class and multi-class probabilistic classifiers using ROC curves. In: Fawcett, T., Mishra, N. (eds.) 20th International Conference on Machine Learning (ICML 2003), pp. 416–423. AAAI Press, Washington, DC, USA (2003)
7. Provost, F., Fawcett, T.: Robust classification for imprecise environments. Machine Learning 42, 203–231 (2001)
8. Flach, P.: The geometry of ROC space: Understanding machine learning metrics through ROC isometrics. In: Fawcett, T., Mishra, N. (eds.) 20th International Conference on Machine Learning (ICML 2003), pp. 194–201. AAAI Press, Washington, DC, USA (2003)
9. Newman, D., Hettich, S., Blake, C., Merz, C.: UCI repository of machine learning databases (1998)
10. Guyon, I.: Data representation discovery workshop of the 20th international joint conference on neural networks (IJCNN 2007) (2007)

Towards Data Mining Without Information on Knowledge Structure

Alexandre Vautier[1], Marie-Odile Cordier[1], and René Quiniou[2]

[1] Irisa - Université de Rennes 1
[2] Irisa - Inria
Campus de Beaulieu 35042 Rennes Cedex, France
{Alexandre.Vautier}@irisa.fr

Abstract. Most knowledge discovery processes are biased since some part of the knowledge structure must be given before extraction. We propose a framework that avoids this bias by supporting all major model structures e.g. clustering, sequences, etc., as well as specifications of data and DM (Data Mining) algorithms, in the same language. A unification operation is provided to match automatically the data to the relevant DM algorithms in order to extract models and their related structure. The MDL principle is used to evaluate and rank models. This evaluation is based on the covering relation that links the data to the models. The notion of schema, related to the category theory, is the key concept of our approach. Intuitively, a schema is an algebraic specification enhanced by the union of types, and the concepts of list and relation. An example based on network alarm mining illustrates the process.

1 Introduction

Frawley et al. [5] have introduced the well-known definition of Knowledge Discovery: "Knowledge discovery is the non trivial extraction of implicit, previously unknown, and potentially useful information from data. Given a set of facts (data) F, a language L, and some measure of certainty C, we define a *pattern* as a statement S in L that describes relationships among a subset F_S of F with a certainty c, such that S is simpler (in some sense) than the enumeration of all facts in F_S".

In most DM (Data Mining) tasks, the language drives the pattern search. This is the case in inductive databases [6], for instance: the user who mines the data has to query a database by using a language L. In other words, he has to define, to some extent, the structure of the "unknown" information. By structure, we mean a decision tree, a clustering, frequent itemsets, etc. So the information is not completely unknown, the structure is at least guessed by the user even if the data inside the structure are unknown.

It is usual to handle data without any idea on their underlying structure. It is the case when you have to mine alarms from a telecommunication network to detect intrusion without a priori knowledge on them. To choose the relevant

J.N. Kok et al. (Eds.): PKDD 2007, LNAI 4702, pp. 300–311, 2007.

DM algorithm to run on a set of alarms is a challenge since even DM specialists are not familiar with the full range of DM algorithms.

The key idea is to propose a framework which provides a specification language called *Schema* in which the user can describe his data and which automatically builds models from various DM algorithms and evaluate them. This point of view is different from the data mining formalization by inductive databases [6] where the DM task is defined as an inductive query to a database. The query contains the "unknown" structure. This inductive database scheme has been instantiated in specific fields: association rules, frequent itemsets, decision trees, etc. However a common framework for inductive databases is still missing.

Our approach is closer to the 3W model [7] that proposes a language to unify DM processes. However, we focus on the automatic computation of models from data whereas the 3W model supports DM as a multi-step process correctly specified. The same authors have proposed later a method [9] to find constrained regions (sets of data cube) that summarize data. The evaluation of these regions relies on the Kolmogorov complexity that we use also to evaluate models.

Bernstein et al. [2] introduce the concept of Intelligent Discovery Assistants (IDAs) "which provide systematic enumeration of valid DM processes and an effective ranking of these valid processes by different criteria to facilitate the choice of DM processes to execute". We use also data description in the form of a specification to find DM algorithms. For the time being, we do not rank DM processes and propose instead an execution of all of them to extract many models. However, we propose a generic rank of extracted models.

The evaluation scores the relevance of a model relatively to the data and a specification (a schema). We do not use classic model evaluation methods since they are not homogeneous and cannot be compared. We introduce a generic evaluation function based on the Kolmogorov complexity and more precisely on MDL [11] (Minimum Description Length). The more a model and a representation of the data in the model are short, the more the model is interesting. The covering relation between a model and the data is used to find different ways to encode data knowing a model.

Back to the example above, the system can extract different structures of models. The network alarms can be viewed as network links in which a DM algorithm looks for frequently used links. Another DM algorithm can extract frequent sequence patterns from an alarm sequence. Finally, alarms can be clustered on their severity and their date. The system has to rank these models in order to present them to the user.

In this paper, we focus on the power and the versatility of the proposed specification language. A prototype implementation exists and is currently applied on the detection of DDoS (Distributed Denial of Service) attacks from an alarm stream. Due to lack of space, we will not develop those computational aspects nor present experiments. The rest of the paper is organized as follows. Section 2 motivates our approach on an example on network alarms that will be used throughout the paper. Section 3 describes our DM framework. Section 4 gives details on the specification of data, models and DM algorithms. It shows how specifications can be unified to execute automatically DM algorithms on data and introduce schema foundations. Furthermore, it presents the specification of covering relations used in the Section 5 to rank models from Kolmogorov complexity. We end with concluding remarks and directions for future work.

2 Motivating Example

To illustrate our method, we show an example where the goal is to extract knowledge from a set of network alarms. An alarm corresponds to a suspicious flow from a source actor to a target actor. An alarm is formally a triple (d, e, s) where d represents the date, e is the link $(source, destination)$ associated to the alarm and s is the severity (1 - low to 3 - high). The graph of Figure 1(a) represents the dataset $\mathcal{A}$. A node represents an actor and an edge labeled by (d, s) represents an alarm occurring at date d with severity s.

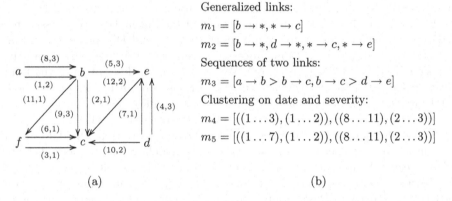

Generalized links:

$m_1 = [b \rightarrow *, * \rightarrow c]$

$m_2 = [b \rightarrow *, d \rightarrow *, * \rightarrow c, * \rightarrow e]$

Sequences of two links:

$m_3 = [a \rightarrow b > b \rightarrow c, b \rightarrow c > d \rightarrow e]$

Clustering on date and severity:

$m_4 = [((1 \ldots 3), (1 \ldots 2)), ((8 \ldots 11), (2 \ldots 3))]$

$m_5 = [((1 \ldots 7), (1 \ldots 2)), ((8 \ldots 11), (2 \ldots 3))]$

(a)   (b)

Fig. 1. (a) Example set $\mathcal{A}$ of network alarms. (b) Models extracted from data $\mathcal{A}$.

Without any information on the structure of knowledge, knowing which DM algorithm to execute on data $\mathcal{A}$ is difficult. In the example of Figure 1, models in the form of generalized links, sequences and clusters could be extracted, as depicted in Figure 1(b). The models m_1 and m_2 are generalized links (links with the symbol $*$). They can be extracted from the alarms in $\mathcal{A}$ by searching nodes with high degree. The model m_3 is composed of 2 sequences of 2 events (an event is a link) and could be extracted by the algorithm of Srikant and Agrawal [12]. The models m_4 and m_5 are two partial clusterings of alarms from $\mathcal{A}$ on the date and the severity. They can be generated by algorithm k-means, for instance.

Firstly, the diversity of DM algorithms and the numerous different ways they can be executed on data make it very difficult for a user to choose DM computation on a given dataset. That is why we provide an automatic way to connect DM algorithms and data. Secondly, the many results of DM algorithm executions on data require a generic evaluation of the resulting models so they could be ranked before being displayed to the user. These two important points are developed in the sequel.

3   A General Data Mining Framework

The framework associated to Data Mining is illustrated in Figure 2. ① The first component is a database that contains the specification of DM algorithms. Each DM algorithm is described by a unique schema. The structure of the model that the algorithm outputs is also described by a schema. ② The user provides also the specification of data in a schema. ③ The system finds the relevant DM algorithms by matching data and algorithm specifications. The corresponding schemas are unified in a new operational schema. ④ Each valid DM algorithm is executed and outputs one or several models. ⑤ Each model is ranked on the MDL principle. This evaluation corresponds to minimize the size of the data according to a model and the covering relation between the model and the data. ⑥ The score of the model is the minimal size found at the previous step.

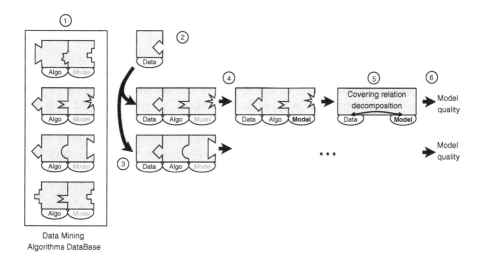

Fig. 2. Framework for mining data without information on knowledge structure

4   A Specification Language Based on Schemas

To begin with, the foundations of schemas are discussed. Instead of giving a formal description of schemas, the grounding elements of the specification language based on schemas are illustrated on the specification of data and the specification of three DM algorithms. Finally, we show how a data schema and a DM algorithm schema can be unified to yield an operational specification.

4.1   Schema Theory

Algebraic specification [4] is based on notions from universal algebras in pure mathematics and on concepts of abstract data types and software specification in computer science. An algebraic specification is composed of a set of *sorts* (also called *type*), a set of *operations* on sorts and a set of *equations* on operations. The domain of an operation is a cartesian product of sorts and its codomain is a unique sort. Three additional concepts are needed to specify a DM problem: powerset, relation and union of types. Powersets are included in the form of lists, relation and union of types will be introduced in the next sections.

The union of types enables a very precise type specification. The concept of *sketch*, later introduced by Ehresmann [3], includes a definition of union of types. Further, this notion of sketch was particularly well described by Barr and Wells [1]. Intuitively a sketch (precisely a finite discrete sketch) is an algebraic specification using the union of types in the form of a graph. The nodes represent the type and the edges represent the operations. To add the notions of relation and powerset, we have extended the concept of sketch to the concept of schema. Intuitively, a schema is an algebraic specification where the concepts of relation, list of types and union of types can be expressed.

4.2 Data Specification

The specification of network alarms is depicted in the schema $\mathcal{S}_d$ of Figure 3. This specification corresponds to the step ② of the general framework. In such a schema, a node represents a type (in the sequel, a type is viewed as a set) and an edge represents a function ($\rightarrow$) or a relation ($\mapsto$). The symbol ($\leftrightarrow$) that is traditionally used to describe a relation is not employed since the relations are represented by lists instead of sets. The green dotted lines ⋯⋯ represent projections and the red dashed lines − − represent inclusions. Functions and relations on some path in the graph can be composed by the operator ∘ . To each node **T** is associated an edge, named *identity*, written $id_T : \mathbf{T} \rightarrow \mathbf{T}$. This edge represents the identity function and is not drawn on a schema, it is implicit.

The type **alarm** is the cartesian product of the types **date**, **link** and **severity**. The edges d, e and s represent projection functions from the type **alarm** to the types **date**, **link** and **severity**, respectively. In the same way, the type **link** is the cartesian product **actor** × **actor** associated to the two projection functions $source : \mathbf{link} \rightarrow \mathbf{actor}$ and $target : \mathbf{link} \rightarrow \mathbf{actor}$.

The type **1** (named terminal object in category theory) is used to define constants. An edge e from the type **1** to the type **T** represents a set of constants of type **T**. For example, the edge $\Sigma_{actor} : \mathbf{1} \mapsto actor$ represents the six constants: a, b, c, d, e and f of the type **actor**. We also assume that each type **T** can be enumerated, i.e. for all type **T**, there is a unique edge $\Sigma_T : \mathbf{1} \mapsto T$. For example, the edge $\Sigma_{link} : \mathbf{1} \mapsto \mathbf{link}$ enumerates the elements $\{(x, y)|x, y \in \mathbf{actor}\}$.

From each node there is an edge, named $\varnothing$, to the terminal object. Thus from every node, one can access any constant by the composition ∘ . For example, the relation $\Sigma_{actor} \circ \varnothing : \mathbf{severity} \mapsto \mathbf{actor}$ gives an access from a severity to any actor. The edges $\varnothing$ are not represented in a schema, they are implicit.

The edge Σ_{actor} represents an inclusion relation. It means that the type **actor** represents a set that contains only the constants defined by the inclusions that arrives to **actor**: the type **actor** contains only the elements a, b, c, d, e and f.

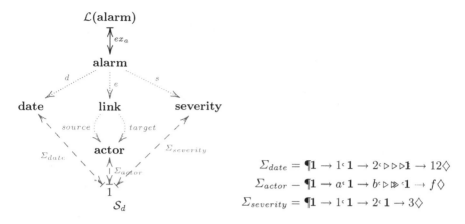

$$\Sigma_{date} = \P\mathbf{1} \to 1^{c}\mathbf{1} \to 2^{c}\triangleright\triangleright\triangleright\mathbf{1} \to 12\lozenge$$

$$\Sigma_{actor} - \P\mathbf{1} \to a^{c}\mathbf{1} \to b^{c}\triangleright\triangleright\cdot\mathbf{1} \to f\lozenge$$

$$\Sigma_{severity} = \P\mathbf{1} \to 1^{c}\mathbf{1} \to 2^{c}\mathbf{1} \to 3\lozenge$$

Fig. 3. The schema $\mathcal{S}_d$ of the network alarms

In the same manner, the type **date** is composed of integers from 1 to 12 and the type **severity** is composed of the integers $1, 2$ and 3. The type $\mathcal{L}(\textbf{alarm})$ represents the set of lists composed of elements of type **alarm**. The relation $ex_a : \mathcal{L}(\textbf{alarm}) \leftrightarrow \textbf{alarm}$ associates a list of alarms with the alarms of the list.

4.3 DM Algorithm and Model Specification

The DM algorithms stored in the database ① are specified in the same specification language as data. In order to show the versatility of schemas, we give the description of three DM algorithms that can extract the five models in the Figure 1(b): the schemas $\mathcal{S}_g, \mathcal{S}_c$ and $\mathcal{S}_s$ of Figures 4, 6 and 8. The specification of covering relations in schemas is particularly emphasized: it explicits the relationship between models and data, and it is useful for model evaluation.

In the schema $\mathcal{S}_g$, $\mathcal{L}(\textbf{edge})$ is the type of graphs which is represented by edge lists. The function *mine_graph* extracts a list of generalized edges (type $\mathcal{L}(\textbf{edgeG})$) from an element of $\mathcal{L}(\textbf{edge})$. The type **node** is considered as *abstract* since the relation Σ_{node} can only partly specified. An element of **nodeG** is either an element of type **node** or the constant $*$ which stands for any element of **node**. This is expressed by the two inclusions $i : \textbf{node} \leftrightarrow \textbf{nodeG}$ and $gen : \mathbf{1} \to \textbf{nodeG}$. These inclusions enable the construction of the relation $c_n = \langle \Sigma_{node}; id_{node} \rangle$. c_n is the covering relation between a generalized node and a node. Precisely, c_n is a *cofactorisation* of the relations $\Sigma_{node} : \mathbf{1} \leftrightarrow \textbf{node}$ and $id_{node} : \textbf{node} \leftrightarrow \textbf{node}$. It corresponds to the construction "if-then-else": for all $x \in \textbf{nodeG}$, if $x \in \textbf{node}$ then $c_n(x) = id_{node}(x)$ else $(x = *)$ $c_n(x) = \Sigma_{node}(x)$.

edgeG is the cartesian product **nodeG** $\times$ **nodeG** and **edge** is the cartesian product **node** $\times$ **node**. The relation $c_e : \textbf{edgeG} \leftrightarrow \textbf{edge}$ is the covering relation between generalized edges and edges. c_e is a *factorisation* of the relation $c_n \circ sourceG : \textbf{edgeG} \leftrightarrow \textbf{node}$ and the relation $c_n \circ targetG : \textbf{edgeG} \leftrightarrow \textbf{node}$. Since an element of **edge** is defined by two elements of **node**, c_e "creates"

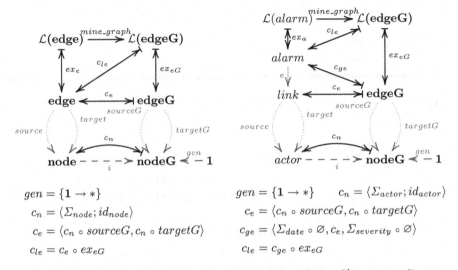

$$gen = \{1 \rightarrow *\}$$
$$c_n = \langle \Sigma_{node}; id_{node} \rangle$$
$$c_e = \langle c_n \circ sourceG, c_n \circ targetG \rangle$$
$$c_{le} = c_e \circ ex_{eG}$$

$$gen = \{1 \rightarrow *\} \qquad c_n = \langle \Sigma_{actor}; id_{actor} \rangle$$
$$c_e = \langle c_n \circ sourceG, c_n \circ targetG \rangle$$
$$c_{ge} = \langle \Sigma_{date} \circ \varnothing, c_e, \Sigma_{severity} \circ \varnothing \rangle$$
$$c_{le} = c_{ge} \circ ex_{eG}$$

Fig. 4. The schema $\mathcal{S}_g$ corresponding to the *mine_graph* algorithm

Fig. 5. The schema $\mathcal{S}'_g$ corresponding to a unification of $\mathcal{S}_g$ with $\mathcal{S}_d$

several elements of **edge** from an element of **edgeG**. Finally the relation c_{le} is the covering relation between a list of generalized edges and the edges.

The schema $\mathcal{S}_c$ (Figure 6) describes a 2-dimensional clustering. In the 2-dimensional space, a cluster can be approximated to a rectangle which can be represented by an horizontal and a vertical interval. A clustering is a list of clusters and the clustering algorithm is represented by the function *mine_cluster* : $\mathcal{L}(\textbf{point}) \rightarrow \mathcal{L}(\textbf{clusterP})$. The considered clustering algorithms are parameter free since we assume that we have no knowledge about the data.

The schema $\mathcal{S}_s$ (Figure 8) describes an algorithm that extract 2-event sequences. An event (**event**) is composed of a time (**time**) and a type (**eventType**). A sequence (**seq**) is composed of two event types. In order to specify the covering relation between a list of events ($\mathcal{L}(\textbf{event})$) and a list of sequences ($\mathcal{L}(\textbf{seq})$) we need to convert an event into a 2-event (two successive events). This encoding expresses the event succession relation by chained 2-events. For example, the event list $(1, A), (2, B), (3, A), (4, A)$ is converted into the 2-event list $(1, A, B), (2, B, A), (3, A, A)$. This way, the covering relation c_s between a sequence and a 2-event list can be specified.

4.4   Schema Unification

In this section, we detail the mechanism of schema unification ③. It is illustrated by unifying each of the three DM algorithm schemas $\mathcal{S}_g, \mathcal{S}_c$ and $\mathcal{S}_s$ to the data schema $\mathcal{S}_d$. The resulting unified schemas $\mathcal{S}'_g, \mathcal{S}'_c$ and $\mathcal{S}'_s$ are presented in Figures 5, 7 and 9. They are used to extract models automatically.

In order to simplify the presentation of unified schemas, all the types are not represented in Figures. For example in the schema $\mathcal{S}'_g$, the type **actor** is not

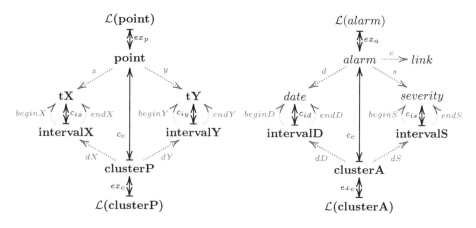

$$c_c = \langle c_{ix} \circ dX, c_{iy} \circ dY \rangle \qquad\qquad c_c = \langle c_{id} \circ dD, c_{is} \circ dS, \Sigma_{link} \circ \varnothing \rangle$$

$$c_{lc} : \mathcal{L}(\mathbf{clusterP}) \longmapsto \mathbf{point} = c_c \circ ex_c \qquad c_{lc} : \mathcal{L}(\mathbf{clusterA}) \longmapsto alarm = c_c \circ ex_c$$

$$mine\_cluster : \mathcal{L}(\mathbf{point}) \to \mathcal{L}(\mathbf{clusterP}) \quad mine\_cluster : \mathcal{L}(\mathbf{alarm}) \to \mathcal{L}(\mathbf{clusterA})$$

Fig. 6. The schema $\mathcal{S}_c$ corresponding to the *mine_cluster* algorithm

Fig. 7. The schema $\mathcal{S}'_c$ corresponding to a unification of $\mathcal{S}_c$ with $\mathcal{S}_d$.

completely defined since the relation Σ_{actor} is not written. By convention, a type that is not written in bold is defined in the schema $\mathcal{S}_d$. This is formally supported by the notion of *morphism of schemas* close to the *morphism of sketches* in category theory. This corresponds intuitively to type inheritance.

The unification $\mathcal{S}_U$ of two schemas $\mathcal{S}_A$ and $\mathcal{S}_B$ is obtained by matching edges and rewriting composed relations (composition, factorisation and cofactorisation). The unification of two schemas is not unique. However, in the DM context of this proposed framework, several constraints have to be respected which decrease the number of potential unifications. Firstly, the type corresponding to the inputs of the DM algorithm and the type corresponding to the type of data specified by the user must be unified. Secondly, the unified schema $\mathcal{S}_U$ should not contain any abstract type. This reflects the fact that mining algorithms works on completely defined inputs and outputs.

The schema $\mathcal{S}'_g$ (Figure 5) is constructed from schemas $\mathcal{S}_d$ and $\mathcal{S}_g$ according to these constraints. *mine_graph*, the graph DM algorithm, is instantiated by the type of data, network alarms in this case. This way, the type **nodeG** represents the set $\{a, b, c, d, e, f, *\}$. The rewriting of the relation c_{le} is a little bit more complex. The types **edge** and $\mathcal{L}(\mathbf{edge})$ are unified with types **link** and $\mathcal{L}(\mathbf{alarm})$ respectively. This way, the *mine_graph* views an alarm as a link. The forgotten attributes date and severity must, however, be taken into account in the covering relation. A new covering relation c_{ge} between **edgeG** and **alarm** is added where c_{ge} is defined by the factorisation $\langle \Sigma_{date} \circ \varnothing, c_e, \Sigma_{severity} \circ \varnothing \rangle$. The old covering relation $c_{le} = c_e \circ ex_{eG}$ is rewritten by replacing c_e by c_{ge}.

The schema $\mathcal{S}'_c$ (Figure 7) is constructed from schemas $\mathcal{S}_d$ and $\mathcal{S}_c$. The function *mine_cluster* views a list of alarms as a list of pairs $(date, severity)$. The relation

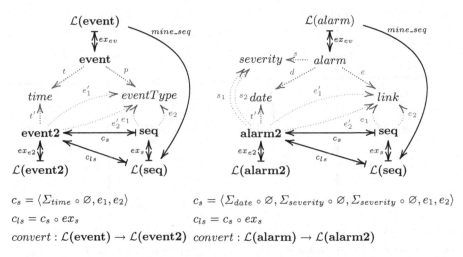

$$c_s = \langle \Sigma_{time} \circ \varnothing, e_1, e_2 \rangle \qquad c_s = \langle \Sigma_{date} \circ \varnothing, \Sigma_{severity} \circ \varnothing, \Sigma_{severity} \circ \varnothing, e_1, e_2 \rangle$$

$$c_{ls} = c_s \circ ex_s \qquad\qquad\qquad c_{ls} = c_s \circ ex_s$$

$$convert : \mathcal{L}(\mathbf{event}) \rightarrow \mathcal{L}(\mathbf{event2}) \quad convert : \mathcal{L}(\mathbf{alarm}) \rightarrow \mathcal{L}(\mathbf{alarm2})$$

Fig. 8. The schema $\mathcal{S}_s$ related to the *mine_sequence* algorithm

Fig. 9. The schema $\mathcal{S}'_s$ corresponding to a unification of $\mathcal{S}_s$ with $\mathcal{S}_d$.

c_c is rewritten in order to include the forgotten information represented by the link element of an alarm. Other unifications could be proposed for clustering network alarms, the only constraint being that the types **Tx** and **Ty** have to be unified with completely ordered types.

The schema $\mathcal{S}'_s$ (Figure 9) is constructed from schemas $\mathcal{S}_d$ and $\mathcal{S}_s$. The event type is the link element of an alarm. The type **alarm2** needs two projections s_1 and s_2 to represent two consecutive alarms.

The unification can result in many schemas. Some of these schemas would turn to be meaningless if the DM algorithm schema is not well specified. In practice, the number of unifications can be reduced by the use of heuristics and constraints. For example, every abstract relations of a DM algorithm schema must be unified with implemented relations of a data schema. Furthermore equations between relations can be expressed in a schema. These equations constrain also the unification. As a result, after stating such constraints, the number of unifications is reduced to two in the case of the generalized edges example.

5   Generic Evaluation

This section shows how to compute a generic evaluation of models relatively to data. First we describe the principles on which this evaluation is grounded. Next, we take as example how to evaluate the model m_1 (Figure 1(b)).

The Kolmogorov complexity [10] measures the complexity of an object from the "absolute" length of a program executed on a "universal" machine (an equivalent of the Turing machine) that outputs this object. In our context of data mining, we use the MDL principle [11]. Some approaches [8] use compression to minimize the length of data description. In our case, given a unified schema, the data complexity relatively to a model is the cost of the model plus the cost of accessing the data from the model by the covering relation.

Definition 1 (Generic measure for model evaluation). *Let S be a schema, $c : M \leftrightarrow D$ the covering relation of S, $d \subseteq D$ a set of data, $m \in M$ a model, and $k(x)$ the cost in bits of representing the input[1] x. The complexity of the data d in the schema S relatively to the model m is:*

$$K(d, m, S) = k(m|M) + k(d|m, c, D)$$

For each model m extracted from the data d, the system has to compute $K(d, m, S)$ where S contains the specification of m. We illustrate the computation of $K(A, m_1, S'_g)$ where $m_1 = [e_{g1}, e_{g2}]$ (Figure 1), $e_{g1} = (b, *)$ and $e_{g2} = (*, c)$. Firstly, $k(m_1|\mathcal{L}(\mathbf{edgeG}))$ is computed. A generalized edge e_{gi} is composed of two generalized actors. The type **actorG** is composed of 7 constants. Thus the size (a size is a number of bits) of an **edgeG** is $2 \times log_2(7)$ and the size of m_1 is $k(m_1|\mathcal{L}(edgeG)) = 4 \times log_2(7) = 11.2$.

Secondly, $k(A|m_1, c_{le}, \mathcal{L}(\mathbf{alarm}))$ is computed. There are many different ways to retrieve the data from the model. Each corresponds to some data encodings according to the model. The covering relation is composed of relations, factorisations and cofactorisations. To each *decomposition* of the covering relation corresponds a data encoding. The evaluation leads to finding the decomposition that minimizes the size of the data encoding. Three among many possible decompositions of c_{le} decompositions are shown. To each decomposition i of c_{le} corresponds a cost k_i. Recall that $c_{le} = c_{ge} \circ ex_{eG}$ where $c_{ge} = \langle \Sigma_{date} \circ \varnothing, c_e, \Sigma_{severity} \circ \varnothing \rangle$ and where $c_e = \langle c_n \circ sourceG, c_n \circ targetG \rangle$.

1. c_{le} is not decomposed. The set $c_{le}(m_1)$ is computed and the twelve alarms of A are found in this set: $log_2(C^{|A|}_{|c_{le}(m_1)|})$, C^k_n is the number of subsets of size k in a subset of size n. The size of $|c_{le}(m_1)|$ is easily computed by enumerating all the edges covered by e_{g1} and e_{g2}. The alarms not covered by m_1 needs also to be represented: $k(A \setminus c_{le}(m_1)|\mathcal{L}(\mathbf{alarm}))$ noted k' in the sequel.
   $\Rightarrow k_1 = log_2(C^{|A|}_{|c_{le}(m_1)|}) + k' = log_2(C^{12}_{432}) + k' = 76.0 + k'$.

2. c_{le} is decomposed into $c_{ge} \circ ex_{eG}$. The cost to find e_{g1} and e_{g2} in m_1 is $2 \times log_2(2)$. The size to find the alarms of A in $c_{ge}(e_{g1})$ and $c_{ge}(e_{g2})$ is:
   - $A_{e_{g1}} = [(9, (b, c), 3), (2, (b, c), 1), (5, (b, e), 3), (11, (b, f), 1)] \subseteq c_{ge}(e_{g1})$
     $\Rightarrow log_2(C^{|A_{e_{g1}}|}_{|c_{ge}(e_{g1})|})$
   - $A_{e_{g2}} = [(6, (f, c), 1), (3, (f, c), 1), (10, (d, c), 2), (12, (e, c), 2), (9, (b, c), 3), (2, (b, c), 1)] \subseteq c_{ge}(e_{g2})$
     $\Rightarrow log_2(C^{|A_{e_{g2}}|}_{|c_{ge}(e_{g2})|})$
   $\Rightarrow k_2 = 2 \times log_2(2) + log_2(C^{|A_{e_{g1}}|}_{|c_{ge}(e_{g1})|}) + log_2(C^{|A_{e_{g2}}|}_{|c_{ge}(e_{g2})|}) + k'$
   $\Rightarrow k_2 = 2 \times log_2(2) + log_2(C^4_{216}) + log_2(C^6_{216}) + k' = 65.3 + k'$

---

[1] $k(d|m, c, D)$ means the cost to represent d knowing m, c and D.

3. c_{le} is decomposed as the second decomposition plus the relation $\Sigma_{severity} \circ \varnothing$. It means that only the elements of $c_{ge}(e_{g1})$ and $c_{ge}(e_{g2})$ that have a severity of at least one element of $\mathcal{A}_{e_{g1}}$ and $\mathcal{A}_{e_{g2}}$ respectively are selected. In other words, the elements $c'_{ge}(e_{g1}) \subseteq c_{ge}(e_{g1})$ of severities 1 and 3 and the elements $c'_{ge}(e_{g2}) \subseteq c_{ge}(e_{g2})$ of severities 1 and 2 are retained. $2 \times 2 \times log_2(3)$ bits are required to represent these severities.

$$\Rightarrow k_3 = 2 \times log_2(2) + 4 \times log_2(3) + log_2(C_{|c'_{ge}(e_{g1})|}^{|\mathcal{A}_{e_{g1}}|}) + log_2(C_{|c'_{ge}(e_{g2})|}^{|\mathcal{A}_{e_{g2}}|}) + k'$$
$$\Rightarrow k_3 = 2 \times log_2(2) + 4 \times log_2(3) + log_2(C_{144}^4) + log_2(C_{144}^6) + k'$$
$$\Rightarrow k_3 = 2 + 6.3 + 24 + 33.4 + k' = 65.7 + k'$$

All the decompositions of the covering relation c_{le} can be enumerated by analyzing its formal expression in the unified schema. Then, the system chooses the decomposition that minimizes the cost of the access to data $\mathcal{A}$. This minimal cost plus the cost of the model gives the evaluation of the model.

6   Conclusion and Future Work

We have presented a framework for mining data without any information on the knowledge structure present in the data. Instead of seeing DM as several ad hoc processes, we have shown that DM algorithms, DM models and data can be unified by the same specification language, based on schemas. The schema expressiveness, due to its grounds in category theory, is the major contribution of this paper. Moreover, several kinds of computations can be performed on specifications, such as unification or evaluation of models. The unification formalizes how to interface data and DM algorithms. This step releases the user from the burden of describing exhaustively this interface in order to run manually DM algorithms on specific data. The formalization of the covering relation is the key concept to evaluate models by using the Kolmogorov complexity.

We chose to illustrate the major aspects through an example. The approach and the language of schema are very general. Until now, we have used schemas to represent two mining processes concerning two very different datasets related to network alarms: DDoS attacks (the example of the paper) and intrusion detection in a virtual private network (VPN). However, the approach should be applied to other domains to fully assess its generality.

The presented framework is being developed in Prolog and Java. Experiments on real network alarms provided by France-Telecom have been undertaken. Furthermore, as DM algorithms are automatically executed in the proposed framework, we are working also on how resulting models can be displayed to the user and, furthermore, how they could be visualized directly on data. Finally, the user should be able to explicit what is the "useful information" of the Frawley et al. Knowledge Discovery definition. Thus, the user should be able to adapt the evaluation method, for instance, by restricting the enumeration of covering relation decompositions or by suggesting other ways to encode data from a model.

Acknowledgements. Special thanks to Dominique Duval from LMC-Grenoble for numerous discussions on Category Theory. This work is part of the CU-RAR Project (CRE #171938) supported by France-Telecom R&D. Thanks to the other project members: Mireille Ducassé from INSA-Rennes and Christophe Dousson and Pierre Le Maigat from France Telecom-Lannion. Thanks also to the anonymous referees.

References

[1] Barr, M., Wells, C.: Category Theory for Computing Science. Prentice-Hall, Englewood Cliffs (1990)

[2] Bernstein, A., Provost, F., Hill, S.: Towards intelligent assistance for a data mining process: An ontology-based approach for cost-sensitive classification. IEEE Transactions on Knowledge and Data Engineering 17(4), 503–518 (2005)

[3] Ehresmann, C.: Catégories et structures. Dunod, Paris (1965)

[4] Ehrig, H., Mahr, B.: Fundamentals of Algebraic Specification I. Springer-Verlag, New York, Inc. Secaucus, NJ, USA (1985)

[5] Frawley, W.J., Piatetsky-Shapiro, G., Matheus, C.J.: Knowledge discovery in databases: an overview. AI Mag. 13(3), 57–70 (1992)

[6] Imielinski, T., Mannila, H.: A database perspective on knowledge discovery. Communications of the ACM 39, 58–64 (1996)

[7] Johnson, T., Lakshmanan, L.V.S., Ng, R.T.: The 3w model and algebra for unified data mining. In: VLDB'00: Proc. 26th International Conference on Very Large Data Bases, pp. 21–32. Morgan Kaufmann Publishers, USA (2000)

[8] Keogh, E., Lonardi, S., Ratanamahatana, C.A.: Towards parameter-free data mining. In: KDD'04: Proc. 10th ACM SIGKDD international conference on Knowledge discovery and data mining, pp. 206–215. ACM Press, USA (2004)

[9] Lakshmanan, L.V.S., Ng, R.T., Xing Wang, C., Zhou, X., Johnson, T.: The generalized MDL approach for summarization. In: Bressan, S., Chaudhri, A.B., Lee, M.L., Yu, J.X., Lacroix, Z. (eds.) CAiSE 2002 and VLDB 2002. LNCS, vol. 2590, pp. 766–777. Springer, Heidelberg (2003)

[10] Li, M., Vitanyi, P.: Introduction to Kolmogorov complexity and its applications. Springer, Heidelberg (1997)

[11] Rissanen, J.: Stochastic Complexity in Statistical Inquiry Theory. World Scientific Publishing Co. Inc. River Edge, NJ, USA (1989)

[12] Srikant, R., Agrawal, R.: Mining sequential patterns: Generalizations and performance improvements. In: Apers, P.M.G., Bouzeghoub, M., Gardarin, G. (eds.) EDBT 1996. LNCS, vol. 1057, pp. 3–17. Springer, Heidelberg (1996)

Relaxation Labeling for Selecting and Exploiting Efficiently Non-local Dependencies in Sequence Labeling

Guillaume Wisniewski and Patrick Gallinari

LIP6 — Université Pierre et Marie Curie
104 avenue du président Kennedy
75016 Paris France
guillaume.wisniewski@lip6.fr, patrick.gallinari@lip6.fr

Abstract. We consider the problem of sequence labeling and propose a two steps method which combines the scores of local classifiers with a relaxation labeling technique. This framework can account for sparse dynamically changing dependencies, which allows us to efficiently discover relevant non-local dependencies and exploit them. This is in contrast to existing models which incorporate only local relationships between neighboring nodes. Experimental results show that the proposed method gives promising results.

1 Introduction

Sequence labeling aims at assigning a label to each element of a sequence of observations. The sequence of labels generally presents multiple dependencies that restrict the possible labels elements can take. For example, for the task of part of speech tagging, the observations that there is only one verb in a sentence and that an article is followed either by a noun or an adjective provide valuable information to the labeling of an element. The aim of *Structured Prediction* is to develop models able to detect and exploit these dependencies so as to improve prediction performance.

Taking dependencies between labels into account entails two main difficulties: parameter estimation and complexity of inference. For the former, the more dependencies we have to consider, the more parameters we have to estimate, which creates a sparse data problem. For the latter, inferring jointly rather than independently a label sequence consistent with the dependencies often proves to be a combinatorial problem and, in the worst case, inference is known to be NP-hard to solve [1].

Several approaches have been developed for many years for sequence labeling. In order to solve the inherent difficulties of both the training and the inference steps, all current methods, like CRFs [2] or SVM'ISO [3], impose a fixed label dependency structure in which only local interactions between labels are considered: they generally incorporate only dependencies between a few successive labels. This Markov assumption limits the number of parameters to be estimated and maintains a tractable exact inference thanks to the use of dynamic programming techniques. While this assumption is critical in preserving computational

J.N. Kok et al. (Eds.): PKDD 2007, LNAI 4702, pp. 312–323, 2007.

efficiency, it is a key limitation since taking non-local dependencies into account is mandatory to achieve good performance for several tasks [4,5,6].

Two main families of approaches have been proposed to take advantage of non-local dependencies. The first one [7,8] relies on a grammar-based formalism to model non-local relationships by introducing a hierarchy of hidden variables, while the second one proposes alternative inference procedures like Gibbs sampling [9] or Integer Linear Programming [10]. In most of these methods, approximate inference algorithms are used to allow tractable inference with long-range dependencies. All these methods suffer a high complexity for both training and inference and rely on an expert knowledge to explicitly define all relevant dependencies involved.

In this paper, we propose a new approach for learning and modeling unknown dependencies among labels. Dependencies are represented using *constraints* which are logical relations among several elements and their value. This approach has several interesting properties. Firstly, it allows the dependency structure to vary according to the actual value of elements, while this structure is fixed in most existing models. Secondly, both local and long-range dependencies can be considered while preserving the computational efficiency of the training and inference steps. More precisely, following [11], we consider a two-parts model, in which, a local classifier predicts the values of variables regardless of their context, while a set of constraints maintains global consistency between local decisions. These constraints are learned and represent dependencies between labels. In this work, we use maximum entropy classifiers [12] to make local decision and relaxation labeling [13] to efficiently build an approximate solution, that satisfies as many constraints as possible.

The paper is organized as follows. We first formalize the task and explain the difficulty of incorporating non-local dependencies in representative existing approaches in Section 2. Our approach is presented in Section 3. Related work is reviewed in Section 2.3 and experimental results are presented in Section 4.

2   Background

2.1   Formalization of the Task

Sequence labeling consists in assigning a label to every element of a sequence of observations. Let $\mathbf{x} = (x_i)_{i=1}^{n}$ be a sequence of n observations and y_i be the label of the ith element of this sequence. The sequence of labels denoted $\mathbf{y} = (y_i)_{i=1}^{n}$ can be seen as a *macro-label* [3] describing a set of labels with possible dependencies between them. Let Λ be the set of all possible labels (the domain of the y_i), and $\mathcal{Y}$, the domain of the macro-label $\mathbf{y}$. Because of the interdependencies between the y_i, some combinations of labels will not be possible, and $\mathcal{Y}$ is only a subset of Λ^n. Intuitively, the smaller $\frac{\#\{\mathcal{Y}\}}{\#\{\Lambda^n\}}$ [1] is, the more regularity in the output there is and the more dependencies can help to predict the label of a variable.

---

[1] We use $\#\{\mathcal{A}\}$ to denote the cardinal of the set $\mathcal{A}$.

2.2   Existing Methods for Sequence Labeling: Local Output Dependencies

Many machine learning models have been proposed to take advantage of the information conveyed by label dependencies. In practice, most of them rely on local hypothesis on the label dependencies. For the sequence labeling task, a popular model is Conditional Random Fields [2]. CRFs will be used for comparison in the evaluation in section 4. More recently, the prediction of structured outputs has motivated a series of new models [3,1,4]. The SVM'ISO family of models [3], for example, is a generalization of Support Vector Machines designed for predicting structured outputs. Both CRFs and SVM'ISO consider sequence labeling as a generalization of multi-class classification: they aim at determining the macro-label $\mathbf{y}$ which is the most *compatible* with a given specific sequence of observations $\mathbf{x}$. The compatibility between the observation and the macro-label is evaluated by a θ-parametrized scoring function $F(\mathbf{x}, \mathbf{y}; \theta)$. The task of sequence prediction then amounts at finding the most compatible output among all legal outputs $\mathcal{Y}$:

$$\mathbf{y}^* = \operatorname*{argmax}_{y \in \mathcal{Y}} F\left(\mathbf{x}, \mathbf{y}; \theta\right) \qquad (1)$$

The argmax operator denotes the search in the space of all possible outputs $\mathcal{Y}$ that takes place during inference. Several methods have been developed to estimate the parameters θ that either optimize the conditional likelihood (in the case of CRFs) or optimize a maximum margin criterion (in the case of SVM'ISO).

In their general formulation, both CRFs and SVM'ISO can describe arbitrary dependencies. But, in practice, due to the complexity of inference and parameter estimation, the scoring function F has to be *decomposable*: a function is said to be decomposable if it can be expressed as a product of local scoring functions. The decomposition used in CRFs for sequence labeling is the following:

$$F(\mathbf{x}, \mathbf{y}; \theta) = \prod_{i=1}^{n} f(y_{i-1}, y_i, \mathbf{x}; \theta) \qquad (2)$$

where f is the local scoring function, which, in the case of CRFs, is chosen to be $f(y_{i-1}, y_i, \mathbf{x}; \theta) = \exp \langle \theta, \phi(y_{i-1}, y_i, \mathbf{x}) \rangle$, where $\langle \cdot, \cdot \rangle$ is the standard dot product and ϕ is the feature vector[2]. In this decomposition, only the interactions between contiguous labels are taken into account. This allows the use of the efficient Viterbi algorithm for inference and limits the number of parameters to be estimated. The SVM'ISO family of algorithms, has been developed for modeling general output dependencies. When used for sequence labeling, however the scoring function is closely related to the one used for CRFs.

While this factorized form is critical in enabling models to work on real data, it precludes any possibility of taking non-local dependencies into account for two

---

[2] Compared to the usual presentation of CRFs [2], we have: $p(\mathbf{y}|\mathbf{x}; \theta) = \frac{1}{Z(\mathbf{x})} \cdot F(\mathbf{y}, \mathbf{x}; \theta)$. $Z(\mathbf{x})$ is a normalizing function that allows us to give a probabilistic interpretation to the CRF.

reasons. Firstly, in the Viterbi algorithm, the output is built by incrementally extending a partial solution towards a complete solution. As a label is often chosen before the other labels involved in the dependency are known (i.e. before the dependency can be "evaluated"), non-local dependencies cannot be exploited. Secondly, in this decomposition, a dependency is modeled by a local scoring function that has as many parameters as elements involved in the dependency. For instance, to describe a second-order dependency, we need a local scoring function with two parameters like $f(y_i, y_{i-1})$. The local scoring function has to be defined for each possible labeling of these elements. Consequently, to describe a dependency between n elements that can take m labels, m^n parameters have to be estimated, possibly resulting in sparse data problems.

Exploiting non-local dependencies therefore requires both an alternative inference algorithm and an alternative modeling of the dependencies.

2.3 Existing Methods for Sequence Labeling: Long-Term Output Dependencies

Different approaches have been proposed for taking advantage of long-range dependencies.

The *N-Best* method combined with reranking of the selected solutions [4] is a general strategy which offers an approximate solution to the structured prediction problem. It allows the use of non-local dependencies by separating the structure prediction in two independent steps. In a first step considering only local features, a limited set of potential candidates are generated with a dynamic programming algorithm. In a second step (reranking), considering arbitrary features, the "best" solution is chosen among all these candidates. Note that a limitation of reranking strategies is that the correct answer is not guaranteed to be contained in the set of potential candidates.

Several works have proposed alternatives to the Viterbi inference algorithm. Popular choices among alternative inference procedures include Gibbs sampling [9] or loopy belief propagation [5]. Another alternative is [10] which replaces the Viterbi algorithm by Integer Linear Programming to tackle the cases of non-local dependencies. In [9] and in [5], long-range dependencies are used to include domain-specific prior-knowledge, namely the so-called *label consistency* which ensures that identical observations in the corpus get the same label (e.g. in as information extraction task, it ensures that Paris is always recognized as a town). In [10] long-range dependencies are described by hand-crafted Boolean functions like "at least one element is assigned a label other than O". In all these works, the non-local dependencies are hand-crafted.

Another popular idea consists in capturing interaction among labels in a hierarchical approach [7,8]. For instance, in an Information Extraction task, [8] proposes to use a Context Free Grammar to escape the "linear tyranny" of chain-models. Within this framework, inference amounts to syntactic parsing. There are two main drawbacks in this approach: firstly, the grammar describing the interactions among labels is constructed by an expert; secondly, inference done by the CKY algorithm (a generalization of the Viterbi algorithm to trees) has a complexity in $\mathcal{O}(n^3)$, where n is the number of elements in the sequence. A CFG has been used for comparison as a baseline non local method in the experiments described in section 4.

3   Proposed Approach

Sequence labeling with interdependent labels amounts to identifying the best assignment to a collection of variables, so that this assignment is *consistent* with a set of dependencies or a *structure*. The dependencies between outputs can be thought of as constraining the output space. We build on this idea and propose to model the dependencies with constraints (logical relations among several elements and their value), rather than by local scoring function as in the approaches discussed in Section 2.2. Typical examples of such constraints are "the label of the ith variable has to be λ, if the $(i-2)^{\text{th}}$ variable is labeled by μ" or "there should be at least one variable labeled with ξ". We will associate to each constraint a weight to be interpreted as a confidence or a level of preference. More precisely, we treat sequence labeling as a *constrained assignment* problem. Let $\mathcal{V} = \{v_1, ..., v_n\}$ denote the variables describing the labels of an input sequence $\{x_1, ..., x_n\}$; each v_i may take its value in the set of labels Λ. We aim at assigning a label to each variable while satisfying a set of constraints, automatically learned from the training set.

To solve this constrained assignment problem, we propose a two-step process as advocated by [11] that relies on a well-known constraint satisfaction algorithm, relaxation labeling [13,14]: firstly, a *local classifier* affects an initial assignment to elements regardless of their context (i.e. without considering any dependencies) and, secondly, the relaxation process applies successively the constraints to propagate information and ensure global consistency.

In the following sections we detail these two steps and describe how constraints can be automatically learned from the training set. Eventually, we explain how this approach offers a solution to the problems discussed in Section 2.

3.1   Local Classifier

The local classifier aims at estimating, for each variable, a probability distribution over the set of labels. To produce these estimates, we adopt here the maximum entropy framework [12]. Note that any classifiers that output a probability distribution over the set of labels could be used as well.

Maximum entropy classifiers model the joint distribution of labels and input features. The probability of labeling a variable v with label λ is modeled by an exponential distribution:

$$p(\lambda|v; \theta) = \frac{1}{Z_\theta(v)} \exp \langle \theta, \phi(v, \lambda) \rangle$$

where $\phi(v, \lambda)$ is the feature vector describing jointly variable v and label λ, $Z_\theta(v)$ is a normalizing factor and θ the parameter vector. To estimate θ, the maximum entropy framework advocates to choose, among all the probability distributions that satisfy the constraints imposed by the training set, the one with the highest entropy, that is to say the one that is maximally noncommittal with regard to missing information [12].

3.2 Relaxation Labeling

Relaxation Labeling (RL) [13,14] is an iterative optimization technique that solves efficiently the problem of assigning labels to a set of variables that satisfy a set of constraints. It aims at reaching an assignment with maximal consensus among the set of labels, that is to say, to assign a label to each variable while satisfying as many constraints as possible. We denote $\mathcal{V} = \{v_1, ..., v_n\}$ the set of n variables, Λ will be the set of m possible labels, and λ and μ two elements of Λ.

In the following, we assume that interactions between labels are described by a *compatibility matrix* $R = \{r_{ij}(\lambda, \mu)\}^3$. Each coefficient $r_{ij}(\lambda, \mu)$ represents a *constraint* and measures to which extent we want to label the ith variable with λ when knowing that the label of the jth variable is μ: the higher $r_{ij}(\lambda, \mu)$, the more we want to label v_i with λ when the label of v_j is μ. These coefficients are estimated from the training set as detailed in Section 3.3.

The iterative algorithm of relaxation labeling works as follows: starting from an initial label assignment computed from the local classifier, the relaxation process iteratively modifies this assignment so that the labeling globally satisfies the constraints described by the compatibility matrix as well as possible. All labels are updated in parallel using the information provided by the compatibility matrix and the current label assignment.

More precisely, the local classifier defines, for each variable $v_i \in \mathcal{V}$ an initial probability vector, $\bar{p}_i^0$, with one component for each label of Λ. Let $\bar{p}_i^{(t)}(\lambda)$ denote the component of $\bar{p}_i^{(t)}$ corresponding to the label λ. Each $p_i^{(t)}(\lambda)$ describes the current confidence in the hypothesis "the label of the ith variable is λ". The set $\bar{p} = \{\bar{p}_1, ..., \bar{p}_n\}$ is called a *weighted label assignment*.

Let us define, for each variable v_i and each label λ a *support function*. This function describes the compatibility of the hypothesis "the label of v_i is λ" and the current label assignment of other variables. It is generally defined by:

$$q_i^{(t)}(\lambda; \bar{p}) = \sum_{j=1}^{n} \sum_{\mu \in \Lambda} r_{ij}(\lambda, \mu) p_j^{(t)}(\mu) \tag{3}$$

Intuitively, the more confident we are in the labelings that support the hypothesis "the label of v_i is λ", the higher the support of this hypothesis (i.e. the higher $q_i(\lambda; \bar{p})$). Hypothesis we are not confident in (i.e. the ones for which $p_i(\lambda)$ is small) have only little influence. A natural way to update the weighted assignment is

---

^{3} In this presentation, for simplification, we only consider pairwise dependencies. The extension to dependencies between an arbitrary number of dependencies is straightforward.

therefore to increase $p_i(\lambda)$ when $q_i(\lambda)$ is big, and decrease it otherwise. More precisely, the update of each $p_i(\lambda)$ is defined by:

$$p_i^{(t+1)}(\lambda) \leftarrow \frac{p_i^{(t)}(\lambda) \cdot q_i^{(t)}(\lambda, \bar{p}^{(t)})}{\sum_{\mu \in \Lambda} p_i^{(t)}(\mu) \cdot q_i^{(t)}(\mu, \bar{p}^{(t)})} \tag{4}$$

The denominator is just a normalizing factor that ensures $p_i^{(t+1)}(\lambda)$ remains a probability. This process (the calculation of the support and the update of the mapping) is iterated until convergence (i.e.:until $\bar{p}^{(t+1)} = \bar{p}^{(t)}$).

It can be proved [13,14] that, under mild assumptions, the relaxation algorithm finds a local maximum of the *average local consistency function* defined as the average support received by each variable. The latter measures the compatibility between each hypothesis "the label of v_i is λ" and all the other assignments: relaxation labeling can be seen as a method that employs the labelings we are the most confident in to disambiguate those with low confidence. The complexity of the relaxation labeling process is linear with respect to the number of variables to label.

3.3   Learning the Constraints

In some applications, the constraints are provided by hand [10] or can be easily derived from the problem specification. Here, they will be learned from the training set. Let us first observe that relevant constraints should reduce the labeling ambiguity and that choosing these constraints can however quickly become computationally intractable as the number of possible dependencies in a sequence grows exponentially with the length of the sequence.

To efficiently select the most relevant constraints, we will take advantage of the following observation: the compatibility coefficients used in the relaxation labeling process can be interpreted as *association rules*. For instance, the compatibility coefficient $r_{ij}(\lambda, \mu)$ can also be interpreted as the rule $v_j = \mu \Rightarrow v_i = \lambda$: both of them mean that, if the label μ appears at the jth position of a sequence, we have a good chance of finding the label λ at the ith position. Higher order dependencies are described by conjunction of label assignments. This connection between the compatibility coefficient used in relaxation labeling and association rules is appealing, since, intuitively, knowing which labels frequently co-occur in the training set, helps reducing the uncertainty of the labeling decisions.

We will draw on this intuition and consider that the compatibility coefficients are to be defined as the conditional entropy [15] of the corresponding association rule. The conditional entropy of an association rule is a combination of the usual *support* and *confidence* used to evaluate the importance of a rule. Combinatorial algorithms, such as Apriori [16] can be used to extract efficiently all the association rules the conditional entropy of which is higher than a user-provided value. This value is a parameter of our algorithm. Note that while Apriori is a combinatorial method, it is used here during training and inference remains linear wrt the sequence size.

For simplicity, relaxation labeling was described using absolute i, j positions. The algorithm is actually implemented using relative variable positions which are more general and flexible. An example of a rule expressed with relative value is $v_{i-3} = \alpha \Rightarrow v_i = \beta$. Relaxation labeling can be easily generalized to handle arbitrary rules. The only difference relies on the definition of the support (Equation 3) that has to be generalized to consider all elements that appear in the rule. For instance, if we consider the previous rule and a sequence of 9 elements in which $v_1 = \alpha$ and $v_5 = \alpha$, the hypothesis $v_4 = \beta$ and $v_8 = \beta$ will be both strengthened. In the same way, association rules are learned with relative values and are then instantiated in any position that fits elements in a sequence.

3.4 Advantages of Our Approach

Our approach solves several of the restrictions that were pointed out in Section 2. First, the most likely labeling is inferred by iteratively reassigning labels based on the current assignment of all other variables, contrary to chain-models in which only previous assignments are taken into account. As a result, each label is chosen according to a global context and dependencies describing the sequence as a whole or involving non-sequential variables can be considered. Relaxation labeling is therefore an interesting alternative to the Viterbi algorithm for sequence labeling.

Secondly, as they are formed by variable-value pairs (v_i, λ), constraints can account for dynamically changing dependencies, which are conditioned on the values of variables. Consequently, sparse data problems are avoided: in our framework only the variable values involved in a constraint have to be considered, while, for the models described in Section 2, the score of all possible assignments to these variables has to be estimated This representation also allows us to efficiently select the relevant dependencies we want to incorporate in the model.

4   Empirical Evaluation

4.1   Tasks and Corpora

We have tested our approach on two different sequence labeling tasks: structured data extraction [17] and chunking [18].

Structured Data Extraction. The first application considered, structured data extraction [19,20,6], is motivated by the development of the *Semantic Web*. Semantic Web aims at providing value-added services by taking advantage of a semantically-rich structured view of HTML or XML documents instead of their traditional bag-of-words representation. The semantic technologies need hard-wiring knowledge of the structure they are using and, therefore, can only deal with documents that comply strictly with a *schema*. This schema, generally expressed by a DTD or an XML Schema, defines, a priori, allowed structures of documents. Because of the lack of standardization, the representation of data varies from source to source, and, consequently, the deployment of the Semantic Web is only possible if we are able to resolve these heterogeneities. One first step towards this resolution consists in *extracting* relevant information from structured documents.

Structured data extraction is closely related to the standard task of information extraction, but it is made easier by the presence of a document structure [6]. We consider this task here as a sequence labeling task: structured data extraction amounts to labeling a sequence of observations defined by the leaves of an HMTL or XML document, where dependencies between the labels are described by a *target schema*.

Our model has been tested on two different corpora. The first one is the collection Courses used in [20] that describes lectures in five different universities. There are more than 12,000 descriptions which are annotated with 17 different labels such as lecturer, title, start time or end time; each description contains between 4 and 552 elements to be extracted. The second corpus, MovieDB is based on the IMDb database [6]. There are 4,483 movie descriptions, annotated with 16 different labels such as actor, director or title.

Considering non-local dependencies between the labels is mandatory to achieve good performances in this task. Indeed, in many cases the local classifier does not have enough information to choose the correct label and dependencies have to be considered to reduce labeling ambiguities. For instance, the only way to distinguish a start time from an end time or actor from a director is by taking into account the context, i.e. the position of the element in the sequence of labels.

Each collection was randomly split in two equal parts for training and testing. Experiments were performed on the two corpora using the same features for the local classifier described in Section 3.1. We used the kind of features generally used in information extraction tasks: typical examples of these features include NumberOfWords, NumberOfCapitalLetter or ContainsHTTP.

Chunking. The second application we considered is the "All-phrase chunking" task of CoNLL 2000 [18]. Text chunking aims at identifying the non-recursive cores of various phrase type in text. There are 11 different chunks in the corpora such as "noun phrase", "adjective phrase" or "subordinated clause". The chunks are represented with three kinds of labels: "B-X" stands for "first word of a chunk of type X", "I-X" for "non-initial word in an X chunk" and "O" means "word outside of any chunk". Using this so-called BIO representation allows to tackle a sequence segmentation task[4] as a sequence labeling task. As pointed out by [10] and [18], the sequences of labels involved in this task present many dependencies, and the BIO representation naturally forbids some combinations of labels. In our experiments we used the features and the train set and test set provided by [18].

---

[4] when several consecutive elements of the observation sequence are mapped to a single label.

4.2 Results

Baseline Models. As baseline models we used a Maximum Entropy classifier (the local classifier described in Section 3.1), standard linear-chain CRF[5] and a grammar-based extraction approach similar to the ones presented by [8] or [22]. Because of its computational complexity, the SVM'ISO approach we described in Section 2.2 cannot be used on our corpora. The local classifier does not incorporate any information about the dependencies between labels, the CRF only considers local and sequential dependencies (see Section 2) while the grammar-based approach can take long-range dependencies into account.

The principle of this latter approach is as follows: a Maximum Entropy Classifier estimates a probability distribution over the set of labels for each observation and a Probabilistic Context Free Grammar is then used to infer a tree structure. This tree structure can be seen as a hierarchy of hidden variables that describes long-range dependencies. The predicted label sequence is defined by the labels of the leaves of the tree with the highest score. The inference complexity of this approach is $\mathcal{O}(n^3)$, where n is the number of elements to be labeled, which should be contrasted with the complexity $\mathcal{O}(n)$ of both CRFs and our method. This model requires us to define a context-free grammar that describes both local and non-local dependencies. This can be done in the structured extraction task by converting the target schema in context-free grammar [22,6], but not in the chunking task, where no general grammar that describes interactions between chunks is known.

Results. Table 1 presents the results of the different experiments. The scores presented correspond to the standard F1 measure.

Results show the importance of taking into account the dependencies: in all the tasks, the score of the local classifier is always the worst. As was explained in Section 4.1, this is mainly due to the fact that in many cases, an observation does not contain enough information to choose the correct label so that the context has to be considered. Exploiting non-local dependencies is also of great help. On the data extraction task both our approach and the grammar-based approach clearly outperform CRFs. In the chunking task, CRFs achieve slightly better performances, likely because the dependencies between labels are less relevant, but both learning and inference are much faster with our method than with CRFs. The grammar-based approach and our approach achieve similar performances, which shows the ability of the proposed method to select relevant dependencies. Note that inference with our method is also an order of magnitude faster than with a grammar-based approach.

In the experiments, CRFs were used with default parameters and better results might be obtained by tuning these parameters for the different tasks. CRFs results on the Course corpus are particularly low. This is due to regularization in learning: to avoid estimation problem, parameters are smoothed so that the weight of each possible transition between two successive nodes is non-zero. All transitions are thus allowed which does not reflect correctly the structure of the data (For course data, most transitions should be 0).

---

[5] In our experiments we used FlexCRF [21].

Table 1. Results of the different experiments. Dashes indicates that the experiments could not be performed. The reported score corresponds to the standard F1 measure.

	MovieDB	Course	Chunking
Local Classifier	90.6%	47.9%	90.3%
Proposed Model	**97.4%**	**88.1%**	**93.2%**
CRF	96.4%	78.7%	**94.6%**
Grammar	**97.5%**	**87.4%**	—

5   Conclusion

We have proposed a general method for efficiently discovering relevant non-local dependencies and exploiting them in the sequence labeling task. Our approach relies on the modeling of relationships between labels by constraints. It is a two steps process: initial label assignment is provided by a local classifier, the dependencies among variables are considered and propagated using an iterative relaxation procedure.

This model can account for dynamically changing dependencies. This is in contrast to most existing approaches that assume that a fixed-sized neighborhood is relevant for predicting each label. The proposed approach has achieved convincing results on different tasks at a low computational cost. A key element in developing such approaches is to define measures to assess the strength of a dependency and the amount of information a dependency provides to reduce labeling ambiguity.

Acknowledgments

We thank Nicolas Usunier and Alexander Spengler for enlightening discussions and insightful comments on early drafts of this paper. This work was supported in part by the IST Programme of the European Community, under the PAS-CAL Network of Excellence, IST-2002-506778. This publication only reflects the authors' views.

References

1. Taskar, B., Lacoste-Julien, S., Jordan, M.I.: Structured prediction, dual extragradient and bregman projections. Journal of Machine Learning Research (2006)
2. Lafferty, J., McCallum, A., Pereira, F.: Conditional random fields: Probabilistic models for segmenting and labeling sequence data. In: ICML'01, pp. 282–289. Morgan Kaufmann, San Francisco, CA (2001)
3. Tsochantaridis, I., Joachims, T., Hofmann, T., Altun, Y.: Large margin methods for structured and interdependent output variables. Journal of Machine Learning Research 6, 1453–1484 (2005)
4. Collins, M., Koo, T.: Discriminative reranking for natural language parsing. Computational Linguistics 31, 25–69 (2005)

5. Sutton, C., McCallum, A.: Collective segmentation and labeling of distant entities in information extraction. In: ICML workshop on Statistical Relational Learning (2004)
6. Wisniewski, G., Gallinari, P.: From layout to semantic: a reranking model for mapping web documents to mediated xml representations. In: 8th RIAO International Conference on Large-Scale Semantic Access to Content (2007)
7. Awasthi, P., Gagrani, A., Ravindran, B.: Image modeling using tree structured conditional random fields. In: IJCAI (2007)
8. Viola, P., Narasimhan, M.: Learning to extract information from semi-structured text using a discriminative context free grammar. In: SIGIR '05 (2005)
9. Finkel, J., Grenager, T., Manning, C.D.: Incorporating non-local information into information extraction systems by gibbs sampling. In: ACL'05 (2005)
10. Roth, D.: Integer linear programming inference for conditional random fields. In: ICML, pp. 736–743. ACM Press, New York (2005)
11. Punyakanok, V., Roth, D., tau Yih, W., Zimak, D.: Learning and inference over constrained output. In: IJCAI'05, pp. 1124–1129 (2005)
12. Berger, A.L., Pietra, V.J.D., Pietra, S.A.D.: A maximum entropy approach to natural language processing. Comput. Linguist. 22(1), 39–71 (1996)
13. Pelillo, M.: The dynamics of nonlinear relaxation labeling processes. J. Math. Imaging Vis. 7(4), 309–323 (1997)
14. Hummel, R.A., Zucker, S.W.: On the foundations of relaxation labeling processes. IEEE PAMI 5(1), 267–287 (1983)
15. Blanchard, J., Guillet, F., Gras, R., Briand, H.: Using information-theoretic measures to assess association rule interestingness. In: ICDM'05., pp. 66–73 (2005)
16. Borgelt, C.: Efficient implementations of apriori and eclat. In: Workshop of Frequent Item Set Mining Implementations (FIMI) (2003)
17. Liu, B., Chen-Chuan-Chang, K.: Editorial: special issue on web content mining. SIGKDD Explor. Newsl. 6(2), 1–4 (2004)
18. Sang, E.F.T.K., Buchholz, S.: Introduction to the conll- 2000 shared task: chunking. In: 2nd workshop on Learning language in logic and the 4th conference on Computational natural language learning, Morristown, NJ, USA pp. 127–132 (2000)
19. Shadbolt, N., Berners-Lee, T., Hall, W.: The semantic web revisited. IEEE Intelligent Systems 21(3), 96–101 (2006)
20. Doan, A., Domingos, P., Halevy, A.: Learning to match the schemas of data sources: A multistrategy approach. Mach. Learn. 50(3), 279–301 (2003)
21. Phan, X.H., Nguyen, L.M.: Flexcrfs: Flexible conditional random field toolkit (2005), http://www.jaist.ac.jp/~hieuxuan/flexcrfs/flexcrfs.html
22. Chidlovskii, B., Fuselier, J.: A Probabilistic Learning Method for XML Annotation of Documents. In: IJCAI (2005)

Bridged Refinement for Transfer Learning

Dikan Xing, Wenyuan Dai, Gui-Rong Xue, and Yong Yu

Department of Computer Science and Engineering
Shanghai Jiao Tong University, Shanghai, China
{xiaobao,dwyak,grxue,yyu}@apex.sjtu.edu.cn

Abstract. There is usually an assumption in traditional machine learning that the training and test data are governed by the same distribution. This assumption might be violated when the training and test data come from different time periods or domains. In such situations, traditional machine learning methods not aware of the shift of distribution may fail. This paper proposes a novel algorithm, namely *bridged refinement*, to take the shift into consideration. The algorithm corrects the labels predicted by a shift-unaware classifier towards a target distribution and takes the mixture distribution of the training and test data as a bridge to better transfer from the training data to the test data. In the experiments, our algorithm successfully refines the classification labels predicted by three state-of-the-art algorithms: the Support Vector Machine, the naïve Bayes classifier and the Transductive Support Vector Machine on eleven data sets. The relative reduction of error rates is about 50% in average.

1 Introduction

Supervised learning requires enough, if not many, high-quality labeled examples to guide the learning progress for a model, by which we predict the labels of newly coming test data. Labeling examples are labor-intensive, and what makes things worse is that more and more labeled data become out of date as time goes by. For example, in the past years, there are a large number of textual data on the Web such as news reports that were written in formal style. But recently, blogs have been emerging, and their owners begin to write their posts in a style increasingly different from what they read in news reports. Past labeled news data thus cannot be used to reliably classify blog articles, since the usage of vocabulary becomes different in blog articles from news articles.

Transfer learning focuses on how to utilize those data from different time periods or domains to help the current learning task. Many previous researches on transfer learning are for so-called "multi-task" learning, where there are K tasks at hand and one wants to complete the K-th task, the mainly-focused task, with the help of the previous $K - 1$ tasks.

In this paper, we focus on such a situation that we only have one task, by which we mean that *the set of target categories are fixed*, but the document-marginal distribution, $P(d)$, of the training data $\mathcal{L}$ and the test data $\mathcal{U}$ are different, for

J.N. Kok et al. (Eds.): PKDD 2007, LNAI 4702, pp. 324–335, 2007.

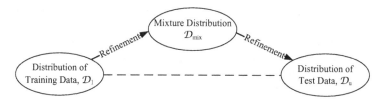

Fig. 1. Bridged Refinement

example being $\mathcal{D}_l$ and $\mathcal{D}_u$ respectively. We only have labeled data governed by $\mathcal{D}_l$, i.e. the training data $\mathcal{L}$, and have *no labeled data governed by* $\mathcal{D}_u$. All test data, $\mathcal{U}$, which are governed by $\mathcal{D}_u$ are to be classified. We want to transfer the knowledge in the training data $\mathcal{L}$ to well classify the test data $\mathcal{U}$.

To achieve this goal, in this work, we propose a bridged refinement algorithm. The algorithm receives the predictions from a shift-unaware classifier, which is trained from data governed by $\mathcal{D}_l$ and is expected to work well on data governed by $\mathcal{D}_l$. But now the predictions produced by it may be far from satisfactory since the distribution of the test data changes from $\mathcal{D}_l$ to $\mathcal{D}_u$. Our algorithm refines the classification labels in a two-step way, firstly towards the mixture distribution $\mathcal{D}_{mix}$ and secondly towards the test distribution $\mathcal{D}_u$. The mixture distribution $\mathcal{D}_{mix}$ governs both the training and test data as a whole. Hence $\mathcal{D}_{mix}$ is more similar both to $\mathcal{D}_l$ and to $\mathcal{D}_u$ than $\mathcal{D}_l$ to $\mathcal{D}_u$, so regarding it as an intermediate step, or a *bridge*, makes the two steps of refinement relatively easy. The bridged transfer process is intuitively depicted in Figure 1.

In each step of refinement, we want to refine the classification labels to make them more consistent under the target distribution (i.e. $\mathcal{D}_{mix}$ in the first step and $\mathcal{D}_u$ in the second). Considering the observation that two identical documents even under two different distributions, e.g. news vs. blogs, are supposed to be in the same category, the conditional distribution of the class label c on the input d, $P(c|d)$, does not vary. Based on this assumption[1], our refinement operation tries to make the labels more consistent under the target distribution.

The algorithm aims at refining classification labels instead of the decision boundary of a model. Refining a model seems more useful since it can be later used for newly coming test data. But it may be difficult in our problem, because what can be transferred from the labeled training data $\mathcal{L}$ totally depends on the test data $\mathcal{U}$. Different test data may require different knowledge to be transferred. Hence it is more realistic to set our goal to better classifying only the currently-given test data rather than learning or refining a universal classifier model, so what we are to solve in this paper is to refine the classification labels instead of the classifier itself.

The experimental results show that our bridged refinement algorithm can successfully refine results predicted from different kinds of shift-unaware classifiers, reducing the error rates by 50% in average compared with those classifiers. We attribute the success of our algorithm to the awareness of distribution

---

[1] More details are available at http://ida.first.fraunhofer.de/projects/different06/.

shift in each refinement step and the utilization of the mixture distribution as a bridge.

The rest of this paper is organized as follows. Some related work is discussed in Section 2. Our bridged refinement algorithm is described in Section 3 and the experiments are reported in Section 4. In Section 5, we conclude our work and mention some future work, as well as several interesting thoughts for extension.

2   Related Work

Utilizing labeled data from different domains, tasks or distributions as auxiliary data for a primary learning task is discussed in the *transfer learning* context [12,11,3,10]. [1] provided a theoretical justification of transfer learning through multi-task learning.

However, there are two slight differences between our work and many previous researches in the transfer learning literature. Firstly they usually utilize the auxiliary data to bias the classifier while our work focuses on refinement of classification results of the test data instead of the classifier itself, which is like what transductive learning does, as Vapnik [15] said,

> When solving a problem of interest, do not solve a more general problem as an intermediate step. Try to get the answer that you really need but not a more general one.

What we really need here is the correct labels for those test data, so we do not refine the model as an intermediate step, which is more general and more difficult, but directly refine the classification labels. Secondly, many researches require confidently labeled data under the new distribution, albeit little in amount, while *ours do not need any*.

The sample selection bias problem was mentioned in e.g. [16]. In that paper, the author concluded four kinds of biases, based on the dependency and independency between whether one example is selected and its feature-label pair (x, y). The problem we solve in this paper is similar to the second kind, which is, whether a training example is picked up (e.g. out-of-date labeled data vs. new data) is independent of the response y (positive or negative) given the input x. The author also suggested a solution to this kind of bias provided that given the input x, the selection criterion can be modeled properly, which is not needed in our work.

The refinement step in our proposed algorithm is suggested by the PageRank algorithm [9]. The PageRank algorithm conveys a mutual reinforcement principle that good pages may also link to some other good pages, thus it yields

$$PR = M^T \times PR \qquad (1)$$

where PR is a vector, each element of which is the score of each page, and M is the adjacent matrix of pages with each row L_1-normalized. Under a random surfer model, the formula is appended with another vector, E, to reflect the fact that the surfer may be bored with clicking through hyper-links in the Web pages and input a completely new URL in the browser:

$$PR = \alpha \, M^T \times PR + (1 - \alpha) \, E \qquad (2)$$

Such modification also involves some computational consideration such as rank sinks [2] and convergence to a unique point. Meanwhile a lot of researches regard the extra item E as a complement to the mutual reinforcement principle. The tradeoff between the two factors are controlled by the teleportation coefficient α. This kind of work includes: Topic-sensitive PageRank [5], TrustRank [4], etc.

3   Bridged Refinement Algorithm

Our algorithm receives the predictions from a shift-unaware classifier and then makes two-step refinements on these predictions, taking the mixture distribution $\mathcal{D}_{mix}$ as a bridge.

3.1   Refinement

One run of refinement is to correct the labels of the documents to make them more consistent with each other under the distribution observed in the documents themselves, which we name as the *target distribution*. In the first (left) refinement in Figure 1, the target distribution is $\mathcal{D}_{mix}$, observed in $\mathcal{L} \cup \mathcal{D}$, and in the second one, it is $\mathcal{D}_u$, observed in $\mathcal{U}$.

In order to clarify what we mean by "consistent", we introduce our assumption first. Formally, we assume that the conditional probability of a specified class given a document d, $P(c|d)$, does not vary among different distributions: $P_{\mathcal{D}_u}(c|d) = P_{\mathcal{D}_{mix}}(c|d) = P_{\mathcal{D}_l}(c|d)$, although the probability of a document $P(d)$ varies. This is based on such an observation that if identical documents appear both in the training data $\mathcal{L}$ and the test data $\mathcal{U}$, the labels should be the same. Taking it a step further, if one's neighbors have high confidence scores to belong to a specified category, the document itself may also receive a high confidence score to that category. This situation constitutes a mutual reinforcement relationship between documents. This kind of influence may occur across the decision boundary found by a shift-unaware classifier and thus can be used to correct the labels.

In a word, by the term "consistent", we mean similar documents should have close confidence values to the same category, where the similarity is measured in some way under the target distribution.

We now introduce the algorithm for refining the confidence score of each document to belong to a specified category under the target distribution. Mathematically, let M denote the adjacent matrix of documents, where M_{ij} is set to 0 if d_j is not a neighbor of d_i and $1/K$ if d_j is a neighbor of d_i, where K is the number of neighbors. The information of distribution governing the documents is, in some way, captured by the matrix M.

Input: The document collection D in bag-of-words representation, the unrefined
confidence score $UConf_{i,j}$ of each document d_i and each category c_j, and
α, K
Output: The refined confidence scores $RConf$
1 **foreach** *Pair of documents d_i, d_j* **do**
2 | $dist(i,j) = 1 - \cos(d_i, d_j)$
3 **end**
4 **foreach** *Document d_i* **do**
5 | Heap sort $\{< j, dist(i,j) >\}$ on $dist(i,j)$ until top $K+1$ are sorted
6 | $N_i \leftarrow K$ nearest neighbors of document d_i (excluding d_i itself)
7 **end**
8 **foreach** *Category c_j* **do**
9 | $UConf_{.,j} \leftarrow UConf_{.,j}/\|UConf_{.,j}\|_1$
10 **end**
11 **repeat**
12 | **foreach** *Document d_i* **do**
13 | | **foreach** *Category c_j* **do**
14 | | | $RConf_{i,j}^{(t+1)} \leftarrow \alpha \sum_{s \in N_i} RConf_{s,j}^{(t)}/|N_i| + (1-\alpha)\, UConf_{i,j}$
15 | | **end**
16 | **end**
17 **until** *RConf converges* ;
18 **return** $RConf$

Algorithm 1. Refinement Algorithm

$UConf_{.,j}$ is a vector, each element of which denotes the confidence score of a document to belong to the category c_j reported by those shift-unaware classifiers. Let $RConf_{.,j}$ denote the refined confidence scores of each document to belong to the category c_j. The mutual reinforcement principle yields the following equation for solving $RConf_{.,j}$:

$$RConf_{.,j} = \alpha\, M^T\, RConf_{.,j} + (1-\alpha)\, UConf_{.,j} \tag{3}$$

where α is the trade-off factor between the refinement process and the original (unrefined) confidence scores.

The equation can be solved in a closed form

$$RConf_{.,j} = (1-\alpha) \times (I - \alpha\, M^T)^{-1} UConf_{.,j} \tag{4}$$

or in an iterative manner as what is done in Algorithm 1 if M is too large to be inversed efficiently.

$RConf_{.,j}$ can also be explained in a random surfer model. Suppose the surfer wants to read documents of the category c_j, he/she starts at a random place and then reads similar documents until he/she decides to switch to a very different document from the one being read, but of the same category. The i-th element in $RConf_{.,j}$, $RConf_{i,j}$, is the probability that the surfer may read d_i. $RConf_{.,j}$ tells the relatedness of each document to the category c_j. In other words, it can be regarded as an estimate to $P(d_i|c_j)$.

Input: The document collection $\mathcal{L}$ and $\mathcal{U}$ in bag-of-words representation, the
     unrefined confidence score $UConf_{i,j}$ of each document d_i and each
     category c_j, and α, K
Output: The refined labels of $\mathcal{U}$

1 Call Algorithm 1 with $D = \mathcal{L} \cup \mathcal{U}$, $UConf$, α, and K. $RConf^1$ is returned.
2 Call Algorithm 1 with $D = \mathcal{U}$, $RConf^1$ projected on $\mathcal{U}$, α, and K. $RConf^2$ is
    returned.
3 **foreach** *Document d_i in $\mathcal{U}$* **do**
4 | $cat_i = \arg\max_j RConf^2_{i,j}$
5 **end**
6 **return** *cat_i for all documents in $\mathcal{U}$.*

Algorithm 2. Bridged Refinement Algorithm

The refinement step can be also regarded as a *thinker*. Given a document,
he/she seeks from his/her memory for similar documents whatever they are new
or old, and transfer the confidence of those similar articles to give a refined
confidence score.

The refinement algorithm is shown in Algorithm 1. Line 1 to 3 compute the
pairwise distances and in Line 4 to 7 neighbor sets are calculated. Line 8 to
10 normalize each $UConf._{,j}$ to unit length. The iterative manner for solving
Equation 3 is implemented in Line 11 to 17. The refined results are returned
in the last line. It should be noticed that the operations in Line 8 to 17 can
be accomplished in very neat codes of a matrix programming language such as
Matlab and that most part of the refinement algorithm can be easily adapted to
be performed in a parallel way.

3.2 Bridged Refinements

The refinement step is mainly controlled by the examples governed by the target
distribution, reflected in the adjacent matrix M in Algorithm 1. Alternatives of
refinements include refining under the mixture distribution $\mathcal{D}_{mix}$, or directly the
target distribution $\mathcal{D}_u$. But in our algorithm, we work in a "bridged" way that
we first refine under $\mathcal{D}_{mix}$ and then $\mathcal{D}_l$. Algorithm 2 gives the full description of
our bridged refinement algorithm.

The mixture distribution $\mathcal{D}_{mix}$ observed by $\mathcal{L} \cup \mathcal{U}$ is similar both to $\mathcal{D}_l$ and
to $\mathcal{D}_u$. Hence a relatively hard transfer from $\mathcal{D}_l$ to $\mathcal{D}_u$ is now bridged by $\mathcal{D}_{mix}$,
by which we decompose a problem into two relatively easy ones. Algorithm 1 is
performed twice. For the first time, we let D be $\mathcal{L} \cup \mathcal{U}$, and for the second time, D
be $\mathcal{U}$. In this way, two different M's will be calculated in the two steps, reflecting
the fact that each refinement is conducted towards a different distribution.

Finally we classify each document to one category that the final confidence in
$RConf._{,j}$ is the highest, which is done in Line 3 to 5 in Algorithm 2:

$$cat_i = \arg\max_{c_j} P(d_i|c_j) = \arg\max_j RConf_{i,j} \qquad (5)$$

4   Experiment

In this section, we want to verify the validity of the refinement step and the bridged way of refinements. We perform the experiments on two categories for the sake of simplicity, though it is straightforward to apply it for multi-categories.

4.1   Data Preparation

We prepare our data sets from three data collections: 20 Newsgroups[2], SRAA[3] and Reuters-21578[4] for evaluating our bridged refinement algorithm.

These three data collections are not originally designed for transfer learning, so we need to make some modifications to make the distribution between the training data and the test data different. Each of these data sets has at least two level hierarchical structure. Suppose A and B are two root categories in one data collection, and A_1, A_2 and B_1, B_2 are sub-level categories of A and B respectively. Now we form the training and test data in this way. Let $A.A_1$, $B.B_1$ be the positive and negative examples in the training data respectively. Let $A.A_2$, $B.B_2$ be the positive and negative examples in the test data respectively.

Thus, the target categories are fixed, being A and B, but the distributions of the training data and the test data are different but still similar enough for the evaluation of our refinement algorithm for transfer learning.

In SRAA, there are four discussion groups: simulated autos (simauto), simulated aviation (simaviation), real autos (realauto), real aviation (realaviation). We compose two data sets from SRAA. In 20 Newsgroups, there are seven top level categories, while three of them have no sub-categories. We compose six (C_4^2) data sets from the remaining four categories. From Retures-21578, we compose three data sets. The detailed composition of these data sets are provided in the Appendix.

We make some preprocessing on the raw data, including turning all letters into lowercases, stemming words by the Porter stemmer [13], removing all stop words. According to [14], DF Thresholding can achieve comparable performance to Information Gain or CHI, but it is much easier to implement and less costly both in time and space complexity. Hence we use it to cut down the number of words/features and thus speed up the classification. The words that occur in less than three documents are removed. Then each document is converted into bag-of-words presentation in the remaining feature space. Each feature value is the term frequency of that word in the document, weighted by its IDF ($\log N/DF$).[5]

4.2   Performance

Working with Different Classifiers. To ensure that our bridged refinement algorithm for transfer learning is robust enough, we perform our algorithm on

---

[2] http://people.csail.mit.edu/jrennie/20Newsgroups/

[3] http://www.cs.umass.edu/ mccallum/data/sraa.tar.gz

[4] http://www.daviddlewis.com/resources/testcollections/reuters21578/

[5] We also conduct experiments on TF without IDF weighting. Both the unrefined and refined results are worse than that of the TFIDF representation.

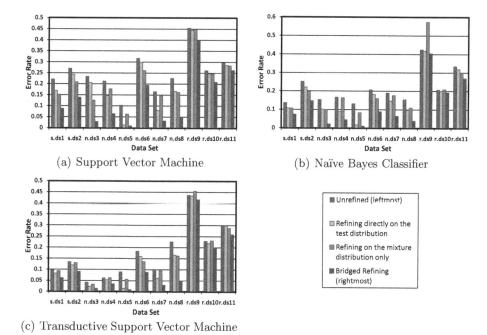

(a) Support Vector Machine   (b) Naïve Bayes Classifier

(c) Transductive Support Vector Machine

Fig. 2. Error Rates on Eleven Data Sets

the labels predicted by three different shift-unaware classifiers, from which we receive the unrefined results: the Support Vector Machine[15], the naïve Bayes classifier[8] and the Transductive Support Vector Machine[7]. The Support Vector Machine is a state-of-the-art supervised learning algorithm. In our experiment, we use the implementation SVM^{light}[6] with a linear kernel and all options set by default. The naïve Bayes classifier, despite its simplicity, performs surprisingly well in many tasks, so we also try to refine the classification results predicted by the naïve Bayes classifier. Last, we select a representative from semi-supervised learning algorithms, the Transductive Support Vector Machine, to see whether our refinement algorithm can reduce the error rate on that kind of learning algorithms. The experimental results show that, in all cases, our bridged refinement algorithm can reduce the error rates.

We first use each of the three models to learn from the training data $\mathcal{L}$ and classify the test data $\mathcal{U}$. Now we have labels for both the training data and the test data. For those positive examples, we set the corresponding element in $UConf._{,+}$ to 1 and for negative examples, we set to 0. In binary classification, $UConf._{,-}$ happens to be $1 - UConf._{,+}$. Then we pass these unrefined confidence scores to our algorithm and receive its outputs as the refined classification results.

Figure 2(a) shows the error rates of the Support Vector Machine on different data sets compared with the error rates of the refined results. In all data sets, our algorithm reduces the error rates. The greatest relative reductions of the error rates are achieved on n.ds5 and n.ds3, being 89.9% and 87.6% respectively. The average relative reduction of the error rates over all the eleven data sets is 54.3%. Figure 2(b) shows the results about the naïve Bayes classifier. The greatest relative reductions of the error rates are achieved on n.ds5 and n.ds3, being 91.2% and 85.7%. The average relative reduction is 51.5%. Figure 2(c) shows the result about the Transductive Support Vector Machine. The greatest relative reductions of the error rates are achieved on n.ds5 and n.ds8, being 89.2% and 78.1% respectively. The average relative reduction is 45.4%.

Does the Bridged Way Help? For each data set and classifier, we also experiment three different refinement strategies. They are (1) directly refining towards $\mathcal{D}_u$, (2) refining towards $\mathcal{D}_{mix}$ only, and (3) refining firstly towards $\mathcal{D}_{mix}$ and then towards $\mathcal{D}_u$.

Taking the Support Vector Machine as an example, each kind of refinement brings improvement in accuracy on all the eleven data sets. Moreover, still on all the eleven data sets, the bridged refinement outperforms another two one-step refinements consistently. Similar outperforming can be observed on another two classifiers. We attribute the success of the bridged refinement to the fact that the mixture distribution $\mathcal{D}_{mix}$ is more similar both to $\mathcal{D}_l$ and to $\mathcal{D}_u$ than $\mathcal{D}_l$ to $\mathcal{D}_u$ and thus the transfer is easy to perform.

4.3   Parameter Sensitivity

Our refinement algorithm includes two parameters, the number of neighbors K, and the teleportation coefficient α. In this section we want to show that the algorithm is not that sensitive to the selection of these parameters.

We only report six series from all eleven data sets and three classifiers, that is in total 33 combinations, to make the figures clear and easy to read. The selected six series are composed of three groups, two in each working with a different classifier. The six series can also be decomposed into another three groups, two in each from a different collection. Such consideration makes the selected six data sets representative.

From Figure 3(a), we can find that the performance is not greatly sensitive to the selection of K as long as K is moderately large, so K is empirically chosen as 70 in this paper. The teleportation coefficient α used in PageRank is reported to be 0.85. From Figure 3(b), we find 0.7 is a better choice. It is interesting to investigate the error rates when α is 0 or 1. In the former case, the error rate is just that of the unrefined labels, while in the latter case, the labels predicted by those shift-unaware classifiers impose no impact on the final labels. In both cases, the performance is not that good.

Although the iterative manner for solving Equation 3 may require many iterations before convergence, in our experiments, we find that in our situation, convergence is reached within five iterations. Thus the iterative way is preferred to directly computing the closed form since inversing a large matrix requires much time. In Figure 3(c), Iteration 1 to 10 are performed for the first refinement towards $\mathcal{D}_{mix}$ and 11 to 20 are for the second refinement towards $\mathcal{D}_u$.

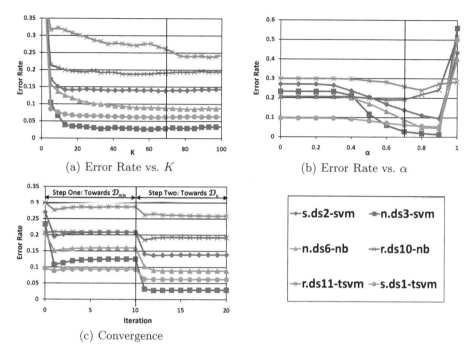

(a) Error Rate vs. K

(b) Error Rate vs. α

(c) Convergence

Fig. 3. Parameter and Iteration. (Points at $K = 70$ in (a), at $\alpha = 0.7$ in (b) and at Iteration= 20 in (c) correspond to the same situations.)

5   Conclusion and Future Work

In this paper, we propose an algorithm to solve a problem in transfer learning, i.e. how to refine the classification labels predicted by a shift-unaware classifier when the distribution of inputs (documents), $P(d)$, varies from training data to test data, under the assumption that the conditional probability of the category given the document, $P(c|d)$, do not vary from training data to test data.

We concentrate on the refinement of the classification labels instead of the classification model, based on the consideration by Vapnik [15] that predicting test data when they are available at hand may be easier than first learning a model and then predicting.

To show the robustness of the algorithm, we refine the results produced by two supervised learning representatives, the Support Vector Machines and the naïve Bayes classifer, and one semi-supervised learning representative, the Transductive Support Vector Machine. In all cases, our refinement algorithm can reduce the error rate.

We also verify that taking the mixture distribution $\mathcal{D}_{mix}$ into account is successful since directly refining the results under the test distribution $\mathcal{D}_u$ is not as good as the two-step way. We attribute this to the fact that $\mathcal{D}_{mix}$ is more similar both to $\mathcal{D}_l$ and to $\mathcal{D}_u$ than $\mathcal{D}_l$ to $\mathcal{D}_u$ directly.

There is a very interesting way to generalize our algorithm that can be tried. In our algorithm, only $(1 - \lambda)\,\mathcal{D}_l + \lambda\,\mathcal{D}_u$, where $\lambda = 0.5$ and 1 are considered, as $\mathcal{D}_{mix}$ and $\mathcal{D}_u$. Given a series: $\lambda_0 = 0$, λ_1, ..., λ_{n-1}, $\lambda_n = 1$, where $\lambda_i < \lambda_j, i, j = 1, \cdots, n$, $i < j$, one can perform an n-step refinement instead of the two-step way in our algorithm, where $n = 2$, and $\lambda_1 = 0.5$, $\lambda_2 = 1$. Thus the transfer process may be carried out in a more smooth way.

There are also some points in this work that can be handled more elegantly. For example, the unrefined confidence scores of both the training examples and the test examples are the same, but the labels of the test data are much less reliable than the training data. Therefore applying an discount to the confidence scores of the test data can be considered. Besides, given the refined confidence scores, the decision rule is based on the maximum likelihood principle. Other alternatives can be tried, including maximizing a posterior after modeling the category a prior probabilities, or directly assigning a document to a category where it ranks highest, etc. We will explore them further as our future work.

Acknowledgements

All authors are supported by a grant from National Natural Science Foundation of China (NO. 60473122). We thank the anonymous reviewers for their great helpful comments.

References

1. Ben-David, S., Schuller, R.: Exploiting task relatedness for multiple task learning. In: Proceedings of the Sixteenth Annual Conference on Learning Theory (2003)
2. Brin, S., Motwani, R., Page, L., Winograd, T.: What can you do with a web in your pocket. In: Bulletin of the IEEE Computer Society Technical Committee on Data Engineering, IEEE Computer Society Press, Los Alamitos (1998)
3. Caruana, R.: Multitask learning. Machine Learning 28(1), 41–75 (1997)
4. Gyongyi, Z., Garcia-Molina, H., Pedersen, J.: Combating web spam with trustrank. In: Proceedings of the 30th International Conference on Very Large Data Bases (2004)
5. Haveliwala, T.: Topic-sensitive PageRank. In: Proceedings of the Eleventh International World Wide Web Conference (2002)
6. Joachims, T.: Making large-scale SVM learning practical. In: Schölkopf, B., Burges, C., Smola, A. (eds.) Advances in Kernel Methods: Support Vector Learning, pp. 169–184. MIT Press, Cambridge (1999)
7. Joachims, T.: Transductive inference for text classification using Support Vector Machines. In: Bratko, I., Dzeroski, S. (eds.) Proceedings of the Sixteenthth Interantional Conference on Machine Learning, San Francisco, CA, USA, pp. 200–209 (1999)
8. Lewis, D.: Naive Bayes at forty: The independence assumption in information retrieval. In: Nédellec, C., Rouveirol, C. (eds.) Proceedings of European Conference on Machine Learning, Chemnitz, DE, pp. 4–15 (1998)
9. Page, L., Brin, S., Motwani, R., Winograd, T.: The PageRank citation ranking: bring order to the web, Stanford Digital Library Technologies Project (1998)

10. Raina, R., Ng, A.Y., Koller, D.: Constructing informative priors using transfer learning. In: Proceedings of the Twentieth International Joint Conference on Artificial Intelligence (2006)
11. Schmidhuber, J.: On learning how to learn learning strategies. Technical Report FKI-198-94, Technische Universität München (1995)
12. Thrun, S., Mitchell, T.M.: Learning one more thing. In: Proceedings of the Fourteenth International Joint Conference on Artificial Intelligence (1995)
13. Porter, M.F.: An algorithm for suffix stripping. Program 14(3), 130–137 (1980)
14. Yang, Y., Pedersen, J.O.: A comparative study on feature selection in text categorization. In: Fisher, D.H. (ed.) Proceedings of Fourteenth International Conference on Machine Learning, pp. 412–420 (1997)
15. Vapnik, V.N.: Statistical learning theory. Wiley, Chichester (1998)
16. Zadrozny, B.: Learning and evaluating classifiers under sample selection bias. In: Proceedings of the Twenty-First International Conference on Machine Learning (2004)

Appendix: Data Set Composition

There are too many sub-categories in Reuters-21578, so we ignore composition details of the last three data sets, which are from Reuters-21578. The detailed composition of the data sets from other two collections are completely shown below:

Source	Data Set	Train/Test	Positive	Negative	# of examples
SRAA (s.)	s.ds1	train	simauto	simaviation	8,000
		test	realauto	realaviation	8,000
	s.ds2	train	realaviation	simaviation	8,000
		test	realauto	simauto	8,000
20 Newsgroups (n.)	n.ds3	train	rec.autos rec.motorcycles	talk.politics.guns talk.politics.misc	3669
		test	rec.sport.baseball rec.sport.hockey	talk.politics.mideast talk.religion.misc	3561
	n.ds4	train	rec.autos rec.sport.baseball	sci.med sci.space	3,961
		test	rec.motorcycles rec.sport.hockey	sci.crypt sci.electronics	3,954
	n.ds5	train	comp.graphics comp.sys.mac.hardware comp.windows.x	talk.politics.mideast talk.religion.misc	4,482
		test	comp.os.ms-windows.misc comp.sys.ibm.pc.hardware	talk.politics.guns talk.politics.misc	3,652
	n.ds6	train	comp.graphics comp.os.ms-windows.misc	sci.crypt sci.electronics	3,930
		test	comp.sys.ibm.pc.hardware comp.sys.mac.hardware comp.windows.x	sci.med sci.space	4,900
	n.ds7	train	comp.graphics comp.sys.ibm.pc.hardware comp.sys.mac.hardware	rec.motorcycles rec.sport.hockey	4,904
		test	comp.os.ms-windows.misc comp.windows.x	rec.autos rec.sport.baseball	3,949
	n.ds8	train	sci.electronics sci.med	talk.politics.misc talk.religion.misc	3,374
		test	sci.crypt sci.space	talk.politics.guns talk.politics.mideast	3,828
Reuters - 21578 (r.)	r.ds9	train	orgs.*	places.*	1,078
		test	orgs.*	places.*	1,080
	r.ds10	train	people.*	places.*	1,239
		test	people.*	places.*	1,210
	r.ds11	train	orgs.*	people.*	1,016
		test	orgs.*	people.*	1,046

A Prediction-Based Visual Approach for Cluster Exploration and Cluster Validation by HOV³*

Ke-Bing Zhang[1], Mehmet A. Orgun[1], and Kang Zhang[2]

[1] Department of Computing, ICS, Macquarie University, Sydney, NSW 2109, Australia
{kebing,mehmet}@ics.mq.edu.au
[2] Department of Computer Science, University of Texas at Dallas
Richardson, TX 75083-0688, USA
kzhang@utdallas.edu

Abstract. Predictive knowledge discovery is an important knowledge acquisition method. It is also used in the clustering process of data mining. Visualization is very helpful for high dimensional data analysis, but not precise and this limits its usability in quantitative cluster analysis. In this paper, we adopt a visual technique called HOV³ to explore and verify clustering results with quantified measurements. With the quantified contrast between grouped data distributions produced by HOV³, users can detect clusters and verify their validity efficiently.

Keywords: predictive knowledge discovery, visualization, cluster analysis.

1 Introduction

Predictive knowledge discovery utilizes the existing knowledge to deduce, reason and establish predictions, and verify the validity of the predictions. By the validation processing, the knowledge may be revised and enriched with new knowledge [20]. The methodology of predictive knowledge discovery is also used in the clustering process [3]. Clustering is regarded as an unsupervised learning process to find group patterns within datasets. It is a widely applied technique in data mining. To achieve different application purposes, a large number of clustering algorithms have been developed [3, 9]. However, most existing clustering algorithms cannot handle arbitrarily shaped data distributions within extremely large and high-dimensional databases very well. The very high computational cost of statistics-based cluster validation methods in cluster analysis also prevents clustering algorithms from being used in practice.

Visualization is very powerful and effective in revealing trends, highlighting outliers, showing clusters, and exposing gaps in high-dimensional data analysis [19]. Many studies have been proposed to visualize the cluster structure of databases [15, 19]. However, most of them focus on information rendering, rather than investigating on how data behavior changes with the parameters variation of the algorithms.

---

* The datasets used in this paper are available from http://www.ics.uci.edu/~mlearn/Machine-Learning.html.

J.N. Kok et al. (Eds.): PKDD 2007, LNAI 4702, pp. 336–349, 2007.
© Springer-Verlag Berlin Heidelberg 2007

In this paper we adopt HOV^3 (*Hypothesis Oriented Verification and Validation by Visualization*) to project high dimensional data onto a 2D complex space [22]. By applying predictive measures (quantified domain knowledge) to the studied data, users can detect grouping information precisely, and employ the clustered patterns as predictive classes to verify the consistency between the clustered subset and unclustered subsets.

The rest of this paper is organized as follows. Section 2 briefly introduces the current issues of cluster analysis, and the HOV^3 technique as the background of this research. Section 3 presents our prediction-based visual cluster analysis approach with examples to demonstrate its effectiveness on cluster exploration and cluster validation. A short review of the related work in visual cluster analysis is provided in Section 4. Finally, Section 5 summarizes the contributions of this paper.

2 Background

The approach reported in this paper has been developed based on the projection of HOV^3 [22], which was inspired from the Star Coordinates technique. For a better understanding of our work, we briefly describe Star Coordinates and HOV^3.

2.1 Visual Cluster Analysis

Cluster analysis includes two major aspects: clustering and cluster validation. Clustering aims at identifying objects into groups, named clusters, where the similarity of objects is high within clusters and low between clusters. Hundreds of clustering algorithms have been proposed [3, 9]. Since there are no general-purpose clustering algorithms that fit all kinds of applications, the evaluation of the quality of clustering results becomes the critical issue of cluster analysis, i.e., cluster validation. Cluster validation aims to assess the quality of clustering results and find a fit cluster scheme for a given specific application.

The user's initial estimation of the cluster number is important for choosing the parameters of clustering algorithms for the pre-processing stage of clustering. Also, the user's clear understanding on cluster distribution is helpful for assessing the quality of clustering results in the post-processing of clustering. The user's visual perception of the data distribution plays a critical role in these processing stages. Using visualization techniques to explore and understand high dimensional datasets is becoming an efficient way to combine human intelligence with the immense brute force computation power available nowadays [16].

Visual cluster analysis is a combination of visualization and cluster analysis. As an indispensable aid for human-participation, visualization is involved in almost every step of cluster analysis. Many studies have been performed on high dimensional data visualization [2, 15], but most of them do not visualize clusters well in high dimensional and very large data. Section 4 discusses several studies that have focused on visual cluster analysis [1, 7, 8, 10, 13, 14, 17, 18] as the related work of this research. Star Coordinates is a good choice for visual cluster analysis with its interactive adjustment features [11].

2.2 Star Coordinates

The idea of Star Coordinates technique is intuitive, which extends the perspective of traditional orthogonal X-Y 2D and X-Y-Z 3D coordinates technique to a higher dimensional space [11]. Technically, Star Coordinates plots a 2D plane into n equal sectors with n coordinate axes, where each axis represents a dimension and all axes share the initials at the centre of a circle on the 2D space. First, data in each dimension are normalized into [0, 1] or [-1, 1] interval. Then the values of all axes are mapped to orthogonal X-Y coordinates which share the initial point with Star Coordinates on the 2D space. Thus, an n-dimensional data item is expressed as a point in the X-Y 2D plane. Fig.1 illustrates the mapping from 8 Star Coordinates to X-Y coordinates.

In practice, projecting high dimensional data onto 2D space inevitably introduces overlapping and ambiguities, even bias. To mitigate the problem, Star Coordinates and its extension iVIBRATE [4] provide several visual adjustment mechanisms, such as axis scaling, axis angle rotating, data point filtering, etc. to change the data distribution of a dataset interactively in order to detect cluster characteristics and render clustering results effectively. Below we briefly introduce the two relevant adjustment features with this research.

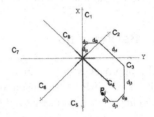

Fig. 1. Positioning a point by an 8-attribute vector in Star Coordinates [11]

- **Axis scaling**

The purpose of the *axis scaling* in Star Coordinates (called α-*adjustment* in iVI-BRATE) is to interactively adjust the weight value of each axis so that users can observe the data distribution changes dynamically. For example, the diagram in Fig.2 shows the original data distribution of Iris (Iris has 4 numeric attributes and 150 instances) with the clustering indices produced by the K-means clustering algorithm in iVIBRATE, where clusters overlap (here k=3).

A well-separated cluster distribution of Iris is illustrated in Fig. 3 by a series of random α-*adjustments*, where clusters are much easier to be recognized than those of the original distribution in Fig 2.

For tracing data points changing in a certain period time, the *footprint* function is provided by Star Coordinates. It is discussed below.

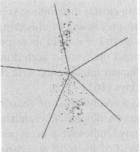

Fig. 2. The initial data distribution of clusters of Iris produced by k-means in iVIBRATE

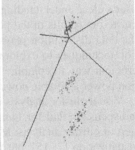

Fig. 3. The separated version of the Iris data distribution in iVIBRATE

- **Footprint**

We use another data set *auto-mpg* to demonstrate the *footprint* feature. The data set *auto-mpg* has 8 attributes and 398 items. Fig. 4 presents the footprints of axis tuning of attributes "weight" and "mpg", where we may find some points with longer traces, and some with shorter footprints.

The most prominent feature of Star Coordinates and its extensions such as iVIBRATE is that their computational complexity is only in linear time. This makes them very suitable to be employed as a visual tool for interactive interpretation and exploration in cluster analysis.

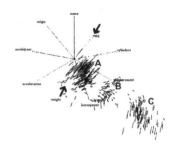

Fig. 4. Footprints of axis scaling of "weight" and "mpg" attributes in Star Coordinates [11]

However, the cluster exploration and refinement based on the user's intuition inevitably introduces randomness and subjectiveness into visual cluster analysis, and as a result, sometimes the adjustments of Star Coordinates and iVIBRATE could be arbitrary and time consuming.

2.3 HOV³

In fact, the Star Coordinates model can be mathematically depicted by the Euler formula. According to the Eular formula: $e^{ix} = \cos x + i \sin x$, where $z = x + i.y$, and i is the imaginary unit. Let $z_0 = e^{2\pi i/n}$; such that $z_0^1, z_0^2, z_0^3, \ldots, z_0^{n-1}, z_0^n$ (with $z_0^n = 1$) divide the unit circle on the complex 2D plane into n equal sectors. Thus, Star Coordinates can be simply written as:

$$P_j(z_0) = \sum_{k=1}^{n}[(d_{jk} - \min d_k)/(\max d_k - \min d_k) \cdot z_0^k] \qquad (1)$$

where $\min d_k$ and $\max d_k$ represent the minimal and maximal values of the kth coordinate respectively. In any case equation (1) can be viewed as mapping from $\mathbb{R}^n \rightarrow \mathbb{C}^2$.

To overcome the arbitrary and random adjustments of Star Coordinates and iVIBRATE, Zhang et al proposed a hypothesis-oriented visual approach called HOV³ to detect clusters [22]. The idea of HOV³ is that, in analytical geometry, the difference of a data set (a matrix) D_j and a measure vector M with the same number of variables as D_j can be represented by their inner product, $D_j \cdot M$. HOV³ uses a measure vector M to represent the corresponding axes' weight values. Then given a non-zero measure vector M in $\mathbb{R}^n$, and a family of vectors P_j, the projection of P_j against M, according to formula (1), the HOV³ model is presented as:

$$P_j(z_0) = \sum_{k=1}^{n}[(d_{jk} - \min d_k)/(\max d_k - \min d_k) \cdot z_0^k \cdot m_k] \qquad (2)$$

where m_k is the kth attribute of measure M.

The aim of interactive adjustments of Star Coordinates and iVIBRATE is to have some separated groups or full-separated clustering result of data by tuning the weight value of each axis, but their arbitrary and random adjustments limit their applicability. As shown in formula (2), HOV³ summarizes these adjustments as a coefficient/measure vector. Comparing the formulas (1) and (2), it can be observed that

HOV^3 subsumes the Star Coordinates model [22]. Thus the HOV^3 model provides users a mechanism to quantify a prediction about a data set as a measure vector of HOV^3 for precisely exploring grouping information.

Equation (2) is a standard form of linear transformation of n variables, where m_k is the coefficient of kth variable of P_j. In principle, any measure vectors, even in complex number form, can be introduced into the linear transformation of HOV^3 if it can distinguish a data set into groups or have well separated clusters visually. Thus the rich statistical methods of reflecting the characteristics of data set can be also introduced as predictions in the HOV^3 projection, such that users may discover more clustering patterns. The detailed explanation of this approach is presented next.

3   Predictive Visual Cluster Analysis by HOV^3

Predictive exploration is a mathematical description of future behavior based on historical exploration of patterns. The goal of predictive visual exploration by HOV^3 is that by applying a prediction (measure vector) to a dataset, the user may identify the groups from the result of visualization. Thus the key issue of applying HOV^3 to detect grouping information is how to quantify historical patterns (or users' domain knowledge) as a measure vector to achieve this goal.

3.1   Multiple HOV^3 Projection (M-HOV^3)

In practice, it is not easy to synthesize historical knowledge about a data set into one vector; rather than using a single measure to implement a prediction test, it is more suitable to apply several predictions (measure vectors) together to the data set, we call this process *multiple HOV^3 projection*, M-HOV^3 in short. Now, we provide the detailed description of M-HOV^3 and its feature of enhanced group separation. For simplifying the discussion of the M-HOV^3 model, we give a definition first.

Definition 1. (poly-multiply vectors to a matrix) The inner product of multiplying a series of non-zero measure vectors M_1, M_2,...,M_S to a matrix A is denoted as

$$A \cdot * \prod_{i=1}^{s} M_i = A \cdot * M_1 \cdot * M_2 \cdot * \ldots \cdot * M_S.$$

Zhang et al [23] gave a simple notation of HOV^3 projection as $\mathcal{D}_p = \mathcal{H}_C(\mathcal{P}, M)$, where $\mathcal{P}$ is a data set; $\mathcal{D}_p$ is the data distribution of $\mathcal{P}$ by applying a measure vector M. Then the projection of M-HOV^3 is denoted as $\mathcal{D}_p = \mathcal{H}_C(\mathcal{P}, \prod_{i=1}^{s} M_i)$. Based on equation (2), we formulate M-HOV^3 as:

$$P_j(z_0) = \sum_{k=1}^{n} [(d_{jk} - \min d_k)/(\max d_k - \min d_k) \cdot z_0^k \cdot \prod_{i=1}^{s} m_{ik}] \qquad (3)$$

where m_{ik} is the kth attribute (dimension) of the ith measure vector M_i, and $s \geq 1$. When $s=1$, the formula (3) is transformed to formula (2).

We may observe that instead of using a single multiplication of m_k in formula (2), it is replaced by a poly-multiplication of $\prod_{i=1}^{s} m_{ik}$ in formula (3). Formula (3) is more general and also closer to the real procedure of cluster detection, because it introduces several aspects of domain knowledge together into the cluster detection.

In addition, the effect of applying M-HOV³ to datasets with the same measure vector can enhance the separation of grouped data points under certain conditions.

3.2 The Enhanced Separation Feature of M-HOV³

To explain the geometrical meaning of M-HOV³ projection, we use the real number system. According to equation (2), the general form of the distance σ (i.e., weighed Minkowski distance) between two points a and b in HOV³ plane can be represented as:

$$\sigma(a,b,m) = \sqrt[q]{\sum_{k=1}^{n} |m_k(a_k - b_k)|^q} \quad (q>0) \tag{4}$$

If $q = 1$, σ is Manhattan (city block) distance; and if $q = 2$, σ is Euclidean distance. To simplify the discussion of our idea, we adopt the Manhattan metric for the explanation. Note that there exists an equivalent mapping (bijection) of distance calculation between the Manhattan and Euclidean metrics [6]. For example, if the distance between points a and b is longer than the distance between points a' and b' in then Manhattan metric, it is also true in the Euclidean metric, and vice versa.

Then the Manhattan distance between points a and b is calculated as in formula (5).

$$\sigma(a,b,m) = \sum_{k=1}^{n} |m_k(a_k - b_k)| \tag{5}$$

According to formulas (2), (3) and (5), we can present the distance of M-HOV³ in Manhattan distance as follows:

$$\sigma(a,b,\prod_{i=1}^{s} m_i) = \sum_{k=1}^{n} |\prod_{i=1}^{s} m_{ki}(a_k - b_k)| \tag{6}$$

Definition 2. (the distance representation of M- HOV³) The distance between two data points a and b projected by M- HOV³ is denoted as $\overset{s}{\underset{i=1}{M}}\sigma ab$. In particular, if the measure vectors in an M-HOV³ are the same, $\overset{s}{\underset{i=1}{M}}\sigma ab$ can be simply written as $M^s\sigma ab$; if each attribute of M is 1 (no measure case), the distance between points a and b is denoted as σab.

Thus, we have $\overset{s}{\underset{i=1}{M}}\sigma ab = \mathcal{H}_C((a,b), \prod_{i}^{s} M_i)$. For example, the distance between two points a and b projected by M-HOV³ with the same two measures can be represented as $M^2\sigma ab$. Thus the projection of HOV³ of a and b can be written as $M\sigma ab$.

We now give several important properties of M- HOV³ as follows.

Lemma 1. In Star Coordinates space, if $\sigma ab \neq 0$ and $M \neq 0$ ($\exists m_k \in M \mid 0 < |m_k| < 1$), then $\sigma ab > M\sigma ab$.

Proof

$$\sigma ab = \sum_{k=1}^{n} |(a_k - b_k)| \text{ and } M\sigma ab = \sum_{k=1}^{n} |m_k(a_k - b_k)|$$

$$\sigma ab - M\sigma ab = \sum_{k=1}^{n} |(a_k - b_k)| - \sum_{k=1}^{n} |m_k(a_k - b_k)| = \sum_{k=1}^{n} |(a_k - b_k)|(1 - |m_k|)$$

$$M \neq 0 \Rightarrow \{\exists m_k \neq 0 \wedge m_k \in M \mid 0 < |m_k| < 1, k=1...n\} \Rightarrow (1 - |m_k|) > 0$$

$$\sigma ab \neq 0 \Rightarrow \sigma ab > (M\sigma ab) \qquad \square$$

This result shows that the distance $M\sigma ab$ between points a and b projected by HOV^3 with a non-zero M is less than the original distance σab between a and b.

Lemma 2. In Star Coordinates space, if $\sigma ab \neq 0$ and $M \neq 0$ ($\forall m_k \in M \mid 0 < |m_k| < 1$), then $M^n \sigma ab > M^{n+1} \sigma ab$, $n \in \mathbf{N}$.

Proof

Let $M^n \sigma ab = \sigma'ab$

Definition 1 $\Rightarrow M^{n+1} \sigma ab = M \sigma'ab$

Lemma 1 $\Rightarrow \sigma'ab > M\sigma'ab \Rightarrow M^n \sigma ab > M^{n+1} \sigma ab \qquad \square$

In general, it can be proved that in Star Coordinates space, if $\sigma ab \neq 0$ and $M \neq 0$ ($\forall m_k \in M \mid |m_k| < 1$), then $M^m \sigma ab > M^n \sigma ab$, $n \in \mathbf{N}$, $m \in \mathbf{N}$ and $m < n$.

Theorem 1. If the measure vector is changed from M to M', ($|m_k| \leq 1, |m_t + \Delta_t| < 1$) and $|M\sigma ab - M\sigma ac| < |M'\sigma ab - M'\sigma ac|$ then

$$\frac{|M'\sigma ab - M'\sigma ac| - |M'^2 \sigma ab - M'^2 \sigma ac|}{|M'\sigma ab - M'\sigma ac|} > \frac{|M\sigma ab - M\sigma ac| - |M'\sigma ab - M'\sigma ac|}{|M\sigma ab - M\sigma ac|}$$

Proof

$$M'\sigma ab = \sum_{k=1}^{n} |m'_k(a_k - b_k)| \text{ and } M'\sigma ac = \sum_{k=1}^{n} |m'_k(a_k - c_k)|$$

$$\Rightarrow M'\sigma ab - M'\sigma ac = \sum_{k=1}^{n} |m'_k| [|(a_k - b_k)| - |(a_k - c_k)|]$$

$$M'^2 \sigma ac - M'^2 \sigma ab = \sum_{k=1}^{n} |m'_k|^2 [|(a_k - b_k)| - |(a_k - c_k)|]$$

Let $|a_k - b_k| = x_k$ and $|a_k - c_k| = y_k$

$$\Rightarrow M'\sigma ac - M'\sigma ab = \sum_{k=1}^{n} |m'_k| [|(a_k - b_k)| - |(a_k - c_k)|] = \sum_{k=1}^{n} |m'_k| (x_k - y_k)$$

$$\Rightarrow M'^2 \sigma ac - M'^2 \sigma ab = \sum_{k=1}^{n} |m'_k|^2 (x_k - y_k)$$

$$\Rightarrow |M'\sigma ac - M'\sigma ab| = M'\sigma xy$$

$$\Rightarrow |M'^2 \sigma ac - M'^2 \sigma ab| = M'^2 \sigma xy$$

Lemma 2 $\Rightarrow M'^2 \sigma xy < M'\sigma xy \Rightarrow \dfrac{|M'^2 \sigma xy|}{|M'\sigma xy|} < 1 \Rightarrow \dfrac{|M'^2 \sigma ab - M'^2 \sigma ac|}{|M'\sigma ab - M'\sigma ac|} < 1$

$$\Rightarrow |M'^2 \sigma ab - M'^2 \sigma ac| < |M'\sigma ab - M'\sigma ac|$$

$$|M\sigma ab - M\sigma ac| < |M'\sigma ab - M'\sigma ac|$$

$$\Rightarrow |M'^2 \sigma ab - M'^2 \sigma ac| . |M\sigma ab - M\sigma ac| < |M'\sigma cab - M'\sigma ac|^2$$

$$\Rightarrow \frac{|M'^2 \sigma ab - M'^2 \sigma ac|}{|M'\sigma ab - M'\sigma ac|} < \frac{|M'\sigma ab - M'\sigma ac|}{|M\sigma ab - M\sigma ac|}$$

$$\Rightarrow 1 - \frac{|M'^2 \sigma ab - M'^2 \sigma ac|}{|M'\sigma ab - M'\sigma ac|} > 1 - \frac{|M'\sigma ab - M'\sigma ac|}{|M\sigma ab - M\sigma ac|}$$

$$\Rightarrow \frac{\mid M'\sigma ab - M'\sigma ac \mid - \mid M'^{2}\sigma ab - M'^{2}\sigma ac \mid}{\mid M'\sigma ab - M'\sigma ac \mid} > \frac{\mid M\sigma ab - M\sigma ac \mid - \mid M'\sigma ab - M'\sigma ac \mid}{\mid M\sigma ab - M\sigma ac \mid}$$

☐

Theorem 1 shows that if the user observes that the difference of the distance between a and b and the distance between a and c are increased relatively (it can be observed by the footprints of points a, b and c, as shown in Fig 4) by tuning weight values of axes from M to M', then after applying M-HOV^3 to a, b and c, the distance variation rate of the distances between pairs of points a, b and a, c is enhanced, as presented in Fig 5.

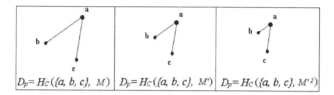

Fig. 5. The contraction and separation effect of M-HOV^3

In other words, if it is observed that several data point groups can be roughly separated visually (there may exist ambiguous points between groups) by projecting a measure vector in HOV^3 to a data set, then applying M-HOV^3 with the same measure vector to the data set would lead to the groups being more condensed, i.e., have a good separation of the groups.

3.3 Predictive Cluster Exploration by M-HOV^3

According to the notation of HOV^3 projection of a dataset $\mathcal{P}$ as $\mathcal{D}_p = \mathcal{H}_C(\mathcal{P}, M)$, the M-HOV^3 is denoted as $\mathcal{D}_p = \mathcal{H}_C(\mathcal{P}, M^n)$ where $n \in \mathbb{N}$.

We use the *auto-mpg* dataset again as an example to demonstrate predictive cluster exploration by M-HOV^3. Fig. 6a illustrates the original data distribution of *auto-mpg* produced by HOV^3 in MATLAB, where it is not possible to recognize any group information. Then we tuned each axis manually and had roughly distinguished three groups, as shown in Fig 6b. The weight values of axes were recorded as a vector M=[0.10, 0, 0.25, 0.2, 0.8, 0.85, 0.1, 0.95]. Fig. 6b shows that there exist several ambiguous data points between groups. Then we employed M^2 (inner dot) as a predictive measure vector and applied it to data set *auto-mpg*. The projected distribution $\mathcal{D}_{p2}$ of *auto-mpg* is presented in Fig 6c. It is much easier to identify 3 groups of *auto-mpg* in Fig 6c than in Fig 6b. To show the contrast between these two diagrams $\mathcal{D}_{p1}$ and $\mathcal{D}_{p2}$, we overlap them in Fig. 6d.

By analyzing the data of these 3 groups, we have found that, group 1 contains 70 items and with "original" value 2 (sourcing Europe); group 2 has 79 instances and with "original" 3 (Japanese product); and group 3 includes 249 records with "original" 1 (from USA). Actually this "natural" grouping based on the user's intuition serendipitously clustered the data set according to the "original" attribute of *auto-mpg*. In the same way, the user may find more grouping information from the interactive cluster exploration by applying predictive measurement.

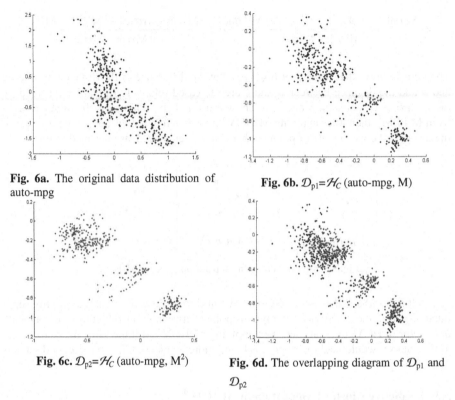

Fig. 6a. The original data distribution of auto-mpg

Fig. 6b. $\mathcal{D}_{p1}=\mathcal{H}_C$ (auto-mpg, M)

Fig. 6c. $\mathcal{D}_{p2}=\mathcal{H}_C$ (auto-mpg, M²)

Fig. 6d. The overlapping diagram of $\mathcal{D}_{p1}$ and $\mathcal{D}_{p2}$

Fig. 6. Diagrams of data set *auto-mpg* projected by HOV³ in MATLAB

3.4 Predictive Cluster Exploration by HOV³ with Statistical Measurements

Many statistical measurements, such as mean, median, standard deviation and etc. can be directly introduced into HOV³ as predictions to explore data distributions. In fact, prediction based on statistical measurements is more purposefully cluster exploration, and give an easier geometrical interpretation of the data distribution.

We use the Iris dataset as an example. As shown in Fig. 3, by random axis scaling, the user can divide the Iris data in 3 groups. This example exhibits that cluster exploration based on random adjustment may expose data groping information, but sometimes, it is hard to interpret such groupings.

We employ the standard deviation of Iris M = [0.2302, 0.1806, 0.2982, 0.3172, 0.4089] as a prediction to project Iris by HOV³ in iVIBRATE. The result is shown in Fig. 7, where 3 groups clearly exist. It can be observed in Fig 7 that, there is a blue point in the pink-colored cluster and a pink point in the green-colored cluster, resulting from the K-means clustering algorithm with k=3. Intuitively, they have been wrongly clustered. We re-clustered them by their distributions, as shown in Fig 8.

The contrast of clusters (C_k) produced by the K-means clustering algorithm and new clustering result (C_H) projected by HOV³ is summarized in Table 1. We can see that the

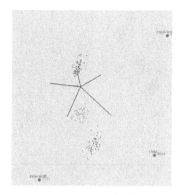
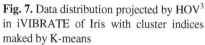
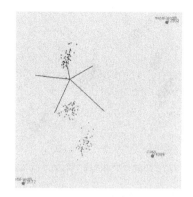

Fig. 7. Data distribution projected by HOV^3 in iVIBRATE of Iris with cluster indices maked by K-means

Fig. 8. Data distribution projected by HOV^3 in iVIBRATE of Iris with the new clustering indices by the user's intuition

quality of the new clustering result of Iris is better than that obtained by K-means according to their *"Variance"* comparison. Each cluster projected by HOV^3 has a higher similarity than that produced by K-means. By analyzing the new grouping data points of Iris, we have found that they are distinguished by the *"class"* attribute of Iris, i.e. *Iris-setosa, Iris-versicolor and Iris-virginica*. The cluster 1 generated by K-means is an outlier.

Table 1. The statistics of the clusters in Iris' produced by HOV^3 with predictive measure

C_k	%	Radius	Variance	MaxDis	C_H	%	Radius	Variance	MaxDis
1	1.333	1.653	2.338	3.306					
2	32.667	5.754	0.153	6.115	*1*	*33.333*	*5.753*	*0.152*	*6.113*
3	33.333	8.196	0.215	8.717	*2*	*33.333*	*8.210*	*0.207*	*8.736*
4	33.333	7.092	0.198	7.582	*3*	*33.333*	*7.112*	*0.180*	*7.517*

With the statistical predictions in HOV^3 the user may even expose the cluster clues that are not easy to be found by random adjustments. For example, we adopted the 8th row of *auto-mpg*'s covariance matrix as a predictive measure (0.04698, -0.07657, -0.06580, 0.00187, -0.05598, 0.01343, 0.02202, 0.16102) to project *auto-mpg* by HOV^3 in MATLAB. The result is shown in Fig 9. We grouped them by their distribution as in Fig 10. Table 2 (right part) reports the statistics of the clusters generated by the projection of HOV^3, and reveals that the points in each cluster have very high similarity.

As we chose the 8th row of *auto-mpg*'s covariance matrix as the prediction, the result mainly depends on the 8th column of *auto-mpg* data, i.e., "origin" (country). Fig. 10 shows that C1, C2 and C3 are closer because they have the same "origin" value 1. The more detailed formation of clusters is given in the right part of Table 2. We believe that a domain expert could give a better and intuitive explanation about this clustering.

Then we chose number 5 to cluster *auto-mpg* by the K-means. Its clustering result is presented in the left part of Table 2. Comparing their corresponding statistics, we can see that according to the *Variance* of clusters, the quality of the clustering result by

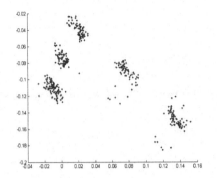

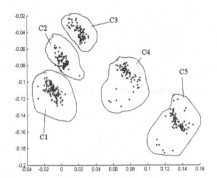

Fig. 9. Data distribution projected by HOV³ in MATLAB of auto-mpg with 8ᵗʰ row of auto-map's covariance matrix as prediction

Fig. 10. Clustered distribution of data in Fig. 8 by the user's intuition

Table 2. The statistical contrast of clusters in *auto-mpg* produced by K-means and HOV³

	Clusters produced by K-means (k=5)				Clusters generated by the user intuition on the data distribution					
C	%	Radius	Variance	MaxDis	Origin	Cylinders	%	Radius	Variance	MaxDis
1	0.503	681.231	963.406	1362.462	1	8	25.879	4129.492	0.130	4129.768
2	18.090	2649.108	0.206	2649.414	1	6	18.583	3222.493	0.098	3222.720
3	16.080	2492.388	0.139	2492.595	1	4	18.090	2441.881	0.090	2442.061
4	21.608	3048.532	0.207	3048.897	2	4	17.588	2427.449	0.142	2427.632
5	25.377	3873.052	0.220	3873.670	3	3	19.849	2225.465	0.093	2225.658
6	18.593	2417.804	0.148	2417.990						

HOV³ with covariance prediction of *auto-mpg* is better than that one produced by K-means (k=5, cluster 1 produced by K-means is an outlier).

3.5 Predictive Cluster Validation by HOV³

In practice, with extremely large sized datasets, it is infeasible to cluster an entire data set within an acceptable time scale. A common solution used in data mining is that, clustering algorithms are first applied to the training (a sampling) subset of data from a database to extract cluster patterns, and then the cluster scheme is assessed to see whether it is suitable for other subsets in the database. This procedure is regarded as *external cluster validation* [21]. Due to the high computational cost of statistical methods on assessing the consistency of cluster structures between large sized subsets, to achieve this goal by statistical methods is still a challenge in data mining.

Based on the assumption that if two same-sized data sets have a similar cluster structure, by applying a linear transformation to the data sets, the similarity of the newly produced distributions of the two sets would still be high, Zhang *et al* proposed a visual external validation approach by HOV³ [23]. Technically, their approach uses a clustered subset and a same-sized unclustered subset from a database as the observation by applying the measure vectors that can separate clusters in the clustered subset by HOV³. Thus each cluster and its geometrically covered data points (called *quasi-Cluster* in their approach) are selected. Finally, the overlapping rate of each cluster-quasicluster pair is calculated; and if the overlapping rate approaches 1, this means that the two subsets have a similar cluster distribution.

Compared with the statistics-based validation methods, their method is not only visually intuitive, but also more effective in real applications [23]. As mentioned above, sometimes, it is time consuming to separate clusters manually in Star Coordinates or iVIBRATE. Thus, separation of clusters from lots of overlapping points is an aim of this research. As we described above, the approaches such as M-HOV³ and HOV³ with statistical measurement can be introduced into external cluster validation by HOV³. In principle, any linear transformation can be employed into HOV³ if it can separate clusters well.

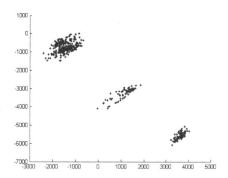

Fig. 11. The data distribution of auto-mpg projected by HOV3 with $cos(M*10i)$ as the prediction

We therefore introduce the complex linear transformation to this process. We again use *auto-mpg* data set as an example. As shown in Fig. 6b, three roughly separated clusters appear there, where the vector M=[0.10, 0, 0.25, 0.2, 0.8, 0.85, 0.1, 0.95] was obtained from the axes values. Then we adopt $cos(M·10i)$ as a prediction, where i is the imaginary unit. The projection of HOV³ with $cos(M·10i)$ is illustrated in Fig. 11, where three clusters are separated very well. In the same way, many other linear transformations can be applied to different datasets to obtain well-separated clusters. With the fully separated clusters, there will be marked improvement of the efficiency of visual cluster validation.

4 Related Work

Visualization is typically employed as an observational mechanism to assist users with intuitive comparisons and better understanding of the studied data. Instead of quantitatively contrasting clustering results, most of the visualization techniques employed in cluster analysis focus on providing users with an easy and intuitive understanding of the cluster structure, or explore clusters randomly.

For instance, Multidimensional Scaling, MDS [14] and Principal Component Analysis, PCA [10] are two commonly used multivariate analysis techniques. However, the relative high computational cost of MDS (polynomial time $O(N^2)$) limits its usability in very large datasets, and PCA first has to find the correlated variables for reducing the dimensionality, which makes it not suitable for unknown data exploration.

OPTICS [1] uses a density-based technique to detect cluster structure and visualizes clusters in "Gaussian bumps", but its non-linear time complexity makes it neither suitable for dealing with very large data sets, nor for providing the contrast between clustering results. H-BLOB visualizes clusters into blob manners in a 3D hierarchical structure [17]. It is an intuitive cluster rendering technique, but its 3D and two stages expression restricts it from interactively investigating cluster structures apart from existing clusters.

Kaski *el. al* [13] uses Self-organizing maps (SOM) to project high-dimensional data sets to 2D space for matching visual models [12]. However, the SOM technique is based on a single projection strategy and it is not powerful enough to discover all the interesting features from the original data set.

Huang *et. al* [7, 8] proposed the approaches based on FastMap [5] to assist users in identifying and verifying the validity of clusters in visual form. Their techniques work well in cluster identification, but are unable to evaluate the cluster quality very well. On the other hand, these techniques are not well suited to the interactive investigation of data distributions of high-dimensional data sets. A recent survey of visualization techniques in cluster analysis can be found in the literature [18].

5 Conclusions

In this paper, we have proposed a prediction-based visual approach to explore and verify clusters. This approach uses the HOV^3 projection technique and quantifies the previously obtained knowledge and statistical measurements about a high dimensional data set as predictions, so that users can utilize the predictions to project the data on 2D plane in order to investigate grouping clues or verify the validity of clusters based on the distribution of the data. This approach not only inherits the intuitive and easy understanding features of visualization, but also avoids the weaknesses of randomness and arbitrary exploration of the existing visual methods employed in data mining.

As a consequence, with the advantage of the quantified predictive measurement of this approach, users can identify the cluster number in the pre-processing stage of clustering efficiently, and also can intuitively verify the validity of clusters in the post-processing stage of clustering.

References

1. Ankerst, M., Breunig, M.M., Kriegel, S.H.P.J.: OPTICS: Ordering points to identify the clustering structure. In: Proc. of ACM SIGMOD Conference, pp. 49–60. ACM Press, New York (1999)
2. Ankerst, M., Keim, D.: Visual Data Mining and Exploration of Large Databases. In: 5th European Conference on Principles and Practice of Knowledge Discovery in Databases (PKDD'01), Freiburg, Germany (September 2001)
3. Berkhin, P.: A Survey of Clustering Data Mining Techniques. In: Jacob, K., Charles, N., Marc, T. (eds.) Grouping Multidimensional Data, pp. 25–72. Springer, Heidelberg (2006)
4. Chen, K., Liu, L.: iVIBRATE: Interactive visualization-based framework for clustering large datasets. ACM Transactions on Information Systems (TOIS) 24(2), 245–294 (2006)
5. Faloutsos, C., Lin, K.: Fastmap: a fast algorithm for indexing, data-mining and visualization of traditional and multimedia data sets. In: Proc. of ACM-SIGMOD, pp. 163–174 (1995)
6. Fleming, W.: Functions of Several Variables. In: Gehring, F.W., Halmos, P.R. (eds.) 2nd edn. Springer, Heidelberg (1977)
7. Huang, Z., Cheung, D.W., Ng, M.K.: An Empirical Study on the Visual Cluster Validation Method with Fastmap. In: Proc. of DASFAA01, pp. 84–91 (2001)
8. Huang, Z., Lin, T.: A visual method of cluster validation with Fastmap. In: Terano, T., Chen, A.L.P. (eds.) PAKDD 2000. LNCS, vol. 1805, pp. 153–164. Springer, Heidelberg (2000)

 9. Jain, A., Murty, M.N., Flynn, P.J.: Data Clustering: A Review. ACM Computing Surveys 31(3), 264–323 (1999)
10. Jolliffe Ian, T.: Principal Component Analysis. Springer Press, Heidelberg (2002)
11. Kandogan, E.: Visualizing multi-dimensional clusters, trends, and outliers using star coordinates. In: Proc. of ACM SIGKDD Conference, pp. 107–116. ACM Press, New York (2001)
12. Kohonen, T.: Self-Organizing Maps, 2nd extended edn. Springer, Berlin (1997)
13. Kaski, S., Sinkkonen, J., Peltonen, J.: Data Visualization and Analysis with Self-Organizing Maps in Learning Metrics. In: Kambayashi, Y., Winiwarter, W., Arikawa, M. (eds.) DaWaK 2001. LNCS, vol. 2114, pp. 162–173. Springer, Heidelberg (2001)
14. Kruskal, J.B., Wish, M.: Multidimensional Scaling, SAGE university paper series on quantitive applications in the social sciences, pp. 7–11. Sage Publications, CA (1978)
15. Oliveira, M.C., Levkowitz, H.: From Visual Data Exploration to Visual Data Mining: A Survey. IEEE Transaction on Visualization and Computer Graphs 9(3), 378–394 (2003)
16. Pampalk, E., Goebl, W., Widmer, G.: Visualizing Changes in the Structure of Data for Exploratory Feature Selection. In: SIGKDD '03, Washington, DC, USA (2003)
17. Sprenger, T.C, Brunella, R., Gross, M.H.: H-BLOB: A Hierarchical Visual Clustering Method Using Implicit Surfaces. In: Proc. of the conference on Visualization '00, pp. 61–68. IEEE Computer Society Press, Los Alamitos (2000)
18. Seo, J., Shneiderman, B.: From Integrated Publication and Information Systems to Virtual Information and Knowledge Environments. In: Hemmje, M., Niederée, C., Risse, T. (eds.) From Integrated Publication and Information Systems to Information and Knowledge Environments. LNCS, vol. 3379, Springer, Heidelberg (2005)
19. Shneiderman, B.: Inventing Discovery Tools: Combining Information Visualization with Data Mining. In: Jantke, K.P., Shinohara, A. (eds.) DS 2001. LNCS (LNAI), vol. 2226, pp. 17–28. Springer, Heidelberg (2001)
20. Weiss, S.M., Indurkhya, N.: Predictive Data Mining: A Practical Guide. Morgan Kaufmann Publishers, San Francisco (1998)
21. Vilalta, R., Stepinski, T., Achari, M.: An Efficient Approach to External Cluster Assessment with an Application to Martian Topography, Technical Report, No. UH-CS-05-08, Department of Computer Science, University of Houston (2005)
22. Zhang, K-B., Orgun, M.A., Zhang, K.: HOV^3, An Approach for Cluster Analysis. In: Li, X., Zaïane, O.R., Li, Z. (eds.) ADMA 2006. LNCS (LNAI), vol. 4093, pp. 317–328. Springer, Heidelberg (2006)
23. Zhang, K-B., Orgun, M.A., Zhang, K.: A Visual Approach for External Cluster Validation. In: Proc. of IEEE Symposium on Computational Intelligence and Data Mining (CIDM2007), Honolulu, Hawaii, USA, April 1-5, 2007, pp. 576–582. IEEE Press, Los Alamitos (2007)

Flexible Grid-Based Clustering

Marc-Ismaël Akodjènou-Jeannin, Kavé Salamatian, and Patrick Gallinari

LIP6 - Université Paris 6 Pierre et Marie Curie
104 avenue du Président Kennedy, Paris, France
{Marc-Ismael.Akodjenou,Kave.Salamatian,Patrick.Gallinari}@lip6.fr

Abstract. Grid-based clustering is particularly appropriate to deal with massive datasets. The principle is to first summarize the dataset with a grid representation, and then to merge grid cells in order to obtain clusters. All previous methods use grids with hyper-rectangular cells. In this paper we propose a flexible grid built from arbitrary shaped polyhedra for the data summary. For the clustering step, a graph is then extracted from this representation. Its edges are weighted by combining density and spatial informations. The clusters are identified as the main connected components of this graph. We present experiments indicating that our grid often leads to better results than an adaptive rectangular grid method.

1 Introduction

With the ever-increasing amount of storage and processing capacities, huge datasets are now common in many areas : earth science, astronomy, or computer networks, just to name a few. The mining of such datasets, and especially the clustering task, calls for robust and efficient techniques. Grid-based clustering methods have been the subject of many recent studies [1,2,3].

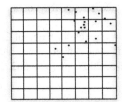
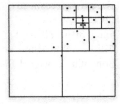

Fig. 1. Summaries of datasets. Left a regular rectangular grid. Right an adaptive hypercubic grid.

Grid-based clustering consists in clustering the space surrounding the datapoints instead of the datapoints themselves [4]. The basic idea is to cover the data space with a grid in order to construct a spatial summary of the data. Each nonempty cell of the grid is weighted by the number of original datapoints it contains (see Figure 1). The clustering is performed by aggregating adjacent dense cells to form clusters. Grid-based methods are similar to density-based clustering, but

J.N. Kok et al. (Eds.): PKDD 2007, LNAI 4702, pp. 350–357, 2007.
© Springer-Verlag Berlin Heidelberg 2007

with local densities and neighborhood relations taking place between cells, and no longer between individual points.

In this paper, we propose a new type of grid to build the dataset summary. The cells of the grid are general polyhedra, and are not axis-aligned hypercubes or hyper-rectangles like in all existing methods. The neighborhood relation between cells is richer; hence the aggregation process (which is the base operation for clustering) is more efficient. The clustering step is performed by extracting a graph from the spatial summary, and identifying clusters as its main connected components. The edges of the graph are weighted by a similarity metric which uses both spatial and density information from the summary.

The remainder of the paper is structured as follows : Section 2 presents related work and motivations. Section 3 describes the construction of the flexible grid. Section 4 describes the clustering step and the similarity metric. Section 5 discusses complexity and sensitivity to dimensionality. Section 6 contains results of experiments and a comparison with a hypercubic adaptive grid method.

2 Related Work and Motivation

Many grid-based clustering approaches [1,3] rely on the traditional regular, hypercubic grid (Figure 1, left). The main drawback of these approaches is that the grid construction requires to cover all the data space with the same precision independently of the data density. Thus a very high resolution could be needed to obtain a satisfying spatial summary. Another class of methods [5,2] uses multi-resolution grids with size-varying hypercubic or hyper-rectangular cells (Figure 1, right). The basic idea is to cover with more precision regions with many points. Usually the clustering step follows the hierarchy of the data structure. Both sets of methods are parametrized by the resolution of the grid. The clustering step usually relies on a density threshold discarding low-density cells. The complexity of these methods is linear in the number of data points $O(N)$. The complexity of the clustering step depends only on the number of (non-empty) cells M.

The aggregation of neighbor cells is the basis for the clustering process. Since the ultimate goal is to find patterns in the original data, one wants to minimize the impact of the particular geometry of the grid on the efficiency of the aggregation process. Classical grids (be they regular or multi-resolution) have their cell borders aligned with the axes of the space; this directional bias has a strong influence on the resulting data summaries.

In this work, we propose a multi-resolution grid whose cells have randomly oriented borders. It is close to the Crack STIT tessellation model of stochastic geometry [6]. The resulting spatial summary has no particular orientation and does not suffer from the rigid geometry of hyper-rectangular tilings. The cells are general polyhedra, allowing a spatially more flexible aggregation process. For the clustering step, we extract a weighted graph from the spatial summary. We propose a similarity metric to weight edges; it takes into account both spatial and density similarities of cells. The clusters are identified as the main connected components of the graph. The complexity of the clustering step is $O(M)$. The parameters of the whole method are the size of the summary M, the number of clusters K and the minimum number of points $MinPts$ per cluster.

3   Flexible Grid

3.1   Hyperplanes and Polyhedra

We recall here simple facts about hyperplanes and polyhedra. A hyperplane in a d dimensional space $H = \{z \mid \langle u \cdot z \rangle = t\}$ is defined by its *orientation vector* $u \in \mathbb{S}^{d-1}$ and *its offset* $t \in \mathbb{R}$ ($\langle \cdot \rangle$ denotes scalar product). For a given u, a hyperplane with offset $t = \langle z_0 \cdot u \rangle$ passes through the point z_0. A uniform random hyperplane can be obtained by taking a random d-dimensional gaussian random vector, and normalizing its norm to 1. A polyhedra $P \subset \mathbb{R}^d$ admits two representations : the H-representation (set of delimiting hyperplanes) and the V-representation (convex hull of vertices). The H-representation describes P as the intersection of halfspaces defined by a set of hyperplanes ($\cap H_i^{\sigma_i}$), where $\sigma = (\sigma_1 \ldots \sigma_m)$ is a binary codeword locating the point in halfspaces defined by the hyperplanes.

3.2   Construction

The principle of the construction of the multi-resolution flexible grid is simple (see Algorithm 1). It begins with the hypercube containing the data. At each step, the cell containing the largest number of points is splitted into two sub-cells by a random hyperplane. This process is iterated until a given number of non-empty cells M (fixed by the user) is reached. The hyperplanes are chosen with a uniform random orientation. The splitting process has a natural binary tree structure, as depicted in Figure 2. The algorithm iteratively encodes the data points into binary codewords. These binary codewords correspond to the H-representation of the cells. At the end of the algorithm, the dataset has been summarized to a set of weighted polyhedra. Each datapoint belongs to a partic-ular cell.

The flexible grid is a particular realization of a stochastic process. It is built iteratively during the cell refinement process and automatically adapts its reso-lution to the local data density. Finer parts of the gird are revealed in regions

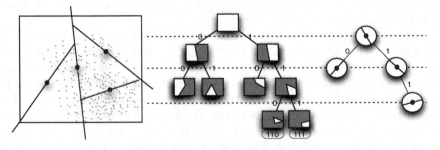

Fig. 2. Data domain, cell tree and hyperplane tree

Algorithm 1. Construction of flexible grid

Inputs
$X = \{x_1, \ldots, x_N\}$ dataset of N points in $\mathbb{R}^d$
D : hyper-rectangle containing X
M : desired size (number of occupied cells) of the summary

Outputs
$\mathcal{S} = \{(S_1, p_1), \ldots, (S_M, p_M)\}$ set of M polyhedra along with the proportion of points they contain
NR =neighborhood relation between the cells

Begin
$\mathcal{S}_0 \leftarrow \{(D, 1, [])\}$ initial region containing all the points
$T \leftarrow$ empty hyperplane binary tree
$NR \leftarrow$ empty list

While $| \mathcal{S}_0 | < M$
    $(C, p, w) \leftarrow$ cell of $\mathcal{S}_0$ with the maximum p and with codeword w
    $H_{split} \leftarrow$ random hyperplane passing through the center of C
    $T \leftarrow$ Add hyperplane H_{split} to hyperplane tree at node of binary index w
    $\{(C_1, p_1, w_1), (C_2, p_2, w_2)\} \leftarrow$ subcells created by splitting C with hyperplane H_{split}
    Replace (C, p, w) in $\mathcal{S}_0$ by non-empty elements of $\{(C_1, p_1, w_1), (C_2, p_2, w_2)\}$
    $NR \leftarrow$ Update neighborhood relations of the new cells replacing C
End

Extract $\mathcal{S}$ from $\mathcal{S}_0$
End

where there are many datapoints. The resulting summary has small, high-density cells in dense regions and big, low-density cells in sparsely populated regions.

4 Clustering

4.1 Graph Clustering

Graph clustering has been the subject of numerous studies (see [7]). The idea is to modelize the clustering problem by a weighted graph; the original clustering problem reduces to find clusters of vertices of the graph. In this paper, we extract the graph from the spatial summary (Figure 3). The graph representation is well suited to our problem since it allows to describe in a compact form the polyhedra (the vertices), their neighborood relation (the edges) and their similarities (edge weights). An edge links two vertices if they correspond to neighbor cells. Two cells are neighbors if they have a $(d - 1)$-dimensional intersection. Edges are weighted with a similarity metric described below. We iteratively remove edges of the graph until we have K connected components, of at least $MinPts$ points each. At each step the edge with the minimum weight is chosen for removal.

 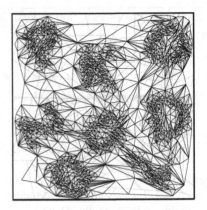

Fig. 3. (a) Flexible data summary (b) Structure of extracted graph

4.2   Similarity Metric

In the majority of previous works, the grouping of two cells is determined (implicitly or explicitly) by the closeness of the densities of the two cells. This stems from the intuitive assumption that cells at the frontier of a cluster will see a large density variation. This is not robust since even in dense regions, important density variations may appear. We propose a more robust cell grouping criterion incorporating spatial closeness between cells. Because of the multi-resolution, the distance between cell centers already conveys much information about the density of the data. The spatial information allows a smoothing of the density variations, thereby allowing better clustering results.

Given two cells with centers c_i, $c_j \in \mathbb{R}^d$, and cell density values D_i and $D_j \in \mathbb{R}$, we set the similarity between cells i and j to be $f_{sim} = f_{dens} \cdot f_{spat}$, with $f_{dens}(i,j) = exp\left(-\frac{\|D_i - D_j\|^2}{2 \cdot \sigma_{dens}^2}\right)$ and $f_{spat}(i,j) = exp\left(-\frac{\|c_i - c_j\|^2}{2 \cdot \sigma_{spat}^2}\right)$ with σ_{dens} being the mean euclidean difference between their densities, and σ_{spat} the average euclidean distance between centers of two neighbors cells of the grid. The exponentiation is the most natural way to express the similarities. The density D_i is the ratio (p_i/V_i) where p_i and V_i are respectively the proportion of points and the volume of the cell.

5   Dimensionality and Complexity

5.1   Dimensionality

Grid-based methods are well suited for small dimensional spaces. For high dimensional data, the number of grid cells and of neighbor cells increases exponentially and the methods cannot be used as such when the number of dimensions iq too high [8]. the exponential number of grid cells, and the high number of neighbor cells are highlighted as the main issues. Compared to regular rectangular grids, the multi-resolution grid and our graph clustering technique partly circumvents this phenomenon. In our case the main limitation comes from the complex structure of polyhedra : computing the volume of the polyhedra and testing which polyhedra are in its neighborhood rapidly becomes prohibitive. We propose here to approximate these two steps: instead of computing the whole neighborhood of cell, we compute the distance between all cell centers. The neighborhood of a specific cell is then defined as the set of cells whose center is among the closest centers according to the distance matrix. The volume of a cell of the grid is then approximated by the volume of a ball, the diameter of which is set to the distance to the nearest cell center. These approximations are reasonable with regard to the multi-resolution nature of the summaries and to our edge-removal graph clustering technique. This approximation does not degrade the performance of the method as will be seen in Section 6.

5.2   Complexity

The construction step has linear complexity in the input data size $O(N)$ (with an analysis similar to [2]). All the other steps depend only on M. Neighborhood check has complexity $O(m \cdot LP(m, d))$, $LP(m, d)$ being the complexity of a linear program with m constraints in a d-dimensional space, m depending on the polyhedra. For the clustering step, each search for connected components has a complexity linear in the size of the graph: $O(V + E)$, V and E being respectively the number of vertices and edges of the graph. It is $O(M)$ for our problem since the number of edges can be bounded by $(n_{max} \cdot M)/2$ with n_{max} the maximum number of neighbors of a node.

6   Experiments

6.1   Experimental Setting

We implemented in C++ the construction of our flexible grid, flexible grid approximation. We also implemented the AMR-like (Adaptive Mesh Refinement) grid (Figure 1 right), which is an adaptive, axis-aligned, hypercubic grid described in [9,2]. Experiments were performed with four datasets : a first complex 2D dataset of 3000 points from [10], a 3D dataset of 8000 points with five non-convex "banana" shapes, the Pageblocks database of 5400 points ($d = 10$), and a subset of 7800 points Letter Recognition database ($d = 16$) from the UCI Machine Learning Repository. For the 2D and 3D datasets, we compared the axis-aligned case ('AMR-like'), the flexible case ('flexible') and the approximation of the flexible case described in Section 5 ('flexible-approx'). For the higher dimensional datasets we compared the flexible approximation and the AMR-like summaries. For flexible approximations, we took respectively 3,4,11 and 17 neighbors per cell for the 2D, 3D, 10D and 16D datasets (following the simple idea that a polyhedron in d dimensions has at least $(d + 1)$ faces). We measured the raw performance with respect to the full original dataset with the Normalized Mutual Information criterion ([11]). Error bars show standard deviation of experiments for the flexible and flexible-approximation cases. Clustering parameters are indicated in the lower right corner of the figures.

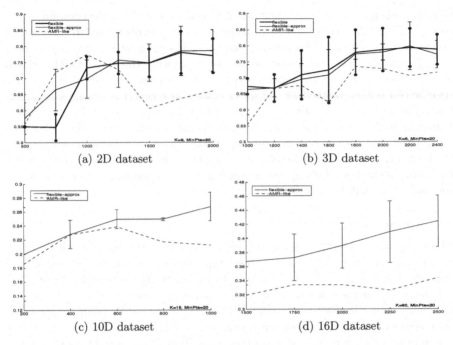

(a) 2D dataset   (b) 3D dataset

(c) 10D dataset   (d) 16D dataset

Fig. 4. Clustering quality for growing summary sizes

6.2 Discussion

The results show that the axis-aligned summary type has a rather unpredictable behavior. The clustering performance does not always grow with the summary size : it may remain approximately constant (16D dataset) or even degrade (2D and 10D datasets). The flexible grid yields better results most of the time. The clustering performance globally grows with the resolution. Note that the flexible approximation has practically the same performance than the full flexible summary. With this approximation, the complexity of the algorithm is greatly reduced so that it could be used reasonably for dimensions up to 50.

7   Conclusion and Future Work

We have proposed a new type of grid for data summaries in the context of grid-based clustering methods. The grid is locally-adaptive and has a flexible geometry. We also proposed an approximation of this method adapted to high dimensional spaces. We have presented results indicating that the proposed grid often yields more accurate clustering results than its axis-aligned counterpart. In future work, we will incorporate the flexible grid into classical variations and improvements for grid-based methods (e.g subspace clustering [12]).

References

1. Peter, W., Chiochetti, J., Giardina, C.: New unsupervised clustering algorithm for large datasets. In: Proceedings of the ninth ACM SIGKDD international conference on Knowledge discovery and data mining, ACM Press, New York (2003)
2. Liao, W.K., Liu, Y., Choudhary, A.: A grid-based clustering algorithm using adaptive mesh refinement. In: 7th Workshop on Mining Scientific and Engineering Datasets of SIAM International Conference on Data Mining (2004)
3. Yu, Z., Wong, H.S.: Gca: A real-time grid-based clustering algorithm for large dataset. In: Proceedings of the 18th International Conference on Pattern Recognition (ICPR) (2006)
4. Schikuta, E.: Grid-clustering: An efficient hierarchical clustering method for very large data sets. In: 13th International Conference on Pattern Recognition (ICPR'96) (1996)
5. Schikuta, E., Erhart, M.: The bang-clustering system: Grid-based data analysis. In: Liu, X., Cohen, P.R., Berthold, M.R. (eds.) Advances in Intelligent Data Analysis. Reasoning about Data. LNCS, vol. 1280, Springer, Heidelberg (1997)
6. Nagel, W., Weiss, V.: Crack stit tessellations: characterization of stationary random tessellations stable with respect to iteration. Advances In Applied Probability 37, 859–883 (2005)
7. Brandes, U., Gaertler, M., Wagner, D.: Experiments on graph clustering algorithms. In: ESA, pp. 568–579 (2003)
8. Hinneburg, A., Keim, D.: Optimal grid-clustering towards breaking the curse of dimensionality in high-dimensional clustering. In: Proceedings of the 25th International Conference on Very Large Databases (VLDB) (1999)
9. Wang, W., Yang, J., Muntz, R.: Sting: a statistical information grid approach to spatial data mining. In: Twenty-Third International Conference on Very Large Databases (1997)
10. Salvador, S., Chan, P.: Determining the number of clusters/segments in hierarchical clustering/segmentation algorithm. In: Proceedings of the 16th IEEE International Conference on Tools with Artificial Intelligence (ICTAI'04), pp. 576–584. IEEE Computer Society Press, Los Alamitos (2004)
11. Strehl, A., Gosh, J.: Cluster ensembles - a knowledge reuse framework for combining multiple partitions. Journal of Machine Learning Research (JMLR) 3 (2002)
12. Agrawal, R., Gehrke, J., Gunopoulos, J., Raghavan, P.: Automatic subspace clustering of high dimensional data for data mining applications. In: Proceedings of the 1998 ACM International Conference on Management of Data (SIGMOD '98), pp. 94–105. ACM Press, New York (1998)

Polyp Detection in Endoscopic Video Using SVMs

Luís A. Alexandre[1,2], João Casteleiro[1], and Nuno Nobre[1]

[1] Department of Informatics, Univ. Beira Interior, Portugal
[2] IT - Networks and Multimedia Group, Covilhã, Portugal

Abstract. Colon cancer is one of the most common cancers in developed countries. Most of these cancers start with a polyp. Polyps are easily detected by physicians. Our goal is to mimic this detection ability so that endoscopic videos can be pre-scanned with our algorithm before the physician analyses them. The method will indicate which part of the video needs attention (polyps were detected there) and hence can speedup the procedures. In this paper we present a method for polyp detection in endoscopic images that uses SVM for classification. Our experiments yielded a result of $93.16 \pm 0.09\%$ of area under the Receiver Operating Characteristic (ROC) curve on a database of 4620 images indicating that the approach proposed is well suited to the detection of polyps in endoscopic video.

1   Introduction

A polyp is an abnormal growth of tissue projecting from a mucous membrane.

In this paper we are concerned with polyps in the colon. An example is presented in figure 1. Polyps are important since they can, with time, turn into colon cancer. The cumulative risk of cancer developing in an unremoved polyp is 2.5% at 5 years, 8% at 10 years, and 24% at 20 years after the diagnosis [7]. If detected on an early stage these polyps can be easily removed.

In Portugal there are six thousands people per year diagnosed with the disease of colon cancer[1]. However, this disease is also the most tractable of all the digestive cancers when diagnosed at an early stage. This cancer is one of the most fatal illness all over the world.

Our interest in the video processing approach comes from the fact that there is a new medical examination, where the patient ingests a capsule (with the form of a pill) that films the digestive tube (video capsule endoscopy). The video is recorded in a device that the patient carries, usually in the belt. This video is then screened by the physician to search for polyps (and possibly other illnesses). Our goal is to develop a method that can be applied to the resultant video and avoid the time necessary to completely screen these videos. The idea is that if our systems detects polyps, the physician will focus on the signalled portions of the video with urgency. Videos where polyps aren't detected will be left for latter processing by the physician (have a lower priority).

J.N. Kok et al. (Eds.): PKDD 2007, LNAI 4702, pp. 358–365, 2007.

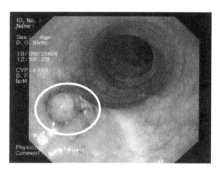

Fig. 1. An endoscopic image with a polyp marked by the white ellipse

Given the importance of colon polyps, some researchers have developed methods for its automatic detection. There are basically two approaches for this: video processing and CT (Computed Tomography) image processing [4,3,5]. Although the approaches based on CT images are able to produce a virtual representation of the colon which can speed up the visual analysis by the physician, they also have some disadvantages: the extensive amount of radiologist time (during CT scanning) involved in the process, the radiation that the patient is subjected to and the cost of such an exam. Given our motivation presented above, we are more interested in the video based approaches. We will now briefly describe some of the work done under this approach.

In [6] a comparative study of texture features for the detection of gastric polyps in endoscopic video was presented. Of the four approaches tested, texture spectrum histogram, texture spectrum and color histogram statistics, local binary pattern histogram and the color wavelet covariance, this last one presented the best results with an area under the ROC curve value of 88.6%.

In [8] the authors presented new approaches for extracting texture- and color-based features from colonoscopic images for the analysis of the colon status. Note that the abnormal status can be due to pathologies other than polyps. They used Principal Component Analysis (PCA) for feature selection and Backpropagation Neural Networks for classification. They found that using texture and color features improved classification results when compared to using only one type of feature.

In [9] the authors were also concerned with abnormality detection from endoscopic images. They use a fusion approach to reach a final decision from sub-decisions made based on associated component feature sets. They report that the overall detectability of abnormalities using the fusion approach is improved when compared with corresponding results from the individual methods.

In this paper we show that, given the SVM's ability to deal with high dimensional input spaces, we can produce very interesting results in terms of polyp detection in endoscopic video images by using only color and pixel position information, without any further feature extraction or selection technique.

The paper is organized as follows: the next section presents our method for polyp detection. Section 4 presents the experiments and the final section contains the conclusions.

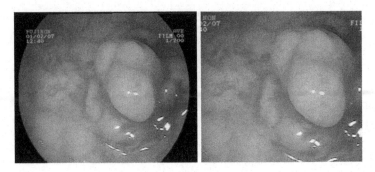

Fig. 2. Image before and after the operation that removes the black frame

2 Polyp Detection

The goal is to detect polyps on colonoscopic images similar to the one in figure 2. Our approach to this problem is the following: first we pre-processed the images to retain only the image portion that contains relevant information; then we subdivided each of the original images into sub-images of 40 × 40 pixels. Then we applied the feature extraction algorithms to these sub-images. Finally a classifier (SVM) was used to make the decision about the existence or not of a polyp in an image.

We will now describe these operations in more detail.

2.1 Pre-processing

The videos were captured with PAL (768 × 576) resolution. The frames have a black frame around the useful region of image as in figure 2. This black frame is removed leaving each image with a resolution of 514 × 469. This approach discards some of the useful area, but since we are working with video, we can recover the lost data from other video frames.

2.2 Image Division and Tagging

Our approach considered the division of the original images into smaller sub-images, that is, we will not classify directly an input image but, we subdivide it and classify each sub-image individually as containing a polyp or not. Then this information is used to classify the original image.

The idea consists in processing sub-images that can sometimes be completely contained within the polyp region. This means that ideally we should use sub-images of the size of a single pixel. Of course this would not produce enough data to have statistically significant results on the sub-image level. So, we define the sub-image area with dimensions of 40 × 40 pixels. This is small enough such that the sub-images are frequently completely contained in the polyp region but are also big enough to produce significant feature results.

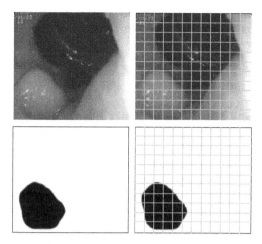

Fig. 3. From left to right and top to bottom: original image, its subdivision, the corresponding manual classification mask (black means polyp) and its subdivision

The sub-images were obtained by sliding a window with a 40 pixels step, both horizontally and vertically through the original image. This means that there is no overlap between the sub-images.

Given the dimensions of the input images after the black frame removal, the sub-division process generates 132 sub-images for each original image.

To simplify (automate) the manual classification of each of these sub-images we produced a binary image that was used as a classification mask. This mask is a manual painted image the size of the original image, that has the polyp region painted black and the remaining portion is white (an example is shown in figure 3). This painting was checked by a grastroentrologist.

The subdivision of the original image into sub-images is done also on the classification mask, yielding a sub-image that contains only black and white pixels. To assign a class label to the original sub-images we look at the corresponding classification mask sub-image and count the number of black pixels it contains. If this number is higher than a given threshold, λ, we consider that the sub-image 'contains' a polyp. This process allows the automatic classification of the sub-images.

2.3   Choosing the Sub-Image Classification Threshold λ

The value of λ can vary between 1 and 1600 (the total number of pixels of the sub-image). Naturally, as λ gets smaller, more sub-images are classified as containing polyps. This might look like a good option so that the system has a smaller false negative rate. But since these images are used for training the system, if we choose to classify a sub-image with few black pixels as a polyp, we are using very little real polyp information in that image to teach what a polyp is. In fact, if we choose λ smaller than 800, we may give more non-polyp than polyp information in a sub-image.

We studied the influence of λ in the classification results. These results are presented in the experiments section but we can say that the best results are obtained for a larger λ.

2.4   Feature Extraction

Out approach to feature extraction is quite simple and produced very interesting results. Given the capabilities of the SVMs in dealing with high-dimensional input data, we chose as features only color and position information for each pixel. Each pixel in a sub-image is represented by five values: its RGB components and its coordinates in the sub-image. So each sub-image is in fact represented by a total of 8000 features: 5 features for each of the 1600 pixels.

These features were reduced and centered in the corresponding training and test sets (see below).

3   Experiments

3.1   Dataset

The dataset consists of 35 video frames obtained with a Fuji 410 video endoscope system at the Hospital Cova da Beira, Portugal, during the year of 2007. The images were subdivided into smaller images after the pre-processing described in section 2.1. Each image produced 132 sub-images. Each sub-image was defined as polyp or not polyp according to the correspondent sub-image obtained from the manual generated classification mask described in section 2.2. The resulting data set contained 4620 images each with a dimension of 40×40 pixels.

The features were centered and reduced such that, for each feature, the mean value is 0 and the standard deviation is 1.

3.2   Classifier

The classifier used was a support vector machine (LIBSVM) [2]. The kernel type used was the radial basis function (RBF):

$$K(\mathbf{x}_i, \mathbf{x}_j) = \exp(-\gamma ||\mathbf{x}_i - \mathbf{x}_j||^2), \ \gamma > 0 \tag{1}$$

where $\mathbf{x}_i$ is an input and γ is a parameter inversely proportional to the kernel width. The SVM with this kernel has two free parameters to be set: $C > 0$ that corresponds to the penalty parameter of the error and γ. The values for these parameters are discussed below.

A different weight can be assign to each class when the prior probabilities for each class are not equal. This is the case in our dataset given that there are more non-polyp images than polyp ones. The exact proportion depends on the threshold λ used. This weighting was done using the svm-train parameters w_0 and w_1. This is also discussed below.

Table 1. Number of images considered polyp for each threshold λ

λ	100	300	500	700	900	1100	1300	1500
N. of polyps	966	900	850	809	770	716	682	609

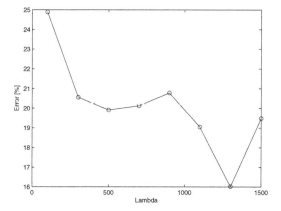

Fig. 4. Error for different values of λ

3.3   Evaluating the Effect of λ

In this section we present the results of experiments made to evaluate the effect of the value of λ used in the automatic classification of the sub-images, as discussed in section 2.3.

We varied λ from 100 to 1500 in steps of 200. Table 1 presents the number of sub-images that are classified as polyp given the value of λ considered.

For each value of λ a grid search was performed using half the dataset to find the best values of C and γ parameters for the SVM classifier. The search was done by varying C from 1 to 64 in integer powers of 2, and γ from 2^{-6} to 2^{-16} also in (negative) integer powers of 2.

We created a subset of with 10% of the images (462) randomly selected from the full image set. The error in this subset was evaluated with 2-fold cross-validation method. The results obtained are shown in figure 4. It can be seen that the smallest error, 16.02%, was obtained for $\lambda = 1300$. The correspondent value of C was 32 and $\gamma = 0.0001$. Given these results we decided to use $\lambda = 1300$ for our subsequent experiments.

3.4   Results on the Full Dataset

We evaluated the error using the 2-fold cross-validation method on the 4630 images, using $\lambda = 1300$ and $C = 32$. We experimentally found that good values for the class weights in this case are $w_0 = 1$ and $w_1 = 5$. The value of γ was varied to produce several points on the ROC curve. Figure 5 contains the ROC plot.

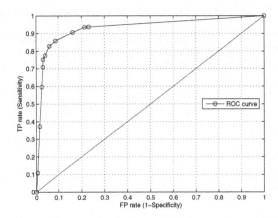

Fig. 5. ROC curve for $\lambda = 1300$. FP stands for False Positive and TP for True Positive.

The area under the Receiver Operating Characteristic (ROC) curve, (AUC), is $93.16 \pm 0.09\%$. The results show that we can get a value of False Negative Rate (FNR) of $6.31 \pm 1.04\%$ at $23.11 \pm 0.93\%$ of False Posite Rate (FPR).

4 Conclusions

In this paper we presented a method for polyp detection in endoscopic videos. The goal was to be able to develop a method that could do a first automatic screening of a endoscopic video and warn the physician of frames where attention is needed. Videos in which the method does not detect any polyp can perhaps be given a smaller priority then others where possible polyps are detected.

Our method subdivides each image into smaller images (with 40×40 pixels). These are the images that are searched for polyps. Of course if a subimage is considered a polyp, its parent image is also considered to have a polyp. (Other approaches can be used like the need for a number of detected subimages with polyp in order to consider that the parent image contains a polyp. Our approach is the most cautious: it perhaps implies the existence of some false positives but will minimize the false negatives). We did not analyse this aspect in this paper though. We focused on the correct classification of the subimages.

We used a very simple approach for feature extraction, relying only on color and pixel position. Although it creates many features per image (8000) the SVM was able to deal with this high dimensionality. It can take longer for the training phase, but may decrease the time taken when processing a new image with the trained system since the feature extraction does not involve many computations.

The results we obtained are quite satisfactory: in a database with 4620 images we were able to obtain an AUC value of $93.16 \pm 0.09\%$, a sensitivity of $93.69 \pm 1.04\ \%$ at $23.11 \pm 0.93\%$ of FPR.

Future work will concern the application of this method to video images obtained from video capsule endoscopy instead of video from a colonoscope. Other future challenges include the identification of the type of polyp and its development stage.

Acknowledgements

We wish to thank the support of Dr. Carlos Casteleiro Alves at providing the images and validating the classification masks.

References

1. Rui Cernadas. Algumas reflexões sobre o cancro colorrectal. In: Endonews (2004)
2. Chang, C.-C., Lin, C.-J.: LIBSVM: a library for support vector machines, Software (2001), available at http://www.csie.ntu.edu.tw/~cjlin/libsvm
3. Chowdhury, T.A., Ghita, O., Whelan, P.F.: A statistical approach for robust polyp detection in CT colonography. In: 27th Annual International Conference of the Engineering in Medicine and Biology Society, pp. 2523–2526. IEEE Computer Society Press, Los Alamitos (2005)
4. Gokturk, S.B., Tomasi, C., Paik, D., Beaulieu, C., Napel, S.: A learning method for automated polyp detection. In: Niessen, W.J., Viergever, M.A. (eds.) MICCAI 2001. LNCS, vol. 2208, pp. 85–92. Springer, Heidelberg (2001)
5. Huang, A., Summers, R.M., Hara, A.K.: Surface curvature estimation for automatic colonic polyp detection. In: Amini, A.A., Manduca, A. (eds.) Medical Imaging 2005: Physiology, Function, and Structure from Medical Images. Proceedings of the SPIE, vol. 5746, pp. 393–402 (2005)
6. Iakovidis, D.K., Maroulis, D.E., Karkanis, S.A., Brokos, A.: A comparative study of texture features for the discrimination of gastric polyps in endoscopic video. In: 18th IEEE Symposium on Computer-Based Medical Systems (CBMS'05), pp. 575–580. IEEE Computer Society, Los Alamitos (2005)
7. Stryker, S.J., Wolff, B.G., Culp, C.E., Libbe, S.D., Ilstrup, D.M., MacCarty, R.L.: Natural history of untreated colonic polyps. Gastroenterology 93(5), 1009–1013 (1987)
8. Tjoa, M.P., Krishnan, S.M.: Feature extraction for the analysis of colon status from the endoscopic images. BioMedical Engineering OnLine 2(9) (2003)
9. Zheng, M.M., Krishnan, S.M., Tjoa, M.P.: A fusion-based clinical decision support for disease diagnosis from endoscopic images. Computers in Biology and Medicine 35(3), 259–274 (2005)

A Density-Biased Sampling Technique to Improve Cluster Representativeness

Ana Paula Appel, Adriano A. Paterlini, Elaine P. M. de Sousa,
Agma J. M. Traina, and Caetano Traina Jr.*

Computer Science Department - ICMC
University of São Paulo at São Carlos – Brazil
{anaappel@,adriano@grad.,parros@,agma@,caetano@}icmc.usp.br

Abstract. The volume and complexity of data collected by modern applications has grown significantly, leading to increasingly costly operations for both data manipulation and analysis. Sampling is an useful technique to support manager a more sensible volume in the data reduction process. Uniform sampling has been widely used but, in datasets exhibiting skewed cluster distribution, biased sampling shows better results. This paper presents the *BBS - Biased Box Sampling* algorithm which aims at keeping the skewed tendency of the clusters from the original data. We also present experimental results obtained with the proposed BBS algorithm.

1 Introduction

Data reduction and sampling techniques have been employed to speed up data mining algorithms, and the more representative a dataset sample, the better the results obtained. Many of the data reduction techniques for multi-dimensional data rely on uniform sampling. However, several tasks have to deal with non-uniform data distribution, in particular clustering activities when the original clusters have distinct properties among themselves, such as the number of elements and/or the density. In such cases, density-biased sampling can provide better results, as the probability of a point to be added to the sample depends on the local density of its neighborhood.

Figure 1 shows nine clusters over uniform noise (fifteen percent), one containing 50,000 points and the others containing 1000 points. Extracting a 1% uniform sampling will produce a sample dataset containing one cluster with roughly five hundred points, and eight containing around ten points each as well as some noise. Since the number of clusters will be discovered only after the clustering has been finished, this information is not available to the sampling process. Therefore, in this example the clustering algorithm will not be able to spot the small cluster. The problem here is that when the representation of a cluster in the dataset is significantly lower than those of the other clusters, the clustering algorithm may miss the small clusters, mixing it with noise. Therefore, the question posed is:

* The authors thank CNPq, Capes and FAPESP for the financial support.

J.N. Kok et al. (Eds.): PKDD 2007, LNAI 4702, pp. 366–373, 2007.

"How to sample a multi-dimensional dataset without missing the clusters, even if they are unbalanced (that is, their number of points are quite different) and without any previous knowledge about the clusters?".

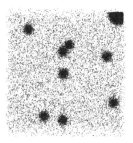

Fig. 1. A 20-dimensional dataset comprising 1 cluster with 50,000 points, 8 with 1000 points and 15% of noise

The problem of uniform sampling over skewed data was first treated in [1]. The authors divided the space into equal-sized cells, storing the points in a set of hash table. Thereafter, the cells with few points are oversampled and the cells with many points are under sampled. In [2], the hash table is substituted by uniform sampling to avoid collision problems. Another density-biased sampling technique was proposed in [3], which samples points according to the local density near each point. The authors use kernel-density based methods to estimate the local density. However, these functions cause a significant time overhead. All of the previously discussed techniques are sensitive to noise and dimensionality. In [4] a density-biased sampling method based on R-Trees that can do sampling in noisy datasets is presented. However, due to the R-Tree shortcomings for high-dimensional data, the authors advised that the technique aimed at sampling data in a Database Management System (DBMS) indexed by a spacial index, and it is also sensitive to the data dimensionality.

We present a new technique for sampling based on local density. It is less sensitive to noise and high dimensionality problems than the most of existing techniques. We also presents the **BBS - Biased Box Sampling** algorithm [5], that implements our technique with linear cost $O(N \cdot E)$ on the number N of dataset points and on the number of attributes E, and can also be integrated into a DBMS. The experimental results show that, even at a lower sampling rate, it can generate samples that allow clustering processes to find clusters more accurately. The remainder of the paper is organized as follows. Section 2 discusses the technique and the algorithm developed. Section 3 discusses experimental studies. Section 4 concludes the paper.

2   Biased-Box Sampling

This section details our main contribution: a technique to sample multi-dimensional datasets to reduce the amount of data to be submitted to clustering processes based on a multi-resolution mapping of the data space, aiming at reducing the computational cost of such processes. We also present the BBS algorithm (**Biased Box Sampling**) based on this technique, which enables clustering processes to find clusters retaining their accuracy even at very low sampling rates.

The main idea of the proposed technique is to divide the data space into 2 regions, so that each attribute splits the space by half, and counting the number of points of the dataset that lies at each region. Each region holding more than a given threshold δ of points is recursively divided, generating a "hyper-quad tree" like structure, the compact multi-resolution grid tree - *MGc-tree*. Whenever a region does not hold the given threshold, it is represented in a leaf node and its points are sampled. The leafs can be at different levels, reflecting the density of the space at each region, so performing the sampling at a constant rate at every leaf will in fact retrieve more points from less dense regions, reflecting accurately the density variation over the full data space.

The BBS algorithm has three parts. The first one creates the multi-level grid, implemented as a tree. It is similar to the LiBOC algorithm [6], the main modification being the need to store the points at each leaf node. As this new requirement has a computational cost constant for each point, this step has the same complexity as LiBOC, which is $O(N \cdot E)$. The resulting structure is a tree (the *MG-tree*) where each non-leaf node stores a counter of the number of points lying on its corresponding region, an identification of the region, and from zero up to 2^E pointers to the next higher-resolution regions that divide the current one. A leaf node only stores the pointers for the data points lying on the corresponding region. Notice that at this step, the tree has every leaf node at the highest resolution $R - 1$. The value R defines the maximum number of resolutions the algorithm must try to obtain a good sample. Its minimum is bounded by how good the sampling must be, and its maximum is bounded by the highest density of the sampled dataset. If R is set too high, the algorithm will require more memory to operate (as the complexity of the memory required by the algorithm is $O(N \cdot E \cdot R)$), but after a threshold it will not improve the sampling anymore. Experimentally we observed that $R = 5$ is suitable for all the datasets evaluated. The first part of BBS is shown as Algorithm 1.

The second part of the BBS algorithm aims at reducing the deep of the multi-resolution grid tree at the less dense regions, transforming the MG-tree generated by Algorithm 1 into a condensed *MGc-tree*. This part, shown as Algorithm 2, looks for cells in the tree where the number of points is lower than the threshold δ. When such cells are found, the index lists from the children cells are concatenated, transforming their parent into a leaf-node. Thus, the resulting condensed *MGc-tree* has leaf nodes of approximately the same number of points.

The threshold δ is set $\delta = E * 100/(2 * Ratio)$, and is evaluated in Step 5 of Algorithm 2. The concatenation of points in Step 7 generates varying number of elements at each leaf node. To assure none of them are under-represented we double δ (using $2 * Ratio$). The next step is to perform the sampling. The third (and last) part of the BBS algorithm effectively performs the multi-resolution sampling. It retrieves each leaf node of the *MGc-tree* and the difference between the level of the node and the maximum level of the tree is used to increase the number of points selected, performing another over sampling in lower resolution regions (Step 7 of Algorithm 3). Step 8 uses the concatenated list to choose points for the sample.

The complete BBS algorithm shown as Algorithm 4 just calls the three parts. It receives the set of points to be sampled and creates an array indexing the sampled points of the original dataset (Step 4). The δ employed in Algorithm 2 and Step 7 of Algorithm 3 contributes to enlarge the number of points in the sample. Therefore, Step 4 of Algorithm 4 generates the final sample, randomly dropping the points selected until the desired number is achieved.

Algorithm 1. Biased Box Sampling: Create Tree - BBSCT

Input: Dataset with N points with the E attributes, number of levels R
Output: Multi-resolution grid Tree - MG-tree
  1: Normalize the dataset points to a unit cube;
  2: **for** each point t in dataset **do**
  3:     **for** $r = 1/2^j$, $j = 1, 2, 3, .., R - 1$ **do**
  4:         select the cell in the next level where t lies as i;
  5:         Increment the counter $C_{i,r}$;
  6:         **if** level $= R - 1$ **then**
  7:             insert point t in the index list of cell i;

Algorithm 2. Biased Box Sampling: Join List - BBSJL

Input: MG-tree and sample ratio $Ratio$
Output: Grid structure Concatenated - MGc-tree
  1: $\delta = E * 100/(2 * Ratio)$
  2: **while** Cell $\neq$ NULL **do**
  3:     **if** Next Level Cell $\neq$ NULL **then**
  4:         **if** Actual Level $<$ R **then**
  5:             Call $BBSJL(ActualLevel + 1, Ratio)$;
  6:         **else if** $C_{r,i} < \delta$ **then**
  7:             Concatenate sibling cells index lists and make parent a leaf node;
  8:     Cell receives next level cell;

Algorithm 3. Biased Box Sampling: Extract Sample - BBSES

Input: Concatenated Grid structure - $MGc - tree$ and sample ratio $Ratio$
Output: vector with indexes of sampling points
  1: **while** Cell $\neq$ NULL **do**
  2:     **if** Next Level Cell $\neq$ NULL and First $=$ NULL **then**
  3:         Call $BBSES(Actual\ Level + 1, Ratio)$
  4:     **else if** First $\neq$ NULL **then**
  5:         $samplesize = Ratio * C_{r,i}/100$;
  6:         **if** $Actual\ Level - R \neq 0$ **then**
  7:             $samplesize = samplesize * 2 * (R - Actual\ Level)$;
  8:         Selected a point in the index list and insert it into the result;
  9:     Cell receives next cell;

Algorithm 4. Biased Box Sampling

Input: Dataset with N points, E attributes, number of levels R and sample ratio $Ratio$
Output: Dataset with the selected points
  1: Call $BBSCT(Dataset, R)$
  2: Call $BBSJL(MG - tree, Ratio)$
  3: Call $BBSES(MGc - tree, Ratio)$
  4: uniformly selected $N * Ratio/100$ points from the sample obtained by BBSES()

3   Experimental Results

In this section we discuss the results of experiments performed using the sampling algorithm BBS on two synthetic and one real datasets. The "OneBig" dataset has twenty attributes and nine clusters, one containing fifty thousand points and the others containing one thousand points each. The remaining ten thousand points are randomly distributed noise (fifteen percent). The "UniformClusters" has two attributes and five clusters forming one big circle, two small ones, two ellipsoids connected by a chain of outliers and random outliers scattered over the entire space, as described in [3]. "UniformClusters" and "OneBig" present, respectively, uniform and Gaussian intra-cluster distribution. The third dataset is the real world dataset "Pendigits" [1] with sixteen attributes and ten thousand and nine hundred ninety two points.

The same experimental methodology was applied to every dataset. In the first step, we ran the BBS algorithm, and for comparison purposes, we also ran the DBS sampling algorithm presented in [3], the GBS presented in [1][2] and the Uniform Sample (US) algorithm, generating four sample sets for several sampling rates of each dataset. In the second step, we evaluated the quality of the samples regarding the preservation of original properties of the full dataset regarding cluster distribution. Every experiment was repeated 10 times, so 10 samples were created with the same parameters. The values presented are the averages of processing each set of 10 samples. We tuned the parameters of DBS according to the indications in [3]. The parameter a was set as follows: for datasets containing noise and clusters with various densities (including small clusters) a was set to -0.25. The other parameter is the number of kernels for DBS, set to 1000. In the GBS algorithm the only parameter is e, set to 0.5 as indicated in [1].

We applied the well-known clustering algorithm DBSCAN [7] to evaluate the precision of our techniques, over the original datasets and their corresponding samples. DBSCAN is based on local density to discover clusters, and it can detect noise as well. We used the WEKA[3] implementation of DBSCAN, which requires the following parameters: the minimum number n of points that a cluster should have, and the radius ϵ that defines the maximum distance to determine if two points are neighbors or not. The value of n was set as a proportion of the size of the smallest cluster in the dataset. The radius ϵ was experimentally set to 0.2 for every sample size for the "OneBig" and "UniformClusters" datasets, and between 0.4 and 0.5 for the "Pendigits" dataset.

Figure 2 shows a visualization of the "UniformClusters" dataset, both complete (Figure 2(a)) and sampled at a 0.5 percent rate by each algorithm (Figure 2(b): BBS, 2(c): DBS, 2(d): GBS and 2(e): US). As we can see, the samples are different but every one allowed DBSCAN to find four clusters. However, visually we can also see that the sample generated with BBS resembles more closely the original

---

[1] UCI http://www.ics.uci.edu/~mlearn/MLSummary.html

[2] We would like to thank also Chris Palmer for putting his biased sampling generator on the Web and Alexandros Nanopoulos for send to us his implementation of DBS.

[3] http://www.cs.waikato.ac.nz/ml/weka/

one data. DBSCAN found four clusters in every sample of "UniformClusters", in spite of the five original clusters, due to the chain of outliers between two of the clusters, which was preserved by all sampling methods.

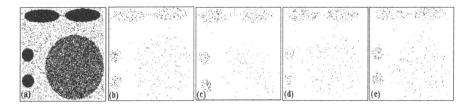

Fig. 2. Visualization of the "UniformClusters". (a) Original dataset, and samples from: (b) BBS, (c) DBS, (d) GBS; (e) US.

As the intra-cluster distribution of datasets "OneBig"(Figure 1) and "Uniform-Clusters" (Figure 2) are distinct, the experiments with "OneBig" led to distinct results. Table 1 shows that the number of clusters found by DBSCAN varies for samples from "OneBig" generated by distinct algorithms. In particular, the DBSCAN was able to find the 9 clusters from the samples generated by BBS, even at a sampling rate as low as 0.5%. All the competing algorithms, on the contrary, were significantly affected by noise and dataset distribution. The DBS in special was strongly affected by the data distribution, and its samples allowed DBSCAN to find just one cluster at any sampling rate. The GBS algorithm is affected by noise and high dimensionality. The US algorithm is also affected by the noise.

Table 3 presents the average amount of noise (in number of points) identified by DBSCAN in samples from "OneBig". As we can see, BBS can filter the noise better than the competing algorithms. Table 4 presents the average error rates produced by DBSCAN when evaluating each sample. The error rates were measured using a class attribute (not submitted to the sampling algorithms nor to DBSCAN), which specifies either the class of the point or whether it is noise. As it can be noted, almost all error rates obtained by DBS is very low. However, looking at Tables 1, 3 and 4 we observe why DBS found just one cluster: DBSCAN correctly classified almost every point, but DBSCAN also found that every point in the DBS samples were from only one cluster, that is, DBS sampled points almost only from the big cluster. The GBS and US algorithms led to bigger error rates than DBS, they allowed finding more clusters than DBS. US and GBS algorithms retrieve points from all clusters in every sample, but several of the sampled clusters do not have enough points to allow DBSCAN to find them. In contrast, samples from BBS allowed DBSCAN to find every cluster in every sample, even at the lowest sampling rate, and always with the lowest error rate.

The last experiment was carried out on "Pendigits" a well-known dataset that illustrates the efficiency of BBS over real data. We applied a different experimental methodology for this dataset, as its number of clusters was not known

Table 1. Number of Clusters Found in the "OneBig" dataset

Algorithm	Sample Size (%)					
	0.5%	1%	1.5%	2%	3%	4%
BBS	9	9	9	9	9	9
DBS	1	1	1	1	1	1
GBS	6	6	7	7	8	7
US	5	7	7	6	8	5

Table 2. Number of Clusters Found in the "Pendigits" dataset

Algorithm	Sample Size (%)			
	3%	4%	5%	10%
BBS	7	7	7	8
DBS	5	5	6	6
GBS	3	4	7	6
Random	2	3	6	2

Table 3. Average noise in the "OneBig" dataset (number of misclassified points)

Algorithm	Sample Size (%)					
	0.5%	1%	1.5%	2%	3%	4%
BBS	3	9	23	50	92	176
DBS	193	292	4	6	10	7
GBS	52	113	194	257	367	479
US	70	113	179	241	320	584

Table 4. Average Error Percentage in samples of the "OneBig" dataset

Algorithm	Sample Size (%)					
	0.5%	1%	1.5%	2%	3%	4%
BBS	0.3	0	0.21	0.08	0	0
DBS	5	43	0.38	0.43	0.48	0.26
GBS	3.24	2.9	1.8	2.05	1.1	2.2
US	5.9	1.6	2.4	2.3	1.2	4.4

beforehand. We applied DBSCAN to the original data, setting the minimum number of points (n) in a cluster as ten percent of the dataset. For ϵ, several values were also evaluated, but DBSCAN was able to find a maximum of 8 clusters in the original dataset only when setting $\epsilon = 0.4$, so this is the value employed in the experiments. The results from the experiments are shown in Table 2, where we can notice that BBS allowed the best clustering accuracy. Although BBS required a sampling rate of 10% to allow DBSCAN to find 8 clusters, the other algorithms did not allow finding more than 7 clusters at any sampling rate.

As expected, almost every sampling algorithms generated the samples sets in low computational time in. Only the DBS took ≈ 15 minutes while every other algorithm took less than 8 seconds to generate a sample. On the other hand, DBSCAN spent 3 hours to process "UniformClusters", 11 hours to process "OneBig" and 18 minutes to process "Pendigits".

The experiments show that BBS is efficient for biased sampling. Furthermore, BBS is tougher to withstand high-dimensionality drawbacks and noise in the datasets than the other techniques. This fact supports us to conclude that the proposed sampling approach is efficient and effective to speed up clustering algorithms, yet having a small impact on their precision.

4   Conclusions

This paper presents a new technique and a corresponding algorithm, the **BBS - Biased Box Sampling**, to perform sampling based on local density. The technique is based on a multi-dimensional multi-resolution grid structure whose depth depends on the local density of the points in the corresponding region. Therefore, even at low sampling rate, the points are selected such that the representativeness of each cluster occurring in the original dataset is preserved. Moreover, BBS is tougher to withstand high-dimensionality drawbacks and it is less sensitive to noise in the datasets than the competing techniques.

The dataset must be read only twice: one when preparing the grid structure, and thereafter to retrieve the points chosen to be in the sample. In fact, the whole process is linear on both the number of points N and on the number E of attributes in the original dataset. Therefore, the proposed technique can handle dimensionality higher than the other methods. We performed extensive experiments on both synthetic and real-world datasets. They highlighted the fact that the BBS algorithm is a very efficient technique to select samples for clustering algorithms with very little impact on their precision, always outperforming the existing techniques, particularly at very low sampling rate.

References

1. Palmer, C.R., Faloutsos, C.: Density biased sampling: An improved method for data mining and clustering. In: ACM SIGMOD, San Diego, pp. 82–92. ACM Press, New York (2000)
2. Kerdprasop, K., Kerdprasop, N., Sattayatham, P.: Density-biased clustering based on reservoir sampling. In: DEXA Workshops, Copenhagen, pp. 1122–1126 (2005)
3. Kollios, G., Gunopulos, D., Koudas, N., Berchtold, S.: Efficient biased sampling for approximate clustering and outlier detection in large data sets. TKDE 15(5), 1170–1187 (2003)
4. Nanopoulos, A., Theodoridis, Y., Manolopoulos, Y.: Indexed-based density biased sampling for clustering applications. DKE 57(1), 37–63 (2006)
5. Appel, A.P., Paterlini, A.A., Sousa, E.P.M.: d., Traina Jr., C., Traina, A.J.M.: Biased box sampling - a density-biased sampling for clustering. In: ACM SAC, Seoul, Korea, pp. 445–446 (2007)
6. Traina, J.C., Traina, A.J.M., Wu, L., Faloutsos, C.: Fast feature selection using fractal dimension. In: Brazilian Symposium on Data Base - SBBD, João Pessoa, PB, Brazil, pp. 158–171 (2000)
7. Ester, M., Kriegel, H.P., Sander, J., Xu, X.: A density-based algorithm for discovering clusters in large spatial databases with noise. In: KDD, pp. 226–231 (1996)

Expectation Propagation for Rating Players in Sports Competitions

Adriana Birlutiu and Tom Heskes

Institute for Computing and Information Sciences,
Radboud University Nijmegen Toernooiveld 1, 6525 ED Nijmegen,
The Netherlands
{adrianab,tomh}@cs.ru.nl

Abstract. Rating players in sports competitions based on game results is one example of paired comparison data analysis. Since an exact Bayesian treatment is intractable, several techniques for approximate inference have been proposed in the literature. In this paper we compare several variants of expectation propagation (EP). EP generalizes assumed density filtering (ADF) by iteratively improving the approximations that are made in the filtering step of ADF. Furthermore, we distinguish between two variants of EP: EP-Correlated, which takes into account the correlations between the strengths of the players and EP-Independent, which ignores those correlations. We evaluate the different approaches on a large tennis dataset to find that EP does significantly better than ADF (iterative improvement indeed helps) and EP-Correlated does significantly better than EP-Independent (correlations do matter).

1 Introduction

Our goal is to develop and evaluate methods for the analysis of paired comparison data. In this paper we illustrate such methods by rating players in sports, in particular in tennis.

We consider the player's strength as a probabilistic variable in a Bayesian framework. Before taking into account the match outcomes, information available about the players can be incorporated in a prior distribution. Using Bayes' rule we compute the posterior distribution over the players' strengths. We take the mean of the posterior distribution as our best estimate of the players' strengths and the covariance matrix as the uncertainty about our estimation.

An exact Bayesian treatment is intractable, even for a small number of players; the posterior distribution cannot be evaluated analytically, and therefore we need approximations for it. Expectation propagation [1] is a popular approximation technique. We will use it in this paper for approximating the posterior distribution over the players' strengths. The question that we want to answer here is: how do different variants of expectation propagation perform for this setting? In particular, does it make sense to perform backward and forward iterations for the approximations and does it help to have a more complicated (full) covariance structure?

J.N. Kok et al. (Eds.): PKDD 2007, LNAI 4702, pp. 374–381, 2007.

The paper is structured as follows: in the next section we introduce the probabilistic framework used to estimate players' strengths; in Section 3 we present algorithms for approximate inference and the way they apply to our setting; in Section 4 we show experimental results for real data, which we use to compare the performance of the algorithms; and in the last section we draw the conclusions.

2   Probabilistic Framework to Estimate Players' Strengths

Let $\boldsymbol{\theta}$ be an n_{players}-dimensional probabilistic variable whose components represent the players' strengths. We define $r_{ij} = 1$ if player i beats player j, and $r_{ij} = -1$ otherwise. For the probability of r_{ij} as a function of the strengths θ_i and θ_j, we take the Bradley-Terry model [2]:

$$p(r_{ij}|\theta_i, \theta_j) = \frac{1}{1 + \exp[-r_{ij}(\theta_i - \theta_j)]} . \tag{1}$$

A straightforward method to approximate the players' strengths is to build the likelihood of $\boldsymbol{\theta}$ given R; where R stands for the outcomes of all played matches. We take the maximum of the likelihood as the estimate for the strengths of the players.

The maximum likelihood approach gives a point estimate, the Bayesian approach, on the other hand, yields a whole distribution over the players' strengths. Furthermore, useful sources of information, like results in previous competitions and additional information about the players, can be incorporated in a prior distribution over the strengths. Using Bayes' rule we compute the posterior distribution over the players' strengths:

$$p(\boldsymbol{\theta}|R) = \frac{1}{d} p(R|\boldsymbol{\theta}) p(\boldsymbol{\theta}) = \frac{1}{d} p(\boldsymbol{\theta}) \prod_{i \neq j} p(r_{ij}|\theta_i, \theta_j) , \tag{2}$$

where $p(\boldsymbol{\theta})$ is the prior, $p(r_{ij}|\theta_i, \theta_j)$ from (1), and d is a normalization constant.

We take the mean or the mode of the posterior as the best estimate for the players' strengths. While computing the mean of the posterior distribution is computationally intractable, its mode (MAP) can be determined using optimization algorithms. For the MAP estimate the computation time is linear in the number of matches, and the number of iterations needed to obtain convergence. Typically, the number of iterations needed scales linearly with the number of players with a state-of-the-art optimization method such as conjugate gradient.

For making predictions and estimating the confidence of these predictions, we need the whole posterior distribution over the players' strengths. The posterior obtained using Bayes' rule in equation (2) cannot be evaluated analytically, hence we need to make approximations for it. For this task, sampling methods are very costly because of the high-dimensionality of the sampling space: the dimension is equal to the number of players. Therefore, for rating players, we here focus on deterministic approximation techniques, in particular expectation propagation and variants of it.

3 Expectation Propagation

Expectation propagation (EP) [1] is an approximation technique which tunes the parameter of a simpler approximate distribution, to match the exact posterior distribution of the model parameters given the data.

Assumed Density Filtering. ADF is an approximation technique in which the terms of the posterior distribution are added one at a time, and in each step the result of the inclusion is projected back into the assumed density. As the assumed density we take the Gaussian, to which we will refer below as q.

The first term which is included is the prior, $q(\boldsymbol{\theta}) = p(\boldsymbol{\theta})$; then we add terms one at a time $\tilde{p}(\boldsymbol{\theta}) = \Psi_{ij}(\theta_i, \theta_j)q(\boldsymbol{\theta})$, where $\Psi_{ij}(\theta_i, \theta_j) = p(r_{ij}|\theta_i, \theta_j)$; and at each step we approximate the resulting distribution as closely as possible by a Gaussian $q^{\text{new}}(\boldsymbol{\theta}) = \text{Project}\{\tilde{p}(\boldsymbol{\theta})\}$. Using the Kullback-Leibler (KL) divergence as the measure between the non-Gaussian $\tilde{p}$ and the Gaussian approximation, projection becomes moment matching: the result q^{new} of the projection is the Gaussian that has the first two moments, mean and covariance, the same as $\tilde{p}$.

After we add a term and project, the Gaussian approximation changes. We call the quotient between the new and old Gaussian approximation a *term approximation*.

Iterative Improvement. EP generalizes ADF by performing backward-forward iterations to refine the term approximations until convergence. The final approximation will be independent of the order of incorporating the terms. The algorithm performs the following steps.

1. Initialize the term approximations $\tilde{\Psi}_{ij}(\theta_i, \theta_j)$, e.g., by performing ADF; and compute the initial approximation

$$q(\boldsymbol{\theta}) = p(\boldsymbol{\theta}) \prod_{i \neq j} \tilde{\Psi}_{ij}(\theta_i, \theta_j).$$

2. Repeat until all $\tilde{\Psi}_{ij}$ converge:

   (a) Remove a term approximation $\tilde{\Psi}_{ij}$ from the approximation, yielding

   $$q^{\backslash ij}(\boldsymbol{\theta}) = \frac{q(\boldsymbol{\theta})}{\tilde{\Psi}_{ij}(\theta_i, \theta_j)}.$$

   (b) Combine $q^{\backslash ij}(\boldsymbol{\theta})$ with the exact factor $\Psi_{ij} = p(r_{ij}|\theta_i, \theta_j)$ to obtain

   $$\tilde{p}(\boldsymbol{\theta}) = \Psi_{ij}(\theta_i, \theta_j)q^{\backslash ij}(\boldsymbol{\theta}). \tag{3}$$

   (c) Project $\tilde{p}(\boldsymbol{\theta})$ into the approximation family

   $$q^{\text{new}}(\boldsymbol{\theta}) = \underset{q \in Q}{\text{argmin}}\, KL[\tilde{p}\|q].$$

   (d) Recompute the term approximation through the division

   $$\tilde{\Psi}_{ij}^{\text{new}}(\theta_i, \theta_j) = \frac{q^{\text{new}}(\boldsymbol{\theta})}{q^{\backslash ij}(\boldsymbol{\theta})}.$$

Computational Complexity. When minimizing the KL divergence in step (c) we can take advantage of the locality property of EP [3]. From equation (3), because the term Ψ_{ij} does not depend on $\boldsymbol{\theta}^{\backslash ij}$, we can rewrite $\tilde{p}$ as:

$$\tilde{p}(\boldsymbol{\theta}) = \tilde{p}(\boldsymbol{\theta}_{\backslash ij}|\theta_i, \theta_j)\tilde{p}(\theta_i, \theta_j) = \tilde{p}(\theta_i, \theta_j)q^{\backslash ij}(\boldsymbol{\theta}_{\backslash ij}|\theta_i, \theta_j).$$

Furthermore we obtain:

$$\begin{aligned} KL[\tilde{p}(\boldsymbol{\theta})||q(\boldsymbol{\theta})] = KL[\tilde{p}(\theta_i, \theta_j)||q(\theta_i, \theta_j)] \\ +E_{\tilde{p}(\theta_i, \theta_j)}[KL[q^{\backslash ij}(\boldsymbol{\theta}_{\backslash ij}|\theta_i, \theta_j)||q(\boldsymbol{\theta}_{\backslash ij}|\theta_i, \theta_j)]]. \end{aligned} \quad (4)$$

The two terms on the right-hand side can be minimized independently. Minimization of the second term gives:

$$q^{\text{new}}(\boldsymbol{\theta}_{\backslash ij}|\theta_i, \theta_j) = q^{\backslash ij}(\boldsymbol{\theta}_{\backslash ij}|\theta_i, \theta_j). \quad (5)$$

Minimizing the KL divergence for the first term in the right-hand side in (4) reduces to matching the moments, mean and covariance, between the 2-dimensional distributions $\tilde{p}(\theta_i, \theta_j)$ and $q(\theta_i, \theta_j)$.

Exploiting this locality property, we managed to go from n_{players}-dimensional integrals to 2-dimensional integrals, which can be further reduced to 1 dimension, by rewriting them in the following way (see e.g., the appendix of [4]):

$$\langle \Psi(\theta_i, \theta_j)\rangle_{\mathcal{N}(\boldsymbol{m},\boldsymbol{C})} = \langle F(\boldsymbol{a}\boldsymbol{\theta}_{ij})\rangle_{\mathcal{N}(\boldsymbol{m},\boldsymbol{C})} = \langle F(\theta\sqrt{\boldsymbol{a}^T\boldsymbol{C}\boldsymbol{a}} + \boldsymbol{a}^T\boldsymbol{m})\rangle_{\mathcal{N}(0,1)}$$

where $\boldsymbol{a}$ is the vector $[-1, 1]$ if player i is the winner, or $\boldsymbol{a} = [1, -1]$ if player j is the winner, $\boldsymbol{\theta}_{ij} = [\theta_i, \theta_j]$, F is defined through equation (1), and $\mathcal{N}(\boldsymbol{m}, \boldsymbol{C})$ stands for a Gaussian with mean $\boldsymbol{m}$ and covariance matrix $\boldsymbol{C}$. Substituting the solution (5), we see that the term approximation, in step (d) of the algorithm, indeed only depends on θ_i and θ_j.

We can simplify the computations by using the canonical form of the Gaussian distribution. Because, when projecting, we need the moment form of the distribution, we go back and forth between distributions in terms of moments and in terms of canonical parameters. For a Gaussian, this requires computing the inverse of the covariance matrix, which is of the order n_{players}^3. Since the covariance matrix, when refining the term corresponding to the game between players i and j, changes only for the elements corresponding to players i and j, we can use the Woodbury formula [5] to reduce the cubic complexity of the matrix inversion to a quadratic one. Thus, the complexity of EP is:

$$\mathcal{C}(\text{EP}) = \mathcal{O}(n_{\text{iterations}} \times n_{\text{players}}^2 \times n_{\text{matches}})$$

where $n_{\text{iterations}}$ is the number of iterations back and forth in refining the term approximations. In practice, the number of iterations to converge seems largely independent of the number of players or matches. In our experiments, we needed $n_{\text{iterations}} \approx 5$ to converge.

We will refer to this version of EP as EP-Correlated: by projecting into a nonfactorized Gaussian, it takes into account the correlations between the players' strengths.

EP-Independent. The complexity of the EP algorithm can be reduced further if we keep track only of the diagonal elements of the covariance matrix, ignoring the correlations. The matrix inversion has in this case linear complexity. The algorithm is faster and requires less memory.

4   Experiments

We applied the approximation algorithms, presented in the previous section, to the analysis of a real dataset. The dataset consists of results of 38538 tennis matches played on ATP events among 1139 players between 1995 and 2006. The goal was to compute ratings for the players based on the match outcomes. The methods described yield a Gaussian distribution of the players' strengths; the mean of the distribution represents our estimate of the players' strengths, the rating, and the variance relates to the uncertainty. Furthermore, we predict results of future games, and estimate the confidence of our predictions. We take as the prior a Gaussian distribution with mean zero and covariance equal to the identity matrix.

Figure 1 shows the empirical distribution of the players' strengths (means of the posterior distribution) in comparison with the average width of the posterior for an individual player. It can be seen that the uncertainty for individual players is comparable to the diversity between players.

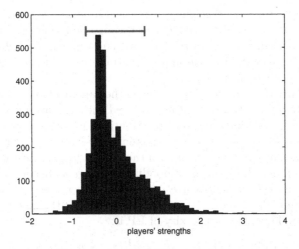

Fig. 1. A histogram of the players' strengths (means of the posterior distribution) for all years. The bar indicates the average width of the posterior distribution for each of the individual players. The results shown are for EP-Correlated.

4.1   Accuracy

We computed the ratings for the players at the end of each year, based on the matches from that year. Furthermore, based on these ratings we made predictions for matches in the next year: in a match we predicted the player with the highest rating to win.

EP-Correlated Versus ADF. We compared the accuracy of the predictions based on EP-Correlated ratings with the ones based on ADF ratings. We divided all joint predictions into 4 categories as shown in Table 1. We applied a binomial test on the matches for which the two algorithms gave different predictions to check the significance of the difference in performance [6]. The p-value obtained for this one-sided binomial test is 3×10^{-14}, which indicates that the difference is highly significant: EP-Correlated performs significantly better than ADF.

EP-Correlated Versus EP-Independent. The same type of comparison was performed between EP-Correlated and EP-Independent, the results are shown in Table 1. As for the previous comparison, the p-value is very small, 3×10^{-7}: the binomial test suggests that the difference between the two algorithms is again highly significant.

Table 1. Comparison between EP-Correlated, ADF and EP-Independent based on the number of matches correctly/incorrectly predicted

	ADF		EP-Independent	
	correct	incorrect	correct	incorrect
EP-Correlated				
correct	16636 (54.48%)	2395 (7.81%)	17857 (58.46%)	1174 (3.83%)
incorrect	1902 (6.21%)	9620 (31.50%)	945 (3.09%)	10577 (34.62%)

EP-Correlated Versus Laplace and ATP Rating. We compared Laplace and EP-Correlated to find out that EP-Correlated does slightly, but not significantly better (p-value is 0.3). They disagree on only 0.2% of all matches.

We also compared the accuracy of the predictions based on the EP ratings with the accuracy of the predictions obtained using the ATP ratings at the end of the year. The ATP rating system gives points to players according to the type of the tournament and how far in the tournament they reached. Averaged over all the years, both EP and ATP ratings, give similar accuracy of predictions for the next, about 62%.

4.2 Confidence

With a posterior probability over the players' strengths we can compute the confidence of the predictions.

The algorithms presented perform about the same in estimating the confidence. However, they all tend to be overconfident, in the sense that the actual fraction of correctly predicted matches is smaller than the predicted confidence, as indicated by the solid line in the left plot of Figure 2. We can correct this by adding noise to the players' strengths, to account for the fact that a player's strength changes over time:

$$\theta_{t+1} = \theta_t + \epsilon$$

where ϵ has mean zero and variance σ^2. To evaluate the confidence estimation, we plot on the right side of Figure 2 the Brier score [7] for different values of σ. The optimum is obtained for $\sigma = 1.4$, which then yields the dashed line in the left plot of Figure 2.

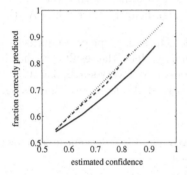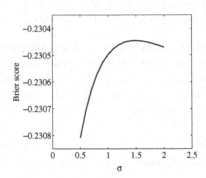

Fig. 2. Left: the actual fraction of correctly predicted matches as a function of the predicted confidence; without added noise (solid line) and with noise of standard deviation 1.4 added (dashed line); the dotted line represents the ideal case and is drawn for reference. Right: the Brier score for the confidence of the predictions as a function of the standard deviation of the noise added to each player's strength.

5 Conclusions

Based on the experimental results reported in this study we draw the conclusion that EP-Correlated performs better in doing predictions for this type of dataset than its modified versions, ADF and EP-Independent. Further experiments should reveal whether this also applies to other types of data.

Our results are generalizable to more complex models, e.g. including dynamics over time, which means that a players rating in the present is related to his performance in the past [8]; and team effects: a player's rating is inferred from team performance [9,10]. Specifically for tennis, the more complex models should also incorporate the effect of surface because the performance of tennis players in a match is influenced by the type of surface they play on (grass, clay, hard court, indoor). In this paper we considered the most basic probabilistic rating model; this model performs as good as the ATP ranking system. We would expect that the more complex models could outperform ATP.

Acknowledgments. The statistical information contained in the tennis dataset has been provided by and is being reproduced with the permission of ATP Tour, Inc., who is the sole copyright owner of such information. We would like to thank

Franc Klaassen for pointing us in the right direction. This research is supported by the Dutch Technology Foundation STW, applied science division of NWO and the Technology Program of the Ministry of Economic Affairs.

References

1. Minka, T.P.: A Family of Algorithms for Approximate Bayesian Inference. PhD thesis, M.I.T (2001)
2. Bradley, R.A, Terry, M.E.: Rank analysis of incomplete block designs: I, the method of paired comparisons. Biometrika (1952)
3. Seeger, M.: Notes on Minka's expectation propagation for Gaussian process classification. Technical report, University of Edinburgh (2002)
4. Barber, D., Bishop, C.: Ensemble learning in Bayesian neural networks. Neural Networks and Machine Learning (1998)
5. Press, W.H., Teukolsky, S.A., Vetterling, W.T., Flannery, B.P.: Numerical Recipes in C: The Art of Scientific Computing. Cambridge University Press, Cambridge (1992)
6. Salzberg, S.L.: On comparing classifiers: Pitfalls to avoid and a recommended approach. Data Mining and Knowledge Discovery 1(3), 317–328 (1997)
7. Brier, G.W.: Verification of forecasts expressed in terms of probability. Monthly Weather Review (1950)
8. Glickman, M.: Paired Comparison Models with Time Varying Parameters. PhD thesis, Harvard University (1993)
9. Herbrich, R., Minka, T., Graepel, T.: TrueSkill: A Bayesian skill rating system. In: Schölkopf, B., Platt, J., Hoffman, T. (eds.) Advances in Neural Information Processing Systems 19, pp. 569–576. MIT Press, Cambridge (2007)
10. Huang, T.K., Lin, C.J., Weng, R.C.: A generalized Bradley-Terry model: From group competition to individual skill. In: Saul, L.K., Weiss, Y., Bottou, L. (eds.) Advances in Neural Information Processing Systems 17, pp. 601–608. MIT Press, Cambridge (2005)

Efficient Closed Pattern Mining in Strongly Accessible Set Systems

(Extended Abstract)

Mario Boley[1], Tamás Horváth[1], Axel Poigné[1], and Stefan Wrobel[1,2]

[1] Fraunhofer IAIS, Schloss Birlinghoven, Sankt Augustin, Germany
[2] Dept. of Computer Science, University of Bonn, Germany
{mario.boley,tamas.horvath,axel.poigne,stefan.wrobel}@iais.fraunhofer.de

Abstract. Many problems in data mining can be viewed as a special case of the problem of enumerating the closed elements of an independence system with respect to some specific closure operator. Motivated by real-world applications, e.g., in track mining, we consider a generalization of this problem to strongly accessible set systems and arbitrary closure operators. For this more general problem setting, the closed sets can be enumerated with polynomial delay if deciding membership in the set system and computing the closure operator can be solved in polynomial time. We discuss potential applications in graph mining.

1 Introduction

Over the past years, a large body of research has been devoted to finding efficient algorithms for the frequent itemset enumeration problem, and it has turned out that by looking at *closed* frequent itemsets, important gains can be made in the design of efficient algorithms (see, e.g., [6]). A closed frequent itemset is a frequent itemset that cannot be further enlarged without changing its support in the database. Unfortunately, similar results do not yet exist for (closed) pattern enumeration tasks in many of the more complex representations that are becoming increasingly popular due to applications in highly structured domains. Consider for example the task of finding closed frequent *connected* subgraphs of movements of people or cars in a street network given a database of GPS-based recordings of spatio-temporal movements (so called *tracks*) [5]. In mining such tracks instead of itemsets, some important properties that are true for the frequent itemset mining problem no longer hold. In particular, it is not true that all subpatterns of a frequent connected pattern must necessarily also be frequent connected, since subpatterns need not be connected.

Technically, for problems like the track mining problem mentioned above, we note that unlike for the simple frequent itemset case, where the underlying set system is an *independence system*, here we are dealing with a weaker property of set systems which is only *strongly accessible*. In this paper, we show that for this generalized problem, it is possible to design an algorithm that enumerates all

J.N. Kok et al. (Eds.): PKDD 2007, LNAI 4702, pp. 382–389, 2007.

closed frequent patterns, for arbitrary closure operators, with polynomial delay (provided deciding membership in the set system and computing the closure operator can be done in polynomial time). To our knowledge, this result gives the first efficient closed pattern enumeration algorithm for this generalized and practically important task.

2   Preliminaries

In this section we define the notions and notations used in this paper. We will sometimes denote a set $\{a_1, \ldots, a_n\}$ by the string $a_1 \ldots a_n$.

Set Systems. A *set system* is an ordered pair $(E, \mathcal{F})$, where E is the ground set and $\mathcal{F} \subseteq 2^E$. A set system is called finite if its ground set is finite. An element X of $\mathcal{F}$ is called *maximal* if there is no $Y \in \mathcal{F}$ such that X is a proper subset of Y. A set system $(E, \mathcal{F})$ with $\emptyset \in \mathcal{F}$ is called

- *accessible* if for all $X \in \mathcal{F} \setminus \{\emptyset\}$ there is an $e \in X$ such that $X \setminus \{e\} \in \mathcal{F}$,
- *strongly accessible* if for every $X, Y \in \mathcal{F}$ satisfying $X \subsetneq Y$, there is an $e \in Y \setminus X$ such that $X \cup \{e\} \in \mathcal{F}$, and
- an *independence system* if $Y \in \mathcal{F}$ and $X \subseteq Y$ implies $X \in \mathcal{F}$.

The definitions imply that (i) every independence system is strongly accessible and (ii) every finite strongly accessible set system is accessible. However, the converse of (i) and (ii) does not hold.

Closure Operators. We now recall some notions related to closure operators. Let $(E, \mathcal{F})$ be a set system. A mapping $\rho : \mathcal{F} \to \mathcal{F}$ is called a *closure operator* if (i) $X \subseteq \rho(X)$ (*extensitivity*), (ii) $X \subseteq Y \Rightarrow \rho(X) \subseteq \rho(Y)$ (*monotonicity*), and (iii) $\rho(X) = \rho(\rho(X))$ (*idempotence*) hold for all $X, Y \in \mathcal{F}$. A set $F \in \mathcal{F}$ satisfying $\rho(F) = F$ is called *ρ-closed*. The *family of ρ-closed sets* of a set system $(E, \mathcal{F})$ with respect to a closure operator ρ is denoted by $\rho(\mathcal{F})$. For a ρ-closed set $C \in \rho(\mathcal{F})$, the family of all ρ-closed *proper* subsets of C is denoted by $\lambda(C)$. Let $(E, \mathcal{F})$ be a set system with $\emptyset \in \mathcal{F}$. Then, because of the monotonicity, for every $C \in \rho(\mathcal{F})$ it holds that $\lambda(C) = \emptyset$ if and only if $C = \rho(\emptyset)$. A set $F \in \mathcal{F}$ is a *generator* of a ρ-closed set $C \in \rho(\mathcal{F})$ if $\rho(F) = C$.

Graphs. An *undirected graph* is a pair (V, E), where $V \neq \emptyset$ is a finite set of *vertices* and $E \subseteq \{e \subseteq V : |e| = 2\}$ is a set of *edges*. Unless otherwise stated, in this paper by graphs we always mean undirected graphs and denote the set of vertices and the set of edges of a graph G by $V(G)$ and $E(G)$, respectively. Let G and G' be graphs. G' is a *subgraph* of G, if $V(G') \subseteq V(G)$ and $E(G') \subseteq E(G)$. Let $X \subseteq E(G)$ for some graph G. Then the graph *induced* by X, denoted $G[X]$, is the subgraph G' of G such that $V(G')$ is the set of vertices occurring in X and $E(G') = X$. A graph G is connected if for every $u, v \in V(G)$, there is a sequence $w_0, w_1, \ldots, w_\ell$ of vertices such that $w_0 = u$, $w_\ell = v$, and $\{w_i, w_{i+1}\} \in E(G)$ for every i $(0 \leq i < \ell)$. A *connected component* of G is a maximal subgraph of G that is connected. For the sets $Y \subseteq X \subseteq E(G)$ of edges such that $G[Y]$ is connected, $\text{CONN}_X(Y)$ denotes the connected component of $G[Y]$ in $G[X]$.

Frequent Pattern Mining. We recall some notions from frequent pattern mining. Let $\mathcal{D}$ be a transaction database over a set I, i.e., $\mathcal{D}$ is a multiset of subsets of I. For a set $X \subseteq I$, the *support* of X with respect to $\mathcal{D}$, denoted $\mathcal{D}[X]$, is the multiset of transactions of $\mathcal{D}$ containing X. The *support count* of X, denoted $\sigma(X)$, is defined by $|\mathcal{D}[X]|$. For an integer *frequency threshold* $t > 0$, a subset $X \subseteq I$ is *t-frequent* if $\sigma(X) \geq t$. The proof of the following proposition is immediate from the definitions.

Proposition 1. *Let $\mathcal{D}$ be a transaction database over a set I and $X \subseteq Y \subseteq I$. Then $X \subseteq Y \Rightarrow \mathcal{D}[X] \supseteq \mathcal{D}[Y] \Rightarrow \bigcap \mathcal{D}[X] \subseteq \bigcap \mathcal{D}[Y]$.*

A subset $X \subseteq I$ is *closed* if for every $X \subsetneq Y \subseteq I$ it holds that $\mathcal{D}[Y] \subsetneq \mathcal{D}[X]$. We note that this notion of "closeness" does not relate necessarily to that used in the definition of closure operators. In Section 4 we present an example, where there is no closure operator defining the above notion of closeness.

For some enumeration problems, the size of the set to be enumerated can be exponential in the size of the input. In such cases, the algorithm cannot enumerate the output in time polynomial only in the size of the input. We consider enumeration with *polynomial delay* (see, e.g., [4]), i.e., the number of steps between the output of two successive elements is bounded by a polynomial in the input size.

3   The General Problem

Many problems in data mining (e.g., closed frequent itemset mining) can be considered as a special case of the following enumeration problem:

> THE CLOSED SET MINING (CSM) PROBLEM: *Given* a finite set E, a membership oracle $M_{\mathcal{F}} : 2^E \to \{0,1\}$ defining a family $\mathcal{F} \subseteq 2^E$ satisfying $\emptyset \in \mathcal{F}$, and a closure operator $\rho : \mathcal{F} \to \mathcal{F}$, *compute* $\rho(\mathcal{F})$.

As an instance of this problem, consider the closed frequent itemset mining problem. For this problem, E and $\mathcal{F}$ correspond to the set of items and the family of frequent itemsets, respectively. Notice that $\mathcal{F}$ is given implicitly by a frequency testing procedure denoted by $M_{\mathcal{F}}$ in the problem definition.

Usually, $\mathcal{F}$ can be enumerated efficiently. Even then the naïve algorithm enumerating each set $S \in \mathcal{F}$ and testing whether S is ρ-closed is inefficient because $|\mathcal{F}|$ can be exponential in $|\rho(\mathcal{F})|$. There are several results on efficient enumeration of ρ-closed sets for the case that the underlying set system is finite and closed under intersection (see, e.g., [2,3]). Among others, formal concept analysis [7] and closed frequent itemset mining (see, e.g., [6]) provide some representative applications of this case.

In contrast to these results, we do *not* require the set system to be closed under intersection. Instead, we consider finite set systems $(E, \mathcal{F})$ associated with closure operators $\rho : \mathcal{F} \to \mathcal{F}$ satisfying the following property: for any ρ-closed element of $\mathcal{F}$, there exists an *inductive generator*. An inductive generator of a ρ-closed element $C \in \rho(\mathcal{F})$ is an element $C' \cup \{e\} \in \mathcal{F}$ such that $C' \in \rho(\mathcal{F})$,

$e \in E \setminus C'$, and $C = \rho(C' \cup \{e\})$. These inductive generators can then be used to enumerate all ρ-closed sets with a DFS algorithm resulting in the following positive result:

Lemma 2. *The CSM problem can be solved with polynomial delay for instances satisfying (i) the membership oracle $M_{\mathcal{F}}$ and the closure operator ρ can be computed in polynomial time and (ii) for every ρ-closed set except $\rho(\emptyset)$, there exists an inductive generator.*

Proof (sketch). The ρ-closed sets can be enumerated by traversing the graph $(\rho(\mathcal{F}), X)$ with

$$X = \{(C, C') : \ C' = \rho(C \cup \{e\}) \text{ for some } e \in E \setminus C \text{ satisfying } C \cup \{e\} \in \mathcal{F}\}$$

in a depth first manner. Using prefix trees for the storage of the enumerated ρ-closed sets, one can decide in time linear in the size of a new ρ-closed set, whether it has already been visited. Condition (ii) implies that every ρ-closed set is reached when starting from $\rho(\emptyset)$. Since a new ρ-closed set is reached or the algorithm terminates after at most $|E|^2$ closure computations and membership queries, condition (i) implies polynomial delay. $\square$

To formulate our main result, we first give a sufficient and necessary condition for generators in arbitrary set systems and closure operators.

Proposition 3. *Let $(E, \mathcal{F})$ be a set system, $\rho : \mathcal{F} \to \mathcal{F}$ a closure operator, and $C \in \rho(\mathcal{F})$ a ρ-closed set. Then for every $F \in \mathcal{F}$ satisfying $F \subseteq C$ it holds that F is a generator of C if and only if there is no $C' \in \lambda(C)$ satisfying $F \subseteq C'$.*

Proof. (*"if"*) Since ρ is a closure operator we have $F \subseteq \rho(F) \subseteq \rho(C) = C$. Now consider the case that $\rho(F) \subsetneq C$. Then $\rho(F) \in \lambda(C)$ contradicting the assumption. Hence, $\rho(F) = C$ must hold.

(*"only if"*) Let $F \in \mathcal{F}$ be a generator of C and suppose for contradiction that $F \subseteq C'$ for some $C' \in \lambda(C)$. Then $F \subseteq C'$, but $\rho(F) = C \not\subseteq C' = \rho(C')$ contradicting the monotonicity of ρ. $\square$

Thus, if C' is a maximal element of $\lambda(C)$ for some ρ-closed set C then C' is not necessarily an *inductive* generator of C together with some element from $E \setminus C'$. For strongly accessible set systems, however, we have a different situation which allows us to state the main result of this section.

Theorem 4. *For any finite strongly accessible set system $(E, \mathcal{F})$ given by a polynomial membership oracle and for any polynomially computable closure operator $\rho : \mathcal{F} \to \mathcal{F}$, the family $\rho(\mathcal{F})$ of ρ-closed sets can be enumerated with polynomial delay.*

Proof. Let $C = \rho(F)$ be a ρ-closed set satisfying $C \neq \rho(\emptyset)$ (or equivalently, $\lambda(C) \neq \emptyset$). Then there is a *maximal* ρ-closed set $C' \in \lambda(C)$ because $(E, \mathcal{F})$ is finite. Since $(E, \mathcal{F})$ is strongly accessible, there is an $e \in C \setminus C'$ such that $C' \cup \{e\} \in \mathcal{F}$. The maximality of C' in $\lambda(C)$ implies that there is no set in $\lambda(C)$ containing $C' \cup \{e\}$. Thus, by Proposition 3, $C' \cup \{e\}$ is an inductive generator of C and the statement follows from Lemma 2. $\square$

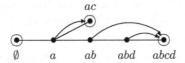

Fig. 1. An accessible set system with the closure operator $\rho(\emptyset) = \emptyset$, $\rho(a) = \rho(ac) = ac$, and $\rho(ab) = \rho(abd) = \rho(abcd) = abcd$. The ρ-closed set $abcd$ has no inductive generator.

We note that accessibility alone is not enough to guarantee the existence of inductive generators. In Figure 1 we give such an example.

4   Applications

As a first application, we state a positive result on *efficient* mining of closed frequent itemsets, which, although it is well-known (see, e.g., [1]), demonstrates the power of Theorem 4.

Theorem 5. *The family of closed frequent itemsets can be enumerated with polynomial delay.*

Proof. Let $\mathcal{D}$ be a transaction database over a set I of items and $t > 0$ be an integer. Let $\mathcal{F}_{\mathcal{D},t}$ denote the family of t-frequent itemsets and $\rho : \mathcal{F}_{\mathcal{D},t} \to \mathcal{F}_{\mathcal{D},t}$ be the function $\rho : F \mapsto \bigcap \mathcal{D}[F]$. Clearly, ρ is a closure operator on $\mathcal{F}_{\mathcal{D},t}$.

Since any subset of a frequent itemset is also frequent, $(I, \mathcal{F}_{\mathcal{D},t})$ is an independence system and thus, it is strongly accessible. It can be decided in time polynomial in the size of $\mathcal{D}$ whether a subset of I is frequent. Finally, for all $F \in \mathcal{F}_{\mathcal{D},t}$, $\rho(F)$ can be computed in time polynomial in the size of $\mathcal{D}$. The statement then follows by Theorem 4. $\square$

For the rest of this section, let $G = (V, E)$ be an undirected graph and $\mathcal{D}$ be a database of subgraphs of G. That is, for every graph $G' \in \mathcal{D}$ we have that $G' = G[E']$ for some $E' \subseteq E$. Notice that the graphs in $\mathcal{D}$ can be represented by subsets of E. We make use of this fact and consider $\mathcal{D}$ as a transaction database over E. Transaction datasets of this type occur e.g. in *track mining* applications. Indeed, consider the application scenario, where we have a network represented by an undirected graph $G = (V, E)$ and points moving in the network within a time interval T (e.g., the network and the points could represent a city and persons moving in the city, respectively). For each point i, let $E_i \subseteq E$ be the set of edges of G visited by point i in T and let $G_i = G[E_i]$ be the subgraph of G induced by E_i. The collection of graphs G_i for every i forms the database $\mathcal{D}$. In contrast to other frequent subgraph mining problems defining the embedding operator by *subgraph isomorphism*, we define it by the *subset* relation.

Given some frequency threshold $t > 0$, one can consider different types of t-*frequent* patterns corresponding to appropriately chosen set systems and closure operators defined by $\mathcal{D}$ and t. As an example, for the set system $(E, \mathcal{F}_{\mathcal{D},t})$ with

$$\mathcal{F}_{\mathcal{D},t} = \{X \subseteq E : \sigma(X) \geq t\}$$

and closure operator $\rho : F \mapsto \bigcap \mathcal{D}[F]$, $\rho(\mathcal{F}_{\mathcal{D},t})$ corresponds to the family of *closed t-frequent subgraphs* of $\mathcal{D}$. Notice that in this case, the underlying set system $(E, \mathcal{F}_{\mathcal{D},t})$ is an independence system and the problem of enumerating $\rho(\mathcal{F}_{\mathcal{D},t})$ is equivalent to the closed frequent itemset mining problem. Hence, by Theorem 5, $\rho(\mathcal{F}_{\mathcal{D},t})$ can be enumerated with polynomial delay.

4.1   Mining Closed Frequent Connected Subgraphs

In the last example, we considered closed frequent subgraphs without any structural restriction, such as, for example, *connectivity*. Connectivity is perhaps the most natural structural property of graphs. In track mining (see, e.g., [5]) for instance, closed frequent connected subgraphs of a network can be considered as *homogeneous* connected subnetworks. From an algorithmic point of view, we consider the following problem:

CLOSED FREQUENT CONNECTED SUBGRAPH MINING PROBLEM: *Given* an undirected graph $G = (V, E)$, a transaction database $\mathcal{D}$ of subgraphs of G, and an integer $t > 0$, *enumerate* the family of closed t-frequent connected subgraphs of $\mathcal{D}$.

Applying Theorem 4, we show that the above problem can be solved with polynomial delay. Let the set system $(E, \mathcal{F}_{\mathcal{D},t})$ and the function $\rho : \mathcal{F}_{\mathcal{D},t} \to \mathcal{F}_{\mathcal{D},t}$ be defined by

$$\mathcal{F}_{\mathcal{D},t} = \{X \subseteq E : \sigma(X) \geq t \ \wedge \ G[X] \text{ is connected}\} \qquad (1)$$

and

$$\rho : X \mapsto \begin{cases} \emptyset & \text{if } X = \emptyset \\ \text{CONN}_{\bigcap \mathcal{D}[X]}(X) & \text{otherwise .} \end{cases} \qquad (2)$$

That is, ρ maps a frequent connected subgraph X to the largest connected supergraph Y of X such that Y is a subgraph of each supergraph of X in $\mathcal{D}$. We note that the set system $(E, \mathcal{F}_{\mathcal{D},t})$ is *not* an independence system because a subgraph of a frequent connected graph is not necessarily connected. In fact, it is *not* even closed under intersection. But, as we note without proof due to space limitation, it is strongly accessible.

Lemma 6. *Let $G = (V, E)$ be an undirected graph, $\mathcal{D}$ be a transaction database over E, and $t > 0$ be an integer. Let $\mathcal{F}_{\mathcal{D},t}$ and ρ be defined as in Equations (1) and (2), respectively. Then $(E, \mathcal{F}_{\mathcal{D},t})$ is a strongly accessible set system and ρ is a closure operator.*

One can easily see that deciding the membership in $\mathcal{F}_{\mathcal{D},t}$ and computing ρ can both be solved efficiently. Combining these properties with the results of Lemma 6, we can apply Theorem 4 and state the main result of this subsection:

Theorem 7. *The closed frequent connected subgraph mining problem can be solved with polynomial delay.*

4.2   Closed Frequent Subpath Mining

The setting defined in this paper actually goes beyond the standard definition of "closeness" usually employed in data mining, as it can be used to resolve an anomaly with this notion described as follows. In the standard case as discussed above, the design of a closure operator was straightforward. Since for all $X \in \mathcal{F}_{\mathcal{D},t}$, the family

$$\max\{X' \in \mathcal{F}_{\mathcal{D},t} : X \subseteq X' \subseteq \bigcap \mathcal{D}[X]\}$$

of *maximal* sets had always exactly one element, the closure operator could just assign this unique maximum element to X. However, it does not hold in general that such a unique maximum element exists. In this subsection, as an illustrative example of how to define a closure operator for such cases, we consider the problem of mining closed frequent *paths*.

We again have a graph $G = (V, E)$ and a database $\mathcal{D}$ of transactions containing subsets of E. The set system of interest is now $(E, \mathcal{F}_{\mathcal{D},t})$ with

$$\mathcal{F}_{\mathcal{D},t} = \{P \subseteq E : \sigma(P) \geq t \wedge P \text{ is a path}\} \ .$$

As in the previous case, $\mathcal{F}_{\mathcal{D},t}$ is not an independence system and also not closed under intersection, but it is strongly accessible. Notice that the membership problem can be solved efficiently.

In data mining, a common definition for closed frequent subpaths is given by "A path P is *closed frequent* if it is frequent and $\mathcal{D}[P'] \subsetneq \mathcal{D}[P]$ for every path P' containing P." Using this definition, let $\mathcal{C}_{\mathcal{D},t}$ denote the set of closed frequent paths in $\mathcal{D}$. However, there is a problem with this definition: $\mathcal{C}_{\mathcal{D},t}$ is *not* induced by a closure operator. In fact, there are cases, for which there is no closure operator $\rho : \mathcal{F}_{\mathcal{D},t} \to \mathcal{F}_{\mathcal{D},t}$ with $\mathcal{C}_{\mathcal{D},t} = \rho(\mathcal{F}_{\mathcal{D},t})$. In the example below, we give such a case.

Example 8. *Let $G = (\{1,2,3,4\}, \{12,23,24\})$, $\mathcal{D} = \{\{12,23,24\}\}$ (i.e., $\mathcal{D}$ is a singleton consisting of G), and $t = 1$. Then $\mathcal{C}_{\mathcal{D},1} = \{\{12,23\},\{12,24\},\{23,24\}\}$. Consider the set system $(\{12,23,24\}, \mathcal{F}_{\mathcal{D},1})$, where $\mathcal{F}_{\mathcal{D},1}$ denotes the set of frequent paths, i.e., $\mathcal{F}_{\mathcal{D},1} = \{\emptyset, \{12\}, \{23\}, \{24\}, \{12,23\}, \{12,24\}, \{23,24\}\}$. Assume that there is a closure operator ρ such that $\rho(\mathcal{F}_{\mathcal{D},1}) = \mathcal{C}_{\mathcal{D},1}$. Then $\rho(\{12\})$ must be either $\{12,23\}$ or $\{12,24\}$, say $\{12,23\}$. But then $\rho(\{12\}) \not\subseteq \rho(\{12,24\}$ contradicting the monotonicity.*

To resolve this anomaly, we consider another natural notion of "closeness" which is induced by a closure operator. As stated above, we have the problem that a path X can be contained in more than one maximal path in $\bigcap \mathcal{D}[X]$. This prevents a closure operator definition in a fashion similar to the connectivity case. A canonical way to overcome this problem is to define it as the intersection of all such maximal elements. For paths, in particular, this results in the definition

$$\rho : P \mapsto \bigcap \max\{M : M \text{ is a path in } \bigcap \mathcal{D}[P] \text{ such that } P \subseteq M\}$$

for every $P \in \mathcal{F}_{\mathcal{D},t}$.

One can show that ρ is a closure operator that can be computed efficiently and thus, by Theorem 4, $\rho(\mathcal{F}_{\mathcal{D},t})$ can be enumerated with polynomial delay. Although $\rho(\mathcal{F}_{\mathcal{D},t})$ is a superset of $\mathcal{C}_{\mathcal{D},t}$, it still can reduce the output significantly and is a semantically more meaningful set of patterns than $\mathcal{F}_{\mathcal{D},t}$. As an example, let G be a path of length n and $\mathcal{D} = \{G\}$. For frequency threshold $t = 1$ we have $|\mathcal{F}_{\mathcal{D},1}| = (n+1)n/2$ and $|\rho(\mathcal{F}_{\mathcal{D},1})| = 1$.

5   Conclusion

In this paper, we have presented a positive result on efficient enumeration of the family of closed sets of strongly accessible set systems with respect to arbitrary closure operators. The significance of our result in the context of data mining is that most of the closed frequent pattern mining algorithms are restricted to the case that the underlying set system corresponding to the set of frequent patterns is an independence system or at least closed under intersection. Strongly accessible set systems, however, are not necessarily independence systems or closed under intersection. We have presented graph mining applications motivated by track mining, where the underlying set systems are strongly accessible, but not closed under intersection.

Although the applications of this paper have resorted to strongly accessible set systems, we note that the algorithm works also for set systems satisfying weaker requirements on accessibility. An interesting question is whether the positive result holds for accessible set systems as well.

References

1. Boros, E., Gurvich, V., Khachiyan, L., Makino, K.: On maximal frequent and minimal infrequent sets in binary matrices. Annals of Mathematics and Artificial Intelligence 39(3), 211–221 (2003)
2. Ganter, B., Reuter, K.: Finding all closed sets: A general approach. Order 8(3), 280–283 (1991)
3. Habib, M., Medina, R., Nourine, L., Steiner, G.: Efficient algorithms on distributive lattices. Discrete Applied Mathematics 110(2-3), 169–187 (2001)
4. Johnson, D.S., Yannakakis, M., Papadimitriou, C.H.: On generating all maximal independent sets. Information Processing Letters 27(3), 119–123 (1988)
5. Kuijpers, B., Nanni, M., Körner, C., May, M., Pedreschi, D.: Spatio-temporal data mining. In: F. Giannotti and D. Pedreschi, editors, Geography, mobility, and privacy: a knowledge discovery vision (to appear)
6. Pasquier, N., Bastide, Y., Taouil, R., Lakhal, L.: Efficient mining of association rules using closed itemset lattices. Information Systems 24(1), 25–46 (1999)
7. Wille, R.: Restructuring lattice theory: an approach based on hierarchies of concept. In: Rival, I. (ed.) Ordered Sets, pp. 445–470. Reidel, Dordecht/Boston (1982)

Discovering Emerging Patterns in Spatial Databases: A Multi-relational Approach

Michelangelo Ceci, Annalisa Appice, and Donato Malerba

Dipartimento di Informatica, Università degli Studi di Bari
via Orabona, 4 - 70126 Bari - Italy
{ceci,appice,malerba}@di.uniba.it

Abstract. Spatial Data Mining (SDM) has great potential in supporting public policy and in underpinning society functioning. One task in SDM is the discovery of characterization and peculiarities of communities sharing socio-economic aspects in order to identify potentialities, needs and public intervention. Emerging patterns (EPs) are a special kind of pattern which contrast two classes. In this paper, we face the problem of extracting EPs from spatial data. At this aim, we resort to a multi-relational approach in order to deal with the degree of complexity of discovering EPs from spatial data (i.e., (i) the spatial dimension implicitly defines spatial properties and relations, (ii) spatial phenomena are affected by autocorrelation). Experiments on real datasets are described.

1   Introduction

Spatial data are collected in a spatial database at a rate which requires automated data analysis methods to extract implicit, unknown, and potentially useful information. Data mining technology provides several data analysis tools for a variety of tasks. However, the presence of a spatial dimension in the data adds complexity to the data mining tasks. First, geometrical representation and positioning of spatial objects implicitly define spatial properties and relations. Second, spatial phenomena are characterized by autocorrelation (observations of spatially distributed variables are not location-independent). This means that when attributes of some units of analysis are investigated, attributes of *any* spatial object in the neighborhood of the unit of analysis may have a certain influence. This leads distinguishing between the *reference objects* of analysis and other *task-relevant spatial objects*, and to represent their spatial interactions.

In this work, we consider the spatial descriptive task of emerging patterns (EPs) discovery. Initially introduced in [3], EPs are kind of patterns (or multivariate features) whose support significantly changes from one class of data to another: the larger the difference of pattern support, the more interesting the patterns. Due to this sharp change in support, EPs can be used to characterize object classes. Several algorithms [8,3,4] have been proposed to discover EPs from data belonging to separate classes (data populations) and stored in a single relational table. But, the challenges posed by spatial dimension in the data makes

J.N. Kok et al. (Eds.): PKDD 2007, LNAI 4702, pp. 390–397, 2007.

necessary to resort to a powerful data representation in order to model properties and interactions possibly involving several spatial object types. In a recent work [5], the system SPADA has been proposed to deal with challenges posed by spatial dimension in the task of association rule discovery. But, although association rules and EPs are both descriptive patterns, they are significantly different: association rules capture *regularities* in data belonging to the same class, while EPs capture *changes* from a data class to another. This adds one source of complexity to the EPs discovery task, i.e., the fact that differently from association rules monotonicity property does not subsist for EPs.

We propose a Multi-Relational Emerging Patterns discovery (Mr-EP) algorithm that deals with the challenges posed by spatial dimension in the data. The class variable is associated with the reference objects, while explanatory attributes refer to either the reference objects or the task-relevant objects which are someway related to the reference objects.

2   Problem Definition

We assume that a spatial database reduces to a relational database D with schema S once implicit spatial relationships between reference objects and task relevant objects have been extracted and stored in separate tables. In this perspective, reference objects, task-relevant objects and (spatial) interactions among them are tuples stored in tables of D. The set R of reference objects is the collection of tuples stored in a table T of D called target table. Each set R_i of task-relevant objects corresponds to a distinct table of D. Reference objects and task-relevant objects are described by means of both spatial and aspatial attributes. Similarly, the (spatial) interactions between different sets of spatial objects (reference objects and task-relevant objects) are stored in tables of D. The inherent "structure" of data, i.e., the (spatial) relations between reference objects and task-relevant objects is expressed in the schema S by the foreign key constraints (FK). By this mapping, the discovery of spatial EPs can be reformulated as the task of discovering (multi-)relational EPs. Before providing a formal definition of the problem to be solved, some definitions need to be introduced.

Definition 1 (Key, Structural and Property predicate)
Let S be a database schema.
− The "key predicate" associated with the target table for the task at hand T in S, is a first order unary predicate $p(t)$ such that p denotes the table T and the term t is a variable that represents the primary key of T.
− A "structural predicate" associated with the pair of tables $\{T_i, T_j\}$ in S such that there exists a foreign key FK in S between T_i and T_j, is a first order binary predicate $p(t, s)$ such that p denotes FK and the term t (s) is a variable that represents the primary key of T_i (T_j).
− A "property predicate" associated with the attribute ATT of the table T_i (which is neither primary nor foreign key) is a binary predicate $p(t, s)$ such that p denotes the attribute ATT, the term t is a variable representing the primary key of T_i and s is a constant representing a value belonging to the range of ATT.

Structural predicates are used to represent spatial relations between spatial objects. A relational pattern over S is a conjunction of predicates consisting of the key predicate and one or more (structural or property) predicates over S. More formally, a relational pattern is defined as follows:

Definition 2 (Relational pattern). *Let S be a database schema. A "relational pattern" P over S is a conjunction of predicates $p_0(t0_1), p_1(t1_1, t1_2), \ldots, p_m(tm_1, tm_2)$, where $p_0(t0_1)$ is the key predicate associated with the target table of the task at hand and $\forall i = 1, \ldots, m$ $p_i(ti_1, ti_2)$ is either a structural predicate or a property predicate over S.*

A spatial pattern is a relational pattern that involves objects and relations which have a spatial nature. Henceforth, we will also use the set notation for relational patterns, that is, a relational pattern is considered a set of atoms.

Definition 3 (Key linked predicate). *Let $P = p_0(t0_1), p_1(t1_1, t1_2), p_2(t2_1, t2_2), \ldots, p_m(tm_1, tm_2)$ be a relational pattern over the database schema S. For each $i = 1, \ldots, m$, the (structural or property) predicate $p_i(ti_1, ti_2)$ is "key linked" in P if $p_i(ti_1, ti_2)$ is a predicate with $t0_1 = ti_1$ or $t0_1 = ti_2$, or there exists a structural predicate $p_j(tj_1, tj_2)$ in P such that $p_j(tj_1, tj_2)$ is key linked in P and $ti_1 = tj_1 \vee ti_2 = tj_1 \vee ti_1 = tj_2 \vee ti_2 = tj_2$.*

Definition 4 (Completely linked relational pattern). *A "completely linked" relational pattern is a relational pattern $P = p_0(t0_1), p_1(t1_1, t1_2), \ldots, p_m(tm_1, tm_2)$ such that $\forall i = 1 \ldots m$, $p_i(ti_1, ti_2)$ is a predicate which is key linked in P.*

Definition 5 (Relational emerging patterns). *Let D be an instance of a database schema S that contains a set of reference objects labeled with $Y \in \{C_1, \ldots, C_L\}$ and stored in the target table T of S. Given a minimum growth rate value (minGR) and a minimum support value (minsup), P is a "relational emerging pattern" in D if P is a completely linked relational pattern over S and some class label C_i exists such that $GR^{\overline{D_i} \rightarrow D_i}(P) > minGR$ and $s_{D_i}(P) > minsup$, where (i) D_i is an instance of database schema S such that $D_i.T = \{t \in D.T | D.T.Y = C_i\}$ and $\forall T' \in S, T' \neq T: D_i.T' = \{t \in D.T' |$ all foreign key constraints FK are satisfied in $D_i\}$ and (ii) $\overline{D_i}$ is an instance of database schema S such that $\overline{D_i}.T = \{t \in D.T | D.T.Y \neq C_i\}$ and $\forall T' \in S, T' \neq T: \overline{D_i}.T' = \{t \in D.T' |$ all foreign key constraints FK are satisfied in $\overline{D_i}\}$.*

The support $s_{D_i}(P)$ of P on database D_i is $s_{D_i}(P) = |O_P|/|O|$, where O denotes the set of reference objects stored as tuples of $D_i.T$, while O_P denotes the subset of reference objects in O which are covered by the pattern P. The growth rate of P for distinguishing D_i from $\overline{D_i}$ is $GR^{\overline{D_i} \rightarrow D_i}(P) = s_{D_i}(P)/s_{\overline{D_i}}(P)$. As in [3], we assume that $GR(P) = \frac{0}{0} = 0$ and $GR(P) = \frac{\geq 0}{0} = \infty$.

The problem of discovering spatial EPs can be formalized as follow.
Given: (i) A spatial database SDB to be reduced to a relational database D, (ii) a set R of reference objects tagged with a class label $Y \in \{C_1, \ldots, C_L\}$, (iii) Some

sets R_i, $1 \leq i \leq h$ of task-relevant objects, (iv) a pair of thresholds, that is, the minimum growth rate ($minGR \geq 1$) and the minimum support ($minsup > 0$). The goal is to discover the set of the *relational emerging patterns* to discriminate between reference objects belonging to contrasting classes in SDB.

In this work, we resort to the relational algebra formalism to express a relational emerging pattern P by means of an SQL query.

3   Relational EPs Discovery

We have adapted the algorithms proposed for frequent pattern discovery to the special case of EPs. The blueprint for the frequent patterns discovery algorithms is the levelwise method [6] that explores level-by-level the lattice of patterns ordered according to a generality relation ($\geqslant$) between patterns. Formally, given two patterns $P1$ and $P2$, $P1 \geqslant P2$ denotes that $P1$ ($P2$) is more general (specific) than $P2$ ($P1$). The search proceeds from the most general pattern and iteratively alternates the candidate generation and candidate evaluation phases.

In this paper, we propose an enhanced version of the aforementioned levelwise method which works on EPs rather than frequent patterns. The space of candidate EPs is structured according to the θ-subsumption generality order [7].

Definition 6 (θ-subsumption). *Let $P1$ and $P2$ be two relational patterns on a data schema S such that both $P1$ and $P2$ are key completely linked patterns with respect to a target table T in S. $P1$ θ-subsumes $P2$ if and only if a substitution θ exists such that $P2 \, \theta \subseteq P1$.*

Having introduced θ-subsumption, we now go to define generality order between completely linked relational patterns.

Definition 7 (Generality order under θ-subsumption). *Let $P1$ and $P2$ be two completely linked relational patterns. $P1$ is more general than $P2$ under θ-subsumption, denoted as $P1 \geqslant_\theta P2$, if and only if $P2$ θ-subsumes $P1$.*

θ-subsumption defines a quasi-ordering, since it satisfies the reflexivity and transitivity property but not the anti-symmetric property. The quasi-ordered set spanned by $\geqslant_\theta$ can be searched according to a downward refinement operator which computes the set of refinements for a completely linked relational pattern.

Definition 8 (Downward refinement operator under θ-subsumption). *Let $\langle G, \geqslant_\theta \rangle$ be the space of completely linked relational patterns ordered according to $\geqslant_\theta$. A downward refinement operator under θ-subsumption is a function ρ such that $\rho(P) \subseteq \{Q \in G | P \geqslant_\theta Q\}$.*

We now define the downward refinement operator ρ' for EPs.

Definition 9 (Downward refinement operator for EPs). *Let P be a relational EP for distinguishing D_i from $\overline{D_i}$. Then $\rho'(P) = \{P \cup \{p(t1, t2)\} | p(t1, t2)$ is a structural or property predicate key linked in $P \cup \{p(t1, t2)\}$ and $P \cup \{p(t1, t2)\}$ is an EP for distinguishing D_i from $\overline{D_i}\}$.*

The downward refinement operator for EPs is a refinement operator under θ-subsumption. In fact, it can be easily proved that $P \geqslant_\theta Q$ for all $Q \in \rho'(P)$. This makes Mr-EP able to perform a levelwise exploration of the lattice of EPs ordered by θ-subsumption. More precisely, for each class C_i, the EPs for distinguishing D_i from $\overline{D_i}$ are discovered by searching the pattern space one level at a time, starting from the most general EP (the EP that contains only the key predicate) and iterating between candidate generation and evaluation phases. In Mr-EP, the number of levels in the lattice to be explored is limited by the user-defined parameter $MAX_M \geq 1$. In other terms, MAX_M limits the maximum number of structural predicates (joins) within a candidate EP. Since joins affects the computational complexity of the method, a low value of MAX_M guarantees the applicability of the algorithm to reasonably large data The monotonicity property of the generality order $\geqslant_\theta$ with respect to the support value (i.e., a superset of an infrequent pattern cannot be frequent) is exploited to avoid the generation of infrequent relational patterns. In fact, an infrequent pattern on D_i cannot be an EP for distinguishing D_i from $\overline{D_i}$.

Proposition 1 (Property of θ-subsumption monotonicity). *Let $\langle G, \geqslant_\theta \rangle$ be the space of relational completely linked patterns ordered according to $\geqslant_\theta$. P1 and P2 are two patterns of $\langle G, \geqslant_\theta \rangle$ with $P1 \geqslant_\theta P2$ then $O_{P1} \supseteq O_{P2}$.*

Therefore, when $P1 \geqslant_\theta P2$, we have $s_{D_i}(P1) \geq s_{D_i}(P2)$ and $s_{\overline{D_i}}(P1) \geq s_{\overline{D_i}}(P2)$ $\forall i = 1, \ldots, L$. This is the counterpart of one of the properties exploited in the family of the Apriori-like algorithms [1] to prune the space of candidate patterns. To efficiently discover relational EPs, Mr-EP prunes the search space by exploiting the θ-subsumption monotonicity of support (*prune1* criterion). Let P' be a refinement of a pattern P. If P is an infrequent pattern on D_i ($s_{D_i}(P) < minsup$), then P' has a support on D_i that is lower than the user-defined threshold (*minsup*). According to the definition of EP, P' cannot be an EP for distinguishing D_i from $\overline{D_i}$, hence Mr-EP does not refine patterns which are infrequent on D_i. Unluckily, the monotonicity property does not hold for the growth rate: a refinement of an EP whose growth rate is lower than the threshold $minGR$ may or may not be an EP. Anyway, as in the propositional case [8], some mathematical considerations on the growth rate formulation can be usefully exploited to define two further pruning criteria.

First (*prune2* criterion), Mr-EP avoids generating the refinements of a pattern P in the case that $GR^{\overline{D_i} \to D_i}(P) = \infty$ (i.e., $s_{D_i}(P) > 0$ and $s_{\overline{D_i}}(P) = 0$). Indeed, due to the θ-subsumption monotonicity of support $\forall P' \in \rho'(P)$: $s_{\overline{D_i}}(P) \geq s_{\overline{D_i}}(P')$ then $s_{\overline{D_i}}(P') = 0$. Thereby, $GR^{\overline{D_i} \to D_i}(P') = 0$ in the case that $s_{D_i}(P') = 0$, while $GR^{\overline{D_i} \to D_i}(P') = \infty$ in the case that $s_{D_i}(P') > 0$. In the former case, P' is not worth to be considered (*prune1*). In the latter case, $P \geqslant_\theta P'$ and $s_{D_i}(P) \geq s_{D_i}(P')$. Therefore, P' is useless since P has the same discriminating ability than P' ($GR^{\overline{D_i} \to D_i}(P) = GR^{\overline{D_i} \to D_i}(P') = \infty$). We prefer P to P' based on the Occams razor principle, according to which all things being equal, the simplest solution tends to be the best one.

Second (*prune3* criterion), Mr-EP avoids generating the refinements of a pattern P which add a property predicate in the case that the refined patterns have the same support of P on $\overline{D_i}$. We denote by:
$SameSupp(P)_{\overline{D_i}} = \{P' \in \rho'(P)|s_{\overline{D_i}}(P) = s_{\overline{D_i}}(P'), \ P' = P \wedge p(t1, t2),$
$\qquad p(t1, t2)$ is a property predicate$\}$.

For the monotonicity property, $\forall P' \in SameSupp_{\overline{D_i}}(P)$: $s_{D_i}(P) \geq s_{D_i}(P')$. This means that $GR^{\overline{D_i} \rightarrow D_i}(P) \geq GR^{\overline{D_i} \rightarrow D_i}(P')$. P' is more specific than P but, at the same time, P' has a lower discriminating power than P. This pruning criterion prunes EPs that could be generated as refinements of patterns in $SameSupp_{\overline{D_i}}(P)$. However, it is possible that some of them may be of interest for our discovery process. Their identification is guaranteed by the following:

Proposition 2. *Let $P' \in SameSupp_{\overline{D_i}}(P)$ such that $P' = P \cup \{p(t1, t2)\}$ with $p(t1, t2)$ being a property predicate. Let $P'' \in \rho'(P')$ such that $P'' = P' \cup \{q(t3, t4)\}$ with $q(t3, t4)$ being a property predicate. If P'' is an EP discriminating D_i from $\overline{D_i}$ and $s_{\overline{D_i}}(P'') \neq s_{\overline{D_i}}(P)$ then $P''' = P \cup \{q(t3, t4)\} \notin SameSupp_{\overline{D_i}}(P)$.*

The proof is reported in [2]. According to proposition 2, we can prune P' (but not P''') without preventing the generation of EPs more specific than P'. It is noteworty to observe that this pruning criterion operates only when $p(t1, t2)$ is a property predicate. Differently, pruning of structural predicates would avoid the introduction of a new variable thus avoiding the discovery of further EPs obtained by adding property or structural predicates involving such variable.

Finally, additional candidates not worth being evaluated are those equivalent under θ-subsumption to some other candidate (*prune4*).

4   Experimental Results

Spatial EPs have been discovered in two spatial databases named North-West England (NWE) Data and Munich Data. EPs have been discovered with *min* $GR = 1.1$, $minsup = 0.1$. MAX_M is set to 3 for NWE Data and to 5 for Munich Data.

NWE data (provided in the European project *SPIN!*) concern both 214 census sections (wards) of Greater Manchester and digital maps data. Census data describe the mortality percentage rate and four deprivation indexes: Jarman (need for primary care), Townsend and Carstairs (health-related analyzes) and DoE (for targeting urban regeneration funds). The higher the index value the more deprived the ward. The mortality rate (target attribute) takes values in the finite set $\{low, high\}$. Vectorized boundaries of the 1998 census wards as well as of other Ordnance Survey digital maps of NWE are available for several layers such as urban area (115 lines), green area (9 lines), road net (1687 lines), rail net (805 lines) and water net (716 lines). The number of "non disjoint" relationships is 5313. Mr-EP discovers 60 EPs to discriminate high from low mortality rate wards and 55 EPs to discriminate low from high mortality rate wards. In the following, some EPs are reported. For the class mortality_rate=high:

$wards(A) \land wards\_rails(A, B) \land wards\_doeindex(A, [6.5..9.2])$

where $wards(A)$ is the key predicate, $wards\_rails(A, B)$ is the structural predicate representing an interaction between the ward A and a ward B (this means that A is crossed by at least one railway) and $wards\_doeindex(A, [6.5..9.2])$ (i.e. A is a deprived zone to be considered as target zone for regeneration fundings) is a property predicate. This pattern has a support of 0.22 and growth rate 3.77. This means that wards crossed by railways and with a relatively high doeindex value present a high percentage of mortality. This could be due to urban decay condition of the area. The pattern corresponds to the Oracle Spatial 10g query:

SELECT distinct W.ID FROM Wards W, Rails R

WHERE RELATE(W.Geometry,R.Geometry)='INTERSECTS'

   AND W.DoEIndex between 6.5 and 9.2

An example of EP discovered for the class mortality_rate=low:

$wards(A) \land wards\_townsendidx(A, [-3.8.. - 2.01] \land wards\_greenareas(A, B)$

This pattern has a support of 0.113 and a growth rate of 2.864. It captures the event that a ward with a low townsend deprivation level (i.e., A is not deprived from the point of view provided by health-related analysis) which overlaps at least one green area discriminates wards with low mortality rate from the others.

Munich data describe the level of monthly rent per square meter for flats in Munich expressed in German Marks. Data describe 2180 flats located in the 446 suquarters of Munich obtained by dividing the Munich metropolitan area up into three areal zones and decomposing each of these zones into 64 districts. The vectorized boundaries of subquarters, districts and zones as well as the map of public transport stops (56 U-Bahn stops, 15 S-Bahn stops and 1 railway station) within Munich are available for this study.The "area" of subquarters is obtained by the spatial dimension of this data. Transport stops are described by means of their type (U-Bahn, S-Bahn or Railway station), while flats are described by means of their "monthly rent per square meter", "floor space in square meters" and "year of construction". The monthly rent per square meter (target attribute) have been discretized into the two intervals $low = [2.0, 14.0]$ or $high =]14.0, 35.0]$. The "close to" relation between districts (autocorrelation on districts) and the "inside" relation between apartments and districts have been considered. Mr-EP discovers 31 (31) EPs to discriminate the apartments with high (low) rent rate per square meters from the class of apartments with low (high) rent rate per square meters. In the following, some EPs are reported. For rate_per_squaremeters=high:

$apartment(A) \land apartment\_inside\_district(A, B) \land$

   $district\_close\_to\_district(B, C) \land district\_ext\_19\_69(B, [0.875..1.0])$

This pattern has a support of 0.125 and a growth rate of 1.723. It represents the event that an apartment A is inside a district B that contains a high percentage (between 87.5% and 100%) of apartments with a relatively low extension (between 19 m^2 and 69 m^2). This pattern distriminates apartments with high rate per square meters form the others. This pattern can be motivated by considering that the rent rate is not directly proportional to the apartment extension but it includes fixed expenses that do not vary with the apartment size.

For the class rate_per_squaremeters=low:

$apartment(A) \wedge apartment\_inside\_district(A, B) \wedge$

$district\_crossedby\_tranStop(B, C) \wedge apartment\_year(A, [1893..1899])$

This pattern has a support of 0.265 and a growth rate of 2.343. It represents the event that an apartment A built between 1893 and 1899 is inside a district B that contains a railway public stop. This pattern discriminates apartments with low rate per square meters form the others. It can be motivated by considering that old buildings do not offer the same facilities of a recently built apartment.

5 Conclusions

In this paper, we present a spatial data mining method that resorts to a MRDM approach to discover a characterization of classes in terms of EPs involving spatial objects and relations thus providing a human-interpretable description of the differences between separate classes of spatially referenced data. The method is implemented in a system that is tightly integrated with a Oracle 10g DBMS. The tight-coupling with the database makes the knowledge on data structure available free of charge to guide the search in the pattern space by taking into account spatial interaction implicit in spatial dimension. Spatial EPs have been used to capture data (spatial) changes among several populations of geo-referenced data.

Acknowledgments

This work is supported by "ATENEO-2007" project "Metodi di scoperta della conoscenza nelle basi di dati: evoluzioni rispetto allo schema unimodale".

References

1. Agrawal, R., Imielinski, T., Swami, A.N.: Mining association rules between sets of items in large databases. In: Buneman, P., Jajodia, S. (eds.) International Conference on Management of Data, pp. 207–216 (1993)
2. Appice, A., Ceci, M., Malgieri, C., Malerba, D.: Discovering relational emerging patterns. In: Basili, R., Pazienza, M. (eds.) AI*IA 2007: Artificial Intelligence and Human-Oriented Computing, LNAI. Springer, (to appear)
3. Dong, G., Li, J.: Efficient mining of emerging patterns: Discovering trends and differences. In: International Conference on Knowledge Discovery and Data Mining, pp. 43–52. ACM Press, New York (1999)
4. Li, J.: Mining Emerging Patterns to Construct Accurate and Efficient Classifiers. PhD thesis, University of Melbourne (2001)
5. Lisi, F.A., Malerba, D.: Inducing multi-level association rules from multiple relations. Machine Learning 55, 175–210 (2004)
6. Mannila, H., Toivonen, H.: Levelwise search and borders of theories in knowledge discovery. Data Mining and Knowledge Discovery 1(3), 241–258 (1997)
7. Plotkin, G.D.: A note on inductive generalization. 5, 153–163 (1970)
8. Zhang, X., Dong, G., Ramamohanarao, K.: Exploring constraints to efficiently mine emerging patterns from large high-dimensional datasets. In: Knowledge Discovery and Data Mining, pp. 310–314 (2000)

Realistic Synthetic Data for Testing Association Rule Mining Algorithms for Market Basket Databases

Colin Cooper[1,*] and Michele Zito[2,*]

[1] Department of Computer Science, Kings' College, London WC2R 2LS, UK
colin.cooper@kcl.ac.uk
[2] Department of Computer Science, University of Liverpool, Liverpool, L69 3BX, UK
michele@liverpool.ac.uk

Abstract. We investigate the statistical properties of the databases generated by the IBM QUEST program. Motivated by the claim (also supported empirical evidence) that item occurrences in real life market basket databases follow a rather different pattern, we propose an alternative model for generating artificial data.

1 Introduction

The ARM problem is a well established topic in KDD. Many techniques have been developed to solve this problem (e.g. [1,3,5,9]), however several fundamental issues are still open. The evaluation of ARM algorithms is a difficult task [12], often tackled by resorting to data generated by the well established QUEST program from the IBM Quest Research Group [1]. The intricacy of this program makes it difficult to draw theoretical predictions on the behaviour of the various algorithms on such databases. Empirical analyses are also difficult to generalise because of the wide range of possible variation, both in the characteristics of the data (the structural characteristics of the synthetic data bases generated by QUEST are governed by a several interacting parameters), and in the environment in which the algorithms are being applied. It has also been noted [3] that data produced using QUEST might be inherently not the hardest to deal with. In fact it seems that the performance of some algorithms on real data is much worse than on synthetic data generated using QUEST [11].

In this paper we first claim that *heavy tail* statistical distributions (see [10] for a survey on the topic) arise naturally in characterizing the item occurrence distribution in market basket databases, but are not evident in data generated by QUEST. Statistical differences have been found before [11] between real-life and QUEST generated databases. We contend that our analysis points to possible differences at a much deeper level. Motivated by the outcomes of our empirical investigation, we then study mathematically the distribution of item occurrences in a typical

* The work of both authors was supported by EPSRC grant EP/D059372/1 *Scale-free structures: models and algorithms*.

J.N. Kok et al. (Eds.): PKDD 2007, LNAI 4702, pp. 398–405, 2007.

large QUEST database. At least in a simplified setting, such study confirms the empirical findings. To the best of our knowledge, this is the first analysis of the structural properties of the databases generated by QUEST. Such properties may well be responsible for the observed [3,11] behaviour of various mining algorithms on such datasets. The final contribution of this paper is the description of an alternative synthetic data generator. Our model is reminiscent of the proposal put forward in the context of author citation networks and the web by Barabási and Albert [2]. The mechanism that leads to the desired properties is the so called *preferential attachment*, whereby successive transactions are filled by selecting items based on some measure of their popularity. We complete our argument by proving mathematically that the resulting databases show an asymptotic heavy tailed item occurrence distribution and giving similar empirical evidence.

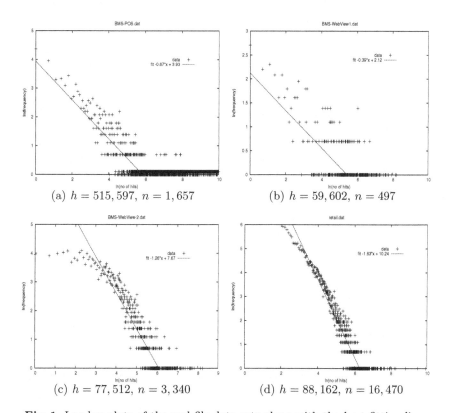

Fig. 1. Log-log plots of the real-file data sets along with the best fitting lines

The rest of the paper has the following structure. Section 2 reports our empirical analysis of a number of real and synthetic databases. Section 3 presents the results of our mathematical investigation of the structural properties of the databases generated by QUEST. Finally in Section 4 we describe our proposal for an alternative synthetic data generator.

2   Analysis of Real Data

From now on a database $\mathcal{D}$ is a collection of h transactions, each containing items out of a set $\mathcal{I}$ of n items. For $r \in \{0, \ldots, h\}$ let N_r be the number of items that occur in r transactions. In this section we substantiate the claim that, at least for market basket data, the sequence $(N_r)_{r \in \{0,\ldots,h\}}$ follows a distribution that has a "fat" tail and, on the contrary, the typical QUEST data shows rather different patterns. To this end we use the real data sets BMS-POS, BMS-WebView-1, BMS-WebView-2, and retail.data and the synthetic QUEST data T10I4D100K.dat and T40I10D100K.dat already used in [11], all available from http://fimi.cs.helsinki.fi/data/. The plots in Figure 1 show the sequence $(N_r)_{r \in \{0,\ldots,h\}}$ in each case, along with the least square fitting lines computed using the fit command of gnuplot, over the whole range of values (slope values are reported in each picture). Figure 2 shows the same statistics obtained using the two synthetic databases.

Although it may be argued that the number of real datasets examined is too small and the test carried out too coarse, our calculations indicate that the sequences $(N_r)_{r \in \{0,\ldots,h\}}$ obtained from real-life databases fit a straight line much better than the sequences obtained from the synthetic QUEST databases. Furthermore this phenomenon leads to the additional conjecture that the studied distributions may be *heavy tailed*, i.e. decay at a sub-exponential rate [10]. In the case of *power law* distributions such decay is proportional to x^{-z} for some fixed $z > 0$. On a doubly logarithmic scale data points having such decay would seem to be clustered around a line, which is exactly what happens in the case of the four market-basket datasets described above.

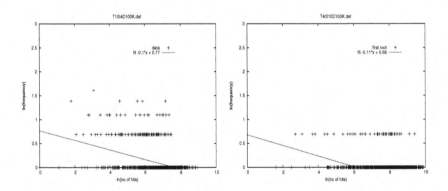

Fig. 2. Log-log plots of the QUEST data sets along with the best fitting lines

3   A Closer Look at QUEST

The QUEST program returns two related structures: the actual database $\mathcal{D}$ and a collection $\mathcal{T}$ of l *potentially large itemsets* or *patterns*, that are used to populate $\mathcal{D}$. The transactions in $\mathcal{D}$ are generated by first determining their size (picked from a Poisson distribution) and then filling the transaction with the items contained in a number of itemsets selected independently and uniformly at random (u.a.r.) from $\mathcal{T}$. Each itemset in $\mathcal{T}$ is generated by first picking its size from a Poisson distribution with mean equal to I. Items in the first itemset are then chosen randomly. To model the phenomenon that large itemsets often have common items, some fraction (chosen from an exponentially distributed random variable with mean equal to the expected correlation level ρ) of items in subsequent itemsets are chosen from the previous itemset generated. The remaining items are picked at random. The use of QUEST is further complicated by its dependence on a number of additional parameters. Agrawal and Srikant claim that the resulting database $\mathcal{D}$ mimics a set of transactions in the retailing environment. Furthermore they justify the use of the structure $\mathcal{T}$ based on the fact that people tend to buy sets of items together, some people may buy only some of the items from a large itemset, others items from many large itemsets. Finally they observe that transaction sizes and the sizes of the large itemsets (although variable) are typically clustered around a mean and a few collections have many items.

To maximise the clarity of exposition we simplify the definition of the QUEST process. A part from $n = |\mathcal{I}|$, h and l, we assume that system parameters will be the number of patterns per transaction, k, the number of items in each pattern, s, and the number of items shared between two consecutive patterns, ρ, with $\rho \in \{0, \ldots, s\}$. Typically $h >> n$ (e.g. $h = n^{O(1)}$) while $l = O(n)$. Parameters k and s are small constants independent of n. Note that the assumption that the size of the transactions is variable as stated in Agrawal and Srikant's work is "simulated" by the fact that different patterns used to form a transaction may share some items. Following Agrawal and Srikant we use a correlation level ρ between subsequent patterns, but assume it takes the fixed constant value $s/2$. Hence $s/2$ items in each pattern (except the first one) belong its predecessor in the generation sequence and the remaining $s - \rho$ are chosen u.a.r. in $\mathcal{I}$. We assume that $\mathcal{D}$ and $\mathcal{T}$ are populated as follows:

1. Generate $\mathcal{T}$ by selecting s random elements in $\mathcal{I}$ and any subsequent pattern by choosing (with replacement) ρ elements u.a.r. from the last generated pattern and $s - \rho$ elements u.a.r. (with replacement) in $\mathcal{I}$.
2. Generate $\mathcal{D}$ by filling each transaction independently with the elements of k patterns in $\mathcal{T}$ chosen independently u.a.r. with replacement.

Let $d_{\mathcal{D}}(v)$ (resp. $d_{\mathcal{T}}(v)$) denote the number of transactions in $\mathcal{D}$ (resp. patterns in $\mathcal{T}$) containing item v. Also, $b(x; n, p) = \binom{n}{x} p^x (1 - p)^{n-x}$. Obviously items occurring in many patterns of $\mathcal{T}$ have a higher chance of occurring in many database transactions. The following result quantify the influence of $\mathcal{T}$ on the item occurrence distribution in $\mathcal{D}$.

Theorem 1. *Let $(\mathcal{D}, \mathcal{T})$ with parameters n, h, l, k, s, and ρ as described above. Then for each $v \in \mathcal{I}$, $d_{\mathcal{D}}(v)$ has binomial distribution with parameters h and $p_{k,l} = \sum_{i=1}^{k} \binom{k}{i}(-1)^{i+1} \frac{\mathrm{E}(d_{\mathcal{T}}(v))^i}{l^i}$, where $\mathrm{E}(d_{\mathcal{T}}(v))^i$ is the i-th moment of $d_{\mathcal{T}}(v)$.*

Proof. By definition the transactions of $\mathcal{D}$ are generated independently of each other. An item v has degree r in $\mathcal{D}$ if it belongs to r fixed transactions. If we assume that each transaction is formed by the union of k patterns chosen independently u.a.r. from $\mathcal{T}$ then $d_{\mathcal{D}}(v)$ has binomial distribution. The result then follows from the binomial theorem after noticing that the probability that v belongs to a given transaction is: $1 - \frac{E(l - d_{\mathcal{T}}(v))^k}{l^k}$. $\square$

The study of the item occurrence distribution in $\mathcal{D}$ is thus reduced to finding the first k moments of $d_{\mathcal{T}}(v)$. Solving the latter is not easy in general. In the remainder of this Section we sketch our analysis under more restricted assumptions.

Item occurrences in $\mathcal{T}$ when $s = 2$. If $s = 2$, $\mathcal{T}$ is a graph and its structure depends on the value of ρ. W.l.o.g. we focus on the case $\rho = 1$. In such case, the resulting graph can be seen as directed, with edges chosen one after the other according to the following process:

1. The first directed edge e_1 is a random pair from $\mathcal{I}$.
2. If the edge chosen as step i is $e_i = (w, z)$, (for $i \geq 1$), then e_{i+1} is chosen by selecting an item u.a.r. in $\mathcal{I}$, and then selecting the second element of the pair at random as either w or z with probability $\frac{1}{2}$.

Define the degree of v in $\mathcal{T}$ as the sum of its *in-degree* $d_{\mathcal{T}}^-(v)$ (number of edges having v as second component) and out-degree $d_{\mathcal{T}}^+(v)$ (number of edges having v as first component).

Theorem 2. *Let $\mathcal{T}$ be given with $s = 2$, $\rho = 1$ and all other parameters specified arbitrarily. Then for each $v \in \mathcal{I}$, $d_{\mathcal{T}}^+(v)$ has binomial distribution with parameters l and $\frac{1}{n}$. Furthermore, the distribution of $d_{\mathcal{T}}^-(v)$ can also be computed exactly. In particular $\lim_{n \to \infty} \frac{E d_{\mathcal{T}}^-(v)}{E d_{\mathcal{T}}^+(v)} = 1$.*

Proof. (Sketch) Under the given assumptions, the first result follows from classical work random allocation of l identical balls in n distinct urns (see for instance [6]).

The in-degrees can also be estimated through a slightly more elaborate argument. Essentially v can occur as second end-point of an edge only if it occurred as first end-point in some previous step. Therefore assuming that $d_{\mathcal{T}}^+(v) = d$, $d_{\mathcal{T}}^-(v)$ can be defined as a sum of d non-negative and independent contributions. The asymptotic result on $E d_{\mathcal{T}}^-(v)$ is a consequence of the fact that for n large $d_{\mathcal{T}}^+(v)$ stays very close to its expected value. $\square$

The occurrence distribution in $\mathcal{D}$ in a very simple case. In this Section we further simplify our model, assuming that $k = 1$, i.e. each transaction in $\mathcal{D}$ is formed by a single random edge of $\mathcal{T}$. The following result shows that, when n becomes large, in such simple setting the item occurrence distribution decays super-polynomially (and therefore it cannot have, asymptotically, a heavy tail).

Theorem 3. *If r is such that $n \cdot b(r; h, \frac{2}{n}) \to \infty$ then $\frac{N_r}{n} \to b(r; h, \frac{2}{n})$ with probability tending to one.*

Proof. (Sketch) If $k = 1$ by Theorem 1 $d_{\mathcal{D}}(v)$ has binomial distribution. By linearity of expectation and Theorem 2 $Ed_{\mathcal{T}}(v)$ is approximately $2l/n$ as n tends to infinity. Thus, for large n, $p_{1,l}$ is approximately $\frac{2}{n}$. Hence EN_r tends to $n \cdot b(r; h, \frac{2}{n})$ and the stated result follows from Chebyshev inequality.   □

We close this section by noticing another peculiar feature of QUEST. Since l is much smaller than h, there is a constant probability that a given item will never occur in a transaction of $\mathcal{D}$. Equivalently a constant fraction of the n available items will never occur in the resulting database. This phenomenon was observed in the two synthetic databases analysed in Section 2: `T40I10D100K.dat` only uses 941 of the 1,000 available items, `T10I4D100K.dat`, only 861. Of course this irrelevant from the practical point of view, but it's a strange artifact of the choice of having a two-component structure in the QUEST generator.

4   An Alternative Proposal

In this Section we describe an alternative way of generating synthetic databases. Our model is in line with the proposal of Barabási and Albert [2], introduced to model structures like the scientific author citation network or the world-wide web. A mechanism, called *preferential attachment*, that allows the process that generates one after the other the transactions in $\mathcal{D}$ to choose their components based on the frequency of such items in previously generated transactions, leads to databases with the desired properties. Instead of assuming an underlying set of patterns $\mathcal{T}$ from which the transactions are built up, the elements of $\mathcal{D}$ are generated sequentially. At the start there is an initial set of e_0 transactions on n_0 existing items. The model can generate transactions based entirely on the n_0 initial items, but in general we assume that new items can also be added to newly defined transactions, so that at the end of the simulation the total number of items is $n > n_0$. The simulation proceeds for a number of steps generating groups of transactions at each step. For each group in the sequence there are four choices made by the simulation at step t:

1. The type of transaction. An OLD transaction (chosen with probability $1 - \alpha$) consists of items occurring in previous transactions. A NEW transaction (chosen with probability α) consists of a mix of new items and items occurring in previous transactions.
2. The number of transactions in a group, $m_O(t)$ (resp. $m_N(t)$) for OLD (resp. NEW) transactions. This can be a fixed value, or given any discrete distribution with mean $\overline{m_O}$ (resp. $\overline{m_N}$). Grouping corresponds to e.g. the persistence of a particular item in a group of transactions in the QUEST model.
3. The transaction size. This can again be a constant, or given by a probability distribution with mean $\overline{\pi}$.
4. The method of choosing the items in the transaction. If transactions of type OLD (resp. NEW) are chosen in a step we assume that each of them is selected using preferential attachment with probability P_O (resp. P_N) and randomly otherwise.

Our main result (its proof, along the lines of similar results given by Cooper in [4], is skipped due to space limitations) is that, provided that the number of transactions is large, with probability approaching one, the distribution of item occurrence in $\mathcal{D}$ follows a power law distribution with parameter $z = 1 + \frac{1}{\eta}$, where $\eta = \frac{\alpha \overline{m_N}(\overline{\pi}-1)P_N+(1-\alpha)\overline{m_O}\,\overline{\pi}P_O}{(\alpha \overline{m_N}+(1-\alpha)\overline{m_O})\overline{\pi}}$. In other words, the number of items occurring r times after t steps of the generation process is approximately Ctr^{-z} for large r and some constant C. Furthermore, for fixed values of t, the expected number of items and transactions after t steps are, respectively, $n_0 + \alpha t$ and $e_0 + t(\alpha \overline{m_N} + (1-\alpha)\overline{m_O})$.

Practical considerations. Turning to examples, in the simplest case, the group sizes are fixed (say $m_N(t), m_O(t) = 1$ always) and the preferential attachment behaviour of the transaction types is the same $P_N = P_O = P$. Thus $\eta = 1 - \frac{\alpha P}{\overline{\pi}}$, and $z = 1 + \frac{\overline{\pi}}{\overline{\pi}-\alpha P}$. The following pseudo-code (whose translation in Java can be found at http://www.csc.liv.ac.uk/~michele/soft.html) describes a specialization of the proposed procedure under the additional assumption that the transaction sizes are given by the absolute value of a normal distribution with parameters μ and σ (this was done only because Java offers a pseudo-random generator of normally distributed real numbers). Initially $\mathcal{D}$ contains one item and one transaction.

Input: μ, σ, α, P, h
Output: A database $\mathcal{D}$ with h transactions
for $t = 1$ to h
    select the size x as the absolute value of a normally distributed number
        with mean μ and deviation σ
    if $(x > 0)$
        with probability α add a NEW transaction to $\mathcal{D}$
        otherwise add an OLD transaction to $\mathcal{D}$.

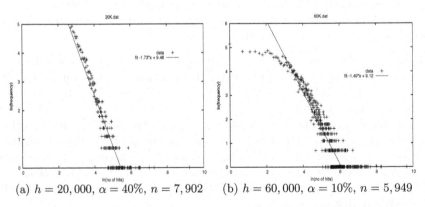

(a) $h = 20,000$, $\alpha = 40\%$, $n = 7,902$ (b) $h = 60,000$, $\alpha = 10\%$, $n = 5,949$

Fig. 3. Log-log plots of the item distributions in databases generated from our model

Figure 3 displays item distribution plots obtained from running the program with parameters $\mu = 7$, $\sigma = 3$, $P = 50\%$ for different values of h and α. While there are many alternative models for generating heavy tailed data (surveys such as [7] or [10] present a rich catalogue) and so different communities may prefer to use alternative processes, we contend that synthetic data generators of this type should be a natural choice for the testing of ARM algorithms.

References

1. Agrawal, R., Srikant, R.: Fast algorithms for mining association rules in large databases. In: Proc. of the 20th Int. Conf. on Very Large Data Bases, pp. 487–499. Morgan Kaufmann Publishers Inc, San Francisco (1994)
2. Barabási, A., Albert, R.: Emergence of scaling in random networks. Science 286, 509–512 (1999)
3. Brin, S., Motwani, R., Ullman, J.D., Tsur, S.: Dynamic itemset counting and implication rules for market basket data. In: Proc. of the ACM SIGMOD Int. Conf. on Management of Data, pp. 255–264. ACM Press, New York (1997)
4. Cooper, C.: The age specific degree distribution of web-graphs. Combinatorics. Probability and Computing 15(5), 637–661 (2006)
5. Han, J., Pei, J., Yin, Y.: Mining frequent patterns without candidate generation. In: Proc. of the ACM SIGMOD Int. Conf. on Management of Data, pp. 1–12. ACM Press, New York (2000)
6. Kolchin, V.F., Sevast'yanov, B.A., Chistyakov, V.P.: Random Allocations. Winston & Sons (1978)
7. Mitzenmacher, M.: A brief history of generative models for power law and lognormal distributions. Internet Mathematics 1(2), 226–251 (2004)
8. Redner, S.: How popular is your paper? an empirical study of the citation distribution. European Physical Journal, B 4, 401–404 (1998)
9. Savasere, A., Omiecinski, E., Navathe, S.B.: An efficient algorithm for mining association rules in large databases. In: Proc. of the 21th Int. Conf. on Very Large Data Bases, pp. 432–444. Morgan Kaufmann Publishers Inc, San Francisco (1995)
10. Watts, D.J.: The "new" science of networks. Annual Review of Sociology 30, 243–270 (2004)
11. Zheng, Z., Kohavi, R., Mason, L.: Real world performance of association rule algorithms. In: Proc. of the 7th ACM SIGKDD Int. Conf. on Knowledge Discovery nd Data mining, pp. 401–406. ACM Press, New York (2001)
12. Zaïane, O., El-Hajj, M., Li, Y., Luk, S.: Scrutinizing frequent pattern discovery performance. In: Proc. of the 21st Int. Conf. on Data Engineering (ICDE'05), pp. 1109–1110. IEEE Computer Society, Los Alamitos (2005)

Learning Multi-dimensional Functions: Gas Turbine Engine Modeling*

Chris Drummond

Institute for Information Technology
National Research Council Canada
Ottawa, Ontario, Canada, K1A 0R6
Chris.Drummond@nrc-cnrc.gc.ca

Abstract. This paper shows how multi-dimensional functions, describing the operation of complex equipment, can be learned. The functions are points in a shape space, each produced by morphing a prototypical function located at its origin. The prototypical function and the space's dimensions, which define morphological operations, are learned from a set of existing functions. New ones are generated by averaging the coordinates of similar functions and using these to morph the prototype appropriately. This paper discusses applying this approach to learning new functions for components of gas turbine engines. Experiments on a set of compressor maps, multi-dimensional functions relating the performance parameters of a compressor, show that it more accurately transforms old maps, into new ones, than existing methods.

1 Introduction

This paper discusses the inductive learning of predictive models where the output is not a label, nor a continuous value but a multi-dimensional function. It proposes using morphological analysis techniques [2] to learn a shape space capturing the common characteristics of a set of existing functions. These functions, and new ones, are points in this space, each produced by morphing a prototypical function located at its origin. The dimensions of the shape space define individual morphing operators. In this paper, the functions represent components of a gas turbine engine, primarily one used to power aircraft. However, the idea of morphing functions to be useful in new situations should readily generalize to other applications. Certainly, the expectation is that it can be used to learn functions that describe the operation of other complex equipment. It should also be useful in applications of reinforcement learning, where functions representing the solution to tasks [1] could be morphed to form solutions to other tasks.

The focus, here, is on the components of gas turbine engines, such as compressors, combustion chambers and turbines. These are described by performance parameter maps, which give the relationship between the input and output parameters. Component models based on these maps can be combined to produce

---

* Copyright ©: National Research Council Canada 2007.

J.N. Kok et al. (Eds.): PKDD 2007, LNAI 4702, pp. 406–413, 2007.
© Springer-Verlag Berlin Heidelberg 2007

an accurate simulation of an engine [12]. This simulation is an important tool
in engine design and diagnostics. An essential, yet time and resource consuming
activity, is to generate such maps, preferably prior to prototypes of the engine
being available. A common approach, as is typical in much of engineering design,
is to refine an existing solution to a similar problem. A current method used for
producing a new compressor map is to take one for a similar compressor, according to expert judgment, and to multiply each of its dimensions by a single factor
[11]. This is satisfactory if the compressor is "sufficiently similar", an ill-defined
concept. More recently more sophisticated methods, with broader applicability
and more accurate approximations, have been proposed [7,6].

Selecting an existing map and then adapting it to apply to a new problem
is very reminiscent of case based learning. This work certainly has many similarities with research into learning and adapting cases representing the real valued parameters of manufacturing equipment [4]. Learning how to morph maps
based on the similarities between existing maps is analogous to learning adaptation rules from cases [5]. What sets this work apart is that the cases are
multi-dimensional functions and more complex adaptive operators are needed.
Overall, this work might best be compared to that combining instance-based
and model-based learning. Quinlan's [9] used model based learning to generalize
across multiple instances, an important component of this work.

The original motivation for this work was to exploit engineering knowledge
within a learning algorithm used to predict, and diagnose, engine faults [8]. But,
here, I argue that as a scaling procedure it stands on its own merits. Not only,
as the experimental section will show, does it produce maps that are more than
competitive with those produced by a recently introduced non-linear scaling
method [7] but it achieves this through generalization of examples rather than a
human analysis of the commonalities of compressors. This gives the approach a
number of clear advantages. More examples can be added to improve accuracy.
Extra information about the compressor can be added to improve accuracy. The
approach should generalize to other components within the engine and to other
complex equipment. The approach should be applicable not only in modeling
new engines but also in accurately modeling older engines.

2   Generating New Functions

To generate a new function, an existing function must be selected and modified.
For a compressor, two properties are of particular importance; the pressure rise
from input to output and the mass flow, the rate of air flow through the compressor. In this work, the most similar map is selected using a distance measure
calculated from these two properties. There are different types of compressor in
a turbofan engine, the most common engine configuration used in commercial
aircraft. At front of the engine, see figure 1, is the fan, this drives a large portion
of the air around the core of the engine providing most of the thrust (at least
in commercial aircraft). The air passing through the core is pressurized by the
high pressure compressor. The core air is driven into the combustion chamber.

It there mixes with the fuel and is ignited, the resultant high velocity gases pass through the two turbines. The turbines extract energy from the flow and drive the fan and the high pressure compressor.

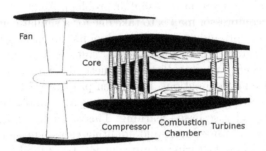

Fig. 1. A Gas Turbine Turbofan Engine

Not only are there different types of compressor, they also vary in the number and complexity of the rows of blades depending on the size and thrust of the engine. To give some feel for the range of variability, figures 2 and 3 show two examples of quite different compressor maps. Figure 2 is for a fan; figure 3 is for a high pressure compressor. The y-axis is the pressure ratio; the x-axis is the mass flow. The fan moves a lot of air, but produces only a small pressure rise. The high pressure compressor has a much larger pressure rise although it moves much less air. The data points, the black dots in the figures, have recorded values of mass flow, pressure ratio and efficiency, temperature rise for a given pressure rise, for the particular engine. They form lines, called speed lines, recorded from operation of the compressor at selected rotational speeds. Two dimensional b-splines are used here to generate the complete function. This produces the three dimensional function shown by the contours, representing the efficiency of the compressor in terms of the pressure ratio and mass flow. As the function also includes information about the compressor speed, it is really four dimensional. This representation, using speed lines, is the conventional mechanical engineering one, as it can be readily projected onto two dimensions making it much more easily interpretable by humans. The standard way to modify an existing map to generate a new one, in use for over 30 years [11], is to multiply each dimension by its own scale factor.

2.1   Non-linear Scaling

Looking back at figures 2 and 3, we can see that the general shape of the speed lines and the efficiency contours are similar but by no means the same. In figure 2 the contours are more spread out and each speed line covers a broader range on the x-axis. The shape of the contours also differs: in figure 2 a line down their center has a quite appreciable curve, in figure 3 this line is much straighter. The range of speed values, as indicated by the values at the end of each speed line, is

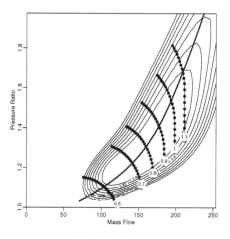

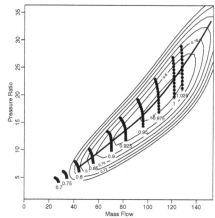

Fig. 2. Fan   **Fig. 3.** High Pressure Compressor

much greater in figure 2 than figure 3. Linear scaling does not account for these differences and research has continued to rectify these shortcomings. Kurzke and Riegler [7] analyzed a large number of compressor maps, to determine in what ways they are similar and in what ways they are different. They proposed an alternative non-linear scaling method.

The heart of this new scaling method is the ridge, or backbone, of the function, a line connecting points of maximum efficiency along each speed line. The axes are first normalized based on a reference point, the specifics of how it is identified can be found in the original paper [7]. The ridges on all the maps were found to be well approximated by an arc of a circle passing through coordinates (0,0) and (1,1). The radius of the circle depends on the pressure ratio at the reference point. Figure 4 shows an example of the scaling procedure in action. The bold solid black line is the ridge of the existing map; the bold dashed line is the ridge of the new one. The two adjacent thin lines are their approximation as arcs of a circle. First the range of each speed line in terms of mass flow is adjusted, according to the ratio of old and new values. The points on the ridge for each speed line are translated horizontally according to the different horizontal distances between the circles. The gray curves in figure 4 indicate one example of a translated and scaled speed line. New values are assigned to each speed line and the efficiency values rescaled, see the paper for details [7].

2.2   Morphological Scaling

The approach taken in this paper is to use morphological image analysis techniques [2] to identify the commonalities and differences between compressor maps. The method relies on identifying common points on all functions, called landmarks, and then finding the morphological operators needed to morph the points on one function to those on another.

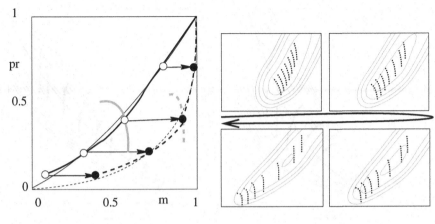

Fig. 4. Non-linear Scaling   **Fig. 5.** Morphing between Maps

The speed lines are a central feature of the compressor maps. Following Kurzke and Riegler [7], a ridge is defined at the maximum efficiency point along these lines. Where the speed lines intersect with the ridge, an initial set of landmarks is defined. Unfortunately, the speed lines are not at the same compressor speeds for each map. To get a consistent set of landmarks interpolation and extrapolation are necessary. A method based on domain knowledge is best at minimizing error. This is particularly true for extrapolation, where many schemes can produce large deviations arising from small errors in values. A commonly accepted, and reasonably accurate, approximation of the relationship described by the speed lines is the cubic polynomial [3]. Polynomials are notoriously ill-suited to extrapolation and this one is only an approximation, albeit a very useful one. So, to incorporate this knowledge, while producing better behaved curves, splines are used. These are the linear combination of smooth functions, here b-spline basis functions. Instead of simply minimizing the squared error, a penalty function is included that penalizes large differentials [10].

Penalizing the fourth order differential means the smoothest possible function is a cubic polynomial. By requiring that the gradient of the function is zero when the mass flow is zero, we get a cubic of the required form. The process fits all the speed lines at the same time so that their common characteristics, defined by the form of the cubic, are included. New speed lines can then be generated at selected speeds. With these new speed lines and their intersection point with the ridge, 8 landmarks are generated along each of 7 speed lines. Splines have been used extensively in this work, with different differential functions to smooth the data in different ways. The ridge is located using an exponential penalty function which produces a curve that fits the ridge somewhat better than the circle discussed in the previous section. Though it is worth noting that for most maps the difference is small. A two dimensional cubic spline, with a third order differential penalty function, is used to generate the complete function. This gives a preference quadratics in two dimensions.

To extract the commonalities between maps, "relative warps" are used [2]. A prototypical map is produced by averaging the coordinates of each landmark separately for all maps. Morphing this will produce any other map in the space. A thin plate spline is used; it is easy to flex but bending it sharply is difficult. This encourages morphological operations that move the landmarks together in one direction and allows the smooth morphing from one map to another. This is controlled by the convex combination of coordinates in the shape space, converted back to the original coordinate system. Figure 5 shows the effect of morphing the maps shown earlier in figures 2 and 3. These were deliberately chosen to be very different. Yet, looking at figure 5, the arrow indicates the morphing direction, the intermediate functions certainly appear to be sensible compressor maps. So, we can produce a new map by averaging over the values of any number of existing maps. The only question is which maps to average over. Sensibly, the nearest neighbor, based on mass flow and pressure ratio, should be one of the maps. If we wish to interpolate, always safer than extrapolation, other points must be found that surround the desired reference point. By using Delaunay triangulation, three points are identified surrounding this point. Their shape space coordinates are averaged and the new map produced using the appropriate morphological operators.

3   Experiments

The experiment uses each of 19 maps as the target. Its aim is to determine which of the three scaling method best approximates the target, knowing only its pressure ratio and mass flow. The least squares distance between the values of pressure ratio, mass flow and efficiency at the landmark points, on the scaled and target maps, is used as the measure of fit. The pressure ratio and the mass flow are normalized, so that the units chosen have no influence on the results and the dimensions contribute in equal part. The nearest neighbor is chosen based on the Euclidean distance of the normalized coordinates. Figure 6 shows how the maps' reference points vary in terms of pressure ratio and mass flow. The black circles are fans, they have very low pressure ratios and a large range of mass flow. The gray circles are low and high pressure compressors, they are designed for much greater pressure ratios but usually with smaller mass flows.

The results are shown in table 7. The best fit, the smallest squared error, is indicated by bold print. The performance using morphological scaling is in most cases the best. This form of scaling is less effective when no surrounding triangle is found. Originally extrapolation was used but this produces unrealistically shaped maps. Finding the nearest point on the black line in figure 6, the convex hull of all the maps (without the target map in the experiments) resulted in sensible maps and produced the results shown in the table. For instance, map 19, at the top, is not only outside the hull formed by the other maps, it is some distance away. Interpolation means averaging only maps 17 and 18. Here, the non-linear scaling is most effective, using human extracted knowledge. Similarly for map 9, on the far right, the nearest point on the convex hull is the nearest

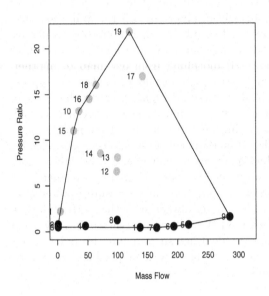

Map	Linear	Non-linear	Morphed
1	0.774	**0.765**	0.808
2	0.880	**0.768**	0.856
3	0.880	1.329	**0.879**
4	1.160	**0.929**	0.969
5	0.109	0.096	**0.054**
6	0.109	0.226	**0.040**
7	0.774	0.805	**0.176**
8	4.188	3.229	**1.456**
9	0.236	**0.225**	0.236
10	**2.062**	2.416	2.522
11	1.051	2.569	**0.798**
12	2.791	**1.014**	1.346
13	**2.791**	2.828	3.057
14	4.177	5.051	**2.233**
15	3.274	3.676	**1.983**
16	0.989	1.174	**0.676**
17	1.996	1.769	**1.079**
18	2.931	5.152	**2.124**
19	1.986	**1.310**	2.177

Fig. 6. Reference Points   **Fig. 7.** Mean Squared Error

neighbor itself, map 5, so the morphing error is identical. Again, non-linear scaling offers some improvement, through for this map it is rather small.

This method produces the best fit for compressor maps, further work is needed to see if the same is true for maps of other components. Maps for turbines are similar in many ways, so offer some promise. The same parameters, mass flow, pressure ratio and efficiency are used, although the shape is quite different. Other components also have maps, such as the combustion chamber. There are likely to be fewer such maps readily obtainable. But, with a few examples and greater use of domain knowledge, generalization should still be possible. Maps for several components of an engines would allow the effective incorporation of knowledge into learning algorithms used to diagnose and predict engine faults. Effectively modeling the normal performance of engines already in service would make identifying deviations associated with problems much easier.

The approach presented here should generalize beyond gas turbine engines and be applicable whenever functions are being learned. The general idea is using knowledge in support of learning functions. This is both theoretical, such as restrictions on the form of speed lines, and empirical, the set of maps for existing compressors. At a more detailed level, constraints in the form of penalty functions for splines, allow the input of theoretical knowledge without the overly strong constraints of a parametric model. The morphology approach does require the identification of landmarks. But if these can be found, the approach over a useful way of extracting the common characteristics of functions. In this author's earlier work [1], on transfer in reinforcement learning, previously learned solutions to subtasks were adapted to fit new ones. It would be worth revisiting this work to test the effectiveness of the approach taken here.

4 Conclusions

This paper introduced a way of learning multi-dimensional functions. Each functions is a point in shape space, produced by morphing a prototypical function located at its origin. The prototypical function and the space's dimensions, which define morphological operations, are learned from a set of existing functions. New ones are generated by averaging the coordinates of similar functions and using these to morph the prototype appropriately. The efficacy of this approach was experimentally demonstrated on a set of compressor maps, multi-dimensional functions relating various parameters of engine compressors.

References

1. Drummond, C.: Accelerating reinforcement learning by composing solutions of automatically identified subtasks. Journal of Artificial Intelligence Research 16, 59–104 (2002)
2. Dryden, I., Mardia, K.: Statistical Shape Analysis. John Wiley and Sons, Chichester, UK (1998)
3. Gravdahl, J.T., Egeland, O.: Compressor Surge and Rotating Stall: Modeling and Control. Springer, London (1999)
4. Griffiths, A.D., Bridge, D.G.: Formalising the knowledge content of case memory systems. In: Watson, I.D. (ed.) Progress in Case-Based Reasoning. LNCS, vol. 1020, pp. 32–41. Springer, Heidelberg (1995)
5. Hanney, K., Keane, M.T.: The adaptation knowledge bottleneck: How to unblock it by learning from cases. In: Leake, D.B., Plaza, E. (eds.) Case-Based Reasoning Research and Development. LNCS, vol. 1266, pp. 359–370. Springer, Heidelberg (1997)
6. Kong, C., Ki, J., Kang, M.: A new scaling method for component maps of gas turbine using system identification. Journal of Engineering for Gas Turbines and Power 125(4), 979–985 (2003)
7. Kurzke, J., Riegler, C.: A new compressor map scaling procedure for preliminary conceptional design of gas turbines. In: Proceedings of the ASME Turbo Expo (2000)
8. Létourneau, S., Famili, F., Matwin, S.: Data mining for prediction of aircraft component replacement. IEEE Intelligent Systems Journal: Special Issue on Data Mining, 59–66 (1999)
9. Quinlan, J.R.: Combining instance-based and model-based learning. In: Proceedings of the Tenth International Conference on Machine Learning, pp. 236–243 (1993)
10. Ramsay, J.O., Silverman, B.W.: Functional Data Analysis. Springer, New York (1997)
11. Sellers, J.F., Daniele, C.J.: DYNGEN-A program for calculating steady-state and transient performance of turbojet and turbofan engines. Technical Report NASA-TN D-7901, NASA (1975)
12. Visser, W., Broomhead, M.: GSP, a generic object-oriented gas turbine simulation environment. In: Proceedings of the ASME Turbo Expo, pp. 8–11 (2000)

Constructing High Dimensional Feature Space for Time Series Classification

Victor Eruhimov, Vladimir Martyanov, and Eugene Tuv

Analysis & Control Technology, Intel,
5000 W Chandler Blvd, Chandler AZ85226, USA

Abstract. The paper investigates a generic method of time series classification that is invariant to transformations of time axis. The state-of-art
methods widely use Dynamic Time Warping (DTW) with One-Nearest-
Neighbor (1NN). We use DTW to transform time axis of each signal in
order to decrease the Euclidean distance between signals from the same
class. The predictive accuracy of an algorithm that learns from a heterogeneous set of features extracted from signals is analyzed. Feature
selection is used to filter out irrelevant predictors and a serial ensemble
of decision trees is used for classification. We simulate a dataset for providing a better insight into the algorithm. We also compare our method
to DTW+1NN on several publicly available datasets.

1 Introduction

The problem of time series classification (TSC) has attracted a lot of attention
from the machine learning society in the past decade. Many domains such as
computer vision, medicine, biology, manufacturing, and others possess time dependencies as natural problem descriptions, as opposed to individual features
extracted from signals. A challenge in working with signals as class predictors
is large amount of features and complex dependence of the signal class on these
features. Advances in supervised learning methods that allow to work with ultra
high dimensional feature space make TSC a very appealing problem. However
the Euclidean metric together with One-Nearest-Neighbor (1NN) classifier has
proven to be one of the most robust TSC methods. A generalization of this
approach that takes into account transformations of time axis has been introduced about a decade ago. [1] suggested a similarity measure called Dynamic
Time Warping (DTW) that is based on matching two signals with dynamic programming. Later [2] showed that the complexity $O(n^2)$ of DTW for matching
two signals of length n can be reduced to $O(n)$ by constraining the search path
without sacrificing accuracy. DTW was proved to be the best state-of-the art
technique in multiple domains, "1NN with DTW is exceptionally hard to beat"
[3]. We will not provide a full review of TSC methods due to limited space, an extensive survey is available in [4]. A large group of papers is devoted to extracting
generic features from signals and transforming a TSC problem into a classical
machine learning problem of predicting signal class from a given feature set. A
list of features includes Singular Value Decomposition features, Discrete Fourier

J.N. Kok et al. (Eds.): PKDD 2007, LNAI 4702, pp. 414–421, 2007.

Transform, coefficients of the decomposition into Chebyshev Polynomials, Discrete Wavelet Transform, Piecewise Linear Approximation, ARMA (AutoRegression Moving Average) coefficients, various symbolic representations. Each of the methods has its own faults. Euclidean/DTW based methods suffer from the curse of dimensionality – 1NN is known to perform poorly on high-dimensional problems (i.e. long signals) [5]. [6] shows superior performance of a boosted tree ensemble learned on a set of generic features compared to 1NN with Euclidean distance on datasets where time warping is not needed. This paper is devoted to a generalization of this method for the case when time warping is essential for classifying signals.

The essense of the method is to transform time axis of both train and test signals so that the same salient points appear at the same time moments. Then we can apply a generic feature extraction method described in [6]. We sample one signal from each class that we call a base signal. Then we use DTW to warp time axis of every time series to each of base signals, resulting in several time series, one per class. A generic set of features – wavelets, coefficients of the decomposition into Chebyshev polynomials, statistical moments – and several DTW-specific features are extracted from each warped signal. A joint set of features is used as predictors. The number of features could be very high – from hundreds to tens of thousands. Such high dimensional representations are hard to learn from. However if we reduce the feature set we run into a risk of loosing information about the signal and increasing classification error. Recent advances in feature selection methods [7,8] allow us to learn a boosted ensemble of trees with a built-in feature weighting method directly in the original high-dimensional feature space. We show that this method is comparable or superior to DTW on several UCR datasets [9]. We also analyze the performance of the method on simulated data to better understand its pros and cons.

The outline of the paper is as follows: Section 2 is devoted to the warped feature extraction algorithm, Section 3 describes our time series generator and Section 4 discusses experimental results on UCR and simulated data.

2   Warping Time for Feature Extraction

Generic features described in [6] such as wavelets and Chebyshev coefficients are not invariant to time warping that changes position and scale of signal features differently for each signal. Statistical moments do not change much with warping as they take into account only the distribution of signal values but in many cases they are weak predictors. We want to build a generic set of features that would work on signals with arbitrarily (with reasonably low loss of information) warped time.

The general idea of the method that we discuss here is to select a base signal and transform time axis of each signal to minimize DTW distance. Then (when all salient features are aligned) we can extract generic features and learn a classifier. But we cannot select a single base signal because transforming all signals to it could cause deformation of class-dependent signal profiles that are crucial

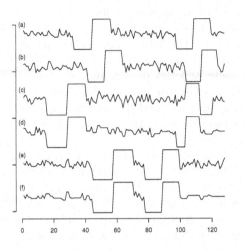

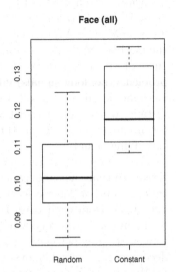

Fig. 1. Example of warped signals from two_patterns [9] dataset: (a) test signal, (b) the DTW-closest training sample, (c) base signal b_2 from class 2, (d) warped test signal wrt b_2, (e) base signal b_4 from class 4, (f) warped test signal wrt b_4

Fig. 2. Distribution of test errors for **Face(all)** dataset for randomly chosen (left boxplot) and fixed (right boxplot) base signals

for classification. So we choose one base signal from each response class. For each signal s and base signal b we find a point-to-point correspondence with DTW and calculate a warped version of s by averaging all values of s corresponding to each point of b. We use DTW algorithm described in [2]. An example of warping is given in Figure 1. We have taken a test signal (a) from the UCR dataset **Two_patterns**, class 4, and warped it to base signals of classes 2 (base signal (c) and warped test signal (d)) and 4 ((e) and (f) correspondingly). The pair of signals (e) and (f) illustrates the alignment of warped signal from the same class. This allows us to extract meaningful features from warped signals characterizing class-dependent signal profiles.

We extract a set of generic features to be used as class predictors from each warped signal. Also, we keep the features from the original unwarped signal in case no time warping is necessary. The essense of the approach is to use as many features that *could be* important as possible, if they are irrelevant, feature selection algorithm will filter them out. The exact description of feature selection method is given in Algorithm 2. We do not use any warping window when transforming signals. We do use a Sakoe-Chiba band [10] when calculating the DTW+1NN (see 2ef of Algorithm 2), the band width is obtained by optimizing DTW+1NN leave-one-out error on the training part of data. Signal warping is described in Algorithm 2. Note that we use the class predicted by DTW+1NN as a feature so the test error can hardly be larger than that of DTW+1NN. We also use DTW distances to base signals as predictors to supply GBT with an additional information about the features from different base signals. The total number of features that we extract is equal to $C \cdot (W + L + Ch + 1) + W + Ch + 6$, where C is the number of classes, L is the signal length and W is minimum power of 2 greater than L. In order to extract wavelets we add $W - L$ zeros to each signal making Discrete Wavelet Transform applicable. Ch is the number of Chebyshev coefficients – we filter out higher coefficients that proved to be too noisy, we use the value $Ch = 20$ throughout the paper.

Algorithm 1. Warped time feature selection

1. For each class c randomly choose a base signal b_c endfor.
2. For each signal s
   a. feature set $F_s = \{\}$
   b. for each class c
       warp the signal wrt base signal $s_c^{(w)} = Warp(s, b_c)$
       add wavelet D8, Chebyshev coefficients and
         raw features (signal values) of $s_c^{(w)}$ to F_s
       calculate DTW distance from $s_c^{(w)}$ to b_c and add to F_s
   c. endfor
   d. calculate statistical moments (mean, variance, skewness, curtosis, and maximum value) and add to F_s
   e. find a signal s_m from the training set D_T such that $s_m = \underset{s_m \in D_T \setminus \{s\}}{\operatorname{argmin}} DTW(s, s_m)$
   f. add the class of s_m as a feature to F_s
   g. add wavelet D8 and Chebyshev coefficients of s to F_s
3. endfor

Algorithm 2. Signal warping $Warp(b, s)$

1. Run DTW for a base signal b and an input signal s.
   Let L_i be the list of elements from s corresponding to the element i from b
2. For each i
   set the i-th value of the warped signal to the average of values in L_i
3. endfor

---

3   Time Series Dataset Generator Description

We used a data generator designed specifically to mimic most of the challenges we face in the real environment (semiconductor manufactuirng signals classification) and to better investigate TSC methods by having insight into signal class nature. Each time series is a trapezoid-like parameterized function (a sample signal is shown in Figure 5). 9 parameters (left node position, horizontal and vertical shifts, right node position, horizontal and vertical shifts, oscillation amplitude, frequency and phase, left and right slope curvatures) are sampled from predefined distributions for each signal. Curvatures and oscillation amplitude are used to generate a numeric response that is a sum of linear and quadratic functions of parameters:

$$y_n = AV + V^T BV + \varepsilon. \tag{1}$$

Here A is a vector $1 \times N$, B is a matrix $N \times N$, V is a vector of $N = 3$ parameters and ε is Gaussian noise. The values of A and B are taken from a uniform distribution $U(0, 1)$ before we start generating any time series. Categorical response is

$$y = 1(y_n - median(y_n)), where$$
$$1(x) = \begin{cases} 1, x \geq 0 \\ 0, x < 0 \end{cases} \tag{2}$$

Note that some parameters (such as the phase of oscillation sampled from the $U(0, 2\pi)$) do not participate in the response but have considerable influence on signals. It is a challenge for any predictive engine to recover this functional relationship due to the complex dependence of time series on V and the problem dimensionality.

In order to make things more complex, we add a random (from 0 to 16) amount of zeros to the beginning of each signal. We will refer to the dataset without such random shifts as to **Quad1**, and to the dataset with random shifts as to **Quad1S16**.

4   Experimental Results

We test our TSC method on several UCR datasets and on the simulated datasets in order to better understand how warping works. We have selected a subset of UCR datasets that have more than 30 samples per class on average. Smaller amount of training samples would produce higher noise and would require a more accurate approach to feature selection. Our implementation of GBT is very close to [11] with feature weighting [7] on top of it. All parameters of GBT learning algorithm were fixed: the number of trees $N = 2000$, shrinkage $\nu = 0.02$, subsampling parameters and probability thresholds. Each tree was trained on a randomly chosen 60% portion of the training dataset, the probability threshold was equal to 0.5.

Figure 2 shows the distributions of test errors on one of UCR datasets when we randomly choose base signals (left boxplot) and when we keep base signals fixed (right boxplot), so the variation of test errors is mostly due to GBT, and one can see that the particular choice of base signals is not crucial.

We run the algorithm on each dataset 10 times with randomly chosen base signals and GBT random seed. The results are summarized in Table 1. One can see that we are almost always superior to DTW+1NN or comparable in the case when the problem is easy enough for DTW+1NN and the absolute number of misclassified samples is very low. In order to check how important our warping features are, we run a set of experiments with zero warping window

Table 1. Test errors

Dataset	DTW+1NN test error	Average Test Error	Standard Deviation of Test Error	p-value for warped features
Quad1	0.108	0.0572	0.0018	1
Quad1S16	0.148	0.0855	0.0027	$4.1 \cdot 10^{-8}$
Wafer	0.005	0.00259	0.000453	$2.2 \cdot 10^{-2}$
Yoga	0.155	0.150	0.00219	0.93
Swedish_Leaf	0.157	0.118	0.00394	1
Face(all)	0.192	0.103	0.0122	$1.3 \cdot 10^{-10}$
Synthetic_Control	0.017	0.00233	0.00260	$5.2 \cdot 10^{-5}$
ECG	0.12	0	0	1
OSU_Leaf	0.384	0.379	0.0130	$2.4 \cdot 10^{-2}$
Two_patterns	0.0015	0.00055	0.000384	$1.9 \cdot 10^{-5}$

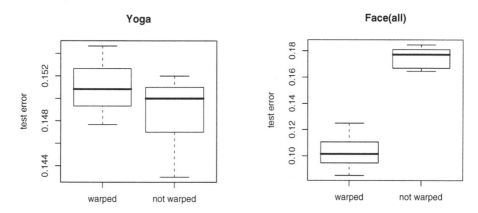

Fig. 3. Distribution of test errors for **Yoga** and **Face(all)** with and without warped features used as predictors

size (keeping the predicted class by DTW+1NN the same) which means that we do not transform signals at all. A one-sided t-test was used to check if test error with warped features is less than test error without warping. The corresponding p-values are given in the last column of Table 1. The improvement in test error is visible for 6 datasets out of 10. There are datasets such as **Quad1** where we did not get any improvement since warping is not important there. Figure 3 illustrates the distributions of test errors with and without warped features for **Yoga** and **Face(all)**. One can see that we get a significant decrease in test error on **Face(all)** when we use warped features. **Yoga**, however, shows a slight increase in test error, most probably due to failure of feature selection to filter out all irrelevant features.

Figure 4 presents the dependence of test error on the width of Sakoe-Chiba band for warping signals (it is important that the feature corresponding to the class predicted by DTW+1NN is kept constant for this experiment corresponding to the optimal width of the band). Note that the dependence of our algorithm

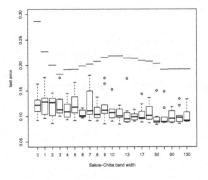

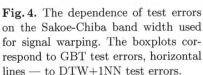

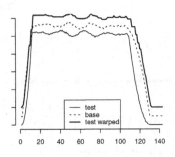

Fig. 4. The dependence of test errors on the Sakoe-Chiba band width used for signal warping. The boxplots correspond to GBT test errors, horizontal lines — to DTW+1NN test errors.

Fig. 5. Example of warped signal from **Quad1S16**. The signals are shifted along the vertical axis 10% from each other due to strong overlapping.

test error on the band width is different from the dependence of DTW+1NN test error. Our interpretation of this effect is that DTW+1NN considers every signal in the training dataset for matching while our approach matches only base signals. So DTW+1NN has higher chances of a correct match with smaller window. This is why we did signal warping without any band restriction. Note that this does not pose computational problems as with DTW+1NN since the latter has to match a test signal with all training time series while we match it with only base signals. The diminishing trend in Figure 4 also shows that the features obtained from warping signals are important for classification. The class predicted by DTW+1NN is also very important – by removing just this one feature from the predictors list of the **Face(all)** dataset we increase the average test error from 0.118 up to 0.179 – a 50% difference!

Quad1 and **Quad1S16** allow us to get an insight of the algorithm weak spots. Going back to Section 3, curvatures of the signal front and back are used to generate response for these datasets. Curvature is not invariant to the transformation that we apply to signals as illustrated by Figure 5. Note the difference in curvatures of test and warped signals in the right part, around time value 120. This means that the information about curvature will be lost in the warped signal and hence will not be reflected in extracted features. Features from the original signal do not help much either since there is a random shift and curvature features are scattered in time corresponding to different wavelet features. The resulting dependence is hard to learn, this is the major reason for 50% difference in test errors for **Quad1** and **Quad1S16**.

5   Conclusion

This work deals with TS classification problems where input signals need to be aligned in time (warped). The proposed approach creates a massive number of features including original signals, by-class warped signals, wavelet and chebychev decomposition coefficients of warped signals, summary statistical moments of warped signals, and even labels predicted by DTW-1-NN used as input features. Gradient boosting of trees with imbedded dynamic feature selection capable of handling hundreds of thousands predictors is then used for classification. A set of experiments on UCR and artificial datasets show that this combination provides a superior learner relative to the well know state of the art approach. No single subset of features by itself carries enough information to achieve the best performance on different classification tasks. The future work will concentrate on refining this approach for important industrial applications with influential curvature-like features not easily detected currently by any method, and porting the methodolgy to time series regression problems.

References

1. Berndtand, D.J., Clifford, J.: Using dynamic time warping to find patterns in time series. In: Working Notes of the Knowledge Discovery in Databases Workshop, pp. 359–370 (1994)
2. Ratanamahatana, C.A., Keogh, E.: Everything you know about dynamic time warping is wrong. In: Third Workshop on Mining Temporal and Sequential Data, in conjunction with the Tenth ACM SIGKDD International Conference on Knowledge Discovery and Data Mining, ACM Press, New York (2004)
3. Xi, X., Keogh, E., Shelton, C., Wei, L., Ratanamahatana, C.A.: Fast time series classification using numerosity reduction. In: International Conference on Machine Learning (2006)
4. Keogh, E.: Data mining and machine learning in time series databases (2004)
5. Hastie, T., Tibshirani, R., Friedman, J.: The elemetns of statistical learning: Data mining, inference, prediction. Springer, Heidelberg (2001)
6. Eruhimov, V., Martyanov, V., Tuv, E.: Feature class selection for time series classification. In: Submitted to the Workshop on Time Series Classification, SIGKDD'07
7. Borisov, A., Eruhimov, V., Tuv, E.: Dynamic soft feature selection for tree-based ensembles. In: Guyon, I., Gunn, S., Nikravesh, M., Zadeh, L. (eds.) Feature Extraction, Foundations and Applications, Springer, New York (2006)
8. Borisov, A., Torkkola, K., Tuv, E.: Best subset feature selection for massive mixed-type problems. In: Corchado, E., Yin, H., Botti, V., Fyfe, C. (eds.) IDEAL 2006. LNCS, vol. 4224, pp. 1048–1056. Springer, Heidelberg (2006)
9. Keogh, E., Xi, X., Wei, L., Ratanamahatana, C.A.: The ucr time series classification/clustering homepage (2006)
10. Sakoe, H., Chiba, S.: Dynamic programming algorithm optimization for spoken word recognition. IEEE Trans. Acoustics, Speech, and Signal Proc. ASSP-26, 43–49 (1978)
11. Friedman, J.H.: Stochastic gradient boosting. Technical report, Dept. of Statistics, Stanford University (1999)

A Dynamic Clustering Algorithm for Mobile Objects

Dominique Fournier[1], Gaële Simon[2], and Bruno Mermet[2]

[1] LITIS EA 4051, 25 rue Philippe Lebon, BP 540 76058 Le Havre cedex
dominique.fournier@univ-lehavre.fr
[2] GREYC CNRS UMR 6072, 6 Boulevard du Maréchal Juin 14050 CAEN cedex
{gaele.simon,bruno.mermet}@univ-lehavre.fr

Abstract. In this paper, a multiagent algorithm for dynamic clustering is presented. This kind of clustering is intended to manage mobile data and so, to be able to continuously adapt the built clusters. First of all, potential applications of this algorithm are presented. Then the specific constraints for this kind of clustering are studied. A multiagent architecture satisfying these constraints is described. It combines an ants algorithm with a cluster agents layer which are executed simultaneously. Finally, the first experimental results of our work are presented.

1 Introduction

1.1 Agents Clustering

In this article, a dynamic technique for clustering mobile objects is proposed. Initially, this work aims at characterizing groups of agents during the execution of a multiagent system. A subset of agents is considered as a group if, for a given period, the internal properties of the agents evolve similarly. Each observed agent can be represented by a vector of properties evolving in time. For example, these properties can represent the number of communications of the agent, its reinforcement value (for agents with training capacities), or its position in a located environment (ants [1] for instance). Thus, groups of agents can be highlighted as evolutionary clusters. More generally, we have to solve a problem of dynamic clustering in which the cardinality of the set of data to cluster is not constant (some agents can appear or disappear during the clustering process). Moreover, already clustered data can be modified or reorganized according to the evolution of the corresponding agents. These characteristics also appear in other kinds of applications like meteorology (detection and follow-up of cyclones), road traffic analysis, animals migrations, ... Thus, a dynamic method of clustering is necessary in order to adapt the set of clusters continuously so that it reflects as accurately as possible the current state of data (of agents in our case).

1.2 Related Works

Recently, some works are focused on particular kinds of clustering where the set of data to cluster is not completely known by the algorithm at the beginning

J.N. Kok et al. (Eds.): PKDD 2007, LNAI 4702, pp. 422–429, 2007.

of the process. This is for instance the case of data stream clustering [2,3]. A data stream is a sequence of numerous data such as the log of actions of a website user. The large volume of data prevents the storage of the stream in main memory; this is why clustering algorithms have to manage the streams gradually. So, most of them work on subsets of streams constituted by several consecutive data. The first disadvantage of these approaches is that the number of clusters built remains generally constant. Moreover data already clustered cannot evolve with time, which thus does not correspond to our specific needs. There also exists evolutionary data streams clustering algorithms [4,5,6]. The main difference with the previous category is that, the underlying distribution of the data stream can evolve significantly with time. Thus, these algorithms have to be able to modify strongly the built clusters, and even to destroy some of them during the clustering process. This point joins one of our concerns. But, as in the first case, these algorithms do not take into account the fact that already clustered data can also evolve. Evolutionary clustering[7] also considers the problem of clustering data over time, even if in this case, at time t the clustering algorithm uses all objects seen so far. The main difference with our work is that, at time t, the set of clusters produced by our algorithm is expected to only characterize the current state of objects.

The closest work to our problem relates to an algorithm for clustering mobile data presented in [8]. This algorithm allows to cluster evolving data. It is based on moving micro-clusters which are adapted from the micro-cluster notion defined in [9]. The location of each object to cluster is described by its location at time t_0 together with a velocity vector. Using this two components, the current location of an object can be evaluated.

One major advantage of this algorithm is its efficiency when updating the set of micro-clusters thanks to their profile. Indeed, if micro-clusters are stable enough, the update can be performed using only profiles which summarize the state of each micro-cluster. Moreover, using time velocities as additional dimensions in the clustering process is very interesting.

The main disadvantage of this approach with respect to our constraints is the need to use a standard clustering algorithm jointly. The authors have chosen K-Means which needs to provide the number of clusters to build. Moreover, K-Means is first used to build initial micro-clusters. This implies that the set of all mobile objects to cluster must be known at the beginning of the clustering process which is not always possible in our context. The other important disadvantage is that the algorithm does not allow to take into account new mobile objects during the clustering process which is necessary in our agents clustering context.

2 Our Approach

In addition to the algorithms previously presented, there also exists in the MAS community, ants algorithms that are able to achieve clustering tasks [10]. In these algorithms, data are distributed in a grid on which ants agents can move.

They can also carry data and gather them in heaps. These algorithms have properties which seem to fit with our needs, in particular to take into account the evolution of data, which is necessary in our context. Unfortunately, they have a slow convergence, which is accentuated in an unstable context.

To improve the convergence of these approaches within a static framework, N. Monmarché proposed the AntClass algorithm [11] which successively associates in four steps an ants algorithm and the K-Means algorithm. This approach is not compatible with the dynamic aspect of our problem as shown in [12]. Nevertheless, we decided to associate a layer of cluster agents with the ants described in AntClass to make a two layers multiagent architecture. In this architecture, each heap created by the ants corresponds to a cluster and is encapsulated in a cluster agent. Thereafter, we describe the grid which constitutes the environment of our agents and the two kinds of agents : ants and clusters.

2.1   The Grid

The grid is divided into cells and allows to store non-clustered objects[1] and heaps. There are two kinds of non-clustered objects: objects which have never been put into a heap; objects which have been rejected from their initial heap by Cluster agents or by Ants (this can occur when objects evolve or when ants build inaccurate clusters).

2.2   Ants

Their behaviour is identical to the one specified in AntClass during the first stage of the algorithm. We summarize this behaviour quickly, more details being available in [11,13]. The ants move on the grid, can carry an object by picking up an isolated object found in a cell, or by taking the most dissimilar object of a heap. They can also drop an object o on a free cell, on an isolated object if they are similar enough (creating a new heap), or on an existing heap if o and the heap centre are similar enough. Similarity is evaluated with a distance measure. These actions are based on probabilities in order to produce a partial random behaviour which is fundamental for the effectiveness of ants algorithms.

Our architecture is based on the same ants algorithm as AntClass, yet we have defined a new measure to evaluate objects with respect to a cluster. This measure takes into account the dispersion of the distances of clustered objects from the gravity centre of their heap. It also allows to reduce the number of parameters used in AntClass while increasing the stability of the clustering. Thus, we suppose that the distribution of data in a cluster follows a Normal law represented by a variable R_C. So, considering the average μ and the standard deviation σ of R_C, by definition, one knows that 99.7% of data are in interval $I_3 = [0, \mu + 3\sigma]$. This led us to redefine the condition for an ant to remove the most dissimilar object o from a heap H: if the distance between o and the centre of H does not belong to the interval $[0, \mu + 3\sigma]$, then the ant removes o

---

[1] From here, when we talk about a piece of data, we will also use the word *object*.

from H. Moreover, we have also modified the mechanism used by ants to add data to heaps. Experiments shows that the aggregation criterion used by ants of AntClass leads sometimes to build inaccurate clusters which are difficult to remove. Thus a more restrictive criterion was defined: our ants drop an object in a heap if its distance from the centre of the heap is in the interval $I_2 = [0, \mu + 2\sigma]$.

2.3   Cluster Agents

In AntClass, after the work of ants, K-Means is used to reduce the number of remaining isolated data. In our approach, we dedicate the management of this problem to cluster agents.

A cluster agent encapsulates a heap C (i.e a cluster) created by ants. This one lives while its heap contains at least two objects. Moreover, just before its death, a cluster agent rejects its remaining objects on the grid. C is defined by a 4_tuple (G_C, R_C, V_C, S_C) where G_C is its centre, R_C its radius, V_C its volume (an hypersphere specified by given by G_C and R_C) and S_C the set of its data. In the following, C refers to a cluster agent and also to its encapsulated heap.

Main Behaviour. As soon as an ant drops an object in a heap or removes an object from a heap, the associated cluster agent must update its 4_tuple. Moreover, cluster agents can also grow by attacking other cluster agents perceived as obstructing their own development. A cluster agent C_{obs} is considered to *obstruct* an other cluster agent C_{att} when $V_{C_{att}}$ intersects $V_{C_{obs}}$. Moreover, we suppose that C_{att} attacks C_{obs} only if its size is at least equal to 20% of the size of C_{obs} (to avoid too many attacks). If such an intersection of volumes occurs, it is probably due to a misconfiguration of the clusters, maybe they could be unified, maybe one of them could be divided into two parts. Thus, the clusters must be updated. That is the reason why, when a conflict between two Cluster agents occurs, one of them attacks the other one which flees. As shown in next section, the flee mechanism allows to update the clusters.

Cluster agents must also detect the evolution of objects contained in their heap. If an object becomes too distant from the centre of its heap, the cluster agent rejects it into the grid and updates its 4_tuple. To determine if an object is too far away from the centre of its heap, a cluster agent uses the same measure as the one used by the ants (§ 2.2).

The Attack/Flee Behaviour. If a cluster agent called C_{att} has an obstructing cluster agent C_{obs}, C_{att} attacks C_{obs} and the former must flee. To describe the attack/flee interaction we use the following notations:

- $\delta(x, y)$ is the Euclidean distance between 2 objects x et y;
- $diss(C)$ is the most dissimilar object of C with respect to δ.

To manage the flight of C_{obs}, two kinds of situations are considered depending on whether the intersection between $V_{C_{att}}$ and $V_{C_{obs}}$ may contain objects or may not. In the first case, all objects located in the intersection are added to the heap of C_{att}. In the second case, C_{obs} drops one by one its $diss(C_{obs})$ until the intersection of volumes disappears. If $\delta(diss(C_{obs}), G_{C_{att}}) < R_{C_{att}} + 2\sigma$ then $diss(C_{obs})$ is put in C_{att} and on the grid if not. This aggregation mechanism of the objects into the attacking cluster during a flight is identical to the aggregation mechanism of data into a heap of the ants behaviour (§ 2.2). Here our aim is to reinforce homogeneity of C_{att} and to avoid the construction of too large heaps which are difficult to dissociate.

3   Experiments

3.1   Scenarios

In order to evaluate our dynamic clustering algorithm, a simulation platform is used to generate sets of mobile objects specified by "scenarios". A scenario allows to describe "populations" and "trajectories". A population is considered to be a set of mobile objects which are supposed to evolve similarly (i.e. to have a similar motion) and distributed inside an hyper-sphere. The description of the motion of this hyper-sphere is called the population's trajectory. As a consequence, at the clustering level, each detected cluster is expected to correspond to an existing population.

The main goal of the first experiments is to show the ability of our algorithm to detect the population trajectories as they are evolving. In this section, we have chosen to present three particular scenarios. The main differences between these scenarios are the number of populations, the kind of population trajectories used and the presence of noise. In the following, each scenario is summarized by a graphical representation (FIG. 1) inside which each population is represented by a circle. Bold and dotted circles represent respectively initial states and intermediate or final states of the populations. The trajectory of each population is illustrated by arrows. In each scenario, a population contains 100 mobile objects distributed according to a normal law. It is important to notice that, for the first experiments, only two attributes have been used for mobile objects in order to simplify the visualisation and the results analysis. The *Noisy* scenario contains a single population (FIG. 1(a)) and 33 additional mobile objects which are randomly distributed to simulate a kind of noise. They also move randomly during the scenario. The *Disjoined* and the *Crossroad* scenarios contains two different populations. These scenarios are rather different because in the *Disjoined* one, trajectories are such that populations converge without any overlapping in a first time and then move in two different directions. In the *Crossroad* scenario, populations not only converge but also overlap, then they also continue on their own ways.

3.2   Results

Evaluating the results produced by a dynamic clustering is a difficult problem because usual criteria defined only on the final results of the clustering process can not be used alone. More precisely, this implies to be able to compare, almost

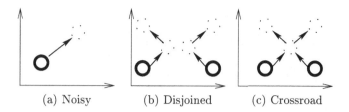

(a) Noisy   (b) Disjoined   (c) Crossroad

Fig. 1. Scenarios schemas

continuously, the state of the different populations of the scenario with the set of detected clusters. Moreover, the temporal gap between what is detected by the clustering algorithm and the different states of the populations in the simulation must be evaluated.

For the moment, three types of temporal graphs have been used which allow to analyse the following measures over time: the number of mobile objects put in clusters, the number of clusters and the mean clusters purity. The purity is a measure used in [5] and defined as follows:

$$purity = \frac{\sum_{i=1}^{K} \frac{|C_i^d|}{|C_i|}}{K} \tag{1}$$

where K is the number of detected clusters at time t, $|C_i|$ is the cardinality of cluster i and $|C_i^d|$ is the number of mobile objects of cluster C_i associated to the population d which is the most represented population in cluster C_i. It allows to evaluate, at time t, the mean quality of clusters with respect to populations in the simulation. Due to a lack of space, these graphs are not detailed in the paper. That's why, in the next section, results evaluation is presented using mean measures of the clustering process. Nevertheless, they take into account a potential evolution of the number of populations during the execution of the scenario.

For each experiment, the following measures have been used:

– #data: number of all mobiles objects in the scenario
– NAvgC: mean ratio between the number of detected clusters and the number of populations in the scenario at the same moment (the closer to 1 this measure is, the better is the clustering result)
– NAvgDC: mean number of aggregated mobile objects (i.e. the number of mobile objects which are put in a cluster).
– TimeNbP: amount of simulation time corresponding to states in which the number of detected clusters is equal to the number of populations in the scenario.
– PM: mean of the mean clusters purity.

It is important to notice that for these measures, a heap containing two mobile objects is considered as a cluster.

Table 1. Experiments results

Test	#data	NAvgC	NAvgDC	TimeNbP	PM
Noisy	133	1.5	93.0	77.0	0.97
Disjoined	200	1.15	180.0	84.9	0.99
Crossroad	200	2.5	132.6	35	0.7

Purity measures show that, generally, clusters are good ones i.e. they contain mobiles objects associated to the same population. NAvgC values are often too high but they show, however, that the number of detected clusters is close to the number of populations in the scenario. Indeed, this is also entailed by the analysis of graphs giving the number of detected clusters with time. This analysis shows also that, at the beginning of the simulation, the clustering algorithm needs a short adaptation period during which the number of clusters is big. The mean number of aggregated objects shows that a little part of the mobiles objects is not clustered at the end of the simulation. The results show that the kind of populations trajectories does not modify the results quality. Moreover, the good results obtained on *Noisy* scenario show that the clustering algorithm has not been disturbed by noisy objects.

However, the algorithm does not provide good results on *Crossroad* scenario. In this scenario, the 2 populations are temporarily superposed. At the clustering level, the consequence is that the two corresponding clusters are merged into a single big cluster SC which is a normal phenomenon. But, when in the simulation the two populations begin to move away, the clustering algorithm keeps the cluster SC instead of splitting it. Indeed, as there exists only one big cluster, it can not be attacked by the new little clusters built by ants. On the contrary, these last ones are attacked by SC and then are merged with SC. This phenomenon is a consequence of the attack constraint described previously which is however useful in most cases.

4 Conclusion

In this paper, a dynamic clustering algorithm for mobile objects (or agents) based on a multiagent architecture has been presented. This last one is made of an ants layer coupled with a cluster agents layer, the two layers being executed simultaneously. The first experiments on scenarios corresponding to various kinds of populations evolutions give good results. Nevertheless, new solutions must be found, particularly to solve the problem of unsplitable "big clusters". A first solution to explore is to use the velocity of mobile objects in the clustering process [8] which could avoid that two clusters corresponding to two populations with different dynamics are merged. Further experiments have also to be performed on new scenarios in which the number of populations and the size of these populations evolve significantly during the simulation. To better evaluate the accuracy of our results we plan to use the MONIC framework[14] which proposes an algorithm for cluster transition detection. Indeed, this framework can help us to compare more precisely scenarios and clustering results.

References

1. Drogoul, A., Corbara, B., Lalande, S.: artificial societies: The Computer Simulation of Social Life. In: Conte, Gilbert (eds.) MANTA: new experimental results on the emergence of (artificial) ant societies (1995)
2. Barbara, D.: Requirements for clustering data streams. SIGKDD Explorations 3(2), 23–27 (2002)
3. Guha, S., Mishra, N., Mortwani, R., O'Callaghan, L.: Clustering data streams. In: IEEE Annual Symposium on Foundations of Computer Science, pp. 359–366. IEEE Computer Society Press, Los Alamitos (2000)
4. Aggarwal, C.C., Han, J., Wang, J., Yu, P.S.: A framework for clustering evolving data streams. In: Aberer, K., Koubarakis, M., Kalogeraki, V. (eds.) Databases, Information Systems, and Peer-to-Peer Computing. LNCS, vol. 2944, Springer, Heidelberg (2004)
5. Cao, F., Ester, M., Qian, W., Zhou, A.: Density-based clustering over an evolving data stream with noise. In: SIAM Conference on Data Mining, p. 11 (2006)
6. Nasraoui, O., Uribe, C.C., Coronel, C.R., Gonzalez, F.: TECNO-STREAMS: Tracking Evolving Clusters in Noisy Data Streams with a Scalable Immune System Learning Model. In: IEEE International Conference on Data Mining, pp. 235–242. Melbourne, Florida (2003)
7. Chakrabarti, D., Kumar, R., Tamkins, A.: Evolutionary clustering. In: KDD'06, USA (2006)
8. Li, Y., Han, J., Yang, J.: Clustering moving objects. In: KDD (Knowledge Discovery in Databases), pp. 617–622 (2004)
9. Zhang, T., Ramakrishnan, R., Livny, M.: Birch: an efficient data clustering method for very large databases. In: SIGMOD (International Conference on Management of Data) (1996)
10. Deneubourg, J.L., Goss, S., Franks, N., Sendova-Franks, A., Detrain, C., Chretien, L.: The dynamics of collective sorting: robot-like ant and ant-like robots. In: Meyer, J.-J., Wilson, S. (eds.) Proceedings of the First International Conference on Simulation of Adaptative Behavior (1990)
11. Monmarché, N.: Algorithmes de fourmis artificielles: applications á la classification et á l' optimisation PhD thesis, Université de Tours, France (2000)
12. Simon, G., Fournier, D.: Agents clustering with ants. In: Proceedings of 5th International Workshop on Agent-Based Simulation (ABS'04, pp. 147–152. SCS Publishing House (2004)
13. Coma, R., Simon, G., Coletta, M.: A multi-agent architecture for agents clustering. In: Proceedings of 4th International Workshop on Agent-Based Simulation (ABS'03), SCS Publishing House (2003)
14. Spiliopoulou, M., Ntoutsi, I., Theodoridis, Y., Schult, R.: The MONIC Framework for Cluster Transition Detection. In: Fifth Hellenic Data Management Symposium, Greece, p. 10 (2006)

A Method for Multi-relational Classification Using Single and Multi-feature Aggregation Functions

Richard Frank, Flavia Moser, and Martin Ester

Simon Fraser University
Burnaby BC, Canada V5A 1S6
{rfrank, fmoser, ester}@cs.sfu.ca

Abstract. This paper presents a novel method for multi-relational classification via an aggregation-based Inductive Logic Programming (ILP) approach. We extend the classical ILP representation by aggregation of multiple-features which aid the classification process by allowing for the analysis of relationships and dependencies between different features. In order to efficiently learn rules of this rich format, we present a novel algorithm capable of performing aggregation with the use of virtual joins of the data. By using more expressive aggregation predicates than the existential quantifier used in standard ILP methods, we improve the accuracy of multi-relational classification. This claim is supported by experimental evaluation on three different real world datasets.

Keywords: multi-relational datamining, multi-relational classification, multi-feature aggregation, existential quantifier.

1 Introduction

Multi-relational (MR) data mining [4] deals with gathering knowledge from multiple related tables by exploring their own features as well as the relationships between them. Classical mining algorithms are not applicable to MR data since tuples linked to the table studied, referred to as the target table, and stored in directly or indirectly related tables have potentially valuable information about target tuples which is not expressible in single-relational (SR) data mining without loss of knowledge [1, 4, 13].

MR classification is inherently different from SR classification because all tables have to be searched for valuable information, and relationships between the features present in the database have to be explored. There are a few techniques extending SR methods into the MR domain. One method of classifying MR data is to adopt the framework of Inductive Logic Programming (ILP) [1, 3] and use it to find rules such that they entail one of the classes. CrossMine [13] is such an ILP based MR classifier using TupleID propagation, propagating data into related tables through foreign key relationships instead of performing a physical-join in the database. As the propagation was expressed in [13], there is not enough data moved to perform aggregations over related tables, other than the target table, since the IDs from those related tables are missing. MDRTL-2 [2] uses selection graphs to represent rules which visually depict the SQL statements used to describe the rules. [6] extends this technique to include

J.N. Kok et al. (Eds.): PKDD 2007, LNAI 4702, pp. 430–437, 2007.
© Springer-Verlag Berlin Heidelberg 2007

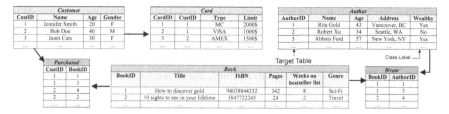

Fig. 1. Customers, books and their authors. (Arrows denote 1-many relationships).

single-feature aggregation functions. The approach of [9] uses virtual features to summarize tuples not contained in the target table. It relies on discretization resulting in information loss [5]. [11] proposes a probabilistic model to learn cluster labels. This model results in information loss as well since the information of non-target table tuples is aggregated into a single value. Decision trees were extended to the MR domain while incorporating single-feature aggregation and probability estimates for the classification labels [8]. [7] introduces a model for MR classification using attribute values and link distributions. It requires a self-join of the target table and hence cannot be applied to datasets used in this paper.

In this paper, we propose CLAMF (*CL*assification with *A*ggregation of *M*ultiple *F*eatures), a method that classifies MR data using aggregation involving single and multiple features without physical joins of the data. As our running example we use the database shown in Fig. 1 consisting of the target table **Author** with attribute 'wealthy' as a class label. There are three other entity tables, **Book**, **Customer** and **Card,** and two relationship tables **Wrote** and **Purchased**. Our contributions are:

- selecting and using appropriate aggregation functions for different numbers of features and different data-types,

- incorporating multi-feature aggregation predicates into MR classification rules,

- the extension of TupleID propagation [13] in order to efficiently perform single- and multi-feature aggregation over related tables, and

- the extensive experimental evaluation of our method on three real life databases, demonstrating that the incorporation of multi-feature aggregation predicates substantially improves classification performance and produces meaningful rules.

2   Classification Rules with Aggregation Predicates

To build classification rules on a dataset, a target table T_t and a class label from T_t are selected by the user. Each tuple in T_t, called a target tuple, has exactly one class label assigned. The goal of the proposed MR classification method is to find rules, using the format of Horn clauses, that predict which class a target tuple belongs to given its own feature values, relationships to other tuples and their features. The tuples in T_t can either be interrelated to other tuples also in T_t or related to tuples in other tables. If they are interrelated, the class label may depend on other tuples in the same table and their class labels [11, 8, 7]. Our algorithm does not deal with this special case. While we assume data to be managed in a relational database management system, we use an ILP format, ILP clauses, to represent multi-relational classification rules.

Definition 1. ILP clauses, referred to as rules, are of the form $R{:}head \leftarrow body$ and are made up of a one-literal *head* specifying a class assignment, and a *body* $L_1 \wedge L_2 \wedge ... \wedge L_n$, represented simply as $L_1, L_2, ..., L_n$, which is a conjunction of literals L_i.

In this paper, a literal is restricted to be either a predicate or a comparison between two terms. As an example, the rule *wealthy(A,'Yes')* ← *author(A), wrote(B,A), book(B)* states that *"if an author has written a book then the author is wealthy"*. Here book *B* is implicitly existentially quantified. Although the rule expresses that there does exist a book for the author, it is very likely that not all authors are rich. Hence this rule is relatively weak. What could help is determining how many books were written by an author but such rules cannot be expressed in classical ILP format.

2.1 Integrating Single-feature Aggregation Functions into ILP Rules

Most state-of-the-art multi-relational classification methods use only the existential quantifier but not aggregation functions to build rules. The authors of [12] extended the ILP formalism that allows for single-feature aggregation (SFA) functions to be used in ILP via *single-feature aggregation predicates* as follows:

Definition 2. A *single-feature aggregation function* maps a bag of elements from the domain of a feature to a single value from another (possibly different) domain.

Definition 3. A *single-feature aggregation predicate Agg* has the form *Agg(input, {conditions}, result)* where *input* specifies the bag of feature values to be aggregated, constrained by the *conditions*, and the *result* is a variable referencing the result of the single-feature aggregation function corresponding to *Agg*.

To use an aggregation predicate, we need an additional literal, called a comparison literal, comparing the result of the SFA predicate against a term *t*, i.e. *result θ t* where $\theta \in \{=, <, \leq, >, \geq\}$. For example, *Avg(A,{purchased(B,C),age(A,C)},N),N<30* formulates that *"the average age of customers C who purchased a book B is less than 30"*. *Avg* is the aggregation function, *Avg(...)* the corresponding aggregation predicate, and *N<30* the comparison literal. The different aggregation functions based on the different input data-types are below:

> **Numerical:** *sum, min, average, median* and *standard deviation.*
> **Date:** *difference* between the earliest and latest date, *earliest* date and *latest* date.
> **Categorical:** Contains a fixed set of unordered values. A category is not 'better' or 'greater' than another category, hence only the equivalency comparison can be used. *Count, most frequent* and *least frequent* can also be performed.
> **Ordinal:** This is ordered categorical data, hence in addition to the aggregation functions for the categorical values, the greater/less-than operator can be applied.

2.2   Integrating Multi-feature Aggregation Functions into ILP Rules

Analyzing and aggregating multiple features of the same table simultaneously can yield valuable information. Using multi-feature aggregation (MFA) functions, dependencies between features can be discovered which then aid in classification. For example, an increasing income of a person over time could indicate wealth. We adapt SFA predicates to MFA predicates by allowing multiple features as arguments.

Definition 4. A *multi-feature aggregation function* maps multiple lists of elements from the domains of the corresponding features to a single value.

Definition 5. A *multi-feature aggregation predicate Agg* has the form *Agg({input₁, ...inputᵢ}, {conditions}, result)* where *{input₁,...inputᵢ}* specifies the lists of feature values to be aggregated and constrained by *conditions*. The *result* is a variable referencing the result of the multi-feature aggregation function corresponding to *Agg*.

As an example, if book *B* has features *pages* and 'weeks on best seller list' (*WoBSL*) then the input to the aggregation function consists of vectors *{pages(B), WoBSL(B)}* with the selection condition *{wrote(B,A),pages(B)≤200, WoBSL(B)≥3}*. The correlation can be calculated by applying the function *corr* to *page* and *WoBSL* to get:

$$\text{corr}(<\text{pages(B)},\text{WoBSL(B)}>, \{\text{wrote(B,A)},\text{pages(B)} \leq 200, \text{WoBSL(B)} \geq 3\}, R), R > 0.5$$

Restrictions are then placed on the result *R* of the multi-feature aggregation, for example, requiring that the correlation be larger than 0.5. In this paper we restrict discussion to the case of aggregating two features only. However, the framework can be generalized to express aggregation of any number of features, for example to see how the *age-gender* distribution changes over time for each book sold.

For 2-dimensional feature analysis, the features are analyzed pair-wise by taking into account both feature-types. Dates and numbers can be binned and analyzed as ordinal data. The MFA functions we use are discussed below:

Numerical vs. Numerical: The slope of line of best fit, correlation, covariance or the T-test can be applied in order to show a relationship between numerical features.

Date vs. Numerical: Slope of line-of-best-fit can illustrate a temporal trend or cycle. Correlation, covariance and T-test can also be calculated by treating dates as numbers (by calculating the number of days from a certain date).

Categorical/Ordinal vs. Categorical/Ordinal: Pair-wise frequency tables can be built and analyzed to find the least and most frequent combination of values. The Chi-Square test indicates whether there is a dependency between two variables.

3   Learning Rules with Aggregation Predicates

In this section, we show how to learn rules with SFA and MFA predicates. Our algorithm, CLAMF (**CL**assification with **A**ggregation of **M**ultiple **F**eatures), is an adaptation of the sequential covering algorithm and is based on the idea of the well-known CrossMine algorithm [13]. The task is to address the two class classification problem by finding rules which predict the class label of a target tuple.

3.1 Learning Rules

The building of rules is done by generating one rule at a time and refining them incrementally until some termination condition applies. When refining a rule, a method (*GetBestLit*) extends the rule by at least one literal at a time. The 'goodness' of a literal is determined via FOILGain [10].

GetBestLit employs a look-ahead strategy extending a given rule by possibly multiple literals at a time. In addition to the standard cases, we allow the extension by SFA or MFA predicates and a corresponding comparison literal. The search-space of all allowed literals is explored by recursively searching all referenced tables T_r in the rule being built. If T_r is referenced in the rule then the tuple IDs are propagated to the linked table which is then explored by recursively applying *GetBestLit* to that table.

To illustrate further our algorithm, we give the following example. Once the IDs have been propagated (Fig. 2), existential quantification (EQ), SFA, and MFA over any previously referenced table can be performed since all necessary IDs are in T_r. Using only the information in T_r, for each previously referenced table over which aggregation is to occur, a new table is created (Fig. 3) and scanned for the best combination of one EQ, SFA, or MFA predicate and a threshold. The overall highest FOILGain is selected and the search for the next best literal is restarted until there are insufficiently many tuples left which are not covered by a rule.

3.2 Extending TupleID Propagation for Aggregation

CrossMine [13] introduced the concept of TupleID propagation to efficiently mine MR classification rules using the existential operator. The propagation appends to each tuple of a non-target table the IDs and class labels of tuples in target table T_t that are related to it. The following example illustrates how TupleID, in the context of aggregation, cannot do what we need. Using our running example, starting with the target table **Author**, the data is propagated to **Book**. During the next iteration of propagations, **Book** becomes the source table for the propagation and the related table **Customer** becomes the destination. For each tuple in **Customer**, the related tuples in **Book** are determined and the IDs, along with the class labels of **Author** are appended to **Customer**. The result is similar to Fig. 2 but without BookID. Aggregation of authors can now be performed to find the '*total number of unique customers each author has sold to*'. Aggregating **Customer** over books to determine the '*number of customers each book sold to*' however is not possible since there is no information from **Book** in **Customer**. Due to this, TupleID propagation does not allow for aggregation over previously referenced tables but only the target table.

Customer					
CustID	Name	...	AuthorID	Class	BookID
1	Jennifer Smith	...	1, 5, 3	Y, Y, Y	1, 1, 3
2	Bob Doe	...	2, 4	N, Y	4, 2
3	Janet Cats	...	1, 5	Y, Y	1, 1
4	Terry Wolfe	...	4	Y	2

Fig. 2. Result of our propagation. Aggregation over *Books* can be done.

Aggregated Customer Table						
BookID	Class	Age				Gender
		Min	Max	Avg	...	Most Common	...
1	Y, Y	20	30	25	...	F	...
2	Y	18	40	29	...	M	...
3	Y	20	20	20	...	M	...

Fig. 3. Aggregated table can now be analyzed for literal selection

To allow for aggregation over all tables we extend TupleID for aggregation over any previously referenced table in the rule being built. We do this by iteratively propagating not just the IDs of T_t, but the IDs of all the previously referenced tables as well (represented as a comma-delimited ordered string in Fig. 2). This allows the calculation of an aggregation involving **Customer** and **Book,** such as: '*the number of unique customers each book sold to*'. Further propagation, e.g. from **Customer** to **Card**, takes the IDs of **Customer** and **Book**, and moves it along with the AuthorID and class label. The resulting table would contain all information required to aggregate over authors, books and customers. We are now able to detect, for example, '*what is the most frequent credit-card type used to purchase a science fiction book*'.

In order to perform aggregation over any previously referenced table, we simply summarize by the ID of that table and can immediately determine which tuples of the current table are associated to each ID. For example, from Fig. 2, book 2 is associated to customers 2 and 4 and book 3 is associated to customer 1. Aggregation can be performed by aggregating over each BookID. The result is shown in Fig. 3. FOILGain is then applied to this table to determine the best combination of aggregation function, feature(s) and threshold value.

4   Experimental Results

We performed extensive experiments on the Financial and Medical datasets from the PKDD'99[1] and the Hepatitis dataset from the PKDD'02[2] Discovery Challenges. The main objective of our experiments was to demonstrate the gain in classifier performance achievable by SFA and MFA. Three classifiers were built per dataset: the first classifier (EQ) used only the existential quantifier, the second (SFA) used single-feature aggregation and EQ, and the third (MFA) used multi-feature aggregation, SFA and EQ. The experiments evaluated the classification accuracy of MFA against SFA and EQ. 5-fold cross validation was performed to evaluate the classifier performance. The results are presented in Fig. 4.

PKDD'99 Financial Dataset. The dataset contains 606 successful and 76 not successful loans along with their information and transactions. Bad loans were chosen as the target class and the *transaction* table was pruned by removing all transactions which occurred after a loan was approved. EQ achieved very poor precision, between 20% and 45%, similar to [5]. Adding SFA resulted in an increase to 90% in the best case, 60% in the worst. With MFA the precision reached 100% and was still above 90% in the worst case. This gain was not at the expense of recall, as can be seen in the F-Measure results. A sample rule that was found is:

R_1: loan(L,'bad') ← loan(L), max(A,{transaction(T),trans_of_loan(T,L),amount(A,T)},M),M<99.6,
        correlation({B,D},{transaction(T),trans_of_loan(T,L),balance(B,T),date(D,T)},corr), corr<0.143.

According to R_1, a loan is bad if the maximum transaction amount corresponding to this loan is smaller than 99.6 (the average transaction amount in the entire dataset is 9,101), and the correlation of the balance and date is less than 0.143. Intuitively, this

---

[1] http://lisp.vse.cz/pkdd99/
[2] http://lisp.vse.cz/challenge/ecmlpkdd2002/

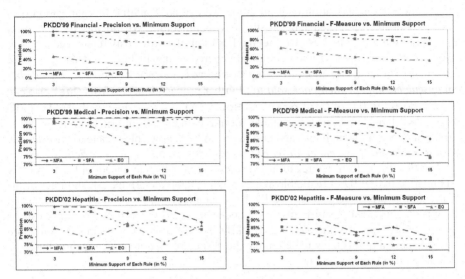

Fig. 4. Precision and F-Measure for the three datasets from PKDD'99 and PKDD'02

indicates that the payments made on the loan are very small and there is no correlation between the balance of the transaction and the date.

PKDD'99 Medical Dataset. Classification was done on the 41 instances of patients with Thrombosis. 56,197 exams, dates, and results, with each exam having 33 numerical results, allowed MFA to perform numerous multi-feature analyses to detect relationships between different features. MFA classification precision was 100% with SFA being above 94% for all minimum support values. EQ was quite competitive for small minimum support values, but for higher values EQ dropped to 82% precision. The gain in classification precision of MFA was not at the expense of recall since MFA consistently also had the highest F-Measure. As anecdotal evidence of the meaningfulness of the rules, for example, the following rule was discovered:

$$R_2: thrombosis(S,'bad') \leftarrow patient(P),slope(\{T,H\},\{exams(E,P),TBIL(T,E),HCT(H,E)\},S),$$
$$S<-4.5,correlation(\{D,B\},\{exams(E,P),RBC(B,E),date(D,E)\},C),C<0.91$$

R_2 states that a person will have thrombosis if the relationship between the results of the TBIL and HCT tests is negatively proportional, and the RBC test values and date are not very highly correlated.

PKDD'02 Hepatitis Dataset. The classifier was built on 206 instances of Hepatitis B (contrasting them against 484 cases of Hepatitis C). The *inhospital* table had to be preprocessed such that each unique test was in a column and all tests for a patient on a given date were in a single tuple. The resulting *inhospital* table had 12,614 tuples and 120 features. Due to the transformation, a lot of columns contained insignificant numbers of non-NULL values, and were removed, leaving only 25 features for classification. MFA consistently outperformed SFA and EQ both in precision and F-Measure. Precision was up to 8% higher than SFA and up to 20% higher than EQ.

5 Conclusions

In this paper we presented a novel method for classification of MR data using an aggregation-based ILP approach. We proposed an ILP framework for representing rules with multi-feature aggregation predicates. The different types of aggregations available for different feature-types, and their combinations, were discussed. For efficient classifier construction, we extended TupleID propagation so that it allows for aggregation over related tables and not only the target table. Experiments on three real-world datasets showed substantial gains in precision and F-Measure compared to existing approaches. Anecdotal evidence was provided to illustrate the rule meaning.

In temporal databases, classification with multi-feature aggregation could provide very interesting rules that are much more meaningful to the end-user by allowing for temporal trends. Another direction is to investigate spatial classification where dependencies between features are prevalent since the spatial relationship of objects impacts their mutual influences. We plan to explore these important applications.

References

1. Appice, A., Ceci, M., Lanza, A.: Discovery of spatial association rules in geo-referenced census data: a relational mining approach. Intelligent Data Analysis 7, 541–566 (2003)
2. Atramentov, A., Leiva, H., Honavar, V.: A multi-relational decision tree learning algorithm -implementation and experiments. In: Horváth, T., Yamamoto, A. (eds.) ILP 2003. LNCS (LNAI), vol. 2835, Springer, Heidelberg (2003)
3. De Raedt, L., Lavrac, N.: Multiple Literal Learning in two Inductive Logic Programming Settings. Journal on Pure and Applied Logic (1996)
4. Dzeroski, S.: Multi-relational data mining: an introduction. SIGKDD Explorations 5(1) (2003)
5. Keim, D.A., Wawryniuk, M.: Identifying Most Predictive Items. In: Workshop on Pattern Representation and Management, Heraklion, Hellas (2004)
6. Knobbe, A.J., Siebes, A., Marseille, B.: Involving Aggregate Functions in Multi-Relational Search. In: Elomaa, T., Mannila, H., Toivonen, H. (eds.) PKDD 2002. LNCS (LNAI), vol. 2431, Springer, Heidelberg (2002)
7. Lu, Q., Getoor, L.: Link-based Text Classification. In: Proceedings of ICML (2003)
8. Neville, J., Jensen, D., Friedland, L., Hay, M.: Learning relational probability trees. In: Proceedings of KDD (2003)
9. Perlich, C., Provost, F.: Aggregation-Based Feature Invention and Relational Concept Classes. In: Proceedings of KDD (2003)
10. Quinlan, J.R., Cameron-Jones, R.M.: FOIL: – A midterm report. In: Brazdil, P.B. (ed.) Machine Learning: ECML-93. LNCS, vol. 667, Springer, Heidelberg (1993)
11. Taskar, B., Segal, E., Koller, D.: Probabilistic classification and clustering in relational data. In: Proceedings IJCAI (2001)
12. Vens, C., Van Assche, A., Blockeel, H., Dzeroski, S.: First order random forests with complex aggregates. In: Camacho, R., King, R., Srinivasan, A. (eds.) ILP 2004. LNCS (LNAI), vol. 3194, Springer, Heidelberg (2004)
13. Yin, X., Han, J., Yang, J., Yu, P.S.: CrossMine: Efficient Classification Across Multiple Database Relations. In: Proceedings of ICDE (2004)

MINI: Mining Informative Non-redundant Itemsets

Arianna Gallo[1], Tijl De Bie[1], and Nello Cristianini[1,2]

[1] University of Bristol, Department of Engineering Mathematics, UK
[2] University of Bristol, Department of Computer Science, UK

Abstract. Frequent itemset mining assists the data mining practitioner in searching for strongly associated items (and transactions) in large transaction databases. Since the number of frequent itemsets is usually extremely large and unmanageable for a human user, recent works have sought to define condensed representations of them, e.g. *closed* or *maximal* frequent itemsets. We argue that not only these methods often still fall short in sufficiently reducing of the output size, but they also output many redundant itemsets. In this paper we propose a philosophically new approach that resolves both these issues in a computationally tractable way. We present and empirically validate a statistically founded approach called MINI, to compress the set of frequent itemsets down to a list of informative and non-redundant itemsets.

1 Introduction

Frequent itemsets (or patterns) mining has been a focused research theme in data mining due to its broad applications at mining association, correlation, sequential patterns, episodes, multidimensional-patterns, max patterns, partial periodicity, emerging patterns, and many other important data mining tasks.

Since their introduction in 1993 in [1], hundreds of new scalable methods have been proposed to solve the mining frequent itemsets problems. Typically, while a too high support threshold leads to generate only commonsense patterns (or none), mining all the itemsets having a low support, or dealing with high correlated data, may generate an explosive number of results, often hard to examine for a user. In order to solve this problem, several methods were proposed to compress (or *summarize*) the set of frequent itemsets, i.e. to find a *concise representation* [2] of the whole collection of patterns. In general, a concise (or condensed) representation must enable to regenerate not only the patterns, but also the values of an evaluation function like the support without accessing the data. If these regenerated values are only approximated, the condensed representation is called *approximate*, *exact* otherwise.

Among these methods, closed [3] [4], non-derivable [5], closed non-derivable itemsets [6] (and 0-Free Sets [7]) have been suggested for finding an exact representation of the data. However, the number of closed itemsets, non-derivable and closed non-derivable itemsets can still be very large, thus an additional effort is essential to allow the user to better understanding the data.

J.N. Kok et al. (Eds.): PKDD 2007, LNAI 4702, pp. 438–445, 2007.

While most of the aforementioned state-of-the-art algorithms make use of the *restoration error* to measure the accuracy of the found patterns, here we develop a new probabilistic and *objective* [8] measure of interestingness. We describe MINI, a new scalable algorithm which discovers interesting and non redundant insights in the data. One of the novelties is that both the computation of the interestingness measure and the redundancy reduction are carried out by considering both the domain of the items and that of the transactions. Moreover, MINI does not require the user to manually choose any parameter, but a value which allows to manage the memory consumption without affecting the quality of the result. The experiments show the efficiency of the MINI algorithm which outputs effectively only the informative and non redundant itemsets that matter.

2 Problem Statement

A transactional database $T = \{t_i\}_{i=1}^n$ consists of a set of n transactions having an unique identifier. Let $\mathcal{I}$ be a set of items $\{i_1, i_2, ..., i_m\}$. A transaction is a couple $t = (tid, X)$ where tid is the transaction identifier and $X \subseteq \mathcal{I}$ is an *itemset*. A transaction $t = (tid, X)$ contains an item (or an itemset) i, if $i \in X$ (or $I \in X$), and we write it $i \in t$ (or $I \in t$) for convenience. For any itemset I, its *tidset* T is defined as the set of identifiers of the transactions containing I. For any itemset I in the database, its *support* is defined as the number of transactions containing I: $supp(I) := |\{t = (tid, X) | I \subseteq X, t \in T\}|$.

Similarly, we can define the support $supp(T)$ of a tidset T being the number of items shared by all the transactions $t \in T$: $supp(T) := |\{i \mid \forall t = (tid, X) \in T, i \in X\}|$.

An itemset is *frequent* in the database if its support in is at least a certain support threshold σ. According to this definition, any subset of a frequent itemset is frequent. This Apriori property leads to an explosive number of frequent patterns. For example, a frequent itemset with c items may generate 2^c sub-itemsets, all of which are frequent. This redundancy can be solved with the introduction of the notion of *closed* itemset, i.e. *an itemset with no frequent superset with the same support.*

Definition 1. *An itemset I is called* closed *if it has no frequent superset with the same support.*

Theorem 1. *The support of non closed itemsets is uniquely determined by the support of the closed itemsets.*

Because of theorem 1, the set of frequent closed itemsets forms a lossless representation of all frequent itemsets. Unfortunately, real applications are often subjected to noise. As long as there is a small noise on the transactions containing an itemset I, hundreds of sub-itemsets can be still generated with different supports. Since their itemsets strongly overlap with each other, these sub-itemsets are considered *redundant*. For this reason, dealing with real domains requires a more sophisticated technique to minimize this redundancy.

3   Measuring Surprise as an Interestingness Measure

Measuring the interestingness of discovered patterns is an active area of data mining research, and there is no widespread agreement on a formal definition. Most itemset summarization methods proposed so far are based on the *restoration error*, which measures the average relative error between the estimated support of a pattern and its true support. In this paper, we introduce a radically different interestingness measure, and show its practical relevance and applicability. In addition to that, we provide an efficient and scalable framework to compute this measure of interestingness of frequent itemsets.

Our interestingness measure is based on hypothesis testing ideas. In particular, we formulate a model for the database that represents the "uninteresting" situation in which no item associations are present, i.e. in which items occur independently from each other in transactions. This model is known as the *null model*. Then, for each itemset discovered, we compute its probability to be "surprising" under the null model. This probability is known as the *p-value* of the itemset. The smaller the p-value, the more surprising (informative, interesting) the itemset is, and the more interesting we consider the itemset to be.

Possible Null Models for the Data. There are two possible null models we consider appropriate. The first possible null model considers all items to be independent random variables. By this we mean that they are contained in any particular transaction with a certain probability $f_{i_k} = \frac{c_{i_k}}{n}$ for item i_k (where $c_{i_k} = supp(i_k)$ is the total item count in the database), independently from the presence or absence of other items. The independence assumption implies that the probability of all items $i \in I$ to be jointly present in a given transaction is equal to $p_I = \prod_{i \in I} f_i$, the product of their individual probabilities. We can then compute the probability that the support of an itemset is exactly $supp(I)$ under the null model by means of the binomial distribution:

$$P(supp(I); n, p_I) = \binom{n}{supp(I)} p_I^{supp(I)} (1 - p_I)^{n - supp(I)}. \tag{1}$$

The p-value for an itemset I under this null model is then computed as the probability to observe an itemset at least as "surprising" I, which we define here as having a support at least as large as $supp(I)$. Given that all items in I occur jointly in a transaction with probability p_I, we can compute the probability of a support larger than or equal to $supp(I)$ out of n transactions by means of the cumulative binomial distribution function, as

$$P_c^I = \sum_{s = supp(I)}^{n} \binom{n}{s} p_I^s (1 - p_I)^{n - s}, \text{ with } p_I = \prod_{i \in I} f_i \tag{2}$$

This is the p-value under the first null model we consider.

A different null model is obtained by considering the transactions as independent random variables, containing each of the items with the same transaction-dependent probability $f_{t_k} = \frac{c_{t_k}}{m}$ for transaction t_k (where $c_{t_k} = supp(t_k)$ is the total number of items in the transaction $\mathbf{t}_k$). Using a reasoning analogous to the above, we obtain the following p-value for a transaction set T:

$$P_c^T = \sum_{s=supp(T)}^{m} \binom{m}{s} p_T^s (1 - p_T)^{m-s}, \text{ with } p_T = \prod_{t \in T} f_{\mathbf{t}} \qquad (3)$$

Note that, because of the Definition 1, $supp(T)$ is equal to the *cardinality* $|I|$ of the itemset I.

The first model takes effective account of the fact that associations between more frequent items are less surprising (even though they may be relevant in terms of restoration error), and hence may not be as interesting to a user (the user would *expect* to see these associations, even by chance). The second model accounts for the fact that two large transactions are likely to share items, even by mere chance, so that again such associations are less interesting. Therefore, for any itemset I supported by transactions T, we define our measure of interestingness as the largest p-value obtained by these null two models:

$$\text{p-value}(I) = \max(P_c^I, P_c^T) \qquad (4)$$

In accordance with Definition 1, it is interesting to note the following fact:

Theorem 2. *For any non-closed itemset there exists a closed itemset that has a lower p-value, i.e. that is more significant.*

Because of this theorem, we can safely restrict our attention to closed itemsets only and sort them by their interestingness instead of considering all the frequent itemsets. Additionally, all the p-values are efficiently and effectively updated in a second step, described in detail in the next sections, further minimizing the redundancy.

4   The MINI Algorithm for Mining Informative Non-redundant Itemsets

The set of closed itemsets represent a concise representation of the set of all frequent itemsets. However, it could still contains redundancy. Several solution to this problem were proposed, such as relaxing the requirement that a supporting transaction contains exactly all the items in the itemset [9], or applying traditional k-means clustering algorithm [10]. In this paper we adopt a completely different approach to address the problem. The list of closed itemsets is sorted by increasing p-value, and then an iterative procedure to this list is applied to updating the p-values ignoring itemsets that are redundant with more highly listed ones. This is done by penalizing itemsets that overlap with highly ranked itemsets by increasing their p-value in a statistically principled way, so that they go down in the sorted list.

Let I_k be the itemset at position k whose p-value is going to be updated. In our framework, all the itemsets $\hat{\mathcal{I}}_{k-1} = \bigcup_k I_k, (k = 1, .., N - 1)$ are said to be *covered*, as well as their respective tidset $\hat{T}_{k-1} = \bigcup_k T_k, (k = 1, .., N - 1)$. Note that for each covered item $\hat{i} \in \hat{I}_{k-1}$, we can always individuate a set of covered tids $\{\hat{t} = (tid, X) | \hat{t} \in \hat{T}_{k-1}, \hat{i} \in X\} = \{\hat{t}\}_{\hat{i}}$, being the tids of the covered itemsets containing i in their set of items.

An itemset I_k could share some items with j covered itemsets. Its success probability p_{I_k} is updated as follows:

$$p'_{I_k} = \sum_j \frac{supp(\hat{I}_j)}{n} \prod_{i \in \hat{I} \backslash I_k} f_i + \frac{\sum_j supp(\hat{I}_j) - supp(I_k)}{n} \prod_{i \in I_k \backslash \hat{I}_{k-1}} f'_i \quad (5)$$

where for each item $i \in I_k$ already covered by some more interesting itemsets, its probabilities of being contained in a transaction (f_i) is updated to be $f'_i = f_i - \frac{|\{\hat{t}\}_{\hat{i}}|}{n}$.

Again, using an analogous to the above, we can update the probability p_T for the other null model discussed so far.

The two p-values in Equations 2 and 3 are then computed using these two updated probabilities as new parameters, and the new interestingness measure for I_k is then chosen accordingly to Equation 4.

An Iterative Algorithm. The sketch of the MINI algorithm is given below. First of all, the set R of closed itemsets is sorted in ascending order by p-values (steps 1-3). Since there are no itemsets more interesting than it, the first itemset is assumed to be covered.

Property 1. The p-value of the itemset I_k is updated considering only the highly listed (i.e. most interesting) $k - 1$ itemsets in the set.

The algorithm starts thus from the *2nd* position (step 4) and, at each step (labeled `current_step`), it updates the p-value of the itemset I_k (accordingly to the Property 1), using the cumulative binomial probability functions in 2 and 3 with the updated probabilities discussed in the last section as parameters. This means that at each step, the iterative algorithm searches for the k-*th* most interesting itemset in the set.

If after the updating of its p-value the itemset I_k still detains the same position k in R, I_k, it is considered to be "covered" as well as its items and tids, and, in the next step, the algorithm will search for the $(k + 1)$-*th* most interesting itemset (steps 13-14). Hence, at the end of the algorithm, the number of covered itemsets is at most $k \le 1 + $ `max_steps`, where `max_steps` is an user defined parameter defining the maximum number of steps to carried out. Usually, the user does not know apriori how many exactly interesting patterns will well summarize the data. In our framework, instead of an exact number of interesting patterns, the user needs to choose only the maximum number `c_max` of itemsets that he/she wants to try to cover, and the number `max_steps` $\ge$ `c_max` of steps to carry out.

```
Algorithm MINI(R, max_steps)
1: for each I ∈ R do
2:      p-value(I) := max{P_c^I(supp(I); n, p_I), P_c^T(supp(T); m, p_T)}
3: sort R by p-values in ascending order
4: k := 2
5: current_step= 1
6: while (current_step<max_steps) do
7:      current_step+=1
8:      I_k := I at position k in R
9:      p-value(I_k) := max{P_c^{I_k}(supp(I_k); n, p'_I), P_c^{T_k}(supp(T); m, p'_T)}
10:     last_updated_itemset:= I_k
11:     update ordering in R
12:     I_k := I at position k in R
13:     if (I_k ==last_updated_itemset) do
14:         k+=1
```

5 Experiments and Results

All the experiments were performed on a 1.60GHz Intel(R) Pentium(R)M with 512 MB of memory. We show the performance of our methods on 5 datasets. The first 2 consist of news articles in different periods: one month (August 2006), one year (Year2006). Another dataset, Iraq was constructed to contain news articles with keyword query "iraq". We will also show experiments performed on two other different kinds of dataset, containing all the titles and titles+abstracts of the papers published at the PKDD conferences of the last years (from 2000 to 2006), labeled TitlesPkdd and AbstractsPkdd. We run a closed itemsets mining algorithm to extract all the frequent closed itemsets with different support thresholds for each dataset. Details of the datasets are shown in Figure 1. In

	rows	different words	min_supp	closeds
Year2006	65754	43100	0.001	5103
August 2006	17085	23630	0.0018	5522
Iraq	5069	8593	0.001	45396
AbstractsPkdd	423	4693	0.012	51494
TitlesPkdd	423	1069	0.012	143

Fig. 1. Details of the datasets used in the experiments

Figure 2 are shown some experimental results. Figure 2(a) shows the number of covered itemsets (on the left) and the number of penalized itemsets at each step (on the right). Looking at the results, we can deduce that the ordered closed itemsets set of August2006 contains several very disjoined (i.e. non redundant) patterns in its first positions. On the other hand, the Year2006 initial list contains a lot of redundancy. Indeed, the slope of the functions of August2006 is higher than that of Year2006 in Figure 2(a) on the left, while it is lower in Figure 2(a) on the right.

We also evaluated the running time of MINI w.r.t. the parameters c_max (with dataset Year2006) and max_steps (with dataset AbstractsPkdd) (Figure 2(b)). As we already pointed out in Section 4, the number of closed itemsets is often very large, so much so that it is hard to store entirely in the memory. One solution

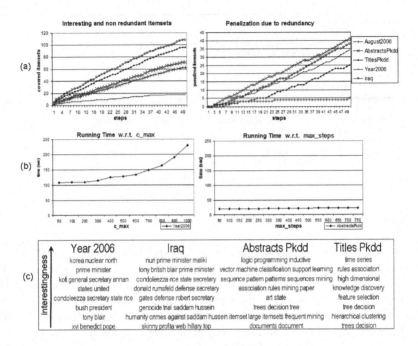

Fig. 2. Experimental Results

could be choosing a higher support threshold σ. However, not only σ is a parameter typically hard to choose a priori by the user, but it could also delete some interesting itemsets. In general, it is desirable to mine all closed itemsets without any support threshold and then perform the summarization with our interesting measure, which we claim is often more appropriate than a support threshold.

However, even if the set of closed itemsets is very large, MINE allows the user to manage the memory consumption, reducing the running time without affecting the summarization quality. In the experiments in Figure 2(b) on the left, the algorithm carries out 50 steps and covers always 17 itemsets, while the c_max parameter is varied. The running time grows exponentially with c_max, while the summarization result does not change.

In the experiments shown in Figure 2(b) (right) we varied max_steps on the dataset AbstractsPkdd. Here, the c_max was chosen to be equal to 1200. Note that, in this case, the summarization result varies, from 9 itemsets covered with 50 steps, to 53 in 750 steps. However, because of the MINI property 1, these results differ from each other only in their size. In each of these results, at each position the same itemset is returned, as well as its measure.

In Figure 2(c) the 8 most interesting and non redundant itemsets found in Year2006, Iraq, AbstractsPkdd, and TitlesPkdd are shown. We included the Iraq example to illustrate an advantage of our method over methods that focus on the restoration error. Such methods would usually be biased towards the *more frequent itemsets*, which in this case are likely to include the word iraq, perhaps even as an itemset of size 1. We argue that, in order to discover *interesting insights*, the results of such techniques are often suboptimal: arguably the itemset `iraq` is not interesting considering the database we are looking at. In our framework, this pattern is automatically considered as *not* interesting and *not* surprising, as it occurs in each row of the dataset.

6   Conclusions

We propose MINI, a two-steps algorithm to mine interesting and non redundant itemsets from a database. In the first step, the cumulative binomial probability of each itemset is computed in both the items and transactions domains independently from the other itemsets in the set. In its second step, the MINI algorithm further reduces the redundancy updating their p-values, penalizing the itemsets which share some items and/or tids with more interesting itemsets. The experiments show that the MINI mines the interesting and non redundant itemsets from a dataset, *without requiring the user to define any parameter*, but `c_max` to reduce the memory consumption and `max_steps`, which can be chosen arbitrarily high without affecting the quality of the summarization. This makes MINI very scalable and thus applicable to many other interesting real world tasks.

Acknowledgment. This work was partially supported by NIH grant R33HG00 3070-01, and the EU Project SMART.

References

1. Agrawal, R., Imieliski, T., Swami, A.: Mining association rules between sets of items in large databases. In: SIGMOD, pp. 207–216. ACM Press, New York (1993)
2. Mannila, H., Toivonen, H.: Multiple uses of frequent sets and condensed representations. In: KDD, Portland, USA, pp. 189–194 (1996)
3. Pasquier, N., Bastide, Y., Taouil, R., Lakhal, L.: Efficient mining of association rules using closed itemset lattices. 24(1), 25–46 (1999)
4. Chi, Y., Wang, H., Yu, P.S., Muntz, R.R.: Moment: Maintaining closed frequent itemsets over a stream sliding window. In: Perner, P. (ed.) ICDM 2004. LNCS (LNAI), vol. 3275, Springer, Heidelberg (2004)
5. Calders, T., Goethals, B.: Mining all non-derivable frequent itemsets, pp. 74–85. Springer, Heidelberg (2002)
6. Muhonen, J., Toivonen, H.: Closed non-derivable itemset. In: Fürnkranz, J., Scheffer, T., Spiliopoulou, M. (eds.) PKDD 2006. LNCS (LNAI), vol. 4213, pp. 601–608. Springer, Heidelberg (2006)
7. Boulicaut, J.F., Bykowski, A., Rigotti, C.: Free-sets: A condensed representation of boolean data for the approximation of frequency queries. Data Min. Knowl. Discov. 7(1), 5–22 (2003)
8. Geng, L., Hamilton, H.J.: Interestingness measures for data mining: A survey. ACM Comput. Surv. 38(3), 9 (2006)
9. Yang, C., Fayyad, U., Bradley, P.S.: Efficient discovery of error-tolerant frequent itemsets in high dimensions. In: SIGKDD, pp. 194–203. ACM Press, New York (2001)
10. Yan, X., Cheng, H., Han, J., Xin, D.: Summarizing itemset patterns: a profile-based approach. In: 11th ACM SIGKDD, pp. 314–323. ACM Press, New York (2005)

Stream-Based Electricity Load Forecast

João Gama[1,2] and Pedro Pereira Rodrigues[1,3]

[1]LIAAD - INESC Porto L.A.
[2]Faculty of Economics, University of Porto
[3]Faculty of Sciences, University of Porto
Rua de Ceuta, 118 - 6 andar, 4050-190 Porto, Portugal
`jgama@fep.up.pt pprodrigues@fc.up.pt`

Abstract. Sensors distributed all around electrical-power distribution networks produce streams of data at high-speed. From a data mining perspective, this sensor network problem is characterized by a large number of variables (sensors), producing a continuous flow of data, in a dynamic non-stationary environment. Companies make decisions to buy or sell energy based on load profiles and forecast. We propose an architecture based on an online clustering algorithm where each cluster (group of sensors with high correlation) contains a neural-network based predictive model. The goal is to maintain in real-time a clustering model and a predictive model able to incorporate new information at the speed data arrives, detecting changes and adapting the decision models to the most recent information. We present results illustrating the advantages of the proposed architecture, on several temporal horizons, and its competitiveness with another predictive strategy.

1 Motivation

Electricity distribution companies usually set their management operators on SCADA/DMS products (Supervisory Control and Data Acquisition / Distribution Management Systems). Load forecast is a relevant auxiliary tool for operational management of an electricity distribution network, since it enables the identification of critical points in load evolution, allowing necessary corrections within available time. In SCADA/DMS systems, the load forecast functionality has to estimate, for different horizons, certain types of measures which are representative of system's load: active power, reactive power and current intensity. Given its practical application and strong financial implications, electricity load forecast has been targeted by innumerous works, mainly relying on the non-linearity and generalizing capacities of neural networks (ANN), which combine a cyclic factor and an auto-regressive one to achieve good results [4]. Nevertheless, static iteration-based training, usually applied to train ANN, is not adequate for high speed data streams. On current real applications, data is being produced in a continuous flow at high speed. In this context, faster answers are usually required, keeping an anytime model of the data, enabling better decisions. Moreover, a predictive system may be developed to serve a set of thousands of load

J.N. Kok et al. (Eds.): PKDD 2007, LNAI 4702, pp. 446–453, 2007.

sensors, but the load demand values tend to follow a restricted number of profiles, considerably smaller than the total set of sensors. Clustering of sensors greatly allows the reduction of necessary predictive models. However, most work in data stream clustering has been concentrated on example clustering and less on variable clustering [7].

The paper is organized as follows. In the next section we present the general architecture of the system, main goals and preprocessing problems with sensors data, the clustering module, and the incremental predictive models. Section 3 presents the evaluation methodology and preliminary results using real data from an electricity network. Last section resumes the lessons learned and future work.

2   General Description

The main objective of this work is to present an incremental system to continuously predict in real time the electricity load demand, in huge sensor networks. The system must predict the value of each individual sensor with a given temporal horizon, that is, if at moment t we receive an observation of all network sensors, the system must execute a prediction for the value of each variable (sensor) for the moment $t + k$. In this scenario, each variable is a time series and each new example included in the system is the value of one observation of all time series for a given moment. Our approach is to first cluster the sensors using an online data stream sensor clustering algorithm, and then associate to each cluster a ANN trained incrementally with the centroid of the cluster. Overall, the system predicts all variables in real time, with incremental training of ANN and continuous monitoring of the clustering structure.

Pre-processing Data. The electrical network spreads out geographically. Sensors send information at different time scales and formats: some sensors send information every minute, others send information each hour, etc.; some send the absolute value of the variable periodically, while others only send information when there is a change in the value of the variable. Sensors act in adversary weather and battery conditions. The available information is noisy. To reduce the impact of noise, missing values, and different granularity, data is aggregated and synchronized in time windows of 15 minutes. This is done in a server, in a pre-processing stage, and was motivated by the fact that it allows to instantiate sensor values for around 80% of the sensors. Data comes in the form of tuples: $< date, time, sensor, measure, value >$. All pre-processing stages (agglomeration and synchronization) require one single scan over the incoming data.

Incremental Clustering of Data Streams. Data streams usually consist of variables producing examples continuously over time at high-speed. The basic idea behind clustering time series is to find groups of variables that behave similarly through time. Applying variable clustering to data streams, requires to incrementally compute dissimilarities. The goal of an incremental clustering system for streaming time series is to find (and make available at any time t) a partition of the streams, where streams in the same cluster tend to be more alike than streams in different clusters. In electrical networks there are clear clusters of demands (like sensors placed near towns or in countryside) which evolve smoothly over time. We believe that a top-down hierarchical approach to the clustering problem is the most appropriate as we do not need to define *a priori* the number of clusters and allow an analysis at different granularity levels.

The system uses the ODAC clustering algorithm which includes an incremental dissimilarity measure based on the correlation between time series, calculated with sufficient statistics gathered continuously over time. There are two main operations in the hierarchical structure of clusters: *expansion* that splits one cluster into two new clusters; and *aggregation* that aggregates two clusters. Both operators are based on the diameters of the clusters, and supported by confidence levels given by the Hoeffding bounds. The main characteristic of the system is the monitoring of those diameters. In ODAC, the dissimilarity between variables a and b is measured by an appropriate metric, the $rnomc(a,b) = \sqrt{(1 - corr(a,b))/2}$, where $corr(a,b)$ is the Pearson's correlation coefficient. More details can be found in [7]. For each cluster, the system chooses two variables that define the diameter of that cluster (those that are less correlated). If a given heuristic condition is met on this diameter, the system splits the cluster in two, assigning each of those variables to one of the two new clusters. Afterwards, the remaining variables are assigned to the cluster that has the closest pivot (first assigned variables). The newly created leaves start new statistics, assuming that only the future information will be useful to decide if the cluster should be split.

A requirement to process data streams is change detection. In electrical networks and for long term conditions, the correlation structure evolves smoothly. The clustering structure must adapt to this type of changes. In a hierarchical structure of clusters, considering that the data streams are produced by a stable concept, the intra-cluster dissimilarity should decrease with each split. For each given cluster C_k, the system verifies if older split decision still represents the structure of data, testing the diameters of C_k, C_k's sibling and C_k's parent. If diameters are increasing above parent's diameter, changes have occurred, so the system aggregates the leaves, restarting the sufficient statistics for that group.

The presented clustering procedure is oriented towards processing high speed data streams. The main characteristics of the system are constant memory and constant time in respect to the number of examples. In ODAC, system space complexity is constant on the number of examples, even considering the infinite amount of examples usually present in data streams. An important feature of this algorithm is that every time a split is performed on a leaf with n variables, the global number of dissimilarities needed to be computed at the next iteration diminishes at least $n - 1$ (worst case scenario) and at most $n/2$ (best case scenario). The time complexity of each iteration of the system is constant given the number of examples, and decreases with every split occurrence. Figure 1 presents the resulting hierarchy of the clustering procedure.

Incremental Learning of ANN. In this section we describe the predictive module of our system. Each group defined by the cluster structure has a feed-forward MLP ANN attached, which was initially trained with a time series representing

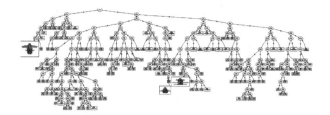

Fig. 1. ODAC hierarchy in the Electrical Network ($\sim$2500 sensors)

the global load of the sensor network, using only past data. The ANN is incrementally trained with incoming data, being used to predict future values of all the sensors in the cluster.

At each moment t, the system executes two actions: one is to predict the moment $t + k$; the other is to back-propagate in the model the error, obtained by comparing the current real value with the prediction made at time $t - k$. The error is back-propagated through the network only once, allowing the system to cope with high speed streams. Although the system builds the learning model with the centroid of the group, the prediction is made for each variable independently. Every time a cluster is split, the offspring clusters inherit the parent's model, starting to fit a different copy separately. This way, a specification of the model is enabled, following the specification of the clustering structure. When an aggregation occurs, due to changes in the clustering structure, the new leaf starts a new predictive model.

The goal of our system is to continuously maintain a prediction for three time horizons: next hour, one day ahead, and one week ahead. This means that after a short initial period, we have three groups of predictions: prediction for the next hour, 24 predictions for the next 24 hours, and 168 predictions for the next week. For the purposes of this application in particular, all predictions are hourly based. For all the horizon forecasts, the clustering hierarchy is the same but the predictive model at each cluster may be different.

The strucure of the MLP consists of 10 inputs, 4 hidden neurons (tanh-activated) and a linear output. The input vector for next hour prediction at time t is t minus $\{1, 2, 3, 4\}$ hours and t minus $\{7, 14\}$ days. As usual [6], we consider also 4 cyclic variables, for hourly and weekly periods (*sin* and *cos*). The choice of the networks topology and inputs was mainly motivated by experts suggestion, autocorrelation analysis and previous work with batch approaches [4]. One implication of the chosen inputs is that we no longer maintain the property of processing each observation once[1]. Thus, we introduce a buffer (window with the most recent values) strategy. The size of the buffer depends on the horizon forecast and data granularity and is at most two weeks. Figure 2 presents a general description of the procedure executed at each new example.

ANNs are powerful models that can approximate any continuous function [2] with arbitrary small error with a three layer network. The *mauvaise reputation*

---

[1] A property that the clustering algorithm satisfies.

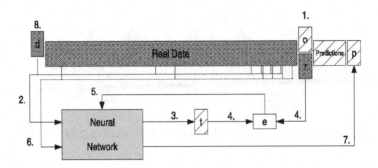

Fig. 2. Buffered Online Predictions: 1. new real data arrives (r) at time stamp i, substituting previously made prediction (o); 2. define the input vector to predict time stamp i; 3. execute prediction (t) for time stamp i; 4. compute error using predicted (t) and real (r) values; 5. back-propagate the error one single time; 6. define input vector to predict time stamp i plus one hour; 7. execute prediction of next hour (p); 8. discard oldest real data (d).

of ANNs comes from slower learning times. Two other known problems of the generalization capacity of neural networks are overfitting and large variance. In our approach the impact of overfitting is reduced due to two main reasons. First we use a reduced number of neurons in the hidden layer. Second, each training example is propagated and the error backpropagated through the network only once, as data is abundant and flow continuously. This is a main advantage, allowing the neural network to process an infinite number of examples at high speed. Another advantage is the smooth adaptation in dynamic data streams where the target function evolves over time. Craven and Shavlik [2] argue that the inductive bias of neural networks is the most appropriate for sequential and temporal prediction tasks. However, this flexibility implies an increase on error variance. In stationary data streams the variance shrinks when the number of examples goes to infinity. In dynamic environments where the target function changes, the variance of predictions is problematic. An efficient variance reduction method is the *dual perturb and combine* [3] algorithm. It consists on perturbing each test example several times, adding white noise to the attribute-values, and predicting each perturbed version of the test example. The final prediction is obtained by aggregating (usually by averaging) the different predictions. The method is directly applicable in the stream setting because multiple predictions only involves test examples. We use the *dual perturb and combine* algorithm to reduce the variance exhibited by neural networks and to estimate a confidence for predictions. For continuous and derivable functions over time one simple prediction strategy, reported elsewhere to work well, consists of predicting for time t the value observed at time $t - k$. A study on the autocorrelation in the time series used to train the scratch neural network reveals that for next hour forecasts $k = 1$ is the most autocorrelated value, while for next day and next week the most autocorrelated one is the corresponding value one week before ($k = 168$). The Kalman filter is widely used in engineering for two main purposes: for combining measurements of the same variables but from different sensors, and for combining an inexact forecast of system's state with an inexact measurement of the state [5]. We use Kalman filter to combine the neural network forecast with the observed value at time $t - k$, where k depends on the horizon forecast as defined above. The one dimensional Kalman filters works by considering $\hat{y}_i = \hat{y}_{i-1} + K(y_i - \hat{y}_{i-1})$, where $\sigma_i^2 = (1 - K)\sigma_{i-1}^2$ and $K = \frac{\sigma_{i-1}^2}{\sigma_{i-1}^2 + \sigma_r^2}$.

3   Experimental Evaluation

The electrical network we are studying contains more than 2500 sensors spreaded out over the network, although some of them have no predictive interest. The measure of interest is current intensity (I). There are 565 High Tension (HT) sensors, 1629 Mean Tension (MT) sensors, and 299 Power Transformers (PT) sensors. We consider around three years of data, aggregated in an hourly basis, unless a fault was detected. The analysis of results were aggregated by month. The system makes forecasts for next hour, one day ahead, and one week ahead. At each time point t, the user can consult the forecast for next hour, next 24 hours and all hours for the next week. The design of the experimental evaluation in streams is not an easy task. For each point in time and sensor we have an estimate of the error. This estimate evolves over time. To have insights about the quality of the model these estimates must be aggregated. In this particular application, there are natural time windows for aggregations: week windows and month windows. For all time horizons, we aggregate the error estimates by month and type of sensor, for a one year test period. The quality measure usually considered in electricity load forecast is the MAPE *(Mean Absolute Percentage Error)* defined as $MAPE = \sum_{i=1}^{n} \frac{|(\hat{y}_i - y_i)/y_i|}{n}$, where y_i is the real value of variable y at time i and $\hat{y}_i$ is the corresponding predicted value. In this work, we prefer to use as quality measure the MEDAPE *(Median Absolute Percentage Error)* to reduce sensibility to outliers [1]. Table 1 presents global results for predicting the load over all horizons and on all sensors. We can stress that the system is stable over time, with acceptable performance. All experiments reported here ran in a AMD Athlon(tm) 64 X2 Dual Core Processor 3800+ (2GHz). The system processes around 30000 points per second with a total running time for all the experiments reported here of about 1 hour. For a 24 hours forecast, electricity load demand has a clear daily pattern, where we can identify day and night, lunch and dinner time. For a single forecast at time t, the historical inputs are: $t - \{24h, (168 - 24)h, 168h, 169h, (168 + 24)h, 336h\}$. The results for the 24 hours ahead forecast are also presented in Table 1. In comparison with the one hour forecast, the level of degradation in the predictions is around 2-3%. For the one week ahead load forecast, the standard profile is also well defined: five quite similar week days, followed by two weekend days. As for the 24 hours forecast, several strategies could be designed for one week ahead forecast. Our lab experiments pointed out consistent

Table 1. Median of MEDAPE for all sensors by month, for three different horizons

1 Hour Ahead					1 Day Ahead					1 Week Ahead			
HT	MT	PT	All		HT	MT	PT	All		HT	MT	PT	All
I, %					I, %					I, %			
Jan 4.34	4.98	4.60	4.63		Jan 6.61	6.52	6.98	6.44		Jan 5.95	6.10	6.50	5.95
Feb 4.24	5.07	4.74	4.73		Feb 6.97	6.83	7.14	6.73		Feb 6.81	6.87	6.95	6.68
Mar 4.24	5.01	4.60	4.66		Mar 7.23	7.09	7.74	7.03		Mar 7.14	7.49	7.69	7.27
Apr 4.46	5.38	5.08	4.98		Apr 8.05	7.71	9.12	7.75		Apr 7.13	7.31	8.10	7.17
May 3.90	4.77	4.45	4.38		May 6.79	6.42	7.46	6.41		May 5.57	6.17	6.32	5.97
Jun 3.93	4.91	4.58	4.55		Jun 7.23	7.21	8.56	7.21		Jun 6.22	6.79	7.02	6.58
Jul 3.87	4.62	4.25	4.26		Jul 7.13	6.98	7.77	6.95		Jul 7.02	7.38	7.40	7.11
Aug 3.68	4.30	3.89	3.98		Aug 6.97	6.20	7.06	6.22		Aug 7.99	8.11	9.10	7.96
Sep 4.33	4.93	4.42	4.59		Sep 6.99	6.83	7.46	6.80		Sep 6.14	6.69	6.86	6.46
Oct 4.50	5.19	4.67	4.84		Oct 8.03	7.38	8.25	7.41		Oct 6.41	6.40	6.94	6.31
Nov 3.89	4.66	4.32	4.37		Nov 7.17	6.87	7.86	6.87		Nov 6.39	5.97	6.49	5.91
Dec 4.34	5.18	4.65	4.84		Dec 8.62	8.02	8.73	7.96		Dec 9.02	8.58	8.85	8.48

advantages using the simplest strategy of a single forecast using the historical inputs $t-\{168h, 169h, (336-24)h, 336h, (336+24)h, (336+168)h\}$. The results for the one week ahead forecast are also presented in Table 1. Again, when comparing these results with one hour ahead forecast, one can observe a degradation of around 2%. At this point we can state that our strategy roughly complies with the requirements presented by the experts. To assess the quality of prediction, we have compared with another predictive system. For the given year, the quality of the system in each month is compared with Wavelets [8][2] on two precise variables each month, chosen as relevant predictable streams (by an expert) but exhibiting either low or high error. Results are shown on Table 2, for the 24 variables, over the three different horizons. The Wilcoxon signed ranks test was applied to compare the error distributions, and the corresponding *p-value* is shown (we consider a significance level of 5%). The relevance of the incremental system using neural networks is exposed, with lower error values on the majority of the studied variables. Moreover, it was noticed an improvement on the performance of the system, compared to the predictions made using Wavelets, after failures or abnormal behavior in the streams. Nevertheless, weaknesses arise that should be considered by future work.

4　Conclusions and Future Issues

This paper introduces a system that gathers a predictive model for a large number of sensors data within specific horizons. The system incrementally constructs a hierarchy of clusters and fits a predictive model for each leaf. Experimental results show that the system is able to produce acceptable predictions for different horizons. Focus is given by experts on overall performance of the complete system. The main contribution of this work is the reduction of the human effort needed to maintain the predictive models over time, eliminating the batch cluster analysis and the periodic ANN training, while keeping the forecast quality at competitive levels. Directions for future work are the inclusion of background knowledge such as temperature, holiday, and special events, into the learning process. Moreover, sensor network data is distributed in nature, suggesting the study of ubiquitous and distributed computation.

---

[2] Wavelets are the standard method used in the company we are working with.

Table 2. MEDAPE for selected variables of current intensity (I), exhibiting low or high error. Comparison with Wavelets is considered for the three different horizons.

	1 Hour Ahead				1 Day Ahead				1 Week Ahead		
	Wav %	NN %	NN-Wav %	p-value	Wav %	NN %	NN-Wav %	p-value	Wav %	NN %	NN-Wav %	p-value
Low Error					**Low Error**				**Low Error**			
Jan	1.69	2.72	**1.03**	<0.001	3.50	3.98	**0.48**	<0.001	7.07	3.81	**-3.26**	<0.001
Feb	2.99	2.79	-0.20	0.196	7.89	5.62	**-2.27**	<0.001	7.18	4.28	**-2.90**	<0.001
Mar	3.63	2.75	**-0.88**	<0.001	6.11	6.38	**0.27**	<0.001	5.33	3.99	**-1.34**	<0.001
Apr	2.05	2.58	**0.53**	0.002	8.04	5.45	**-2.60**	<0.001	14.68	4.74	**-9.94**	<0.001
May	2.69	2.28	**-0.41**	<0.001	19.47	7.63	**-11.84**	<0.001	14.16	4.05	**-10.11**	<0.001
Jun	2.33	2.52	0.29	0.051	3.68	4.26	**0.58**	0.002	4.41	3.40	**-1.01**	<0.001
Jul	2.14	2.12	-0.02	0.049	5.83	5.61	-0.22	<0.001	7.13	4.45	**-2.68**	<0.001
Aug	2.59	2.54	-0.05	0.537	6.14	3.64	**-2.50**	<0.001	4.73	5.96	**1.23**	0.008
Sep	2.65	2.64	-0.01	0.374	7.57	7.65	0.08	0.835	10.03	3.73	**-6.30**	<0.001
Oct	2.28	2.36	0.08	0.127	7.05	8.77	**1.73**	0.001	6.28	7.34	**1.06**	0.010
Nov	2.41	2.14	-0.27	0.085	4.08	4.52	**0.44**	0.047	3.15	4.06	**0.91**	0.003
Dec	3.56	2.97	**-0.59**	0.029	9.92	5.70	**-4.23**	<0.001	14.02	7.02	**-7.00**	<0.001
High Error					**High Error**				**High Error**			
Jan	9.04	10.34	**1.30**	<0.001	9.04	10.34	**1.30**	<0.001	19.73	14.91	**-4.82**	<0.001
Feb	8.51	9.82	**1.31**	0.002	8.51	9.82	**1.31**	0.002	9.95	10.54	0.59	0.053
Mar	11.52	11.28	-0.24	0.166	11.52	11.28	-0.24	0.166	32.18	28.95	**-3.23**	<0.001
Apr	9.36	12.74	**1.38**	<0.001	9.36	12.74	**1.38**	<0.001	18.22	17.93	-0.30	0.074
May	12.89	10.54	**-2.35**	0.035	12.89	10.54	**-2.35**	0.035	14.65	10.43	**-4.22**	<0.001
Jun	6.68	8.10	**1.42**	<0.001	6.68	8.10	**1.42**	<0.001	8.96	8.11	-0.86	0.373
Jul	14.52	10.68	**-3.84**	<0.001	14.52	10.68	**-3.84**	<0.001	32.68	21.12	**-11.56**	<0.001
Aug	11.11	12.27	**1.16**	0.034	11.11	12.27	**1.16**	0.034	13.19	14.28	1.09	0.062
Sep	10.52	9.81	-0.71	0.656	10.52	9.81	-0.71	0.656	30.58	21.71	**-8.87**	<0.001
Oct	12.45	11.25	**-1.20**	0.002	12.45	11.25	**-1.20**	0.002	29.44	24.65	**-4.79**	0.009
Nov	8.85	7.71	-1.14	0.356	8.85	7.71	-1.14	0.356	17.19	12.46	**-4.72**	<0.001
Dec	11.76	10.91	**-0.85**	0.040	11.76	10.91	**-0.85**	0.040	38.26	45.08	6.82	0.056

Acknowledgement. Thanks to Paulo Santos from EFACEC, the financial support of projects ALES II (POSC/EIA/55340/2004) and RETINAE (PRIME/ IDEIA/70/00078). PPR thanks FCT for a PhD Grant (SFRH/BD/29219/2006).

References

1. Armstrong, J.S., Collopy, F.: Error measures for generalizing about forecasting methods: empirical comparisons. Int. Journal of Forecasting 8, 69–80 (1992)
2. Craven, M., Shavlik, J.W.: Understanding time-series networks: a case study in rule extraction. International Journal of Neural Systems 8(4), 373–384 (1997)
3. Geurts, P.: Dual perturb and combine algorithm. In: Proceedings of the Eighth International Workshop on Artificial Intelligence and Statistics pp. 196–201 (2001)
4. Hippert, H.S., Pedreira, C.E., Souza, R.C.: Neural networks for short-term load forecasting: a review and evaluation. IEEE Transactions on Power Systems 16(1), 44–55 (2001)
5. Kalman, R.E.: A new approach to linear filtering and prediction problems. In: Transaction of ASME - Journal of Basic Engineering, pp. 35–45 (1960)
6. Khotanzad, A., Afkhami-Rohani, R., Lu, T.-L., Abaye, A., Davis, M., Maratukulam, D.J.: ANNSTLF – A neural-network-based electric load forecasting system. IEEE Transactions on Neural Networks 8(4), 835–846 (1997)
7. Rodrigues, P.P., Gama, J., Pedroso, J.P.: ODAC: Hierarchical clustering of time series data streams. In: SIAM Int. Conf. Data Mining pp. 499–503 (2006)
8. Rauschenbach, T,: Short-term load forecast using wavelet transformation. In: Proceedings of Artificial Intelligence and Applications (2002)

Automatic Hidden Web Database Classification

Zhiguo Gong, Jingbai Zhang, and Qian Liu

Faculty of Science and Technology
University of Macau
Macao, PRC
{fstzgg,ma46597,ma46620}@umac.mo

Abstract. In this paper, a method for automatic classification of Hidden-Web databases is addressed. In our approach, the classification tree for Hidden Web databases is constructed by tailoring the well accepted classification tree of DMOZ Directory. Then the feature for each class is extracted from randomly selected Web documents in the corresponding category. For each Web database, query terms are selected from the class features based on their weights. A hidden-web database is then probed by analyzing the results of the class-specific query. To raise the performance further, we also use Web pages which have links pointing to the hidden-web database (HW-DB) as another important source to represent the database. We combine link-based evaluation and query-based probing as our final classification solution. The experiment shows that the combined method can produce much better performance for classification of hidden Web Databases.

1 Introduction

With the explosive growth of the World Wide Web, the traditional Crawlers fail to satisfy the users' demand for information searching yet. Many recent studies [1, 2] have observed that a significant fraction of Web content known as the Hidden-Web (HW) [3], the Invisible Web [4], or the Deep Web [2], lies outside the PIW. In fact, these pages can only be dynamically generated in response to users' queries, which the traditional Crawlers cannot handle. However, we cannot simply ignore them, because some recent studies claim that the size of the Hidden-Web pages are as many as 500 billion pages, comparing to "only" two billion pages of the ordinary web [5]. Furthermore, the information on the HW is usually generated from structured databases, which are referred to as Hidden-Web Databases (HW-DB) [6]. In [7], the study has estimated that there are 250,000 private databases, and the access of 95% of them is free. These databases represent 54% of the Hidden Web.

In this paper, in order to effectively guide users to find the relevant information from such databases, we present a prototype system for classifying the HW-DB into a predefined category hierarchy which is tailored from some existing classification tree for Web documents. The feature for each class is extracted from randomly selected Web documents in corresponding Web class. For each Web database, query terms are selected from such class features based on their weights. A hidden-web database is

J.N. Kok et al. (Eds.): PKDD 2007, LNAI 4702, pp. 454–461, 2007.
© Springer-Verlag Berlin Heidelberg 2007

then probed by analyzing the results of the class-specific query to the hidden database. To raise the performance further, we also use Web pages which have links pointing to the hidden database as another important source to represent the database. We combine link-based evaluation and query-based probing as our final classification solution for hidden database classification. In addition, our focus is on text databases, since 84% of all searchable databases on the web are estimated to provide access to text documents [2], and other kinds of databases like image or video databases are out of the scope of this paper.

The contributions presented in this article are organized as follows. We present the details of our HW-DB classification system based on query probing and based on link evaluation in Section 2. A system evaluation is conducted and important experimental results are discussed in Section 3. And finally section 4 provides conclusions.

2   Hidden Databases Classification

Our system aims to automatically assign each Hidden-Web Database to the "best" category or categories of the classification scheme. Instead of constructing a new classification scheme manually, similar to the approaches proposed by [5, 7], we exploit a category hierarchy for HW-DB classification from the popular DMOZ Directory. Fig.1. shows a fraction of the category hierarchy used in our system.

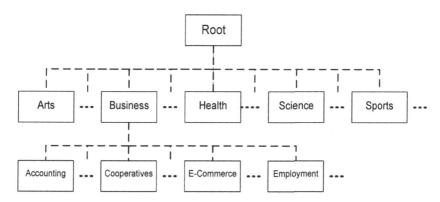

Fig. 1. A fraction of the category hierarchy for HW-DB used in our system

2.1   Classification Models

In order to assign a Web document to corresponding categories, a classifier algorithm is needed. Several classifier models exist in literature, such as SVM (Support Vector Machine) [8], kNN (key nearest Neighbor) [9], LLSF (Linear Least Square Fit) [10], NNet (Neural Network) [11], NB (Naïve Bayes) [11], and RIPPER [12]. Though [6] shows that RIPPER can provide good overall performance for HW-DB classification, the correctness of the rules for each category are critical for the precision of the classifications. However, it is a hard work for correct rule extractions. Take the rule ("ibm" AND "computer" →"Computer") as an example, even though some document may contain both "ibm" and "computer", it may not belong to the category of "Computer" in many cases. Furthermore, the classification needs to extensively interact with a HW-DB. That means, for each rule the system needs to interact at least one time with the database. Y. Yang and X. Liu compared other classifiers and pointed out that model SVM, kNN and NB can always produce better performance for the document classifications over LLSF and NNet. Considering both effectiveness and efficiency as the important factors, in our work, we employ kNN as the classifier for HW-DB classifications.

To use kNN, training documents for each category are needed. In our implementation, for any category c_j, we select N Web documents $(d_{j,1}, d_{j,2}, ..., d_{j,N})$ from the corresponding DMOZ directory. Then, for any Web document x, the classification rule in kNN can be written as:

$$Similarity(x, c_j) = \sum_{1 \le i \le N} sim(x, d_{j,i}) - b_j \qquad (1)$$

where $sim(x, d_{j,i})$ is the metric of similarity between x and $d_{j,i}$, b_j is the category specific threshold for the binary classification.

In order to calculate $sim(x, d_{j,i})$, we represent each Web document d as a vector $(w_1, w_2, ... w_M)$, where w_l is the weight of term t_l in d. And w_l is defined as:

$$w_l = \frac{tf(t_l, d)}{Max_{\{n\}}\{tf(t_n, d)\}} \qquad (2)$$

where $tf(t_l, d)$ is the frequency of term t_l in d. With this definition, for any two web documents $d = (w_1, w_2, ... w_M)$ and $d' = (w'_1, w'_2, ... w'_M)$, $sim(d, d')$ is defined as the cosine value between them:

$$sim(d, d') = \frac{\sum w_l * w'_l}{\sqrt{\sum (w_l)^2} \sqrt{\sum (w'_l)^2}}. \qquad (3)$$

The original kNN classifier is designed for document classification. For a hidden Web database HD, let $\{hd_1, hd_2, ..., hd_R\}$ be all the documents contained in HD. We concatenate all the documents of HD into one document, still denoted as HD. That is, $HD = \cup hd_j$. Then, HD is assigned to category c if and only if $Similarity(HD, c) > 0$. However, we do not have the knowledge about the documents within HD. To solve the problem, we use two techniques to approximate it in this study. Firstly, we detect the HW-DB through probing; secondly, link structure of the Web is used.

2.2 Hidden Database Probing

In each category, some queries are needed for probing hidden databases. [6] uses extensive number of rules or queries for probing. As mentioned before, multiple query probing is expensive for both rule extracting and database probing. For such reasons, in our approach, we only use one query for probing in each category. Our one-query probing is based on the assumption that it does not affect the classification too much because every category uses the same number of queries (one query in our paper). We extract candidate query terms for each category from the concatenation of its all training documents selected from the corresponding DMOZ Directory. Those terms, called category feature, are ordered with their weights. We chose several terms according to their weights as a query to probe hidden databases.

After sending the request message including form filled-out information to the server, our proposed system will receive the result pages. Perhaps the most common case is that a web server returns results page by page consecutively, with a fixed number, say ten or twenty, result matches per page.

To classify the HW databases effectively, we need to analyze the content of each result document. However, full-text of results from some HW-DB cannot be obtained for some reasons like copyright. So the system handles differently for these two situations.

2.2.1 Result Documents Without Full-Text

In this situation, only the number of returned documents for the query can be used for analysis. Let c_1, c_2, ..., c_K be K categories under the same parent node in the hierarchy, the returned number for each category by a hidden database HD are L_1, L_2, ..., L_K respectively. Then, we approximate HD as:

$$HD \approx \sum_{1 \leq j \leq K} f_j * L_j \qquad (4)$$

where f_j is the category feature of class c_j. In fact, f_j is the centroid of the training documents in c_j.

2.2.2 Result Documents with Full-Text

For the hidden databases whose full-texts can be accessed, our system can analyze the document content further to get more accurate approximation for HD. In such case, not only the number of the results for a category can be got, but also the relevance of the documents can be used. To save the cost, we only access documents in several positions along the result list. For example, the positions can be set to the first result and the last result, or more complex to 0%, 25%, 50%, 75%, and 100% of the result list.

Suppose we only access the first document $hd_{j,1}$ and the last document hd_{j,L_j} for category c_j. Then, the hidden database HD can be approximated as:

$$HD \approx \sum_{1 \leq j \leq K} \frac{1}{2} (hd_{j,1} + hd_{j,L_j}) * L_j \qquad (5)$$

where L_j is defined as before. It can be easily extended to support more document accesses.

With the probed result for a hidden database HD (equation 4, or 5), HD can always be classified using equation (1).

2.3 HW-DB Classification Based on Link Structure

In last subsection, we introduced the methods for the Hidden-Web databases classification based on probing, which produces good experimental results. However, it does not make use of the properties of Web structure, especially the links among the Web documents. Actually, link structure of the Web provides another important clue for HW-DB classifications. In fact, as in fig.2, a hidden database may be referenced by many Web pages. Those pages can also be used to derive the semantics of the hidden database.

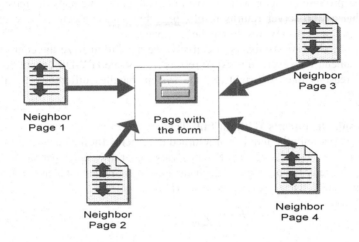

Fig. 2. A Hidden database linked by other pages (neighbor pages)

Web pages, which have links to the hidden database *HD*, are called neighbor pages for this database. To use them for the classification of *HD*, we concatenate all of the neighbor pages into one document called *NP*. Then, the semantics of *HD* is represented with a vector of terms extracted from *NP*. Therefore, *HD* is also can be classified by the values of *Similarity(NP,c)*.

2.4 Combined Classifier for Hidden Databases

To raise the performance of the classification, we try to combine probing model with link-based model. In fact, new hidden databases often have less neighbor pages to be referenced. Therefore, probing method is the only way for the classification in such situation. To avoid outlier, we use link-based classifiers only for hidden databases which have at least 20 neighbor pages. The combined classifier is defined as:

$$C\text{-}Similarity(HD,\ c)=W*Similarity(HD,c)+(1-W)*Similarity(NP,c) \qquad (6)$$

where *W* is used to balance this two classifiers.

3 Experiment

Our objective functions for system performance are based on two basic metrics—precision and recall [5].

When evaluating the result of classification, there are three important values for each category:

A ---- Number of documents which are classified into the category correctly;

B ---- Number of documents which are classified into the category wrongly;

C ---- Number of documents which are classified into other category wrongly;

Recall is the ratio of the number of documents classified into a category correctly to the total number of relevant documents in the same category. Precision is the ratio of the number of documents classified into a category correctly to the total number of irrelevant and relevant documents classified into the same category. Both of them can be represented with the above values, A, B, and C.

$$precision = \frac{A}{A + B} \times 100\% \quad , \quad recall = \frac{A}{A + C} \times 100\%$$

To condense precision and recall into one number, we use the F_1-measure metric [5]:

$$F_1 = \frac{2 \times precision \times recall}{precision + recall}$$

which is only high when both precision and recall are high, and is low for design options that trivially obtain high precision by sacrificing recall or vice versa. Recall and precision are evenly weighted.

3.1 Determining the Number of the Feature Terms for Form Filling-Out

There are two types of form-elements in general, A-Element (support Boolean 'AND') and O-Element (support Boolean 'OR'). We must choose a proper number of query terms for filling out form elements for these two types. A-Element and O-Element models occupy 41% and 59% respectively in our testing hidden databases. We fill-out those hidden databases with changing number of query terms.

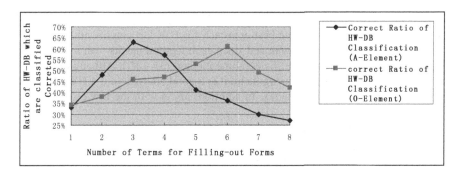

Fig. 3. The ratio of HW-DB which are classified correctly using different quantities of terms

Fig.3. shows the correct ratio of HW-DB classification using different numbers of feature terms. The horizontal axis shows the number of terms to fill-out the forms and the vertical axis shows the ratio of HW-DB which are classified correctly. It can be seen from the figure, for A-Element, the correct radio reaches its summit 63% when we choose 3 terms to fill-out the forms. For O-Element, the optimal number of terms is 6, which leads to 61% correct radio. That is, we should choose 6 terms to fill-out the forms for O-Element, in order to receive the maximal correct ratio.

3.2 Evaluating Results over Different Classification Approaches

In our system, three basic models for classification of hidden databases are addressed, including full-text probing M_1, result-number only probing M_2, and link-based classifying M_3. By combining M_1 with M_3, M_2 and M_3, we get two combined classification models.

Fig.4. shows the classification performances for basic model M_1 and M_2, as well as the two combined models. It is clear from the figure, by far the combined method ($M_1 + M_3$) receives the best performance when balance perimeter W=0.4. And the second combined mode ($M_2 + M_3$) reaches its optimal performance when W=0.3.

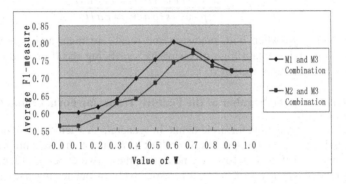

Fig. 4. The Classification Performance of Combine Methods by Different Weight

The average F_1-measures of those methods are shown in Table 1. By far, the combined method (M_1+M_3) is the best approach for HW-DB classification. However, other methods should not be abandoned since each method has its own merit. Method M_1 and M_3 are the basic ones for the combine method (M_1+M_3). Although method M_2 shows the worst performance among them, it is a good alternative if a HW-DB cannot returns full-text of result documents. In addition, M_2 is with the low cost comparing with M_1.

Table 1. Average F_1-measure of M_1, M_2, M_3 and the Combine Methods

Methods	M_1	M_2	M_2+M_3	M_1+M_3
Average F_1-measure	0.60	0.56	0.72	0.80

4 Conclusions

In this paper, we have proposed a novel and efficient approach for classification of Hidden-Web Databases. We have introduced a category hierarchy for HW-DB and described the process to extract the feature for each category. With terms of the features, we probe the hidden databases and analyze the results documents in order to classify the HW-DB. To raise the performance further, we also use Web pages which have links pointing to the hidden-web database as another important source to represent the databases. We combine link-based evaluation and query-based probing as our final classification solution. Our experiment shows the combined approach can generate a much better performance for the HW-DB classification.

Acknowledgement. This Work was supported in part by the University Research Committee under Grant No. RG069/05-06S/07R/GZG/FST and by the Science and Technology Development Found of Macao Government under Grant No. 044/2006/A.

References

[1] Lawrence, S., Giles, C.L.: Accessibility of Information on the Web. Nature 400, 107–109 (1999)
[2] Bergman, M.K.: The Deep Web: Surfacing Hidden Value Latest Access: 11/1/2007 (September 2001), http://www.brightplanet.com/resources/details/deepweb.html
[3] Raghavan, S., Garcia-Molina, H.: Crawling the Hidden Web. In: Proceedings of the 27th International Conference on Very Large Data Bases (VLDB) (2001)
[4] Lin, K.I., Cheng, H.: Automatic Information Discovery form the Invisible Web. In: Proceedings of the International Conference on Information Technology: Coding and Computing (ITCC) (2002)
[5] Ipeirotis, P.G., Gravano, L., Sahami, M.: Probe, Count, and Classify: Categorizing Hidden-Web Databases. In: Proceedings of the 20th ACM SIGMOD International Conference on Management of Data, ACM Press, New York (2001)
[6] Gravano, L., Ipeirotis, P.G., Sahami, M.: QProber: A System for Automatic Classification of Hidden-Web Databases. ACM Transactions on Information Systems (TOIS) 21(1), 1–41 (2003)
[7] Bergholz, A., Chidlovskii, B.: Crawling for Domain-Specific Hidden Web Resources. In: Proceedings of the 4th International Conference on Web Information Systems Engineering (WISE '03) (2003)
[8] Vapnik, V.: The Nature of Statistic Learning Theory. Springer, New York (1995)
[9] Dasarathy, B.V.: Nearest Neighbor (NN) Norms: NN Pattern Classification Techniques. In: McGraw-Hill Computer Science Series, IEEE Computer Society Press, Las Alamitos, California (1991)
[10] Yang, Y., Chute, C.G.: An example-based mapping method for text categorization and retrieval. ACM Transaction on Information Systems (TOIS) 12(3), 252–277 (1994)
[11] Mitchell, T.: Machine Learning. McGraw Hill, New York (1996)
[12] Cohen, W.W.: Learning trees and rules with set-valued features. In: Proceedings of the Thirteenth National Conference on Artificial Intelligence, pp. 709–716 (1996)

Pruning Relations for Substructure Discovery of Multi-relational Databases

Hongyu Guo[1], Herna L. Viktor[1], and Eric Paquet[2]

[1] School of Information Technology& Engineering, University of Ottawa, Canada
{hguo028, hlviktor}@site.uottawa.ca,
[2] National Research Council of Canada, Ottawa, Canada
eric.paquet@nrc-cnrc.gc.ca

Abstract. Multirelational data mining methods discover patterns across multiple interlinked tables (relations) in a relational database. In many large organizations, such a multi-relational database spans numerous departments and/or subdivisions, which are involved in different aspects of the enterprise such as customer profiling, fraud detection, inventory management, financial management, and so on. When considering multirelational classification, it follows that these subdivisions will express different interests in the data, leading to the need to explore various subsets of relevant relations with high utility with respect to the target class. The paper presents a novel approach for pruning the uninteresting relations of a relational database where relations come from such different parties and spans many classification tasks. We aim to create a pruned structure and thus minimize predictive performance loss on the final classification model. Our method identifies a set of strongly uncorrelated subgraphs to use for training and discards all others. The experiments performed demonstrate that our strategy is able to significantly reduce the size of the relational schema without sacrificing predictive accuracy.

1 Introduction

Knowledge discovery from relational databases poses a unique opportunity for the data mining community. In many large organizations, such a relational database spans numerous departments and/or subdivisions and these subdivisions will express different interests in the data, leading to the need to explore various subsets of relevant relations with high utility with respect to the target class. Furthermore, acquiring such data is often expensive. Also, it becomes harder to preserve data privacy when data in a database is from multiple sources. These problems can be mitigated by pruning uninteresting relations before constructing a model. Few attempts to address this issue have been made so far [5,12].

This paper presents the Subgraph Ensemble Structure Pruning (SESP) method for pre-pruning relational databases. The SESP approach aims to create a pruned relational schema that models only the most informative substructures, while maintaining satisfactory predictive performance. This is achieved

J.N. Kok et al. (Eds.): PKDD 2007, LNAI 4702, pp. 462–470, 2007.

by removing either *irrelevant* or *redundant* substructures from the database. The SESP algorithm assumes that strongly correlated substructures contain redundant information. Those which are weakly correlated to the class are said to be of low relevance. The SESP approach initially *decomposes* the relational domain into subgraphs. From this set, it subsequently identifies a subset of subgraphs which are strongly uncorrelated with each other, but correlated with the target class (*uncorrelated subgraphs*). All other subgraphs are discarded. We compare the classifiers constructed from the original schema with those constructed from the pruned database. Our experiments on both real-world and synthetic databases demonstrate that the SESP approach is able to significantly reduce the complexity of the relational schema without sacrificing the predictive accuracy.

The paper is organized as follows. Section 2 describes the problem setting. Next, a detailed discussion of the SESP algorithm is provided in Section 3. Section 4 presents a comparative evaluation. Section 5 concludes the paper.

2   Problem Setting

The problem setting for our SESP strategy is defined by a relational database schema $\Re$, which is described by a set of tables $\{R_1, \cdots, R_n\}$. Each table R_i consists of a set of tuples T_R, a primary key, and a set of foreign keys (in this paper, we refer to primary key and foreign keys as *key attributes*). Foreign key attributes[1] link to primary keys of other tables. This type of linkage defines a *join* (relationship) between the two tables involved. A set of joins with n tables $R_1 \bowtie \cdots \bowtie R_n$ describes a join path, where the length of the path is defined as the number of joins it contains.

In a relational classification setting a database $\Re$ contains a target relation R_t, a set of background relations R_b, and a set of joins (J). Each tuple $x \in T_{R_t}$ includes a unique primary key attribute $x.k$ (*tuple identifier*) and a categorical variable y. The task in this setting is to find a function $F(x)$ which maps each tuple x of the target table R_t to a category label y. That is, $y = F(x, R_t, R_b, J)$.

The SESP method aims to pre-prune the relational structure for the task of multirelational classification without loss of predictive accuracy. The SESP pruning strategy is closely related to our Multiple View Relational Classification (MRC) method presented by Guo and Viktor in [4]. The MRC method aims at constructing an accurate relational classifier directly from a relational database by using multiple views on the data.

3   The SESP Pruning Approach

The core idea of the SESP method is to identify a small set of strongly uncorrelated subgraphs given a database schema. As presented in Algorithm 1, the process consists of two key steps: 1) to initially decompose the relational database schema into subgraphs (*subgraph construction*); and 2) to identify the subset of strongly uncorrelated subgraphs (*subgraph evaluation*).

---

[1] For simplicity, we only consider key attribute as a single attribute here.

Algorithm 1. The SESP Pruning Approach

Input: A relational database $\Re = \{R_t, R_b, J\}$
Output: A pruned database $\Re' = \{R_t, R_b', J'\}$

1: Divide $\Re$ into training data set $\mathcal{T}_t$ and evaluation set $\mathcal{T}_m$;
2: Convert $\Re$ into undirected graph $\mathcal{G}$;
3: Decompose $\mathcal{G}$ into a set of subgraphs $\{\mathcal{S}\}$;
4: Build a classifier for each subgraph in $\{\mathcal{S}\}$, using $\mathcal{T}_t$;forming $\{C_d^1, \cdots, C_d^m\}$;
5: Let subgraph set $C = \emptyset$;
6: Generate an evaluation examples set $\mathcal{T}_m'$, using $\mathcal{T}_m$ and $\{C_d^i\}_1^m$;
7: Select a subgraph feature set $\mathcal{A}'$ from $\mathcal{T}_m'$;
8: For each C_d^i with at least one attribute in $\mathcal{A}'$: $C.add(C_d^i)$; forming $\{C^i\}_1^k$ $(k \leq m)$;
9: Remove duplicate relations/relationships from $\{C^i\}_1^k$; forming $\Re' = \{R_t, R_b', J'\}$.

3.1   Subgraph Construction

The *subgraph construction* process aims to build a set of subgraphs given a relational database schema where each subgraph corresponds to a unique join path. The construction process initially converts the relational database schema into an undirected graph, using the tables as the nodes and joins as edges.

Two heuristic constraints are imposed on each constructed subgraph. The first is that each subgraph must start at the target relation. This constraint ensures that each subgraph will contain the target relation and, therefore, be able to construct a classification model (details to be discussed in Section 3.2). The second constraint is for relations to be unique for each candidate subgraph. Typically in a relational domain, the number of possible join paths given a large number of relations is usually *very* large, making it too costly to exhaustively search all join paths [7]. Also, join paths with many relations may decrease the number of entities related to the target tuples. Therefore, we propose to this restriction for the SESP algorithm as a tradeoff between accuracy and efficiency.

Using these constraints, the subgraph construction process proceeds initially by finding unique join paths with two relations, i.e. join paths with a length of one. These join paths are progressively lengthened, one relation at a time. We use the length of the join path as the stopping criterion, preferring subgraphs with shorter length. The reason for preferring shorter subgraphs is that semantic links with too many joins are usually very weak in a relational database [13]. Thus we specify a maximum length for join paths. When this number is reached, the entire join path extraction process stops. Note that a special subgraph, one that is comprised solely of the target relation, is created as well.

3.2   Subgraph Evaluation

In order to select which subgraphs will be best for classification, subgraphs are evaluated according to the following methodology (Algorithm 1). Firstly, each of the created subgraphs is used to construct a separate classifier (*subgraph-based relational classification*). Secondly, the constructed classification models are used to generate an evaluation data set in which each instance is described by sets of feature sets (*subgraph features*). Each feature set describes the knowledge contained in one of the classification models. Thirdly, the set of *subgraph features* which are strongly uncorrelated with each other, but correlated with the target class, are identified. These subgraph feature sets *correspond*, therefore, to a set of subgraphs. Lastly, subgraphs which are not selected are discarded.

Subgraph-based Relational Classification. Each subgraph created in Section 3.1 can be used to build a relational classifier using traditional single-table learning algorithms. These methods require "flat" data presentations. In order to employ these "flat" data methods, aggregation operators are usually used to squeeze a bag of tuples into one attribute-based entity. Often, different aggregation functions are employed on different types of attributes. Here, for a *Nominal Attribute*, the aggregation *COUNT* function is applied to calculate the number of times the attribute occurs within the data set. For a *Binary Attribute* for which there are only two possible values, the *COUNT* function is applied separately on each of the two values. For a *Numeric Attribute*, the aggregation functions *SUM, AVG, MIN, MAX, STDDEV and COUNT* are applied. In this way, each subgraph separately creates a set of attribute-based training instances.

Subgraph Features. *Subgraph features* are used to describe the knowledge held by subgraph-based classifiers and represent the corresponding subgraphs. The subgraph features are generated as follows. Let $\{C^1, \cdots, C^n\}$ be n classifiers (each is built using a different subgraph). Let $\mathcal{T}_m$ be an *evaluation data set* with m labels $\{y_1, \cdots, y_m\}$. For each instance t (with label L) in $\mathcal{T}_m$, each classifier is called upon to produce predictions $\{f_{C^i}^{y_k}(t)\}$ ($i \in \{1, \cdots, n\}$ and $k \in \{1, \cdots, m\}$) for it. Here, $f_{C^i}^{y_k}(t)$ denotes the probability that instance t belongs to class y_k, as predicted by classifier C^i. In this way a set of *evaluation examples* is constructed where each consists of sets of prediction set $\{P_{C^i}^{y_k}(t)\}$ and the original class label L. For instance, C^1 is described and represented by features $\{f_{C^1}^{y_k}(t)\}$. We define $\{f_{C^1}^{y_i}(t)\}$ to be the *subgraph features* of classifier C^1.

   As an example, consider a two class (y_1 and y_2) problem which has two subgraph-based classifiers C^1 and C^2 constructed. For each tuple t in the evaluation data set, each classifier C^i assigns for each class y_i a prediction p_i^i. For each such prediction a new attribute is thus generated, which we refer to as a subgraph feature. In our example, for the subgraph corresponding to classifier C^1, each evaluation example contains two new attributes (subgraph features) p_1^1 and p_2^1. Similarly, two subgraph features p_1^2 and p_2^2 are generated for the subgraph corresponding to classifier C^2. This evaluation example also contains the original class label of the tuple t.

Correlation Measurement of Subgraphs. The SESP strategy uses a heuristic measure to calculate the correlation score of a set of subgraphs (represented by subgraph features). This measure considers the correlation information both between subgraphs and between those subgraphs and the class to be learned. A similar heuristic principle has previously been applied in test theory by Ghiselli [3] and in feature selection by Hall [6]. This heuristic assigns each subset of features a level of "goodness." The score is formalized by Equation 1 [6]:

$$Q = \frac{K\overline{R_{cf}}}{\sqrt{K + K(K-1)\overline{R_{ff}}}} \tag{1}$$

Here, K is the number of features in the subset, $\overline{R_{cf}}$ is the average feature-to-class correlation, and $\overline{R_{ff}}$ represents the average feature-to-feature dependence.

To measure the degree of correlation between features and the target class and between the features themselves, we use the notion of *Symmetrical Uncertainty* (*U*) [10] to calculate $\overline{R_{cf}}$ and $\overline{R_{ff}}$. This score is a variation of the *Information Gain* (*InfoGain*) measure [11]. It compensates for *InfoGain*'s bias toward attributes with more values, and has been successfully applied by Ghiselli [10] and Hall [6]. *Symmetrical Uncertainty* is defined as follows:
Given features X and Y,

$$U = 2.0 \times \left[\frac{InfoGain}{H(Y) + H(X)} \right]$$

where $H(X)$ and $H(Y)$ are the entropies of the random variables X and Y, perspectively. Note that, these measures need all of the features to be nominal, so numeric features are first discretized properly.

Subgraph Pruning. In order to identify a set of uncorrelated subgraphs, the evaluation procedure searches all of the possible subgraph feature subsets, and constructs a ranking on them. The best ranking subset will be selected, i.e. the subset with the highest Q value.

To search the subgraph feature space, the SESP method uses a best first search strategy [8]. The method starts with an empty set of features, and keeps expanding, one feature at a time. In each round of the expansion, the best feature subset, namely the subset with the highest "goodness" value Q is chosen. In addition, the SESP algorithm terminates the search if a preset number of consecutive non-improvement expansions occurs. Based on our experimental observations we empirically set the number to five (5).

Subgraphs are selected based on the final best subset of subgraph features. If a subgraph has no features that are strongly correlated to the class, the knowledge possessed by this subgraph can be said to be unimportant for the task at hand. Thus, it makes sense to prune this subgraph. The SESP algorithm, therefore, keeps a subgraph *if* and *only if* any of its subgraph features appears in the final best ranking subset.

4   Experimental Results

In our evaluation, we compare the accuracy of a relational classifier constructed from the original schema with the accuracy of one built from a pruned schema.

We perform our experiments using the MRC, RelAggs [9], TILDE [2], and Cross-Mine algorithms. The MRC and RelAggs approaches are aggregation-based algorithms where C4.5 decision trees [11] were applied as the single-table learner. The C4.5 decision tree learner was used due to its de facto standard for empirical comparisons. In contrast, the CrossMine and TILDE methods are two benchmark logic-based strategies. In addition, we only consider join paths which contain less than four tables. This number was empirically determined and provides a good trade off between accuracy and execution time. Also, C4.5 decision tree was applied as the subgraph-based classifiers of the SESP strategy. All experiments were conducted using ten-fold cross validation. We report the average running time of each fold (run on a 3 GHz Pentium4 PC with 1 GByte of RAM).

A) Real Databases

Financial Database: Our first experiment uses the financial database from the PKDD 1999 discovery challenge [1]. This database consists of eight tables. This database provides us with two different learning problems. Our first learning task (F234) is to learn if a loan is good or bad from the 234 finished tuples. The second experimental task (F400) uses the Financial database as prepared in [13], which has 400 examples in the target table.

ECML98 Database: Our second experiment uses the database from the ECML 1998 Sisyphus Workshop. The learning task (ECML) is to categorize the 7,329 households into classes 1 and 2 [9]. Eight background relations are provided for this learning task. We here used the new star schema prepared in [9].

Experimental Results and Discussion: The predictive accuracy we obtained, using MRC, RelAggs, TILDE, and CrossMine is presented in Table 1. The results obtained with the respective original and pruned schemas are shown side by side. We also present the compression rates achieved by the SESP approach in the last column of Table 1. The compression rate considers the number of relations of the original schema ($N_{original}$) and the number of relations pruned (N_{pruned}) and is calculated as ($N_{original} - N_{pruned})/N_{original}$. In table 2, we also provide the execution time of the pruning process, as well as the running time required for the four tested algorithms against the original and pruned schemas.

From Table 1, one can see that the SESP algorithm not only significantly reduces the size of the relational schema, but also produces compact pruned schemas that provide comparable multi-relational classification models in terms of the accuracy obtained. The results shown in Table 1 provide us with two

Table 1. Accuracies obtained using methods MRC, RelAggs, TILDE, and CrossMine against the original and pruned schemas, along with the compression rate

Schema	MRC		RelAggs		TILDE		CrossMine		Compress.
	Original	Pruned	Original	Pruned	Original	Pruned	Original	Pruned	Rate
F400	88.0 %	88.0 %	89.0 %	86.8 %	81.3 %	81.0%	85.8 %	87.3 %	50.0 %
F234	92.3 %	92.3 %	90.2 %	90.2 %	86.8 %	86.8%	88.0 %	89.4 %	25.0 %
ECML	88.2 %	87.5 %	88.0 %	86.2 %	53.7 %	52.0%	85.3 %	83.7 %	55.5 %

Table 2. Execution time (seconds) required using the four tested methods against the original and pruned schemas, along with the computational time of the SESP method

Schema	MRC		RelAggs		TILDE		CrossMine		Pruning Time
	Original	Pruned	Original	Pruned	Original	Pruned	Original	Pruned	
F400	2.83	2.25	60.00	51.83	650.00	132.32	8.10	6.76	1.97
F234	1.60	1.17	40.80	34.13	568.30	80.36	5.00	3.41	1.07
ECML	424.43	220.99	1703.58	1206.39	1108.60	167.76	570.90	366.78	356.24

meaningful observations. The first is that the SESP strategy is capable of pruning the database schemas meaningfully. The compression rates for these three learning schemas are 50%, 25%, and 55.5%, respectively. The second finding is that the pruned schemas produce comparable accuracies, when compared to the results obtained with the original schemas. Results as obtained by the aggregation-based methods show that, for two of the three databases (F400 and F234), the MRC algorithm obtained the same predictive results when pruned. Only against the ECML database, did the pruned MRC algorithm obtain a slightly lower accuracy than the original (lower by only 0.7%). When considering the RelAggs algorithm, the accuracy produced by the RelAggs method against both the pruned and full schemas were comparable. Against the F234 data set, the RelAggs algorithm achieved the same predictive results. Only against the F400 and ECML data sets, did the RelAggs method yield slightly lower accuracy than the original.

When testing with the logic-based strategies, Table 1 shows that, the TILDE algorithm obtained almost the same accuracy against two of the three tested data sets (F400 and F234). Only against the ECML database, did the TILDE algorithm obtain a slightly lower accuracy than the original (lower by only 1.7%). When considering the CrossMine method, the accuracy produced by this method against both the pruned and full schemas was also very close.

In terms of computational cost of the SESP method, results presented in Table 2 show that the pruning processes were fast. The fast pruning time is especially relevant when considering the time required when training all four methods against the original schemas. This result is promising, especially when considering large, complex schemas where the cost to process the entire schema is prohibitive. Also, the results indicate that meaningful execution time reductions may be achieved when building the models against the pruned schemas.

B) Synthetic Databases

To further test the SESP algorithm, we generated six synthetic databases with 10, 20, 50, 80, 100, and 150 relations (denoted as SynR10, SynR20, SynR50, SynR80, SynR100, and SynR150), respectively. The database generator was obtained from Yin et al. in [13]. For each database in this paper, we set the expected number of tuples and attributes to 1000 and 15, respectively. The accuracies obtained by the MRC and CrossMine methods are shown in Figures 1(a) and 1(b), respectively. The compression rates obtained are provided in Figure 1(c). We also provide the execution time needed using the MRC and CrossMine algorithms against the original and pruned schemas in Figure 1(d).

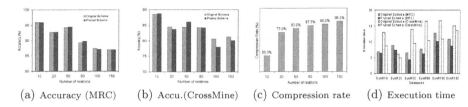

(a) Accuracy (MRC)   (b) Accu.(CrossMine)   (c) Compression rate   (d) Execution time

Fig. 1. Accuracies obtained and execution time (seconds) required by the MRC and CrossMine methods, as well as compression rates achieved by the SESP

From these Figures one can again see that the SESP algorithm not only significantly reduces the complexity of the structural schemas, but also produces very comparable classification models in terms of the accuracy obtained. The MRC algorithm, for example, produced equal or higher accuracies for all databases, except for a slight decrease of 0.3% with the SynR80 database. When using the CrossMine method, the results also convince us that the pruned schemas produce comparable classifiers in terms of accuracies obtained. In terms of compression rate, the results in Figure 1(c) show that the compression rates were more than 80% for databases with more than 50 relations. In addition, results as presented in Figure 1(d) show that the execution time needed for constructing relational models using the two tested algorithms was meaningful reduced when pruned.

5   Conclusion and Future Work

Multirelational data mining application usually involves a large number of relations. This paper presents a novel algorithm to pre-prune uninteresting relations of relational learning tasks. Our experiments demonstrate that the strategy is able to significantly reduce the size of the relational schema while still maintaining the accuracy of the final model. This research suggests that one can build an accurate relational classification model using only a small subset of the original schema. Our future work will include research on extending the subgraph definition to include graphs with more than one slot chain. We also aim to empirically evaluate the optimum choice of maximum subgraph length.

References

1. Berka, P.: Guide to the financial data set. In: Siebes, A., Berka, P. (eds.) PKDD2000 Discovery Challenge (2000)
2. Blockeel, H., Raedt, L.D.: Top-down induction of first-order logical decision trees. Artificial Intelligence 101(1-2), 285–297 (1998)
3. Ghiselli, E.E.: Theory of Psychological Measurement. McGrawHill Company, New York (1964)
4. Guo, H., Viktor, H.L.: Mining relational data through correlation-based multiple view validation. In: KDD '06, pp. 567–573, New York, USA (2006)

5. Habrard, A., Bernard, M., Sebban, M.: Detecting irrelevant subtrees to improve probabilistic learning from tree-structured data. Fundamenta Informaticae: Special Issue on Mining Graphs, Trees and Sequences (2005)
6. Hall, M.: Correlation-based feature selection for machine learning, Ph.D thesis, department of computer science, university of waikato, new zealand (1998)
7. Hamill, R., Martin, N.: Database support for path query functions. In: Williams, H., MacKinnon, L.M. (eds.) Key Technologies for Data Management. LNCS, vol. 3112, pp. 84–99. Springer, Heidelberg (2004)
8. Kohavi, R., John, G.H.: Wrappers for feature subset selection. Artificial Intelligence 97(1-2), 273–324 (1997)
9. Krogel, M.-A.: On Propositionalization for Knowledge Discovery in Relational Databases. PhD thesis, Otto-von-Guericke-Universität Magdeburg (2005)
10. Press, W.H., Flannery, B.P., Teukolsky, S.A., Vetterling, W.T.: Numerical recipes in C: the art of scientific computing. Cambridge University Press, Cambridge (1988)
11. Quinlan, J.R.: C4.5: programs for machine learning. Morgan Kaufmann, San Francisco (1993)
12. Singh, L., Getoor, L., Licamele, L.: Pruning social networks using structural properties and descriptive attributes. In: ICDM '05, pp. 773–776 (2005)
13. Yin, X., Han, J., Yang, J., Yu, P.S.: Crossmine: Efficient classification across multiple database relations. In: ICDE '04, Boston (2004)

The Most Reliable Subgraph Problem

Petteri Hintsanen

HIIT Basic Research Unit, Department of Computer Science,
PO Box 68, FI-00014 University of Helsinki, Finland
petteri.hintsanen@cs.helsinki.fi

Abstract. We introduce the problem of finding the most reliable subgraph: given a probabilistic graph G subject to random edge failures, a set of terminal vertices, and an integer K, find a subgraph $H \subset G$ having K fewer edges than G, such that the probability of connecting the terminals in H is maximized. The solution has applications in link analysis and visualization. We begin by formally defining the problem in a general form, after which we focus on a two-terminal, undirected case. Although the problem is most likely computationally intractable, we give a polynomial-time algorithm for a special case where G is series-parallel. For the general case, we propose a computationally efficient greedy heuristic. Our experiments on simulated graphs illustrate the usefulness of the concept of most reliable subgraph, and suggest that the heuristic for the general case is quite competitive.

1 Introduction

Many contemporary domains in data mining have heterogeneous objects linked together by various relations. Graphs are natural models for data arising from such domains; for example, social networks and the World Wide Web can be naturally described as graphs. In this article we consider probabilistic graphs, whose edges are unreliable and can fail with specified probabilities. Telecommunications and electrical networks are classical examples of real-world structures often modeled as probabilistic graphs.

Informally, given a probabilistic graph and a set of terminal vertices, the reliability of the graph is the probability that there exists at least one path between all pairs of terminals at the time of inspection. It is easy to see that some edges can be more important for the existence of a connection than others. For example, edges forming a cut between two terminals cannot fail simultaneously without breaking connections between those terminals. A natural question to ask is which edges contribute most to the reliability, or equivalently, which edges are safest to remove without a significant loss of reliability? We formulate this question as the problem of finding the most reliable subgraph: given a probabilistic graph, a set of terminal vertices, and an integer K, what is the optimal way to remove K edges from the graph, such that the remaining graph has maximum reliability?

A solution to this problem can be readily applied to a variety of network problems. Consider, for example, a telecommunications network, where edges

J.N. Kok et al. (Eds.): PKDD 2007, LNAI 4702, pp. 471–478, 2007.

represent links between communicating parties, and are subject to random malfunctions. The most reliable subgraph between two communicating terminals describes the most reliable channels for exchanging messages. In social networks, where relative mutual acquaintances could be represented as edge probabilities, one can discover most important relationships between specified individuals by finding a reliable subgraph connecting them. Reliable subgraphs are also useful when visualizing large graphs: they can be highlighted in a picture, or extracted altogether for visual inspection.

A closely related concept of *connection subgraph* was recently introduced by Faloutsos and others [1]. They formalized the *connection subgraph problem*: given a weighted graph, two vertices s and t, and an integer k, find a k-vertex subgraph containing s and t which maximizes a given goodness function. Our framework has a similar goal, but is based on probabilistic reasoning and is defined for multiple terminal vertices.

Overall, little has been published on the extraction and analysis of general connection subgraphs. Lin and Chalupsky use rarity of simple paths and cycles for evaluating the novelty and interestingness of links [2]. Following Faloutsos et al., Ramakrishnan and others propose a method for extracting informative connection subgraphs from RDF graphs [3]. There are a couple of methods utilizing network reliability. One is described by Asthana et al., who predict protein complex memberships in a network of protein interactions [4]. Sevon et al. use network reliability for evaluating the connection strength between entities in biological graphs [5]. Finally, De Raedt et al. consider compression of probabilistic first-order theories and their uses for link discovery in biological networks [6].

Network reliability, on the other hand, has been under extensive research. A canonical summary is given by Colbourn [7]. However, we have been unable to find any references to our problem from the vast literature on reliability theory. Closest effort into this direction seems to be Birnbaum's classical text on *reliability importance*, which measures the importance of a single edge for the reliability of a graph [8]. It has been extended for pairs of edges by Hong and Lie [9]. Finally, Page and Perry consider the reliability importance for ranking the edges of a given graph [10].

2   Problem Definition and Complexity

We use a standard probabilistic graph model. Let $G = (V, E)$ be a graph with a vertex set V and an edge set E. Edges are unreliable: each edge $e \in E$ has an associated probability p_e for functioning; conversely, each edge can fail with probability $1 - p_e$. Edge failures are assumed to be independent. On the other hand, vertices are expected to be fully reliable, that is, they do not fail.

Let G be an undirected probabilistic graph, and let $U \subset V$ be a set of k terminal vertices or nodes. We review the six classical reliability measures, following Colbourn [7]. First, k-*terminal reliability* Rel_k is defined as the probability that each of the k terminal nodes in U can communicate in G; equivalently, it is the probability that there exists a path between any pair of terminals.

When $k = 2$, this measure is referred to as *two-terminal network reliability*, while the case $k = |V|$ is known as *all-terminal network reliability*. We denote these measures by Rel_2 and Rel_A, respectively. (We omit explicit references to G and U whenever they are clear from the context.)

These measures have natural counterparts for directed probabilistic graphs. One vertex $s \in U$ is chosen as the source node, and the rest of the vertices of U are target nodes. The directed version of Rel_k, known as s,T-*connectedness* or Conn_k, is the probability that there exists a (directed) path from s to all target nodes. When $k = 2$, this measure is called s,t-*connectedness* or Conn_2. Finally, the directed analogue of Rel_A is known as *reachability* or Conn_A.

The objective in the *most reliable subgraph problem* (MRSP) is to find the most reliable subgraph obtained from G by removing exactly K edges:

Definition 1 (The Most Reliable Subgraph Problem). *Let $G = (V, E)$ be a probabilistic graph, and let $U \subset V$ be a set of k terminal vertices, where $2 \leq k \leq |V|$. Let $f \in \{\text{Rel}_2, \text{Rel}_k, \text{Rel}_A, \text{Conn}_2, \text{Conn}_k, \text{Conn}_A\}$ be the corresponding reliability measure with respect to U, and let $K \in \mathbb{N}$ with $0 \leq K \leq |E|$. The objective is to find a subgraph $H \subset G$ with $|E| - K$ edges, such that $f(H) \geq f(H')$ for all subgraphs $H' \subset G$ having $|E| - K$ edges.*

Given the fact that exact calculations of $\text{Rel}_{\{2,k,A\}}$ and $\text{Conn}_{\{2,k,A\}}$ are #P-complete problems [11], it is not surprising that the MRSP is likely to be computationally hard as well. The problem does not ask for the value of $f(H)$ for the chosen f and an optimal subgraph H, so the MRSP could be in that sense easier than computing the reliability. Despite this relaxation, it is easy to see that the k-terminal undirected MRSP is NP-hard:

Theorem 1. *MRSP with $f = \text{Rel}_k$ is NP-hard.*

Proof. We give a polynomial time reduction from the NP-complete STEINER TREE problem [12] to the MRSP. Let (G, U, B) be an instance of STEINER TREE, where $G = (V, E)$ is a graph with positive edge weights, $U \subset V$ is a set of terminals, and $B \in \mathbb{N}$ is a bound for the size of the tree.

Without a loss of generality we assume that all edge weights are equal to 1. We transform G into a probabilistic graph $H = (V, E)$ by setting $p_e = 1/2$ for each $e \in E$. Next, we find the smallest (that is, having the least number of vertices and edges) optimal subgraph $H^* \subset H$ connecting the terminals, by solving the MRSP for $K = 0, \ldots, |E| - |U| + 1$ and checking the results in polynomial time. Obviously H^* is a tree; it is also a minimal Steiner tree. Assume to the contrary that there exists a minimal Steiner tree T such that $\|T\| < \|H^*\|$, where $\| \cdot \|$ denotes the number of edges. By construction, we have $\text{Rel}_k(T) = 1/2^{\|T\|} > 1/2^{\|H^*\|} = \text{Rel}_k(H^*)$, which contradicts the optimality of H^*, since T is also a subgraph of H connecting the vertices in U.

To complete the reduction, we simply check if $\|H^*\| \leq B$ holds.   □

The complexity of cases where $f \in \{\text{Rel}_2, \text{Rel}_A\}$ remains open, but we conjecture that they are also NP-hard. The directed variants of the problem are probably hard too, considering the fact that the directed reliability problems are as hard as the corresponding undirected problems [13].

3   Algorithms

It is most likely that there is no efficient algorithm for solving the MRSP in a general case. However, we next describe a polynomial-time algorithm for solving the two-terminal undirected MRSP in an important special case, where the graph is series-parallel. We then give a computationally efficient greedy heuristic for the MRSP in a general case.

3.1   Series-Parallel Graphs

The class of (edge) series-parallel graphs is usually defined using series and parallel composition rules [14]. For our purposes, the following equivalent definition is better: a probabilistic graph G with specified terminals s and t is *series-parallel*, if it can be reduced into a single edge (s, t) by repeatedly applying the following reductions:

- *Series reduction*: If G has a vertex $v \notin \{s, t\}$ of degree two, v and its adjacent edges $e = (u, v)$ and $f = (v, w)$ can be replaced with a single edge $g = (u, w)$ with $p_g = p_e p_f$.
- *Parallel reduction*: If G has two parallel edges $e = (u, v)$ and $f = (u, v)$, they can be replaced with a single edge $g = (u, v)$ with $p_g = 1 - (1 - p_e)(1 - p_f)$.

The specific sequence of reductions is irrelevant, that is, if reductions are applied in any order until no reduction is possible, the result is the single edge (s, t) [14].

Before we describe the algorithm, let us introduce some terminology and notation. For an arbitrary edge set $F \subset E$, let $G[F]$ be the subgraph edge-induced by F. We denote the set of edges reduced into an edge e by $S(e)$; i.e. $f \in S(e)$, if f occurs in the sequence of series-parallel reductions that produced e. Initially, we let $S(e) = \{e\}$ for each $e \in E$.

Let $e = (u, v) \in E$. An i-edge subset $S(e, i) \subset S(e)$ is said to be an *optimal solution for* $G[S(e)]$, if $G[S(e) - S(e, i)]$ is the most reliable subgraph of $G[S(e)]$ with $|S(e)| - i$ edges and terminals u and v. In other words, $G[S(e) - S(e, i)]$ is a solution to the MRSP for $G[S(e)]$ with $K = i$ and $U = \{u, v\}$. Let $S_R(e, i)$ be the reliability of an optimal solution $S(e, i)$, i.e. $S_R(e, i) = \mathrm{Rel}_2\big(G[S(e) - S(e, i)]\big)$.

The iterative definition of series-parallel graphs suggests an iterative, dynamic programming algorithm for solving the MRSP, given that an optimal solution can be constructed from optimal solutions to smaller subgraphs. The following lemma states that this is indeed the case.

Lemma 1. *Let e and f be two edges in series or parallel, and let $S(e, i)$, $S(f, i)$ be optimal solutions for $G[S(e)]$ and $G[S(f)]$, where $0 \le i \le K$. Optimal solutions $S(g, i)$, for all i, can be formed in $O(K^2)$ time, where g is the edge produced by the reduction of e and f.*

Proof. Let i be fixed. Since $S(g) = S(e) \cup S(f)$ and $S(e) \cap S(f) = \emptyset$, an optimal solution $S(g, i)$ has exactly j edges in $S(e)$ and $i - j$ edges in $S(f)$, where $0 \leq j \leq i$. If e and f are in series, we have $S_R(g, i) = S_R(e, j) \cdot S_R(f, i - j)$. Otherwise e and f are parallel, and we have $S_R(g, i) = 1 - (1 - S_R(e, j)) \cdot (1 - S_R(f, i - j))$. An optimal solution can be found by simply enumerating all i possible combinations of edge assignments and choosing one which maximizes $S_R(g, i)$:

$$k = \arg\max_{0 \leq j \leq i} \begin{cases} S_R(e, j) \cdot S_R(f, i - j) & \text{if } e \text{ and } f \text{ are in series} \\ 1 - (1 - S_R(e, j)) \cdot (1 - S_R(f, i - j)) & \text{if } e \text{ and } f \text{ are parallel} \end{cases}$$

$$S(g, i) = S(e, k) \cup S(f, i - k) \ .$$

The solution can be found in $O(i)$ time. By repeating the procedure for all i, $0 \leq i \leq K$, we obtain the solutions $S(g, i)$ in $O(K^2)$ time. □

To solve the MRSP for a series-parallel graph G, we repeatedly apply series and parallel reductions until the graph is reduced into a single edge. As initialization, let $S(e, 0) = \emptyset$, $S_R(e, 0) = p_e$, $S(e, 1) = \{e\}$, and $S_R(e, 1) = 0$ for each $e \in E$. This establishes an invariant: each e has optimal solutions $S(e, i)$ for $G[S(e)]$, where $i = 0, \ldots, \min\{|S(e)|, K\}$. We maintain the invariant by keeping track of optimal solutions $S(e, i)$ and their reliabilities, for each remaining edge e. The invariant, with the definition of series-parallel graphs, guarantees that in the end we have an optimal solution to the MRSP for G.

At the beginning of each iteration, we identify a pair $\{e, f\}$ of reducible edges. This can be done in constant time by suitably augmenting the graph data structure. These edges are replaced with a new edge g; by Lemma 1 it is straightforward to form optimal solutions $S(g, i)$, thus maintaining the invariant. (Note that some combinations stated in Lemma 1 are undefined when $S(e, i)$ or $S(f, i)$ are available only for small i, however, these special cases can be easily detected.) Since each reduction effectively removes one edge from G, after $|E| - 1$ iterations only a single edge e remains, and $S(e, K)$ contains an optimal solution for G. Putting the pieces together, we have established the following theorem.

Theorem 2. *Let $G = (V, E)$ be a series-parallel probabilistic graph. The MRSP for G can be solved in $O(K^2 |E|)$ time, where $1 \leq K \leq |E|$.*

3.2   General Graphs

The set of series-parallel graphs is a very restricted class of graphs; in general, graphs are lot more complex. Unfortunately, as suggested in Sect. 2, the computational effort required for an exact solution quickly becomes excessive. It is most likely that one must content with approximate or heuristic solutions.

We next describe a simple, greedy heuristic for solving the MRSP on general graphs. The heuristic is based on a well-known Monte-Carlo (MC) simulation procedure, which is in many cases sufficient for approximating the reliability of a probabilistic graph G: one just simulates random edge failures in G by flipping a suitably biased coin for each edge, and checks if the terminals are connected in the resulting graph. By counting the number of positive outcomes (i.e. there is a connection between the terminals in that particular outcome) over many repetitions, the reliability estimate is then the fraction of positive outcomes out of the total number of simulations. If the reliability is not very low, then with a reasonable number of simulations we have a good estimate with high probability [15].

The MC procedure is not directly suitable for the MRSP, due to the large number of possible solutions (subgraphs) to consider. However, we use it to estimate $\mathrm{Rel}_2(G-e)$ for each $e \in E$, where $G-e$ denotes G with an edge e removed. Intuitively, edges with large values of $\mathrm{Rel}_2(G-e)$ are less critical, so we iteratively remove K edges with the highest $\mathrm{Rel}_2(G-e)$ values. If, at the beginning of an iteration, there are edges with an endpoint of degree one, we remove those edges first. Such edges are irrelevant from the reliability's standpoint, since they do not occur on any acyclic path between terminals. This heuristic can be implemented in a straightforward manner to run in $O\big(N|E|^2 + K(|E| + \log|E|)\big)$ time. We emphasize the computational efficiency of the heuristic, suggesting that it is suitable for interactive use such as visualization.

4   Experiments

Series-parallel graphs are practicable for evaluating the usefulness of the concept of most reliable subgraph and the relative performance of the proposed heuristic, since they are easy to generate with controlled parameters (size and reliability). Furthermore, we can efficiently calculate optimal solutions with the algorithm described in Sect. 3.1.

We generated nine datasets of random series-parallel multigraphs by repeatedly applying the series-parallel composition rules. Each set consists of 50 graphs with the same number of edges and the same reliability. The sizes are 50, 100 and 200 edges, and the reliabilities are 0.25, 0.5 and 0.75. These parameter choices give nine possible combinations, one for each dataset. The given sizes are averages: there is a slight random variation in the number of edges, because we reduced the parallel edges from the generated multigraphs in order to obtain proper probabilistic graphs.

To evaluate the approximation quality of the heuristic, we used it to solve the MRSP on the generated graphs, with different values of K. After this, the reliabilities of the results were estimated. Optimal solutions were calculated using the algorithm of Sect. 3.1. In order to assess the effect of using $\mathrm{Rel}_2(G-e)$ values to control the heuristic, we implemented a baseline heuristic. This heuristic is identical to the proposed heuristic with the exception that instead of using $\mathrm{Rel}_2(G-e)$ values to decide which edges to remove, it simply considers edge probabilities in ascending order. All estimates were done with 1,000,000 MC simulations. To control random variation, we report the average performance of each method over the 50 graphs in each dataset.

The results for two datasets are depicted in Fig. 1; in the remaining cases the results were comparable. From Fig. 1, we see that the relative performance (the estimated reliability of the subgraph produced by the heuristic divided by the reliability of the optimal subgraph) is fairly stable over different values of K. This is in contrast to the baseline heuristic, whose performance gets significantly worse as K grows. The results suggest that the proposed heuristic is quite competitive; in most cases, the mean relative performance over the different values of K is close to 85%.

The usefulness of most reliable subgraph is also observable in Fig. 1. In both cases, over half of the edges could be removed without a significant loss of reliability. The resulting small, reliable subgraphs contain the most critical edges for the connection, and could be used as a starting point for further analysis or visualization. Our preliminary experiments with real biological graphs show similar or even more accentuated effect (results not shown).

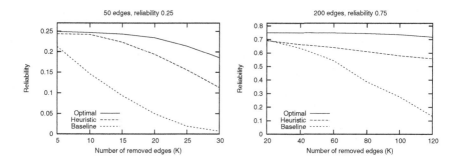

Fig. 1. Results for two generated datasets

5   Conclusions

As more and more domains of interest are best described as interlinked heterogeneous objects, we can expect graphs to become the data models of choice in many situations [16]. Applications may demonstrate a degree of randomness on links, e.g. technical unreliability, subjective uncertainty, or relevance with respect to a specific task. Probabilistic graphs are useful models in such cases.

In the light of these observations, novel graph mining concepts and methods are essential for coping with the increasing number of graph mining problems. The concept of most reliable subgraph, the associated most reliable subgraph problem, and the analysis of the problem are novel additions to this setting. We believe that the concept is useful in many data mining challenges on probabilistic graphs.

We described efficient methods for solving the MRSP, and demonstrated their usefulness with experimental results on synthetic probabilistic graphs. Future work will include improving the methods and assessing their performance with more varied and extensive experiments. There are also open questions on the complexity, and on the other variants of the MRSP.

Despite the apparent usefulness of the concept of most reliable subgraph, we were surprised to found out that (to the best of our knowledge) there is practically no previous research on the subject. Therefore, our proposed methods for solving the MRSP could be seen as the first steps toward more efficient and robust algorithms.

Acknowledgements. The author is grateful to Hannu Toivonen, Kimmo Kulovesi, Petteri Sevon, and Lauri Eronen for constructive discussions on the subject, and for numerous valuable improvements to the manuscript. This research has been supported by Tekes, Jurilab Ltd., Biocomputing Platforms Ltd., and GeneOS Ltd.

References

1. Faloutsos, C., McCurley, K.S., Tomkins, A.: Fast discovery of connection subgraphs. In: Proceedings of the Tenth ACM SIGKDD International Conference on Knowledge Discovery and Data Mining, pp. 118–127. ACM Press, New York (2004)
2. Lin, S., Chalupsky, H.: Unsupervised link discovery in multi-relational data via rarity analysis. In: Proceedings of the Third IEEE International Conference on Data Mining, pp. 171–178. IEEE Computer Society Press, Los Alamitos (2003)
3. Ramakrishnan, C., Milnor, W.H., Perry, M., Sheth, A.P.: Discovering informative connection subgraphs in multi-relational graphs. SIGKDD Explorations 7, 56–63 (2005)
4. Asthana, S., King, O.D., Gibbons, F.D., Roth, F.P.: Predicting protein complex membership using probabilistic network reliability. Genome Research 14, 1170–1175 (2004)
5. Sevon, P., Eronen, L., Hintsanen, P., Kulovesi, K., Toivonen, H.: Link discovery in graphs derived from biological databases. In: Proceedings of Data Integration in the Life Sciences, Third International Workshop, pp. 35–49 (2006)
6. De Raedt, L., Kersting, K., Kimmig, A., Revoredo, K., Toivonen, H.: Compressing probabilistic Prolog programs (Submitted)
7. Colbourn, C.J.: The Combinatorics of Network Reliability. Oxford University Press, Oxford (1987)
8. Birnbaum, Z.W.: On the importance of different components in a multicomponent system. In: Multivariate Analysis - II, pp. 581–592 (1969)
9. Hong, J., Lie, C.: Joint reliability-importance of two edges in an undirected network. IEEE Transactions on Reliability 42, 17–33 (1993)
10. Page, L.B., Perry, J.E.: Reliability polynomials and link importance in networks. IEEE Transactions on Reliability 43, 51–58 (1994)
11. Valiant, L.G.: The complexity of enumeration and reliability problems. SIAM Journal on Computing 8, 410–421 (1979)
12. Garey, M.R., Johnson, D.S.: Computers and Intractability: A Guide to the Theory of NP-Completeness. W. H. Freeman and Company (1979)
13. Ball, M.O.: Complexity of network reliability computations. Networks 10, 153–165 (1980)
14. Valdes, J., Tarjan, R.E., Lawler, E.L.: The recognition of series-parallel digraphs. SIAM Journal on Computing 11, 298–313 (1982)
15. Karp, R.M., Luby, M., Madras, N.: Monte-Carlo approximation algorithms for enumeration problems. Journal of Algorithms 10, 429–449 (1989)
16. Getoor, L., Diehl, C.P.: Link mining: A survey. SIGKDD Explorations 7, 3–12 (2005)

Matching Partitions over Time to Reliably Capture Local Clusters in Noisy Domains

Frank Höppner[1] and Mirko Böttcher[2]

[1] University of Applied Sciences Braunschweig/Wolfenbüttel
Robert Koch Platz 10-14, D-38440 Wolfsburg, Germany
[2] BT Group, Intelligent Systems Research Centre,
Adastral Park, Orion Bldg. pp1/12, Ipswich, IP5 3RE, UK

Abstract. When seeking for small clusters it is very intricate to distinguish between incidental agglomeration of noisy points and true local patterns. We present the PAMALOC algorithm that addresses this problem by exploiting temporal information which is contained in most business data sets. The algorithm enables the detection of local patterns in noisy data sets more reliable compared to the case when the temporal information is ignored. This is achieved by making use of the fact that noise does not reproduce its incidental structure but even small patterns do. In particular, we developed a method to track clusters over time based on an optimal match of data partitions between time periods.

1 Introduction

Clustering is the partitioning of data into groups such that similar data objects belong to the same and dissimilar objects to different groups (according to some similarity measure). In this paper, we focus on one important topic in clustering: how to reliably detect small clusters in the presence of heavy noise.

We are particularly interested in so-called *local patterns* rather than the global structure of the data, because (a) the global structure is much better known by domain experts and thus not considered as interesting and (b) small structures may indicate niches or upcoming trends and as such are potentially of high value to businesses. Two difficulties are connected with local pattern discovery: Firstly, within the multitude of local patterns discovered most prove to be uninteresting (since in line with the global structure) or incidental on closer inspection (cf. [1]). Secondly, local patterns are easily obscured by noise. In particular, cluster algorithms often fail in discovering small clusters in noisy domains since they aim at the detection of large and well separated clusters.

In our approach, we divide up the data (based on time stamps) into several slices and analyze the data of consecutive slices. This idea and its consequences are discussed in Section 2 and the clustering algorithm we are going to use (modified OPTICS [2]) in Section 3. By matching clusters in consecutive partitions, a cluster history is obtained (cf. Sec. 4). We exploit this historical information to distinguish incidental from substantial small clusters (Sec. 5). Finally, some results are shown in Section 6.

J.N. Kok et al. (Eds.): PKDD 2007, LNAI 4702, pp. 479–486, 2007.
© Springer-Verlag Berlin Heidelberg 2007

Related Work. Clustering in the presence of time-stamped and changing data has been studied in the context of moving cluster detection and data streams. Although it is possible to track the change of clusters with our proposed algorithm, we will focus on the detection of local patterns here only. In contrast to most stream mining approaches [3,4] we do not keep a condensed representation of the data stream in main memory, but analyze subsequent parts of the stream one after another. The algorithm we propose is not an online algorithm.

We will compare the partitions obtained from data of consecutive time periods, therefore *clustering ensembles* is also a related area. In the clustering ensembles literature, however, the partitions are usually obtained from identical data sets but different clustering algorithms, e.g. [5], or from the same algorithm using different subsamples of the original data set, e.g. [6]. In this paper, however, we compare partitions obtained at different points in time in order to gain information about the temporal stability of a cluster.

Brief Review of OPTICS. Since we are going to use the OPTICS algorithm we want to briefly review it here. OPTICS [2] is a density-based approach, which means that for a data point x to belong to a cluster, the density around x must exceed some density threshold ϱ. The density ϱ is given implicitly by the size of a hypersphere with radius ε (defining the neighborhood $N_\varepsilon(x)$ of volume V_ε) and a required number MinPts of data objects within this neighborhood. The condition for x to establish a cluster is $|N_\varepsilon(x)| \geq MinPts$. In terms of density this is equivalent to $\frac{|N_\varepsilon(x)|}{V_\varepsilon} \geq \frac{MinPts}{V_\varepsilon} =: \varrho$. OPTICS requires only the parameter MinPts and an upper bound for the neighborhood size. The outcome of the algorithm is an ordering of all data points together with a reachability distance for each data point (cf. Fig. 2). Simplifying, the reachability distance of x is the smallest neighborhood radius ε such that x belongs to a cluster. In the reachability plot all data objects are ordered on the horizontal axis with their reachability distance on the vertical axis, such that consecutive data objects in the plot having all their reachability distances below some given threshold belong to the same cluster at the corresponding density level.

2   Selection of the Clustering Algorithm

We start by dividing our existing data into slices S_i of approximately the same size and consider the sequence of partitions obtained from clustering a window of n consecutive slices. By comparing the clusters of consecutive partitions we trace the clusters over time and thereby also measure their stability. The rough sketch of our proposed PAMALOC algorithm (**pa**rtition **ma**tching to detect **loc**al clusters) is shown in Fig. 1. The data does not have to be timestamped, but all data in S_i should have been observed before the data in any S_j, $j > i$.

Given this coarse idea of how to attack the problem, we may use a variety of clustering algorithms in line 2 and 6. They should, however, meet at least the following four requirements:

- *Flexibility.* The notion of a cluster supported by the clustering algorithm must be flexible in order to cope with any arbitrary cluster shape. This rules out

INPUT: t time interval, n number of slices to use, S_t data slice collected during t	
Set $t = n$	1
Run clustering algorithm with slices $S_{t-n+1} \cup S_{t-n+2} \cup ... \cup S_t$	2
Extract clusters, denote the set of clusters (partition) by $P^t = \{C_1^t, C_2^t, ..\}$	3
While there is another data slice S_{t+1} available {	4
t=t+1	5
Run clustering algorithm with slices $S_{t-n+1} \cup S_{t-n+2} \cup ... \cup S_t$	6
Extract clusters, denote the set of clusters (partition) by $P^t = \{C_1^t, C_2^t, ..\}$	7
Match clusters in P^t to those in P^{t-1}	8
Evaluate the stability of the cluster in the cluster's history}	9
OUTPUT: partition of stable clusters	

Fig. 1. Sketch of the proposed PAMALOC algorithm

all clustering algorithms that assume a certain model for a cluster (e.g. k-Means, mixture of Gaussians).

- *Robustness wrt. parameters.* At the time of setting up the algorithm, we do not know what clusters will show up in the future. So we cannot tailor our parameters to a 'representative dataset', but it must work with a wide range of noise and cluster densities.

- *Stability wrt. input data.* The algorithm should be as stable as possible (similar input should lead to similar output). This is important since we want to trace and compare clusters obtained from different runs of the algorithm. Lack of stability might be caused by iterative optimization, where a different initialization may already lead to a different partition (k-means and derivatives, mixture of Gaussians), or by the choice of the distance measure (in hierarchical clustering with single-linkage a single new data object may have dramatic effects).

There is probably no *truly optimal* choice, but the OPTICS algorithm [2] is at least considered being a very good candidate. The cluster's shape is highly flexible due to the concept of *density-connected regions*, that is, clusters are made up by data points close to each other with their individual data point density exceeding some threshold. Regarding robustness, the MinPts parameter of the OPTICS algorithm can be used to control the robustness of the results: With MinPts=2 small perturbations may have large effects on the connectivity (similar to single-linkage clustering), but by increasing MinPts the response to small changes is reduced. Regarding stability, as long as the clusters are clearly separated and the noise level is low, the clusters exhibit themselves quite prominently in the reachability plot and OPTICS produces stable results.

However, stability becomes a major problem if the noise level increases. The top right image in Fig. 2 shows an artificial example consisting of about 500 uniformly distributed noise objects and about 200 data objects that belong to artificially superimposed clusters. It is really hard to tell whether some of the data agglomerations are small but substantial local patterns or just occurred by chance. At this noise level the random noise points lump together into spurious clusters (e.g. shaded region in the image). If we now remove some data and add some other (replace oldest with a new data slice) this will cause considerable changes in the *shape* of the *incidental data agglomerations* we see in this figure – but to a lesser extent to the shape of true clusters. The main idea for the distinction between (small local) patterns and agglomeration by chance is therefore the *stability of a cluster*, i.e. whether the cluster reappears in the next iteration. Even if the random points spontaneously agglomerate, it is very unlikely that this happens in the same way over multiple analyses. The longer an agglomeration – even if it is of low density – remains stable over time, the more likely it is that we observe some *true pattern* in the data.

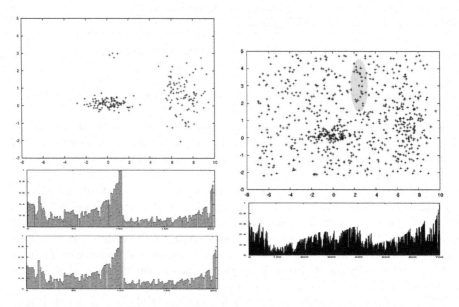

Fig. 2. The outcome of the OPTICS algorithm (reachability plot, bottom line) for the clean and a noisy data set

But still, the procedure in Fig. 1 requires that we come up with some clusters. As we have argued, we *cannot* be sure about the partition, so we are condemned to make errors. This imposes a fourth requirement:

• *Ability to preserve ambiguity in the partition.* Ambiguity in the partition can help in the sense, that we can consider more *cluster candidates* than actually expected clusters. We keep several possible clusters under consideration to be able to defer the decision making, whether a data agglomeration is incidental or significant, to a later point in time. Ambiguity can be achieved, for instance, by means of a hierarchical clustering (which is ambiguous in the sense that a particular data object may belong to several clusters). Since we can not think of a clustering algorithm that is capable of detecting *local patterns* without having to discover more global structure first (for a discussion see [7]), it seems that generally hierarchical approaches are better suited than partitional approaches.

3   Extraction of Clusters

OPTICS finds an ordering of the data objects such that the reachability plot
has the following property: consecutive data objects in the plot having all their
reachability distances below some given threshold ε belong to the same cluster (at density level $\frac{MinPts}{V_\varepsilon}$). Therefore, clusters are represented by valleys in
the reachability plot and some user-specified thresholds on the steepness of the
flanks of the valleys are used in [2] to decide whether a valley qualifies as a
deep valley and thus as a cluster. Fixing such thresholds is easy, if the cluster
density differs clearly from the background density. However, with noisy data
and comparatively small clusters of non-uniform density, clusters do not show
up that clearly in the reachability plot and fixing the thresholds is difficult if at
all possible (cf. Fig. 2, left vs. right).

We therefore propose a different approach that still uses the reachability plot
as its basis. If we draw a straight line in this plot, we thereby mark a certain data
density level. In practice we are only interested in subclusters with significantly
increased density, say, a factor of f_d higher. The same is true for the subcluster of
a subcluster. Thus we set up a cascade of density levels (given by $\varrho_i = \varrho_0 \cdot f_d{}^i$ with
initial density ϱ_0 and levels $i = 1...L$) and whenever the density of some subset of
the data drops below such a density level we consider it as a cluster *candidate*. To
depict the cluster candidates we draw a rectangle in the graphical representation
(cf. Fig. 3) whenever this happens. A deep valley in the reachability plot will
cross multiple lines and therefore indicate hierarchically nested rectangles in
Fig. 3. By considering all these sets as cluster *candidates*, we tolerate for the
moment that many of them may be incidental.

Given a data density ϱ_i, how can we match it to a vertical line in the reachability plot? Within a cluster, the reachability of a data point x corresponds
to the core distance of x, which is the smallest distance such that MinPts data
objects can be found in a hypersphere of this radius around x. Given the volume
V of the hypersphere ($V = \frac{\sqrt{\pi^d}}{\Gamma(d/2)}\varepsilon^d$ where Γ denotes the Gamma function) we
may estimate the density at x by $\varrho = MinPts/V$. Using this equation we can
transform our sequence of densities into a sequence of hypersphere radii.

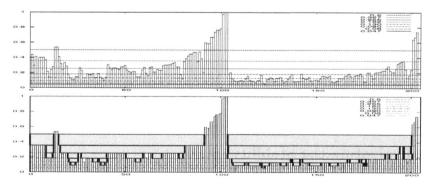

Fig. 3. Top: Reachability plot with density lines. Bottom: Whenever the reachability
drops below a density line, a new cluster candidate (shown as a rectangle) is induced

4   Matching of Clusters

Once we have two hierarchies of clusters we match the clusters against each
other to see if a cluster survives over time. As the rationale for this matching we
need some measure that indicates for each pair of clusters (one cluster candidate
from the old, one from the new hierarchy) whether it would be a good match or
not. When both partitions were obtained from two completely different data sets
(and the clusters cannot be represented by some model) such a match becomes
somewhat difficult, because we have no means to compare the clusters directly
against each other. Therefore we perform clustering on more than one data slice
per clustering, such that data from some slices are contained in *both* clusterings.

Figure 4 illustrates this: For every cluster analysis we use a data set that consists of n data slices. Then, between any two consecutive runs, we have a common
basis of $n-1$ slices for matching. Given clusters C_i^{t-1} and C_j^t from consecutive
data sets D^{t-1} and D^t, we adopt the Jaccard measure for this purpose:

$$J_{i,j} = \frac{|C_i^{t-1} \cap C_j^t|}{|(C_i^{t-1} \cup C_j^t) \setminus (S_t \cup S_{t-n})|}$$

$J_{i,j}$ becomes 1 if both, old and new cluster, contain the same data objects in the
shared part of their dataset. We have to exclude the set $S_t \cup S_{t-n}$ from the usual
Jaccard denominator, because this set does not contain shared data objects.

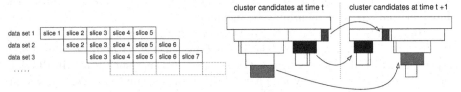

Fig. 4. Data slices used for clustering

Fig. 5. Matching of rectangles in two hierarchies

We construct a $|P^{t-1}| \times |P^t|$ cost matrix M with $M_{i,j} = 1 - J_{i,j}$. To match
the clusters pairwise, we use the bipartite graph matching algorithm by Munkres
(also known as the Hungarian algorithm) [8]. We accept an assignment of two
clusters only in case they share at least some percentage p_m of the data, otherwise
most of the data has obviously been scattered among other clusters and the
original cluster does no longer exist. We have selected a relatively low value of
$p_m = 30\%$ to account for effects such as shrinking or varying noise levels.

5   Evaluating the Stability

Whenever a new data slice arrives, we perform a new cluster analysis, extract new
clusters and match them with the old ones. To find *good clusters*, usually cluster

validity measures are employed that account for the compactness or separation of clusters (or, in case of OPTICS, the steepness of the flanks). In the presence of much noise, these measures tend to break down. We do not want to investigate clusters that are likely to vanish in the next period of time, that is, have a very short *lifetime*. By *lifetime of a cluster* we denote the number of data sets over which a cluster survives. Whenever a cluster is successfully matched into the next hierarchy, its lifetime is increased.

But not all matched clusters are matched equally well. The Jaccard measure tells us precisely how much data is shared between the old and the new cluster. Rather than requiring, e.g., three matches in a row, we may directly aggregate the matching values in the most recent r periods to a single stability value. If a cluster has been matched several times, each time with a matching value of m_i, we define its stability as: $S_t = \sum_{i=0}^{r} \beta_i \cdot m_i$. We choose β_i heuristically such that S_t becomes a weighted average. Since recent matches are more important than matches in the past, we use normalized coefficients from a Gaussian ($\beta_i = \exp(-i^2/r^2)$ normalized to $\sum_{i=1}^{r} \beta_i = 1$). The best stability value is 1.0 for a cluster that has been matched r times in the past with a Jaccard measure of 1.0. For a cluster to be stable at t, we require $S_t > 0.5$. Note that this measure does not explicitly account for the size or density of the cluster, and thus represents **no** bias towards large and dense clusters (which is important for the discovery of *local* patterns).

6   Example and Conclusion

Due to lack of space, we discuss the performance of PAMALOC for an artificial data set. We consider a sequence of 11 data slices S_t with $t = 0...10$. Every slice contains up to four (normally distributed) moving clusters, two of them are moving (at different speed) and the others (denoted by A and D) are static. The size of the clusters is also varying: The moving clusters are expanding and shrinking, respectively (by 5 data points per time step t). Cluster D will appear at $t = 7$ for the first time and its initial size is 13; with every time step t its size will increase by 3 data points. Cluster A is a very small one, it consists of 4 data points per time slice only. In the final slice ($t = 10$) the clusters consist of about 250 data points in total. We add 500 uniformly distributed data points, such that we have at least twice as much noise as substantial data (cf. Figure 2 (right) for the first data slice). In particular, cluster A, which consists of 4 data points per slice only, is impossible to detect in the right image.

Using five time slices for each OPTICS run, PAMALOC reaches the last time slice ($t = 10$) after 6 iterations, but only the last iteration is shown in Fig. 6 ($MinPts = 3$, $f_d = 1.3$). The stable clusters ($S_t > 0.5$) have been shaded: the darker the rectangle the more stable the cluster. Compared to experiments with less noise (not shown) two observations can be made: (1) due to the high noise level, the whole data set is density-connected for the first 5-6 data density levels and (2) we have more 'small clusters' (small rectangles), which is due to spontaneous data agglomerations in the noise. Most of these agglomerations vary from time slice to time slice that much, that they do not become stable.

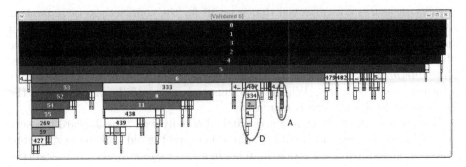

Fig. 6. Final hierarchy for example 2 (67 % noise)

At the final hierarchy only very few of these small rectangles are stable with the rectangles belonging to cluster A being the most stable among them. Cluster D is also difficult to detect, because its initial small size ($t = 7$) must compete against all the noise around it. Nevertheless, in the final hierarchy of Fig. 6 it has been recognized as a stable cluster (light shaded rectangle).

Conclusion. We have presented the PAMALOC algorithm for detecting small patterns in very noisy data. Our initial experiments show, that even in cases where we have much more noise than substantial data points, we are capable of identifying very small local patterns. We consider the fact that only a very small number of false positives were flagged as very encouraging.

References

1. Hand, D.J., Adams, N.M., Bolton, R.J. (eds.): Pattern Detection and Discovery. LNCS (LNAI), vol. 2447. Springer, Heidelberg (2002)
2. Ankerst, M., Breunig, M.M., Kriegel, H.P., Sander, J.: OPTICS: Ordering points to identify the clustering structure. In: Proc. of ACM SIGMOD Int. Conf. on Management of Data, Philadelpha, pp. 49–60. ACM, New York (1999)
3. Zhang, T., Ramakrishnan, R., Livny, M.: BIRCH: an efficient data clustering method for very large databases. In: Proc. of ACM SIGMOD Int. Conf. on Management of Data, Montreal, pp. 103–114. ACM Press, New York (1996)
4. Guha, S., Mishra, N., Motwani, R., O'Callaghan, L.: Clustering data streams. In: Proc. Ann. Symp. Foundations of Computer Science, pp. 359–366 (2000)
5. Ghosh, J., Strehl, A., Merugu, S.: A consensus framework for integrating distributed clusterings under limited knowledge sharing. In: Proc. NSF Workshop on Next Generation Data Mining, pp. 99–108 (2002)
6. Topchy, A., Minaei-Bidgoli, B., Jain, A.K., Punch, W.F.: Adaptive clustering ensembles. In: Proc. of the 17th Int. Conf. on Pattern Recognition, pp. 272–275 (2004)
7. Höppner, F.: Local pattern detection and clustering – are there substantive differences? In: Morik, K., Boulicaut, J.-F., Siebes, A. (eds.) Local Pattern Detection. LNCS (LNAI), vol. 3539, pp. 53–70. Springer, Heidelberg (2005)
8. Munkres, M.: Algorithms for the assignment and transportation problems. Journal of the Society of Industrial and Applied Mathematics 5(1), 32–38 (1957)

Searching for Better Randomized Response Schemes for Privacy-Preserving Data Mining

Zhengli Huang, Wenliang Du, and Zhouxuan Teng

Department of Electrical Engineering and Computer Science,
Syracuse University, Syracuse, NY 13244, U.S.A.
{zhuang,wedu}@ecs.syr.edu,zteng@syr.edu

Abstract. To preserve user privacy in Privacy-Preserving Data Mining (PPDM), the randomized response (RR) technique is widely used for categorical data. Although various RR schemes have been proposed, there is no study to systematically compare them in order to find optimal RR schemes. In the paper, we choose the R-U (Risk-Utility) confidentiality map to compare different randomization schemes. Using the R-U map as our metric, we present an optimal RR scheme for binary data, which helps us find an optimal class of RR matrices. From this optimal scheme, we have discovered several heuristic rules among the elements in the optimal class. We generalize these rules to find optimal class of RR matrices for categorical data. Based on these rules, we propose an RR scheme to find a class of RR matrices for categorical data. Our experimental results have shown that our scheme has much better performance than the existing RR schemes.

1 Introduction

Data mining is a technology to extract useful patterns and trends from a large sum of data. If the original information is obtained directly, many techniques, e.g. statistics, artificial intelligence, can be utilized to analyze the information. Generally, the data are collected from different individuals or data owners. The data owners, however, might not be willing to share their data to others because of privacy concerns. This motivates a field called Privacy-Preserving Data Mining (PPDM). In PPDM, the original data are disguised; with the disguised data, the data patterns included in the original data can still be extracted. This is becoming more and more important in current society, especially when privacy is becoming a prominent concern.

When data are categorical, randomized response (RR) technique can be used to achieve the privacy-preserving data mining. The RR technique for binary data was first introduced by Warner [13], which can be easily extended to general categorical data [4]. In the generalized RR scheme, there is a transformation matrix M (called RR matrix) of $n \times n$, for n of categories of the data. With a RR matrix M, the original data are randomized and produce disguised data. Different M's will achieve different accuracy level (regarding data mining results) and different privacy level (regarding how much of the original data is disclosed). How to

J.N. Kok et al. (Eds.): PKDD 2007, LNAI 4702, pp. 487–497, 2007.
© Springer-Verlag Berlin Heidelberg 2007

choose the elements of M has been studied in [5]. But when the paper compares different schemes, it only compares accuracy, which is just one important aspect of a RR scheme. Because the two aspects - accuracy and privacy - are all important in an RR scheme, using any single one of them to compare different schemes is incomplete. The first goal of this paper is to find a way to compare different RR matrices. Once we can fairly compare different matrices, our next question is whether there is a way to find an optimal class of RR matrices; namely, given an accuracy level, we would like to generate a matrix with optimal privacy.

To compare different randomized response schemes, we use a Risk-Utility confidentiality (R-U) map, proposed by Duncan et. al [7]. R-U map is a curve describing the relationship of accuracy versus privacy when the parameters of a RR scheme change. Armed with the R-U map which can be used to compare different RR schemes, we would like to find an optimal class of RR matrices, forming a R-U map that is consistently better than the R-U maps of other RR schemes.

To find the optimal class of RR matrices, we present a deterministic algorithm to find an optimal class of RR matrices for binary data (two categories) which have the highest utility given any risk value. According to the deterministic approach for binary data, we have identified several heuristic rules from the optimal matrices. By using the rules and a constraints possibly added by the data owners, we propose an algorithm that can find a better class of RR matrices for categorical data. Our experiments show that the R-U map of the class of RR matrices generated by our algorithm is much better than the existing schemes.

The rest of this paper is organized as follows. Section 2 reviews the related work. The problem formulation is presented in Section 3. A deterministic approach to finding the optimal RR matrices for binary data is presented in Section 4. Our algorithm to find a class of better RR matrices for categorical data is described in Section 5. Section 6 experimentally evaluates our scheme. Section 7 concludes the paper.

2   Related Work

Randomization is one of the major approaches in privacy-preserving data mining. There are two different randomization methods: the Random Perturbation scheme and the Randomized Response scheme.

Agrawal and Srikant propose a scheme for privacy-preserving data mining using randomized perturbation [3]. This work has been extended by Agrawal and Aggarwal [2], who propose an Expectation Maximization (EM) algorithm for distribution reconstruction. Kargupta et al. point out an important issue, that is, arbitrary randomization is not safe [10]. We study why and how correlations affect privacy [9].

The randomized response is mainly used to deal with categorical data and originally proposed by Warner [13]. Other works based on the randomized response can be found in [6,8,5,11,12]. The work closest to our study is that of Agrawal and Haritsa [5].

3 Problem Statement

3.1 General Randomized Response Technique for Categorical Data

The randomized response technique is mainly applied to each individual attribute of a data set (although it can be applied to multiple attributes); therefore, in this paper, we only study one attribute. Suppose that an attribute of the original data has n different categories, e.g. $\{c_1, c_2, ..., c_n\}$. The randomized response technique replaces each c_i with c_j with certain probability. We use $\theta_{j,i}$ to denote the probability that category c_i is randomized to c_j, where $i, j = 1, \ldots, n$. Based on the randomized response technique, we have the following equation:

$$\boldsymbol{P^*} = M\boldsymbol{P}, \quad where \ \boldsymbol{P^*} = \begin{pmatrix} P^*(c_1) \\ \vdots \\ P^*(c_n) \end{pmatrix}, M = \begin{pmatrix} \theta_{1,1} \ \ldots \ \theta_{1,2} \\ \vdots \ \ddots \ \vdots \\ \theta_{n,1} \ \ldots \ \theta_{n,n} \end{pmatrix}, \boldsymbol{P} = \begin{pmatrix} P(c_1) \\ \vdots \\ P(c_n) \end{pmatrix}.$$

Based on maximum likelihood estimation, we can derive the following theorem (the proofs for all the theorems are omitted due to page limitation):

Theorem 1. *The MLE (Maximum Likelihood Estimate) of the probabilities $\boldsymbol{P}$ or $(\widehat{P(c_1)}, ..., \widehat{P(c_n)})^T$ is $\widehat{\boldsymbol{P}} = M^{-1}\widehat{\boldsymbol{P^*}}$, where $\widehat{\boldsymbol{P^*}} = (\frac{N_1}{N}, \frac{N_2}{N}, \ldots, \frac{N_n}{N})^T$, N_i (i = 1, ..., n) is the number of c_i in the randomized data and N is the total number of the records in the data set. It is an unbiased estimate.*

3.2 Problem Formulation

When applying an RR scheme to the data set, both privacy and accuracy are desirable, but the two factors cannot be optimized at the same time; there are tradeoffs. If we want high accuracy, privacy must be disclosed to some extent. When different schemes have different tradeoffs, which tradeoff is better? i.e., which scheme is better? This is an optimization problem: *Given a data set, find the optimal randomization scheme which can achieve the optimal performance in terms of privacy and accuracy.*

There are two main challenges in solving this optimization problem. First, when comparing two different RR schemes, what does the "better performance" means? Second, in a straightforward brute-force approach, the search space for finding an optimal solution is large.

3.3 R-U Confidentiality Map

We choose to use Risk-Utility confidentiality map (R-U map) [7] to compare RR schemes. An R-U map is a curve describing the impact on disclosure risk and data utility when the parameters of a scheme changes. It uses risk as the x-axis and utility as the y-axis.

By drawing the R-U maps, the comparisons between two schemes become easy. We can easily see when one scheme has better performance than the other, for example, when the utility is higher for one scheme than the other under certain risk range. Based on the maps, users can choose different parameter for different schemes under their risk and utility requirements. Therefore, we mainly use the R-U maps to compare different randomization schemes, and based on the maps to examine our scheme.

3.4   Utility

Since the probability distribution of the data set is used to conduct most data mining computations, it is reasonable to study the distribution when analyzing the utility of the data set. We use the mean squared errors to measure the difference between the estimated distribution and the original distribution.

Mean Squared Error. Let the probability of each data value is $P(c_k), k = 1, 2, ..., n$ and the corresponding estimated probability is $\widehat{P(c_k)}$. The mean squared error of each probability estimator is defined as follows.

Definition 1. *The Mean Squared Error (MSE) of an estimator $\widehat{P(c_k)}$ of the probability $P(c_k)$ is the function of $P(c_k)$, which is $E_{P(c_k)}(\widehat{P(c_k)} - P(c_k))^2$, where $E_{P(c_k)}(.)$ is the expected value.*

The MSE can be calculated from the following theorem.

Theorem 2. *The Mean Squared Error (MSE) of an estimator $\widehat{P(c_k)}$ from the probability $P(c_k)$ is*

$$\sum_{i=1}^{n} \beta_{k,i}^2 Var(\frac{N_i}{N}) + 2 \cdot \sum_{i=1,j=1,i\neq j}^{n} \beta_{k,i}\beta_{k,j} Cov(\frac{N_i}{N}, \frac{N_j}{N}), \tag{1}$$

where $\beta_{g,h}$ is the gth row and hth column element of the inverse of M, $Var(\frac{N_i}{N})$ is the variance of $\frac{N_i}{N}$, $Cov(\frac{N_i}{N}, \frac{N_j}{N})$ is the covariance of $\frac{N_i}{N}$ and $\frac{N_j}{N}$ and they are

$$Var(\frac{N_i}{N}) = \frac{1}{N} \cdot P^*(c_i)(1 - P^*(c_i)), \quad Cov(\frac{N_i}{N}, \frac{N_j}{N}) = -\frac{1}{N} \cdot P^*(c_i) \cdot P^*(c_j).$$

We use the average MSE of $\widehat{P(c_k)}$ as the utility of the estimator of the probability distribution.

Definition 2. *The Average Mean Squared Error (AMSE) of an estimator $\widehat{P}$ of the probability distribution P is $\frac{1}{n}\sum_{k=1}^{n} E_{P(c_k)}(\widehat{P(c_k)} - P(c_k))^2$.*

When AMSE is large, the difference between the two probability distributions are large. In the Randomized Response scenario, large AMSE means that the estimation error is high so that the accuracy of the estimated probability distribution is low. We will use it as the utility criteria to study different RR schemes.

3.5   Risk

There are many different risk measurements. Based on different risk measurements, the R-U maps will be different and the optimal RR scheme will be different. To measure the risk, we propose a new approach of privacy quantification, which is considered as the following estimation problem:

Problem 1. Given a disguised data set $Y_s = \{y_1, \ldots, y_N\}$ and certain prior knowledge of the original data X, adversaries would like to estimate the original value x_i from y_i, for $i = 1, \ldots, N$. How accurate can their estimates be?

The less accurate their estimates are, the higher privacy is. Therefore, to quantify privacy, we just need to quantify the accuracy of the estimates. We define an accuracy function G to represent the accuracy score of an estimate (denoted by $\widehat{x}_i$) against the actual value x_i. With this function, the average accuracy score of all the estimates can be computed in the following:

$$A = \frac{1}{N} \sum_{i=1}^{N} G(\widehat{x}_i, x_i). \tag{2}$$

We will derive the optimal value for A, which represents the best estimates that can be achieved by adversaries. We will then use this optimal result to quantify privacy.

We use $\widehat{X_Y}$ to represent this estimate. However, according to the randomized response technique, the same value Y might be the disguised results of different values from $C = \{c_1, \ldots, c_n\}$. To maximize accuracy, given a specific Y, adversaries would like to find an estimate $\widehat{X_Y}$, such that the expected accuracy score is maximized. The expected accuracy score can be computed using the following formula:

$$E_X[G(\widehat{X_Y}, X) \mid Y] = \sum_{X \in C} G(\widehat{X_Y}, X) \cdot P(X \mid Y). \tag{3}$$

This is actually the *Bayes Estimate*, the theory of which not only provides optimal estimates for a variety of accuracy functions G, but also provides a methodology to derive optimal estimates for an arbitrary G. In this paper, we study one specific accuracy function:

$$G(\widehat{X}, X) = \begin{cases} 1, & \text{when } \widehat{X} = X; \\ 0, & \text{otherwise.} \end{cases} \tag{4}$$

The intuitive meaning of the above function says that when an estimate is correct, the score is 1; otherwise, the score is 0.

Theorem 3. *For the accuracy function G defined in Equation (4), the optimal estimate $\widehat{X_Y}$ for a given Y is the MAP (maximum a posteriori) estimate, i.e.,*

$$\widehat{X_Y} = argmax_{X \in C} P(X \mid Y) \tag{5}$$

From Theorem 3, we know that the MAP estimate is optimal. In other words, MAP estimate gives an upper bound on what adversaries can achieve. Therefore, MAP estimate can be used to quantify the privacy.

In Equation (3), the expected accuracy value is computed for a particular Y over all $X \in C$. To consider all $Y \in C$ from the disguised data set, we compute another expected value, but this time, over Y. This is the expected accuracy value for the entire disguised data. Namely, the value A defined in Equation (2) can be computed using the following formula, rather than using the sample means:

$$A = E_Y\{E_X[G(\widehat{X_Y}, X) \mid Y]\} = \sum_{Y \in C} P(\widehat{X_Y} \mid Y) \cdot P(Y) = \sum_{Y \in C} P(Y \mid \widehat{X_Y}) \cdot P(X_Y),$$

where $\widehat{X_Y}$ is the MAP estimate for an given Y and the last equality is from Bayes rules.

The value of A represents the estimation accuracy from adversary's perspective; the larger A is, the worse for privacy. Therefore, we define the BE privacy as $1 - A$, i.e.,

$$\mathcal{P}_{BE} = 1 - \sum_{Y \in C} P(Y \mid \widehat{X_Y}) \cdot P(X_Y). \tag{6}$$

4   Optimization of RR for Binary Data

With the R-U maps, we can evaluate different RR schemes. Starting from the simplest, we take into account them for binary data (only two value c_1, c_2 in data), and present an approach to find the optimal RR matrices with highest accuracy given certain privacy level. We denote the RR matrix M to be

$$M = \begin{pmatrix} \theta_{1,1} & \theta_{1,2} \\ \theta_{2,1} & \theta_{2,2} \end{pmatrix}.$$

Because the sum of each column is 1, a combination of the two elements $\theta_{1,1}$ and $\theta_{2,2}$ will determine the M.

We analyze distribution of the BE privacy along $\theta_{1,1}$ and $\theta_{2,2}$, which can be obtained from the following:

Theorem 4. *When the BE privacy for binary data is less than $min\{P(X = c_1), P(X = c_2)\}$, it is $\mathcal{P}_{BE} = (1 - \theta_{2,2})P(X = c_2) + (1 - \theta_{1,1})P(X = c_1)$ or $\mathcal{P}_{BE} = \theta_{1,1}P(X = c_1) + \theta_{2,2}P(X = c_2)$*

If there is a coordinate system composed of variables $\theta_{1,1}$ and $\theta_{2,2}$, a BE privacy value ($< min\{P(X = c_1), P(X = c_2)\}$) is represented by a straight line in the coordinate system. Actually, 2 corresponding straight lines correspond to a privacy value and they are centrosymmetric with respect to the center point $(0.5, 0.5)$.

After conducting many experiments, we find, in the optimal matrix with the highest BE privacy $\mathcal{P} = P(c_1)$ ($P(c_1) < P(c_2)$), some interesting rules, or equations or inequality between different elements:

- $\theta_{1,1} * P(c_1) = \theta_{1,2} * P(c_2)$. This means that the proportion of c_1 randomized from original c_1 is the same as that from original c_2.
- $\theta_{1,1} > \theta_{2,2}$ and $\theta_{1,1}$ is as large as possible (here is 1). This results in the proportion of c_1 and that of c_2 in the randomized data being closer to each other.

The above two rules exist in the optimal RR matrices for binary data and probably also exist in those for categorical data.

5   Improved RR Scheme for Categorical Data

After finding the optimal matrices for binary data which can achieve much better R-U map than Warner's scheme and observing the rules in an optimal matrix, we generalize the rules for categorical data. Based on the generalized rules together with another one obtained from other work, and under some constraint probably imposed by the data owners, we propose a heuristic approach which can find a RR matrix for categorical data achieving much better performance than the existing approaches. We introduce the rules and the constraint below.

5.1   Rules and Constraint

Rule 1. The first rule is from that used for binary data in the previous section. Suppose that the probabilities are sorted in an increasing order, i.e. $P(X = c_1) \leq ... \leq P(X = c_n)$. We have

$$\theta_{r,r} * P(X = c_r) = \theta_{r,r+1} * P(X = c_{r+1}) = ... = \theta_{r,n} * P(X = c_n). \quad (7)$$

Rule 2. The second rule is also from that used for binary data. If there are n different values $c_1, ..., c_n$ in a data set and their probabilities are in an incremental order, we have $\theta_{1,1} \geq \theta_{2,2} \geq ... \geq \theta_{n,n}$ in M.

However, when there are more than two categories in the data, the two rules can not be satisfied simultaneously. To control the tradeoff between the two rules, we introduce a variable, the upper bound of the difference between two neighbouring diagonal elements,

$$E_{dif}, \quad s.t. \quad \theta_{i,i} \leq \theta_{i+1,i+1} + E_{dif},$$

where $i = 1, ..., n - 1$. The variable will be used in our algorithm to find the optimal M.

Rule 3. From [5], the condition number is an important parameter to decide the estimation errors of the original probability distribution; the smaller the number is, the higher the estimation accuracy level. We wish the condition number of the optimal M has an upper bound. Since the eigenvalues of an upper triangular matrix are the diagonal elements of the matrix, the lower bound of the condition number of a optimal M is $\frac{max_i(\theta_{i,i})}{min_j(\theta_{j,j})} < \delta$, where $i, j = 1, ..., n$, δ is a variable (> 1).

Because the 1st diagonal element is usually 1, we simplify the above inequality as

$$\theta_{j,j} > E_{min}, \; for \; j = 1, ..., n,$$

where E_{min} is a positive variable to control the condition number.

Constraint From Data Owners. Besides the rules for the optimal matrix, there could be a constraint imposed by the data owners. If any element in M is 0, e.g. $\theta_{1,3} = 0$, the adversaries can immediately know that c_1 in randomized data must not come from c_3. This situation should be prevented. So the data owners might want to add a constraint which is $P(Y = c_j | X = c_i) > \sigma$, where $i, j = 1, ..., n$ and σ is a small number decided by the users.

Based on the rules and constraint, we propose an algorithm to find improved RR matrices for categorical data of a given probability distribution. We will show that the RR matrices achieve much better performance than Warner's scheme and FRAPP scheme [5] with experiments.

5.2   Algorithm

Suppose that the data set has n categories and the probability distribution $\boldsymbol{P}$. For 2 specified variables E_{min} and E_{dif}, both of which can change from 0 to 1, our algorithm is designed to find better RR matrices for the categorical data. The algorithm uses a brute force method, discretizing E_{min} and E_{dif} due to certain step length and traversing all the combination of both variables to find the matrices. The brute force method has only 2 variables of the ranges from 0 to 1 so that it has much smaller search space than the brute force method used on all elements of an RR matrix M which has n^2 variables of range from 0 to 1. The algorithm is shown below.

Algorithm. *Finding − Improved − M($\boldsymbol{P}, E_{dif}, E_{min}$)*

Input: the probability distribution of the data $\boldsymbol{P}$,
the maximal difference between two consecutive diagonal elements E_{dif},
the minimal value of the diagonal elements E_{min}
Output: the RR matrix M of $n \times n$
1. sort the probability distribution $\boldsymbol{P}$ in an increasing order
   //The following is **Step 1**. The elements are determined column by column
2. **for** $c \leftarrow 2$ **to** n
3.   *Phase 1*: set the first $c - 1$ elements of a column to met Rule 1
      //Tradeoff between Rule 1&2 is obeyed in *Phase 2*. So is Rule 3
4.   *Phase 2*: subtract values from the non-diagonal elements if needed, set
      the diagonal element of the column
5. **end for**
6. **Step 2**: Modify small elements in M to satisfy the Constraint
7. restore the matrix M with its original order
8. return M

Complexity Analysis. Since in the main loop of the algorithm Phase 1 and Phase 2 all have computational overhead of $O(n)$, the loop has computational overhead of $O(n^2)$. The process of modifying the small elements has computational overhead of $O(n^2)$. So the total computational overhead is still $O(n^2)$. The space complexity is also $O(n^2)$ because of the used space for storing the matrix M. Suppose that E_{dif} and E_{min} are all discretized into s different intervals, the whole algorithm has the computational overhead of $O(s^2 \cdot n^2)$. But for the space overhead, if we want to store all the optimal matrix M, the total space overhead is $O(s^2 \cdot n^2)$. If s is not too large and data sets do not have too many categories (n is not very large), the algorithm is fast and does not consume much space.

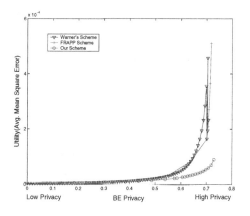

Fig. 1. Experiment 1: Utility(Average Mean Square Error) vs. Risk(BE privacy)

6   Experiments

We design a series of experiments to compare our RR scheme and other RR schemes, i.e. the Warner's scheme and FRAPP [5]. Our experiments use real data, *Adult* data set in the UCI Machine Learning Repository [1] which has 14 attributes, including some continuous attributes. We discretize those continuous attributes in order to utilize any RR scheme. Because of the space limitation, we only show the results for the 1st attribute, but the results for other attributes have shown the same trends as the attribute.

The results in Figure 1 all show that our scheme has almost the same utility as Warner's scheme and the FRAPP scheme when BE privacy level is low while ours has much better utility (lower AMSE) than them when the BE privacy level becomes higher. This means that our scheme has much better performance than them especially when the BE privacy level is high.

The reason why our scheme achieves better performance than them is the following. An optimal class of RR matrices possess many rules, which are any equations or inequalities between different elements. The more rules the RR matrices generated with a scheme possess, the more similar to the optimal the generated matrices are. The more similar the RR matrices generated by a scheme is to the optimal, the better performance the scheme has. In our scheme, by following the rules and constraint, our generated RR matrices possess more rules than the matrices generated by other schemes, e.g. Warner's and FRAPP. So our scheme achieves better performance than them.

7   Conclusion

In this paper, using R-U maps as a metric to systematically compare different RR schemes we present an approach to find an optimal class of RR matrices for binary data. In the optimal class, we discover some heuristic rules. By using the rules and a constraint probably imposed by the data owners, we propose a deterministic approach to finding a better class of RR matrices for categorical data. The experiments show that our scheme achieves much better performance than the existing RR schemes according to their R-U maps. Our future work is to continue searching for the optimal RR schemes for categorical data. We will also apply our scheme to some data mining computations to examine how utility is affected.

References

1. http://www.ics.uci.edu/~mlearn/MLRepository.html
2. Agrawal, D., Aggarwal, C.: On the design and quantification of privacy preserving data mining algorithms. In: Proccedings of the 20th ACM SIGACT-SIGMOD-SIGART Symposium on Principles of Database Systems, Santa Barbara, California, USA, ACM Press, New York (2001)
3. Agrawal, R., Srikant, R.: Privacy-preserving data mining. In: Proceedings of the 2000 ACM SIGMOD on Management of Data, pp. 439–450, Dallas, TX USA, May 15–8 (2000)
4. Agrawal, R., Srikant, R., Thomas, D.: Privacy preserving olap. In: Proceedings of the ACM SIGMOD Conference (SIGMOD 2005), Baltimore, Maryland, USA, June 14-16, 2005., ACM Press, New York (2005)
5. Agrawal, S., Haritsa, J.: A framework for high-accuracy privacy-preserving mining. In: Proceedings of 21st IEEE Intl. Conf. on Data Engineering (ICDE), Tokyo, Japan (2005)
6. Du, W., Zhan, Z.: Using randomized response techniques for privacy-preserving data mining. In: Proceedings of The 9th ACM SIGKDD International Conference on Knowledge Discovery and Data Mining, Washington, DC, USA, August 24-27, 2003 (2003)
7. Duncan, G.T., Keller-McNulty, S.A., Stokes, S.L.: Disclosure risk vs. data utility: The r-u confidentiality map. In: Technical Report Number 121, National Institute of Statistical Sciences (December 2001)
8. Evfimievski, A., Srikant, R., Agrawal, R., Gehrke, J.: Privacy preserving mining of association rules. In: Proceedings of 8th ACM SIGKDD International Conference on Knowledge Discovery and Data Mining, Edmonton, Alberta, Canada (July 2002)
9. Huang, Z.: Deriving private information from randomized data. In: Proceedings of the ACM SIGMOD Conference (SIGMOD 2005), Baltimore, Maryland, USA, June 14-16, 2005 (2005)
10. Kargupta, H., Datta, S., Wang, Q., Sivakumar, K.: On the privacy preserving properties of random data perturbation techniques. In: the IEEE International Conference on Data Mining, Melbourne, Florida, November 19 - 22, 2003, IEEE Computer Society Press, Los Alamitos (2003)

11. Rizvi, S., Haritsa, J.R.: Maintaining data privacy in association rule mining. In: Proceedings of the 28th VLDB Conference, Hong Kong, China (2002)
12. Krishnan, V., Agrawal, S., Haritsa, J.R.: On addressing efficiency concerns in privacy-preserving mining. In: Lee, Y., Li, J., Whang, K.-Y., Lee, D. (eds.) DASFAA 2004. LNCS, vol. 2973, Springer, Heidelberg (2004)
13. Warner, S.L.: Randomized response: A survey technique for eliminating evasive answer bias. The American Statistical Association 60(309), 63–69 (1965)

Pre-processing Large Spatial Data Sets with Bayesian Methods

Saara Hyvönen, Esa Junttila, and Marko Salmenkivi

Helsinki Institute for Information Technology, Department of Computer Science
University of Helsinki, Finland
{saara.hyvonen,esa.junttila,marko.salmenkivi}@cs.helsinki.fi

Abstract. Binary data appears in many spatial applications such as dialectology and ecology. We demonstrate that a simple Bayesian modeling approach can be used in pre-processing large spatial data sets with missing or uncertain data. Our experiments on real and synthetic data show that conducting the pre-processing phase before applying conventional data mining methods, such as PCA, clustering or NMF, improves the results significantly.

Keywords: spatial data, pre-processing, Bayesian methods.

1 Introduction

While Bayesian methods have been used in confirmatory data analysis in various application areas, they have not been commonly applied to data mining tasks with relatively modest prior knowledge. In this paper we demonstrate that it is feasible to pre-process large spatial binary data sets with Bayesian methods and – what is the main point – to obtain good results in subsequent analyses.

In real applications the quality of raw data is often unsatisfactory. Missing data, noise, and uncertainty may distort the results of a straightforward application of many data mining methods. Bayesian methods provide tools for modeling, e.g., the missing data explicitly. Our experiments show that a relatively simple Bayesian model improves considerably the quality of the results achieved by different data mining methods. The applied model turns out to be practically "parameter-free": the specified prior distributions have very little influence on the results. The model is essentially based on the well-known Ising model [9], and it was first introduced in [7].

In this paper we investigate a large collection of geographical distributions of Finnish dialect words, and synthetic data sets. In particular, we study the influence of modeling the missing data with Bayesian methods on the results of three kinds of subsequent analyses: principal components analysis, k-means clustering and nonnegative matrix factorization.

Bayesian methods have been employed in spatial confirmatory data analysis in, e.g., ecology, epidemiology, and image reconstruction [1,2,5,9]. Hyvönen et al. analyze a part of the dialect data set used in this paper with several multivariate

J.N. Kok et al. (Eds.): PKDD 2007, LNAI 4702, pp. 498–505, 2007.
© Springer-Verlag Berlin Heidelberg 2007

methods [6]. MDL-principle and association analysis methods have been applied to spatial presence-absence data [8,10]. These approaches ignore the problem of missing data.

The rest of the paper is organized as follows. We introduce the modeling approach and the dialect data set in Section 2. Section 3 compares the results of the subsequent analysis on the original and the pre-processed data. A general discussion is presented in Section 4. Section 5 is a conclusion.

2   Spatial Modeling with Markov Random Fields

Observations at two locations close to one another are often relatively similar, that is, they are spatially *autocorrelated*. Instead of trying to find out all the reasons for the similarity, spatial models often make assumptions of autocorrelation to cover the influence of these unobserved factors. Markov random fields (MRF) are typically employed to model autocorrelation.

Given a neighbor graph of regions, a probability distribution is an MRF, if $\text{PR}(X_i \mid \mathcal{X} \setminus \{X_i\}) = \text{PR}(X_i \mid \mathcal{N}(X_i))$, where $\mathcal{X}$ is the set of all variables, and $\mathcal{N}(X_i)$ is the set of variables associated with the regions that are neighbors of i. Thus, a random variable associated with region i is independent of the variables in all the other regions, given the values of variables in the neighbor regions.

Bayesian modeling requires setting up a joint distribution of the model parameters and data. It is not trivial to specify a valid distribution function that meets the Markov property in spatial domain. The function is valid (*Hammersley-Clifford theorem* [9]) if and only if it is of the form $\text{PR}(\boldsymbol{\omega}) = \frac{1}{Z} \cdot \exp((\sum_{C \in \mathcal{C}} V_C(\boldsymbol{\omega})))$. Here $\boldsymbol{\omega}$ is a vector of values of all variables in the MRF, and Z is a normalizing constant. Further, V_C is a *potential function* of clique C, and $\mathcal{C}$ is the set of all cliques in the neighbor graph. Given that $V_C(\boldsymbol{\omega})$ depends only on the values of the vertices in C, functions V_C may be chosen arbitrarily.

2.1   Dialect Word Data

During the process of writing a comprehensive dictionary of Finnish dialects, a large set of maps describing the regional distribution of the dialect words have been compiled in electronic form. Combining these distributions yields a 17,100 × 563 binary matrix (words × municipalities), the proportion of 1s being 4%.

The overall collection is far from uniform. Roughly 15 % of the municipalities have been systematically surveyed, but even here the number of recorded words varies remarkably. The collections from the rest of the municipalities are often much smaller. For the purpose of compiling the dictionary the data is generally quite good, but the spotty coverage has proved to be one of the main issues in data analysis of the collections [6]. Social relationships spread dialect words, primarily through neighboring areas. Hence, it is likely that an assumption of autocorrelation can be utilized in modeling missing data.

Denote by $y_{m,d}$ the data item indicating, whether word d was recorded in municipality m, and by $x_{m,d}$ the unknown actual status of usage of d in m. We specify a Bayesian model that estimates for each zero in the data the probability that the correct value is actually one. The likelihood $\mathrm{PR}(y_{m,d} = 1 \mid x_{m,d}, r_m) = x_{m,d} \cdot r_m$ of observing a word in m depends on the unobserved quantity r_m that can be interpreted as being related to the research activity in m. The greater the value of r_m, the greater the probability of observing a word in m, given that the word is used in m. We set the uniform prior distribution $r_m \sim \mathrm{Unif}(0,1)$.

An MRF (Ising model) is defined for each word to model the spatial dependencies: the larger the proportion of neighbors n of m having $x_{n,d} = 1$, the more evidence we have for $x_{m,d} = 1$. Two municipalities are defined to be neighbors, if they have at least one common point in their borders. Word-specific variables β_d control the strength of autocorrelation. If $\beta_d = 0$, the neighboring municipalites are ignored. Intuitively, the greater the value of β_d the more probability mass is given to the configurations of $x_{m,d}$ with coherent areas of zeros and ones.

Fig. 1 shows a graphical representation of the model. Denote by $\boldsymbol{x_d} = (x_{1,d}, x_{2,d}, \ldots)$ all the variables in the MRF associated with dialect word d. The joint distribution of the model $\mathcal{M}$ is $\mathrm{PR}(\mathcal{M}) = (\prod_m \mathrm{PR}(r_m)) \cdot \prod_d \mathrm{PR}(\beta_d) \cdot \mathrm{PR}(\boldsymbol{x_d} \mid \beta_d) \cdot \prod_m \mathrm{PR}(y_{m,d} \mid x_{m,d}, r_m)$. We next specify $\mathrm{PR}(\boldsymbol{x_d} \mid \beta_d)$ in detail.

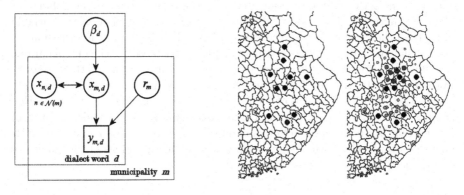

Fig. 1. Left: graph representation of a model with MRF dependencies between neighbor municipalities ($\mathcal{N}(m)$ is the set of municipalities that are neighbors of m). Right: observations of word *korahtaa*, and the approximated posterior probabilities of actual occurrences for the same word.

Each municipality forms a clique, each pair of neighboring municipalities forms a clique of size two etc. In order to achieve the desired interaction between municipalities the cliques of size two are essential. Thus, for the cliques of the other sizes we set $V_C = 0$. For the cliques of size two we assign a word-specific potential function $V_{\{m,n\}}(\boldsymbol{x_d}) = \beta_d$, if $x_{m,d} = x_{n,d}$, and 0 otherwise. Values of β_d are treated as unknown parameters. We set the prior distribution $\beta_d \sim \mathrm{Unif}(0,10)$, which allows no autocorrelation as well as very high correlation. We obtain [9] conditional probabilities for $x_{m,d}$ as $\mathrm{PR}(x_{m,d} = 1 \mid \beta_d, x_{s,d}, s \neq m) = Q/(1+Q)$, where $Q = \exp((\beta_d \cdot \sum_{j \in \mathcal{N}(m)} (2x_{j,d} - 1)))$ and $\mathcal{N}(m)$ is the set of municipalities that are neighbors of m.

Based on a priori knowledge of dialect words, we know that missing observations are, in practice, potential occurrences only within a reasonable distance
from some observation. Ignoring this prior information leads to incorrect simulated occurrences, particularly in edge areas. Thus, we set the probability of
occurrence to zero in the remote municipalities, that is, far from any observation. After conducting trials with different reasonable criteria for remoteness,
we defined a municipality to be *remote* with respect to word d, if there are no
observations of d within k steps in the neighbor graph. This practice eliminated
the edge effects, while the results in other respects showed no significant changes
when a value of $k \geq 3$ was used.

We employed MCMC methods (see, e.g., [4]) to approximate the posterior distribution and to obtain probabilities (expectations of variables x) of word usages
in municipalities. A single run (110,000 sweeps, Linux 3 GHz) on the whole data
set took approx. five days. We tested the convergence with several well-known
methods (Gelman–Rubin, Geweke, Heidelberger–Welch, Raftery–Lewis, see, e.g.
[3]). The exact computation of $\text{PR}(\boldsymbol{x_d} \mid \beta_d)$ is intractable, and we applied the
common pseudo-likelihood approximation, see [5]. Fig. 1 illustrates the marginal
posterior distributions of word occurrences of a single word.

3   Subsequent Analysis

Next we compare the performance of principal components analysis (PCA), nonnegative matrix factorization (NMF) and clustering on the original and pre-
processed dialect data and a synthetic data set.

Principal Components Analysis. The aim of PCA is to capture the intrinsic
variability in the data. Figs. 2a,d show PC 1 for the original and pre-processed
data; it essentially tells about the number of words in each municipality. We
discuss Figs. 2a,d in more detail in Sec. 4. The next component in Figs. 2b,e
captures the east-west variation, which is known to be the dominant direction
of Finnish dialect variation. PC 3 in Figs. 2c,f shows the north-south variation.

Pre-processing clearly improves the results: the plots for the original data are
grainy while the plots for the pre-processed data exhibit a smooth variation.
The graininess is due to the uneven sampling of the original data. The divisions of western/eastern and northern/southern dialects are much clearer in the
preprocessed data than in the original data. A similar effect is present for the
subsequent components as well.

Nonnegative Matrix Factorization. Given a data matrix $\mathbf{D} \in \mathbb{R}^{m \times n}$ NMF finds
nonnegative matrices $\mathbf{W} \in \mathbb{R}^{m \times k}$ and $\mathbf{H} \in \mathbb{R}^{k \times n}$ such that $\mathbf{D} \approx \mathbf{WH}$. This
means that each data vector is expressed as a linear combination of k nonnegative
factors (columns of $\mathbf{W}$). These factors can be interpreted as corresponding to
different dialect regions. A geographical distribution of a word is then expressed
as a weighted combination of these factors. A large k yields relatively local dialect
regions. We show the results for $k = 3$ in Fig. 3. Indeed we observe the eastern,
northern and western dialect regions in Figs. 3a,d, 3b,e and 3c,f respectively.

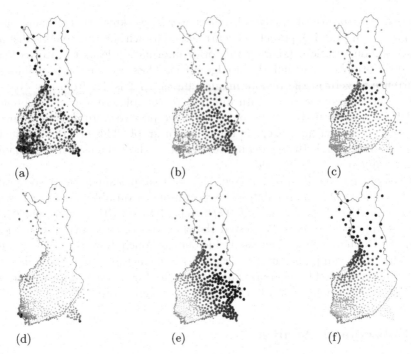

Fig. 2. The first 3 PCA components for the original (a-c) and pre-processed (d-f) data

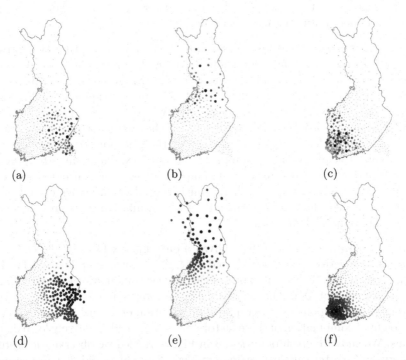

Fig. 3. NMF components ($k = 3$) for the original (a-c) and pre-processed (d-f) data

Again the factors computed on the pre-processed data are smoother and easier to interpret. This phenomenon is also apparent for different choices of k.

Clustering. Dialects are traditionally divided into specific dialect regions, which makes clustering a natural approach to dialect data. The k-means clustering using the Euclidean distance fails for the raw data, because municipalities with few words resemble each other. Fig. 4 shows that the problem of clearly incorrectly clustered points is treated by the pre-processing ($k = 6$). This also holds for other choices of k. Employing the Cosine distance will get rid of the "scatter cluster" ($+$): in those cases the clusterings on both data sets look almost similar.

Synthetic Data. We have also experimented on different synthetic data sets. We present the results for the simple but illustrative case, in which we generate a rectangular map consisting 18 by 20 grid cells. These are divided into three distinct regions as shown in Fig. 5a. Each of the 3000 features is assigned to a single region. Now we remove a fraction f_j of the data in each cell s_j, where $f_j \sim \text{Unif}(0.1, 0.9)$. We then use the pre-processing approach to recover the missing data. We present the results for NMF in Figs. 5b-c. The original data can be expressed in terms of three factors, such that in each factor the elements corresponding to a particular region are equal to one and all others are zero. Introducing missing data makes the components noisy (Fig. 5b), but pre-processing effectively removes this noise (Fig. 5c). In the latter case the missing cells in the factors as well as those that have moved to another factor are cells near the corners of the region borders. We next discuss this issue.

4 Discussion

The edge areas are problematic for the plain Ising model: a small number of neighbors yields unrealistically high probabilities. In order to handle the problem we introduced the notion of remoteness in Sec. 2.1. In some cases this leads to the opposite effect, particularly in corner areas.

The first PC for the pre-processed data (Fig. 2d) correlates very strongly with the (expectation of the) number of words in each municipality. There are some municipalites with only a single neighbor and few recorded words. For instance, in the whole south-eastern region the research activity is low. Thus, the MRF structure cannot spread a lot of probality mass of new words into the southeasternmost municipality, since the nearest municipality with a large number of recorded words is remote. Bottlenecks also appear in the southwestern archipelago, where the number of neighbors is small, and the northernmost areas.

We compared our results to those of earlier linguistic research (see references in [6]). Our findings are in good agreement with them. For instance, the dark points in the north (Fig. 2f, Fig. 3e) agree with the earlier research, unlike the analyses on raw data distorted by low research rates in the most northern municipalities. Yet, one should be careful with large areas having few observations.

Fig. 6 (right) shows the posterior expectations of r_m against the number of recorded words in each municipality. These factors correlate particularly strongly,

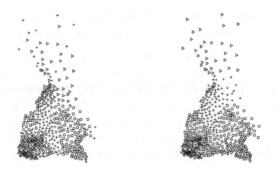

Fig. 4. Clustering results using k-means and squared Euclidean distance for original data (left), and pre-processed data (right) for 6 clusters

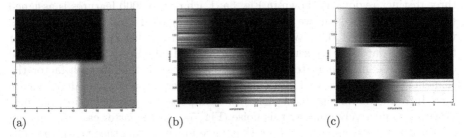

(a)   (b)   (c)

Fig. 5. (a) Synthetic data set, (b) NMF components with missing data introduced, and (c) NMF components for (pre-processed) recovered data set

when the number of recorded words is small. Still, three outliers are discerned: the southeasternmost municipality, and two munipalities in the southwest. The use of remoteness restricts the influence of autocorrelation and decreases the probability of word being used but not observed. Thus, it leads to increasing the probabilities of great values of r_m, since r_m is identical to the probability of observing a word in m, given that the word is used in m.

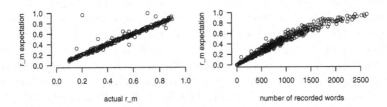

Fig. 6. Actual values vs. posterior expectations of r_m in the synthetic data (left); number of recorded words vs. posterior expectations of r_m in the dialect data (right)

In the case of the synthetic data Fig. 6 (left) shows how the model can very accurately estimate the known values of r_m for almost every cell m. There are a few outliers, always residing next to the border of two or three dialect areas.

Several neighbors of the ourlier cells have very low research rates, which increases the uncertainty. Our experiments indicated that the number of outliers decreased when the amount of data was increased. In real data the connection between the number of recorded words and values of r_m is not as straightforward as in the synthetic data, since the distributions of words are very heterogeneous.

5   Conclusion

We have demonstrated how Bayesian methods can be used as a pre-processing step in spatial data analysis. A relatively simple approach is sufficient to significantly reduce the effects of missing data. The method is practically "parameter-free", since the prior distributions have very little influence on the results. We have investigated synthetic data, and a Finnish dialect data set, that suffers from uneven sampling. We have applied the principal components analysis, nonnegative matrix factorization and clustering to both the original and pre-processed data sets. Some regions suffer slightly from edge effects. Importantly, those local effects cannot distort the results of the subsequent, global analyses.

References

1. Besag, J., York, J., Mollie, A.: Bayesian image restoration with two applications in spatial statistics. Ann. Institute of Statistical Mathematics 43(1), 1–59 (1991)
2. Best, N., Richardson, S., Thomson, A.: A comparison of Bayesian spatial models for disease mapping. Statistical Methods in Medical Research 14(1), 35–59 (2005)
3. Cowles, M., Carlin, B.: Markov chain Monte Carlo convergence diagnostics: a comparative review. J. of the American Statistical Association 91, 883–904 (1996)
4. Gamerman, D.: Markov Chain Monte Carlo: Stochastic Simulation for Bayesian Inference. Chapman–Hall, Great Britain (1997)
5. Heikkinen, J., Full, H.H.: Bayesian approach to image restoration with an application in biogeography. Applied Statistics 43(4), 569–582 (1994)
6. Hyvönen, S., Leino, A., Salmenkivi, M.: Multivariate analysis of Finnish dialect data – an overview of lexical variation. Literary and Linguistic Computing (2007)
7. Junttila, E., Salmenkivi, M.: Modeling missing data with Markov random fields in large data sets. In: Proc. of IADIS European Conference on Data Mining, Lisbon (2007)
8. Papadimitrou, S., Gionis, A., Tsaparas, P., Väisänen, R., Mannila, H., Faloutsos, C.: Parameter-free spatial data mining using MDL. In: Proc. of the 5th IEEE Int. Conf. on Data Mining (ICDM 2005), pp. 346–353. IEEE Computer Society Press, Los Alamitos (2005)
9. Winkler, G.: Image Analysis, Random Fields and Dynamic Monte Carlo Methods. Springer, Berlin (1995)
10. Yoo, J., Shekhar, S.: A joinless approach for mining spatial colocation patterns. IEEE Transactions on Knowledge and Data Engineering 18(10), 1323–1337 (2006)

Tag Recommendations in Folksonomies

Robert Jäschke[1,2], Leandro Marinho[3,4], Andreas Hotho[1],
Lars Schmidt-Thieme[3], and Gerd Stumme[1,2]

[1] Knowledge & Data Engineering Group (KDE), University of Kassel,
Wilhelmshöher Allee 73, 34121 Kassel, Germany
http://www.kde.cs.uni-kassel.de
[2] Research Center L3S, Appelstr. 9a, 30167 Hannover, Germany
http://www.l3s.de
[3] Information Systems and Machine Learning Lab (ISMLL), University of Hildesheim,
Samelsonplatz 1, 31141 Hildesheim, Germany
http://www.ismll.uni-hildesheim.de
[4] Brazilian National Council Scientific and Technological Research (CNPq) scholarship holder

Abstract. Collaborative tagging systems allow users to assign keywords—so called "tags"—to resources. Tags are used for navigation, finding resources and serendipitous browsing and thus provide an immediate benefit for users. These systems usually include tag recommendation mechanisms easing the process of finding good tags for a resource, but also consolidating the tag vocabulary across users. In practice, however, only very basic recommendation strategies are applied.

In this paper we evaluate and compare two recommendation algorithms on large-scale real life datasets: an adaptation of user-based collaborative filtering and a graph-based recommender built on top of FolkRank. We show that both provide better results than non-personalized baseline methods. Especially the graph-based recommender outperforms existing methods considerably.

1   Introduction

Folksonomies are web-based systems that allow users to upload their resources, and to label them with arbitrary words, so-called *tags*. The systems can be distinguished according to what kind of resources are supported. Flickr, for instance, allows the sharing of photos, del.icio.us the sharing of bookmarks, CiteULike[1] and Connotea[2] the sharing of bibliographic references, and Last.fm[3] the sharing of music listening habits. *BibSonomy*,[4] allows to share bookmarks and BibTeX based publication entries simultaneously.

To support users in the tagging process and to expose different facets of a resource, most of the systems offered some kind of tag recommendations already at an early stage. Del.icio.us, for instance, had a tag recommender in June 2005 at the latest,[5] and also included resource recommendations.[6] As of today, nobody has empirically shown

---

[1] http://www.citeulike.org
[2] http://www.connotea.org
[3] http://www.last.fm
[4] http://www.bibsonomy.org
[5] http://www.socio-kybernetics.net/saurierduval/archive/2005_06_01_archive.html
[6] http://blog.del.icio.us/blog/2005/08/people_who_like.html

J.N. Kok et al. (Eds.): PKDD 2007, LNAI 4702, pp. 506–514, 2007.

the quantitative benefits of recommender systems in such systems. In this paper, we will quantitatively evaluate a tag recommender based on collaborative filtering (introduced in Sec. 3) and a graph based recommender using our ranking algorithm FolkRank (see Sec. 4) on the two real world folksonomy datasets BibSonomy and Last.fm. We make the BibSonomy dataset publicly available for research purposes to stimulate research in the area of folksonomy systems (details in Section 5).

The results we are able to present in Sec. 6 are very encouraging as the graph based approach outperforms all other approaches significantly. As we will see later, this is caused by the ability of FolkRank to exploit the information that is pertinent to the specific user together with input from other users via the integrating structure of the underlying hypergraph.

2   Recommending Tags—Problem Definition and State of the Art

Recommending tags can serve various purposes, such as: increasing the chances of getting a resource annotated, reminding a user what a resource is about and consolidating the vocabulary across the users. In this section we formalize the notion of folksonomies, formulate the tag recommendation problem and briefly describe the state of the art on tag recommendations in folksonomies.

A Formal Model for Folksonomies. A folksonomy $\mathbb{F}$ describes the users U, resources R, and tags T, and the user-based assignment of tags to resources by the ternary relation $Y \subseteq U \times T \times R$. We depict the set of all posts by P. The model of a folksonomy we use here is based on the definition in [9].

Tag Recommender Systems. Recommender systems (RS) in general recommend interesting or personalized information objects to users based on explicit or implicit ratings. Usually RS predict ratings of objects or suggest a list of new objects that the user hopefully will like the most. In tag recommender systems the recommendations are, for a given user $u \in U$ and a given resource $r \in R$, a set $\tilde{T}(u, r) \subseteq T$ of tags. In many cases, $\tilde{T}(u, r)$ is computed by first generating a ranking on the set of tags according to some quality or relevance criterion, from which then the top n elements are selected.

Related work. General overviews on the rather young area of folksonomy systems and their strengths and weaknesses are given in [7,11,12]. In [13], Mika defines a model of semantic-social networks for extracting lightweight ontologies from del.icio.us. Recently, work on more specialized topics such as structure mining on folksonomies— e. g. to visualize trends [5] and patterns [16] in users' tagging behavior—as well as ranking of folksonomy contents [9], analyzing the semiotic dynamics of the tagging vocabulary [3], or the dynamics and semantics [6] have been presented.

The literature concerning the problem of tag recommendations in folksonomies is still sparse. The existent approaches [2,10,14] usually adapt methods from collaborative filtering or information retrieval. The standard tag recommenders, in practice, are services that provide the most-popular tags used for a particular resource by means of tag clouds, i.e., the most frequent used tags are depicted in a larger font or otherwise emphasized. These approaches address important aspects of the problem, but they still diverge on the experimental protocol, notion of tag relevance and metrics used, what makes further comparisons difficult.

3   Collaborative Filtering

Due to its simplicity and promising results, collaborative filtering (CF) has been one of the most dominant methods used in recommender systems. In the next section we recall the basic principles and then present the details of the adaptation to folksonomies.

Basic CF principle. The idea is to suggest new objects or to predict the utility of a certain object based on the opinion of like-minded users [15]. In CF, for m users and n objects, the user profiles are represented in a user-object matrix $\mathbf{X} \in \mathbb{R}^{m \times n}$. The matrix can be decomposed into row vectors:

$$\mathbf{X} := [\vec{x}_1, ..., \vec{x}_m]^\top \text{ with } \vec{x}_u := [x_{u,1}, ..., x_{u,n}], \text{ for } u := 1, \ldots, m,$$

where $x_{u,o}$ indicates that user u rated object o by $x_{u,o} \in \mathbb{R}$. Each row vector $\vec{x}_u$ corresponds thus to a user profile representing the object ratings of a particular user. This decomposition leads to user-based CF (see [4] for item-based algorithms).

Now, one can compute, for a given user u, the recommendation as follows. First, based on matrix $\mathbf{X}$ and for a given k, the set N_u^k of the k users that are most similar to user $u \in U$ are computed: $N_u^k := \arg \max_{v \in U}^k \mathrm{sim}(\vec{x}_u, \vec{x}_v)$ where the superscript in the arg max function indicates the number k of neighbors to be returned, and sim is regarded (in our setting) as the cosine similarity measure. Then, for a given $n \in \mathbb{N}$, the top n recommendations consist of a list of objects ranked by decreasing frequency of occurrence in the ratings of the neighbors (see Eq. 1 below for the folksonomy case).

CF for Tag Recommendations in Folksonomies. Because of the ternary relational nature of folksonomies, traditional CF cannot be applied directly, unless we reduce the ternary relation Y to a lower dimensional space. To this end we consider as matrix $\mathbf{X}$ alternatively the two 2-dimensional projections $\pi_{UR}Y \in \{0,1\}^{|U| \times |R|}$ with $(\pi_{UR}Y)_{u,r} := 1$ if there exists $t \in T$ s.t. $(u,t,r) \in Y$ and 0 else and $\pi_{UT}Y \in \{0,1\}^{|U| \times |T|}$ with $(\pi_{UT}Y)_{u,t} := 1$ if there exists $r \in R$ s.t. $(u,t,r) \in Y$ and 0 else. The projections preserve the user information, and lead to log-based like recommender systems based on occurrence or non-occurrence of resources or tags, resp., with the users. Notice that now we have two possible setups in which the k-neighborhood N_u^k of a user u can be formed, by considering either the resources or the tags as objects.

Having defined matrix $\mathbf{X}$, and having decided whether to use $\pi_{UR}Y$ or $\pi_{UT}Y$ for computing user neighborhoods, we have the required setup to apply collaborative filtering. For determining, for a given user u, a given resource r, and some $n \in \mathbb{N}$, the set $\tilde{T}(u,r)$ of n recommended tags, we compute first N_u^k as described above, followed by:

$$\tilde{T}(u,r) := \arg \max_{t \in T}^n \sum_{v \in N_u^k} \mathrm{sim}(\vec{x}_u, \vec{x}_v)\delta(v,t,r) \tag{1}$$

where $\delta(v,t,r) := 1$ if $(v,t,r) \in Y$ and 0 else.

4   A Graph Based Approach

The seminal PageRank algorithm reflects the idea that a web page is important if there are many pages linking to it, and if those pages are important themselves. In [9], we employed the same underlying principle for Google-like search and ranking in folksonomies. The key idea of our FolkRank algorithm is that a resource which is tagged with important tags by important users becomes important itself. The same holds, symmetrically, for tags and users, thus we have a graph of vertices which are mutually reinforcing each other by spreading their weights.

For generating a tag recommendation for a given user/resource pair (u, r), we compute the ranking as described in [9], and then restrict the result set $\tilde{T}(u, r)$ to the top n tag nodes.

5   Evaluation

In this section we first describe the datasets we used, how we prepared the data, the methodology deployed to measure the performance, and which algorithms we used, together with their specific settings.

Datasets. To evaluate the proposed recommendation techniques we have chosen datasets from two different folksonomy systems: *BibSonomy* and *Last.fm*. Table 1 gives an overview on the datasets. For both datasets we disregarded if the tags had lower or upper case.

BibSonomy. Since three of the authors have participated in the development of Bib-Sonomy, [7] we were able to create a complete snapshot of all users, resources (both publication references and bookmarks) and tags publicly available at April 30, 2007, 23:59:59 CEST.[8] From the snapshot we excluded the posts from the DBLP computer science bibliography[9] since they are automatically inserted and all owned by one user and all tagged with the same tag (*dblp*). Therefore they do not provide meaningful information for the analysis.

Last.fm. The data for Last.fm[10] was gathered during July 2006, partly through the web services API (collecting user nicknames), partly crawling the Last.fm site. Here the resources are artist names, which are already normalized by the system.

Core computation. Many recommendation algorithms suffer from sparse data or the "long tail" of items which were used by only few users. Hence, to increase the chances of good results for all algorithms (with exception of the most popular tags recommender) we will restrict the evaluation to the "dense" part of the folksonomy, for which

---

[7] http://www.bibsonomy.org

[8] On request to bibsonomy@cs.uni-kassel.de a snapshot of BibSonomy is available for research purposes.

[9] http://www.informatik.uni-trier.de/~ley/db/

[10] http://www.last.fm

Table 1. Characteristics of the used datasets

dataset	$|U|$	$|T|$	$|R|$	$|Y|$	$|P|$	date	k_{max}
BibSonomy	1,037	28,648	86,563	341,183	96,972	2007-04-30	7
Last.fm	3,746	10,848	5,197	299,520	100,101	2006-07-01	20

Table 2. Characteristics of the p-cores at level k

dataset	k	$|U|$	$|T|$	$|R|$	$|Y|$	$|P|$
BibSonomy	5	116	412	361	10,148	2,522
Last.fm	10	2,917	2,045	1,853	219,702	75,565

we adapt the notion of a p-core [1] to tri-partite hypergraphs. The p-core of level k has the property, that each user, tag and resource has/occurs in at least k posts.

An overview on the p-cores we used for our datasets is given in Table 2. For BibSonomy, we used $k = 5$ instead of 10 because of its smaller size. The largest k for which a p-core exists is listed, for each dataset, in the last column of Table 1.

Evaluation methodology. To evaluate the recommenders we used a variant of the leave-one-out hold-out estimation [8] which we call *LeavePostOut*. In all datasets, we picked, for each user, one of his posts p randomly. The task of the different recommenders was then to predict the tags of this post, based on the folksonomy $\mathbb{F} \setminus \{p\}$.

As performance measures we use precision and recall which are standard in such scenarios [8]. With r being the resource from the randomly picked post of user u and $\tilde{T}(u, r)$ the set of recommended tags, recall and precision are defined as

$$\text{recall}(\tilde{T}(u, r)) = \frac{1}{|U|} \sum_{u \in U} \frac{|\text{tags}(u, r) \cap \tilde{T}(u, r)|}{|\text{tags}(u, r)|} \tag{2}$$

$$\text{precision}(\tilde{T}(u, r)) = \frac{1}{|U|} \sum_{u \in U} \frac{|\text{tags}(u, r) \cap \tilde{T}(u, r)|}{|\tilde{T}(u, r)|}. \tag{3}$$

For each of the algorithms of our evaluation we will now describe briefly the specific settings used to run them.

Most popular tags. For each tag we counted in how many posts it occurs globally and used the top tags (ranked by occurence count) as recommendations.

Most popular tags by resource. For a given resource we counted for all tags in how many posts they occur together with that resource. We then used the tags that occured most often together with that resource as recommendation.

Adapted PageRank. With the parameter $d = 0.7$ we stopped computation after 10 iterations or when the distance between two consecutive weight vectors was less than 10^{-6}. In $\vec{p}$, we gave higher weights to the user and the resource from the post which was chosen. While each user, tag and resource got a preference weight of 1, the user and resource from that particular post got a preference weight of $1 + |U|$ and $1 + |R|$, resp.

FolkRank. The same parameter and preference weights were used as in the adapted PageRank.

Collaborative Filtering UT. For this collaborative filtering algorithm the neighborhood is computed based on the user-tag matrix $\pi_{UT}Y$. The only parameter to be tuned in the CF based algorithms is the number k of best neighbors. For that, multiple runs where performed where k was successively incremented until a point where no more improvements in the results were observed. For this approach the best values for k were 20 for the BibSonomy and 60 for the Last.fm dataset.

Collaborative Filtering UR. Here the neighborhood is computed based on the user-resource matrix $\pi_{UR}Y$. For this approach the best values for k were 30 for the BibSonomy and 100 for the Last.fm dataset.

6 Results

In this section we present and describe the results of the evaluation. We will see that both datasets show the same overall behavior: 'most popular tags' is outperformed by all other approaches; the CF-UT algorithm performs slightly better than and the CF-UR approach approx. as good as the 'most popular tag by resource', and FolkRank uniformly provides significantly better results.

The diagrams 1 and 2 show precision-recall plots as usual. A datapoint on a curve stands for the number of tags used for recommendation (starting with the highest ranked tag on the left of the curve and ending with ten tags on the right). Hence, the steady decay of all curves in both plots means that the more tags of the recommendation are regarded, the better the recall and the worse the precision will be.

BibSonomy. Figure 1 shows the precision and recall of the chosen algorithms. The top-rightmost curve depicts the performance of FolkRank and it can clearly be seen that the graph based algorithm outperforms the other methods in both precision and recall. With ten recommended tags the recall reaches up to 80%, while the second best results only reach around 65% with a comparable precision. While CF-UT, CF-UR and the 'most popular tags by resource' algorithms have a quite similiar performance, the adapted PageRank is significantly worse, especially with its dropdown of precision already after the third recommended tag. Finally, using the most popular tags as recommendation gives very poor results in both precision and recall.

Let us now look at Table 3. We will focus here on a phenomenon which is unique for this dataset. With an increasing number of suggested tags, the precision decrease is steeper for FolkRank than for the collaborative filtering and the 'most popular tags by resource' algorithm such that the latter two approaches for ten suggested tags finally overtake FolkRank. The reason is that the average number of tags in a post is around 4 for this dataset and while FolkRank can always recommend the maximum number of

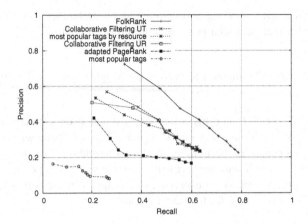

Fig. 1. Recall and Precision for BibSonomy p-core at level 5

Table 3. Precision for BibSonomy p-core at level 5

Number of recommended tags	1	2	3	4	5	6	7	8	9	10
FolkRank	0.724	0.586	0.474	0.412	0.364	0.319	0.289	0.263	0.243	0.225
Collaborative Filtering UT	0.569	0.483	0.411	0.343	0.311	0.276	0.265	0.257	0.243	0.235
most popular tags by resource	0.534	0.440	0.382	0.350	0.311	0.288	0.267	0.250	0.241	0.234
Collaborative Filtering UR	0.509	0.478	0.408	0.341	0.311	0.285	0.267	0.252	0.241	0.234

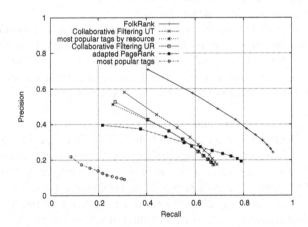

Fig. 2. Recall and Precision for Last.fm p-core at level 10

tags, for the other approaches there are often not enough tags available for recommendation. Hence, less tags are recommended. This is because in the p-core of order 5, for each post, often tags from only four other posts can be used for recommendation with these approaches. Consequently this behaviour is even more noticeable in the p-core of order 3 (which is not shown here).

Last.fm. For this dataset, the recall for FolkRank is considerably higher than for the BibSonomy dataset, see Figure 2. Even when just two tags are recommended, the recall is close to 60 %. Again, the graph based approach outperforms all other methods (CF-UT reaches at most 76 % of the recall of FolkRank). An interesting observation can be made about the adapted PageRank: its recall now is the second best after FolkRank for larger numbers of recommended tags. This shows the overall importance of general terms in this dataset—which have a high influence on the adapted PageRank (cf. Sec. 4).

The results clearly show that the graph based FolkRank algorithm outperforms base line algorithms like 'most popular tags' and collaborative filtering approaches.

Acknowledgement. Part of this research was funded by the EU in the Nepomuk[11] (FP6-027705), Tagora[12] (FP6-2005-34721), and the X-Media[13] (IST-FP6-026978) projects.

References

1. Batagelj, V., Zaversnik, M.: Generalized cores, cs.DS/0202039 (2002), `http://arxiv.org/abs/cs/0202039`
2. Benz, D., Tso, K., Schmidt-Thieme, L.: Automatic bookmark classification: A collaborative approach. In: Proceedings of the Second Workshop on Innovations in Web Infrastructure (IWI 2006), Edinburgh, Scotland (2006)
3. Cattuto, C., Loreto, V., Pietronero, L.: Collaborative tagging and semiotic dynamics (May 2006), `http://arxiv.org/abs/cs/0605015`
4. Deshpande, M., Karypis, G.: Item-based top-n recommendation algorithms. ACM Trans. Inf. Syst. 22(1), 143–177 (2004)
5. Dubinko, M., Kumar, R., Magnani, J., Novak, J., Raghavan, P., Tomkins, A.: Visualizing tags over time. In: Proc. of the 15th International WWW Conference, Edinburgh, Scotland (2006)
6. Halpin, H., Robu, V., Shepard, H.: The dynamics and semantics of collaborative tagging. In: Proceedings of the 1st Semantic Authoring and Annotation Workshop (SAAW'06) (2006)
7. Hammond, T., Hannay, T., Lund, B., Scott, J.: Social Bookmarking Tools (I): A General Review. D-Lib Magazine 11(4) (2005)
8. Herlocker, J.L., Konstan, J.A., Terveen, L.G., Riedl, J.T.: Evaluating collaborative filtering recommender systems. ACM Trans. Inf. Syst. 22(1), 5–53 (2004)
9. Hotho, A., Jäschke, R., Schmitz, C., Stumme, G.: Information retrieval in folksonomies: Search and ranking. In: Sure, Y., Domingue, J. (eds.) ESWC 2006. LNCS, vol. 4011, pp. 411–426. Springer, Heidelberg (2006)
10. Kleinberg, J.M.: Authoritative sources in a hyperlinked environment. Journal of the ACM 46(5), 604–632 (1999)
11. Lund, B., Hammond, T., Flack, M., Hannay, T.: Social Bookmarking Tools (II): A Case Study - Connotea. D-Lib Magazine, 11(4) (2005)
12. Mathes, A.: Folksonomies – Cooperative Classification and Communication Through Shared Metadata (December 2004), `http://www.adammathes.com/academic/computer-mediated-communication/folksonomies.html`
13. Mika, P.: Ontologies Are Us: A Unified Model of Social Networks and Semantics. In: Gil, Y., Motta, E., Benjamins, V.R., Musen, M.A. (eds.) ISWC 2005. LNCS, vol. 3729, pp. 522–536. Springer, Heidelberg (2005)

---

[11] http://nepomuk.semanticdesktop.org

[12] http://www.tagora-project.eu

[13] http://www.x-media-project.org,

14. Mishne, G.: Autotag: a collaborative approach to automated tag assignment for weblog posts. In: WWW '06: Proceedings of the 15th international conference on World Wide Web, pp. 953–954. ACM Press, New York (2006)
15. Sarwar, B.M., Karypis, G., Konstan, J.A., Reidl, J.: Item-based collaborative filtering recommendation algorithms. In: World Wide Web, pp. 285–295 (2001)
16. Schmitz, C., Hotho, A., Jäschke, R., Stumme, G.: Mining association rules in folksonomies. In: Batagelj, V., Bock, H.-H., Ferligoj, A., Žiberna, A. (eds.) Data Science and Classification: Proc. of the 10th IFCS Conf. Studies in Classification, Data Analysis, and Knowledge Organization, pp. 261–270. Springer, Berlin, Heidelberg (2006)

Providing Naïve Bayesian Classifier-Based Private Recommendations on Partitioned Data

Cihan Kaleli and Huseyin Polat

Computer Engineering Department, Anadolu University, Eskisehir, 26470 Turkey
{ckaleli,polath}@anadolu.edu.tr

Abstract. Data collected for collaborative filtering (CF) purposes might be split between various parties. Integrating such data is helpful for both e-companies and customers due to mutual advantageous. However, due to privacy reasons, data owners do not want to disclose their data. We hypothesize that if privacy measures are provided, data holders might decide to integrate their data to perform richer CF services. In this paper, we investigate how to achieve naïve Bayesian classifier (NBC)-based CF tasks on partitioned data with privacy. We perform experiments on real data, analyze our outcomes, and provide some suggestions.

1 Introduction

With the evolution of the Internet, the number of users accessing the Internet and the number of products available online are rapidly increasing. To find the most interesting and useful information is imperative. Collaborative filtering (CF) techniques are used for filtering and recommendation purposes. The goal in CF is to predict the preferences of one user (an active user, a), based on a database consisting of a set of votes corresponding to the ratings of users on items [1,6].

To provide more truthful and dependable referrals, data collected for CF purposes should be large enough. It is impossible to produce recommendations from insufficient data. With increasing available data, it is more likely to have enough neighbors and matchings between users. Many online vendors, especially those newly established ones, might not have enough data for CF purposes. If there is a limited number of users, it becomes a challenge to form a large enough neighborhood. Moreover, some vendors might own ratings for a limited number of items; and that makes it harder to compute the similarities between users.

Data might be partitioned horizontally or vertically between various parties, even competing companies. In horizontal partitioning, data owners hold disjoint sets of users' preferences for the same items. In vertical partitioning, they own disjoint sets of items' ratings collected from the same users. Combining horizontally partitioned data (HPD) is helpful when CF systems own a low number of users. Integrating vertically partitioned data (VPD) is advantageous when data holders have ratings for a limited number of items. A referral computed from the joint data is likely more accurate and reliable than the one calculated from one of the disjoint data sets alone. But, due to privacy, legal, and financial reasons, data owners do not want to collaborate and disclose their data to each other.

J.N. Kok et al. (Eds.): PKDD 2007, LNAI 4702, pp. 515–522, 2007.
© Springer-Verlag Berlin Heidelberg 2007

We study how to provide CF services from partitioned data between two parties without greatly exposing their privacy, using the NBC-based CF algorithm proposed by [5]. Our goals are, as follows: First, data holders should not be able to learn the true ratings and rated items in each other's databases. Second, the referrals calculated from partitioned data with privacy concerns should be close to those referrals computed from combined data without privacy concerns. Finally, additional costs introduced due to privacy concerns should be negligible and make it possible to provide referrals to many users in an acceptable time.

2 Related Work

Privacy-preserving data mining on partitioned data has been receiving increasing attention. Vaidya and Clifton present privacy-preserving methods for association rule mining [9], K-means clustering [10], and NBC [11] on VPD. Although such approaches are based on VPD, we study both VPD- and HPD-based CF with privacy using NBC. Privacy-preserving collaborative filtering (PPCF) on VPD problem is discussed in [7]. Unlike their study in which they show how to achieve predictions from numerical ratings, we investigate how to provide CF tasks based on VPD and HPD using binary ratings employing NBC.

Privacy-preserving NBC for HPD is discussed in [4]. They show that using secure summation and logarithm, they can learn distributed NBC securely. Kantarcioglu and Clifton [3] discuss privacy-preserving association rules on HPD. They address secure mining of association rules over HPD while incorporating cryptographic techniques to minimize the shared data. Polat and Du [8] discuss PPCF on HPD using item-based algorithms. Unlike these works, we explore partitioned data-based CF with privacy employing NBC, where users' preferences are represented with binary ratings. Moreover, our schemes can be easily extended to multi-party schemes. We investigate both VPD and HPD-based CF services using NBC, where users' preferences are represented with binary ratings.

In [5], it is proposed to employ NBC for producing recommendations. The users rate items as *like* (1) or *dislike* (0) based on their preferences. a's ratings for items are class labels of the training examples. In the user ratings matrix, other users correspond to features and the matrix entries correspond to feature values. The probability of an item belonging to $class_j$ (c_j), where $j \in \{like, dislike\}$, given its n feature values, can be written, as follows:

$$p(c_j|f_1, f_2, \ldots f_n) \propto p(c_j) \prod_i^n p(f_i|c_j), \qquad (1)$$

where both $p(c_j)$ and $p(f_i|c_j)$ can be estimated from training data and f_i corresponds the feature value of the target item (q) for user i. To assign q to a class, the probability of each class is computed, and the example is assigned to the class with the highest probability. Only known features and the data that both users commonly rated are used for predictions.

3   Partitioned Data-Based PPCF Using NBC

Due to privacy concerns, data owners do not want to reveal their data. We propose PPCF schemes to achieve CF tasks using NBC from partitioned data. We assume that the parties communicate through a during providing recommendations online and one of the parties acts as a master site. To derive information, data holders can employ different attacks and the proposed schemes should be secure against such attacks, which can be explained, as follows:

Acting as an active user in multiple scenarios. The party acting as an a employs the same ratings vector during the all recommendation computation processes, manipulating only one rating value each time. Since it gets some conditional probability values computed using its ratings and the users' ratings in the other party's database, the party can easily figure out the differences between such probabilities computed successively. Based on such differences, it is able to find out the ratings of the item for which the rating was manipulated or it can learn whether such item is rated or not.

Bribing. Data holders can offer some incentives or bribery to the users who provided data for filtering services. They then can obtain some data from users and try to derive more information about each other's databases. Since both parties can bribe the same users to derive data or to manipulate each other's data, the required data through such bribed users may not be true or trusted. These users can employ such offers against the other party to get more discounts or coupons. This kind of attack becomes expensive and the derived data through this attack become questionable and doubtful. Therefore, we only consider the "acting as an a in multiple scenarios" attack.

Privacy-Preserving HPD-based Schemes. Two vendors, A and B, hold n_A and n_B users' ratings, respectively, of the same m items, where $n = n_A + n_B$ and n is the number of users. They perform CF tasks using the joint data, which is an $(n_A + n_B) \times m$ matrix, while preserving their privacy. It would be difficult to find out whether two users from different online vendors refer to the same person or not. This can be solved by using some unique identities, which can be exchanged off-line. Since data is partitioned, Eq. (1) can be written, as follows:

$$p(c_j|f_1, \ldots, f_n) \propto p(c_j) \times P_{Aj} \times P_{Bj} = p(c_j) \times \prod_{i=1}^{n_A} p(f_i|c_j) \times \prod_{i=n_A+1}^{n} p(f_i|c_j) \quad (2)$$

where P_{Aj} and P_{Bj} represent the products of conditional probabilities computed from data belonging to A and B, respectively. When B acts as a master site, A computes the required data, P_{Aj} values, and sends it to B through a. HPD-based scheme with privacy can be explained, as follows:

   1. a sends her data to both A and B. B computes $p(c_j)$ values.

   2. Since both A and B own the feature ratings of q, they can compute the conditional probabilities for classes *like* and *dislike*.

   3. A then computes P_{Aj} values and sends them to B through a, while B computes P_{Bj} values.

4. Finally, B can find the probabilities of q belonging to c_j using Eq. (2).

B will not learn the true ratings and the rated items in A's database, because it only gets P_{Aj} values, which are products of n_A values, from A. To further improve privacy, before sending P_{Aj} values to B, A multiplies such values with the same value r_A, where r_A is a random number generated by A. Since both values are multiplied by the same number, the comparison between $p(c_j|f_1, f_2, \dots f_n)$ values will not be changed for j being *like* or *dislike*.

Privacy-Preserving VPD-based Schemes. A and B hold m_A and m_B items' ratings, respectively, where $m = m_A + m_B$. To make the data sharing possible, the identity of the products should be established across the data holders' databases. This data exchange can be achieved between vendors off-line.

In VPD-based schemes, a sends the corresponding data to A and B. Ratings of q is held by one of the vendors because data is split vertically. Therefore, the party, which does not have q, should conduct the required computations and send the results to the company that owns q; and such party acts as a master site. The party not having q should be able to compute corresponding results required to find the conditional probabilities in such a way to prevent the master site deriving information from its data set. Since class probabilities are known by the master site, it needs to compute the conditional probabilities, as follows:

$$p(f_i|c_j) = \frac{\#(f_i|c_j)}{\#(c_j)}, \tag{3}$$

where $\#(f_i|c_j)$ shows the total number of similarly rated items of c_j as the feature value of q for corresponding user; and $\#(c_j)$ represents the total number of commonly rated items as j, where $j \in \{like, dislike\}$. Since data is partitioned between A and B vertically, the master site gets the results from other party to find the conditional probabilities. Therefore, Eq. (3) can be written, as follows:

$$p(f_i|c_j) = \frac{\#_A(f_i|c_j) + \#_B(f_i|c_j)}{\#_A(c_j) + \#_B(c_j)}, \tag{4}$$

where A and B compute the corresponding parts of $\#(f_i|c_j)$ and $\#(c_j)$ values. Suppose that B owns q. A then should compute $\#_A(f_i|c_j)$ and $\#_A(c_j)$ values for all $i = 1, 2, \dots, n$ and j being like (1) or dislike (0); and sends them to B. VPD-based scheme with privacy can be explained, as follows:

1. a sends her corresponding data to A and B. a also computes $p(c_j)$ values and sends them to the master site, B.

2. Since A does not know which features of q are known, it computes the corresponding parts of conditional probabilities for all features. Moreover, it computes such values twice, for $f_i = 1$ and $f_i = 0$, because it does not know the feature values of q. However, since $p(f_i = 1|c_j) + p(f_i = 0|c_j) = 1$, it needs to compute $\#_A(f_i|c_j)$ values for each classes for only f_i being 1 or 0. After receiving such values, B selects and/or finds the required data to find the conditional probabilities because it knows the known features of q and their values.

3. Since B gets $p(c_j)$ values from a, it then can figure out how many 1s, 0s, and empty cells are in a's vector. Moreover, B can act as an a in multiple scenarios.

Therefore, A should compute $\#_A(f_i|c_j)$ and $\#_A(c_j)$ values in such a way to prevent B deriving data from its database. To do so, A employs the following steps: It first finds the number of empty cells (m_{ae}) in corresponding part of a's vector. A then uniformly randomly selects a value, R_A, over the range $[1, 100]$. A then can fill randomly selected R_A percent of these m_{ae} empty cells ($f = m_{ae} \times R_A/100$) with random ratings (1s and 0s). However, with increasing randomness, accuracy diminishes. Instead of filling empty cells with random ratings, A can fill them with default votes (v_ds) of items it holds. Therefore, A finds the v_ds for m_A items it holds. Both parties own the all ratings for items they hold. Therefore, they can compute non-personalized votes for the items they hold without the help of each other. For each item's column, they find the total number of 1s (l) and 0s (d). They then compare l and d values for each item. If $l > d$, then v_d for that item is 1, it is 0 otherwise. Such ratings are computed off-line. A finally fills empty cells with the corresponding v_ds. The number of empty cells to be filled depends on how much privacy and accuracy the parties want. With increasing numbers of filled cells, randomness increases; thus, accuracy diminishes.

4. A then computes the corresponding parts of conditional probability values ($\#_A(f_i|c_j)$ and $\#_A(c_j)$ values) based on a's new or filled ratings vector.

5. Since B does not know how many and which empty cells are selected to be filled, it cannot derive information from the received data. Moreover, since empty cells are filled with non-personalized ratings, which are only known by A, B does not know such values, either.

6. After B gets the required data, it finds the final conditional probabilities, the probabilities for q belonging to c_j, and finally sends the prediction to a.

4 Overhead Costs and Privacy Analysis

The extra storage cost is negligible because A and B need to store v_ds into $1 \times m_A$ and $1 \times m_B$ matrices, respectively. For single predictions, in HPD- and VPD-based schemes, additional number of communications is only 3. Moreover, the amount of data sent also increases. Our HPD-based schemes do not introduce additional computation costs. However, VPD-based schemes introduce extra computation costs due to randomly inserted non-personalized ratings. Computing v_ds is done off-line, which is not critical for overall performance. Our HPD-based schemes are secure. B will not be able to learn the true ratings and the rated items, because it receives two aggregate values, which are products of n_A values. VPD-based schemes are also secure. Even if the master site knows a's ratings, since only commonly rated items between a and other users are used for recommendation computations, it will not be able to derive data from other party's data. Finding v_ds is secure because the parties do not need each other's data to find them. Due to randomly inserted v_ds, B will not be able to derive data from the corresponding parts of conditional probability values.

The master site can guess the randomly selected unrated items. The probabilities of guessing the correct R_A and m_{ae} are 1 out of 100 and 1 out of m_A, respectively. After guessing them, it can compute f. The probability of guessing the f

randomly selected cells among m_{ae} empty cells is 1 out of $C_f^{m_{ae}}$, where C_h^g represents the number of ways of picking h unordered outcomes from g possibilities. Since the master site does not know the v_ds, the probability of guessing the inserted v_d for one item is 1 out of 2. Thus, the probability of guessing the randomly selected empty cells and their ratings is 1 out of $\left(100 \times m_A \times (1/2)^f \times C_f^{m_{ae}}\right)$.

5   Experiments

We performed experiments using Jester and EachMovie (EM) data sets. Jester is a web-based joke recommendation system [2]. It includes continuous ratings ranging from -10 to 10. There are 100 jokes and 17,988 users. The DEC Systems Research Center (www.research.compaq.com/SRC/eachmovie/) collected EM. It contains ratings of 72,916 users for 1,628 movies. User ratings were recorded on a numeric six-point scale, ranging from 0 to 1. We employed classification accuracy (CA) and F-measure (FM) to measure accuracy. We also used coverage as a metrical indicator to show the effectiveness of the NBC-based CF algorithm with combining various amounts of data.

We first transformed the numerical ratings into binary ones as done in [5]. We randomly selected 3,000 and 2,000 users for train and test set, respectively, among those users who have rated at least 50 and 60 items from Jester and EM, respectively. We randomly selected 5 rated items from test users' ratings vectors as test items. The number of users and/or items to be selected varies for various experiment sets. We performed CF tasks using the training sets to provide referrals to test users for test items. We withheld the selected rated items' votes, replaced their entries with null, and tried to predict their values. We compared referrals for them with their withheld votes. We ran the experiments for split sets alone and combined data; and found average CAs and FMs.

We hypothesize that accuracy, privacy, and efficiency depend on various factors. We performed experiments using the disjoint data sets alone and the integrated data. We then compared their outcomes. We varied the number of items (m) and users or features (n) to show how various sizes of disjoint and integrated data sets affect our results. Moreover, since v_ds are inserted randomly selected cells, we performed trials to show how different numbers of randomly selected cells (f) affect accuracy. We also computed computation times. We ran our experiments using MATLAB 7.3.0 on a computer, which is Pentium 4, 3.00 GHz with 1 GB RAM. We performed the following experiments:

It is expected to increase the coverage by integrating split data. Since number of users involving in recommendation process increases, integrating HPD improves coverage. We found coverage values for data owners on data they owned and the combined data. For Jester, when n is 50, the coverage is 99.5% and 100% for split and combined data, respectively. When n is bigger than 100, coverage is 100% for both split and integrated data. For EM, we varied n from 50 to 1,250 to show how coverage changes with combining different sizes of split data. When n is 125, coverage is 63% and 85% for split and combined data, respectively.

We performed experiments with varying n values to show how combining different amounts of HPD affect accuracy and recommendation computation time (CT) in seconds. We randomly selected training users while varying n from 50 to 1,250 and randomly selected 1,000 test users from train and test sets, respectively. Using our scheme, we found referrals for randomly selected 5 rated items from each test user's ratings vector on disjoint data sets alone and combined data. We compared predictions with true ratings, calculated the average outcomes and displayed them in Table 1, where combined data contains $2 \times n$ users' data. As seen from Table 1, the accuracy of the referrals becomes better both with combined data and increasing n. Although we improve accuracy by combining HPD, time to provide recommendations increases. CTs represent the times to produce 5,000 referrals based on various amounts of data.

Table 1. Overall Performance with Combining Varying Amounts of HPD

		Jester					EM				
n		50	125	300	750	1,250	50	125	300	750	1,250
Split Data	CA (%)	64.86	66.55	67.37	68.07	69.73	70.96	72.95	74.29	74.88	75.14
	FM (%)	63.42	64.77	65.81	66.40	66.64	78.04	79.77	80.85	81.23	81.46
	CT (secs)	15	35	104	345	706	48	127	315	909	1,302
Combined Data	CA (%)	66.14	67.22	69.16	70.15	71.40	73.12	74.62	75.28	75.50	75.86
	FM (%)	64.50	65.76	66.08	67.57	68.12	79.74	81.02	81.56	81.69	81.79
	CT (secs)	21	82	277	926	1,930	83	224	582	1,680	2,986

To show how overall performance changes with integrating varying amounts of VPD, we conducted experiments while varying m. We randomly selected 1,000 training and test users from EM. We computed referrals for randomly selected 5 test items for each test user. We compared the referrals we found with true ratings. We calculated the CAs,the FMs, and the CTs; and showed only the CAs and the CTs in Table 2. As seen from Table 2, accuracy improves with both combining VPD and increasing m. Our results also change with different f values because randomness increases with increasing f. Inserting v_ds into randomly selected cells affects accuracy and the times required to provide referrals. Although accuracy worsens by inserting v_ds, our results are still promising even if all empty cells are filled with non-personalized ratings. With increasing f, accuracy becomes worse and CTs increase. On the other hand, data owners protect their privacy by adding randomness to the private data. Data holders can adjust f to achieve required levels of privacy, accuracy, and efficiency.

Table 2. Overall Performance with Combining Varying Amounts of VPD

	Split Data					Combined Data				
m	200	350	500	650	814	400	700	1,000	1,300	1,628
CA (%)	63.27	65.12	66.16	67.16	67.33	65.96	67.52	68.04	70.94	71.26
CT (secs)	218	452	582	667	811	561	655	896	1,093	1,260

6   Conclusions and Future Work

We have shown that it is still possible to provide accurate recommendations efficiently based on partitioned data between online vendors without greatly jeopardizing their privacy. Our schemes can be easily extended to provide top-N recommendations. We evaluated our schemes in terms of accuracy and computation costs. The experiment results have shown that our outcomes are promising and the proposed schemes allow online vendors to provide accurate referrals efficiently on partitioned data. The proposed schemes for both HPD and VPD can be easily extended to multi-party schemes. We will investigate multi-part schemes in detail. We will explore more attacks and look for solutions to them.

References

1. Breese, J.S., Heckerman, D., Kadie, C.: Empirical analysis of predictive algorithms for collaborative filtering. In: Proceedings of the 14th Conference on Uncertainty in Artificial Intelligence, pp. 43–52, Madison, WI, USA (July 1998)
2. Gupta, D., Digiovanni, M., Narita, H., Goldberg, K.: Jester 2.0: A new linear-time collaborative ltering algorithm applied to jokes. In: Proceedings of the Workshop on Recommender Systems: Algorithms and Evaluation, 22nd Annual International ACM SIGIR Conference, Berkeley, CA, USA (August 1999)
3. Kantarcioglu, M., Clifton, C.: Privacy-preserving distributed mining of association rules on horizontally partitioned data. Transactions on Knowledge and Data Engineering 16(9), 1026–1037 (2004)
4. Kantarcioglu, M., Vaidya, J.S.: Privacy-preserving naive Bayes classifier for horizontally partitioned data. In: Proceedings of the IEEE ICDM Workshop on PPDM, pp. 3-9, Melbourne, FL, USA (November 19-22, 2003)
5. Miyahara, K., Pazzani, M.J.: Improvement of collaborative filtering with the simple Bayesian classifier. IPSJ Journal 43(11) (2002)
6. Pennock, D.M., Horvitz, E., Lawrence, S., Giles, C.L.: Collaborative filter ing by personality diagnosis: A hybrid memory- and model-based approach. In: Proceedings of the 16th Conference on Uncertainty in Artificial Intelligence, pp. 473–480, Stanford, CA, USA, June 30-July 3 (2000)
7. Polat, H., Du, W.: Privacy-preserving collaborative filtering on vertically partitioned data. In: Proceedings of the 9th European Conference on Principles and Practice of Knowledge Discovery in Databases, Porto, Portugal (October 3-7, 2005)
8. Polat, H., Du, W.: Privacy-preserving top-N recommendation on horizontally partitioned data. In: Proceedings of the 2005 IEEE/WIC/ACM International Conference on Web Intelligence, Paris, France (September 19–22, 2005)
9. Vaidya, J.S., Clifton, C.: Privacy-preserving association rule mining in vertically partitioned data. In: Proceedings of the 8th International ACM SIGKDD Conference, pp. 639–644 (July 23-26, 2002)
10. Vaidya, J.S., Clifton, C.: Privacy-preserving k-means clustering over vertically partitioned data. In: Proceedings of the 2004 SIAM Conference on Data Mining, Lake Buena Vista, FL, USA (April 22–24, 2003)
11. Vaidya, J.S., Clifton, C.: Privacy-preserving naïve Bayes classier for vertically partitioned data. In: Proceedings of the 9th International ACM SIGKDD Conference, pp. 206–215, Washington, DC, USA (August 24-27, 2003)

Multi-party, Privacy-Preserving Distributed Data Mining Using a Game Theoretic Framework

Hillol Kargupta[1,3], Kamalika Das[1], and Kun Liu[2]

[1] University of Maryland, Baltimore County, Baltimore MD 21250, USA
[2] IBM Almaden Research Center, San Jose CA 95120, USA
[3] Agnik, LLC, USA
{hillol,kdas1}@cs.umbc.edu, kun@us.ibm.com

Abstract. Analysis of privacy-sensitive data in a multi-party environment often assumes that the parties are well-behaved and they abide by the protocols. Parties compute whatever is needed, communicate correctly following the rules, and do not collude with other parties for exposing third party's sensitive data. This paper argues that most of these assumptions fall apart in real-life applications of privacy-preserving distributed data mining (PPDM). This paper offers a more realistic formulation of the PPDM problem as a multi-party game where each party tries to maximize its own objectives. It develops a game-theoretic framework to analyze the behavior of each party in such games and presents detailed analysis of the well known secure sum computation as an example.

1 Introduction

Advanced analysis of privacy-sensitive data plays an important role in many multi-party, cross-domain applications. For example, the US Department of Homeland Security-funded PURSUIT project[1] involves analysis of network traffic data from different organizations. Network traffic is usually privacy sensitive and no organization would be willing to share their information with a third party. PPDM offers one possible solution which would allow comparing and matching multi-party network traffic to detect common attacks and compute various statistics for a group of organizations that are not willing to share the raw data. However, many of the existing PPDM algorithms make strong assumptions about the behavior of the participants, *e.g.*, they are semi-honest and not colluding with others. Unfortunately, participants of a real application like PURSUIT may not all be ideal. Some may try to exploit the benefit of the system without contributing much; some may try to sabotage the computation; and some may try to collude with other parties for exposing the private data.

This paper suggests an alternate perspective for relaxing some of the assumptions of PPDM algorithms. It argues that large-scale multi-party PPDM can be thought of as a game where each participant tries to maximize its benefit

---

[1] http://www.agnik.com/DHSSBIR.html

J.N. Kok et al. (Eds.): PKDD 2007, LNAI 4702, pp. 523–531, 2007.
© Springer-Verlag Berlin Heidelberg 2007

by optimally choosing the strategies during the entire PPDM process. Applications of game theory in secure multi-party computation and privacy preserving distributed data mining is relatively new [1,4,2]. This paper develops a game-theoretic framework for analyzing the rational behavior of each party in such a game, and presents detailed equilibrium analysis of the well known secure sum computation [7,3] as an example. A new version of the secure sum is proposed. It works based on well known concepts from game theory and economics such as "cheap talk" and mechanism design. This paper also describes experiments on large scale distributed games and illustrates the validity of the formulations.

The remainder of this paper is organized as follows. Section 2 describes multi-party PPDM from a game theoretic perspective. Section 3 illustrates the framework using multi-party secure sum computation as an example. Section 4 gives the optimal solution using a distributed penalty function mechanism. Section 5 presents the experimental results. Finally, Section 6 concludes this paper.

2   Multi-party PPDM as Games

A game is an interaction or a series of interactions between players, which assumes that 1) the players pursue well defined objectives (they are *rational*) and 2) they take into account their knowledge or expectations of other players' behavior (they *reason strategically*). For simplicity, we start by considering the most basic game - the *strategic game*.

Definition 1 (Strategic Game). *A strategic game consists of*
- *a finite set P: the set of players,*
- *for each player $i \in P$ a nonempty set A_i: the set of actions available to player i,*
- *for each player $i \in P$ a preference relation $\succeq_i$ on $A = \times_{j \in P} A_j$: the preference relation of player i.*

The preference relation $\succeq_i$ of player i can be specified by a utility function $u_i : A \to \mathbb{R}$ (also called a payoff function), in the sense that for any $a \in A, b \in A$, $u_i(a) \geq u_i(b)$ whenever $a \succeq_i b$. The values of such a function is usually referred to as utilities (or payoffs). Here a or b is called the *action profile*, which consists of a set of actions, one for each player. Therefore, the utility (or payoff) of player i depends not only on the action chosen by herself, but also the actions chosen by all the other players. Mathematically, for any action profile $a \in A$, let a_i be the action chosen by player i and a_{-i} be the list of actions chosen by all the other players except i, the utility of player i is $u_i(a) = u_i(\{a_i, a_{-i}\})$.

One of the fundamental concepts in game theory is the Nash equilibrium:

Definition 2 (Nash Equilibrium). *A Nash equilibrium of a strategic game is an action profile $a^* \in A$ such that for every player $i \in P$ we have*

$$u_i(\{a_i^*, a_{-i}^*\}) \geq u_i(\{a_i, a_{-i}^*\}) \text{ for all } a_i \in A_i.$$

Therefore, Nash equilibrium defines a set of actions (an action profile) that captures a steady state of the game in which no player can do better by unilaterally changing her action (while all other players do not change their actions).

When the game involves a sequence of interactive actions of the players, and each player can consider her plan of action whenever she has to make a decision, the *strategic game* becomes an *extensive game*. In that situation, the *action* a_i for player i, is replaced by σ_i, the *strategy* for that player, which is a complete algorithm for playing the game, implicitly including all actions of that player for every possible situation throughout the game. The utility function also assigns a payoff to player i for each joint strategies of all the players, *i.e.*, $u_i(\{\sigma_i, \sigma_{-i}\})$.

Armed with the basic knowledge of game theory, we are now ready to formulate multi-party PPDM as a game. In a multi-party PPDM environment, each node has certain responsibilities in terms of performing their part of the computations, communicating correct values to other nodes and protecting the privacy of the data. Depending on the characteristics of these nodes and their objectives, they either perform their duties or not, sometimes, they even collude with others to modify the protocol and reveal others' private information. Let M_i denote the overall sequence of computations node i has performed, which may or may not be the same as what it is supposed to do defined by the PPDM protocol. Similarly, let R_i be the messages node i has received, and S_i the messages it has sent. Let G_i be a subgroup of the nodes that would collude with node i. The strategy of each node in the multi-party PPDM game prescribes the actions for such computations, communications, and collusions with other nodes, *i.e.*, $\sigma_i = (M_i, R_i, S_i, G_i)$. Further let $c_{i,m}(M_i)$ be the utility of performing M_i, and similarly we can define $c_{i,r}(R_i)$, $c_{i,s}(S_i)$ and $c_{i,g}(G_i)$. Then the overall utility of node i will be a linear or nonlinear function of utilities obtained by the choice of strategies in the respective dimensions of computation, communication and collusion. Without loss of generality, we consider an utility function which is a weighted linear combination of all of the above dimensions:

$$u_i(\{\sigma_i, \sigma_{-i}\}) = w_{i,m}c_{i,m}(M_i) + w_{i,r}c_{i,r}(R_i) + w_{i,s}c_{i,s}(S_i) + w_{i,g}c_{i,g}(G_i),$$

where $w_{i,m}, w_{i,r}, w_{i,s}, w_{i,g}$ represent the weights for the corresponding utility factors. Note that we omitted other nodes' strategies in the above expression just for simplicity. In the next section, we would illustrate our formalizations using one of the most popular PPDM algorithms, the secure sum computation.

3   Case Study: Multi-party Secure Sum Computation

Secure sum computation [7,3] computes the sum of n different nodes without disclosing the local value of any node. It has been widely used in privacy preserving distributed data mining as an important primitive, *e.g.*, privacy preserving association rule mining on horizontally partitioned data [5], k-means clustering over vertically partitioned data [8] and many else.

Secure Sum Computation. Suppose there are n individual nodes organized in a ring topology, each with a value $v_j, j = 1, 2, \ldots, n$. It is known that the sum

$v = \sum_{j=1}^{n} v_j$ (to be computed) takes an integer value in the range $[0, N-1]$. The basic idea of secure sum is as follows. Assuming nodes do not collude, node 1 generates a random number R uniformly distributed in the range $[0, N-1]$, which is independent of its local value v_1. Then node 1 adds R to its local value v_1 and transmits $(R + v_1) \bmod N$ to node 2. In general, for $i = 2, \ldots, n$, node i performs the following operation: receive a value z_{i-1} from previous node $i-1$, add it to its own local value v_i and compute its modulus N. In other words,

$$z_i = (z_{i-1} + v_i) \bmod N = (R + \sum_{j=1}^{i} v_j) \bmod N,$$

where z_i is the perturbed version of local value v_i to be sent to the next node $i+1$. Node n performs the same step and sends the result z_n to node 1. Then node 1, which knows R, can subtract R from z_n to obtain the actual sum. This sum is further broadcasted to all other sites.

Collusion Analysis. It can be shown that [6] any z_i has an uniform distribution over the interval $[0, N-1]$ due to the modulus operation. Further, any z_i and v_i are statistically independent, and hence, a single malicious node may not be able to launch a successful privacy-breaching attack. Then how about collusion?

Let us assume that there are k ($k \geq 2$) nodes acting together secretly to achieve a fraudulent purpose. Let v_i be an honest node who is worried about her privacy. We also use v_i to denote the value in that node. Let v_{i-1} be the immediate predecessor of v_i and v_{i+1} be the immediate successor of v_i. The possible collusion that can arise are:

- If $k = n - 1$, then the exact value of v_i will be disclosed.
- If $k \geq 2$ and the colluding nodes include both v_{i-1} and v_{i+1}, then the exact value of v_i will be disclosed.
- If $n - 1 > k \geq 2$ and the colluding nodes contain neither v_{i-1} nor v_{i+1}, or only one of them, then v_i is disguised by $n - k - 1$ other nodes' values.

The first two cases need no explanation. Now let us investigate the third case. Without loss of generality, we can arrange the nodes in an order such that $v_1 v_2 \ldots v_{n-k-1}$ are the honest sites, v_i is the node whose privacy is at stake and $v_{i+1} \ldots v_{i+k}$ form the colluding group. We have

$$\underbrace{\underbrace{\sum_{j=1}^{n-k-1} v_j}_{\text{denoted by X}} + \underbrace{v_i}_{\text{denoted by Y}}}_{} = v - \underbrace{\sum_{j=i+1}^{i+k} v_j}_{\text{denoted by W}},$$

where W is a constant and is known to all the colluding nodes. Now, it is clear that the colluding nodes will know v_i is not greater than W, which is some extra information contributing to the utility of the collusions. To take a further look, the colluding nodes can compute the posteriori probability of v_i and further use that to launch a maximum a posteriori probability (MAP) estimate-based attack. It can be shown that, this posteriori probability is:

$$f_{posterior}(v_i) = \frac{1}{(m+1)^{(n-k-1)}} \times \sum_{j=0}^{r} (-1)^j C_{(n-k-1)}^j C_{(n-k-1)+(r-j)(m+1)+t-1}^{(r-j)(m+1)+t},$$

where $v_i \leq W$, $r = \lfloor \frac{W-v_i}{m+1} \rfloor$ and $t = W - v_i - \lfloor \frac{W-v_i}{m+1} \rfloor (m+1)$. When $v_i > W$, $f_{posterior}(v_i) = 0$. Due to space constraints, we have not included the proof of this result here. Interested readers can find a detailed proof in [6].

Note that, when computing this posteriori probability, we model the colluding nodes' belief of each unknown v_j ($j = 1, \ldots, n-k-1$) as a uniform distribution over an interval $\{0, 1, \ldots, m\}$. This assumption has its roots in the principle of maximum entropy, which models all that is known and assumes nothing about what is unknown, in that case, the only reasonable distribution would be uniform.

Overall Utilities. The derived posteriori probability can be used to quantify the utility of collusion, e.g., $g(v_i) = Posteriori - Prior = f_{posterior}(v_i) - \frac{1}{N}$. We see here that this utility depends on $W - v_i$ and the size of the colluding group k. Now we can put together the overall utility function for the game of multi-party secure sum computation:

$$u_i(\{\sigma_i, \sigma_{-i}\}) = w_{i,m}c_{i,m}(M_i) + w_{i,r}c_{i,r}(R_i) + w_{i,s}c_{i,s}(S_i) + w_{i,g} \sum_{j \in P-G_i} g(v_j),$$

where P is the set of all nodes and G_i is the set of nodes colluding with node i.

Let us now consider a special instance of the overall utility where the node performs all the communication and computation related activities as required by the protocol. This results in a function: $u_i(\{\sigma_i, \sigma_{-i}\}) = w_{i,g} \sum_{j \in P-G_i} g(v_j)$, where the utilities due to communication and computation are constant and hence can be neglected for determining the nature of the function. Figure 1(Left) shows a plot of the overall utility of multi-party secure sum as a function of the distribution of the random variable $W - v_i$ and the size of the colluding group k. It shows that the utility is maximum for a value of k that is greater than 1. Since the strategies opted by the nodes are dominant, the optimal solution corresponds to the Nash equilibrium. This implies that in a realistic scenario for multi-party secure sum computation, nodes will have a tendency to collude. Therefore the non-collusion ($k = 1$) assumption of the classical secure multi-party sum is sub-optimal. Next section describes a new mechanism that leads to an equilibrium state corresponding to no collusion.

4 Achieving Nash Equilibrium with No-Colluding Nodes

To achieve a Nash equilibrium with no collusions, the game players can adopt a punishment strategy to threaten potential deviators. One may design a mechanism to penalize colluding nodes in various ways:

1. Policy I: Remove the node from the application environment because of protocol violation. Although it may work in some cases, the penalty may be too harsh since usually the goal is to have everyone participate in the process and faithfully contribute to the data mining process.
2. Policy II: Penalize by increasing the cost of computation and communication. For example, if a node suspects a colluding group of size k' (an estimate of

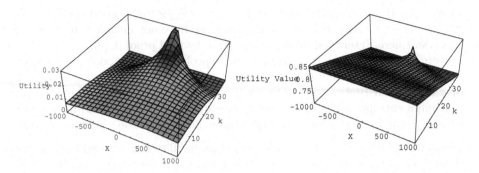

Fig. 1. Overall utility for classical secure sum computation (Left) and secure sum computation with punishment strategy (Right). The optimal strategy takes a value of $k > 1$ in the first case and $k = 1$ in the second case.

k), then it may split the every number used in a secure sum among $\alpha k'$ different parts and demand $\alpha k'$ rounds of secure sum computation one for each of these $\alpha k'$ parts, here $\alpha > 0$ is a constant factor. This increases the computation and communication cost by $\alpha k'$-fold. This linear increase in cost with respect to k', the suspected size of colluding group, may be used to counteract possible benefit that one may receive by joining a team of colluders. The modified utility function is given by $\tilde{u}_i(\{\sigma_i, \sigma_{-i}\}) = u_i(\{\sigma_i, \sigma_{-i}\}) - w_{i,p} * \alpha k'$. The last term in the equation accounts for the penalty due to excess computation and communication as a result of collusion.

Figure 1(Right) shows a plot of the modified utility function for secure sum with policy II. It shows that the globally optimal strategies are all for $k = 1$. The strategies that adopt collusion always offer a sub-optimal solutions which would lead to moving the global optimum to the case where $k = 1$.

As a toy example, consider a three-party secure sum computation with the payoff listed in Table 1. When there is no penalty, all the scenarios with two bad nodes and one good node offer the highest payoff for the colluding bad nodes. So the Nash equilibrium in the classical secure sum computation is the scenario where the participating nodes are likely to collude. However, in both cases with penalty, no node can gain anything better by deviating from good to bad when all others remain good. Therefore, the equilibrium corresponds to the strategy where none of the nodes collude. Note that, the three-party collusion is not very relevant in secure sum computation since there are all together three parties and there is always a good node (the initiator) who wants to only know the sum.

Implementing the Penalty Mechanism without Having to Detect Collusion: In order to implement the penalty protocol, one may use a central mediator who can monitor the behavior of all nodes (see, *e.g.*, [2]). However, this is usually very difficult, if not impossible in a real application environment. Moreover, it requires global synchronization which might create a bottleneck in the distributed system. Instead, we borrow the concept of *cheap talk*, a pre-play

Table 1. Payoff table for three-party secure sum computation

A	B	C	Payoff (No Penalty)	Payoff (Policy I)	Payoff (Policy II)
Good	Good	Good	(3, 3, 3)	(3, 3, 3)	(3, 3, 3)
Good	Good	Bad	(3, 3, 3)	(2, 2, 0)	(2, 2, 2)
Good	Bad	Good	(3, 3, 3)	(2, 0, 2)	(2, 2, 2)
Good	Bad	Bad	(3, 4, 4)	(0, 0, 0)	(2, 2, 2)
Bad	Good	Good	(3, 3, 3)	(0, 2, 2)	(2, 2, 2)
Bad	Good	Bad	(4, 3, 4)	(0, 0, 0)	(2, 2, 2)
Bad	Bad	Good	(4, 4, 3)	(0, 0, 0)	(2, 2, 2)
Bad	Bad	Bad	(0, 0, 0)	(0, 0, 0)	(0, 0, 0)

communication concept from game theory, to realize an asynchronous distributed control. The idea is based on the assumption that collusion requires consent from multiple parties. So a party with intention of collusion might get caught while sending out collusion invitation randomly in the network if those invitations reach some honest parties. The new protocol will therefore have a pre-play phase where "lobbying agents" (well-behaved nodes or advocacy groups) will make participants aware of the fact that one will be penalized if any collusion is detected. This "lobbying" does not affect the utility function. It simply makes everyone aware of that. *It does not require a perfect collusion detection.* A real threat with an estimated high-enough value of the collusion-size (k') will do. The threat of a good node introducing penalty using a perceived value of k' will push everyone toward proper behavior.

The new secure sum with penalty (SSP) protocol we proposed is as follows. Consider a network of n nodes where a node can either be *good* (honest) or *bad* (colluding). Before the secure sum protocol starts, the good nodes set their estimate of bad nodes in the network $k' = 0$ and bad nodes send invitations for collusions randomly to nodes in the network. Every time a good node receives such an invitation, it increments its estimate of k'. Bad nodes respond to such collusion invitations and form collusions. If a bad node does not receive any response, it behaves as a good node. To penalize nodes that collude, good nodes split their local data into $\alpha k'$ random shares. This initial phase of communication is cheap talk in our algorithm. The secure sum phase consists of $O(\alpha k')$ rounds of communication for every complete sum computation. This process converges to the correct sum in $O(n\alpha k)$ time. Note that, the SSP protocol does not require detecting all the colluding parties. Raising k' based on a perception of collusion will do. If the threat is real, the parties are expected to behave as long they are acting rationally to optimize their utility.

5   Experimental Results

We empirically verify our claim that the SSP protocol leads to an equilibrium state where there is no collusion. The utility function used for the experiments

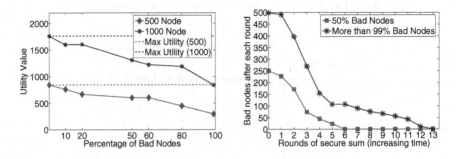

Fig. 2. (Left) Utility vs. Collusion-size. (Right) Rate of decrease of bad nodes.

is the one described in Policy II. The penalty in this case is the excess amount of communication and computation needed. In the first experiment we demonstrate for different sizes of the network (500 nodes and 1000 nodes) that the utility is maximum when the collusion is minimum, see Figure 2 (Left). The maximum utility in the figure corresponds to the classical secure sum computation without collusion. The second experiment shows that the number of bad nodes decreases with successive rounds of SSP, see Figure 2 (Right). Each bad node has a random utility threshold that is assigned during the setup. If the computed utility falls below a node's threshold, the node decides to change its strategy and becomes a good node for the subsequent rounds. The time taken to have a no collusion scenario depends on the initial number of bad nodes in the network.

6   Conclusions

This paper questions some of the common assumptions in multi-party PPDM and shows that if nobody is penalized for cheating, rational participants tends to behave dishonestly. This paper takes a game-theoretic approach to analyze this phenomenon and presents Nash equilibrium analysis of a well-known multi-party secure sum computation. A cheap-talk based protocol to implement a punishment mechanism is proposed to offer a more robust process. The paper illustrates the idea using the secure sum problem as an example. Future work includes theoretical analysis of the existence of Nash equilibrium, as well as the relationship between the amount of penalty and the payoff from collusion.

References

1. Abraham, I., Dolev, D., Gonen, R., Halpern, J.: Distributed computing meets game theory: Robust mechanisms for rational secret sharing and multiparty computation. In: PODC'06, Denver, CO, pp. 53–62 (2006)
2. Agrawal, R., Terzi, E.: On honesty in sovereign information sharing. In: Grust, T., Höpfner, H., Illarramendi, A., Jablonski, S., Mesiti, M., Müller, S., Patranjan, P.-L., Sattler, K.-U., Spiliopoulou, M., Wijsen, J. (eds.) EDBT 2006. LNCS, vol. 4254, pp. 240–256. Springer, Heidelberg (2006)

3. Clifton, C., Kantarcioglu, M., Vaidya, J., Lin, X., Zhu, M.Y.: Tools for privacy preserving distributed data mining. ACM SIGKDD Explorations 4(2), 1–7 (2003)
4. Jiang, W., Clifton, C.: Transforming semi-honest protocols to ensure accountability. In: PADM'06, Hong Kong, China, pp. 524–529 (2006)
5. Kantarcioglu, M., Clifton, C.: Privacy-preserving distributed mining of association rules on horizontally partitioned data. In: DMKD'02, pp. 24–31 (2002)
6. Kargupta, H., Das, K., Liu, K.: A game theoretic approach toward multiparty privacy-preserving distributed data mining. Technical Report TR-CS-0701, UMBC (April 2007)
7. Schneier, B.: Applied Cryptography, 2nd edn. John Wiley & Sons, Chichester (1995)
8. Vaidya, J., Clifton, C.: Privacy-preserving k-means clustering over vertically partitioned data. In: ACM SIGKDD'03, Washington, D.C, pp. 206–215. ACM Press, New York (2003)

Multilevel Conditional Fuzzy C-Means Clustering of XML Documents

Michal Kozielski

Silesian University of Technology, Akademicka 16, 44-100 Gliwice
michal.kozielski@polsl.pl

Abstract. XML documents are the special kind of data having hierarchical structure. Typical clustering algorithms do not meet requirements which may be stated for analysis of such data. A novel, dedicated for XML documents clustering method called *Multilevel clustering of XML documents* (*ML*) is presented in the paper. The method clusters feature vectors encoding XML documents on the different structure levels. Application of *Conditional Fuzzy C-Means* algorithm to *ML* method is proposed in the paper and the advantage of this fuzzy method over hard approach to *ML* algorithm is discussed and proved. An application of *ML* method to accelerating query execution on XML documents is discussed in the paper. The experimental results performed on two data sets having different characteristics show that the proposed method of multilevel conditional fuzzy clustering of XML documents outperforms hard multilevel clustering.

Keywords: clustering, clustering XML documents.

1 Introduction

Popularity of XML standard (*eXtensible Markup Language*) and a large number of its applications triggered an intense development of the new database systems called *native XML databases*. Also the functionality of relational systems supporting XML storage is continuously extended in order to effectively store, query and process XML documents.

Due to flexibility of the XML document structure it is possible that some queries will address only few documents in a large database. In order to accelerate execution of such selective queries on XML documents it is possible to consider a method reducing number of documents which have to be analysed. The reduced document set must contain all the documents which are addressed by a query and shall contain as little number of other documents as possible. It is possible to apply clustering of XML documents according to their structure in order to determine such groups of documents.

There are plenty of clustering algorithms which may be applied to the task presented above [4]. These algorithms however, are not dedicated to the hierarchical structure of XML documents and do not perform in acceptable way concerning the specified task. A new approach called *Multilevel clustering of XML documents* (*ML* algorithm) was proposed in [5]. *ML* algorithm is dedicated to a hierarchical structure

J.N. Kok et al. (Eds.): PKDD 2007, LNAI 4702, pp. 532–539, 2007.

of XML documents and may take advantage of any clustering algorithm. The paper presented defines formally *ML* algorithm and describes the shortcomings of using hard clustering in *ML* approach and proposes the solution of the problem by means of *Conditional Fuzzy C-Means* algorithm (CFCM) [11].

The paper is organised as follows. Section 2 describes the approach to accelerating XML queries by clustering XML documents. Section 3 presents *Multilevel clustering of XML documents* algorithm and application of *CFCM* to *ML* algorithm. Datasets which were used in the analysis and the results of the experiments which were performed are presented in section 4. The final conclusions are drawn in section 5.

2   Accelerating XML Query Execution

Elements and attributes of XML documents which are addressed in the queries are defined by means of path expressions. Flexibility of the structure of XML documents stored in a database causes that occurrence of an element or attribute may be optional and not all the documents in a collection match a path specified in a query. Assuming that an execution of a query on a subset of documents is less time consuming then querying the whole collection it is worth verifying the methods which could determine the collection subsets addressing the given queries.

Occurrence of an element or an attribute is a feature of a structure of XML document. It is possible therefore, to apply methods of clustering XML documents according to their structure to determine such document subsets. Having a cluster of documents it should be possible to calculate a signature of the cluster representing all the features (elements and attributes) existing in the cluster. It should be also possible to calculate a signature of a query representing all the features which are addressed by the query. Comparison of the two signatures (of a cluster and a query) should show whether the query addresses any documents in the cluster and whether the XML documents in a cluster should be processed by the query.

3   Clustering XML Documents

In order to determine the clusters of XML documents having similar structure within the clusters and different structure between the clusters it is needed to:

- apply one of the methods calculating similarity or distance between the XML document structures,
- apply one of the clustering algorithms determining the clusters.

In the presented work an approach calculating structural similarity or distance on the bases of feature vectors encoding the structure features of the documents was used. The analysis of two encoding methods: signal encoding [3] and bit encoding [7], [13] was performed [5] and bit encoding was chosen as the only acceptable approach. An assumption was taken in a work presented that the query paths are fully defined and indicate all the elements starting from a root element up to a target node.

3.1 Clustering Algorithms Review

There is a large number of clustering algorithms [4] which can be applied to the task of clustering bit feature vectors and supporting proposed method of accelerating XML queries execution. Algorithms of different types like hierarchical algorithms, e.g. *Complete* or *Single Link* [4] or partitional algorithms, e.g. *Hard C-Means* [4] can be used in the presented task. However, these algorithms are not dedicated to the data representing a structure of XML documents.

The algorithms mentioned above perform clustering in a full feature vector space. Bit encoding produces very long feature vectors what decreases a clustering quality [6], [8]. Additionally, the queries which may be performed on XML documents do not traverse through the whole tree structure to the leafs very often. Concerning the application presented reduction of the number of features cannot be performed by any of the known methods [6], [8] because they operate on the whole feature space and they do not differentiate features according to the document structure level.

It is also a common observation that the most general and therefore important information is enclosed nearby the root element concerning the structure of XML document. The features which are placed on the levels neighbouring a root element should have therefore, a greater influence on the clustering results then the leaf nodes what cannot be achieved by means of the algorithms mentioned above.

There are approaches to clustering XML documents concerning their structure which take under consideration tree-like structure of XML documents and the significance of the features depending on their level in this structure [3], [10]. These algorithms however, introduce methods dedicated to XML structure on a level of calculating similarity between document structures. They do not operate on bit feature vectors encoding XML document structure which were shown to be very effective in the presented method of accelerating XML queries.

There was therefore, a need to introduce a new clustering algorithm which would be dedicated to XML documents and which would address all the requirements which are not met by the clustering algorithms mentioned above. The new clustering algorithm giving promising results was called *Multilevel Clustering of XML Documents (ML)* [5].

3.2 Multilevel Clustering of XML Document Structure

Multilevel Clustering of XML Documents (ML) [5] is a method dedicated to XML documents. Multilevel approach starts clustering at a root level and continues the process at the following levels. In this way it differentiates features treating the elements placed in the neighbourhood of a root element as more significant. It is possible to stop the algorithm at a certain level of the document structure tree reducing a number of features which are processed.

Defining a set of XML documents as $D = \{ d_1, \ldots , d_N \}$ and a feature vector B encoding each document as a string of bits as $B = \{ b_1{}^l, \ldots , b_{nl}{}^l , \ldots , b_1{}^l, \ldots , b_{nl}{}^l \}$ where $l = 1, \ldots , L_T$ is a number of a level at which occurs a given feature. A hard clustering result on a level l is a partition of a set D to a set of clusters $C^l = \{ C_1{}^l, \ldots , C_{Kl}{}^l \}$, where K_l is a number of clusters determined on a level l, $\bigcup_{i=1}^{K_l} C_i^l = D$, and $C_i^l \cap$

$C_j^l = \Phi$, where $i \neq j$. Each cluster C_i^l may be partitioned on a level $l+1$ giving a set of new clusters. A final clustering result is a set of clusters $C = C^L$, where L is a level of XML structure tree which is defined by a user as a stop condition. The other input parameters are a user defined final number of clusters K_L and a distribution of the features among the document structure levels. Clustering on each level can be performed by means of any clustering algorithm. The definitions presented above concern the process of hard clustering which was used in the previous analysis [5] and which may be illustrated by the figure presented below.

Fig. 1. Illustration of a feature space of XML documents (a) and the partition created by hard Multilevel clustering of XML documents (b)

Figure 1. illustrates a feature space of a set of XML documents (Fig. 1. a)) captured at the three levels of XML document structures. Documents are placed on axis X, structure levels are placed on axis Y and the existing clusters relevant to different feature values are depicted by different shades. The result of applying hard Multilevel clustering of XML documents (Fig. 1. b)) is marked by thick black lines separating the clusters. The final number of clusters was set to three. Figure 1. illustrates also a problem which may be encountered when applying hard clustering (e.g. *HCM* algorithm) to *ML* method. The hard partitioning on the second level of the document trees is directly transferred to the next iteration and determines the clusters on the third level making one of them inconsistent. In order to solve the presented problem it is needed to introduce an algorithm affecting a clustering on a level $l+1$ by the results of a clustering performed on a level l but not determining them in a hard way. Therefore, *Conditional Fuzzy C-Means* [11] algorithm is proposed and applied into *ML* method in the presented work.

Conditional Fuzzy C-Means algorithm was introduced in [11] and generalised in [9]. It extends *Fuzzy C-Means* [9], [11] algorithm by introducing conditional variable f_k. Conditional variable f_k specifies what is the impact of a data object x_k on the created partition. The result of an algorithm is a pair of iteratively modified matrices: a fuzzy partition matrix $U = [u_{ik}]$ defining what is the membership value of each data object x_k to each cluster C_i and a prototype matrix $V = [v_i]$.

3.3 Multilevel Conditional Fuzzy C-Means

ML algorithm where *CFCM* algorithm is applied to perform partitioning on each level of XML document structure trees was named *Multilevel Conditional Fuzzy C-Means* (*MLCFCM*). The clustering results at the level l defined in paragraph 3.2 for hard clustering must be modified and defined as a fuzzy partition having a form of a matrix $U^l = [u_{ik}^l]$ of size $K_l \times N$, where K_l was defined as a number of clusters determined on a level l, N is a number of all documents which are analysed. The created clusters are not separated and partitioned in the consecutive iterations as it was performed when hard clustering was used. The partition matrix U^{l-1} calculated on a previous level $l-1$

of the document structure trees becomes a bases of condition matrix $F_{Kl}^{\ l}$ containing the values of condition parameter f_k used in *CFCM* algorithm. Condition parameters impact a new fuzzy partition U^l on a level l. Condition matrix $F_{Kl}^{\ l} = g(U^{l-1})$ may be calculated by means of different forms of function g. A final partition U^L is binarized what gives a hard partition defining a set of clusters C^L as defined in paragraph 3.2.

The approach presented above ensures that clustering on a level closer to a root element will impact the partition of the features placed further in the XML document tree. At the same time, a direct transfer of cluster borders, which is performed when hard clustering is used, is avoided what enables the algorithm to produce a correct results for the case presented on figure 1.

Another advantage of this method is a possibility of detection of the documents having a strongly different structure comparing to the cluster prototypes. Concerning acceleration of the XML queries it is profitable to receive as compact clusters as possible. The characteristic feature of the documents of this type is a very small variance of the membership values in a partition matrix U^l. The proposed *MLCFCM* algorithm assigns documents strongly distinct from the cluster prototypes on each level to the "others" cluster.

4 Experiments

Performance of *ML* algorithm which was presented in the previous sections was analysed on two sets of XML documents.

Level3 dataset [2] consists of 112 XML documents which were generated by means of ToXgene tool [1]. This dataset was used in order to verify the differences between hard version of *ML* algorithm and *MLCFCM* method. Therefore, the characteristics of the generated documents refer to the data illustrated on figure 1. Additionally, twelve documents having distinct structure from the others were added to the dataset what should enable to verify how fuzzy approach can deal with the data of very different characteristic. Encoding the documents creating *Level3* dataset gave a feature vector containing 48 bits.

Wiki dataset consists of 989 XML documents randomly chosen from a large extract of Wikipedia to XML format [12]. This dataset is a real life dataset of unknown structure and it is used in order to verify how a new *MLCFCM* algorithm performs in general conditions. Encoding the documents from *Wiki* dataset gave a feature vector containing 6557 bits distributed among 36 levels of XML document structure trees.

4.1 Partition Consistency Verification

Section 3.2 presented a problem which may occur when hard clustering is implemented in multilevel method and which may lead to inconsistent clusters. In order to confirm the expected advantage (described in section 3.3) of *MLCFCM* algorithm over hard multilevel approach, *Level3* dataset was clustered by both kinds of algorithms. *Hard C-Means* algorithm was implemented in *ML* method as a hard multilevel algorithm (*MLHCM*). The figures presented below show the structure of *Level3* dataset (fig. 2.) and the partitions created by both algorithms (fig. 3.).

Fig. 2. Designed structure of *Level3* dataset

a)   b)

Fig. 3. Partition created by *MLHCM* algorithm (a) and by *MLCFCM* algorithm (b)

Each figure presents the results of clustering of 112 XML documents placed on X axis on three levels of XML structure placed on Y axis. Different clusters are marked by different shades. The colour of a cluster is not important (except black colour) and was used in order to distinguish the clusters. Documents having a strongly different structure and assigned to "others" cluster were marked in black. Figure 2. presents the expected result of clustering according to designed structure of XML documents.

In this experiment the following function g transforming partition matrix U^{l-1} to condition matrix F_{Kl}^{l} was used in *MLCFCM* algorithm:

$$g = 0.5 \cdot U^{l-1} + 0.5 \qquad (1)$$

The figures presented above show that only *MLCFCM* algorithm revealed the clusters existing on the third level of document structures correctly. Hard multilevel clustering assigned documents numbered from 53 to 112 into one inconsistent cluster.

4.2 Dataset Reduction Verification

Another experiment shows the results of accelerating XML query execution by means of *ML* method. As it was presented in the section 2, it is assumed that a query should be faster executed on a reduced set of documents containing all the documents addressed by the query. Therefore, a reduction degree of the datasets which was received by means of clustering algorithm was compared in the experiment.

In case of *Level3* dataset an average degree of reduction was calculated for all possible query paths. Comparison of fuzzy (*MLCFCM*) and hard (*MLHCM*) implementation of *ML* method was performed and the results of the analysis are presented in table 1.

Table 1. Average reduction degree of *Level3* dataset

MLHCM	MLCFCM ($v=0$)	MLCFCM ($v=0.01$)
59.3	62.9	72.4

Two values of parameter v determining the maximal value of variance of partition matrix which assigns a document as "others" were used in the implementation of

MLCFCM algorithm. A value of $v=0$ means that no document was assigned to "others" cluster. Function g was defined in the same way as in equation (1).

Presented in table 1. average numbers of documents which were reduced as not being addressed by a query show that *MLCFCM* algorithm performs better concerning the presented application to acceleration of queries on XML documents. The difference between the reduction values for *MLHCM* and *MLCFCM* where $v=0$ is not very large but the analysis of the particular paths addressing the documents which were incorrectly clustered by *MLHCM* algorithm (documents numbered from 53 to 83 on fig. 3.) show that the difference in reduction degree for these paths may be significant (71 reduced documents in case of *MLCFCM* instead of 30 documents reduced in case of *MLHCM*). It is also visible that the documents having strongly different structure may decrease the quality of clustering. It is important therefore, to assign that kind of documents to a separate cluster what may be performed by means of *MLCFCM* algorithm.

In case of *Wiki* dataset a reduction degree for four queries was compared. Table 2. presents what is a number of documents which are addressed and which are not addressed by the queries. Table 3. presents the results of the analysis performed on different levels of the document structure trees. In this experiment, a value of parameter v was set to $v = 10^{-13}$ and g function was defined as $g = U^{l-1}$.

Table 2. Characteristics of the queries defined on *Wiki* dataset

Query	Query path	Path length (no. of levels)	Documents addressed by a query	Documents which may be reduced
q1	/article/body/section/title	4	620	369
q2	/article/body/definitionlist	3	3	986
q3	/article/body/normallist	3	56	933
q4	/article/body/figure	3	155	834

Table 3. Number of documents reduced by means of *ML* method

Query	Level 3		Level 4	
	MLHCM	MLCFCM	MLHCM	MLCFCM
q1	355	0	319	0
q2	347	553	211	508
q3	14	93	0	0
q4	0	13	0	127

The results presented in table 3. show that *MLCFCM* algorithm performed better concerning queries *q2*, *q3* and *q4*, where the difference received for query *q2* is significant. *MLHCM* algorithm however, performed significantly better concerning query *q1*.

5 Conclusions

The new method of multilevel clustering of XML documents (*ML*) is presented in the paper. The discussion presented in the paper and the results of the experiments which were performed show that application of conditional fuzzy clustering to *ML* method (*MLCFCM*) is able to produce more consistent clusters comparing to hard clustering algorithm (*MLHCM*).

An application of clustering XML documents according to their structure to accelerating query execution on XML document set was also presented in the paper. An analysis of the methods encoding a structure of XML document which could be applied to this task showed that only bit encoding meets the requirements which were stated. The experiments comparing fuzzy (*MLCFCM*) and hard (*MLHCM*) implementations of *ML* method show that multilevel conditional fuzzy c-means clustering gives better results then *HCM* implementation on both generated and real life data.

References

1. Barbosa, D., Keenleyside, J., Lyons, K., Mendelzon, A.: ToXgene - the ToX XML Data Generator (2007), http://www.cs.toronto.edu/tox/toxgene
2. Dataset used in the experiments: http://dydaktyka.polsl.pl/ZTiPSK/ IndywidualnePlanyZajec/ Michal_Kozielski_dane/mkoz_www.html
3. Flesca, S., et al.: Fast Detection of XML Structural Similarity. IEEE Transactions on Knowledge and Data Engineering 17(2) (2004)
4. Jain, A.K., Murty, M.N., Flynn, P.J.: Data Clustering: A review. ACM Computing Surveys 31(3) (1999)
5. Kozielski, M.: Przyspieszanie realizacji zapytań na dokumentach XML z wykorzystaniem grupowania względem ich struktury, Bazy Danych, Nowe Technologie: Architektura, metody formalne i zaawansowana analiza danych, WKŁ, pp. 305–314 (2007)
6. Kozielski, M.: Improving the Results and Performance of Clustering Bit-encoded XML Documents. In: Proc. of ICDM Workshops 2006, pp. 60–64. IEEE Computer Society Press, Los Alamitos (2006)
7. Lian, W., et al.: An Efficient and Scalable Algorithm for Clustering XML Documents by Structure. IEEE Transactions on Knowledge and Data Engineering 16(1) (2004)
8. Liu, J., et al.: XML Clustering by Principal Component Analysis. In: Proceedings of the 16th IEEE International Conference on Tools with Artificial Intelligence (ICTAI 2004), IEEE Computer Society Press, Los Alamitos (2004)
9. Łęski, J.: Generalized Weighted Conditional Fuzzy Clustering. IEEE Transactions on Fuzzy Systems 11(6) (2003)
10. Nayak, R.: Fast and Effective Clustering of XML Data Utilizing their Structural Information, Under publication in KAIS: Knowledge and Information Systems - An International Journal
11. Pedrycz, W.: Conditional Fuzzy C-Means. Pattern Recognition Letters 17, 625–631 (1996)
12. (2006), http://xmlmining.lip6.fr
13. Yoon, J.P., Raghavan, V., Chakilam, V.: Bitmap Indexing-based Clustering and Retrieval of XML Documents. In: Proceedings of ACM SIGIR Workshop on Mathematical/Formal Methods in Information Retrieval, New Orleans, LA, ACM Press, New York (2001)

Uncovering Fraud in Direct Marketing Data with a Fraud Auditing Case Builder

Fletcher Lu

Department of Math and Computer Science
University of Maryland Eastern Shore
Princess Anne, MD, 21853, USA
flu@umes.edu

Abstract. This paper illustrates an automated system that replicates the investigative operation of human fraud auditors. Human fraud auditors often utilize fraud detection methods that exploit structure in database tables to uncover outliers that may be part of a fraud case. From the uncovered outliers, an auditor will build a case of fraud by searching data related to the outlier possibly across many different databases and tables within these different databases. This paper illustrates an industrial implementation of an adaptive fraud case building system that uses machine learning to conduct the search and decision-making process with an automated outlier detection component. This system was successfully applied to uncover fraud cases in real marketing data.

Keywords: Fraud Detection, Benford's Law, Reinforcement Learning.

1 Introduction

A common definition of fraud requires two components: (1) deception and (2) an unjustified gain or loss [1]. Therefore, fraud is generally hidden to some degree and a party must obtain an unjustified benefit or loss. The system we are proposing differs markedly from what are commonly know as fraud detection tools in that fraud detection tools deal with the first component of fraud by uncovering some hidden structure/anomaly. Determining an *unjustified* gain or loss due to this deception is generally left to human auditors. Thus most tools used in fraud detection could more accurately be described as anomaly detectors since they do not ascribe any loss or gain to their uncovered structures. Our system in contrast may best be described as a fraud auditing *case builder* rather than a fraud detector. Our system will use the fraud detection tools to find the anomalies, and then perform the human auditor task of linking the hidden anomaly to other data by searching possibly vast amounts of records across different database tables for related information that demonstrates said loss or gain. This search thus relates interconnected evidence of fraud into a fraud case.

We propose to use a reinforcement learning (RL) method to conduct the search component. RL models all database records as states in a networked environment. Often there is structure within a database table that popular fraud

J.N. Kok et al. (Eds.): PKDD 2007, LNAI 4702, pp. 540–547, 2007.
© Springer-Verlag Berlin Heidelberg 2007

detection tools exploit. Any data that deviates significantly from the modeled structures is then marked as possibly fraudulent [2,3].

Just as human auditors may use any and all possible fraud/anomaly detectors our automated system may do so as well. The utility of our mechanism is in:

1. Automating the task of human auditors who have to run anomaly detectors and then *manually* search to build the fraud case.
2. Speeding up the searching of possibly vast amounts of data related to an outlier in order to link records together that build a case for fraud.

Under a small enough finite search space, our machine learning approach may degenerate to a simple dynamic programming problem that may be solved completely with sufficient search time. For cases with vast numbers of records and databases, a reinforcement learning approach will be employed. Our system combines a reinforcement learning approach with an outlier detection method that exploits structures within tables to produce a new fraud auditing case building mechanism.

In this paper, we briefly illustrate the technique for combining these two methods and then illustrate issues that may be used to enhance this new technique. We address the appropriateness of using reinforcement learning for fraud auditing by considering the:

1. Objective of both reinforcement learning and fraud auditing.
2. Enviromental requirements of RL (specifically, the Markov requirement).
3. Reward structure and how it relates to fraud outlier detection.

We conduct experiments on our method that demonstrate accuracy improvement over an anomaly detector alone and against a competing fraud case builder that uses a greedy search method. Our tests use real direct retail marketing data with two types of outliers: a Normal distribution and a Benford's Law outlier technique.

2   Background

2.1   Fraud Detection

As Bolton and Hand [4] noted, fraud detection methods may be divided into both supervised and unsupervised methods. For supervised methods, both fraudulent and non-fraudulent records are used to train a system, which then searches and classifies new records according to the trained patterns. Supervised methods require pre-identified fraudulent and non-fraudulent records to train on. Thus, it is limited to only previously known methods of fraud.

Unsupervised methods, in contrast, typically identify records that do not fit expected norms. The advantage of this approach is that one may identify new instances of fraud. The common approach to this method is to use forms of outlier detection. The main limit to this approach is that we are essentially identifying anomalies that may or may not be fraudulent behaviour. Anomalous behaviour is not necessarily fraudulent behaviour. Instead they can be used as indicators of possible fraud, whereby the strength of the anomalous behaviour (how much it deviates from expected norms), may be used as a measure of one's confidence in how likely the behaviour may be fraudulent. Audit investigators are then typically employed to analyze these anomalies. Outlier detection fits data to some statistical distribution isolating any outliers from the fitted data. Another approach common in fraud detection is to use a digital analysis technique known as Benford's Law.

2.2 Benford's Law

Benford's Law specifies the distribution of the digits for *naturally* occurring phenomena. This technique, commonly used in areas of taxation and accounting, describes the frequency with which individual and sets of digits for naturally growing phenomena such as population measures should appear [5]. Such natural growth has been shown to include areas such as spending records and stock market values [6]. One may therefore use significant deviations from Benford's Law expected values as an indicator for possible fraud in these areas. Much of the research on Benford's Law for fraud detection has been in areas of statistics [7,8] as well as auditing [9,2].

2.3 Reinforcement Learning

In reinforcement learning, an environment is modeled as a network of states, $\{s \in S\}$. Each state is associated with a set of possible actions, $a_s \in A_s$ and a reward for entering that state $\{r_s \in R_s\}$. All states are required to be Markov Decision Processes. We can transition from one state s_i to another s_j by choosing an action a_{s_i} and with a certain probability $P(s_j|s_i, a_{s_i})$ we transition to another state. A policy is a mapping of states to actions. The objective is to find an optimal policy that maximizes the long-term rewards one may obtain as one navigates through the network. One may find an optimal policy using an approach known as the temporal differencing method [10] [11].

3   Why Use Reinforcement Learning?

3.1 Fraud Auditing Objective

As a motivation for using reinforcement learning as a tool to help auditor's build a case for fraud or eliminate cases, consider the objective in reinforcement learning (RL). RL builds policies that are designed to make action choices based on the state an agent is in. The RL attempts to build an 'optimal' policy that returns best possible rewards over a long term 'travel' through the environment. An auditor's task is to build a case for fraud by linking suspicious records to some gain or loss. Since this gain or loss is unjustified, it is likely anomalous data. An auditor can be thought of as an agent exploring through tables of databases of records, where each record is a state in our database environment. Just as a robot using reinforcement learning develops a policy to move from state to state collecting high rewards, an auditor agent navigates from database record to database record relating anomalous records together to build it's fraud case. Two highly suspicious records may be linked through the attributes of several intermediate records that are not in themselves suspicious. Thus, the *long-term* rewards nature of RL lends itself well to such multiple state linking where we are not concerned with only the immediate high rewards between two directly linked states that for instance a greedy search approach would relate. Assuming we can equate high rewards with significant outliers, we can utilize an RL approach to link these outliers together.

3.2   The Markov Property

Now let us consider the Markov property. In order to apply reinforcement learning the next state must be determined solely by the current state and current action of an agent. When auditors build their case for fraud they do so by linking database records together through the record attributes. Therefore, to apply an RL approach, we will require attributes for the current record being explored by an agent to be completely determined by the current record. Therefore our algorithm will only look at next records/states that have attributes in common with the *current* record. No previous attributes encountered during our fraud case linking of different records may be considered.

3.3   Fraud Outliers and Rewards

Finally, in order for this approach to work, we need to relate high reward values to a strong indictator of fraud. To do so, we will use the popular outlier detection methods noted in section 2.1. The larger the deviation from expected values that a record's data contains, the larger the reward value we will assign to it. In section 4 we will illustrate a few methods to associate such rewards with records.

One trait of a reinforcement learning approach that makes it particularly useful for the large number of records that are being continuously added to in real business systems is its ability to be applied in an online form. We can use an online form of temporal differencing which allows for continuous updating of our policy based on previous information bootstrapped to new records that are encountered.

4   Algorithm

The best way to illustrate our fraud detection algorithm is through an example. Figure 1 is an example of purchase records for some consumer.

We begin by first deciding what type of outlier detection method we wish to utilize. If we use a standard statistical distribution outlier detection approach, we compute a reward by the deviation of actual frequencies, af_i, from the expected

States	Actions/Attributes				Digit
	Purchase Item	Store	Location	Form of Payment	Sequences
1	shoes	storeA	street15	credit	$52
2	hat	storeB	street12	cash	$38
3	hat	storeC	street17	debit	$22
4	TV	storeB	street11	cheque	$640

Fig. 1. Sample Application: Purchase Records

States	Action/Attributes				Digit	Rewards/Magnitude
	Purchase Item	Store	Location	Form of Payment	Sequences	of Anomalies
1	shoes	storeA	street15	credit	$52	1.6
2	hat	storeB	street12	cash	$38	3.2
3	hat	storeC	street17	debit	$22	0.2
4	TV	storeB	street11	cheque	$340	1.1

Fig. 2. Sample Application: Calculating Rewards & Choosing a Record/State

frequencies, e_i, of our purchase values at state i according to our given statistical distribution using:

$$Reward(i) = \frac{af_i}{e_i}. \tag{1}$$

If we use a Benford's Law outlier approach then we compute the frequency with which each digit sequence from 1 to 999 appears in our purchase value records.[1] We compute a measure of how much any given purchase value deviates from expected Benford value by:

$$Reward(i) = \frac{f_{1i}}{b_{1i}} + \frac{f_{2i}}{b_{2i}} + \frac{f_{3i}}{b_{3i}}, \tag{2}$$

where f_{ji} is the frequency that a digit sequence of length j for state i appears in the dataset and b_{ji} is the expected Benford's Law distribution frequency that the digit sequence of length j for state i should appear.

Once the reward values have been computed, we can now explore our environment as an RL network. We do so by choosing a start state. In figure 2 we chose state 2. This results in a reward value of 3.2. We then need to choose an action. Our actions are any unused attributes of our record. In this case we have four possible actions. There are numerous methods for choosing an action. See [10] for various techniques.

Choosing action/attribute 'Store', the specific instance of this action in state 2 is 'storeB'. We therefore search the store column, in all tables containing this attribute, for any other states/records with 'storeB' as an entry. Every possible record with such an entry is a possible next state. In our example, state 4 is a possible next state which, as figure 3 illustrates, will be our next state. We

---

[1] Benford's Law works with digit sequences of any length. For most practical purposes, the frequencies of sequences of three digits or less are evaluated. For longer digit lengths, the probabilities become so small that they are of little practical value.

| States | Action/Attributes | | | | Digit | Rewards/Magnitude |
	Purchase Item	Store	Location	Form of Payment	Sequences	of Anomalies
1	shoes	storeA	street15	credit	$52	1.6
2	hat	storeB	street12	cash	$38	3.2
3	hat	storeC	street17	debit	$22	0.2
4	TV	storeB	street11	cheque	$340	1.1

Fig. 3. Sample Application: State to Action to Next State Transition

use a uniform random distribution to choose which of our possible next state candidates will be selected. With this method of exploring our environment, we can now apply an RL algorithm to find an optimal policy to our system.

To summarize our approach we,

1. Identify attributes in a database that fit a statistical or Benford distribution.
2. Set reward values based on the amount of deviation from expected distribution values.
3. Use the remaining attributes as action choices in an RL context.
4. Run an RL approach until an optimal policy is found.
5. Navigate through the environment using the found optimal policy with a start state of one with a high reward value produced from part 1. Return all states encountered.

5 Experiments

In our experiments we compare two outlier methods for our reward system, the Benford's Law and standard Normal distribution outlier mechanisms. Our implementation uses a SARSA form of temporal differencing (TD) reinforcement learning. A stop state of any state that was already previously visited in the current trajectory was used. Start states were states with the largest rewards. When multiple states all had the same largest reward value, we uniformly randomly selected one of those states.

5.1 Experiment 1

In this experiment we compare our outlier with RL against an outlier detector alone. The database consisted of a total of 136,929 records with data partitioned across various criteria to produce twelve different test sets.

Our TD algorithm uses a decreasing $\alpha = 1/t$ where t is the number of steps taken during a trajectory and $\gamma = .5$. For comparison purposes with Benford's Law alone, a 95% confidence interval bound such that any digit sequences exceeding our confidence interval would flagged as possibly fraudulent.

Table 1 summarizes the precision results of the records that the Benford's Law alone and the Benford's Law with reinforcement learning flag as possibly fraudulent. Overall, the Benford's Law with reinforcement learning performs better with a higher true positive fraud performance in 10 of the 12 data sets. For the two cases where Benford's Law alone outperformed our method, the 95%

Table 1. Benford's Law Alone versus Benford's Law with Reinforcement Learning, Fraud Precision on Purchasing Data

Set	Size	Benford Alone	Benford with RL
1	40303	18.40%	11.11%
2	40302	18.88%	41.67%
3	28820	17.04%	66.67%
4	28095	13.09%	11.11%
5	28468	14.74%	50.00%
6	28437	15.15%	33.33%
7	14076	13.56%	25.00%
8	42829	15.76%	34.78%
9	14633	14.15%	22.22%
10	42272	14.57%	30.77%
11	14013	16.51%	33.33%
12	42892	14.77%	25.00%

confidence interval possibly allowed for the inclusion of more cases with lower reward value that were still cases of fraud.

5.2   Experiment 2

In this experiment we illustrate the utility of reinforcement learning's long-term reward approach over a greedy search approach which seizes only immediate rewards. A Normal and a Benford's Law distribution was used for our rewards. Tests were performed on 227,156 retail record's containing 1526 fraudulent records. For comparison purposes, our table of data includes the theoretical accuracy rate if states were randomly selected as fraudulent.

A greedy search returns the largest immediate reward deviations, and as table 2 illustrates the such deviations can be poor indicators of fraud. The best greedy results were with the Benford method which still only produced an accuracy of 0.48%, which is below even a random selection of records that yields 0.67% return. In contrast, the RL mechanism, which links multiple deviations together in a

Table 2. RL vs. Greedy Search

Method Search, Reward	# of States Correct	# of States Recommended	Percent Accuracy
Random (Theoretical)	1526	227,156	0.67%
Greedy, Normal	0	105	0.00%
Greedy, Benford	51	10,581	0.48%
RL, Normal	166	1679	9.89%
RL, Benford	126	623	20.22%

long-term pattern, obtained significantly better results, with the Normal distribution accuracy at over 9% and the Benford distribution accuracy at over 20%.

6   Discussion and Conclusions

In this paper we have implemented a machine learning approach that replicates the fraud case investigating and building task of human fraud auditors. We illustrated why a reinforcement learning method may be used to perform this task. We supported this assertion by comparing RL with a competing greedy search method to build fraud cases. We also demonstrated through real retail marketing data how our system enhances the accuracy of a simple outlier fraud detector with a direct comparison between a Benford outlier alone against our Benford outlier with reinforcement learning.

In terms of future work, we wish to explore methods for combining results from multiple fraud detectors. In addition, since determining whether a built case is actually fraud requires interpretation, some automated interpreting method may also be explored.

References

1. Dictionary.com (2007), http://www.dictionary.com
2. Crowder, N.: Fraud Detection Techniques. Internal Auditor, 17–20 (1997)
3. Fawcett, T.: AI Approaches to Fraud Detection & Risk Management. Technical Report WS-97-07, AAAI Workshop: Technical Report (1997)
4. Bolton, R.J., Hand, D.J.: Statistical Fraud Detection: A Review. Statistical Science 17(3), 235–255 (1999)
5. Nigrini, M.J.: Digital Analysis Using Benford's Law. Global Audit Publications, Vancouver, B.C., Canada (2000)
6. Nigrini, M.J.: Can Benford's Law Be Used In Forensic Accounting? The Balance Sheet pp. 7–8 (1993)
7. Pinkham, R.S.: On the Distribution of First Significant Digits. Annals of Mathematical Statistics 32, 1223–1230 (1961)
8. Hill, T.P.: A Statistical Derivation of the Significant-Digit Law. Statistical Science 4, 354–363 (1996)
9. Carslaw, C.A.: Anomalies in Income Numbers: Evidence of Goal Oriented Behaviour. The Accounting Review 63, 321–327 (1988)
10. Sutton, R.S., Barto, A.G.: Reinforcement Learning: An Introduction. MIT Press, Cambridge, Massachusetts (1998)
11. Sutton, R.S.: Learning to predict by the method of Temporal Differences. Machine Learning 3, 9–44 (1988)

Real Time GPU-Based Fuzzy ART
Skin Recognition

Mario Martínez-Zarzuela, Francisco Javier Díaz Pernas,
David González Ortega, José Fernando Díez Higuera,
and Míriam Antón Rodríguez

Higher School of Telecommunications Engineering
University of Valladolid (Spain)
mario.martinez@tel.uva.es
http://gti.tel.uva.es

Abstract. Graphics Processing Units (GPUs) have evolved into powerful programmable processors, becoming increasingly used in many research fields such as computer vision. For non-intrusive human body parts detection and tracking, skin filtering is a powerful tool. In this paper we propose the use of a GPU-designed implementation of a Fuzzy ART Neural Network for robust real-time skin recognition. Both learning and testing processes are done on the GPU using chrominance components in TSL color space. Within the GPU, classification of several pixels can be made simultaneously, allowing skin recognition at high frame rates. System performance depends both on video resolution and number of neural network committed categories. Our application can process 296 fps or 79 fps at video resolutions of 320x240 and 640x480 pixels respectively.

1 Introduction

Graphics Processing Units (GPUs) have been used for many years as CPU co-processors, helping them in the task of rendering complex images onto the computer screen. Inside this special purpose processors, data downloaded from the CPU is transformed along the *graphics pipeline*. Inside this pipeline, several *vertex* and *fragment* processors can be programmed to perform user-defined calculations over the elements of a *stream of data* in a parallel fashion.

Human body parts detection has important applications as a first step in many high-level computer vision tasks such as personal identification, video indexing systems and human-machine interfaces (HMI). HMI needs a real-time video processing while consuming as few system resources as possible. Skin color segmentation on the GPU can be a first step towards robust and efficient HMIs.

James Fung et al. [1] was one of the first researchers who used GPUs in the field of computer vision. James Fung's open source OpenVidia project implements computer vision algorithms on computer graphics hardware. OpenVidia is able to make skin tone segmentation at 30 fps at a resolution of 320x240 pixels, using RGB to HSV color conversion and threshold filtering.

In this paper, we propose a real time GPU-based Fuzzy ART Neural Network for skin recognition. The rest of this paper is organized as follows: for

J.N. Kok et al. (Eds.): PKDD 2007, LNAI 4702, pp. 548–555, 2007.
© Springer-Verlag Berlin Heidelberg 2007

completeness, Section 2 summarizes Fuzzy ART architecture and training algorithm. Section 3 talks about the TSL color space and justify its convenience for skin color segmentation. Section 4 explains our GPU implementation of a Fuzzy ART skin recognizer. Section 5 summarizes the experimental results that were done for training the neural network and measuring the performance of the skin recognizer. Finally, Section 6 draws the main conclusions and outlines future research tasks.

2   Fuzzy ART Neural Networks

Fuzzy ART systems are able to categorize analog or binary input patterns, defined by M components. Each output neuron in the network represents a category and has an associated weight vector or long-term memory (LTM) trace $\boldsymbol{w}_j = (w_{j1}, \cdots, w_{jM})$. Initially, all weights are set to one, so each category is said to be *uncommitted*. When a category is first selected it becomes *committed*. For each input pattern $\boldsymbol{I}$ and output node j, the *choice function*, T_j, is defined by:

$$T_j(\boldsymbol{I}) = \frac{|\boldsymbol{I} \wedge \boldsymbol{w}_j|}{\alpha + |\boldsymbol{w}_j|} \tag{1}$$

where the fuzzy MIN operator $\wedge$ is defined by $(\boldsymbol{p} \wedge \boldsymbol{q}) \equiv \min(p_i, q_i)$ and the norm $|\cdot|$ is defined by $|\boldsymbol{p}| \equiv \sum_{i=1}^{M} |p_i|$ for any M-dimensional vectors $\boldsymbol{p}$ and $\boldsymbol{q}$. The category choice is indexed by J, where $T_J = \max(T_j : j = 1 \cdots N)$. Then, $\boldsymbol{w}_J$ is said to be a *fuzzy subset* of $\boldsymbol{I}$ and it is fed down in order to measure its resemblance to the input pattern $\boldsymbol{I}$. The system enters in *resonance* if the *match function* meets the *vigilance criterion* ρ:

$$\frac{|\boldsymbol{I} \wedge \boldsymbol{w}_J|}{|\boldsymbol{I}|} \geq \rho \tag{2}$$

When this occurs and learning is enabled, vector $\boldsymbol{w}_J$ is updated:

$$\boldsymbol{w}_J^{new} = \beta(\boldsymbol{I} \wedge \boldsymbol{w}_J^{old}) + (1 - \beta)\boldsymbol{w}_J^{old} \tag{3}$$

Otherwise, node J is inhibited and another node is selected. In case no node j is found to meet the vigilance criterion, a new output neuron is committed. Proliferation of categories is avoided in Fuzzy ART by normalizing inputs. A useful rule for achieving normalization while preserving amplitude information is *complement coding*. If $\boldsymbol{a}$ represents the on-response of the pattern, each component of the off-response $\boldsymbol{a}^c$ is defined as $a_i^c \equiv 1 - a_i$. Then, the complement coded input comes $\boldsymbol{I} = (\boldsymbol{a}, \boldsymbol{a}^c) \equiv (a_1 \cdots a_M, a_1^c \cdots a_M^c)$ and $|\boldsymbol{I}| = M$ for every input pattern.

3   Skin Color Segmentation TSL Space Color

Color can be decomposed into one luminance and two chrominance components. Several researches have proved that skin colors have a certain invariance regarding chrominance components and in consequence skin tone and lighting mainly affect the luminance value [3]. Furthermore, normalized probability of skin colors can be modeled by a Gaussian distribution in the chromatic color space [3].

Different color spaces separating chrominance and luminance components have been used for skin color segmentation (YIQ, YCbCr, CIE-Lab, CIE-Luv, HSV, IHS and TSL). In TSL color space [4], colors are specified in terms of Tint (T), Saturation (S) and Luminance (L) values. TSL has been selected as the best color space to extract skin color from complex backgrounds [5] because it has the advantage of extracting a given color robustly while minimizing illumination influence.

4 GPU Implementation for Real Time Fuzzy ART Skin Recognition

4.1 Training

An exhaustive description of a GPU-based Fuzzy ART Neural Network implementation for generic data learn and test can be found in [6]. For specific skin recognition application, complement coded TS features were chosen, so every input pattern is described as $I = (a, a^c) = (T, S, 1-T, 1-S)$. By using *complement coding* we drastically reduce proliferation of categories and force $|I|$ to be constant ($|I| = M = 2$) for every input pattern, thus avoiding extra computing for calculating the *match criteria* (2). During learning process, each four-dimension input pattern I is replicated along a RGBA texture I^T. Four-dimension patterns can be stored in a single texel using its four different channels. Neural network weights are initialized to 1 and stored in another RGBA texture W^T.

Generalized size reduction operations on the GPU through a *ping-pong* technique [7], are used to obtain $|I \wedge w_j|$ and *Multiple Render Targets* support allows to calculate the module of newer committed categories $|w_j|$ at the same time. The use of RGBA textures allows to run MIN and SUM operations on 4-component vectors in one clock cycle on every fragment shader unit, making the process faster.

Reduced textures are then used to store the activity of each neuron, satisfying the match criteria, on the R channel of a texture T^T; G channel is used to store the category index; *alpha* channel takes the value of 1 if the match criteria is satisfied and 0 otherwise; finally, channel B can be used for printing the matching rate, which we found very useful for debugging purposes.

The J_{th} neuron is found using a row reduction operation over texture T^T, where those fragments not satisfying the match criteria are discarded. If the system enters in resonance, the weights of the selected category are updated by rendering into the corresponding texel of the texture W^T using *scissoring* [7]. If not, the new pattern is learned by rendering into a unused texel using equation (3).

4.2 Testing

For the real time skin recognition application, a GPU-based Fuzzy ART Neural Network implementation was developed. The system is able to read data from conventional USB Webcams using the DirectShow library. Webcam captures are directly downloaded to the GPU memory and stored in a texture, avoiding slow CPU-GPU data transfers. Each new frame is stored in the GPU as a bidimensional RGBA texture. Bidimensional textures are more amenable to be used on the GPU than one dimensional textures and this kind of organization maps specially well for skin pattern recognition in our computer vision application.

On each frame, several computations take place in order to recognize skin tone pixels. Before the neural network can start recognition process, a pre-processing stage has to be configured in order to translate 3-color dimension RGB coded frames into 4 color dimension complement coded ones. A shader in this stage uses R,G and B components to calculate T and S chrominance components and its corresponding complements 1-T and 1-S, and renders them to a RGBA texture TS^T. Each texel on the new texture has the whole information we need to categorize each pixel of the original image. Weights learned during training process are loaded into a unidimensional W^T RGBA texture. Each texel on this texture stores a long-term-memory trace w_j.

During neural network categorization, several input patterns stored in texture TS^T can be categorized in a parallel fashion using every fragment processor available on the GPU. Category choice occurs through the execution of a shader for N times, being N the number of categories in the neural network. In each pass, the activation of the j_th output neuron and the match function are computed for every input pattern. This values are rendered into a RGBA output texture, which is used as input for the next iteration, using a *ping-pong* technique. The other two channels in the output texture are used for indexing the selected category of the pattern and counting the number of categories the input pattern has been tested against.

If the activation in pass j is bigger than the computed activation in pass $j-1$ and the match criterion is satisfied, then the index category is updated on the output texture. Rendering both the index of the selected category and the match function to the output texture allows the expert to visually analyze results. In our implementation, different gray levels on channel R belong to different skin categories committed during training process, *alpha* channel represents the level of resemblance of that pixel in the original image to the selected skin category.

For real-time demonstration of the system, another post-processing stage is used. In this last stage a shader is configured to render onto the screen just those pixels in the original image which have crossed all the tests and have been recognized as skin. Final result is an image were only skin recognized regions are drawn over a uniform color background.

5   Experimental Results

5.1   Statistical Skin Color Selection

From two different image databases, regions containing skin pixels were selected. Normalized frequency of TSL chrominance components of every pixel in these regions was graphically represented.

Afterwards, we estimated the skin color distributions as a normal distribution through the minimum covariance determinants estimator (MCD) [8]. A color was classified to belong to skin if its mahalanobis distance to the mean color given by the mean vector of the modeled distribution is lower than a threshold defined through the critical value of α % of a χ^2 distribution with two degrees of freedom. From previous works [9], we concluded that the best results are achieved with the threshold value of 0.99.

We created 2 different Training Sets of colors for the neural network. Training Set 1 is formed by all the colors selected as skin colors from the skin regions in the 3056-image Faces96 database [10], a well-known face database and Training Set 2 from the skin regions in a 4176-image private database (taken by us with different cameras and lighting conditions and including not only faces but hands and arms).

5.2 Fuzzy ART Off-Line Training

From the colors selected in the two training sets, the input features to train the Neural Network were the T and S components and their complementary values. Performance of our GPU-based Fuzzy ART implementation relays on several factors: length of the input pattern I, number of input patterns P presented to the network and number of created categories N. During the learning process, the number of committed categories varies depending both on the grade of similarity between patterns and the *vigilance criterion* ρ. The number of input features used during training were 671438 for Training Set 1 and 559640 for Training Set 2. The network was trained using $\alpha = 0.001$ and different values for ρ in the range $[0.90; 0.97]$. Table 1 subsumes the number of committed categories in each Training Set with different ρ values. The bigger the value of ρ is, the bigger the number of categories created by the network is. Although the number of colors is very big, few categories are created proving both training colors and selection of TS features were correct.

Table 1. Number of committed categories with different ρ

ρ	0,90	0,93	0,95	0,97
Training Set 1	9	15	33	71
Training Set 2	8	17	24	73

5.3 Fuzzy ART Real-Time Testing

In order to measure the performance of our implementation, several tests were done on a dual-core 3.2 GHz Pentium 4 with 1GB RAM, GeForce 7800GTX 256 MB GPU (containing 24 fragment processors) and a generic webcam able to capture at resolutions of 640x480 pixels at a rate up to 90 fps.

Figure 1 shows different skin recognition results for the same image using the weights learned from Training Set 1. Figure 1(a) shows the original captured RGB image. Figure 1(f) represents the input features introduced to the network.

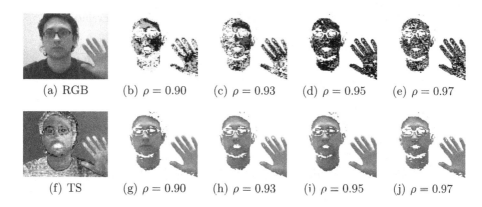

Fig. 1. Selected categories and segmented skin regions for a single image using Training Set 1 and different ρ values

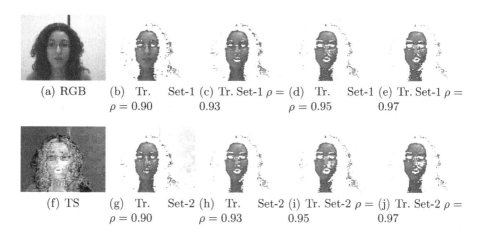

Fig. 2. Testing results with Training Sets 1 and 2

The image was created by writing on every RGB channel the tone (T) value in each pixel and choosing A channel to be the value of the saturation characteristic (S). Figures from 1(b) to 1(e) show regions identified as different skin categories by the network with ρ increasing from 0.90 to 0.97 respectively. Finally, figures from 1(g) to 1(j) show the final filtered image in which only skin recognized pixels are drawn on a white background.

Figure 2 depicts the different behavior of the network with the two different Training Sets. For low ρ values false positive rate is bigger with Training Set 2, making the recognition unacceptable for $\rho = 0.9$. For increasing values of ρ false positive rate decreases on both Training Sets, while hit rate significantly worsen only in Training Set 2.

Table 2 shows frame rate accomplished by the neural network on the GPUs for different video resolutions. As value of ρ and resolution increase, frame rate

Table 2. Frames per second for different resolutions and ρ values

Resolution/fps	320x240 pixels			352x288 pixels			640x480 pixels		
ρ	0.90	0,95	0,97	0.90	0,95	0,97	0.90	0,95	0,97
Color set 1	270	89	42	212	68	32	71	23	11
Color set 2	206	120	43	230	92	31	79	30	10

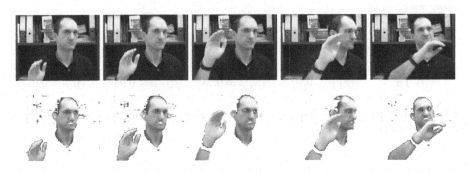

Fig. 3. Skin recognition on a video sequence at 270 fps with Training Set 1

decreases. Best performance, which was achieved using a resolution of 320x240 pixels and $\rho = 0.90$, is 296 fps. The number of frames that can be processed by the network strongly depends on the number of committed categories every pixel has to be tested to.

Figure 3 shows the performance of the neural network with 5 frames extracted from a video sequence, using Training Set 1 and $\rho = 0.90$. With this value of ρ some background pixel are wrongly categorized but the hit rate is very high, achieving an excellent real-time skin recognition.

6   Conclusions

An implementation of a GPU-based Fuzzy ART Neural Network for real time skin recognition was introduced in this paper. Our design successfully faces the problem of using a neural network for pattern classification when time is a major requirement. A robust and complete set of skin colors and a good selection of input features (chrominance components of TSL color space) is necessary to train the network so that it can recognize skin in real changing conditions.

Experimental results show our system achieves excellent skin tone pixels recognition at high frame rates with an nVIDIA 7800GTX GPU equipped video card, which includes 24 fragment shaders in the pipeline. GPUs are quickly evolving and every few months a new generation of improved processors is made publicly available. Forward compatibility of our design is guaranteed and we can expect an incredible performance with newer cards, such as the recently appeared NVIDIA GeForce 8800, which incorporates 128 unified shaders.

Fuzzy ART skin recognition using a fast GPU can be the first high-computational-cost but low-time-consuming step in a much more complex system for computer vision. In this sense, our design is potentially applicable in a wide range of applications, such as human-machine interfaces or video indexing.

References

1. Fung, J.: Computer vision on the gpu. In: Pharr, M. (ed.) GPU Gems 2, pp. 649–665. Addison-Wesley, Reading (2005)
2. Carpenter, G.A., Grossberg, S., Rosen, D.B.: Fuzzy ART: Fast stable learning and categorization of analog patterns by an adaptive resonance system. Neural Networks 4(6), 759–771 (1991)
3. Hsieh, I.S., Fan, K.C., Lin, C.: A statistic approach to the detection of human faces in color nature scene. Pattern Recognition 35(7), 1583–1596 (2002)
4. Terrillon, J., David, M., Akamatsu, S.: Automatic detection of human faces in natural scene images by use of a skin color model and of invariant moments. In: Proceedings of the IEEE Conference on Automatic Face and Gesture Recognition, pp. 112–117. IEEE Computer Society, Washington, DC, USA (1998)
5. Duan-sheng, C., Zheng-kai, L.: A novel approach to detect and correct highlighted face region in color image. In: AVSS '03: Proceedings of the IEEE Conference on Advanced Video and Signal Based Surveillance, pp. 7–12. IEEE Computer Society, Washington, DC, USA (2003)
6. Martínez-Zarzuela, M., Díaz Pernas, F., Díez Higuera, J., Antón Rodríguez, M.: Fuzzy ART neural network parallel computing on the gpu. In: Sandoval, F. (ed.) IWANN 2007. LNCS, vol. 4507, pp. 463–470. Springer, Heidelberg (2007)
7. Pharr, M. (ed.): GPU Gems 2 (Programming Techniques for High-Performance Graphics and General-Purpose Computation). Addison-Wesley, Reading (2005)
8. Rousseeuw, P.J., Driessen, K.V.: A fast algorithm for the minimum covariance determinant estimator. Technometrics 41(3), 212–223 (1999)
9. González Ortega, D., Díaz-Pernas, F., Díez-Higuera, J., Martínez-Zarzuela, M., Boto Giralda, D.: Computer vision architecture for real-time face and hand detection and tracking. In: Bres, S., Laurini, R. (eds.) VISUAL 2005. LNCS, vol. 3736, pp. 35–49. Springer, Heidelberg (2006)
10. Spacek, L.: Faces96 database:
   http://cswww.essex.ac.uk/mv/allfaces/faces96.html

A Cooperative Game Theoretic Approach to Prototype Selection

N. Rama Suri, V.S. Srinivas, and M. Narasimha Murty

Electronic Commerce Laboratory, Dept. of Computer Science and Automation,
Indian Institute of Science, Bangalore, India
{nrsuri,srini,mnm}@csa.iisc.ernet.in

Abstract. In this paper we consider the task of prototype selection whose primary goal is to reduce the storage and computational requirements of the Nearest Neighbor classifier while achieving better classification accuracies. We propose a solution to the prototype selection problem using techniques from cooperative game theory and show its efficacy experimentally.

1   Introduction

Though nearest neighbor (NN) [1] rule is a well-known supervised non-parametric classifier, it demands huge memory and computational requirements as the entire training data set needs to be stored in the main memory. There have been several efforts to alleviate these problems associated with the NN classifier. *Prototype selection is the problem* of finding representative patterns from a given data set to reduce the storage and computational requirements of a classifier while achieving better classification accuracies. In the literature, two different families of prototype selection methods exist [4]. The first method is called *condensing algorithm* that aims at selecting a minimal subset of prototypes that leads to the same performance as using the whole training set. The difficulty [4] with condensing algorithms is that noisy examples are preferred to be selected into prototype set, which harms the accuracy of the final classifier. The second type of algorithms for prototype selection are called *editing algorithms*. The main difficulty [4] with the editing algorithms is that they may not reduce the training set size.

Motivation: In this paper, we use cooperative game theoretic techniques to address the prototype selection problem. Cooperative game theory is a branch of game theory and we refer the reader to [6] for a detailed discussion on game theory and its variants. To analyze cooperative games, there are a set of solution concepts such as Shapley value, core, bargaining sets [6], etc. Among these, Shapley value is a fair solution concept in the sense that it divides the collective (i.e. total) value of the game among the players according to their marginal contribution in achieving that collective value. That is, the higher the Shapley value of a player, the more important that player among the other players. We can make use of this notion of importance among the available patterns to pick a small set of prototypes from the given data set, if we can formulate the prototype

J.N. Kok et al. (Eds.): PKDD 2007, LNAI 4702, pp. 556–564, 2007.

selection problem as a cooperative game. Hence the techniques from cooperative game theory can be useful in proposing a solution approach to the prototype selection problem.

Contributions and Outline of the Paper: To the best of our knowledge, *it is the first time that the techniques from (cooperative) game theory are used in proposing an algorithm to the prototype selection problem.* In Section 2, we present basic concepts underlying cooperative game theory that are useful in understanding the rest of the paper. We formulate a convex game [8], which we call *PS-GAME*, that helps in solving the prototype selection problem in Section 3.1. We propose an exact algorithm for prototype selection in Section 3.2 and also investigate the intractability of this algorithm. Then in Section 3.3, we propose an approximate prototype selection algorithm (APSA), which runs in polynomial time. We finally show the efficacy of the APSA using empirical studies on different data sets in Section 4. We conclude the paper in Section 5 along with a few pointers to the future work.

2   Preliminaries

A cooperative game [9] with transferable utility is defined as the pair (N, v) where $N = \{1, 2, \ldots, n\}$ is the set of players and $v : 2^N \to R$ is a characteristic function, with $v(\phi) = 0$.

We can analyze a cooperative game using solution concept, which is a method of dividing the total value of the game to the individual players. We now briefly discuss two such solution concepts, namely core and Shapley value.

The Core of a Cooperative Game: A payoff allocation $x = (x_1, x_2, \ldots, x_n)$ is any vector in R^n. In this payoff allocation, x_i is the utility of player i, $i \in N$. An allocation $x = (x_1, x_2, \ldots, x_n)$ is said to *feasible* if $\sum_{i \in C} x_i \leq v(C)$, $\forall C \subseteq N$. A payoff allocation is said to be *individual rational* if $x_i \geq v(\{i\})$, $\forall i \in N$. It is said to be collectively rational if $\sum_{i \in N} x_i = v(N)$. A payoff allocation is said to *coalitionally rational* if $\sum_{i \in C} x_i \geq v(C)$, $\forall C \subseteq N$. A payoff allocation vector that is individually rational, collective rational and coalitionally rational is said to be an element in the *core* of the game (N, v). The allocation vectors in the core are stable in the sense that no subset of players can deviate from them.

Shapley Value of a Cooperative Game: Let us consider a cooperative game, (N, v) as defined above. Shapley value suggests a fair allocation scheme to the players in the cooperative game in the sense that it takes the marginal contribution of each player into account. The Shapley value, $\Phi_i()$, for each player i is defined [6] as,

$$\Phi_i(v) = \frac{1}{n!} \sum_{\pi \in \Pi} [v(S_i(\pi) \cup i) - v(S_i(\pi))] \tag{1}$$

where Π is the set of permutations over N, and $S_i(\pi)$ is the set of players appearing before the ith player in permutation π. The Shapley value of a player is a weighted mean of its marginal value, averaged over all possible permutations of players.

Note: In special class of cooperative games called convex games, the core is non-empty and the Shapley value belongs to the core.

Convex Games: Now we introduce a special class of cooperative games called *convex games*. The cooperative game (N, v) is called *superadditive* [8] if

$$v(S) + v(T) \leq v(S \bigcup T), \quad \forall S, T \subseteq N \quad \text{with} \quad S \bigcap T = \phi. \tag{2}$$

It is called *convex game* [8] if

$$v(S) + v(T) \leq v(S \bigcup T) + v(S \bigcap T), \quad \forall S, T \subseteq N. \tag{3}$$

It is called *strictly convex* if the inequality holds in (3) whenever neither $S \subseteq T$ nor $T \subseteq S$. Another condition [8] that is equivalent to (3) (provided N is finite) is,

$$v(S \bigcup \{i\}) - v(S) \leq v(T \bigcup \{i\}) - v(T) \tag{4}$$

for all $i \in N$ and all $S \subseteq T \subseteq N - \{i\}$. This expresses a sort of increasing marginal utility for coalition membership, and is analogous to the *increasing returns to scale* associated with convex production functions in economics [8].

Shapley Value of a Convex Game: We have explained briefly the Shapley value of a cooperative game in the previous section. If the cooperative game turns out to be a convex game, then it is possible to give a nice interpretation for the Shapley value using the *core* of the cooperative game [8].

Let $|N| = n$ in the cooperative game (N, v) under consideration. Any solution concept suggests a vector with n tuples in an $n-$dimensional linear space E^N with coordinates indexed by the elements of N. If $x \in E^N$ and $S \subseteq N$ we shall often write $x(S)$ to represent $\sum_{i \in S} x_i$. The hyperplane in E^N defined by the equation $x(S) = v(S)$, $0 \subseteq S \subseteq N$, will be denoted by H_S.

A payoff vector $x \in E^N$ is said to be feasible (for $v(.)$) if $x(N) \leq v(N)$. Recall that the core of (N, v) is defined as the set C of all feasible $x \in E^N$ such that $x(S) \geq v(S)$, $\forall S \in N$. The core is obviously a subset of the hyperplane H_N and since the inequalities $x_i \geq v(i)$ are included in the previous set of inequalities, i.e., $x(S) \geq v(S)$, $\forall S \in N$, the core is bounded. Thus, C is a compact convex polyhedron, possibly empty, of dimension at most $n - 1$.

Let ω represent a simple ordering of the players. Specifically, let ω be one of the $n!$ functions that map N onto $\{1, 2, \ldots, n\}$. Define

$$S_{\omega,k} = \{i \in N : \omega(i) \leq k\}, \quad k = 0, 1, \ldots, n. \tag{5}$$

these are called the *initial segments* of the ordering. Thus $S_{\omega,0} = 0$ and $S_{\omega,n} = N$. Consider the equations $x^\omega(S_{\omega,k}) = v(S_{\omega,k})$, $k = 0, 1, \ldots, n$ and it is easy to solve the set of equations to find the coordinates of the intersection of the hyper planes $H_{S_{\omega,k}}$, namely

$$x_i^\omega = v(S_{\omega,\omega(i)}) - v(S_{\omega,\omega(i)-1}), \quad \forall i = 1, 2, \ldots, n \tag{6}$$

Thus each ordering ω defines a payoff vector x^ω.

Theorem 1. In convex games, the extreme points of the core are precisely the points x^{ω} [8].

Theorem 2. In convex games, the Shapley value is the center of gravity of the extreme points of the core [8]. That is, $\forall i \in N$,

$$\Phi_i = \frac{1}{n!} \sum_{\omega \in \Omega} x_i^{\omega}. \tag{7}$$

where Ω is the set of all permutations of N.

This gives a nice interpretation of the Shapley value for convex games.

3   Our Solution Approach to Prototype Selection

In this section, we propose our solution approach to the prototype selection problem. We first start with modeling this problem as a convex game.

3.1   Modeling Prototype Selection as a Convex Game (PS-GAME)

Here we consider *the data sets where the clusters corresponding to different classes of patterns are non-overlapping.* In one of the following sections, namely Section 3.4, we consider the *general scenario* where the clusters of patterns corresponding to different classes may be overlapping.

We now give interpretations for the two components of a cooperative game, (N, v), in the context of prototype selection. Let the set of players, N, be the patterns in the given training data set. Now we define the characteristic function, $v(.)$, in the following way. $\forall\ S \subseteq N$, $v(S)$ is *the error rate of the training data set* when only the patterns in S are used as the prototypes. Now it is easy to check that $v(.)$ satisfies the desired condition (4) for convexity for the data sets where the clusters corresponding to different classes are non-overlapping. Hence prototype selection problem can be modelled as convex game and we refer this *PS-GAME*.

3.2   Algorithm-1: An Exact Algorithm for Prototype Selection

Here we propose an exact algorithm for prototype selection that arises naturally as a consequence of the framework that we have developed so far. Due to *Theorem-1*, it is easy to see that *PS-GAME* has non-empty core (since PS-GAME is a convex game). Hence the Shapley value of *PS-GAME* is the center of gravity of the extreme points of the core by means of *Theorem-2*. There are $n!$ extreme points of the core. Now, we can compute the Shapley value (7) of each pattern (i.e. player) with the help of the extreme points of the core. By sorting the patterns in the non-decreasing order of their Shapley values, we can pick and add the patterns one by one to the desired set of prototypes until there is an increase in the error rate. Since this algorithm needs to work with all possible $n!$ permutations of the patterns, its time complexity is $O(\frac{n}{e})^n$ due to Stirling's approximation. To get around this difficulty, we present an approximate algorithm that runs in polynomial time for prototype selection.

3.3   An Approximate Polynomial Time Algorithm for Prototype Selection

Recall that the Shapley value of *PS-GAME* is the center of gravity of the extreme points of the core. We can compute approximate Shapley values of the patterns by randomly sampling a set of extreme points. Here the cardinality of the sampled set is polynomial in the number of given patterns.

Let us represent the randomly sampled set of extreme points with Ψ. Also let t be the cardinality of Ψ, i.e., $t = |\Psi|$. So we approximate the Shapley value using Ψ. This approach is called *multi perturbation Shapley value analysis* [3]. Note that the size of each permutation in the set Ψ is n. In general the size of a permutation is n. For our approximation, it is sufficient even if we randomly pick d patterns out of available n patterns to form each element of the set Ψ, where d is a positive integer constant and $d << n$. We call each element of Ψ a *partial permutation*. Using such a sampled set Ψ consisting of t number of partial permutations where the size of each partial permutation is d, we present an approximate algorithm to solve the prototype selection problem. We call this algorithm *approximate prototype selection algorithm (APSA)*. Following is the description of *APSA*.

Algorithm 2: APSA

1. Construct a set, Ψ, of size t by sampling from available $n!$ permutations. The size of each partial permutation in Ψ is bounded by size d instead of n. Here $d << n$. The elements of Ψ are partial permutations.
2. For each partial permutation in Ψ of size d, compute the contribution of each of the d patterns using expression (6).
3. Compute the Shapley value of each pattern using expression (7).
4. Sort the patterns in the non-decreasing order of their Shapley values.
5. Pick and add patterns one by one, in the sorted order, to the desired set of prototypes until there is an increase in the error rate of the classification.

Time Complexity of APSA: We now discuss the running time of *APSA*. Here we have to compute the marginal contribution of each pattern corresponding to each partial permutation in Ψ. It takes $O(td)$ time. Note that $t << n$ and $d << n$. Sorting the patterns in non-decreaseing order of their Shapley values requires $O(n \, log(n))$ time. The overall running time of *APSA* is $O(td+n \, log(n))$. Since $t << n$ and $d << n$, note that it runs much much faster than a quadratic time algorithm.

3.4   The General Scenario

In Section 3.1 we assumed that the data sets where the clusters corresponding to different classes are non-overlapping. Here we relax this assumption. That is, the clusters of patterns corresponding to different classes may be over-lapping.

In such scenarios, if we want to formulate a convex game as in Section 3.1, the characteristic function $v(.)$ may not satisfy the convexity condition (4). To get around this problem, we have to relax the conditions (4) for convexity. *Average convex games [2]* and *partially average convex games [2]* essentially address this issue.

The class of average convex games strictly includes the class of convex games [2] and the Shapley value of an average convex is in the core [2]. In case of partially average convex games, they include the average convex games. Moreover partially average convex games need not be super-additive [2]. It is also true that Shapley value for these partially average convex games belongs to the core [2]. Hence in the general scenario of modelling prototype selection problem either as average convex games or as partially average convex games, we can use *APSA* to solve the prototype selection problem.

4   Experimental Results

In this section, we carry out two types of experiments to illustrate the performance of *APSA* on 6 bench mark data sets from UCI repository of machine learning databases [5]. Characteristics of the data sets are shown in Table 1.

Table 1. Description of Experimental Data Sets

Data Set	No. of Features	No. of Classes	Size of Data Set
Glass	9	6	214
Iris	4	3	150
Liver (BUPA)	6	2	345
Pima	8	2	768
Wine	13	3	178
Ionosphere	34	2	351

Experiment Type-1: Comparison with Editing Algorithms: Here we compare the performance of *APSA* with some popular editing algorithms [7] for prototype selection. These algorithms and *APSA* are respectively applied to the same data sets and 5-fold cross validation has been applied on the training data set to obtain the performance results. Each data set has randomly been divided into training and test samples as shown in Table 2.

Two main aspects of our interest to carry out these experiments are *classification accuracy*, and *number of prototypes selected* from the training set. The number of prototypes selected is a direct measure of the computational savings and storage space. Classification accuracy represents the ability of the algorithms to select the most representative prototypes. We present the comparison of *the classification accuracies* of APSA with the editing algorithms [7] in Table 3. Similarly, Table 4 presents the comparison of *the number of prototypes selected* by APSA and by the editing algorithms [7]. In both these tables, the numbers provided for editing algorithms are directly reproduced from [7] (since our experimental setup is the same as that reported in [7]).

Table 2. Data Sets with Corresponding Training and Test Set Sizes for Experimental Comparison of APSA with Editing Algorithms

Data Set	Training Set Size	Test Set Size
Glass	174	40
Iris	120	30
Liver (BUPA)	276	69
Pima	615	153
Wine	144	34

Table 3. Comparison of Classification Accuracies of *APSA* and the Editing Algorithms

	Glass	Iris	Liver	Pima	Wine
APSA	65.5 (3.67)	95.0 (3.08)	69.85 (1.58)	70.5 (3.7)	**76.2** (2.72)
Best k-NCN	**68.0** (5.79)	96.0 (1.34)	70.1 (6.71)	74.1 (2.64)	71.8 (3.99)
Wilson (k = 3)	63.0 (6.20)	**96.7** (2.11)	69.3 (6.24)	72.0 (2.59)	71.8 (8.02)
Repetition (k=3)	64.5 (7.58)	**96.7** (2.37)	**71.4** (8.50)	73.2 (3.67)	72.9 (7.61)
Edited k-NCN (k=3)	67.0 (5.34)	**96.7** (2.11)	68.1 (4.85)	72.2 (3.47)	70.0 (6.81)
Edited k-NCN (k=5)	65.0 (7.07)	**96.7** (2.11)	66.4 (6.94)	73.9 (2.45)	70.0 (9.0)
GG	67.0(5.79)	95.3 (1.64)	69.3 (4.71)	74.1 (3.27)	70.0 (5.39)
RNG	67.5 (6.52)	90.2 (1.34)	68.1 (4.85)	72.0 (2.35)	67.88 (6.86)
All k-NN (k=3)	64.2 (6.29)	**96.7** (2.36)	68.1 (7.39)	71.7 (3.84)	67.7 (5.5)
Depuration (k=3, k'=2)	67.0 (5.12)	**96.7** (1.52)	70.3 (7.15)	**75.9** (2.58)	70.6 (11.76)

Table 4. Comparison of the Number of Prototypes Selected by *APSA* and the Editing Algorithms

	Glass	Iris	Liver	Pima	Wine
Original Training Set Size	174	120	276	615	144
APSA	**12.0** (7.48)	**15.2**(8.68)	**26.20**(10.37)	**84**(25.53)	**9.0**(5.36)
Best k-NCN	115.8(5.49)	115.6(0.80)	190.8(4.45)	456.0(10.14)	105.6(2.06)
Wilson (k = 3)	114.8(6.43)	115.4(1.36)	176.0(7.16)	427.8(6.43)	105.6(2.06)
Repetition (k=3)	76.8(43.44)	33.0(47.54)	146.8(54.38)	338.2(47.74)	86.2(15.36)
Edited k-NCN (k=3)	114.6(5.75)	115.6(0.80)	176.8(8.06)	424.0(3.74)	103.8(2.93)
Edited k-NCN (k=5)	109.8(6.68)	116.2(0.75)	185.6(4.45)	438.6(3.07)	104.2(1.94)
GG	105.6(4.32)	115.2(1.17)	190.4(3.93)	472.8(8.42)	110.8(2.23)
RNG	132.2(6.18)	115.0(0.63)	195.2(6.73)	474.6(10.07)	118.0(2.61)
All k-NN (k=3)	111.0(6.24)	114.6(0.89)	145.4(12.95)	363.2(10.13)	92.8(4.15)
Depuration (k=3, k'=2)	142.4(26.34)	35.0(47.53)	232.0(48.39)	403.6(49.55)	109.8(16.66)

In Table 3, and Table 4, the values in brackets correspond to the respective standard deviation. We can observe from Table 4 that *APSA* is performing better than the editing algorithms [7] in terms of the number of prototypes selected. Results from Table 3 suggests that *APSA* can achieve competitive classification accuracies compared with the editing algorithms. Hence we can conclude that our proposed algorithm can select small size prototype set while maintaining comparable classification accuracies.

There are two parameters in the proposed algorithm. They are t and d. Recall that $t << n$ and $d << n$. The values of t and d in these experiments also convey the same message. For Glass data set $t = 10$, $d = 16$, and $n = 174$. Here we can observe that $t << n$ and $d << n$. This observation is true for the remaining data sets as well. For completeness, we provide the values of t and d for the remaining data sets also. For IRIS data set $t = 14$, $d = 21$, and $n = 120$. For WINE data set $t = 19$, $d = 20$, and $n = 144$. For BUPA (Liver) data set $t = 21$, $d = 49$, and $n = 276$. For PIMA data set $t = 31$, $d = 48$, and $n = 615$.

Experiment Type-2: Comparison with Condensing Algorithms: Here we compare the performance of the $APSA$ with some popular condensing algorithms [10] for prototype selection over the four bench-mark data sets reported in Table 1. 10-fold cross validation is used for each experiment. Each data set is divided into 10 sets and each algorithm is given a training set consisting of 9 of the partitions (i.e. 90% of the data), from which it picks the prototype set, S. The other 10% of the data is classified using the prototypes in S. Ten such trails are run for each data set with each algorithm, using a different one of the 10 partitions as the test set for each trail. The average classification accuracy and the average percentage of the number of prototypes selected in each trail are reported in Table 5. The numbers in Table 5 corresponding to the condensing algorithms are directly reproduced from [10] (since our experimental setup is same as that reported in [10]).

Table 5. Comparison of Classification Accuracies of $APSA$ and the Condensing Algorithms. The Column (%) indicates percentage of number of prototypes selected.

	Ionosphere	%	Liver (Bupa)	%	Pima	%	IRIS	%
APSA	85.75	5.73	68.58	7.66	73.49	7.02	90.66	11.81
k-NN	84.62	100	65.57	100	73.56	100	94.0	100
CNN	82.93	21.62	56.8	40.87	65.76	36.89	90.00	12.74
SNN	81.74	19.21	57.70	52.59	67.97	42.95	83.34	14.07
IB2	82.93	21.62	56.80	40.87	65.76	36.89	90.00	12.74
IB3	85.75	14.59	58.24	10.66	69.78	10.97	94.67	19.78
DEL	86.32	12.88	61.38	38.36	71.61	12.64	93.33	9.56
DROP1	79.77	3.23	58.24	10.92	65.23	6. 50	84.67	8.59
DROP2	86.6	7.79	67.77	24.77	70.44	17.59	94.67	14.22
DROP3	87.75	7.06	60.84	24.99	75.01	16.90	95.33	14.81
DROP4	86.90	10.60	62.60	32.56	72.53	21.76	95.33	14.89
DROP5	86.90	9.78	65.50	31.08	73.05	21.95	94.0	12.15
ENN	84.04	84.24	61.12	68.15	75.39	76.37	95.33	94.74
RENN	84.04	82.27	58.77	63.13	75.91	74.52	95.33	94.67
All k-NN	84.05	82.18	60.24	52.34	74.88	64.61	95.33	93.78
ELGROW	73.77	0.63	56.74	0.55	67.84	0.29	88.67	2.30
EXPLORE	80.89	0.63	57.65	0.64	75.27	0.29	92.67	2.30

We can see from Table 5 that the percentage of the number of prototypes selected by $APSA$ is better than most of the condensing algorithms except EL-GROW and EXPLORE. The classification accuracies obtained by $APSA$ are better than the condensing algorithms for some data sets and comparable with the remaining algorithms for some other data sets.

5   Conclusions and Future Work

We considered the prototype selection problem in this paper and proposed a cooperative game theoretic based solution. This is a new direction to address the prototype selection problem which is not yet considered in the literature. Experimental results showed the efficacy of the presented approach.

It would be interesting to probe the use of the game theoretic ideas to address the prototype selection problem in case of incremental data sets.

Acknowledgements

We are very thankful to Prof. Y. Narahari for getting valuable advices and cooperation in carrying out this work.

References

1. Cover, T.M.: Nearest neighbor pattern classification. IEEE Transactions of Information Theory 13, 21–27 (1967)
2. Inarra, E., Usategui, J.M.: The shapley value and average convex games. International Journal of Game Theory 22, 13–29 (1993)
3. Keinan, A., Sandbank, B., Hilgetag, C., Meilijson, I., Ruppin, E.: Fair attribution of functional contribution in artificial and biological networks. Journal of Neural Computation 16(9), 1887–1915 (2004)
4. Li, Y., Hu, Z., Cai, Y., Zhang, W.: Support Vector Based Prototype Selection Method for Nearest Neighbor Rules. In: Advances in Natural Computation. LNCS, Springer, Berlin (2005)
5. Murphy, P.M., Aha, D.W.: UCI repository of machine learning databases (1994)
6. Myerson, R.B.: Game Theory: Analysis of Conflict. Harvard University Press (1997)
7. Sanchez, J.S., Barandela, R., Alejo, R., Marques, A.I.: Performance evaluation of prototype selection algorithms for nearest neighbor classification. In: Proceedings of 14th Brazilian Symposium on Computer Graphics and Image Processing (SIBGRAPI'01) (2001)
8. Shapley, L.S.: Cores of convex games. International Journal of Game Theory 1(1), 11–26 (1971)
9. Straffin, P.D.: Game Theory and Strategy. The Mathematical Association of America (1993)
10. Wilson, D.R., Martinez, T.R.: Reduction techniques for instance-based learing algorithms. Machine Learning 38(3), 257–286 (2000)

Dynamic Bayesian Networks for Real-Time Classification of Seismic Signals

Carsten Riggelsen, Matthias Ohrnberger, and Frank Scherbaum

University of Potsdam, Institute of Geosciences
Karl-Liebknecht-Str. 24/25, 14476 Golm, Potsdam, Germany
{riggelsen,mao,fs}@geo.uni-potsdam.de

Abstract. We present a novel method for automatic classification of seismolog-ical data streams, focusing on the detection of earthquake signals. We consider the approach as being a first step towards a generic method that provides for classifying a broad range of seismic patterns by modeling the interrelationships between essential features of seismograms in a graphical model. Through a con-tinuous Wavelet transform the features are extracted, yielding a time-frequency-amplitude decomposition. The extracted features obey certain Markov properties, which allows us to form a joint distribution in terms of a Dynamic Bayesian Net-work. We performed experiments using real seismic data recorded at different stations in the European Broadband Network, for which we achieve an average classification accuracy of 95%.

1 Introduction

Last decade's developments in data acquisition, data storage and data transmission tech-nologies via internet allow seismologists to collect new data at an unprecedented rate. Still, the impact of recent large earthquakes (e.g. Kobe 1995, Sumatra 2004) demon-strates that the need for establishing even denser seismological network infrastructures will continue [10] in order to allow for detailed research on earthquake source processes and to enable the implementation of early warning systems for critical infrastructure fa-cilities. Consequently, robust automatic algorithms have to be used for scanning data streams automatically in order to aid seismologists in extracting the relevant portions for further analysis.

The main purpose of existing seismic signal detection algorithms is the automatic detection and timing of body phase arrivals in seismogram recordings of tectonic earth-quakes and artificial explosions (chemical or nuclear)—for a review, see [12]. Pattern recognition approaches, which aim to jointly detect and classify the complete seismo-gram, have rarely been used in the context of seismic signal detection and classification. In [6] a robust seismic event detector has been developed which is based on the com-parison of spectral images (sonogram) to a set of reference templates. More recently, in [5] continuous seismogram recordings have been represented by using the discrete wavelet transform and de-noising techniques and then classified into local, regional and teleseismic events categories.

Within this study, we aim to demonstrate the use of Dynamic Bayesian Networks as a very general and symbolic approach to seismic signal classification.

J.N. Kok et al. (Eds.): PKDD 2007, LNAI 4702, pp. 565–572, 2007.

2   Dynamic Bayesian Networks

The graphical models framework is an attractive and intuitively appealing statistical formalism for modeling a broad range of real-life situations and systems. Bayesian networks (BN), a particular kind of graphical model, has been used extensively for modeling *static* systems. When modeling *dynamic* systems that produce data as time passes, it is beneficial to explicitly account for the temporal element in the model. This gives rise to so-called Dynamic Bayesian Networks (DBN), a generalistion of the less powerful Hidden Markov Models.

A BN [11] is a concise way of representing a joint probability distribution over variables $X = \{X_1, \ldots, X_n\}$. Formally a BN consists of directed acyclic graph (DAG), where every vertex coincides with a variable X_i, and a set of conditional probability distributions (CPDs) $p(X_i | \Pi(X_i))$, where $\Pi(X_i)$ is the set of parents of X_i in the DAG. We use (conditional) multinomials and Gaussians as CPDs. The DAG encodes conditional independence restrictions for a joint distribution $\Pr(X)$ via the so-called Markov properties, such that we may write $\Pr(X) = \prod_i p(X_i | \Pi(X_i))$.

In a DBN [4] we are interested in $X = \{X_1, \ldots, X_n\}$ at every point in time; to distinguish times, we write X_i^t to refer to the ith variable at time t. DBNs are time-invariant, such that the DAG of the overall DBN is a repeating structure, and the CPDs do not change with time. The joint distribution is then $\Pr(X^0, \ldots, X^T) = \prod_{t=0}^{T} \prod_i p(X_i^t | \Pi(X_i^t))$ where the parents of the variables in slice t may occur in slice t (intra links) or $t - 1$ (inter links) except for slice $t = 0$, where only intra links exist. Figure 2 shows an example of a DBN. The repeating structure and the CPDs (shaded region) is unrolled until T; the structure and CPDs defined at time $t = 0$ appears only once at the beginning. The variables $X = U \cup O$ may be observed, O, while other variables may be hidden, U (for unobserved).

3   Model Specification

In the next sections we develop a generic generative DBN for seismic waves. In Section 3.1 we start off by defining the feature space of wave patterns. The features are defined such that we explicitly can specify their (in)dependences via intra- and inter-frame links in the qualitative sense—this we discuss in Section 3.2.

3.1   Continuous Wavelet Transform

For the so-called feature extraction we depart from a Continuous Wavelet Transform (CWT) (see for instance [3]), a mathematical analysis technique perfectly suitable for describing cyclic sequences, multiscale features and in particular seismic signals [8].

In contrast to a normal Fourier transform, the Wavelet transform tells us not only what happens, but also when it happens, i.e., we obtain information localized in time and frequency. The CWT coefficients are the inner product of signal $s(\tau)$, and a dilated, shifted and normalized version of a mother wavelet $\psi(\tau)$, a complex valued function with particular properties (e.g. oscillating): $CWT(t, f) = \int s(\tau) f^{-1/2} \psi_{t,f}^*(\tau) d\tau$, where $*$ denotes complex conjugation. For our purpose, we use a Morlet wavelet. The

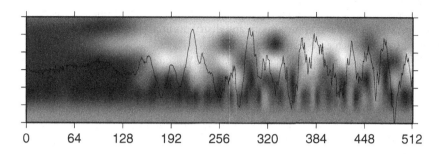

Fig. 1. Part of a seismogram (512 samples) showing a typical P-phase and its CWT coefficients. The coefficients have been normalized, and the level of gray corresponds to the size of the resulting coefficients. At the top band with the lowest center frequency is shown (0.25Hz) and at the bottom the highest frequency (4Hz).

CWT is a measure of similarity between the basis functions and the signal itself. The CWT coefficients refer to the closeness of the signal to the basis function at the current scale f.

In practice, the continuous wavelet and integrals are approximated at discrete points in time and frequency. We employ the tool GWL [7] to produce the coefficients, i.e., for every (discrete) combination of t and f, we obtain $CWT(t, f)$. The time scale for the CWT coefficients is discretized according to the original sampling rate (20Hz) of the recorded time series $s(\tau)$. In the frequency domain, we consider $i = 1 \ldots 5$ frequency bands on a dyadic scale, with center frequency 0.25Hz, 0.5Hz, 1Hz, 2Hz, 4Hz. The choice of those frequencies is due to the signal to noise ratio of seismic signals, which is fairly good in that range.

Figure 1 is an example of how the CWT of a seismic signal looks. We may observe the following: looking at single bands, there is no abrupt change in the level of gray, but rather there is a "bleeding"-effect—the gray transitions are smooth in the time dimension. Hence, there is an obvious dependence between the CWT coefficients in time. Across bands (scales) we observe a similar effect—large coefficients influence neighboring bands at the same time segment. This is the key to our main idea: **1)** Consider the coefficients $CWT(t, f)$ as random variables **2)** Exploit the Markov properties of those variables **3)** Represent those Markov assumptions in a DBN.

3.2   Relating the CWT Coefficients in a DBN

Let $O_i^t = CWT(t, i)$ and introduce discrete *state variables* U_i^t with binary states, $v = 1, 2$, corresponding to a *large* or *small* value of O_i^t. For each state of U_i^t, the *within*-distribution is defined to be a Gaussian. In graphical model terms, we may think of U_i^t as a single binary parent node of O_i^t, that is $U_i^t \to O_i^t$, where $v = 1, 2 : p(O_i^t|U_i^t = v)$ are Gaussian distributions. Hence, for each state the "members thereof" is subject to variation according to a Gaussian. Notice that the state variables U_i^t are considered hidden. Marginalizing out this variable according to $U_i^t \to O_i^t$ yields a mixture with two Gaussian components, $\Pr(O_i^t) = \sum_{u_i^t} \Pr(O_i^t, u_i^t) = \sum_{v=1,2} p(O_i^t|U_i^t = v)p(U_i^t =$

v). For increasing $|\Omega_{U_i}|$ and adding the same number of conditional Gaussians to the mixture, we can, in fact, approximate any distribution arbitrarily close at the price of adding extra parameters to our model.

The observation about the dependence of the CWT coefficients is formulated in terms of U_i^t, i.e., the mixture weights. Hence, we may for any moment in time encode the dependence across bands in a BN with the graph $\forall t : U_1^t \to U_2^t \to U_3^t \to U_4^t \to U_5^t$ which means that U_i^t is independent of all other variables given its neighbors in the graph, for all times. On the other hand, the time dependence between CWT coefficients along bands also gives rise to the graph $\forall i : U_i^0 \to U_i^1 \to U_i^2 \to \ldots \to U_i^T$. We now create the graph of a DBN by combining the two graphs. The unrolled DBN graph of the (conditional) mixture weights becomes:

$$U_1^0 \to U_1^1 \to U_1^2 \to \ldots \to U_1^T$$
$$\vdots \quad\quad \vdots \quad\quad \vdots \quad\quad\quad\quad \vdots$$
$$\downarrow \quad\quad \downarrow \quad\quad \downarrow \quad\quad\quad\quad \downarrow$$
$$U_5^0 \to U_5^1 \to U_5^2 \to \ldots \to U_5^T$$

The CPDs $p(U_i^t|U_i^{t-1}, U_{i-1}^t)$ are the same for every time segment $t > 0$, amounting to a total of 18 unique parameters. For time segment $t = 0$, $p(U_i^0|U_{i-1}^0)$ are unique, amounting to 9 parameters. We take $\forall i : p(O_i^0|U_i^0) = \ldots = p(O_i^T|U_i^T)$ to be conditional Gaussians $\mathcal{N}(O_i^t|\mu_{i,v}, \sigma_{i,v})$, for which we need to determine 20 parameters, $v = 1, 2, \forall i : \{\mu_{i,v}, \sigma_{i,v}\}$.

The dynamic range of the CWTs is very large, leading to numerous problems for parameter estimation and on-line classification. We propose to produce "more bounded" coefficients as follows. Extract the so-called *average log-amplitude* from the CWT coefficients, and associate O_i^t with the *residuals* instead of the CWT coefficients $O_i^t = \log CWT(t, i) - \bar{b}_t$ with $\bar{b}_t = \frac{1}{5}\sum_i^5 \log CWT(t, i)$. By only retaining O_i^t without the base-line $\bar{b}_t$ there is loss of information with respect to the original coefficients $CWT(t, i)$. We therefore exploit the expert-knowledge [1] captured in *source spectra* relating residuals to the average log-amplitude, i.e., a scaling relation between the frequency of the spectral maximum and the average log-amplitude. We model this domain knowledge characteristic explicitly in the DBN.

Add two extra binary variables to the structure, $U_{lo}^t = U_6^t$ and $U_{hi}^t = U_7^t$ and an extra continuous variable $O_6^t = O_{amp}^t = \bar{b}_t$. We add *deterministic arcs* from the low frequency vertices U_1^t, U_2^t and U_3^t to U_{lo}^t and from the high frequency vertices U_3^t, U_4^t and U_5^t to U_{hi}^t. These arcs define a mapping such that when any of U_1^t, U_2^t or U_3^t is in the state *large* then U_{lo}^t is in the state *large*, otherwise it is in the state *small*; for high frequencies the same mapping to U_{hi}^t holds. Finally add arcs from U_{lo}^t and U_{hi}^t to O_{amp}^t. The associated CPDs are conditional Gaussians, one per parent set configuration $z = 1, \ldots, 4 : \mathcal{N}(O_6^t|\mu_{6,z}, \sigma_{6,z})$, increasing the number of model parameters with 8. Figure 2 shows the resulting DBN.

The DBN in figure 2 along with the (conditional) multinomial, Gaussian and deterministic CPDs was implemented in GMTK [2]. This tool allows for efficient inference and learning in DBNs.

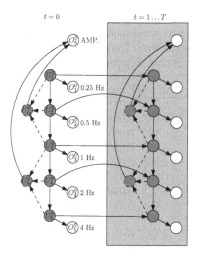

Fig. 2. The DBN capturing the relationships between the residuals, states, and the amplitude of a seismic pattern of length T. The dashed lines are deterministic relationships.

4 Learning, Classification and Confidence

The proposed DBN captures the intrinsic characteristics of a seismic pattern in a non-deterministic way by a joint distribution. We learn C different patterns by quantifying C different DBNs, that is, DBNs with the same model structure but different parameters, $\boldsymbol{\theta}_1, \ldots, \boldsymbol{\theta}_C$. A previously unseen pattern $\boldsymbol{y} = (\boldsymbol{y}^0, \ldots, \boldsymbol{y}^T)$ is classified as $I = \arg\max_i \Pr(\boldsymbol{\theta}_i|\boldsymbol{y}) = \arg\max_i \Pr(\boldsymbol{y}|\boldsymbol{\theta}_i) = \arg\max_i \log \Pr(\boldsymbol{y}|\boldsymbol{\theta}_i)$ where the 1st equality follows from $\Pr(\boldsymbol{\theta}_i|\boldsymbol{y}) \propto \Pr(\boldsymbol{y}|\boldsymbol{\theta}_i)$ under the assumption of a uniform prior over the different DBNs. Since $\boldsymbol{y}$ consists of observed variables, computing $\Pr(\boldsymbol{y}|\boldsymbol{\theta}_i)$ entails marginalizing the hidden state variables out of the joint distribution represented as our DBN, $\Pr(\boldsymbol{y}^0, \ldots, \boldsymbol{y}^T|\boldsymbol{\theta}_i) = \sum_{\boldsymbol{u}} \Pr(\boldsymbol{y}^0, \boldsymbol{u}^0 \ldots, \boldsymbol{y}^T, \boldsymbol{u}^T|\boldsymbol{\theta}_i)$. Rather than summing out the hidden variables directly, which would be intractable in general, the algorithm for efficient DBN inference as described in [13] is called via GMTK.

Assigning $\boldsymbol{y}$ to the "winning" class unfortunately gives no indication about the confidence of the assignment: Is the decision close to the boundary of the other classes? For a binary classification problem the following ratio provides a measure of confidence of assigning $\boldsymbol{y}$ to I rather than to the alternative $\bar{I}$, $\Pr(\boldsymbol{y}|\boldsymbol{\theta}_I)/\Pr(\boldsymbol{y}|\boldsymbol{\theta}_{\bar{I}})$. For multiclass problems, we define this to be the geometric mean of every alternative (inferior) model $\sqrt[C-1]{\prod_{i \neq I}^C \frac{\Pr(\boldsymbol{y}|\boldsymbol{\theta}_I)}{\Pr(\boldsymbol{y}|\boldsymbol{\theta}_i)}} = \Pr(\boldsymbol{y}|\boldsymbol{\theta}_I) \cdot \prod_{i \neq I}^C \Pr(\boldsymbol{y}|\boldsymbol{\theta}_i)^{-\frac{1}{C-1}}$. We take the logarithm of this expression to avoid numerical problems. A value of 0 then means that the classifier is indifferent; a large value means that with high confidence I is correct.

Training of the DBNs amounts to computing the Maximum Likelihood Estimates (MLE) of $\boldsymbol{\theta}_1, \ldots, \boldsymbol{\theta}_C$. C training sets are provided, each set with examples of patterns belonging to the class in question. MLEs are computed via (G)EM [9] using GMTK.

Classification of continuous time-series $s(\tau)$ starting at time τ is done in a sliding-window fashion. A pattern captured by a DBN may last for T time slices (T may vary

Table 1. Results using 4-fold cross-validation. The columns "Acc. -Amp" and "Acc. +Amp" are the accuracy results with and without the amplitude feature. The remaining columns are statistics for the DBNs with the amplitude feature. The number between brackets is the number of "signal" examples for the station. The number of "noise" examples is the same.

Station	Acc. - Amp.	Acc. + Amp.	FP rate	Conf. FP	Conf. TP	FN rate	Conf. FN	Conf. TN
anto (27)	0.87	0.91	0.071	102	2004	0.107	92	453
aqu (30)	0.80	0.92	0.031	106	1744	0.125	163	411
arg (35)	0.86	0.94	0.065	227	3754	0.025	75	658
cey (32)	0.86	0.96	0.055	103	3314	0.056	232	585
css (27)	0.81	0.98	0.050	96	2333	0.028	104	554
itm (35)	0.91	0.93	0.051	346	7546	0	-	506
kek (32)	0.86	0.91	0.098	590	4504	0.032	124	435
psz (33)	0.92	0.98	0.031	40	1238	0	-	635
rdo (35)	0.98	1	0	-	5168	0	-	590
	0.87	0.95	0.050	201	3511	0.041	131	536

per pattern), i.e., the CWT coefficients are computed from $s(\tau + 1), \ldots, s(\tau + T)$, the residuals and the average log-amplitude are extracted, DBN inference is performed, and the window is moved to $\tau := T$, and so on. The computational burden is insignificant: computing the CWT can be done very quickly, and the actual classification, which entails inference in C (sparse) DBNs, can be done very fast as well.

5   Experiments

We performed experiments using seismic data recorded by the European broadband network. We considered 9 individual stations. For now, we are primarily interested in distinguishing *seismic signal* from *noise* (ambient, urban, etc). We learn the typical pattern of *signal* from the transient signal onset of a seismogram, i.e., the so-called *P-phase*, lasting for 512 time slices. For each station these phases were picked by hand by a seismologist (domain expert). Similarly, several examples of *noise* were selected before the P-phase (also lasting for 512 slices).

The performance was tested using 4-fold cross-validation for each station using single patterns. In order to see if the amplitude feature makes a difference when classifying, we ran tests with and without this attribute. Table 1 shows the results. As we may see, including the amplitude indeed improves classification accuracy. In general, the accuracy is high, especially when taking into consideration the relatively small training instances provided. We are primarily concerned with the false positives (presented as noise, but classified as signal) for which the false positive rate has been computed too. How far are these false alarms from being classified correctly as negatives? To get an impression of that, we compute the average confidence of the false positives. We see that they are close to the decision boundary, especially when we compare to the average confidence of the true positives. This means that when the classifier is issuing a false alarm, it has been triggered by a pattern which is far from being a "typical" non-signal. We do the same for the false negatives (presented as positives, classified as negatives). The

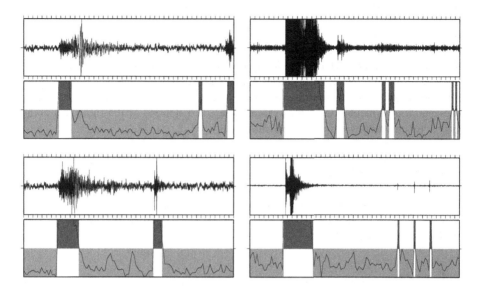

Fig. 3. Four seismograms lasting for 30 minutes. At the top the unfiltered traces and below the real-time classifications are shown together with the plot referring to the confidence of the class assignment with respect to the alternative.

average confidence of a false negative is pretty close to the decision boundary compared to the average true negatives.

The detection of weak seismic signals is difficult, even for the trained seismologist. Presenting the measure of confidence during real-time classification allows the seismologist to intervene: weak (uncertain) false alarms of missed events can be corrected.

Figure 3 shows four examples of seismograms and below each trace the classification and confidence plots are given. The horizontal line in the middle of the confidence plots is the decision boundary between classes. If the plot is above the threshold, the pattern is classified as signal, if it is below it is noise. Every 15th second a classification and confidence is given of the previous 15 seconds. To the untrained eye there seems to be several misclassifications, but in fact all patterns detected were correct signal classifications according to the domain expert. Although the examples might indicate that the amplitude of the seismic signals provides enough information for classification, this is *not* the case; certainly the amplitude plays a role, but other factors are important too as suggested by the accuracy results in table 1 (with and without the amplitude).

6   Conclusion

We have presented a novel method for real-time classification of seismic patterns using DBNs. The symbolic DBN modeling approach gives rise to a model with a limited number of parameters which is crucial because only a small number of training samples exist per station. Experimentally the accuracy and the error rates are good, taking into consideration that the decision problem is very fuzzy in nature. Moreover, the confidence measure provides the seismologist with information about the closeness of the signal to the assigned class.

Since our method is very generic, future works goes in the direction to extend the classification to more detailed modeling of the "anatomy of seismograms", i.e., the distinction of individual seismic phases within an earthquake signal that allow seismologists to study earthquake source processes and effects of seismic wave propagation. The actual implementation of the system at selected stations of the European broadband network will be realized in near future.

Acknowledgments

We would like to acknowledge Andreas Köhler for preparing the data and Michail Kulesh for letting us use GWL. This work has been made possible under the EC-Project NERIES contract no. 026130.

References

1. Aki, K., Richards, P.G.: Quantitative Seismology. University Science Books (2002)
2. Bilmes, J., Zweig, G.: The graphical models toolkit: An open source software system for speech and time-series processing. In: Intl. Conf. on Acoustics, Speech and Signal Proc (2002)
3. Chui, C.K.: An Introduction to Wavelets. Academic Press, London (1992)
4. Dean, T., Kanazawa, K.: A model for reasoning about persistence and causation. Artificial Intelligence 93(1–2), 1–27 (1989)
5. Gendron, P., Ebel, J., Manolakis, D.: Rapid Joint Detection and Classification with Wavelet Bases via Bayes Theorem. Bull. Seism. Soc. Am. 90(3), 764–774 (2000)
6. Joswig, M.: Pattern recognition for earthquake detection. Bull. Seism. Soc. Am. 80(1), 170–186 (1990)
7. Kulesh, M., Holschneider, M., Diallo, M.S.: Geophysics wavelet library: Applications of the continuous wavelet transform to the polarization and dispersion analysis of signals. Computers & Geoscience (Submitted 2007)
8. Kumar, P., Foufoula-Georgiou, E.: Wavelet analysis for geophysical applications. Reviews of Geophysics 35(4), 385–409 (1997)
9. Lauritzen, S.L.: The EM algorithm for graphical association models with missing data. Computational Statistics and Data Analysis 19, 191–201 (1995)
10. Okada, Y., Kasahara, K., Hori, S., Obara, K., Sekiguchi, S., Fujiwara, H., Yamamoto, A.: Recent progress of seismic observation networks in Japan—Hi-net, F-net, K-NET and KiK-net. Earth, Planets, and Space, 56:D15+ (August 2004)
11. Pearl, J.: Probabilistic Reasoning in Intelligent Systems: Networks of Plausible Inference. Morgan Kaufmann, San Francisco (1988)
12. Withers, M., Aster, R., Young, C., Beiriger, J., Harris, M., Moore, S., Trujillo, J.: A comparison of select trigger algorithms for automated global seismic phase and event detection. Bull. Seism. Soc. Am. 88(1), 95–106 (1998)
13. Zweig, G., Russell, S.: Speech Recognition with Dynamic Bayesian Networks. In: AAAI, pp. 173–180 (1998)

Robust Visual Mining of Data with Error Information

Jianyong Sun[1,2], Ata Kabán[1], and Somak Raychaudhury[2]

[1] School of Computer Science, The University of Birmingham, Edgbaston,
Birmingham, U.K. B15 2TT
[2] School of Physics and Astronomy, The University of Birmingham, Edgbaston,
Birmingham, U.K. B15 2TT

Abstract. Recent results on robust density-based clustering have indicated that the uncertainty associated with the actual measurements can be exploited to locate objects that are atypical for a reason unrelated to measurement errors. In this paper, we develop a *constrained* robust mixture model, which, in addition, is able to nonlinearly map such data for visual exploration. Our robust visual mining approach aims to combine statistically sound density-based analysis with visual presentation of the density structure, and to provide visual support for the identification and exploration of 'genuine' peculiar objects of interest that are not due to the measurement errors. In this model, an exact inference is not possible despite the latent space being discretised, and we resort to employing a structured variational EM. We present results on synthetic data as well as a real application, for visualising peculiar quasars from an astrophysical survey, given photometric measurements with errors.

1 Introduction

Providing the users with a qualitative understanding of multivariate data, and automatically detecting atypical objects are among the most important tasks of data mining. Data visualisation techniques aim to capture the underlying structure of the dataset and preserve the structure in the low-dimensional space which can be readily visualised. However, care must be taken with appropriately handling outliers in the data set, to avoid biased parameter estimates [8] [1] and consequently to avoid misleading visual representations [11]. Moreover, outliers may occur for different reasons. Some of them are of interest in certain domain-specific applications, while others are not.

For example, one common reason for the occurrence of outliers or atypical data instances is due to measurement errors. These are inevitable in many areas and may arise from physical limitations of measuring devices and / or measurement conditions. In scientific areas, such as in Astrophysics [3], these errors are recorded, are available, and should be made use of. It is therefore essential to develop methods that are able to incorporate the existing knowledge of measurement errors in order to hunt for the potentially interesting peculiar objects. We have recently proposed a robust mixture modelling method that takes the

J.N. Kok et al. (Eds.): PKDD 2007, LNAI 4702, pp. 573–580, 2007.

errors into account [9]. Here we constrain this model in the spirit of a generative topographic mapping [2], to enable a visual presentation of the data density structure and hence to provide visual support for the identification and exploration of 'genuine' peculiar objects that are not due to the measurement errors.

2   Robust GTM in the Presence of Measurement Errors

GTM. The generative topographic mapping (GTM) [2] is a latent variable model for data visualisation. It expresses the probability distribution $p(\mathbf{t})$ of the data $\mathbf{t} \in \Re^D$ in terms of a small (typically two) number of continuous valued and uniformly distributed latent variables $\mathbf{x} \in \Re^d$ where $d \ll D$. To achieve this, GTM supposes that the data lies on a manifold, that is, a data point can be expressed as a mapping from the latent space to the data space as $\mathbf{t} = y(\mathbf{x}; \mathbf{W}) + \epsilon$; where $y(\mathbf{x}; \mathbf{W}) = \mathbf{W}\phi(x)$ is defined as a generalised linear regression model, and the elements $\phi(\mathbf{x})$ consist of M fixed basis functions $\phi_j(\mathbf{x})$, i.e. $\phi(\mathbf{x}) = (\phi_1(\mathbf{x}), \cdots, \phi_M(\mathbf{x}))$ and $\mathbf{W}$ is the weight matrix. The basis functions may be chosen as RBF-s and ϵ is a Gaussian noise $\mathcal{N}(0, \sigma^{-1}\mathbf{I})$. The distribution $p(\mathbf{t})$ can then be obtained by integrating out the latent variable $\mathbf{x}$: $p(\mathbf{t}) = \int p(\mathbf{x})p(\mathbf{t}|\mathbf{x})d\mathbf{x}$. For tractability reasons, the continuous latent space is discretised in GTM, which results in a latent density $p(\mathbf{x})$ expressed as a finite sum of delta functions, each centered on the nodes of a regular grid in the latent space: $p(\mathbf{x}) = \frac{1}{K}\sum_{k=1}^{K}\delta(\mathbf{x} - \mathbf{x}_i)$ and $p(\mathbf{t}|\mathbf{x}) = \mathcal{N}(\mathbf{t}|y(\mathbf{x}; \mathbf{W}), \sigma^{-1}\mathbf{I})$. This is convenient, since then the formalism of unconstrained mixture models is directly applicable, while a powerful nonlinear mapping is obtained. A robust extension of GTM, using Student t density components has been developed (t-GTM) in [11], for applications where outliers exist in the data set. However, the existing t-GTM cannot make use of the measurement error information, so it cannot differentiate between outliers that are due to a known cause as opposed to those that are of potential interest. Our purpose here is to address this issue.

The Proposed Model. Each individual measurement $\mathbf{t}_n$ is given with an associated error. It is conceptually justified to assume a Gaussian error model (e.g.[10]), where the square of the (known) errors are arranged on the diagonals of variance matrices denoted by $\mathbf{S}_n$, so we have:

$$p(\mathbf{t}_n|\mathbf{w}_n) = \mathcal{N}(\mathbf{t}_n|\mathbf{w}_n, \mathbf{S}_n) \tag{1}$$

The unknown mean values $\mathbf{w}_n$ represent the clean, error-free version of the data. The genuine outliers, which we are interested in, must be those of the density of $\mathbf{w}_n$ rather than those of the density of $\mathbf{t}_n$. We also assume that it is the clean data rather than the contaminated data, who lies on a manifold. We will therefore model the hidden clean density as a GTM with Student t components $p(\mathbf{w}|k) = S_t(\mathbf{w}; \mathbf{W}, \sigma, \nu_k)$ which can be re-written as a convolution of a Gaussian with a Gamma ($\mathcal{G}(u|a,b) = b^a u^{a-1}e^{-bu}/\mathbf{\Gamma}(a)$) placed on the Gaussian precisions [7]. So in our model, the distribution of $\mathbf{t}$ can be obtained by integration over $\mathbf{w}$:

$$p(\mathbf{t}; \mathbf{W}, \sigma, \nu) = \frac{1}{K}\sum_{k}\iint \mathcal{N}(\mathbf{t}|\mathbf{w}, \mathbf{S})\mathcal{N}(\mathbf{w}|\mathbf{W}\phi(\mathbf{x}_k), \frac{1}{u\sigma})\mathcal{G}(u|\frac{\nu_k}{2}, \frac{\nu_k}{2})dud\mathbf{w} \tag{2}$$

3   Structured Variational EM Solution

Since the integration in Eq. (2) is not tractable, we develop a generalised EM (GEM) algorithm (see e.g. [6]), with approximate E-step. In general terms, for each data point $\mathbf{t}_n$, its log-likelihood can be bounded as follows:

$$\log p(\mathbf{t}_n|\theta) \geq \int q(h_n) \log \frac{p(h_n, \mathbf{t}_n|\theta)}{q(h_n)} dh \equiv \mathcal{F}(\mathbf{t}_n|q, \theta) \tag{3}$$

where q is the free-form variational posterior, $\mathcal{F}$ is called variational free energy function, h_n is the set of latent variables associated with $\mathbf{t}_n$, and θ is the set of parameters of the model. In our case, $h_n = (z_n, \mathbf{w}_n, u_n)$ and $\theta = (\mathbf{W}, \sigma, \{\nu_k\})$. Some tractable form needs to be chosen for q, e.g. a fully factorial form [6] is the most common choice. However, under our model definitions, it is feasible to keep some of the posterior dependencies by choosing the following tree-structured variational distribution: $q(\mathbf{w}, u, z = k) = q(z = k)q(\mathbf{w}|z = k)q(u|z = k) \equiv q(k)q(\mathbf{w}|k)q(u|k)$. The free energy function $\mathcal{F}(\mathbf{t}|q, \theta)$ can be evaluated as:

$$\mathcal{F}(\mathbf{t}|q, \theta) = \sum_k q(k) \left[\langle \log p(\mathbf{t}, \mathbf{w}, u, k) - \log (q(u|k)q(\mathbf{w}|k)q(k)) \rangle_{\mathbf{w}, u|k} \right] \equiv \sum_k q(k) A_{\mathbf{t}, k}.$$

Variational E-step. Due to discretisation of the latent space, the E-step expressions follow analogously to the unconstrained case [9], and we obtain the following variational posteriors:

$$q(\mathbf{w}|k) = \mathcal{N}(\mathbf{w}|\langle \mathbf{w} \rangle_k, \mathbf{\Sigma}_{\mathbf{w}|k}); \quad q(u|k) = \mathcal{G}(u|a_k, b_k); \quad q(k) = \frac{\exp(A_{\mathbf{t}, k})}{\sum_{k'} \exp(A_{\mathbf{t}, k'})} \tag{4}$$

$$\mathbf{\Sigma}_{\mathbf{w}|k} = \left[\sigma \langle u \rangle_{u|k} + \mathbf{S}^{-1} \right]^{-1}; \quad \langle \mathbf{w} \rangle_k = \mathbf{\Sigma}_{\mathbf{w}|k} \left[\langle u \rangle_{u|k} \sigma \mathbf{W} \phi(\mathbf{x}_k) + \mathbf{S}^{-1}\mathbf{t} \right] \tag{5}$$

$$a_k = \frac{\nu_k + D}{2}; b_k = \frac{\nu_k + C_k}{2}; C_k = \sigma \left(\|\langle \mathbf{w} \rangle_k - \mathbf{W}\phi(\mathbf{x}_k)\|^2 + \mathrm{Tr}\left(\mathbf{\Sigma}_{\mathbf{w}|k} \right) \right)$$

Using these, the likelihood bound is straightforward to evaluate (omitted for space constraints).

M-step. Taking derivatives of $\mathcal{F}$ w.r.t $\mathbf{W}$, and solving the stationary equations, we obtain the following equation in matrix notation:

$$\mathbf{\Phi}^T \mathbf{G} \mathbf{\Phi} \mathbf{W}^T = \mathbf{\Phi}^T \mathbf{A} \tag{6}$$

where $\mathbf{\Phi}$ is a $K \times M$ matrix with element $\mathbf{\Phi}_{ij} = \phi_j(x_i)$, $\mathbf{A}$ is a $K \times d$ matrix, its (k, i) element is $\sum_n q(z_n = k)\langle u_n \rangle_k \langle \mathbf{w}_n \rangle_{ki}$, $\mathbf{G}$ is a diagonal $K \times K$ matrix with elements $\mathbf{G}_{kk} = \sum_n q(z_n = k)\langle u_n \rangle_k$. Eq. (6) can be solved by using standard matrix inversion techniques. Similarly, maximising the likelihood bound w.r.t σ, we can re-estimate the inverse variance σ as:

$$\frac{1}{\sigma} = \frac{1}{ND} \sum_n \sum_k q(z_n = k)\langle u_n \rangle_k \left\langle \|\mathbf{w}_n - \mathbf{W}\phi(x_k)\|^2 \right\rangle_{\mathbf{w}_n|k}.$$

Time Complexity. The E-step scales as $\mathcal{O}(\max\{D, M\}KN)$, the M-step is $\mathcal{O}(\max\{D, M\}NK + MKd + M^3)$. In conclusion, the theoretical complexity per iteration is the same as that of the t-GTM.

4 Deriving Interpretations from the Model

Outlier Detection Criteria. In addition to mapping the density of non-outliers to 2D, our main goal includes the ability to visualise the genuine outliers. Similarly to the case of unconstrained density modelling [9], the posterior expectation of u is of interest:

$$e \equiv \sum_k q(k) \frac{\nu_k + D}{\nu_k + \sigma \left(\|\langle \mathbf{w} \rangle_k - \mathbf{W}\phi(\mathbf{x}_k)\|^2 + \mathrm{Tr}\left(\Sigma_{\mathbf{w}|k} \right) \right)} \tag{7}$$

where a_k and b_k are given in Eq. (6). A data point is considered to be an outlier not due to errors, if its e value is sufficiently small, or equivalently, the value

$$v \equiv \sum_k q(k)\sigma \left(\|\langle \mathbf{w} \rangle_k - \mathbf{W}\phi(\mathbf{x}_k)\|^2 + \mathrm{Tr}\left(\Sigma_{\mathbf{w}|k} \right) \right) \tag{8}$$

is sufficiently large. By contrast, in the t-GTM [11], the outlier criterion is

$$m \equiv \sum_k p(k|\mathbf{t})\sigma \, \|\mathbf{t} - \mathbf{W}\phi(\mathbf{x}_k)\|^2 . \tag{9}$$

Comparing Eq. (9) with Eq. (8), we can see that the degree of outlierness will differ when measurement errors are taken into consideration.

Data Visualisation. Analogously to the original GTM [2], we can derive 2D coordinates of each multivariate data point $\mathbf{t}_n$, by computing its posterior expectation in the latent space. This is calculated as follows:

$$\langle \mathbf{x}|\mathbf{t}_n, \mathbf{W}^*, \sigma^* \rangle = \int p(\mathbf{x}|\mathbf{t}_n; \mathbf{W}^*, \sigma^*)\mathbf{x}d\mathbf{x} = \int \frac{p(\mathbf{t}_n|\mathbf{x}; \mathbf{W}^*, \sigma^*)p(\mathbf{x})}{p(\mathbf{t}_n)}\mathbf{x}d\mathbf{x} = \sum_k q(k)\mathbf{x}_k$$

where $\mathbf{W}^*$ and σ^* are the parameters after training. Each multivariate data point $\mathbf{t}_n$ will have its image in the latent space (in case $d = 2$), as $\langle \mathbf{x}|\mathbf{t}_n, \mathbf{W}^*, \sigma^* \rangle$.

It should be noted that in the case of an atypical object, outside the manifold defined by the bulk of data density, the posterior responsibilities, i.e. the $q(z_n = k)$, will tend to be approximately equal for all $k, 1 \leq k \leq K$. Hence these outliers will tend to lie in the center of the latent space. It is therefore not particularly convenient to visualise these outliers in the 2D space together with the non-outliers. To deal with this problem, while also trating outlierness as a continuous notion, in this paper we propose to use a third dimension for representing outlierness on the visual image. In addition, we can use markers of different sizes to indicate the extent of measurement error for each object. These pieces of information together have the potential to provide a both comprehensive and intuitive visual representation of complex realistic data sets, which combines theoretical soundness and rigour with intuitive expressiveness.

Visualising a Held-out Set. Since the model is fully generative, it can also be applied to new, previously unseen data from the same source. For a given held-out data set, we need to calculate the posterior distributions of $\mathbf{w}_n$ and u_n associated with each test point $\mathbf{t}_n$. To this end, we fix the parameters $\mathbf{W}$ and σ, obtained from the training set and perform the E-step iterations until convergence. This typically converges at least an order of magnitude faster than the full training procedure.

5   Experiments and Results

Illustrative Experiments on Synthetic Data. First, we create a synthetic clean data set, sampled from a mixture of three well separated Gaussians in $\Re^{10}$ and a uniform distribution simulates the presence of genuine outliers. Then we add Gaussian noise to all sample points, in order to simulate measurement errors. The resulting data set is the input of our algorithm. The aim is to recover and display the genuine outliers, along with the density of non-outliers, despite the added errors. Moreover, we demonstrate the advantage of our 3D display. The leftmost plot of Fig. 1 shows a conventional 2D visualisation of the input data based on posterior means $\langle \mathbf{x} \rangle$. The rightmost plot of Fig.1 provides a 3D image with the outlierness (cf. eq. (7)) being displayed on a third, vertical dimension. On both plots, different markers are used for points in different Gaussian density-components, the outliers are highlighted by star markers and the marker sizes are proportional to the size of errors. We see, since outlierness is a continuous quantity, it is hard to distinguish the outliers based on 2D posterior means only. In the 3D display in turn they are nicely revealed.

In order to evaluate the accuracy of the data representation created in some objective manner, as well as to assess the benefits of including measurement error information, we evaluate k-nearest neighbour (KNN) classifiers on the obtained visual representation of a held-out set, sampled in the same way as the training set. The leftmost plot of Fig.2 shows the visual image of the held-out set, as obtained with our approach. The rightmost plot presents the correct KNN classification rates in comparison with those achieved with t-GTM (i.e. without knowledge of the measurement errors), varying the neighbourhood size as shown on the horisontal axis. At all neighbourhood sizes tested, the proposed algorithm achieves better performance compared with t-GTM. This demonstrates

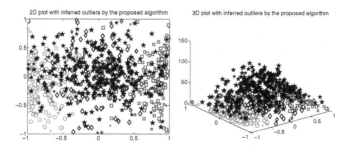

Fig. 1. Synthetic data sets with cluster structure and outliers

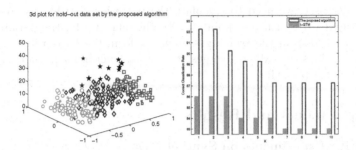

Fig. 2. Left: Visualisation of held-out data. Right: Comparison of KNN classification accuracies in the latent space, computed on held-out data.

that inclusion of measurement error information is beneficial for achieving a more accurate representation of the data, and hence the visual exploration of this representation is more meaningful. We notice that KNNs with small neighbourhoods perform better in both cases, probably because of the high degree of compression from the original data space and since the smoothness of the generative mapping preserves the *local* topology.

Visualising High-redshift Quasars from the SDSS Quasar. In this section, we apply the proposed algorithm on a well-studied data set in astrophysics – the SDSS quasar photometric catalogue [12]. The initial features are five observed magnitudes (measured brightness with a particular filter, in logarithmic units), u_{mag}, g_{mag}, i_{mag}, r_{mag} and z_{mag}, from which we created four features relating to colour, by subtracting r_{mag} from each. The reason for this is to avoid discriminating on the basis of brightness and the choice of r_{mag} as a reference magnitude is because that is most reliably measured. Further, the measurement errors are known for each feature and each object. In addition, spectroscopic redshift estimates are available. These are not used within the algorithm, but are useful for relating our results to known physical quantities, as a way of validating the interpretability of our results. The redshift relates to the distance of the object from us. As there are fewer distant quasars in the catalogue than closer ones, and given that with higher redshift the entire spectral pattern is shifted more towards lower frequencies, this provides a physical reason for high redshift quasars to be perceived as outliers in the overall density of quasars in the colour space. This observation was exploited in a number of previous studies by astronomers for finding high redshift quasars in various 2D projections of the data [4]. However through our developments, in addition to having a model based approach which takes the multivariate feature space and also takes principled account of the measurement errors [9], we are now also able to visualise the structure of the data so that domain experts may explore and understand the automated analysis such as the detection of atypical objects. In this visualisation, an optional step is to determine cluster assignments using the robust mixture modelling method for data with errors [9], in order to display the objects in the various clusters found in different colours or with different markers.

Here we apply our method to a sample of 2,000 quasars. The 3D visualisation produced by the proposed algorithm on the SDSS quasars subset is displayed in Fig. 3. In this plot, we detail two selected outliers and two prototypical objects. The error bars represent their measurement errors, and the numbers shown on the plot are the redshift estimates for the selected outliers. As mentioned, we would expect the genuine outliers to be related to an increased redshift, in average, so we compute the area under the ROC curve (AUC) [5] of our outlierness estimates against a varying redshift threshold. The resulting relation between these two quantities is shown on Fig. 4. The y-coordinate of each point indicates the probability of detecting quasars of redshift greater than its x-coordinate. As expected, in a similar manner to the unconstrained analogue of the model presented [9], our principled method in four-colour space, using errors, can identify as outliers an overwhelming fraction of quasars already at a redshift of 2.5 (or higher). The main advantage of the constrained model over the unconstrained one is allowing us to visually understand these detections in the context of the

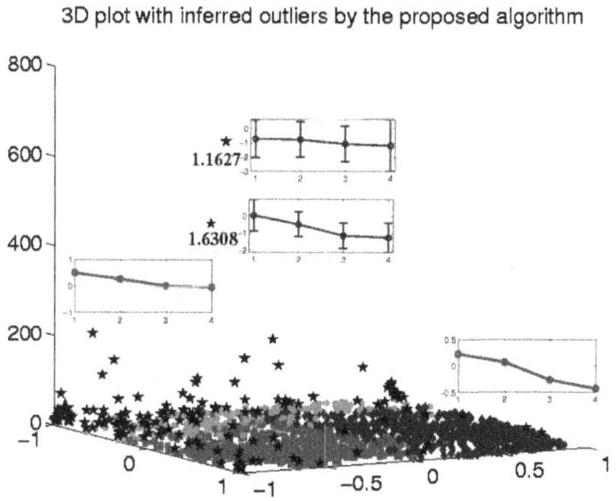

3D plot with inferred outliers by the proposed algorithm

Fig. 3. The 3D plot of the proposed algorithm on the subset of SDSS quasar catalogue

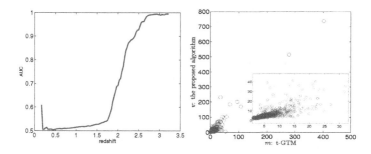

Fig. 4. Left: AUC vs. possible redshift thresholds. Right: The comparison of the distribution of outlierness values with and without taking into account measurement errors.

structure of the entire data set. The rightmost plot on Fig. 4 shows the distribution of outlierness estimates with and without taking measurement errors into account. Upon zooming the denser region we notice the relative ranking produced by the two approaches is indeed quite different.

To relate our developments to existing methods in this application domain, it should be noted the prevalence of 2D projection methods plotting two features at a time against each other, e.g. [4]. Such a method can only manage to detect objects whose redshift is $z > 3.5$, which are extremely rare, and obvious from naive projections. In turn, by being able to map the multivariate data density to a visualisation space in a principled and robust manner, and taking into account measurement error information, we provide domain experts with a framework and methodology for further exploring realistic data sets in a more comprehensive and goal-directed manner.

6   Conclusions

We presented a robust visual mining method for detecting and visualising interesting outliers in the presence of known measurement errors. A generalised EM algorithm was derived for inference and parameter estimation. The resulting algorithm was then demonstrated on both synthetic data and a real application.

References

1. Archambeau, C., Delannay, N., Verleysen, M.: Robust probabilistic projections. In: Proceedings of the 23rd International Conference on Machine Learning (2006)
2. Bishop, C.M., Svensén, M., Williams, C.K.I.: GTM: The generative topographic mapping. Neural Computation 10(1), 215–235 (1998)
3. Djorgovski, S., Mahabal, A., Brunner, R., Gal, R., Castro, S.: Searches for rare and new types of objects. Virtual Observatories of the Future, ASP Conference Series, 225 (2001)
4. Fan, X.: A survey of $z > 5.7$ quasars in the Sloan Digital Sky Survey. IV. discovery of seven additional quasars. The Astronomical Journal 131, 1203–1209 (2006)
5. Fawcett, T.: ROC graphs: Notes and practical considerations for researchers. Machine Learning (2004)
6. Jordan, M., Ghahramani, Z., Jaakkola, T., Saul, L.: An introduction to variational methods for graphical models. Machine Learning 37, 183–233 (1999)
7. Liu, C., Rubin, D.: ML estimation of the t distribution using EM and its extensions: ECM and ECME. Statistica Sinica 5, 19–39 (1995)
8. Peel, D., McLachlan, G.: Robust mixture modelling using the t distribution. Statistics and Computing 10, 339–348 (2000)
9. Sun, J., Kabán, A., Raychaudhury, S.: Robust mixtures in the presence of measurement errors. In: Proceedings of The 24th International Conference on Machine Learning (2007)
10. Taylor, J.: An introduction to error analysis. University Science Books (1996)
11. Vellido, A., Lisboa, P., Vicente, D.: Handling outliers and missing data in brain tumor clinical assessment using t-GTM. Computers in Biology and Medicine (2006)
12. York, D.: The sloan digital sky survey: Technical summary. The Astronomical Journal 120, 1579–1587 (2000)

An Effective Approach to Enhance Centroid Classifier for Text Categorization

Songbo Tan and Xueqi Cheng

Information Security Center, Institute of Computing Technology, China
tansongbo@software.ict.ac.cn, tansongbo@gmail.com

Abstract. Centroid Classifier has been shown to be a simple and yet effective method for text categorization. However, it is often plagued with model misfit (or inductive bias) incurred by its assumption. To address this issue, a novel Model Adjustment algorithm was proposed. The basic idea is to make use of some criteria to adjust Centroid Classifier model. In this work, the criteria include training-set errors as well as training-set margins. The empirical assessment indicates that proposed method performs slightly better than SVM classifier in prediction accuracy, as well as beats it in running time.

Keywords: Text Categorization, Information Retrieval, Data Mining.

1 Introduction

With the rapid growth of texts in the Internet, text categorization has received more and more attention in information retrieval community. Numerous machine learning approaches have been introduced to deal with text classification, including Centroid Classifier [1], Naive Bayes [2], Winnow or Perceptron [3], Voting [4] and Support Vector Machines (SVM) [5].

Despite simplicity and straightforwardness, Centroid Classifier has proved to be an effective and yet robust method for text categorization. However, it is often plagued with inductive bias [6] or model misfit [7]. For example, Centroid Classifier makes a simple assumption that a given document should be assigned a particular class if the similarity of this document to the centroid of the class is the largest. However, this supposition is often violated when there exists a document from class A sharing more similarity with the centroid of class B than that of class A. The more serious the model misfit is, the poorer the classification performance will be.

In this work, we proposed a novel Model Adjustment (MA) algorithm to cope with model misfit problem of Centroid Classifier. The basic idea is to pick out some training examples to adjust Centroid Classifier model. The most popular and simple method is to select misclassified examples. From the perspective of machine learning, however, low training-error-rate does not indicate low error-rate for unseen examples. To address this issue, another measure, i.e., training-set margin, was incorporated. In other words, misclassified examples as well as small-margin examples were picked out to update the classifier model.

J.N. Kok et al. (Eds.): PKDD 2007, LNAI 4702, pp. 581–588, 2007.

The experimental results show that proposed technique is able to improve classification performance of Centroid Classifier significantly. Furthermore, the resulting classifier performs slightly better than SVM classifier in classifying accuracy, as well as beats it in running time.

2 Related Work

In this section, we briefly review the related researches and compare them with proposed method.

Voting [4] is a famous strategy for correction of inductive bias. It works by taking a classifier and training set as input and training the classifier multiple times on different versions of the training set. Compared to this method, MA has three particularities. First, MA does not need to retrain the classifier multiple times on the different versions of the entire training set. Second, MA produces only one refined classifier. Furthermore, Voting utilizes only training-set errors while MA employs training-set errors as well as training-set margins.

A "dragpushing" strategy for Centroid and Naïve Bayes Classifier is proposed by Tan et al[8]. This method takes advantage of misclassified training examples to successively refine classification model by online-modification. Different from this method, MA employs training error as well as training margin.

The weight adjustment scheme [1] is to use a measure of the discriminating power of each term to gradually adjust the weights of all features concurrently. In contrast to this scheme, MA employs training error and training margin rather than discriminating power as adjustment criteria.

3 Centroid Classifier

The idea behind the centroid classification algorithm [1] is extremely simple and straightforward. First we compute the weighted representation of each training document using normalized TFIDF [4]. Then we calculate the prototype vector or centroid vector C_i for each training class c_i:

$$C_i = \frac{1}{|c_i|} \sum_{d \in c_i} d \cdot$$

(1)

where $|z|$ denotes the cardinality of set z.

Thirdly we calculate the similarity of one document d to each centroid by inner-product measure,

$$Sim(d, C_i) = d \cdot C_i \cdot$$

(2)

Lastly, based on these similarities, we assign d the class label corresponding to the most similar centroid.

4 Proposed Technique

In this section we first illustrate the rationale to deal with inductive bias and then present the detailed algorithm of proposed method.

Let us take a two-class text data as an example. The data distribution is illustrated as Fig 1. Class "*A*" spread as grey is elliptically populated; while class "*B*" packed as white is roundly distributed. C_A and C_B are the centroids of class *A* and class *B* respectively. *Middle Line* is the perpendicular bisector of the line between C_A and C_B. From another perspective, *Middle Line* serves as a decision hyper-plane that separate class *A* and class *B*. Obviously, the examples of category *A* on the right of *Middle Line* share more similarity with centroid C_B rather than C_A, so they will be misclassified into class *B*. This is the case that the supposition of Centroid Classifier is violated by data distribution.

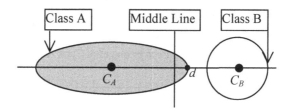

Fig. 1. The outline of Original Centroids

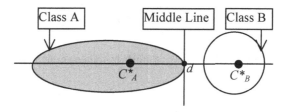

Fig. 2. The outline of Refined Centroids

In order to reduce this model bias, we make use of training errors to adjust its prototype vectors. For example, if document *d* of class *A* is misclassified into class *B*, both centroid C_A and C_B should be moved right by the following formulas (3) and (4) respectively,

$$C_A^* = C_A + \eta \cdot d . \tag{3}$$

$$C_B^* = C_B - \eta \cdot d . \tag{4}$$

where η denotes the "*LearnRate*" that is used to control the strength of update. With this so-called move operation, C_A and C_B are both moving right gradually. At the end of this kind of move operation (see Fig 2), no example of class *A* locates at the right of *Middle Line* so no example will be misclassified.

However, above adjustment approach employs only one criterion, i.e., training-set error. From the point of view of machine learning, training-set error based method cannot guarantee the generalization capability of base classifiers for unseen examples. To fully demonstrate this problem, we revisit the aforementioned two-class dataset. Without loss of generality, we can construct the future distribution of class *A* and class *B*. Obviously the training examples are only a small portion of unseen examples of class *A* and class *B* (as illustrated Fig 3). The unseen examples of class *A* are denoted by "." or grey; the unseen examples of class *B* are denoted by "-" or white.

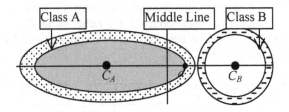

Fig. 3. The distribution of unseen examples of Class A and Class B

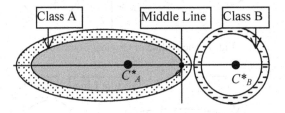

Fig. 4. Refining the centroids by training examples

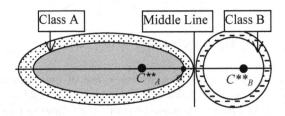

Fig. 5. Refining the centroids by training examples and unseen examples

After adjusting classifier model by misclassified training examples, the *Middle Line* moves right to the border of class *A* (see Fig. 4). In this case, all training examples can be correctly classified, but not all unseen examples can be correctly classified. For example, we can observe that the unseen examples of class *A* on the right of *Middle Line* will be misclassified into class *B*. This observation indicates that training-set error based model update cannot guarantee the classification performance of base classifiers for unseen documents.

With the aim to improve the classification ability of classifier for unseen examples, the *Middle Line* should be moved right again. That is to say, centroid C_A and C_B should be both moved right. To accomplish this goal, some correctly classified examples near *Middle Line* in class A should be employed to adjust centroid C_A and C_B. That is, for each training example in class A, we not only require its $Sim(d, C_A)$ is bigger than $Sim(d, C_B)$; but also demand $Sim(d, C_A)$ exceeds $Sim(d, C_B)$ by a wide margin.

To further illustrate this kind of margin (denoted by ρ), we take document d in Fig 4 as an example. Although document d can be correctly classified because the *Middle Line* has moved to the border of class A, its margin is very close to zero since it lies exactly on the *Middle Line*. Hence in order to enlarge the margin, both centroid C_A and C_B should be moved right again by formulas (3) and (4) respectively. After a number of this kind of moving operations, the *Middle Line* moves to the border of unseen examples of class A (as demonstrated in Fig 5). In this case, all unseen examples can be correctly categorized. This is the mechanism that margin can boost the classification ability of classifier for unseen examples.

To concentrate our attention on small-margin examples, we introduce a small positive margin threshold, *MinMargin* (denoted by θ). If the margin of document d is smaller than *MinMargin*, it should be employed to adjust the classifier model. Furthermore, to balance training errors and training margins, we introduce a constant parameter "*Weight*" (denoted by ω). As a result, we can write down the batch-update formula as following,

$$C_A^* = C_A + \eta \times \left\{ \sum_{\substack{d \in c_A \\ \rho(d) < 0}} d - \sum_{\substack{d \notin c_A \\ \rho(d) < 0}} d + \omega \times \left(\sum_{\substack{d \in c_A \\ 0 < \rho(d) < \theta}} d - \sum_{\substack{d \notin c_A \\ 0 < \rho(d) < \theta}} d \right) \right\} . \quad (5)$$

The detailed algorithm is presented in Fig 6. First it loads training data and parameters (including *MaxIteration*, *Weight* and *LearnRate*), and then calculates one centroid for each category. In one iteration of the updating phase, it needs to categorize all training documents, and then makes use of these misclassified examples and small-margin examples to adjust centroids by formula (5). For the sake of brevity, we refer to the model-adjustment algorithm as MA.

```
1 Load training data and parameters;
2 Calculate C_i for each class c_i;
3 For iter=1 to MaxIteration Do
    3.1 Classify all training documents;
    3.2 Update centroids by formula (5);
```

Fig. 6. The outline of Model Adjustment for Centroid Classifier

5   Empirical Assessment

5.1   Experimental Design

In our experiment, we use five corpora: Reuters-21578[1], 20NewsGroup[2], Industry Sector[3], OHSUMED[4] and RCV1[5]. For **Reuters-21578**, we used its subset: one consisting of 92 categories and in total 10,346 documents; for **20NewsGroup,** we use a subset consisting of total categories and 19,446 documents; for **Industry Sector,** we use a subset called as Sector-48 consisting of 48 categories and in all 4,581 documents; for **OHSUMED,** we use a subset (called ohscal[6] in [1]) that contains 11,162 documents and in total 10 categories; for **RCV1,** we use a subset consisting of 56 categories and 41,320 documents. Typically, we use 2/3 of documents for training and 1/3 for testing; exceptionally, in order to reduce the training time for RCV1, we use 10% of its total documents for training and 90% for testing.

To evaluate the performance of a text classifier, we use MicroF1 and MacroF1 scores [4]. To remove redundant features and save running time, we employ Information Gain as feature selection method. Algorithms are coded in C++ and running on a Pentium-4 machine with a single 3.0GHz CPU and 512M memory. For SVM classifier, we employed LibSvm that can directly deal with multi-class classification problems. (www.csie.ntu.edu.tw/~cjlin/). We left all parameters as default. For Winnow classifier, we run Balanced Winnow [3]. The promotion parameter α and the demotion β (learning rates) were fixed as 1.2 and 0.8 respectively.

5.2   Comparison with Other Methods

Table 1 shows performance comparison in MicroF1 and MacroF1. Feature number is set to 10,000; For MA, *MaxIteration*, *Weight*, *LearnRate*, and *MinMargin* are set to 10, 0.2, 0.5, and 0.1 respectively. From this table, we can see that MA improves the performance of Centroid Classifier dramatically, and the improvement is especially significant on Sector-48. For example, MA improves Centroid Classifier by about 9% on Sector-48. In a word, Model Adjustment is an effective and robust method to boost the performance of Centroid Classifier.

MA outperforms all the other methods on OHSUMED, Reuters, Sector-48 and RCV1. Especially on Reuters, the MicroF1 of MA is one percent lower than LibSvm but its MacroF1 is 12 percent higher than LibSvm. In total MA performs slightly better than LibSvm. Consequently we can say that MA is an efficient and competitive algorithm for text classification.

Table 2 reports the training time of five methods on five text collections. Feature number is set to 10,000; For MA, *MaxIteration*, *Weight*, *LearnRate*, and *MinMargin* are set to 10, 0.2, 0.5, and 0.1 respectively. As we can observe from this table, the

---

[1] http://www.daviddlewis.com/resources/testcollections/reuters21578/.
[2] http://www-2.cs.cmu.edu/afs/cs/project/theo-11/www/wwkb.
[3] http://www-2.cs.cmu.edu/afs/cs.cmu.edu/project/theo-20/www/data/.
[4] ftp://medir.ohsu.edu/pub/OHSUMED/.
[5] http://www.daviddlewis.com/resources/testcollections/rcv1/.
[6] http://www.cs.umn.edu/~han/data/tmdata.tar.gz.

Table 1. The performance of different methods

	MA		Centroid		Winnow		Perceptron		LibSvm	
	MicroF1	MacroF1	MicroF1	MacroF1	MicroF1	MacroF1	MicroF1	MacroF1	MicroF1	MacroF1
OHSUMED	**0.8049**	**0.7940**	0.7676	0.7600	0.7193	0.7110	0.7100	0.6996	0.7906	0.7800
Reuters	0.8565	**0.6061**	0.7820	0.5617	0.8263	0.4891	0.7918	0.4553	**0.8694**	*0.4875*
Sector-48	**0.8970**	**0.9000**	0.8055	0.8152	0.8003	0.8389	0.7845	0.7943	0.8732	0.8780
NewsGroup	0.8892	0.8859	0.8429	0.8389	0.8105	0.8161	0.8089	0.8081	**0.9040**	**0.9029**
RCV1	0.7166	**0.4912**	0.6778	0.4883	0.6142	0.4138	0.6335	0.3379	**0.7213**	0.4126

Table 2. Training Time in seconds

	MA	Centroid	Winnow	Perceptron	LibSvm
OHSUMED	1.39	0.40	1.72	1.95	62.28
Reuters	18.41	0.40	7.75	11.01	80.77
Sector-48	11.91	0.50	4.92	9.51	38.31
NewsGroup	7.56	0.48	4.90	6.34	160.11
RCV1	10.46	0.42	4.64	6.79	38.54

CPU time required by LibSvm is about 40 times larger than that of MA on OHSUMED and about 20 times larger on NewsGroup. In contrast to LibSvm, as a result, the time saving of MA is very obvious.

5.3 Training Margin and Performance vs. MaxIteration

Fig. 7 shows training margin and prediction performance curves of MA vs. *MaxIteration* on five datasets. *Weight*, *LearnRate*, and *MinMargin* are set to 0.2, 0.5, and 0.1 respectively; feature number is set to 10,000.

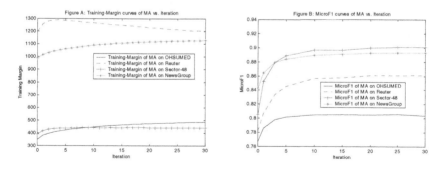

Fig. 7. Training margin and prediction performance of MA vs. *MaxIteration*

We can observe that proposed Model Adjustment can enlarge margin and boost prediction performance. This figure demonstrates that increasing the *MaxIteration* increases training margin and prediction performance. However, the increase in two measures is not directly proportional to increase in *MaxIteration*. As the *MaxIteration* is getting larger, the curves of two measures are starting to level off.

6 Conclusion Remarks

In this work, a novel Model Adjustment (MA) algorithm was proposed to deal with model bias problem of Centroid Classifier. The basic idea is to pick out some training examples to adjust Centroid Classifier model. The main contributions are:

Firstly, in order to avoid over-train problem, we combine two measures for Model Adjustment: training-set errors and training-set margins. That is to say, misclassified examples as well as small-margin examples are picked out to update the classifier model.

Secondly, extensive experiments are conducted on five benchmark evaluation collections. The results show that Model Adjustment could make a significant difference on the performance of Centroid Classifier. Furthermore, the experimental results indicate margin can further improve the performance of Model Adjustment for Centroid Classifier.

References

1. Shankar, S., Karypis, G.: Weight adjustment schemes for a centroid based classifier. Technical report, Dept. of Computer Science, University of Minnesota
2. McCallum, A., Nigam, K.: A Comparison of Event Models for Naive Bayes Text Classification. In: AAAI/ICML-98 Workshop on Learning for Text Categorization, pp. 41–48 (1998)
3. Mun, P.P.T.M.: Text Classification in Information Retrieval using Winnow, http://citeseer.ist.psu.edu/cs
4. Sebastiani, F.: Machine learning in automated text categorization. ACM Computing Surveys 34(1), 1–47 (2002)
5. Joachims, T.: Text categorization with support vector machines: learning with many relevant features. In: ECML, pp. 137–142 (1998)
6. Liu, Y., Yang, Y., Carbonell, J.: Boosting to Correct Inductive Bias in Text Classification. In: CIKM, pp. 348–355 (2002)
7. Wu, H., Phang, T.H., Liu, B., Li, X.: A Refinement Approach to Handling Model Misfit in Text Categorization. In: SIGKDD, pp. 207–216 (2002)
8. Tan, S., Cheng, X., Ghanem, M.M., Wang, B., Xu, H.: A Novel Refinement Approach for Text Categorization. In: ACM CIKM, Bremen, Germany, pp. 469–476 (2005)

Automatic Categorization of Human-Coded and Evolved CoreWar Warriors

Nenad Tomašev, Doni Pracner, Miloš Radovanović, and Mirjana Ivanović

University of Novi Sad
Faculty of Science, Department of Mathematics and Informatics
Trg D. Obradovića 4, 21000 Novi Sad, Serbia
tomasev@nspoint.net, doni@neobee.net, {radacha,mira}@im.ns.ac.yu

Abstract. CoreWar is a computer simulation devised in the 1980s where programs loaded into a virtual memory array compete for control over the virtual machine. These programs are written in a special-purpose assembly language called *Redcode* and referred to as *warriors*. A great variety of environments and battle strategies have emerged over the years, leading to formation of different warrior types. This paper deals with the problem of automatic warrior categorization, presenting results of classification based on several approaches to warrior representation, and offering insight into ambiguities concerning the identification of strategic classes. Over 600 human-coded warriors were annotated, forming a training set for classification. Several major classifiers were used, SVMs proving to be the most reliable, reaching accuracy of 84%. Classification of an evolved warrior set using the trained classifiers was also conducted. The obtained results proved helpful in outlining the issues with both automatic and manual Redcode program categorization.

1   Introduction

CoreWar was introduced by A. K. Dewdney in 1984 in an article published in Scientific American [1]. It was based on a game called *Darwin* developed in Bell Labs in 1960, devised by Victor Vyssotsky, Robert Morris Sr. and Dennis Richie. In CoreWar, several programs, referred to as *warriors*, attempt to survive in a looping memory array, avoiding attacks and at the same time trying to eliminate the opposition. The warrior that takes complete control of the process queue wins the battle. A match between two warriors consists of a number of such battles, each time varying the initial positioning in the memory array which is referred to as *the core*.

Several automatic warrior generators have been created so far, utilizing evolutionary algorithms to create functioning warriors out of sets of randomly generated code sequences. Performance evaluation in warrior generation is usually done via testing against some predetermined benchmark warrior set of considerable size. In evolvers, when a lot of new warriors are being constructed in each generation, determining fitness becomes a very time demanding process.

Automatic warrior categorization could be very useful in CoreWar evolutionary software in order to design control mechanisms for mutation rate adjustment. High mutation rates allow the creation of a greater variety of forms, while the low mutation

J.N. Kok et al. (Eds.): PKDD 2007, LNAI 4702, pp. 589–596, 2007.

rates instigate convergence of the generation pools to the fittest among the generated types. However, the current lack of automatic categorization renders diversity supervision practically impossible. Some warriors may be easy to categorize manually by the combination of strategic components in their code, but finding clear distinctions between warrior types is generally not an easy task.

The goal of the research presented in this paper is to explore the possibilities for automatic warrior categorization using representations based on syntax analysis and benchmark scores. The former may quickly and easily be calculated, while the latter are an essential part of fitness evaluation in CoreWar evolvers. The addressed issues include choosing warrior types from a plethora of possible distinctions depending on the desired level of abstraction, specifying the representations, manually categorizing a warrior set, conducting automatic classification, and testing it on both human-coded and evolved warrior sets. Our main intention was to spot the obstacles in the categorization process, so that further improvements to the representations could be made and, more importantly, to assess the feasibility of automatic warrior categorization. This project is the first attempt to achieve the above mentioned goal using supervised machine learning methods, and also the first to introduce a fully labeled warrior dataset.

The rest of the paper is organized as follows. Section 2 explains the essentials of CoreWar and the Redcode language, while Section 3 discusses possibilities for representing warriors in a form suitable for analysis. Section 4 deals with the issues related to both human-coded and evolved warrior datasets used in this research. The analysis of categorization itself is given in Section 5. The last section provides a summary of the conclusions together with plans for future work.

2   CoreWar

CoreWar is a computer simulation where programs compete in a virtual cyclic memory array. These programs, referred to as *warriors*, are written in an assembly language called *Redcode*. Warrior confrontation takes place in a virtual memory array called *the core* which is wrapped around so that the successor of the last address in the core is the first one. Execution of instructions and management of threads is performed by the *memory array redcode simulator* (MARS).

A warrior's goal in most competitions is to take complete control of the core by forcing all its opponents to eliminate their threads of execution from the process queue. Warriors can read from the core, write to the core, perform basic arithmetic instructions, create new threads, mutate, go through numerous stages in their ontogeny, copy themselves, actively search for their opponents, etc.

Competitions are held regularly on several Internet servers. Corewar leagues are commonly referred to as *hills*. There are several important parameters defining these standard competitions, namely: core size, maximal battle duration, number of threads allowed per warrior, warrior size restrictions, etc. The most popular hill is certainly the 94nop hill, which is the presumed setting for warriors considered in this paper.

The Redcode language. Redcode is currently the default language for writing CoreWar warriors. It contains 19 instructions, 7 instruction modifiers and 8 addressing modes. Each command consists of an instruction name, followed by the instruction modifier,

A-field addressing mode, A-field value, B-field addressing mode and B-field value. The more important instructions include DAT, which is used both to store data and remove the thread executing it from the process queue; the copying instruction MOV; arithmetic instructions ADD, SUB, MUL, DIV and MOD; unconditional jump instruction JMP and conditional jumps JMZ, JMN and DJN; and the thread-creating instruction SPL. There are many combinations of the mentioned elements, more precisely $8512 \cdot \text{CORESIZE}^2$. The set of all possible Redcode programs in the standard setting is of cardinality $100 \cdot (544768000000^{100})$. A description of Redcode is available in [2].

Warrior types. Contemporary warriors are highly sophisticated, a result of over two decades of continuous improvements over the basic ideas, and occasional ascension of new concepts. Most of them represent combinations of several strategic elements. Some common *strategic concepts* are summarized below.

Imps are among the simplest of components, yet quite often used due to the fact that disposing of them is costly in terms of time and space. Imps consist only of MOV instructions, copying themselves through the core, commonly forming structures known as *rings* and *spirals*. *Bombing* is a process of copying some predetermined instructions throughout the core with the intention of overwriting a part of the opponent's code. *Replication* is a process performed by warrior components constantly copying themselves and creating new threads to run the copies. *Core clearing* is a process of sequential overwriting of the core with some predetermined instruction. *Scanning* denotes heuristically searching for opponent's code by comparing pairs of instructions or instruction fields. *Bootstrapping* is a process of quickly copying essential components away from the original code to avoid detection. *Quickscan* is a component performing exceptionally fast scanning at the start of the simulation, trying to locate enemy code in an early stage and disable it before it bootstraps and activates its components.

For the purposes of categorization in this paper, a relatively modest number of 13 warrior types has been selected to represent the strategic abundance of CoreWar. The considered warrior types are given in Table 1.

3   Warrior Representation

One of the main issues in automatic CoreWar warrior categorization is certainly representing warriors in a form suitable for analysis. The code itself can be viewed as a genotype, while the associated behavior in a certain core corresponds to the phenotype of a warrior. Same warrior code can display different properties in different environments and even belong to different warrior types in the respective core settings! Therefore, it is the emergent behavior that outlines the category generalizing the strategic concepts of a warrior in a certain environment. In the rest of this paper the concept of warrior types will be regarded in this manner, only relative to the 94nop setting.

The question arises whether it is possible to draw conclusions about warrior phenotype given the parameters of the considered environment, based on observations of code alone. If it were possible to classify a warrior based on a representation derived from syntax analysis, such a process would be favorable in terms of execution time, and therefore preferable for use in systems performing a lot of calculations, e.g. evolvers.

Table 1. Warrior categories

Type	Description
cds	Clear-directing scanners
clr	Warriors basing their activity on clearing the core
clrwi	Core-clearing warriors using imp components
evo	Evolved warriors, a category denoting all automatically created warriors that do not resemble human-coded warriors enough to be considered one of the other types
onesh	Oneshots – a special class of scanners, focusing on the first potential threat
pap	Replicators, also called *papers*, according to the stone/paper/scissor analogy
pwi	Replicators that also use imps
pws	Replicators that also use stones
sabi	Stones accompanied by both A-field and B-field imps
sai	Stones accompanied by A-field imps
sbi	Stones accompanied by B-field imps
scn	Scanners other than clear-directing scanners and oneshots. All three classes together are referred to as *scissors*
stn	*Stones* are warriors utilizing a bombing strategy. Their name is derived from the stone/paper/scissor analogy

In the first phase, a simple syntax-based representation was evaluated comprised of frequencies of appearance of instruction names in warrior code, and also of some characteristic instruction pairs, namely SPLMOV, MOVJMP, MOVDJN, MOVADD, MOVSUB, SEQSNE, SNEJMP, and SEQSLT. A boolean flag *Impspec* was added to carry information about the potential imp presence within a warrior. We shall denote this representation *static*. It total, it consists of 26 continuous, and one boolean attribute.

In the second phase, a representation formed by benchmark scores was used. These data are usually available in automatic warrior generators, due to necessity of performing fitness estimation. This representation consists of win and loss percentages of the tested warrior against each of the benchmark warriors, and will be referred to as the *score* representation. The benchmark was comprised of 30 carefully chosen warriors, accounting for 60 continuous attributes for this representation.

After both mentioned representations were tested, a hybrid representation combining the former was used (denoted *combined*). It was additionally attempted to extend the representations with another boolean feature named *Qspec*, indicating the presence of a quickscan component within a warrior. This attribute will be treated separately since we do not yet have a satisfactory means of automatically determining its value.

The static representation suffers heavily from its inability to distinguish code that will be executed during the simulation from decoys. There was an attempt in the past to overcome this problem by observing frequencies of command executions during simulations [3]. This may seem to be a good solution, but it has its own pitfalls. In particular, when a warrior is placed alone in the core there is a strong possibility that some parts of its code will never be executed because no enemy is ever detected. If, on the other hand, a warrior is set to confront some other warriors, it will also be executing commands that other warriors might copy over its code. Syntax analysis does not fall prey to the described problem.

Table 2. Class distribution of the h1c dataset

cds	clr	clrwi	evo	onesh	pap	pwi	pws	sabi	sai	sbi	scn	stn	Total
47	20	12	66	73	100	40	40	6	33	39	73	117	666

4   The Datasets

Two warrior sets have been used in this research. The first one, denoted h1c, represents a subset of **94nop** Königstuhl set [4]. All warriors were manually categorized. The dataset consists of 666 warriors (no pun intended), which are unfortunately not evenly distributed among the considered categories (Table 2). The average warrior length in the dataset disregarding data storing instructions was 35, far below the 100 instruction limit imposed by the **94nop** setting. Massive use of quickscanners as an early stage strategic component led to the minimalistic approach in coding to avoid early detection.

The considered evolved warrior set is a subset of the output generated by the CCAI evolver [5], which was written by Barkley Vowk from the University of Alberta in summer 2003. We have already subjected the complete CCAI output to clustering using the static representation in [2]. The respective size of that dataset is 4389 warriors.

5   Warrior Categorization

Classification was performed using the WEKA machine learning workbench [6]. The following classifiers were included in the experiments: SMO – an implementation of the sequential minimal optimization algorithm for training support vector machines [7], performing multi-class classification using a binary classifier for each pair of classes; MultilayerPerceptron (MLP) – a neural network classifier trained using backpropagation; J48 – a decision tree learner based on revision 8 of the C4.5 algorithm; Naive-Bayes; BayesNet – a Bayesian network with automatically determined structure as the maximum weight spanning tree [8]; and IBk – which implements the classical k-nearest neighbor algorithm. We report results for SMO with the linear kernel and $C = 1$, MLP trained in 500 epochs with one hidden layer containing $1/2$ of the total number of input and output nodes, NaiveBayes with supervised discretization used for handling continuous attributes, and IBk with $k = 5$ neighbors and reciprocal distance weighing.

Categorization of human-coded warriors. Figure 1 summarizes the performance of the considered classifiers in experiments involving 10 runs of 10-fold cross-validation on the h1c dataset. The highest accuracy was exhibited by SMO, 84.26%, on the representation including both static and score components, including specification of quickscanner presence. Generally, the use of the *Qspec* attribute only slightly improved the performance of classification with all representations, and its contribution could not be statistically verified using the corrected resampled t-test ($p = 0.05$).

Performance of all classifiers except for MLP was statistically verified as worse when compared to SMO on the complete (combined+Qspec) representation. Naive-Bayes proved to be the worst among the tested classifying methods, reaching accuracy of only 69.7%. Apparently, it had problems coping with the dependencies between

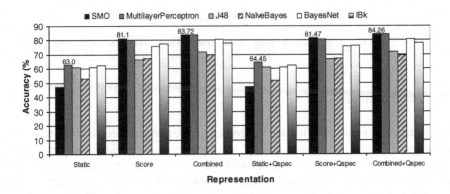

Fig. 1. Performance of various classifiers on the h1c warrior dataset, by representation

attributes, both static (concerning co-occurrences of instructions) and score-based (regarding the pairwise and other dependencies between win and loss percentages), which may be observed on the partial representations. This is corroborated by the bad performance of BayesNet when configured to work with the "naive" network structure (its accuracy is only slightly better than that of the the reported NaiveBayes).

The classification success rate varied respective to warrior categories in question. Replicators were easily recognized. Some scanners belonging to the scn class had been confused with stones. The scanners in question were mostly the less efficient ones, thus differing in crucial benchmark scores from the rest of their group, instead scoring similarly to a warrior subtype known as *incendiary stones*. The lowest accuracy was present in classification of warriors belonging to those categories represented by a small number of instances in the dataset, namely clr, clrwi, and sabi. It is also worth mentioning that most of the evolved warriors in h1c were correctly classified as evo.

Classifier accuracy on the complete representation was superior over isolated use of static features. Also, adding the static feature vector to the score representation improved classification significantly in cases of SMO, MLP, BayesNet and J48 classifiers.

Even though SMO performed best on score-based and combined representations, its accuracy of 47.15% on the static representation was significantly inferior to the performance of J48, BayesNet, IBk and MLP. This can be partially remedied by employing polynomial kernels of higher degree (4–6), with the performance being able to reach that of the best classifiers, but at the expense of worsening accuracy on score-based representations. The highest accuracy of classification on the static representation was achieved by MLP – 64.45%.

Regardless of the apparent advantages of both score-based and combined representations over the static representation, the obtained results indicate that classification according to static features alone might be possible in the future if some modifications were made. First of all, according to the current static representation, it is absolutely impossible to distinguish A-field imps from B-field imps. That can easily be solved by adding new imp presence indicators. Also, it appears that the use of characteristic instruction pairs was insufficient for carrying information about the context in which instructions were used. Some solutions to that problem will be considered in Section 6.

Categorization of the evolved dataset. Generation 4 of the CCAI dataset was already clustered [2], so the categorization of that warrior set was meant both as a test of the reliability of trained classifiers and also to provide insight into the structure of the dataset and its clusters. Clustering had been done using the static representation, without *Qspec*. For purposes of classification in the current research, classifiers were trained on h1c with the score and combined representations, and tested on a random 400-warrior sample from the CCAI dataset which was manually labeled.

Evolved warriors usually differ from their human-coded adversaries. One of the main characteristics of evolved warriors is the presence junk code, i.e. instructions in the source code which actually never get executed. Evolved datasets consist mainly of mutation resistant forms – core-clearing warriors and replicators being the dominant types [2]. Also, such warriors rarely utilize advanced strategic tricks, which distinguishes them from analogous human-coded warrior types. However, the last generation of the CCAI set exhibits somewhat different properties, as the evolver was aimed at generating strong, competitive warrior forms. The final product was a famous warrior that defeated many human-coded opponents, appropriately named *Machines Will Rule*. Hence, warriors in the CCAI set tend to be stronger than typical evolved warriors and bear more resemblance to their human-coded counterparts.

Manual inspection detected only two classes in the 400-warrior sample of the CCAI set: pap and pwi. Papers were mostly using anti-imp core-clearing techniques, and achieved great scores against imp-type warriors in the benchmark. Imp-containing papers, however, were not nearly as well optimized and rarely benefited from the presence of defensive imp structures.

The following classifiers were tested: SMO, MLP, BayesNet, and IBk, with the results summarized in Table 3. It can be seen that the introduction of static features to the score-based representation does not yield consistent improvements as with the h1c dataset. On the contrary, it significantly degraded the performance of MLP and IBk classifiers. This is the result of noise introduced by junk code in the evolved warriors which contained mostly imp-specific and arithmetic instructions, helping MLP confuse many pap warriors with pws. The misclassification rate of most classifiers originated from pap being interpreted as sbi. Besides junk code, this can be attributed to the high resistance of the warriors in question to some common scanner attack techniques. On the other hand, static features continue to carry useful information, which is evident from the introduced improvement to the accuracy of SMO and BayesNet, and also from the increase in the ability of most classifiers to detect the minority pwi class.

All of the warriors from the sample misclassified by either of the classifiers as sbi or pwi using the combined representation were additionally examined. In 37% of instances an interesting structure was discovered, used by the replicators as a strong anti-imp feature, but also forming a sort of imp-like structure, thus enhancing defensive capabilities. Such pseudo-spirals were set up in a similar fashion to imp spirals, the difference being in the instruction used, namely MOV.I #1169, }2667. Apart from this interesting replicator subtype, core-clearing papers were quite frequent in the considered sample.

As for the syntactic clustering described in [2], we can now conclude that it was unable to detect the subtle differences between the two classes in the dataset, being misled by junk code within the warriors. However, this does not mean that the clustering

Table 3. Categorization of a 400-warrior sample from generation 4 of the CCAI dataset, with class counts and accuracy for the score and combined representations

Classifier	evo		pap		pwi		pws		sabi		sbi		stn		Accuracy (%)	
	scr	cmb	scr	cmb	scr	cmb	scr	cmb	scr	cmb	scr	cmb	scr	cmb	scr	cmb
SMO	18	0	345	359	1	0	19	0	0	0	17	41	0	0	85.75	88.50
MLP	0	11	322	217	1	14	63	136	2	0	12	21	0	1	79.25	53.25
BayesNet	0	0	391	392	1	7	0	0	0	0	8	1	0	0	96.25	96.50
IBk	0	0	386	359	2	3	0	0	0	0	12	38	0	0	95.00	88.50
Manual	0		392		8		0		0		0		0		100.00	

was not a good indicator of the diversity of warrior genotype, since junk code can also be combined in subsequent generations to produce working warriors.

6   Conclusions and Future Work

Recently, the attention of the CoreWar community shifted from devising new tricks in the existing strategies to exploring new settings, parameter optimization [9], and automatic warrior generation [5]. Quick and reliable warrior categorization would be of great importance in many such automated optimizing systems. The results obtained from this research indicate that automatic categorization is indeed possible to achieve, at least in the standard 94nop environment. However, further research is necessary in order to improve classification accuracy and possibly form a more universal categorization model, applicable to a wider range of environments. We believe that changing the static part of the representation alone may be enough to ensure the desired increase of accuracy. Adding new n-grams to the representation, as well as modifying and decomposing the *Impspec* feature would certainly improve static-based categorization, and probably the combined representation as well.

References

1. Dewdney, A.K.: Computer recreations: In the game called core war hostile programs engage in a battle of bits. Scientific American 250(5), 14–22 (1984)
2. Pracner, D., Tomašev, N., Radovanović, M., Ivanović, M.: Categorizing evolved CoreWar warriors using EM and attribute evaluation. In: Proc. MLDM'07, 5th Int. Conf. on Machine Learning and Data Mining in Patt. Recognition. LNAI, vol. 4571, Springer, Heidelberg (2007)
3. Varfar, W.: Wilfiz scores of warriors on the 94nop.
   redcoder.sourceforge.net/?p=kepler-wilfiz
4. Birk, C.: CoreWar Koenigstuhl,
   www.ociw.edu/~birk/COREWAR/koenigstuhl.html
5. Vowk, B.: CCAI., www.math.ualberta.ca/~bvowk/corewar.html
6. Witten, I.H., Frank, E.: Data Mining: Practical Machine Learning Tools and Techniques, 2nd edn. Morgan Kaufmann Publishers, San Francisco (2005)
7. Platt, J.: Fast training of support vector machines using sequential minimal optimization. In: Advances in Kernel Methods – Support Vector Learning, MIT Press, Cambridge (1999)
8. Friedman, N., Geiger, D., Goldszmidt, M.: Bayesian network classifiers. Machine Learning 29(2–3), 131–163 (1997)
9. Zapf, S.: Optimax, www.corewar.info/optimax/

Utility-Based Regression

Luis Torgo[1,2] and Rita Ribeiro[1,3]

[1] LIAAD-INESC Porto LA, R. Ceuta, 118, 6., 4050-190 Porto, Portugal
[2] FEP, University of Porto, R. Dr. Roberto Frias, 4200-464 Porto, Portugal
[3] FC, University of Porto, R. Campo Alegre, 1021/1055, 4169-007 Porto, Portugal
{ltorgo,rita}@liacc.up.pt

Abstract. Cost-sensitive learning is a key technique for addressing many real world data mining applications. Most existing research has been focused on classification problems. In this paper we propose a framework for evaluating regression models in applications with non-uniform costs and benefits across the domain of the continuous target variable. Namely, we describe two metrics for asserting the costs and benefits of the predictions of any model given a set of test cases. We illustrate the use of our metrics in the context of a specific type of applications where non-uniform costs are required: the prediction of rare extreme values of a continuous target variable. Our experiments provide clear evidence of the utility of the proposed framework for evaluating the merits of any model in this class of regression domains.

1 Introduction

In many real world applications the costs and benefits of using prediction models are non-uniform. These observations have motivated the work on cost-sensitive learning (e.g. [5]) and more generally on utility-based mining [9,11]. In the context of applying the discovered knowledge under a non-uniform cost setup, most works have focused on classification tasks (e.g. [3,4,5,6]). Still, within numeric prediction problems, also know as regression, similar problems arise. As mentioned by Crone et. al. [2] most works on regression assume uniform costs and use some form of average error statistic. In this context, several authors (e.g. [1,2]) have proposed new cost of error functions that try to address these issues. However, most of these works only consider one particular type of non-uniform costs of errors: the difference between under- and over-predictions, i.e. situations where the predicted values are above or below the true values, respectively.

This paper proposes a framework for evaluating regression models in the context of arbitrarily shaped costs and benefits across the domain of the numeric target variable of regression tasks. We propose two new evaluation metrics that incorporate the notions of costs and benefits and thus are able to provide better feedback on the merits of regression models in the context of the specific biases of any numeric prediction task. These metrics use cost and benefit surfaces that we also formalize, which can be regarded as continuous versions of the well-know notion of misclassification cost matrices. We illustrate the use of our

J.N. Kok et al. (Eds.): PKDD 2007, LNAI 4702, pp. 597–604, 2007.

proposed metrics in a particular class of non-uniform costs/benefits application: the prediction of rare extreme values of a continuous variable.

2   Problem Formulation

Predictive learning tries to obtain an approximation of an unknown function $f : \chi \rightarrow \gamma$, based on a training data set D drawn from a distribution with domain $\chi \times \gamma$, where χ is the domain of the set of predictor variables and γ is either a discrete domain in the case of classification tasks, or $\Re$ in the case of regression. The obtained approximation, $\hat{f}_\beta$, is a model with a set of parameters, β, that are obtained by optimizing some preference criterion. For classification, this is usually the error rate, while in the case of regression the most frequent are the mean squared error, $MSE = \frac{1}{n}\sum_{i=1}^{n}(y_i - \hat{y}_i)^2$, or the mean absolute deviation, $MAD = \frac{1}{n}\sum_{i=1}^{n}|y_i - \hat{y}_i|$.

Many authors (e.g. [3,5]) have noticed the problems arising from the uniform cost assumption of the error rate evaluation criterion, which is unacceptable for many real world domains. The cost matrix formulation overcomes these limitations by allowing the specification of the cost of misclassifying class i by class j, and leads to the criterion of expected cost minimization, $\frac{1}{n}\sum_{i=1}^{n}C(\hat{y}, y)$, where $C(\hat{y}, y)$ is an entry on the pre-specified cost matrix. Regards regression few authors have addressed the issue of differentiated costs. Most of the existing works on having non-uniform costs for regression have been addressing the issue of differentiating the cost of under-predictions ($\hat{y} < y$), from the cost of over-predictions ($\hat{y} > y$) (e.g. [1,2]). Although these approaches address several important application-specific requirements, they fail to provide a means to specify a cost function across all domain of the target variable, which was shown to be of key importance for this type of applications [8]. In this paper we address this issue by associating to each prediction a cost that is dependent on an user-defined relevance of both the true and predicted values.

3   Utility-Based Regression

As mentioned by Zadrozny [10], research on cost-sensitive learning has traditionally been formalized in terms of costs as opposed to benefits or rewards. However, evaluating a model in terms of benefits is generally preferable because there is a natural baseline from which to measure all benefits whether positive (real benefits of a prediction) or negative (that are in effect costs) [5]. Our proposal follows these lines, by measuring the utility of a regression model through the total balance between the costs and benefits originated by its predictions.

3.1   Relevance Functions

We assume that for some applications the relevance (importance) of the values of the target variable is not uniform across its domain. This domain-dependent information shall be provided through the specification of a relevance function,

$\phi(Y) : \Re \rightarrow 0..1$, that maps the domain of the target variable into a $0..1$ scale of relevance, where 1 represents maximum relevance. Our proposal is independent of the shape of the $\phi()$ function. We assume this function is specified by the user using his/her domain knowledge. The specification of the relevance function is the step of our proposal that is most challenging for the user. Given the large range of applications where the relevance of the target variable is non-uniform, it is virtually impossible to describe reasonable default relevance functions for all these applications. Still, in many applications relevance is often associated with rarity (e.g. highly profitable customers; high variations on stock prices; extreme weather conditions, etc.). For these applications relevance can be defined as a function that is inversely proportional to the probability density function (*pdf*) of the target variable. Although obtaining the functional form of these *pdf*'s is generally non trivial, reasonable approximations based on the available data sample can be obtained with techniques like kernel density approximators. In Section 4 we propose an even simpler strategy to derive a relevance function for a class of applications where relevance is associated with rarity: the prediction of rare extreme values of a numeric variable.

3.2   Cost and Benefit Surfaces

Generally, the cost of a prediction depends not only on the relevance of the test case value but also on the relevance of the predicted value. In effect, all three following situations are penalizing in a cost-sensitive application:

1. Predict a relevant value for an irrelevant test case (false alarm);
2. Predict an irrelevant value for a relevant test case (opportunity cost);
3. Predict a relevant but very different value for a relevant test case (the most serious mistakes: confusing relevant events).

We capture this notion of relevance of the prediction for a given test case by means of the definition of a bi-variate relevance function, $\Phi(\hat{Y}, Y)$, that depends on the relevance of both the true and predicted values,

$$\Phi(\hat{Y}, Y) = (1 - m) \cdot \phi(\hat{Y}) + m \cdot \phi(Y) \tag{1}$$

This function is a weighted average of the individual relevances of $\hat{Y}$ and Y. It is maximum when both are highly relevant and these are the cases where the cost of the predictions may reach the maximum if they are not accurate enough. The m parameter ($0 \leq m \leq 1$) differentiates between situations 1 (false alarms) and 2 (opportunity costs). Setting $m > 0.5$ makes the latter more important.

The cost of a prediction should also depend on its precision, i.e. how near are $\hat{Y}$ and Y from each other. Moreover, it should also be possible for the user to establish some kind of application-specific measure of cost in whatever units make sense for the domain. In this context, we define the cost of a prediction as,

$$c(\hat{Y}, Y) = \Phi(\hat{Y}, Y) \times C_{\max} \times L(\hat{Y}, Y) \tag{2}$$

where $C_{\max}$ is the maximum cost that is only assigned when the relevance of the prediction is maximum (i.e. $\Phi(\hat{Y}, Y) = 1$); and $L(\hat{Y}, Y)$ is a loss function that measures the prediction error.

The term $\Phi(\hat{Y}, Y) \times C_{\max}$ can be seen as a kind of case-specific maximum cost value. This is the maximum penalty we get if $\hat{Y}$ is the "worst possible" prediction for the test case under consideration. With respect to the loss function we could use any metric function, e.g. the absolute deviation $|\hat{Y} - Y|$. However, in order to make the meaning of the value $c(\hat{Y}, Y)$ more intuitive, we recommend the use of a percentage-type loss function that ranges from 0 to 1. Such function will then represent the proportion of the case-specific maximum cost we get due to our prediction. For maximum error ($L(\hat{Y}, Y) = 1$) we get the full penalty of the particular test case ($\Phi(\hat{Y}, Y) \times C_{\max}$), while a perfect prediction ($L(\hat{Y}, Y) = 0$) would entail no cost as expected. This means the value of $c(\hat{Y}, Y)$ will be expressed in the same units as $C_{\max}$, which is provided by the user, and thus it is more intuitive for him/her. In this context, we propose the following loss function that ranges from 0 to 1:

$$L(\hat{Y}, Y) = | \max_{i \in \hat{Y}..Y} \phi(i) - \min_{i \in \hat{Y}..Y} \phi(i)| \tag{3}$$

The use of the maximum and minimum functions is due to the fact that we want to let the user specify any arbitrarily shaped $\phi()$ function. This means that we can have two quite different Y values with the same value of $\phi()$, which would look like a perfect prediction if we had used the difference of relevances directly. However, these cases are exactly the most serious mistakes we want to avoid (the 3rd case on the list presented before). With our proposal, if both values have high relevance but are quite different then surely there will be values in between with lower relevance and this will result in a higher value of the loss function.

The function $c()$ can be seen as a continuous version of cost matrices, i.e. a cost surface. The total cost of the predictions of a model is defined as,

$$TC = \sum_{i=1}^{n} c(\hat{y}_i, y_i) \tag{4}$$

Our proposal also considers the benefits of the predictions of a model, with the goal of asserting its ability to accurately predict most of the relevant values in a test set. In the case of benefits it is only the relevance of the true value that counts, i.e. we are interested in asserting how well a model predicts the relevant test cases. In this context, the benefit surface is defined as,

$$b(\hat{Y}, Y) = \phi(Y) \times B_{\max} \times (1 - L(\hat{y}_i, y_i)) \tag{5}$$

where $B_{\max}$ is a user-defined maximum reward that is measured in the same units as the $C_{\max}$ constant; and $L()$ is a loss function as before.

Our definition of benefits associates higher rewards with higher relevance. The term $\phi(Y) \times B_{\max}$ calculates the case-specific benefit, while the last term is the proportion of this reward we get. The total benefits are given by,

$$TB = \sum_{i=1}^{n} b(\hat{y}_i, y_i) \qquad (6)$$

Finally, we can define the utility of the predictions of a model as the net balance between its total costs and benefits,

$$U = TB - TC \qquad (7)$$

4   An Illustrative Application

Modeling extreme data is very important in several application domains, like finance, meteorology, ecology, etc.. Several of these applications involve predicting a continuous variable. For these domains the extreme (high or low) values of the target variable are much more important than the others. Moreover, these extremes are generally quite rare, which turns this into a very hard prediction problem with very clear non-uniform costs and benefits of predictions. In this section we illustrate the use of our proposed framework for utility-based regression, by using it to compare quite diverse modelling techniques on a real world data set where the prediction of rare extreme values is of primary importance.

The application we use to illustrate our proposal concerns stock market forecasting. Namely, the data are about the task of trying to predict the future daily variation in closing prices of the IBM stock, using information regarding the values of these variations on the 10 previous market sessions. The data set consists of information on 8166 daily market sessions (roughly 30 years), each being described by 10 predictor variables (the variations on the 10 previous days) and a target variable (the variation on the next day). This application is a very clear example of non-uniform costs (and benefits) of predictions. In effect, any model that is extremely accurate at predicting small price variations (the most common) is essentially useless for a trader. Profitable trading is based on being able to capitalize on large price changes. Trades carried out over small price changes are usually not able to cover the trading costs and thus are non-profitable or even represent a loss of money. As such, in these applications the accuracy on the relevant (i.e. extreme high or low) changes of prices is the key criterion.

In order to apply our evaluation method we need to specify a relevance function for this domain. In this class of applications relevance is strongly associated with extreme and rare values of the target variable. The distribution of the target variable has a normal-like shape with very marked tails (the rare extreme price variations). From the description of the goals of this application it should be clear that the relevance function should have a shape that is inverse of the *pdf* of the price variations. We propose to use a sigmoid-like function for establishing a smooth relevance function. In order to define this function we use some of the statistics provided by boxplots that summarize the distribution of the target variable. With this strategy we are able to obtain a relevance function without having to deal with computationally complex approximations of the unknown *pdf*. Figure 1 provides a graphical illustration of the quantities involved in the

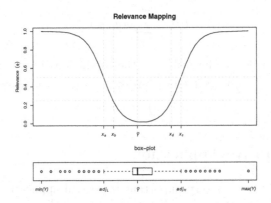

Fig. 1. A sigmoid-based relevance function for rare extreme values prediction

derivation of the relevance function we use. This figure shows the box plot of an arbitrary normal-like distribution and the respective sigmoid-based relevance function. The relevance function is defined using distribution properties of the target variable ($\min(Y)$, $adj_L(Y)$, $\tilde{Y}$, $adj_H(Y)$ and $\max(Y)$) that can be easily estimated from the available data sample. This approach can be generally applied on problems where the target variable has a normal-like shape and where relevance is associated to rare extremes.

The other three parameters necessary to use our U metric are $C_{\max}$, $B_{\max}$ and m. Given our absence of domain expertise on stock market trading we have set these parameters using what seemed to us reasonable settings. Namely, we decided that the maximum benefit should be clearly higher than the maximum cost to try to reward proactive models. In our experiments we have used $C_{\max} = 10$ and $B_{\max} = 20$. With respect to the m parameter we have set it to 0.5, i.e. equal importance to false alarms and opportunity costs.

In order to test our proposed metrics under different experimental setups we have applied 3 quite different modeling techniques to the IBM data set. Namely, regression trees, neural networks and support vector regression. For all 3 methods we have used their implementations freely available on the R software environment [7], more specifically the function `rpart()` of the package **rpart**, the function `nnet()` of the package **nnet** and the function `svm()` of the package **e1071**. All 3 methods were used without any extensive parameter tuning as the goal was not to achieve the best possible accuracy but instead to test an evaluation metric under different setups.

All models were evaluated using the MAD, U, TC and TB statistics, which were described in Sections 2 and 3. The MAD statistic was selected as a "representative" of a standard evaluation metric. The values of all statistics were estimated using a 10-fold cross validation process. Statistical significance (95% level) of the differences when compared to the best ranked model were asserted by means of the non-parametric Wilcox test and signaled by "*". The results are shown on Table 1. For each statistic, we provide the ranking of the models and indicate the median and inter-quartile value measured over the 10 repetitions.

Table 1. The results/rankings on the IBM data set

MAD		U		TB		TC	
dummy		dummy		svm		dummy	
0.01205	(0.00038)	108.18427	(37.76051)	278.41357	(27.94408)	134.12127	(17.29376)
cart	*	rand.forest		dummy	*	cart	*
0.01206	(0.00038)	108.18248	(36.24961)	243.10776	(36.9331)	134.12645	(17.29339)
nnet	*	cart	*	cart	*	nnet	*
0.01209	(0.00042)	108.1688	(37.75762)	243.09674	(36.92891)	134.12665	(17.29379)
rand.forest	*	nnet	*	nnet	*	rand.forest	*
0.01235	(0.00029)	108.1674	(37.75963)	243.09523	(36.93183)	134.14339	(13.72446)
svm	*	svm	*	rand.forest	*	svm	*
0.01447	(0.00047)	21.07661	(30.65326)	242.07468	(36.97128)	248.66908	(32.11022)

The goal of these experiments is not to check if the models are good according to our U metric, as they were obtained optimizing other criteria. Our objective is to check whether by using a metric tunned for giving more weight to rare extreme values, we can spot a method that is better at predicting these cases, particularly if that would not be found by using only standard statistics in the comparitive study.

The results on Table 1 unveil some interesting information that could not be observed from looking at the MAD scores. In effect, we can see that the SVM achieves a much higher score in terms of benefits, clearly indicating that it is able to capture more extreme values. However, this approach also has led to a higher value of TC, resulting of its more risky approach to this prediction problem. This results in a poor score in terms of net balance (U score). Still, given the fact that there was no particular tuning of the model parameters, we can say that the SVM is probably a model where more time should be invested in the context of this application, so that the signals it is producing get more precise. This sort of information is only available due to the use of an evaluation metric that is tunned towards the application goals.

5   Conclusions

This paper has described a new evaluation framework for regression tasks with non-uniform costs and benefits of the predictions. Our proposal is based on the specification of a relevance function over the domain of the target continuous variable. This function is the basis of the definitions of cost and benefit surfaces that can be regarded as continuous versions of cost/benefit matrices used in classification tasks. The use of the relevance function relieves the user from the heavy burden of having to specify a cost (and benefit) for all points in the bi-dimensional space of the predicted and true target values. The total cost and benefit of the predictions of a model provide, either individually or aggregated on an utility measure, important insights on the predictive performance of a model. Moreover, these insights are related to the application goals in terms of what is really relevant.

We have illustrated the use of our evaluation framework in the context of a particular class of applications: the prediction of rare extreme values of a continuous variable. Namely, we have used a data set from stock market prediction and have introduced a general relevance function for rare extremes prediction tasks. The results of our experiments have confirmed that our proposed metric provides a better insight on the ability of the models to accurately predict the cases that are more important for this class of applications.

Acknowledgments

The work of the second author is supported by a PhD scholarship of the Portuguese government (SFRH/BD/1711/2004).

References

1. Christoffersen, P., Diebold, F.: Further results on forecasting and model selection under asymmetric loss. Journal of Applied Econometrics 11, 561–571 (1996)
2. Crone, S., Lessmann, S., Stahlbock, R.: Utility based data mining for time series analysis - cost-sensitive learning for neural networks. In: Weiss, G., Saar-Tsechansky, M., Zadrozny, B. (eds.) Proceedings of the 1st International Workshop on Utility-Based Data Mining, pp. 59–68 (2005)
3. Domingos, P.: Metacost: A general method for making classifiers cost-sensitive. In: Proceedings of the 5th International Conference on Knowledge Discovery and Data Mining (KDD-99), pp. 155–164. ACM Press, New York (1999)
4. Drummond, C., Holte, R.: Exploiting the cost of (in)sensitivity of decision tree splitting criteria. In: Proc. 17th International Conf. on Machine Learning, pp. 239–246. Morgan Kaufmann, San Francisco, CA (2000)
5. Elkan, C.: The foundations of cost-sensitive learning. In: Proceedings of 7th IJCAI'01, pp. 973–978 (2001)
6. Fan, W., Stolfo, S., Zhang, J., Chan, P.: AdaCost: misclassification cost-sensitive boosting. In: Proc. 16th International Conf. on Machine Learning, pp. 97–105. Morgan Kaufmann, San Francisco, CA (1999)
7. R Development Core Team R: A Language and Environment for Statistical Computing. R Foundation for Statistical Computing (2006) ISBN 3-900051-07-0.
8. Torgo, L.: Regression error characteristic surfaces. In: Grossman, R., Bayardo, R., Bennett, K., Vaidya, J. (eds.) Proc. of the 11th ACM SIGKDD Intern. Conf. on Knowledge Discovery and Data Mining, pp. 697–702. ACM Press, New York (2005)
9. Weiss, G., Saar-Tsechansky, M., Zadrozny, B. (eds.): Proceedings of the 1st International Workshop on Utility-Based Data Mining (2005)
10. Zadrozny, B.: One-benefit leaning: Cost-sensitive learning with restricted cost information. In: 1st Intern. Work. on Utility-Based Data Mining, pp. 53–58 (2005)
11. Zadrozny, B., Weiss, G., Saar-Tsechansky, M(eds.): Proceedings of the 2nd International Workshop on Utility-Based Data Mining (2006)

Multi-label Lazy Associative Classification*

Adriano Veloso[1], Wagner Meira Jr.[1], Marcos Gonçalves[1], and Mohammed Zaki[2]

[1] Computer Science Department, Universidade Federal de Minas Gerais, Brazil
{adrianov,meira,mgoncalv}@dcc.ufmg.br
[2] Computer Science Department, Rensselaer Polytechnic Institute, USA
zaki@cs.rpi.edu

Abstract. Most current work on classification has been focused on learning from a set of instances that are associated with a single label (i.e., single-label classification). However, many applications, such as gene functional prediction and text categorization, may allow the instances to be associated with multiple labels simultaneously. Multi-label classification is a generalization of single-label classification, and its generality makes it much more difficult to solve.

Despite its importance, research on multi-label classification is still lacking. Common approaches simply learn independent binary classifiers for each label, and do not exploit dependencies among labels. Also, several small disjuncts may appear due to the possibly large number of label combinations, and neglecting these small disjuncts may degrade classification accuracy. In this paper we propose a multi-label lazy associative classifier, which progressively exploits dependencies among labels. Further, since in our lazy strategy the classification model is induced on an instance-based fashion, the proposed approach can provide a better coverage of small disjuncts. Gains of up to 24% are observed when the proposed approach is compared against the state-of-the-art multi-label classifiers.

1 Introduction

The classification problem is to build a model, which, based on external observations, assigns an instance to one or more labels. A set of examples is given as the training set, from which the model is built. A typical assumption in classification is that labels are mutually exclusive, so that an instance can be mapped to only one label. However, due to ambiguity or multiplicity, it is quite natural that most of the applications violate this assumption, allowing instances to be mapped to multiple labels simultaneously. For example, a movie being mapped to *action* or *adventure*, or a song being classified as *rock* or *ballad*, could all lead to violations of the single-label assumption.

Multi-label classification consists in learning a model from instances that may be associated with multiple labels, that is, labels are not assumed to be mutually exclusive. Most of the proposed approaches [7,1,3] for multi-label classification employ heuristics, such as learning independent classifiers for each label, and employing ranking and thresholding schemes for classification. Although simple, these heuristics do not deal with important issues such as *small disjuncts* and *correlated labels*.

---

* This research was sponsored by UOL (www.uol.com.br) through its UOL Bolsa Pesquisa program, process number 20060519184000a.

J.N. Kok et al. (Eds.): PKDD 2007, LNAI 4702, pp. 605–612, 2007.

In essence, small disjuncts are rules covering a small number of examples, and thus they are often neglected. The problem is that, although a single small disjunct covers only few examples, many of them, collectively, may cover a substantial fraction of all examples, and simply eliminating them may degrade classification accuracy [4]. Small disjuncts pose significant problems in single-label classification, and in multi-label classification these problems are worsened, because the search space for disjuncts increases due to the possibly large number of label combinations. Also, it is often the case that there are strong dependencies among labels, and such dependencies, when properly explored, may provide improved accuracy in multi-label classification.

In this paper we propose an approach which deals with small disjuncts while exploring dependencies among labels. To address the problem with small disjuncts, we adopt a lazy associative classification approach. Instead of building a single set of *class association rules* (CARs) that is good on average for all predictions, the proposed lazy approach delays the inductive process until a test instance is given for classification, therefore taking advantage of better qualitative evidence coming from the test instance, and generating CARs on a demand-driven basis. Small disjuncts are better covered, due to the highly specific bias associated with this approach. We address the label correlation issue by defining *multi-label class association rules* (MCARs), a variation of CARs that allows the presence of multiple labels in the antecedent of the rule. The search space for MCARs is huge and to avoid an exhaustive enumeration. which would be necessary to find the best label combination, we employ a novel heuristic called *progressive label focusing*, which makes feasible the exploration of associations among labels.

The proposed approach was evaluated using two different applications: text categorization and gene functional prediction. It consistently achieves better performance than the state-of-the-art multi-label classifiers, showing gains up to 24%.

2   Related Work

Typical approaches for multi-label classification are based on training an independent binary classifier for each label. These independent classifiers are used to assign a probability of membership to each label, and then an instance is classified into the labels that rank above a given threshold. Examples of this approach include ADTBOOST.MH [2] (decision trees that can directly handle multi-label problems), a multi-label generalization of SVMs [3], and a a multi-label lazy learning based on the kNN approach [7]. In [6] an approach based on independent associative classifiers was proposed. However, this approach was only evaluated in single-label problems, and thus, the performance of multi-label associative classifiers for multi-label problems is still unknown.

The main problem with the binary approach is that it does not consider correlation among labels. The direct multi-label approach explores this correlation by considering a combination of labels as a new, separate label [1]. For instance, a multi-label problem with 10 labels will be transformed to a single-label problem composed of potentially 1,024 labels. The problem now is that a relatively small number of examples may be associated with those new labels, specially if the combination contains many labels. While these approaches are able to capture dependencies among labels, the poor coverage of small disjuncts may degrade overall accuracy.

3   Single-Label Associative Classification

A typical associative classifier, suitable for single-label classification problems, is described in this section. An associative classification model is composed of *class association rules* (CARs), which are defined in the following.

DEFINITION 1. [CLASS ASSOCIATION RULES] CARs are rules of the form $\mathcal{X} \xrightarrow{\sigma,\theta} c_i$, where the set $\mathcal{X}$ is allowed to contain only features (i.e., $\mathcal{X} \subseteq \mathcal{I}$, where $\mathcal{I}$ is the set of all possible features), and c_i is one of the n labels (i.e., $c_i \in \mathcal{C}$, where $\mathcal{C}$ is the set of all possible labels). A valid CAR has support (σ) and confidence (θ) greater than or equal to the corresponding thresholds, σ_{min} and θ_{min}.

Common approaches for associative classification employ a slightly modified algorithm for mining valid CARs directly from the training data. When a sufficient number of valid CARs are found, the model (denoted as $\mathcal{M}$) is finally completed, and it is used to predict the label of the test instances. Due to class overlapping, and since labels are mutually exclusive, CARs may perform contradictory predictions. For example, let $\mathcal{T}$ be a test instance, and let $\mathcal{X}$ and $\mathcal{Y}$ be two subsets of $\mathcal{T}$. Also suppose that the valid CARs $\mathcal{X} \rightarrow c_i$ and $\mathcal{Y} \rightarrow c_j$ (with $i \neq j$) are in $\mathcal{M}$. These CARs are contradictory, since they predict different labels for the same test instance, $\mathcal{T}$. To address this problem, the rule-set $\mathcal{M}$ is interpreted as a poll, in which CAR $\mathcal{X} \xrightarrow{\sigma,\theta} c_i \in \mathcal{M}$ is a vote of weight $\sigma \times \theta$ given by $\mathcal{X}$ for label c_i (note that other criteria for weighting the votes can be used). Weighted votes for each label are then summed, and the score of label c_i is given by the real-valued function s showed in Equation 1. In the end, the label associated with the highest score is finally predicted.

$$s(c_i) = \sum_{\mathcal{X} \xrightarrow{\sigma,\theta} c_i \in \mathcal{M}} \sigma \times \theta \tag{1}$$

Consider the set of instances shown in Table 1, used as a running example in this paper. Each instance corresponds to a movie, and to each movie is assigned a set of labels (but for this example, which refers to single-label classification, only the first label will be considered). If we set σ_{min} to 0.20 and θ_{min} to 0.66, then the model $\mathcal{M}$ will be composed of the following CARs:

1. actor=T. Hanks $\xrightarrow{0.30,0.75}$ label=Drama
2. ~~actor=L. DiCaprio $\xrightarrow{0.20,0.67}$ label=Drama~~
3. actor=M. Damon $\xrightarrow{0.20,0.67}$ label=Crime

Now, suppose we want to classify instance 11. In this case, only first and third CARs are applicable, since feature *actor=L. DiCaprio* is not present in instance 11 (thus, second CAR is crossed out). According to Equation 1, s(Drama)=0.225 and s(Crime)=0.134, and thus *Drama* will be predicted.

4   Multi-label Lazy Associative Classification

In this section we extend the basic classifier described in the previous section, allowing it to predict multiple labels. We also propose an approach for exploring correlated labels, while dealing with small disjuncts, improving the classification model.

4.1 Independent Classifiers

A heuristic employed for multi-label classification is to build an independent classifier for each label. This extension is natural, and it is based on assigning a probability of membership to each label, $f(c_i)$. The probabilities are computed using the proportion of scores associated with each label normalized by the highest score (i.e., max):

$$f(c_i) = \frac{s(c_i)}{max}. \tag{2}$$

Once all probabilities are computed, labels are inserted into a ranking $\mathcal{L}=\{l_1, ..., l_n\}$, so that $f(l_1) \geq f(l_2) \geq ... \geq f(l_n)$. Those labels that rank above a threshold δ_{min} (i.e., $l_k | f(l_k) \geq \delta_{min}$) are assigned to the test instance. To illustrate this process, consider again the example shown in Table 1, but now each movie has multiple labels, which are all considered when mining the CARs. If we set σ_{min} to 0.20 and θ_{min} to 0.66, then $\mathcal{M}$ will be composed of the following CARs:

1. actor=M. Damon $\xrightarrow{0.30,1.00}$ label=Action
2. ~~actor=L. DiCaprio~~ $\xrightarrow{0.30,1.00}$ ~~label=Crime~~
3. ~~actor=T. Hanks~~ $\xrightarrow{0.30,0.75}$ ~~label=Drama~~
4. actor=M. Damon $\xrightarrow{0.20,0.67}$ label=Crime
5. ~~actor=L. DiCaprio~~ $\xrightarrow{0.20,0.67}$ ~~label=Drama~~

Now, suppose we want to classify instance 12 and δ_{min} is set to 0.66. Following Equation 2, f(Action)=1.00 and f(Crime)=0.45, and therefore label *Action* is predicted. Note that, although there is a strong association between feature *actor=B. Pitt* and label *Romance*, CAR *actor=B. Pitt→Romance* is considered a small disjunct, and is neglected by the classifier, even being important to classify instance 12. We refer to this classifier as IEAC (*independent eager associative classifier*).

Table 1. Training and Test Instances

	Id	Label	Title	Actors
Training Set	1	Comedy/Romance	Forrest Gump	T. Hanks
	2	Drama/Romance	The Terminal	T. Hanks
	3	Drama/Crime	Catch Me If You Can	T. Hanks and L. DiCaprio
	4	Drama/Crime	The Da Vinci Code	T. Hanks
	5	Drama/Crime	Blood Diamond	L. DiCaprio
	6	Crime/Action	The Departed	L. DiCaprio and M. Damon
	7	Crime/Action	The Bourne Identity	M. Damon
	8	Action/Romance	Syriana	M. Damon
	9	Romance	Troy	B. Pitt
	10	Drama/Crime	Confidence	E. Burns
Test Set	11	? [Drama/Action]	Saving Private Rian	T. Hanks, M. Damon and E. Burns
	12	? [Action/Romance]	Ocean's Twelve	B. Pitt and M. Damon
	13	? [Crime/Drama]	The Green Mile	T. Hanks

Like most of the eager classifiers, IEAC does not perform well on complex spaces. This is because it generates CARs before the test instance is even known, and the difficulty in this case is in anticipating all the different directions in which it should attempt to generalize its training examples. In order to perform more general predictions, common approaches usually prefer to generalize more frequent disjuncts. This can reduce the performance in complex spaces, where small disjuncts may be important to classify specific instances. Lazy classifiers, on the other hand, generalize the examples exactly as needed to cover a specific test instance.

In lazy associative classification, whenever a test instance is being considered, that instance is used as a filter to remove irrelevant features and examples from the training data. This process automatically reduces the size of the training data, since irrelevant examples are not considered. As a result, disjuncts that are not frequent in the original training data, may become frequent in the filtered training data, providing a better coverage of small disjuncts. To illustrate this process, suppose we want to classify instance 12. As shown in Table 2 only four examples are relevant to this instance. If we set σ_{min} to 0.20 and θ_{min} to 0.66, then $\mathcal{M}$ will be composed of the following CARs:

1. actor=M. Damon $\xrightarrow{0.75, 1.00}$ label=Action
2. actor=M. Damon $\xrightarrow{0.50, 0.67}$ label=Crime
3. actor=B. Pitt $\xrightarrow{0.25, 1.00}$ label=Romance

According to Equation 2, $f(\text{Action})=1.00$, $f(\text{Crime})=0.45$ and $f(\text{Romance})=0.33$, and for $\delta_{min} = 0.66$, label combination *Action/Romance* is predicted. We refer to this classifier as ILAC (*independent lazy associative classifier*).

4.2 Correlated Classifiers

Labels in multi-label problems are often correlated, and as we will see in our experiments, this correlation can be helpful for improving classification performance. In this section we describe CLAC (*correlated lazy associative classifier*), which, unlike IEAC and ILAC, explicitly explores interactions among labels. The classification model is composed of *multi-label class association rules* (MCARs), which are defined next.

DEFINITION 2. [MULTI-LABEL CLASS ASSOCIATION RULES] MCARs are a special type of association rules of the form $\mathcal{X} \cup \mathcal{F} \xrightarrow{\sigma, \theta} c_i$, where $\mathcal{F} \subseteq (\mathcal{C} - c_i)$. A valid MCAR has σ and θ greater than or equal to the corresponding thresholds, σ_{min} and θ_{min}.

The model is built iteratively, following a greedy heuristic called *progressive label focusing*, which tries to find the best label combination by making locally best choices. In the first iteration, $\mathcal{F} = \emptyset$, and a set of MCARs ($\mathcal{M}_1$) of the form $\mathcal{X} \rightarrow c_i$ is generated. Based on $\mathcal{M}_1$, label l_1 is assigned to the test instance. In the second iteration, l_1 is treated as a new feature and thus $\mathcal{F} = \{l_1\}$. A set of MCARs of the form $\mathcal{X} \cup \{l_1\} \rightarrow c_i$ ($\mathcal{M}_2$) is generated, and $\mathcal{M}_2$ is then used to assign label l_2 to the test instance. This process iterates until no more MCARs are generated. The basic idea is to progressively narrow the search space for MCARs as labels are being assigned to the test instance.

Consider again the example in Table 1, and suppose that we want to classify instance 13. The first step is to filter the training data according to the features in instance 13. The filtered training data is shown in Table 3, and if we set σ_{min} to 0.20 and θ_{min} to 0.66, then the corresponding model (i.e., $\mathcal{M}_1$) is composed of the following MCAR:

Table 2. Filtering according to Instance 12

Id	Label	Actors
6	Crime/Action	M. Damon
7	Crime/Action	M. Damon
8	Action/Romance	M. Damon
9	Romance	B. Pitt

Table 3. Filtering according to Instance 13

Id	Label	Actors
1	Comedy/Romance	T. Hanks
2	Drama/Romance	T. Hanks
3	Drama/Crime	T. Hanks
4	Drama/Crime	T. Hanks

Table 4. Filtering according to Instance 13 and Label *Drama*

Id	Label	Actors
2	Drama/Romance	T. Hanks
3	Drama/Crime	T. Hanks
4	Drama/Crime	T. Hanks

1. actor=T. Hanks $\xrightarrow{0.75,0.75}$ label=Drama

Label *Drama* is assigned to instance 13, and now this label is considered a new feature. The training data is filtered again, as shown in Table 4. The corresponding model (i.e., $\mathcal{M}_2$) is composed of the following MCAR:

1. actor=T. Hanks $\wedge$ label=Drama $\xrightarrow{0.67,0.67}$ label=Crime

Thus, label *Crime* is also assigned to instance 13, and since no more MCARs can be generated, the process stops. In summary, labels *Romance* and *Crime* are equaly related to feature *actor=T. Hanks* (see Table 3). Therefore it may be difficult to distinguish these two labels based solely on this feature. However, if we are confident that a movie starred by *T. Hanks* should be classified as *Drama*, then it is more likely that this movie should be classified as *Crime*, rather than *Romance* (as seen in Table 4).

5   Experimental Evaluation

Three datasets were used in our experiments. The first dataset, which is called ACM-DL (first level), was extracted from the first level of the ACM Computing Classification System (http://portal.acm.org/dl.cfm/), comprising a set of 81,251 documents labeled using the 11 first level categories of ACM. The second dataset, ACM-DL (second level) contains the same set of documents of ACM-DL (first level), but these documents are labeled using the 81 second level categories. In both datasets, each document is described by its title and abstract, citations, and authorship, resulting in a huge and sparse feature space. The third dataset, YEAST [5], is composed of a set of 2,417 genes. Each gene is described by the concatenation of micro-array expression data and phylogenetic profile, and is associated with a set of functional classes. There are 14 possible class labels, and the average number of labels for each gene is 4.24.

Figure 1 shows the number of instances associated with each label combination size for each dataset. The YEAST dataset presents very large combinations of labels (combinations of 11 labels). Figure 2 shows the association of each pair of labels for each dataset (an association level of 0.8 between labels $\mathcal{A}$ and $\mathcal{B}$, means that 80% of the instances that belong to $\mathcal{A}$, also belong to $\mathcal{B}$).

The experiments were performed on a Linux-based PC with a INTEL PENTIUM III 1.0 GHZ processor and 1.0 GB RAM. In all experiments with the aforementioned

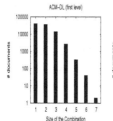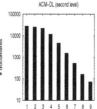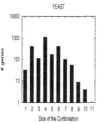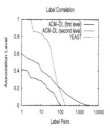

Fig. 1. Number of Instances Associated with each Combination Size

Fig. 2. Association of each Pair of Labels

Table 5. Results for Different Classifiers using the YEAST Dataset

Evaluation	Classifier					
Criterion	BOOSTEXTER	ADTBOOST.MH	RANK-SVM	IEAC	ILAC	CLAC
h	0.220	0.207	0.196	0.203	0.191	**0.179**
r	0.186	–	0.163	0.178	0.164	**0.150**
o	0.278	0.244	0.217	0.232	**0.213**	**0.213**

Table 6. Results for Different Classifiers using the ACM-DL Datasets

	First Level				Second Level			
Evaluation	Classifier							
Criterion	RANK-SVM	IEAC	ILAC	CLAC	RANK-SVM	IEAC	ILAC	CLAC
h	0.225	0.295	0.222	**0.187**	0.327	0.419	0.319	**0.285**
r	0.194	0.276	0.216	**0.179**	0.299	0.378	0.294	**0.273**
o	**0.244**	0.304	**0.238**	**0.238**	0.348	0.427	**0.331**	**0.331**

datasets, we used 10-fold cross-validation and the final results of each experiment represent the average of the ten runs. We used three evaluation criteria that were proposed in [5]: Hamming Loss (h), Ranking Loss (r) and One-Error (o). All the results to be presented were found statistically significant based on a t-test at 5% significance level.

The proposed classifiers, IEAC, ILAC and CLAC are compared against boosting-style classifiers BOOSTEXTER [5] and ADTBOOST.MH [2], and the multi-label kernel method RANK-SVM [3]. We believe that these approaches are representative of some of the most effective multi-label methods available. For BOOSTEXTER and ADT-BOOST.MH, the number of boosting rounds was set to 500 and 50, respectively. For RANK-SVM, polynomial kernels of degree 10 were used. For IEAC, ILAC and CLAC, σ_{min}, θ_{min} and δ_{min} were set to 0.01, 0.90 and 0.25, respectively.

Best results (including statistical ties) on each criterion are shown in bold face. Table 5 shows results obtained using the YEAST dataset, which is considered complex, with strong dependencies among labels. CLAC provide gains of 24% in terms of one-error, considering BOOSTEXTER as the baseline. The reason is that the simple decision function used by BOOSTEXTER is not suitable for this complex dataset. Also, the classification models employed by ILAC and CLAC are able to explore many more associations than the model induced by ADTBOOST.MH. CLAC performs much better than RANK-SVM since CLAC is able to explore dependencies between labels.

In the next set of experiments we compare IEAC, ILAC and CLAC, against RANK-SVM using the ACM-DL dataset (first and second levels). As can be seen, CLAC and ILAC are always superior than their eager counterpart, IEAC. RANK-SVM and ILAC shown competitive performance, and CLAC is the best performer in the ACM-DL datasets. To verify if the association between labels was properly explored by CLAC, we checked if the explicitly correlated categories shown in the ACM Computing Classification System (http://www.acm.org/class/1998/overview.html) were indeed used. We verified that some of these explicitly correlated categories often appear together in the predicted label combination (i.e., *Files* and *Database Management*, or *Simulation/Modeling* and *Probability/Statistics*). We further verified that some of the associated labels appear more frequently in the predictions performed by CLAC than was observed in the predictions of the other classifiers.

6   Conclusions and Future Work

In this paper we propose a novel associative classification approach for multi-label classification. The model is induced in an instance-based fashion, in which the test instance is used as a filter to remove irrelevant features from the training data. Then a specific model is induced for each test instance, providing a much better coverage of small disjuncts. Also, the proposed approach properly explores the correlation among labels by employing a greedy heuristic called progressive label focusing, which allows the presence of multiple labels in the antecedent of the rule. Experimental results underscore the benefits of covering small disjuncts (i.e., lazy model induction) and exploring correlated labels (i.e., progressive label focusing). As future work, we intend to further explore correlated labels by also allowing the presence of multiple labels in the consequent of the rule.

References

1. Boutell, M., Luo, J., Shen, X., Brown, C.: Learning multi-label scene classification. Pattern Recognition 37, 1757–1771 (2004)
2. Comité, F., Gilleron, R., Tommasi, M.: Learning multi-label alternating decision trees from texts and data. In: Proc. of the Intl. Conf. on Machine Learning and Data Mining in Pattern Recognition, pp. 35–49 (2003)
3. Elisseeff, A., Weston, J.: A kernel method for multi-labelled classification. Advances in Neural Information Processing Systems 14, 681–687 (2001)
4. Holte, R., Acker, L., Porter, B.: Concept learning and the problem of small disjuncts. In: Proc. of the Intl. Joint Conf. on Artificial Intelligence, pp. 813–818 (1989)
5. Schapire, R., Singer, Y.: Boostexter: A boosting-based system for text categorization. Machine Learning 39, 135–168 (2000)
6. Thabtah, F., Cowling, P., Peng, Y.: Multiple labels associative classification. Knowledge and Information Systems 9, 109–129 (2006)
7. Zhang, M., Zhou, Z.: Ml-knn: A lazy learning approach to multi-label learning. Pattern Recognition 40, 2038–2048 (2007)

Visual Exploration of Genomic Data

Michail Vlachos[1], Bahar Taneri[2], Eamonn Keogh[3], and Philip S. Yu[1]

[1] IBM T.J. Watson Research Center, Hawthorne, NY, USA
[2] Scripps Institute of Oceanography, UCSD, CA, USA
[3] University of California Riverside, CA, USA

Abstract. In this study, we present methods for comparative visualization of DNA sequences in two dimensions. First, we illustrate a transformation of gene sequences into numerical trajectories. The trajectory visually captures the nucleotide content of each sequence, allowing for fast and easy visualization of long DNA sequences. Then, we project the relative placement of the trajectories on the 2D plane using a spanning-tree arrangement method, which allows the efficient comparison of multiple sequences. We demonstrate with various examples the applicability of our technique in evolutionary biology and specifically in capturing and visualizing the molecular phylogeny between species.

1 Introduction

Identification of evolutionary distances among species has always been a topic of interest to researchers. Several different methods have been used to identify the evolutionary relationships between species, including taxonomic, phylogenetic analyses, geometric morphometric data analysis. In the post-genome era, more accurate evolutionary views have been reached using DNA sequence analyses of species [6,9].

In this work, we also provide a molecular vision of evolution through comparison and visualization of DNA sequences. Using comparative mitochondrial DNA analysis, we illustrate the evolutionary distances among various mammalian species. Mitochondrial DNA (mtDNA) analyses have been proven useful in establishing phylogeny among a wide range of species [2,7,8]. We achieve our goal by mapping the DNA nucleotide sequences into 2-dimensional trajectories. The purpose of this conversion is to facilitate the quick visual comparison between long DNA sequences. We evaluate the affinity between the resulting DNA trajectories by employing an elastic warping distance function. Our empirical results on mitochondrial DNA from various species, suggest that the utilized distance measure can reflect with great accuracy the divergence point between species. Finally, for visually comparing the evolutionary distance between the DNA trajectories we present a spanning-tree-based mapping technique. The technique arranges the objects on the 2-dimensional space, while retaining as much of the original structure as possible. We depict the enhanced visualization power that can be induced from the proposed mapping technique. All our results

J.N. Kok et al. (Eds.): PKDD 2007, LNAI 4702, pp. 613–620, 2007.
© Springer-Verlag Berlin Heidelberg 2007

are validated with freely available genomic data obtained from Genbank [1], and corroborate the current prevalent views on evolutionary biology.

Previous work on DNA visualization has appeared in [3,4], but the techniques pose limitations regarding the visual comparison between multiple DNA sequences. A technique that allows the comparison between different sequences in terms of their common subsequences has been presented in [1]. Our method is unique in that, it not only provides a visual representation of the nucleotide sequences, but also it deciphers the comparative phylogenetic distances among different species.

In the sections that follow we present a DNA conversion technique into trajectories and later on we demonstrate the spanning-tree mapping technique. The final section contains the empirical evaluation of both methods using mammalian DNA sequences.

2   Converting DNA to Trajectories

Visual comparison of DNA symbol strings can be particularly troublesome to perform, because typical DNA datasets contains thousands of symbols. Humans cannot easily compare or visually represent bulk of text; our brains are much more efficient at comparing lines or shapes. Therefore, a technique for converting a DNA string into a low dimensional shape, can significantly enhance our ability of interpreting and comparing very long DNA sequence data. Given a string of length n drawn from the alphabet A,T,C,G, which we will denote as DNA, we wish to convert it to a two-dimensional trajectory of length $n + 1$, which we denote as T. We can use the following rule to build the trajectory vector: $T(i) = T(i - 1) + \mathbf{V}$, where $\mathbf{V}$ is a basis vector constructed as follows:

$$\mathbf{V} = \begin{cases} [0\ 1], \text{ if } DNA(i) = \text{A} \\ [1\ 0], \text{ if } DNA(i) = \text{T} \\ [0\ \text{-}1], \text{ if } DNA(i) = \text{C} \\ [\text{-}1\ 0], \text{ if } DNA(i) = \text{G}. \end{cases}$$

That is, starting from an initial reference point we will direct the trajectory on the relevant direction (up, down, left or right) based on the currently examined symbol. For example, if the sequence contains many A symbols then it will demonstrate a predominantly upward movement. Below, we demonstrate an arithmetic example of the trajectory construction.

Example: Suppose that the starting position $T(1) = [0\ 0]$. Then, for the DNA string AATCG, we get the trajectory vector $\{[0\ 0],[0\ 1],[0\ 2],[1\ 2],[1\ 1],[0\ 1]\}$.

2.1   Comparing Trajectories

In order to quantify the similarity between the resulting trajectories we utilize a warping distance, which can allow for a flexible matching between the DNA

---

[1] http://www.ncbi.nlm.nih.gov/Genbank/

trajectories, supporting local compressions and decompressions. The warping distance can be seen as a real-valued counterpart of the *Edit Distance*, which is customarily used for comparing DNA transcripts.

Suppose that Q and T are the trajectories that we wish to compare. If $Q = (Q_1, Q_2, \ldots, Q_n)$ and $Head(Q) = (Q_1, Q_2, \ldots, Q_{n-1})$ (and similarly for a sequence T) then the recursive equation to provide then warping distance between Q and T is:

$$DTW(Q,T) = D(Q_n, T_n) + min \begin{cases} DTW(Head(Q), Head(T)) \\ DTW(Head(Q), T) \\ DTW(Q, Head(T)) \end{cases}$$

where $D(\cdot, \cdot)$ is the distance between two points of the sequence. Typically, D is the Euclidean distance. The warping distance can be computed using a well known dynamic programming [12]. In Fig. 1 we can see the flexibility of matching that can be achieved between trajectories when utilizing the warping distance. On the left side we demonstrate the mapping between the human and the chimpanzee trajectories, which were derived from their respective mitochondrial DNA. On the right side, the matching between the human and the bear mtDNA is illustrated.

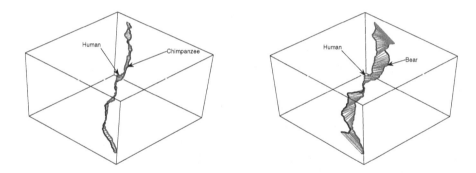

Fig. 1. Matching of DNA trajectories using DTW. Left: Human vs Chimpanzee, Right: Humans vs Bear.

Even though the Warping distance can accommodate a flexible matching between the resulting DNA trajectories, it does not obey the triangle inequality (unlike the Euclidean distance). We will utilize this fact to motivate extensions on the triangulation mapping technique that is presented subsequently.

3   Spanning-Tree Visualization

Given a set of pairwise distances between objects we are seeking a way of visualizing their relationship on two dimensions, while retaining as much of the original structure as possible. We revisit a mapping technique proposed by Lee, *et al.*, in [5], which utilizes the Minimum-Spanning-Tree (MST) and a triangulation method for preserving 2 distances per object on the two-dimensional space. The first distance preserved is the distance to the nearest neighbor of every object. The second distance can either be different for every object (e.g. its 2NN), or it can be the distance to a reference point. The latter option is the one that we adapt, which creates a powerful visualization technique that not only allows for preservation of Nearest Neighbors distances (local structure), but additionally retains distances with respect a single reference point, giving the option for global data view using that object as a pivot.

Once the MST is calculated the mapping on the 2D space can commence from any point/object that the user designates and the MST tree is mapped either in a breadth-first-search (BFS) or depth-first-search (DFS) manner. In this work we utilize a BFS mapping. We illustrate how the mapping works with a running example.

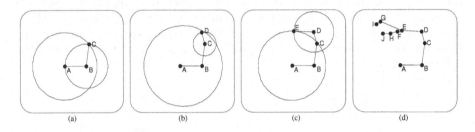

Fig. 2. 2D mapping of objects using spanning-tree and triangulation

Suppose the first two points (A and B) of the MST are already mapped, as shown in Fig. 2 (a). Let's assume that the second distance preserved per object is the distance with respect to a reference point which in our case is the first point. The third point is mapped at the intersection of circles centered at the reference points. The circles are centered at A and B with radii of $D(A, C)$ - the distance between points A and C- and $D(B, C)$, respectively. Due to the triangle inequality, the circles either intersect at 2 positions or at tangent. Any position on the circles' intersection will retain the original distances towards the two reference points. The position of point C is shown in Fig. 2 (a). The fourth point is mapped at the intersection of circles centered at A and C (Fig. 2 (b)) and the fifth point is mapped similarly (Fig. 2 (c)). The process continues until all the points of the MST at positioned on the 2D plane and the final result is shown in Fig. 2 (d).

3.1   Extensions for Non-metric Distances

The triangulation method proposed by Lee, *et al.*, is only applicable for metric distances when the circles around the reference points are guaranteed to intersect. Recall that, the Warping Distance used to quantify the distance between the DNA trajectories does no obey the triangle inequality. This means that the reference circles may not necessarily intersect. We highlight necessary extensions to the triangulation method that allows its proper usage under non-metric distances.

We can identify two cases for the non-intersecting circles:

1. **Case 1:** One circle encloses each other,
2. **Case 2:** The two circles are disjoint and not enclosed within one another

For each of these cases we need to identify the position where to position an object with respect to the two circles, so that the object is mapped as close as possible to the circumference of both circles. In order words, we need to identify the locus of points that minimize the sum of distances to the perimeters of two circles.

One can show that the desired locus always lies on the line connecting the centers of the two circles. Case 1 is shown in Figure 3, and we can identify two sub-cases.

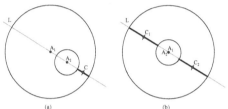
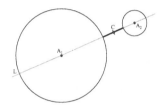

Fig. 3. Circles enclosed within one another   **Fig. 4.** Circles that are disjoint

 – When the two circles have disjoint centers, then the point that minimizes the sum of distances to both perimeters, is point C on Fig. 3 (a), which lies on the line L connecting the two centers, and midway on the line segment with vertices the intersection of L with the circles' perimeters.
 – In the case when the two circles have common centers, then there exist two points that satisfy the distance minimization property as shown in Fig. 3 (b).

Case 2 can be resolved in a similar way, which is shown in Fig. 4.

With the addition of the above rules, we can now discover the mapping positions of the objects on the two-dimensional plane, so that the original pairwise distances are satisfied as well as possible using the spanning-tree triangulation method.

4   Experimental Results

We demonstrate the usefulness of the proposed techniques on comparative molecular phylogenetic studies via visualization of mitochondrial DNA sequences. We utilize publicly available mtDNA obtained from Genbank (see Table 1). All datasets used in this paper along with supplementary material can be found at the project website [2].

---

[2] http://www.cs.ucr.edu/~mvlachos/VizDNA/

Table 1. Example from subset of the mitochondrial DNA data used for our visualizations

Name	Species	mtDNA bps
Indian Elephant	Elephas Manimus Indicus	16800
African Elephant	Loxodonta Africana	16859
Blue Whale	Balaenoptera Musculus	16402
Finback Whale	Balaenoptera Physalus	16398
Hippopotamus	Hippopotamus Amphibius	16407
Human	Homo Sapiens	16571
Chimpanzee	Pan Troglodytes	16554
Pygmy chimpanzee	Pan Paniscus	16563
Dog	Canis familiaris	16727
American Bear	Ursus americanus	16841
Polar Bear	Ursus maritimus	17017

Mitochondrial DNA is passed on only from the mother during sexual reproduction, making the mitochondria clones. This means that there are minor changes in the mtDNA from generation to generation, unlike nuclear DNA which changes by 50% each generation. Therefore, mtDNA has a long *memory*. Each mtDNA string consists of approximately 16000 symbols (with mtDNA of humans being 16,571 nucleotides long, and all other mammals mtDNA are within plus or minus 1-3% of this).

For our first experiment we utilize mtDNA from *Homo sapiens* and other primates. Figure 5 illustrates the spanning-tree mapping for 8 species. Our results are in general agreement with current evolutionary views. We also observe that not only the mapping is very accurate with regard to the evolutionary distance of the species, but the mapping preserves the clustering between the original groups that the various primates belong to. Specifically, Human, Pygmy chimpanzee, Chimpanzee and Orangutan belong to the *hominidae* group, the Gibbon to the *hylobatae* group and the Baboon and the Macaque to the *cercopithicae*.

Adjacent to this mapping, we provide a spanning-tree visualization that utilizes the most commonly used Euclidean distance instead of the Warping distance. One can observe that the use of the Euclidean distance introduces errors, such as mapping the gibbon closer to the human rather than to the orangutan, which is incorrect. Human and orangutan divergence took place approximately 11 million years ago. Whereas, gibbon and human divergence occurred approximately 15 million years ago [10]. According to the same source, gorilla divergence occurred about 6.5 million years ago and chimpanzee divergence took place about 5.5 million years ago.

For our second example in Figure 6 we utilize a larger mammalian dataset and again take the human as the referential point. On this plot we use the formal names of the species (instead of their common names) and we also overlay on the figure the DNA trajectory of the respective mtDNA sequence. Again, the spanning-tree technique exhibits a very strong visualization capacity, particularly in unveiling the similarities and connections between the different species. For example, one can notice the great similarity of the hippopotamus with the

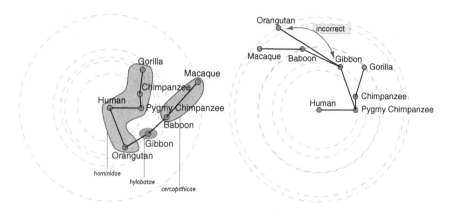

Fig. 5. Visualization of humans and other primates. Left: Using the Warping distance, Right: Using Euclidean distance to compare the DNA trajectories. Various mapping errors are indicated on the figure.

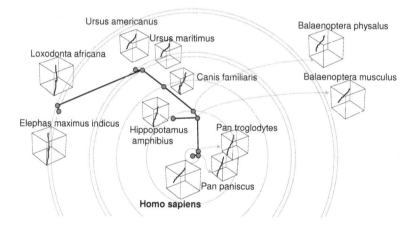

Fig. 6. Evolutionary visualization of mammalian species with respect to the human

whales. The hippopotami are indeed closely related to whales than to any other mammals. Whales and hippopotami diverged 54 million years ago, whereas the whale/hippopotamus group parted from the elephants about 105 million years ago. The group that includes hippopotami and whales/dolphins, but excludes all other mammals above is called Cetartiodactyla [11].

5   Conclusions

We presented techniques that allow the effective visualization and comparison between DNA sequences, by transforming them into trajectories and mapping them on the two-dimensional plane. The mapping technique can have many biomedical applications, including advancement of diagnostic techniques for cancer data. This technology could both be applied for distinguishing cancer transcripts from normal ones, and for the identification of different cancer stages. Future direction of this work, includes expansion of our technique to transcriptome-wide screens of cancer transcripts in human and mouse transcriptomes.

References

1. Apostolico, A., Gong, F., Lonardi, S.: Verbumculus and the discovery of unusual words. Journal of Computer Science and Technology 19(1), 22–41 (2003)
2. Auch, A., Henz, S., Holland, B., Goker, M.: Genome BLAST distance phylogenies inferred from whole plastid and whole mitochondrion genome sequences. BMC Bioinformatics 7, 350 (2006)
3. Chang, H.T., Lo, N.-W., Lu, W.C., Kuo, C.J.: Visualization and Comparison of DNA Sequences by Use of Three-Dimensional Trajectories. In: First Asia-Pacific Bioinformatics Conference (APBC) (2003)
4. Herisson, J., Ferey, N., Gros, P., Gherbi, O.M.R.: 3D visualization and virtual exploration of genomic sequences. Data Science Journal 4, 82–91 (2005)
5. Lee, R., Slagle, J., Blum, H.: A Triangulation Method for the Sequential Mapping of Points from N-Space to Two-Space. IEEE Transactions on Computers C-26(3), 288–292 (1977)
6. Lockwood, C., Kimbel, W., Lynch, J.: Morphometrics and hominoid phylogeny: Support for a chimpanzee-human clade and differentiation among great ape subspecies. Proc. Natl. Acad. Sci. USA 101(13), 4356–4360 (2004)
7. Orrell, T., Carpenter, K.: A phylogeny of the fish family Sparidae (porgies) inferred from mitochondrial sequence data. Mol Phylogenet Evol. 32(2), 425–434 (2004)
8. Royyuru, A.K., Alexe, G., Platt1, D., Vijaya-Satya, R., Parida, L., Rosset, S., Bhanot, G.: Inferring Common Origins from mtDNA. In: Research in Computational Molecular Biology pp. 246–247 (2006)
9. Ruvolo, M.: Comparative primate genomics: the year of chimpanzee. Curr. Opin. Genet Dev. 14(6), 650–656 (2004)
10. Stauffer, R., Walker, A., Ryder, O., Lyons-Weiler, M., Hedges, S.: Human and Ape Molecular Clocks and Constraints on Paleontological Hypotheses. The Journal of Heredity 92(6), 469–474 (2001)
11. Ursing, B.M., Arnason, U.: Analyses of mitochondrial genomes strongly support a hippopotamus-whale clade. In: Proc. of the Royal Society of London, Series B, vol. 265, pp. 2251–2255 (1998)
12. Vlachos, M., Hadjieleftheriou, M., Gunopulos, D., Keogh, E.: Indexing Multi-Dimensional Time-Series with Support for Multiple Distance Measures. In: Proc. of SIGKDD (2003)

Association Mining in Large Databases: A Re-examination of Its Measures*

Tianyi Wu[1], Yuguo Chen[2], and Jiawei Han[1]

[1] Department of Computer Science, UIUC
[2] Department of Statistics, UIUC
{twu5,yuguo,hanj}@uiuc.edu

Abstract. In the literature of data mining and statistics, numerous interestingness measures have been proposed to disclose succinct object relationships of association patterns. However, it is still not clear when a measure is truly effective in large data sets. Recent studies have identified a critical property, *null-(transaction) invariance*, for measuring event associations in large data sets, but many existing measures do not have this property. We thus re-examine the null-invariant measures and find interestingly that they can be expressed as a generalized mathematical mean, and there exists a total ordering of them. This ordering provides insights into the underlying philosophy of the measures and helps us understand and select the proper measure for different applications.

1 Introduction

Despite more than a decade of study over association mining, it has been well recognized that traditional association rules may not disclose truly interesting event relationships [2]. For example, mining a market basket data set may find a rule, *"coffee → milk"*, with nontrivial support and high confidence (*e.g.*, 80%), but this does not imply that *buying coffee* and *buying milk* are strongly associated because *milk* itself might be popular. Thus, researchers have proposed various measures as constraints to mine true relationships among events [3,11,10,4].

Many association, correlation, and similarity measures have been proposed for analyzing the relationships among discretized events [6,12]. For example, χ^2 is a typical measure for analyzing correlations among discretized events in statistics [7]. However, it may not be an appropriate measure for analyzing event associations in large transaction databases. Notice in a typical transaction database, a particular item i (*e.g.*, *coffee*) appearing in a transaction T (*i.e.*, $i \in T$) is often a small probability event. Since most transactions do not contain item i, they are *null transactions w.r.t. i*. If the association among a set of events being analyzed is affected by the transactions that contain none of them (*i.e.*, null-transactions), such a measure is unlikely to be of high quality. Recent studies

---

* The work was supported in part by the U.S. National Science Foundation NSF IIS-05-13678, NSF BDI-05-15813, and NSF DMS-05-03981. Any opinions, findings, and conclusions or recommendations expressed in this paper are those of the authors and do not necessarily reflect the views of the funding agencies.

Table 1. Two-event contingency table

	$milk$	$\overline{milk}$	Σ_{row}
$coffee$	mc	$\overline{m}c$	c
$\overline{coffee}$	$m\overline{c}$	$\overline{m}\,\overline{c}$	$\overline{c}$
Σ_{col}	m	$\overline{m}$	total

[12,9,10] have shown that null-(transaction) invariance is critically important for an interestingness measure. We use the following example to illustrate this.

Example 1. In a typical shopping transaction database, a product appearing in a transaction is called an **event**, and a set of products appearing in a transaction is called an **event-set**. Association analysis is to identify interesting (positive or negative) associations among a set of events. It is expected that a particular event happens with a very low probability.

In Table 1, the purchase history of two events *milk* and *coffee* are summarized by their support counts, where, for instance, mc denotes the support of the event-set "*coffee and milk*", *i.e.*, the occurrences of transactions containing them. Table 2 enumerates six data sets in terms of a "flattened" contingency table. We then select six measures: χ^2, *Lift*, *AllConf*, *Coherence*, *Cosine*, *Kulczynski* (denoted as *Kulc* hereafter), and *MaxConf*, and show their results for each data sets. The definitions for the six measures are given in Table 3. *AllConf*, *Coherence*[1], *Cosine*, and *MaxConf*[2] are the only ones that are not sensitive to the number of null-transactions (hence called *null-invariant measures*) among over 20 interestingness measures studied in [12]. *Kulc* is another null-invariant measure proposed in [1]. Two popular but not null-invariant measures, χ^2 and *Lift* [3,6], are listed for comparison.

Let's examine D_1 and D_2, where *milk* and *coffee* are positively associated because mc is considerably greater than $\overline{m}c$ and $m\overline{c}$. The results of the five null-invariant measures show that m and c are strongly positively associated in both data sets. However, *Lift* and χ^2 generate dramatically different values, due to their sensitivity to $\overline{m}\,\overline{c}$. In such cases, since $\overline{m}\,\overline{c}$ is usually huge and unstable, a good interestingness measure should not be affected by it. Similarly, in D_3, the five null-invariant measures correctly show that m and c are strongly negatively associated; whereas *Lift* and χ^2 judge it in an incorrect or controversial way. For D_4, both *Lift* and χ^2 indicate a highly positive association between m and c, whereas the others[3] a neutral association, because $mc : \overline{m}\,\overline{c} = mc : m\overline{c} = 1 : 1$. This means that given the event *coffee*, the probability of the event *milk* is exactly 50% and vice versa. Note that *milk* and *coffee* are statistically independent if and only if $P(mc) = P(m)P(c)$, where $\overline{m}\,\overline{c}$ cannot be ignored. In fact,

---

[1] Notice that *Coherence*(a, b), though introduced lately [10] and defined differently, is essentially the popularly used Jaccard Coefficient [12].

[2] We use *MaxConf* instead of *Confidence* as in [12] to avoid any confusion with the directional "confidence" measure in traditional association rule mining.

[3] The neutral point of *Coherence* is at 0.33 instead of 0.5 (see [9]).

Table 2. Example data sets

Data set	mc	$\overline{m}c$	$m\overline{c}$	$\overline{m}\overline{c}$	χ^2	$Lift$	$AllConf$	$Coherence$	$Cosine$	$Kulc$	$MaxConf$
D_1	10,000	1,000	1,000	100,000	90557	9.26	0.91	0.83	0.91	0.91	0.91
D_2	10,000	1,000	1,000	100	0	1	0.91	0.83	0.91	0.91	0.91
D_3	100	1,000	1,000	100,000	670	8.44	0.09	0.05	0.09	0.09	0.09
D_4	1,000	1,000	1,000	100,000	24740	25.75	0.5	0.33	0.5	0.5	0.5
D_5	1,000	100	10,000	100,000	8173	9.18	0.09	0.09	0.29	0.5	0.91
D_6	1,000	10	100,000	100,000	965	1.97	0.01	0.01	0.10	0.5	0.99

Table 3. Interestingness measure definitions

Measure	Definition	Range	Null-Invariant
$\chi^2(a,b)$	$\sum_{i,j=0,1} \frac{(c(a_i,b_j)\ o(a_i,b_j))^2}{e(a_i,b_j)}$	$[0,\infty]$	No
$Lift(a,b)$	$\frac{P(ab)}{P(a)P(b)}$	$[0,\infty]$	No
$AllConf(a,b)$	$\frac{sup(ab)}{max\{sup(a),sup(b)\}}$	$[0,1]$	Yes
$Coherence(a,b)$	$\frac{sup(ab)}{sup(a)+sup(b)-sup(ab)}$	$[0,1]$	Yes
$Cosine(a,b)$	$\frac{sup(ab)}{\sqrt{sup(a)sup(b)}}$	$[0,1]$	Yes
$Kulc(a,b)$	$\frac{sup(ab)}{2}\left(\frac{1}{sup(a)}+\frac{1}{sup(b)}\right)$	$[0,1]$	Yes
$MaxConf(a,b)$	$max\{\frac{sup(a)}{sup(ab)},\frac{sup(b)}{sup(ab)}\}$	$[0,1]$	Yes

statistical independence requires $\overline{m}\overline{c}$ to be 1000. Therefore, the neutrality of the null-invariant measures does not necessarily suggest statistical independence. ∎

Our subsequent discussions will be focused only on the five null-invariant measures in Table 3. Although a comparative study of interestingness measures has been done in [12], there are still many important yet unanswered questions on the null-invariant measures, such as *"Are there inherent relationships among them?"*, and *"Which measure is better for evaluating interesting associations among small probability events?"*. This motivates us to conduct an in-depth study to weave a well-organized picture. Specifically, we make the following contributions: (i) We show that there exists a total ordering among these measures based on a mathematical analysis. This not only explains their inherent relationships and underlying philosophy, but also provides a unified view of association analysis in large transaction datasets. And, (ii) we propose a generalized measure to handle multiple events under the unified framework. The rest of the paper is organized as follows. Section 2 presents the re-examination and generalization. Section 3 overviews the related work. Finally, Section 4 concludes the study.

2 A Re-examination of Null-Invariant Measures

2.1 Inherent Ordering Among the Measures

Given two events a and b, the support of a is denoted as $sup(a)$, and we use a for $sup(a)$ when there is no ambiguity. Let M be any of the five null-invariant

measures. From Table 3, we immediately have the following fundamental properties: **(P1)** $M \in [0, 1]$; **(P2)** M monotonically increases with $sup(ab)$ when $sup(a)$
and $sup(b)$ remain unchanged; and it monotonically decreases with $sup(a)$ (or
$sup(b)$) when $sup(ab)$ and $sup(b)$ (or $sup(a)$) stay the same; **(P3)** M is symmetric
under event permutations; and **(P4)** M is invariant to scaling, *i.e.*, multiplying
a scaling factor to $sup(ab)$, $sup(a)$, and $sup(b)$ will not affect the measure. These
four properties justify the "conventional wisdom" about association analysis in
large databases, and therefore are desirable. Specifically, $P1$ states that the value
domain of M is normalized so that 0 indicates no event co-occurrence and 1 indicates that events always appear together[4]. $P2$ is consistent with the basic
intuition that more co-occurrences would result in greater measure values, and
vice versa. $P3$ and $P4$ show the robustness of M.

Despite such "conventional wisdom", there are subtle cases that cannot be
resolved by our common sense. Turn to D_5 and D_6 in Table 2, where m and c have
unbalanced conditional probabilities – $P(m|c) > 0.9$ and $P(c|m) < 0.1$. *AllConf*,
Coherence, and *Cosine* view both as negatively associated, *Kulc* is neutral, and
MaxConf claims strongly positive associations. One may ask, "*Which measure
intuitively reflects the true relationship?*" Unfortunately, there is no commonly
agreed judgment for such cases due to the "*balanced*" skewness of the data.

Interestingly, we show that there is a total ordering that discloses the inherent
relationships among the measures and thus may help the user's decision-making.
To begin with, we rewrite the definitions in Table 3 into the form of conditional
probabilities in Table 4 ($P(a|b) = sup(ab)/(sup(ab)+sup(\bar{a}b))$). The rewriting of
Coherence need the assumption that $sup(ab) \neq 0$, and for simplicity, we assume
that all these measures are equal to 0 when $sup(ab) = 0$.

Table 4. Null-invariant measures defined using conditional probabilities

Measure	Definition	Exponent
$AllConf(a, b)$	$\min\{P(a|b), P(b|a)\}$	$k \to -\infty$
$Coherence(a, b)$	$(P(a|b)^{-1} + P(b|a)^{-1} - 1)^{-1}$	$k = -1$
$Cosine(a, b)$	$\sqrt{P(a|b)P(b|a)}$	$k \to 0$
$Kulc(a, b)$	$(P(a|b) + P(b|a))/2$	$k = 1$
$MaxConf(a, b)$	$\max\{P(a|b), P(b|a)\}$	$k \to +\infty$

Following from the rewritten definitions, we generalize all five measures using
the mathematical generalized mean [7]. Each of them can be represented by the
generalized mean of the two conditional probabilities $P(a|b)$ and $P(b|a)$ as

$$\mathbb{M}^k(P(a|b), P(b|a)) = \left(\frac{P(a|b)^k + P(b|a)^k}{2} \right)^{1/k}, \tag{1}$$

---

[4] The only exception is that $MaxConf(a, b) = 1$ may not indicate that a and b always
co-occur.

where $k \in (-\infty, +\infty)$ is the exponent of the generalized mean. As in Table 4, each measure can be generalized to Eq. (1) with the corresponding exponent.

Proof of Correctness. We first prove $AllConf(a,b) = \lim_{k \to -\infty} \mathbb{M}^k(P(a|b),$ $P(b|a))$. Without loss of generality, let's assume that $P(a|b) \leq P(b|a)$. The proof follows from $\lim_{k \to -\infty} \mathbb{M}^k(P(a|b), P(b|a)) = \lim_{k \to -\infty} \left(\frac{1+(P(b|a)/P(a|b))^k}{2} \right)^{1/k} P(a|b) =$ $P(a|b) = AllConf(a,b)$. The proof for $MaxConf$ has a similar argument. For $Cosine$, let $x = P(a|b)/P(b|a)$. We have $\lim_{k \to 0} \ln \left(\frac{x^k+1}{2} \right)^{1/k} = \ln(x^{1/2})$, because

$\lim_{k \to 0} \ln \left(\frac{x^k+1}{2} \right)^{1/k} = \lim_{k \to 0} \frac{\ln \left(\frac{x^k+1}{2} \right)}{k} = \lim_{k \to 0} \frac{\frac{1}{2} x^k \ln x}{(x^k+1)/2} = \frac{1}{2} \ln x$. Therefore, $Cosine(a,b)$ $= \lim_{k \to 0} \mathbb{M}^k(P(a|b), P(b|a))$. The proof for $Kulc(a,b) = \mathbb{M}^1(P(a|b), P(b|a))$ is trivial. For $Coherence$ however, the equation $Coherence(a,b) = \mathbb{M}^{-1}(P(a|b), P(b|a))$ does not hold. In fact, we have $Coherence(a,b) = (2/\mathbb{M}^{-1} - 1)^{-1}$. For simplicity, we define a new measure $Coherence' = \mathbb{M}^{-1}(P(a|b), P(b|a))$ as a replacement of $Coherence$ in our following discussions. This is a reasonable replacement because $Coherence'$ preserves the ordering of $Coherence$; that is, $Coherence'(a_1, b_1) \leq$ $Coherence'(a_2, b_2) \Leftrightarrow Coherence(a_1, b_1) \leq Coherence(a_2, b_2)$. ∎

All five measures can be expressed nicely as the generalized mean of $P(a|b)$ and $P(b|a)$ except that $Coherence$ (or $Jaccard\ Coefficient$) need a order-preserving transformation. The generalization to $\mathbb{M}^k(P(a|b), P(b|a))$ (note that these measures only differ in terms of the exponent k) gives us two implications, summarized into the following lemmas.

Lemma 1. For any $k \in (-\infty, +\infty)$, $\mathbb{M}^k(P(a|b), P(b|a))$ always satisfies the fundamental properties P1–P4 and the null-invariance property.

PROOF. Both $P(a|b)$ and $P(b|a)$ have range $[0,1]$, so their mean must have the same range. Also, $\mathbb{M}^k(P(a|b), P(b|a))$ is monotone $w.r.t.$ $sup(a)$, $sup(b)$, and $sup(ab)$, and is invariant to event permutation, scaling, and null-transactions. ∎

Lemma 2. Given any two events a and b, we have

$$AllConf(a,b) \leq Coherence'(a,b) \leq Cosine(a,b) \leq Kulc(a,b) \leq MaxConf(a,b).$$
(2)

PROOF. Given any exponents k and k' $(k < k')$, we have $\mathbb{M}^k(a,b) \leq \mathbb{M}^{k'}(a,b)$ [7], where the equality holds if and only if $P(a|b) = P(b|a)$. ∎

These two lemmas provide insights into both sides of the coin. The first one provides a general justification to the *common*, desirable properties of the null-invariant association measures, whereas the second lemma presents an organized picture of the *differences* between them. The total ordering of the measures clearly exhibits their relationships. First, higher-order (*i.e.*, with larger k) measures provides an upper-bound to lower-order (*i.e.*, with smaller k) measures. Therefore, given a fixed interestingness threshold (*e.g.*, 0.9), the patterns output by a higher-order measure must be a superset of those by a lower-order one. This is helpful to association pattern mining, in that computationally expensive measures such as *Cosine* that involves square root computation, is bounded by computationally cheaper measures like *Kulc*, which can be pushed deep into the mining process. Intuitively, a lower-order measure is more strict (*i.e.*, prune more patterns), because a small k tends to mitigate the impact of the larger one of the two conditional probabilities, whereas a large k tends to aggravate it.

While the generalized mean represents a family of null-invariant measures, there is no universally accepted one for association analysis in large databases, because no particular value of k is generally better. Thus, an appropriate value of k should be determined on a case-by-case basis. It is worth noticing that each of the measures being examined is a special case in the whole spectrum of exponent k. That is, *AllConf* $(k \to -\infty)$ and *MaxConf* $(k \to +\infty)$ correspond to the minimum and maximum of the conditional probabilities, whereas *Coherence'* $(k = -1)$, *Cosine* $(k \to 0)$, and *Kulc* $(k = 1)$ correspond to the *harmonic mean*, *geometric mean*, and *arithmetic mean* of the conditional probabilities.

2.2 Multiple Events

In this subsection, we extend these measures to multiple events. In order to preserve the fundamental properties and take the "generalized mean" approach for balancing conditional probabilities, we have the following definition.

DEFINITION 1. (**Generalized Association Measure**) Let X be an event-set containing n $(n \geq 2)$ events $\{a_1, a_2, \cdots, a_n\}$. The generalized measure is

$$\mathbb{M}^k(X) = \mathbb{M}^k\left(P(a_2 \cdots a_n | a_1), \cdots, P(a_1 \cdots a_{n-1} | a_n)\right)$$
$$= \sqrt[k]{\frac{sup(X)^k}{n}\left(\frac{1}{sup(a_1)^k} + \cdots + \frac{1}{sup(a_n)^k}\right)}.$$

∎

The generalized association measure is the generalized mean of the conditional probabilities of all events. The total ordering still applies to this extension in that the smaller k is, the smaller result the measure will produce. It is worth mentioning that *AllConf* has been defined [10,9] on more than two events, which also matches this definition.

2.3 Empirical Evaluation

We choose the DBLP[5] data set for our empirical evaluation. We extract papers from several selected data mining and database conferences including *KDD*, *SIGMOD*, and *VLDB* in recent 10 years and generate a transaction database. We show in Table 5 ten typical skewed pairs of productive authors with at least 10 papers and rank them according to their number of joint papers (*i.e.*, $sup(ab)$). While *AllConf* and *MaxConf* have a straightforward philosophy, their results are omitted and we list the measure value of the other three measures,

---

[5] http://www.informatik.uni-trier.de/~ley/db/

Table 5. Experiment on DBLP data set

ID	Author a	Author b	$sup(ab)$	$sup(a)$	$sup(b)$	Coherence	Cosine	Kulc
1	Hans-Peter Kriegel	Martin Ester	28	146	54	0.163 (2)	0.315 (7)	0.355 (9)
2	Michael Carey	Miron Livny	26	104	58	0.191 (1)	0.335 (4)	0.349 (10)
3	Hans-Peter Kriegel	Joerg Sander	24	146	36	0.152 (3)	0.331 (5)	0.416 (8)
4	Christos Faloutsos	Spiros Papadimitriou	20	162	26	0.119 (7)	0.308 (10)	0.446 (7)
5	Hans-Peter Kriegel	Martin Pfeifle	18	146	18	0.123 (6)	0.351 (2)	0.562 (2)
6	Hector Garcia-Molina	Wilburt Labio	16	144	18	0.110 (9)	0.314 (8)	0.500 (4)
7	Divyakant Agrawal	Wang Hsiung	16	120	16	0.133 (5)	0.365 (1)	0.567 (1)
8	Elke Rundensteiner	Murali Mani	16	104	20	0.148 (4)	0.351 (3)	0.477 (6)
9	Divyakant Agrawal	Oliver Po	12	120	12	0.100 (10)	0.316 (6)	0.550 (3)
10	Gerhard Weikum	Martin Theobald	12	106	14	0.111 (8)	0.312 (9)	0.485 (5)

Coherence, Cosine, and *Kulc*, for each pair and its rank in the parenthesis to demonstrate their similarities and differences.

It can be seen from the support in the table that at least 3 pairs of authors (ID = 5, 7, 9) demonstrate a relationship of the "advisor-advisee" style because $sup(a) \gg sup(b)$ and b *always* coauthors with a, but conversely, a, as an advisor, only coauthors a small portion of his/her papers with b. While *Kulc* shows relative preferences for such very skewed patterns by ranking them the top-3 most strongly associated pairs, *Cosine* and *Coherence* rank relatively balanced data higher. On the other hand, the author pairs ranked top 3 (ID = 1, 2, 3) by *Coherence* are considered to be the bottom 3 by *Kulc*, because these 3 pairs have relatively large $sup(ab)$ but the conditional probabilities are more balanced. The *Cosine* measure, as expected, stands in the middle of the other two: the top *Cosine* patterns (ID = 5, 7, 8) are ranked by *Coherence* as 4th, 5th, and 6th, and by *Kulc* as 1st, 2nd, and 6th. The same observation can be made to the bottom 3 patterns of *Cosine*. In conclusion, *Kulc* tends to give more credits to skewed patterns (*e.g.*, advisor-advisee relationships), *Coherence* prefers balanced patterns (*e.g.*, two comparable collaborators), and *Cosine* lies in-between.

3   Related Work

Both association and correlation mining are essential to the discovery of interesting, inherent relationships among large sets of events in a wide spectrum of applications. Various existing metrics and newly proposed measures have been studied to facilitate such analysis [11,3,12,10,6]. There are statistical correlation analysis methods. χ^2 [2] is borrowed from statistics [7] to identify correlations, considering both the absence and presence of items for interesting rules. TAPER [13], an algorithm for efficiently finding strongly correlated pairs of items, is grounded on the *Pearson*'s coefficient. Another class of work belongs to constraint-based association mining [4], where measures like *Confidence* and *Lift* are used to assist in rule generation. In [10,9] a new interestingness measures *AllConf* is defined based on a few desired properties. There are also measures widely used in other scenarios. For instance, *H-Measure* [5] is tailored for correlation analysis of deep Web-based query templates. Similarity metrics like *Cosine* distance function and

Coherence (or *Jaccard Coefficient*) are also popularly used. *Kulc*, proposed in [8], has been used in chemistry research [1].

An extensive investigation of the implications and connections between different measures has been conducted in [12], which compares a list of twenty-one interestingness metrics. The study describes three desired properties and five other key properties to compare different measures, and claims that no measure is generally better. Thus, one should match the application background against the intrinsic measure properties. Our paper can be viewed as a continued study of [12] in the context of small probability events.

4   Conclusions

We have presented a comprehensive study of null-invariant interestingness measures for mining small probability events. We show a generalization of the measures in one mathematical framework and a total ordering among them that provides an organized view. We also extend their definitions to support multiple events. For future research, it would be interesting to see how this work may influence real-world problems, such as social network analysis and clustering.

References

1. Bradshaw, J.: YAMS - Yet another measure of similarity. EuroMUG (2001), http://www.daylight.com/meetings/emug01/bradshaw/similarity/YAMS.html
2. Brin, S., Motwani, R., Silverstein, C.: Beyond market basket: Generalizing association rules to correlations. In: SIGMOD, pp. 265–276 (May 1997)
3. Brin, S., Motwani, R., Ullman, J.D., Tsur, S.: Dynamic itemset counting and implication rules for market basket analysis. In: SIGMOD, pp. 255–264 (May 1997)
4. Grahne, G., Lakshmanan, L., Wang, X.: Efficient mining of constrained correlated sets. In: ICDE, pp. 512–521 (February 2000)
5. He, B., Chang, K.C.-C., Han, J.: Discovering complex matchings across web query interfaces: A correlation mining approach. In: KDD, pp. 148–157 (2004)
6. Hilderman, R.J., Hamilton, H.J.: Knowledge Discovery and Measures of Interest. Kluwer Academic Publishers, Dordrecht (2001)
7. Kachigan, S.: Multivariate Statistical Analysis: A Conceptual Introduction. Radius Press (1991)
8. Kulczynski, S.: Die pflanzenassoziationen der pieninen. Bulletin, 57–203 (1927)
9. Lee, Y.-K., Kim, W.-Y., Cai, Y.D., Han, J.: CoMine: Efficient mining of correlated patterns. In: ICDM, pp. 581–584 (November 2003)
10. Omiecinski, E.: Alternative interest measures for mining associations. IEEE Trans. Knowledge and Data Engineering 15, 57–69 (2003)
11. Savasere, A., Omiecinski, E., Navathe, S.: Mining for strong negative associations in a large database of customer transactions. In: ICDE, pp. 432–443 (1998)
12. Tan, P.-N., Kumar, V., Srivastava, J.: Selecting the right interestingness measure for association patterns. In: KDD, pp. 32–41 (2002)
13. Xiong, H., Shekhar, S., Tan, P.-N., Kumar, V.: Exploiting a support-based upper bound of Pearson's correlation coefficient for efficiently identifying strongly correlated pairs. In: KDD, pp. 334–343 (2004)

Semantic Text Classification of Emergent Disease Reports

Yi Zhang and Bing Liu

Department of Computer Science, University of Illinois at Chicago,
851 S. Morgan Street, Chicago IL 60607, USA
{yzhang3,liub}@cs.uic.edu

Abstract. Traditional text classification studied in the information retrieval and machine learning literature is mainly based on topics. That is, each class represents a particular topic, e.g., sports and politics. However, many real-world problems require more refined classification based on some *semantic perspectives*. For example, in a set of sentences about a disease, some may report outbreaks of the disease, some may describe how to cure the disease, and yet some may discuss how to prevent the disease. To classify sentences at this semantic level, the traditional bag-of-words model is no longer sufficient. In this paper, we study semantic sentence classification of disease reporting. We show that both keywords and sentence semantic features are useful. Our results demonstrated that this integrated approach is highly effective.

Keywords: Semantic text classification.

1 Introduction

In traditional topic-based text classification, the bag-of-words representation of text documents is often sufficient because a topic can usually be characterized by a set of keywords [12, 13]. However, for semantic classification, the n-gram representation is no longer sufficient because the texts from different classes may be on the same topic. To classify texts at such a level, the system needs to capture some semantic characteristics in order to perform more accurate classification.

In this paper, we propose to integrate the bag-of-words scheme and semantic features extracted from sentences for classification. As a case study, we investigate the disease reporting domain. We want to classify sentences that report disease outbreaks, and sentences that do not. For example, the following sentence reports a possible disease outbreak "*the district hospital reported today that 10 people were diagnosed with cholera this morning*". However, the following sentence does not report an outbreak, "*the district hospital reported today that they have successfully tested a new cholera treatment procedure*". Both sentences are on the topic of cholera. However, they are entirely different semantically. The problem is how to separate sentences based on the required semantic categories, i.e., reporting a possible outbreak or not in this case. We note that sentences rather than documents are used in this work because a document contains a large number of sentences and the sentences have quite different semantic meanings. For example, a piece of disease outbreak

J.N. Kok et al. (Eds.): PKDD 2007, LNAI 4702, pp. 629–637, 2007.

news may contain many pieces of other related information, e.g., symptoms, treatment, vaccine, and past disease history. We show that both the words used in sentences and the sentence semantic characteristics are important. Our experimental results confirm that this integrated approach produces more accurate classifiers than each of them alone. Another original work is the investigation of date representations in natural language.

2 On Semantic Classification

The setting of semantic classification is the same as traditional topic-based text classification. Given a set of documents D and each document $d \in D$ is labeled with a class $c \in C$, where C is a set of known classes. However, semantic classification usually has more refined classes, which are hard to be separated based on bag-of-words or n-grams alone. For example, the sentence, "*the district hospital reported that 10 people were diagnosed with cholera early today*", reports a possible cholera outbreak. It is easy to observe that "reported" and "diagnosed" are indicative of an outbreak. The time "today" indicates a new event. Using the words alone, however, is insufficient as this sentence illustrates: "*10 people from the district hospital submitted a report early today on cholera diagnosis.*" This sentence uses very similar words, but has a completely different semantic meaning, i.e., no disease outbreak is reported.

In this paper, we define *semantic information* as any information extracted from the sentence that is not based on keywords/n-grams. There are multiple levels of semantic information: at the highest level it is the full understanding of the text, which is still not possible with the state-of-art technology; at lower levels, we have features with different amounts of semantic contents, extracted from sentences using current NLP techniques. The exact features used in this work will be discussed in the next section.

We note that the bag-of-words scheme is still useful because to express a particular semantic meaning, certain keywords are more likely to be used. Semantic feature extraction is still not perfect. Keywords can help offset some of the errors. Fig. 1 illustrates the difference between traditional text classification and semantic text classification as described in this work. Note that we do not make a difference of the types of classes or texts used in a classification task because classification of different types of categories may be assisted by some level of semantic information.

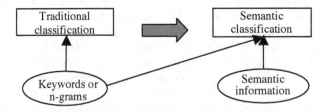

Fig. 1. Traditional classification and semantic classification

3 Important Semantic Features

Our aim is to classify sentences that report possible disease outbreaks and those that do not, which is a classification problem. We will use a supervised machine learning algorithm, e.g., naïve Bayesian or support vector machines. Thus, we only need to design and construct features. As we mentioned above, we use both keywords and semantic features. Keyword features are obtained in the same way as in traditional text classification. Here, we only focus on semantic features.

3.1 Noun Phrase Containing a Disease Word

Center word of a noun phrase: The center word in a noun phrase has a more direct influence on the overall semantic meaning than any other word in the noun phrase. For example, the center word of the noun phrase "*their new cholera treatment procedure*" is "*procedure*". We only use noun phrases containing a disease word because such phrases are more likely to be relevant to our classification task.

Negation modifier and determiner word: Other important features in a noun phrase include negation modifiers e.g., "no", and determiner words such as "every". Their importance can be illustrated by the following examples: "*No case of cholera has been found yesterday*" indicates no disease, and "*For every case of mad cow disease in Switzerland, 100 animals may carry the infection silently*" gives a study result on the disease rather than a specific case.

3.2 Verb Phrase

Verb and adjective: The verb serves as the main skeleton of a sentence and thus is an important feature. Sometimes, a verb is too common to have a specific meaning. In that case, the adjective word after the verb becomes important. For example, the verbs "is" and "become" are not specific, but "is ill" and "become sick" are.

Tense: The verb tense is important for the semantic meaning of a sentence because it shows the time or the subjunctive mood. Past perfect tense usually means something happened in the long past. E.g., "*West Nile Virus had plagued US*" refers to an old event. Subjunctive mood expressed by the past-future perfect tense is often used for conjectures. E.g., "*a bird flu outbreak could have killed millions of people*" is a conjecture of the disease's impact rather than a report of an actual outbreak.

Other features extracted from a verbal phrase include: **auxiliary word, verb phrase being an if-whether clause, negation word, verb phrase being an adjective clause, and subject/object.**

3.3 Dates

Dates are important for disease reporting. We focus on the common ones in this work. Thus, our description below is by no means complete, but is quite sufficient for our data. Date information is usually expressed by a prep word (maybe omitted) followed by a *date phrase*. We call the date expressed in the text as the *expressed date*, and the date of the context as the *context date* (e.g., publication date of a news report).

Prep word: A prep word decides the relationship between an *expressed date* and the date phrase that follows. We summarize the prep words and the corresponding relationships in Table 1. If a prep word is omitted, in most cases it's the same as the first relationship in the table. For example, *"The alert was given last Tuesday"*.

Table 1. Prep words and relationships between *expressed dates* and date phrases

Relationship	Prep word	Example
expressed date is the date phrase	in, at, on, during	on Monday
expressed date is before the date phrase	before	before winter
expressed date is after the date phrase	after	after May 1, 2006
expressed date ends within the date phrase	in, within	in two days
expressed date ends by the date phrase	by, as of, until/till, no later than	by today
expressed date spans the two date phrases	between … and, from…to…	From Jan to Feb
expressed date starts from the date phrase	since	since last year

Adjective and adverb: Adjectives and adverbs may be associated with dates, e.g., "ago" as in "three months ago". In general, a *date phrase* expresses either an *absolute date* or a *relative date*. We will not discuss time in this paper as it can be dealt with in a similar way.

Absolute date: As its name suggests, an absolute date expresses a specific date without ambiguity regardless when it is seen. There are two main types:

- *Historic period*: It is a time period in history usually with a very long duration, and it has a specific name, e.g., "Stone Age".
- *Formal date*: It specifies an absolute time period that can be: a century, a decade, a year, a season, a month, a day, a time period of a specific day, etc.
  Relative date: Its absolute date can only be determined based on the *context date*.
- *Recurrent named date*: Such a relative date occurs repetitively, e.g., annual festival, season, month of year, day of month, and day of week. A restrictive modifier is mandatory, although sometimes it can be omitted based on convention.
- *Other named dates*: Such dates include "today" and "tomorrow" or *special words* (e.g., "now" and "recently") that are dedicated to some relative dates.
- *Number-unit*: This is also popularly used in date phrases, e.g., "three months" in "three months ago". Similar to a recurrent date, a modifier is also required.

A date phrase may have a refiner, such as *"early* 2007" and "the *end* of last month". Now we give a formal definition of date phrases in Backus–Naur form. Due to space limitations, some rules use suspension points in place of similar entries, and definitions of self-explanary items (e.g., <DateUnit> and <Number>) are omitted.

```
<DatePhrase> ::= [<Refiner>]<FormalDate>
  | [<Refiner>]<HistoricPeriod> | [<Refiner>]<FormalPeriod>
  | [<Refiner>]<Modifer><DateName> |   <SpecialWord>
  | [<Refiner>]<Modifier>[<Number>]<Date Unit>
  | [<Number>]<Date Unit>[<PostModifier>] | [<Refiner>]<SpecialDay>
<Refiner> ::= late | early | end of | beginning of| middle of|……
<Modifier> ::= last | previous | next | coming | past|……
<PostModifier> ::= ago | later | early | ……
```

```
<SpecialDay> ::= today | tomorrow | the day before yesterday | ......
<SpecialWord> ::= now | recently | ......
<FormalDate> ::= [<Month>] <Year> | [<Season>] <Year>
| [<Festival>] <Year> | [<DayofWeek>]<Month>[/]<Day>[/][<Year>]
<HistoricPeriod> ::= stone age | ......
<FormalPeriod> ::= <OrdinalNumber> century | <year>[']s | ......
<DateName> ::= <Festival> | <DayOfWeek> [<TimeOfDay>] | ......
```

4 Feature Extraction

Now we describe how to extract features from a sentence. Our first task is to recognize named entities, because most features given in Section 3 can only be found based on correct recognition of named entities, i.e., disease names or dates.

4.1 Dependency Tree

A dependency tree describes the syntactical and semantic relationships in a sentence. Fig. 2 gives a dependency tree generated from the sentence *"Danish health authorities on Friday confirmed the Scandinavian country's fourth case of mad cow disease"*.

Fig. 2. An example of dependency tree. Preposition words and the word "'s" are put on the edge to save space. Arrows point from a parent node to children nodes. Note that "mad cow disease" is a named entity, so it is a single node in spite of that it has three literal words.

4.2 Feature Extraction and Construction

After named entities have been recognized and the dependency tree has been built, features are extracted in the following way: Starting from any infectious disease, we find the center word of the noun phrase containing the disease. Negation modifier and determiner words can be found among children nodes of the center word.

The verb is always the nearest ancestral verb node of the noun phrase. If there is an adjective node between the verb and the noun phrase, it is taken as the adjective word feature. The tense of a verb is determined by the form of the verb and the forms of auxiliary words, which are children nodes of the center word. So are negation modifiers and subject/object words. Other features of a verb phrase can be obtained from sibling or parent nodes of the verb node. If a verb phrase is an adjective clause, the verb node normally has a sibling node of "wh-" word and a relationship of complementary to its parent node. If a verb phrase is an if-whether clause, there will be a sibling node of "if" or "whether". Date feature can be recognized using the definitions given in Section 3.3. To normalize dates, we treat a relative date in current year as "recent", any date before that as "old", and any date after as "future". Thus, the date feature has three possible values, "recent", "old" and "future".

Implementation: We use MINIPAR [7] for dependency tree generation and named entity recognition. In order to recognize diseases, we supplemented the standard MINIPAR database with disease names extracted from Centers for Disease Control and Prevention (http://www.cdc.gov/ncidod/). The feature construction algorithm then reads the generated dependency trees and outputs features.

5   Experiments

Experimental data: Our corpus consists of sentences related to infectious diseases. Some of them are emergent disease reports (EDR), and others are not (non-EDR) but still contain the disease names. The sentences are extracted from disease report documents in ProMED-mail (http://www.promedmail.org). We labeled the sentences into two classes: EDR(1660 sentences) and nonEDR (682 sentences).

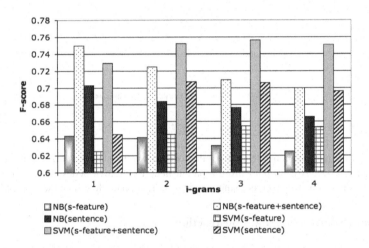

Fig. 3. Experimental results

Experimental settings: Two popular supervised learning algorithms are used to build models, Support Vector Machines (SVM) and naïve Bayesian (NB). Both algorithms are provided in the latest version of the Rainbow package [8], which is used in our experiments. Different types of features are employed and compared:

- **sentence:** bag-of-words representation with 1-gram, 2-gram, 3-gram and 4-gram.
- **s-features:** semantic features (including the date feature).
- **s-features+sentence:** s-features and sentence features in a sentence combined.

To ensure reliable results, we run each technique 10 times. In each run, 80% of the data (randomly selected) is used for training and 20% of the data is used for testing. The results are then averaged and reported below. The evaluation measure is F-score on EDR sentences. F-score is the harmonic mean of precision (p) and recall (r), i.e., $F = 2pr/(p+r)$, which is commonly used in text classification.

Results: Fig. 3 shows the average F-scores of all methods. We observe that semantic features (denoted by s-features) are very helpful. Both SVM and NB produce much better results when sentences and semantic features are both used. SVM (s-features+sentence) with 3-gram for sentences gives the best F-score and it also performs the best for 2-gram and 4-gram, except for 1-gram, in which NB (s-features+sentence) is better. Please note that we are applying i-grams for all three feature sets, so i-grams are also relevant for "s-feature" feature set, as a result, F-scores for NB (s-feature) and SVM(s-feature) differ across different orders of i-grams.

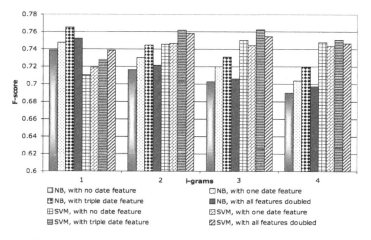

Fig. 4. Experimental results: with different weights of date feature

We also single out the date feature to see how it effects classification since intuitively the dates are important for EDR sentences. The date feature is indeed helpful (Fig. 4). For NB, the F-scores with date features are always better than without date features. For SVM, the results are also better for 1-gram and 2-gram. All the results here use both s-features and sentences. Since the date feature has shown its importance, it may help more if its weight is increased. We found that multiplying each date feature by 3 gives the best results. Fig. 4 shows that "triple date feature" gives better F-scores for both NB and SVM in almost all cases. NB with 1-gram produces the best result. Due to this success, we also tried to increase the weights of all s-features ("with all features doubled") but without improvements (Fig. 4).

In summary, we can conclude that combining bag-of-words and semantic features indeed improve classification. The date feature is also very helpful.

6 Related Work

There are several works on using linguistic information for text classification. Most of them are based on the idea of carefully choosing keywords or phrases. Noun phrases have been used in [2], and terms features extracted from part-of-speech tags were used in [1]. [11] improved the text classification by using bag-of-concepts. [10] investigated adding complex nominals as features. [3] reported classification results based on several keyword extraction methods. [4] assigned feature weights based on the importance of each sentence determined by a text summarization system. Similar works were also reported in [6, 9]. Sub-trees of dependency trees were shown helpful in classification in [5]. Our work is related but also different from these text classification works in several ways. First, they still focus on classifying whole text documents, but we focuses on sentence level classification, which requires more delicate semantic features. We also construct date features, which is also new.

7 Conclusion

In this paper, we studied the problem of classifying disease reporting at the semantic level. It is shown that both keywords and semantic features are valuable for the task. The paper also investigated the representation of dates, which will be useful to other applications. Experimental results demonstrated that the proposed integrated approach significantly outperforms each individual approach alone.

Acknowledgments. This work is funded by the Great Lakes Protection Fund. We also thank Karl Rockne for helpful discussions.

References

1. Aizawa, A.: Linguistic techniques to improve the performance of automatic text categorization. NLPRS-01 (2001)
2. Furnkranz, J., Mitchell, T., Riloff, E.: A case study using linguistic phrases for text categorization on the WWW. In: AAAI-98 Workshop on Learning for Text Categorization
3. Hulth, A., Megyesi, B.: A Study on Automatically Extracted Keywords in Text Categorization. In: ACL-06 (2006)
4. Ko, Y., Park, J., Seo, J.: Improving text categorization using the importance of sentences. Info. Proc. and Manag. 40(1), 65–79 (2004)
5. Kudo, T., Matsumoto, Y.: A boosting algorithm for classification of semi-structured text. In: EMNLP-2004 (2004)
6. Li, C., Wen, J.-R., Li, H.: Text classification using stochastic keyword generation. In: ICML-03 (2003)
7. Lin, D., Pantel, P.: Discovery of Inference Rules for Question Answering. Nat. Lang. Eng. 7(4) (2001)
8. McCallum, A.: Bow: A toolkit for statistical language modeling, text retrieval, classification and clustering (1996), http://www.cs.cmu.edu/ mccallum/bow

9. Mihalcea, R., Hassan, S.: Using the essence of texts to improve document classification. In: RANLP-2005 (2005)
10. Moschitti, A., Basili, R.: Complex linguistic features for text classification: A comprehensive study. In: ECIR-04 (2004)
11. Sahlgren, M., Coster, R.: Using bag-of-concepts to improve the performance of support vector machines in text categorization. In: COLING 2004 (2004)
12. Sebastiani, F.: Machine learning in automated text categorization. ACM Computing Surveys 34(1), 1–47 (2002)
13. Yang, Y., Liu, X.: A re-examination of text categorization methods. In: SIGIR-99 (1999)

Author Index

Lecture Notes in Artificial Intelligence (LNAI)

Vol. 4496: N.T. Nguyen, A. Grzech, R.J. Howlett, L.C. Jain (Eds.), Agent and Multi-Agent Systems: Technologies and Applications. XXI, 1046 pages. 2007.

Vol. 4483: C. Baral, G. Brewka, J. Schlipf (Eds.), Logic Programming and Nonmonotonic Reasoning. IX, 327 pages. 2007.

Vol. 4482: A. An, J. Stefanowski, S. Ramanna, C.J. Butz, W. Pedrycz, G. Wang (Eds.), Rough Sets, Fuzzy Sets, Data Mining and Granular Computing. XIV, 585 pages. 2007.

Vol. 4481: J. Yao, P. Lingras, W.-Z. Wu, M. Szczuka, N.J. Cercone, D. Ślęzak (Eds.), Rough Sets and Knowledge Technology. XIV, 576 pages. 2007.

Vol. 4476: V. Gorodetsky, C. Zhang, V.A. Skormin, L. Cao (Eds.), Autonomous Intelligent Systems: Multi-Agents and Data Mining. XIII, 323 pages. 2007.

Vol. 4456: Y. Wang, Y.-m. Cheung, H. Liu (Eds.), Computational Intelligence and Security. XXIII, 1118 pages. 2007.

Vol. 4455: S. Muggleton, R. Otero, A. Tamaddoni-Nezhad (Eds.), Inductive Logic Programming. XII, 456 pages. 2007.

Vol. 4452: M. Fasli, O. Shehory (Eds.), Agent-Mediated Electronic Commerce. VIII, 249 pages. 2007.

Vol. 4451: T.S. Huang, A. Nijholt, M. Pantic, A. Pentland (Eds.), Artifical Intelligence for Human Computing. XVI, 359 pages. 2007.

Vol. 4441: C. Müller (Ed.), Speaker Classification. X, 309 pages. 2007.

Vol. 4438: L. Maicher, A. Sigel, L.M. Garshol (Eds.), Leveraging the Semantics of Topic Maps. X, 257 pages. 2007.

Vol. 4434: G. Lakemeyer, E. Sklar, D.G. Sorrenti, T. Takahashi (Eds.), RoboCup 2006: Robot Soccer World Cup X. XIII, 566 pages. 2007.

Vol. 4429: R. Lu, J.H. Siekmann, C. Ullrich (Eds.), Cognitive Systems. X, 161 pages. 2007.

Vol. 4428: S. Edelkamp, A. Lomuscio (Eds.), Model Checking and Artificial Intelligence. IX, 185 pages. 2007.

Vol. 4426: Z.-H. Zhou, H. Li, Q. Yang (Eds.), Advances in Knowledge Discovery and Data Mining. XXV, 1161 pages. 2007.

Vol. 4411: R.H. Bordini, M. Dastani, J. Dix, A.E.F. Seghrouchni (Eds.), Programming Multi-Agent Systems. XIV, 249 pages. 2007.

Vol. 4410: A. Branco (Ed.), Anaphora: Analysis, Algorithms and Applications. X, 191 pages. 2007.

Vol. 4399: T. Kovacs, X. Llorà, K. Takadama, P.L. Lanzi, W. Stolzmann, S.W. Wilson (Eds.), Learning Classifier Systems. XII, 345 pages. 2007.

Vol. 4390: S.O. Kuznetsov, S. Schmidt (Eds.), Formal Concept Analysis. X, 329 pages. 2007.

Vol. 4389: D. Weyns, H.V.D. Parunak, F. Michel (Eds.), Environments for Multi-Agent Systems III. X, 273 pages. 2007.

Vol. 4386: P. Noriega, J. Vázquez-Salceda, G. Boella, O. Boissier, V. Dignum, N. Fornara, E. Matson (Eds.), Coordination, Organizations, Institutions, and Norms in Agent Systems II. XI, 373 pages. 2007.

Vol. 4384: T. Washio, K. Satoh, H. Takeda, A. Inokuchi (Eds.), New Frontiers in Artificial Intelligence. IX, 401 pages. 2007.

Vol. 4371: K. Inoue, K. Satoh, F. Toni (Eds.), Computational Logic in Multi-Agent Systems. X, 315 pages. 2007.

Vol. 4369: M. Umeda, A. Wolf, O. Bartenstein, U. Geske, D. Seipel, O. Takata (Eds.), Declarative Programming for Knowledge Management. X, 229 pages. 2006.

Vol. 4343: C. Müller (Ed.), Speaker Classification I. X, 355 pages. 2007.

Vol. 4342: H. de Swart, E. Orłowska, G. Schmidt, M. Roubens (Eds.), Theory and Applications of Relational Structures as Knowledge Instruments II. X, 373 pages. 2006.

Vol. 4335: S.A. Brueckner, S. Hassas, M. Jelasity, D. Yamins (Eds.), Engineering Self-Organising Systems. XII, 212 pages. 2007.

Vol. 4334: B. Beckert, R. Hähnle, P.H. Schmitt (Eds.), Verification of Object-Oriented Software. XXIX, 658 pages. 2007.

Vol. 4333: U. Reimer, D. Karagiannis (Eds.), Practical Aspects of Knowledge Management. XII, 338 pages. 2006.

Vol. 4327: M. Baldoni, U. Endriss (Eds.), Declarative Agent Languages and Technologies IV. VIII, 257 pages. 2006.

Vol. 4314: C. Freksa, M. Kohlhase, K. Schill (Eds.), KI 2006: Advances in Artificial Intelligence. XII, 458 pages. 2007.

Vol. 4304: A. Sattar, B.-h. Kang (Eds.), AI 2006: Advances in Artificial Intelligence. XXVII, 1303 pages. 2006.

Vol. 4303: A. Hoffmann, B.-h. Kang, D. Richards, S. Tsumoto (Eds.), Advances in Knowledge Acquisition and Management. XI, 259 pages. 2006.

Vol. 4293: A. Gelbukh, C.A. Reyes-Garcia (Eds.), MICAI 2006: Advances in Artificial Intelligence. XXVIII, 1232 pages. 2006.

Vol. 4289: M. Ackermann, B. Berendt, M. Grobelnik, A. Hotho, D. Mladenič, G. Semeraro, M. Spiliopoulou, G. Stumme, V. Svátek, M. van Someren (Eds.), Semantics, Web and Mining. X, 197 pages. 2006.

Vol. 4285: Y. Matsumoto, R.W. Sproat, K.-F. Wong, M. Zhang (Eds.), Computer Processing of Oriental Languages. XVII, 544 pages. 2006.

Vol. 4274: Q. Huo, B. Ma, E.-S. Chng, H. Li (Eds.), Chinese Spoken Language Processing. XXIV, 805 pages. 2006.

Vol. 4265: L. Todorovski, N. Lavrač, K.P. Jantke (Eds.), Discovery Science. XIV, 384 pages. 2006.

Vol. 4264: J.L. Balcázar, P.M. Long, F. Stephan (Eds.), Algorithmic Learning Theory. XIII, 393 pages. 2006.